Photonic Instrumentation

Photonic Instrumentation: Sensing and Measuring with Lasers is designed as a source for university-level courses covering the essentials of laser-based instrumentation, and as a useful reference for working engineers. Photonic instruments have very desirable features like non-contact operation and unparalleled sensitivity. They have quickly become a big industrial success, passing unaffected through the bubble years and, not any less important, well-established methods in measurement science. This book offers coverage of the most proven instruments, with a balanced treatment of the optical and electronic aspects involved. It also attempts to present the basic principles, develop the guidelines of design and evaluate the ultimate limits of performances set by noise.

The instruments surveyed include: alignment instruments, such as wire diameter and particle size analyzers, telemeters, laser interferometers and self-mixing interferometers, and speckle pattern instruments, laser doppler velocimeters, gyroscopes, optical fiber sensors and quantum sensing. A few appendices offer convenient reference material for key principles on lasers, optical interferometers, propagation, scattering and diffraction.

Silvano Donati is emeritus professor of optoelectronics at the University of Pavia, Italy. A graduate of physics at the University of Milan, Italy, he has conducted research in photonic instrumentation for more than 30 years, covering virtually all the material treated in this book. He has given seminal contributions to interferometry with the invention of the self-mixing scheme, and to telemetry, gyroscopes and current fiber sensors with new approaches and noise evaluations. He has authored *Photodetectors: Devices, Circuits and Applications, Second Edition,* 2021, and more than 350 papers in international journals, plus a dozen patents. He has chaired several international conferences including Optoelectronic Distance Measurements and Applications (ODIMAP). Prof. Donati is a Life Fellow member of IEEE and Optica. He was the founder and chairman of the IEEE LEOS Italian chapter, and Chair of the IEEE Italy section.

Photonic Instrumentation

Sensing and Measuring with Lasers

Silvano Donati

CRC Press
Taylor & Francis Group
Boca Raton London New York

CRC Press is an imprint of the
Taylor & Francis Group, an **informa** business

Second edition published 2023
by CRC Press
6000 Broken Sound Parkway NW, Suite 300, Boca Raton, FL 33487-2742

and by CRC Press
4 Park Square, Milton Park, Abingdon, Oxon, OX14 4RN

CRC Press is an imprint of Taylor & Francis Group, LLC

First edition published by Prentice Hall 2000

Library of Congress Cataloging-in-Publication Data

Names: Donati, Silvano, author.
Title: Photonic instrumentation : sensing and measuring with lasers / Silvano Donati.
Other titles: Electro-optical instrumentation
Description: Second edition. | Boca Raton : CRC Press, 2023. | Revised edition of: Electro-optical instrumentation : sensing and measuring with lasers / Silvano Donati. c2004. | Includes bibliographical references and index. | Summary: "Photonic Instrumentation is designed as the sourcebook for university-level courses covering the essentials of laser-based instrumentation, and is also a useful reference for working engineers. Photonic instruments have very desirable features like non-contact operation and unparalleled sensitivity. They have soon become a big industrial success, passing unaffected through the bubble years and, not less important, well-established methods in measurement science. This book offers coverage of the most proven instruments, with a balanced treatment of the optical and electronic aspects involved. It also attempts to present the basic principles, develop the guidelines of design and evaluate the ultimate limits of performances set by noise. The instruments surveyed include: Alignment instruments -Wire diameter and Particle Size Analyzers - Telemeters - Laser Interferometers -Self-Mixing Interferometers - Speckle Pattern Instruments- Laser Doppler Velocimeters- Gyroscopes - Optical Fiber Sensors -Quantum Sensing A few appendices offer convenient reference material for key principles on lasers, optical interferometers, propagation and scattering, and diffraction."-- Provided by publisher.
Identifiers: LCCN 2022060555 (print) | LCCN 2022060556 (ebook) | ISBN 9781032469324 (hardback) | ISBN 9781032489063 (paperback) | ISBN 9781003391357 (ebook)
Subjects: LCSH: Optoelectronic devices. | Electrooptical devices. | Lasers. | Measuring instruments. | Engineering instruments. | Scientific apparatus and instruments.
Classification: LCC TA1750 .D66 2023 (print) | LCC TA1750 (ebook) | DDC 681/.2--dc23/eng/20230405
LC record available at https://lccn.loc.gov/2022060555
LC ebook record available at https://lccn.loc.gov/2022060556

ISBN: 978-1-032-46932-4 (hbk)
ISBN: 978-1-032-48906-3 (pbk)
ISBN: 978-1-003-39135-7 (ebk)

DOI: 10.1201/9781003391357

Typeset in Times Roman
by KnowledgeWorks Global Ltd.

Access the Support Material at: www.routledge.com/9781032469324

Contents

Preface

When I started to write and update this book, the intended title was *Electro-Optical Instrumentation, 2nd edition*. But, twenty years had passed from the first edition of this book, published by Prentice Hall, an excellent publisher that meanwhile had ceased its activity on such books, so I had to find another excellent publisher and luckily found it soon in CRC Press.

Thus, I realized that twenty years had been a long time-span, not only for the many new topics I had in mind to add in the book, but also for the most appropriate name identifying the subjects of the book. It was easy to get out of this semantic problem, as the IEEE Society to which I belong and is my reference, had changed his name in 2009, shortly after the 30th anniversary of existence as LEOS (Laser and Electro-Optical Society), preferring the new name PhoS (Photonics Society) to identify their cultural territory. So, I had no choice, and I had to turn the name of my book into *Photonic Instrumentation* to better reflect the discipline it covers.

If the virus lockdown was slowing or impeding normal scientific activity, the front of the medal was the availability of much time to write books, and I took ample advantage of it in rethinking my book from scratch.

Two aspects are intermingled in the book: the pedagogical value, which I wanted to privilege in a book mainly intended as a textbook, and the survey of the field that can be a stimulus for young PhDs and researchers in their activity. I kept constant reference to the first issue, also thanks to the feedbacks from the foreign language translations, in the Chinese (Guang Dien Yi Chi) and Indian edition, as well as from colleagues using the Student Edition as a textbook. So, the main frame of the book is unaltered, and only two new chapters were added on Self-mixing Interferometry and Quantum Sensing. But, in any other of the ten chapters, new sections now cover scientific novelties like long-wave Lidar, Optical Coherent Tomography, several measurement and consumer applications of Self-mixing, Particle Image Velocimetry, Flowmeters, and the Shupe Gyroscope, just to cite a few. In total, about 20% new pages have been added, and more than half of the material has been corrected, updated, or improved for better readability.

Yet, the character of the book is the same as the first edition, i.e., a textbook for MSc and PhD courses in Photonic Engineering, that hopefully will be well received by the scientific community as a useful tool for teaching the exciting subject of photonic (or electro-optical) instrumentation, like it has been for me for twenty years of courses given in my own University and some more years when I have been a visiting professor, mostly in Taiwan universities.

An important novelty of this edition, following the suggestion of colleagues teaching the subject, are the selected problems that I added at the end of each chapter, for a total of about one hundred. I will provide the booklet of solutions to those adopting the book for their classes.

I constantly tried to keep the character of the book appropriate for students of the Faculty of Engineering where I teach, focusing not only on the principles, but also and especially on the design and the engineering applications. Also, I deliberately left unaltered a few figures of the first edition, now definitely vintage, to give the reader an idea of the by now well-established field of activity and discipline of Photonic Instrumentation.

So, I start each new argument with a description of the working principle of devices, then develop in some detail the circuits for their use, evaluate performances, and finally illustrate the applications. Another distinctive feature of the book, reflecting my research interests, is the constant reference to the noise limits of measurements, those that determine the ultimate performances that are achievable.

Thus, I try to leverage the pedagogic value of the book based on developments of concepts, rather than offering a collection of arguments grouped by technicality.

I think that this book, even better than its first edition, is well suited for forging the young minds of MSc and PhD students in courses in Photonics Engineering, and also in Measurement Science and Electronics.

To teachers and instructors using the book as the official course textbook, in addition to the booklet of solved problems, I will be glad to supply the set of about 400 slides useful to deliver classroom lessons.

Silvano Donati, University of Pavia
email: silvano.donati@unipv.it
website: www.unipv.it/donati

Pavia, October 2022

Preface to the First Edition

This book is an outgrowth of the lecture notes for a semester course that the author has given at the University of Pavia for several years. The course was tailored for electronic engineer graduates in their fifth and final year of the curriculum (the 18th grade). The course was designed as a compulsory course for students of the optoelectronic engineering major, and was also offered as an elective course to students of the instrumentation and microelectronics engineering sections.

During the years, I have gone through several versions of the text, trying to improve and expand the material. I have also added new topics taken from the literature or from my own research on interferometers, laser telemeters, speckle-pattern and optical fiber sensors.

About terminology, electro-optics is a very interdisciplinary science that receives contributions from researchers whose cultural roots originate in electronics, as well in optics, laser physics, and electromagnetism.

Other researchers may refer to the content of this book as measurements by lasers, coherent techniques, optoelectronic measurements, optical metrology, and perhaps some more names equally acceptable.

I think that electro-optical instrumentation is preferable to give this field a name of its own and to underline the connection between optics and electronics, a very fruitful synergy we have actually observed in this field through the years.

In distributing the material between text and appendixes, the rationale used here is that core arguments of the book are text, whereas arguments either complementary, common to other disciplines, or recalls of basics are reviewed in the appendixes.

As a general scope, in this book I have tried to (i) illustrate the basic principles behind the application; (ii) outline the guidelines for the design; and (iii) discuss the basic performance achievable and the ultimate limit set by noise.

I have attempted to make the chapters and the accompanying appendixes self-contained, so that a selection of them can be used for a shorter course as well.

As a guideline, in 42 hours of classroom lessons I cover most of the material presented in the text, starting from basic ideas and going on to clarify performance limits and give development hints. To lessons, I add some 10 hours of lab and exercises. In the classes, I usually skip mathematical details and some advanced topics when they are most difficult. I have included them anyway in the text because they illustrate the state of the art of electro-optical instrumentation and are useful for advanced study.

Based on my experience, students with a limited background in optoelectronics can follow the course profitably if they are given a primer of an additional 6 to 8 hours on the fundamentals of lasers and fibers that are collected in the appendixes.

About derivations, I have tried to keep the mathematical details to a minimum and report just the very straight derivations. Purposely, to report just simple expressions, I have typed the equations all on one line to save precious typographical space. This kind of typing may look odd at first, but will be easier as the reader becomes acquainted with it.

Throughout the text, special attention has been paid to complement the understanding of the basic principle with the treatment of development issues. Thus, as compared with other books on the subject, the reader will find a lot of electronic schematics about signal handling, discussions on the impact of practical components, comments about the ultimate limits of performance, etc. Of course, this is just the attitude we try to develop in an electronic engineer, but it is also interesting for the generic reader to grasp the engineering problems of instrumentation.

Presently, the book does not contain problems. Actually, I prefer to publish the book first and follow up with problems afterwards. Problems will be made available to the reader a few months after publication at the author's website, unipv.it/donati. At the same location, instructors adopting the book for their course will be able to find a selection of viewgraphs in PowerPoint.

Many individuals have helped through the years in collecting the lecture notes from which the book has started.

It is also a pleasure to thank Risto Myllyla, Gordon Day, Peter deGroot, Jesse S. Greever, and Thierry Bosch for reading the manuscript and providing useful feedback.

I hope this book will be useful to the young student as a guide and motivation to work in the exciting field of electro-optical instrumentation and to the designer as a reference to the state of the art in this field. If the book will stimulate new ideas or development of products, my effort in writing it will be amply rewarded.

Silvano Donati
Pavia, January 2004
unipv.it/donati

CHAPTER 1

Introduction

The coming of age of photonic instrumentation dates back to a few years after the invention of the laser, that is, to the mid-1960s. The enormous potential of lasers in measurements was soon recognized by the scientific community as one capable of providing new approaches with unparalleled performances. Plenty of successful examples of new instruments and sensors started to appear in the 1970s, including gyroscopes, laser interferometers, pulsed telemeters, and laser Doppler velocimeters.

These examples were at first just bright scientific ideas, but, after several years of research and development, they matured to industrial products of great success. Initially, military applications provided a big push to the research and development effort, and, in the system balance, the laser played the role of the enabling component.

After a few decades, parallel with the growth of a civil application market, photonic instrumentation became to stand on its own feet. The focus of attention then moved from the source to the measurement, thus boosting the synergy of optics and electronics and merging the new technology in the frame of measurement science.

1.1 Looking Back to Milestones

The album of successes of photonic instrumentation soon became very rich and record-breaking.

About *telemeters*, the first astonishing experiment was the LUnar Ranging Experiment (LURE) carried out in 1969 (Fig.1-1) when the astronauts of Apollo 11 left on the moon a 1m × 1m array of corner-cubes [1].

Several telescopes of astronomical observatories on earth aimed their ruby laser, a Q-switched pulsed beam, on the array. The task was difficult, because that small target was clearly invisible and only the lunar coordinates were available. Three out of five telescopes succeeded in hitting the target and getting the very small return (about 10 photons). The pulse delay, or time-of-flight (about 2 seconds), was measured with nanosecond resolution, thus sampling the 384,000-kilometer distance to about 30 centimeters, or about 1 part in 10^9, an all-time record in physical measurements.

This was not just a mere big-science exhibition, as this early laser ranging experiment was the forerunner of the modern earth-to-satellite telemeter network. Soon, long-distance measurement by laser telemeters became an established, powerful tool for geodesy survey.

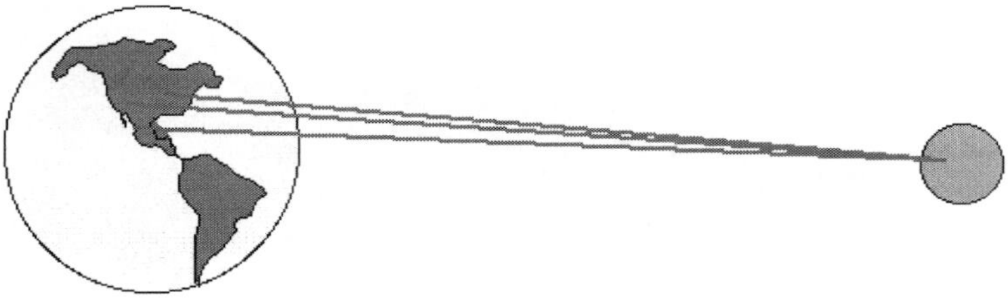

Fig.1-1 The LURE experiment set a record of long-distance pulse telemeter in 1969. The baseline was the 384,000 km earth-to-moon distance, and the resolution of the measurement about 30 cm.

The experiment also opened the way to airborne telemeters and altimeters (Fig.1-2), based mainly on the Nd:YAG Q-switched laser and occasionally on CO_2 lasers for better transmission through haze and fog.

Twenty years later, in the '90s, thanks to semiconductor laser becoming reliable, compact and inexpensive, telemeters started to be considered for automotive and consumer applications. An early experiment (Fig.1-3) [2] aimed at fitting the telemeter in the headlight compartment of the car to develop an anti-collision system, fostered the subsequent years' autonomous–drive systems, which is not a dream of science fiction anymore, but a technical reality being actively pursued.

The application to the consumer market included a hand-held low-cost (typ. US$ 200) telemeter for measuring the size of residential premises and, curiously, for flag distance aid to golf players.

Fig.1-2 Pulsed telemeters (based on Q-switched solid-state lasers) have been routinely mounted aboard military aircraft since the 1970s, with several kilometers operation range and meter accuracy ...

Fig.1-3 ... and their compact variants (based on semiconductor diode lasers) have been experimented in automobiles as the sensor of anti-collision systems as early as 1992 by CRF-Fiat, Turin (Italy) [2]. They are the ancestors of the modern automatic-drive and intelligent lighting systems, see Figs.1-11 and 1-12.

The sine-wave-modulated diode-laser telemeter introduced in the early 1970s is nowadays an instrument of widespread use in construction works and has marked the retirement of the old theodolite and associated set of rulers (Fig.1-4) [3]. As a device evolved from the early concept, the topographic telemeter provides 1-cm accuracy over several hundred-to thousand-meter distance and accounts for a healthy US$ 300 million per year market segment. On the other hand, much cheaper and very compact solutions were demanded by the automotive and consumer products industries, and the answer has been the long-wave, gated detector telemeter (see Ch. 3.3).

Fig.1-4 Meanwhile, in the early 1970s, sine-wave modulated telemeters started to enter the consumer market to replace the old theodolite in civil engineering constructions [3]. Distance measurement is carried out instantly on ranges up to several hundred meters with 1-cm resolution.

The LURE sensational experiment was later renewed and amplified by the Mars Orbiter Laser Altimeter (MOLA) flown aboard the Mars Global Surveyor in 1997, which orbited the red planet for over a year to collect height data on the Mars surface. The MOLA altimeter sent the amazing map in Fig.1-5 back to earth.

In this map, pixels are 100 m $\times$100 m wide, and their relative height has been mapped with a 10-m accuracy [4] from the average 400 km distance flown by the satellite from the planet surface [5]. The telemeter aboard the Mars Global Surveyor used as the source a diode-laser pumped Cr-Nd:YAG laser operating in Q-switching regime and supplying 8-nanosecond (ns) pulses at 10-Hertz (Hz) repetition rate.

The receiver was a 50-cm diameter telescope focusing the collected light on a Si-avalanche photodiode detector and included a 2-nanometer (nm) band pass filter to reject solar background.

Similar to planetary telemeters, a number of Laser Ranger Observatories (LRO) have been built around the world to target geodetic satellites in earth orbit. One of these, the Matera LRO [6], reaches 25 ps of timing accuracy, or approximately 3-millimeter range accuracy.

Another version of the basic telemetry scheme is provided by Laser Identification, Detection, and Ranging (LIDAR, see Ch.3). This instrument is again a pulsed telemeter, but has a λ-tunable laser as the source. By analyzing the echoes collected from the atmosphere, according to different conceptual schemes (DIAL, CARS, etc.), pollutant concentrations can be measured down to part-per-billion (ppb) concentration with distance-resolved plots of the pollutant concentration (Fig.1-6).

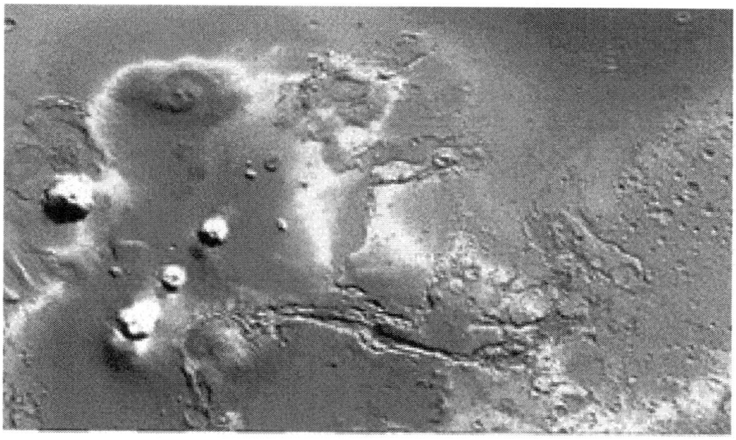

Fig.1-5 Map of the planet Mars as provided by the MOLA orbiting telemeter (1998). Individual pixels are 100 m in size, and their height is measured to a 10-m accuracy by the orbiting telemeter (by courtesy of NASA and MOLA Science Team).

The *optical gyroscope* is another enormous success, both scientific and industrial, of photonic instrumentation. The laser gyroscope proof-of-concept was demonstrated as early as 1962, just one year after the discovery of the He-Ne laser, based on a 1-m square ring configuration just capable of detecting the earth rotation rate (15 degree/hour), see Fig.1-7.

However, from this encouraging start, the way to a commercial gyroscope suitable for field deployment was then long and full of obstacles, like the locking effect and sign detection, which required new ideas as well as new improved technologies.

Only after a decade of worldwide research efforts by the scientific community, clever conceptual solutions and improved technology finally evolved, which led to the modern top-class RLG with the amazing accuracy of 0.001 deg/h.

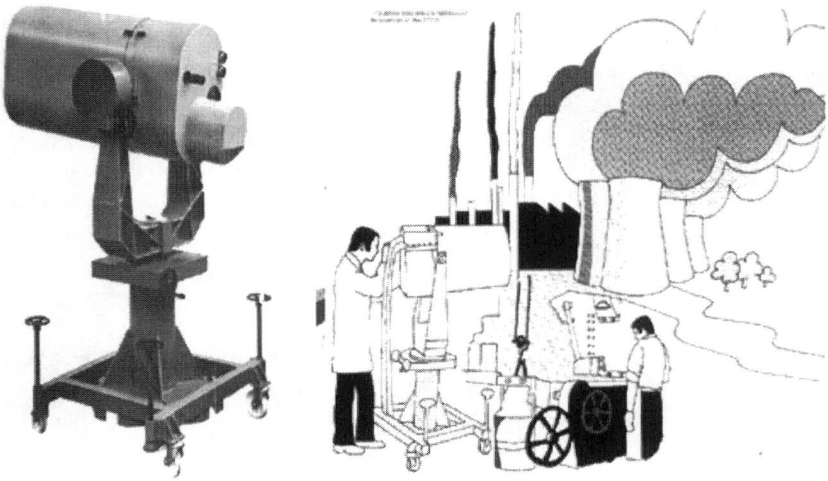

Fig.1-6 As early as 1975, Ferranti (UK) [7] introduced a pulsed telemeter equipped with a λ-tunable laser as the source, called LIDAR at the time, to generate echoes at different wavelengths from the atmosphere and allow measuring pollutants down to ppb concentration and get distance-resolved plots.

Fig.1-7 The demonstration of the Sagnac effect in a square 1-m by side cavity He-Ne laser by Macek at Sperry Corp. in 1962 [8] (from [9], ©IEEE, reprinted by permission).

Such a device (Fig.1-8) has been used in all newly fabricated airliners since the 1980s. It is the heart of modern Inertial Navigation Units (INUs), an electronics box about 10 inches on a side that is tagged at about US$ 4 million. The INU is capable of telling the actual position of the aircraft with 1-mile accuracy after a few hours of flight.

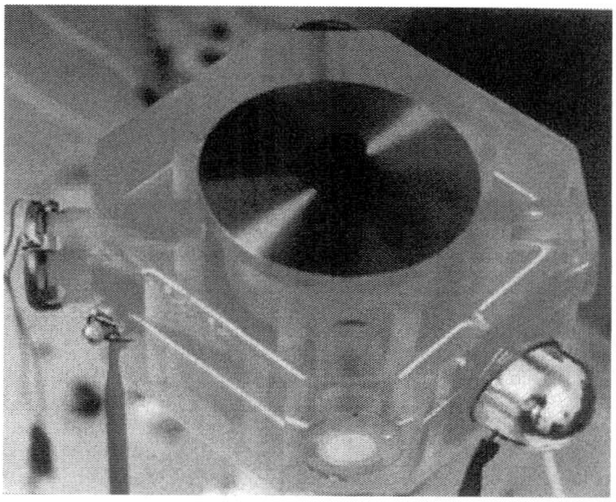

Fig.1-8 A modern 4-inch side ring laser gyro (RLG) is the heart of the Inertial Navigation Unit (INU) of today's airliners (from [9], ©IEEE, reprinted by permission).

The gyro story did not come to an end, however. In 1978, by taking advantage of the newly developed single-mode fibers, the Fiber Optic Gyroscope (FOG) was proposed, and the quest restarted with another technology. The aim was to produce a medium-class light, cheap, and more reliable device especially for space applications.

Again, about a decade of efforts was necessary to cure a series of small idiosyncrasies of the new approach. Units were finally ready in the 1990s (Fig.1-9) for use as a reference aboard telecommunication satellites as an attitude system as well as for military applications.

Thanks to potentially low cost, the FOG was also initially considered as a good candidate in the automotive industry, to demonstrate the car navigation concept (Fig.1-10). Today, it has become clear that, despite the attempts to squeeze the production cost, the FOG is too complex and expensive to meet the targets of performance and price. Yet, continuous efforts to improve performance and the discovery of thermal transient effects (the Shupe effect, see Sect.8.5.3) led the FOG to reach the inertial grade performance.

In addition, because of the interest in the automotive market, a last generation effort of the gyroscope has been pursued using Micro-Optical-Electro-Mechanical System (MOEMS) technology [10]. Starting from accuracy of 1 deg/h adequate for airbag control but far from inertial navigation, this technology is rapidly progressing and likely to offer new breakthroughs in the next years.

Going back to automotive applications, the holy grail is automatic drive (AD), based on a telemeter sensing the distance of targets on the road, coupled to a CCD camera aimed at the field-of-view and feeding an AI (artificial intelligence) recognition system to identify the target within a number of classes (e.g., pedestrian, bike, motorcycle, preceding car/truck).

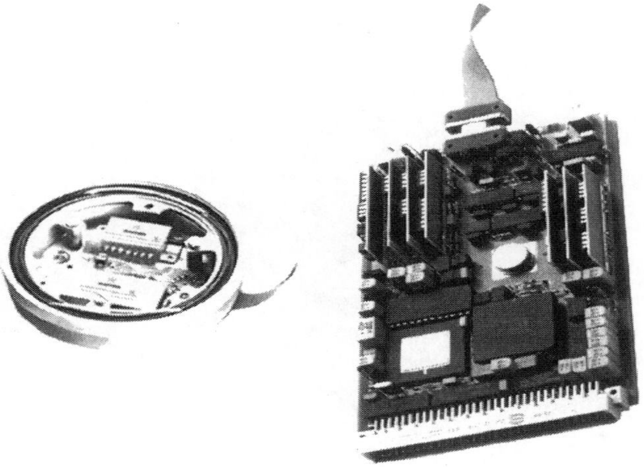

Fig.1-9 One of the first 3-inch diameter fiber optic gyroscope (left) and its printed circuit board for signal processing (right) (model FOG-1B introduced by SEL Alcatel in 1990).

Fig.1-10 The gyroscope has opened the way to car navigators and modern adaptive cruise control, like in this early unit developed by Magneti Marelli Sistemi Elettronici in 1995.

The AD system also includes side detection sensors for tracking the road. This technology has been pursued by a European Community STREP program (2003–08), resulting in products that have reached satisfactory performances as class 3 systems, the one allowing hands off, eyes off.

Manufacturers offering a legally approved Level 3 vehicle are Honda (from 2020), and Mercedes-Benz (from 2021). In addition, Waymo-Chrysler Pacifica minivan was the first commercial raid hailing service, opened to public in Arizona in 2020 of (Fig.1-11), although with a remote human supervising.

Fig.1-11 The Waymo-Chrysler taxi cab has been the first example of a totally automatic, driver-less vehicle for the general public that started service in Phoenix, AZ, in 2020 (from [11], by courtesy of the Waymo).

Central to automatic driving is the laser telemeter, also called *LiDAR* (Light Detection and Ranging, see Chapter 3), which is integrated in the headlight compartment (Fig.1-12) together with the high- and low-beam lights, conveniently obtained by a blue-laser and phosphor combination.

Adding a MEM-mirror device for beam steering, a number of functions can be implemented, like deflecting the beams and switching from high to low when coming across another vehicle, and this new technology is called the intelligent headlight.

Interferometry is another important flagship in photonics instrumentation.

As soon as the first He-Ne lasers were stabilized in wavelength at $\Delta\lambda/\lambda = 10^{-9}$ or better in the mid-1960s, several scientific as well as industrial applications were developed, taking advantage of the unprecedented long coherence length made available by this source.

Well-known examples, and huge commercial successes, have been the so-called *Laser Interferometer* and the *Laser Doppler Velocimeter* (or LDV).

These instruments soon penetrated and captured the markets of mechanical metrology and machine-tool calibration and the fluidics and anemometry engineering segments, respectively. At sale levels of 10,000 units per year, they would justify a place for photonic instrumentation by themselves.

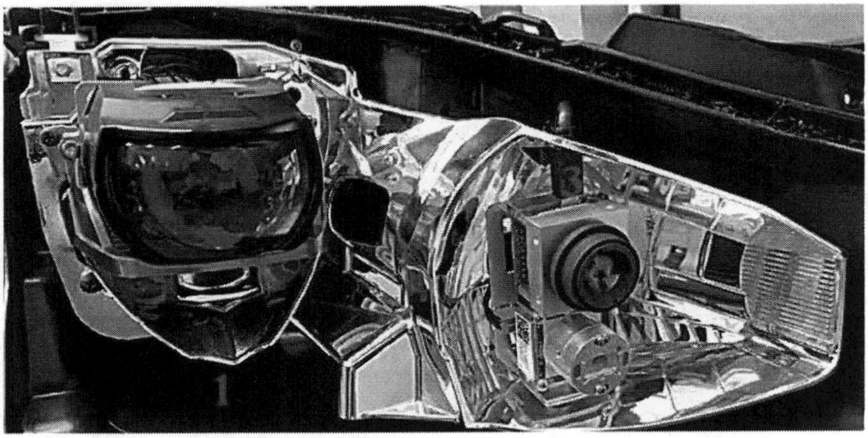

Fig.1-12 Incorporated in the car high- and low-beam light package (1 and 2) of an intelligent headlight, we find a telemeter or LiDAR (3) for target distance measurement (from [12], ©Optica Publ., reprinted by permission).

In addition, the synergy of optics and electronics has pushed the field of interferometry well beyond the fringe (or micrometer) performance of the classical eighteenth-century optics. Picometers are well resolved in displacement, as well as 10^{-9} strains, or relative Dl/l variation.

After the early detection of earth-crust tides in the 1960s at a level of 10^{-9} strain, a big scientific enterprise was completed in these years, namely the interferometer detection of gravitational waves coming from remote galaxies and massive collapsing stars.

A number of teams are developing gravitational antennas around the world (in the United States, Europe and Japan) under the names of Laser Interferometer Gravitational Observatory (LIGO, see Fig.1-13) [13], Laser Interferometer Gravitational Antenna (LIGA), etc.

After several years refinements and tuning of the instrument, finally in 2015 the gravitational collapse of two orbiting black holes was observed, with their gradual increase of the orbital frequency up to some kilohertz before the final merging (see Fig.1-14).

The discovery was awarded the Nobel Prize in Physics in 2015.

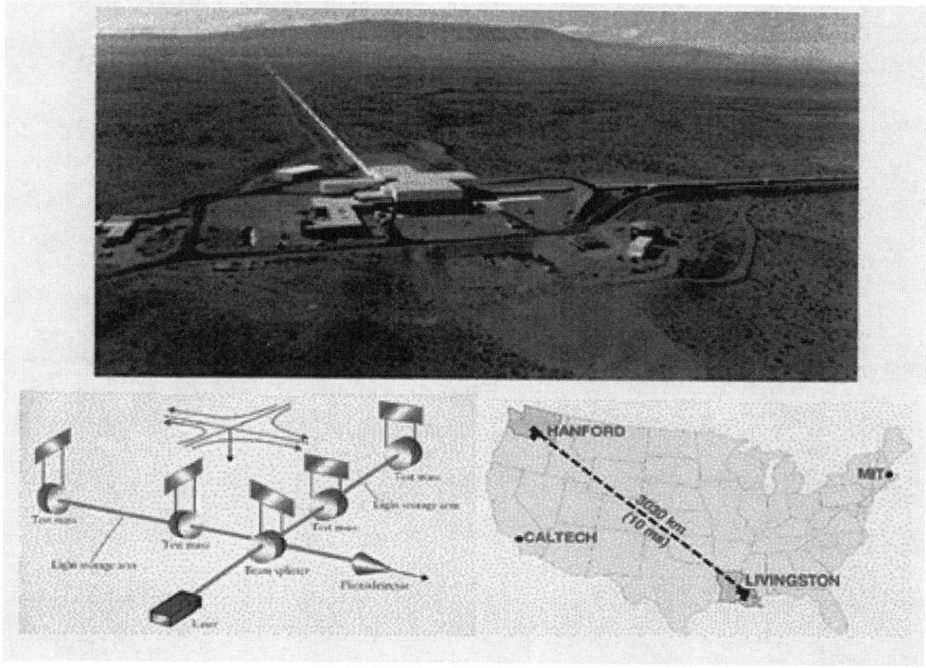

Fig.1-13 The LIGO interferometer has 2-km arms (top: the facility; bottom left: the optical layout). The instrument can detect the gravitational collapse of a double star or black hole a Megaparsec away. To get rid of spurious pulses due to local perturbations, two interferometers are located 3000 km apart and operate in coincidence (bottom right) [by courtesy of Caltech/LIGO].

Since then, many more gravitational collapse events have been recorded by LIGO and other gravitational antenna instruments, totaling 90 in 2021 (see Fig.1-15).

So, The LIGO instrument is opening a new window to probe the cosmos that is comparable in importance of observational results, perhaps, with the advent of the radio telescope.

Next steps to exploit interferometers for astronomy are under way [14]. The ESA Darwin project is aimed at discovering and analyzing earth-size planets orbiting stars up to 50 light-years away.

It is based on an interferometer placed in orbit around the earth in the stable Lagrange points, using a constellation of four to nine spacecraft carrying three or four 2- to 3-m diameter mirrors and an optical beam combiner for the synthetic aperture operation of the assembly, which is then equivalent to have a 30- to 50-m diameter telescope.

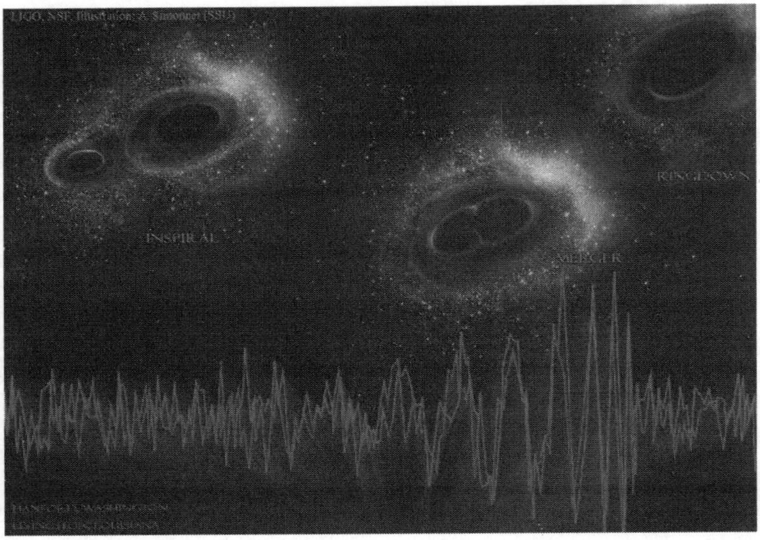

Fig.1-14 In September 2015 LIGO detected in both locations (traces at bottom) the gravitational waves emitted by the collapse of a binary black hole [by courtesy of Caltech/LIGO].

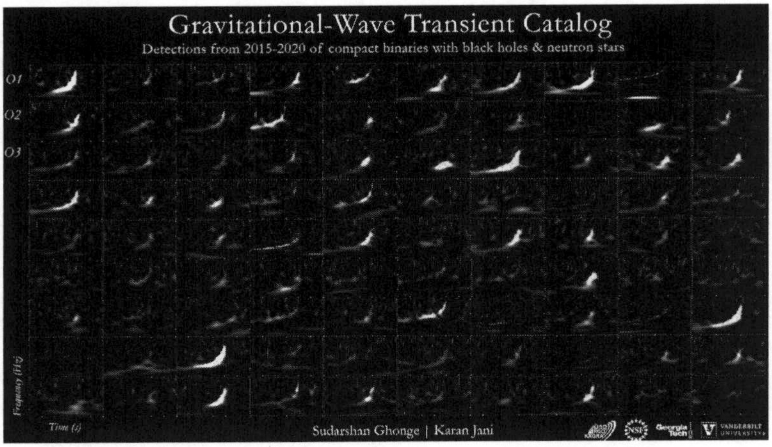

Fig.1-15 Spectrograms of 90 gravitational-collapse events recorded by LIGO in the period 2015 to 2020. In each panel, frequency of revolution is plotted versus time, and trace brightness is proportional to wave amplitude. Common to all panels is the sharp increase in frequency as the two bodies approach closely, followed by sudden disappearance when collapse is complete [by courtesy of Caltech/LIGO].

Another successful application of interferometry is in the biomedical area, with the *Optical Coherence Tomography* (OCT) technology (see Ch.4). OCT is based on a broadband spectrum source, that is, one with very short coherence length l_C (down to micrometers). The interference signal can only be obtained when the interferometer arms are balanced, or differ by less than l_C. Thus, by scanning the reference mirror we obtain a depth-resolved profile of the object under test. Illuminating the object with a light spot, each pixel of the image provides the depth profile at the collecting photodetector (typically a CCD) and thus we have a kind of three-dimensional microscope with depth (z-axis) resolution l_C and spatial (x and y axes) resolution set by the diffraction of the observing objective.

OCT was initially developed for surface diagnostics in the manufacturing industry, then found even more important application in medicine [15] for observing skin melanoma, eye cornea, and blood vessel walls; an example of a raw image is reported in Fig.1-16.

Thus, though only a few decades old, photonic instrumentation has already achieved remarkable scientific and technical successes, has proven the advantages of synergy and of optics and electronics, and continues to grow in ideas and achievements.

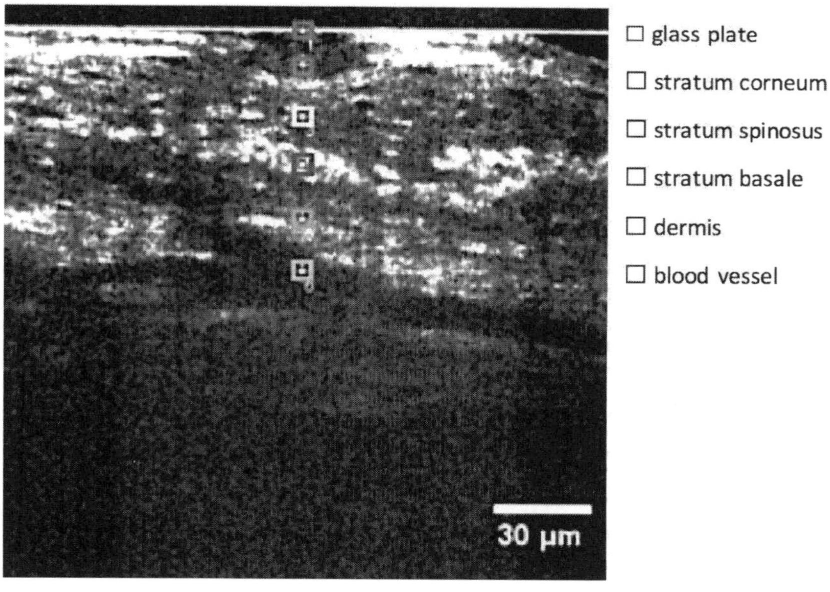

Fig.1-16 A typical *in-vivo* OCT imaging of the forearm skin. Details of skin structure (strata, dermis and blood vessel) are evidenced down to a depth of about 150 μm. Spatial resolution is 0.9 μm and 1.25 μm in the transversal and axial (or depth) direction (from [16], ©IEEE, reprinted by permission).

In the years to come, this field will certainly continue to offer significant potential to researchers and engineers from the point of view of advancement in science, as well as of engineering development of new products.

REFERENCES

[1] C.O. Alley et al.: "*The LURE Experiment: Preliminary Results*", Science, vol.167 (1970), p.368.

[2] L. Ampane, E. Balocco, E. Borrello, and G. Innocenti: "*Laser Telemetry for Automotive Applications*", in: Proc. LEOS Conf. ODIMAP II, ed. by S. Donati, Pavia, 20-22 May 1999, pp.179-189.

[3] AGA Geodimeter model 710, Publication 571.30006 2k8.71 (1971).

[4] M.T. Zuber and D.E. Smith: "*The Mars Orbserver Laser Altimeter Investigation*", J. Geophys. Res., vol.97 (1992), pp.7781-7797.

[5] D.E. Smith and M.T. Zuber: "*The MOLA Investigation of the Shape and Topography of Mars*", Proc. LEOS Conf. ODIMAP III, ed. by S. Donati, Pavia, 20-22 Sept. 2001, pp.1-4.

[6] G. Bianco and M.D. Selden: "*The Matera Ranging Observatory*", in: Proc. LEOS Conf. ODIMAP II, ed. by S. Donati, Pavia, 20–22 May 1999, pp.253-260.

[7] Ferranti 700 Series Lidar System, Publication DDF/524/675 (June 1975).

[8] W.M. Macek and D.I.M. Davis: "*Rotation Rate Sensing with a Travelling Wave Laser*", Appl. Phys. Lett., vol.2 (1963), p.67-68.

[9] G.J. Martin: "*Gyroscopes May Cease Spinning*", IEEE Spectrum, vol.23 (Oct.1986), pp.48-53.

[10] E.A. Brez: "*Technology 2000: Transportation*", IEEE Spectrum, vol.37 (Jan. 2000), pp.91-96; see also: L. Ulrich: "*2014 Top Ten Cars*", IEEE Spectrum, vol.51 (Apr. 2014). pp. 38-47.

[11] T.B. Lee: "*Waymo finally launches an actual public, driverless taxi service*", Ars Technica, Oct.2, 2020, arstechnica.com/cars/2020/10/waymo.

[12] S. Donati, W.-H. Cheng, C.-N. Liu, H.-K. Shih, and Z. Pei: "*Integration of LiDAR and Smart lLaser Headlight in a Compact Module for Autonomous Driving*", OSA Continuum, vol.4 (2021) pp.1587-1597.

[13] B.C. Barish and R. Weiss: "*LIGO and the Detection of Gravitational Waves*", Physics Today (Oct.1999), pp.44-50.

[14] C.V.M. Fridlund: "*Darwin -The Infrared Space Interferometry Mission*", ESA Bulletin 103, Aug.2000, pp.20-25.

[15] D. Huang, E.A. Sampson, C.P. Lin, J.S. Schuman, W.G. Stinson, W. Chang, M.R. Hee, T. Flotte, K. Gregory, C.A. Puliafito, J.G. Fujimoto: "*Optical Coherence Tomography*", Science, vol. 254(5035), (1991), pp.1178-1181.

[16] R. Soundararajan, T.-W. Hsu, M. Calderon-Delgado, S. Donati, S.-L. Huang: "*Spectroscopic Full-field Optical Coherence Tomography in Dermatology*", Proc. ICB2019, 4th Intl. Conf. on Biophotonics, Taipei, 15-28 Sept. 2019, paper P9.

CHAPTER 2

Alignment, Pointing, and Sizing Instruments

A specific property of laser sources is the ability to supply a well-collimated light beam. Indeed, in free propagation, the beam of a common laser can be visualized at a distance much larger than that of an ordinary flashlight. This is a consequence of the low angular divergence of the laser beam and is exploited in the application of laser pointers and instruments used for alignment operation, like an optical version of the plumb for the vertical direction, but now extended to any desired direction.

The divergence of the laser beam is usually not far away from the small diffraction limit, typically 1 milliradian for a 0.2-mm beam size (at λ=633 nm). The diffraction limit is reached when the source works close to the single-transverse mode distribution, which is particularly true for the He-Ne laser (see Appendix A1.1). In semiconductor diode lasers (DL), (Appendix 1.3) the divergence is M times the diffraction limit, with M ranging typically from 2 to 5, depending on the specific DL structure. Thus, while initially He-Ne lasers were preferred despite the much larger size, diode lasers are now entering in new design because of their compact size, ruggedness and low supply voltage, at a divergence comparable to or not much worse than the He-Ne laser, thanks to improved structures and eventually a spatial filtering of the beam (however at the expense of an M^2 loss in available power).

Alignment is used for positioning objects along a desired direction indicated by the propagation direction of the laser beam. Applications range from construction work and laying pipelines, to assembly of parts of large tool-machines, to leveling heights in construction works and to checking terrain planarity in agriculture. In the last two applications, the concept is extended by making a laser level that uses a rotating fan beam and defines a reference plane for alignment.

Another class of instruments exploiting the good spatial quality of the laser beam, as obtained by single transverse-mode operation (or M^2 close to 1), is dimensional measurement by diffraction. The object under measurement is positioned in the beam waist of the laser beam so that to get a plane wavefront of illumination. Light scattered by the object has a far-field diffracted profile precisely related to the dimensions of the object, so the measurement of it can be developed by analyzing the signal collected by a detector scanning the far-field distribution. Examples of noncontact measurement discussed later in this chapter are wire diameter and fine-particle sizing.

2.1 ALIGNMENT

In an alignment instrument, we want to project the beam at a distance of interest and be able to keep its size the smallest possible along the longest possible path.

Because the single-transversal TEM_{00} mode distribution obtained by a He-Ne laser has the least diffraction in propagation, this has been the preferred source traditionally. Wavelength of operation is chosen in the visible at λ=633 nm for direct sight of the beam or detection with a Si-photodiode.

Semiconductor lasers are used as well, and we shall correct the elliptical near-field spot by means of an anamorphic objective lens circularizing the output beam.

Let us now consider the propagation of the Gaussian beam emitted by the laser (see Appendix A1.1), as in Fig.2-1.

The Gaussian beam keeps its distribution unaltered in free propagation and in imaging through lenses, and the spot-size w is the parameter that describes it.
In free propagation (Fig.2-1, top), w evolves with the distance z from the beam waist w_0 as:

$$w^2(z) = w_0^2 + (\lambda z/\pi w_0)^2 \tag{2.1}$$

In optical conjugation through a lens, the spot size is multiplied by the magnification factor $m=r_1/r_2$, with r_1 and r_2 being the radii of curvature of the wave fronts at the lens surfaces. The radii obey a formula similar to Newton lens formula $1/p+1/q=1/f$, that is:

$$1/r_1 + 1/r_2 = 1/f \tag{2.2}$$

On its turn, the radius of curvature of the wave front, as a function of the distance z from the beam waist, is given by:

$$r = z\,[1+ (\pi w_0^2/\lambda z)^2] \tag{2.3}$$

From Eq.2.3, we have r=∞ for z=0 (at the beam waist), and r = z for $z \gg \pi w_0^2/\lambda$ (in the far field, where the Newton formula holds).

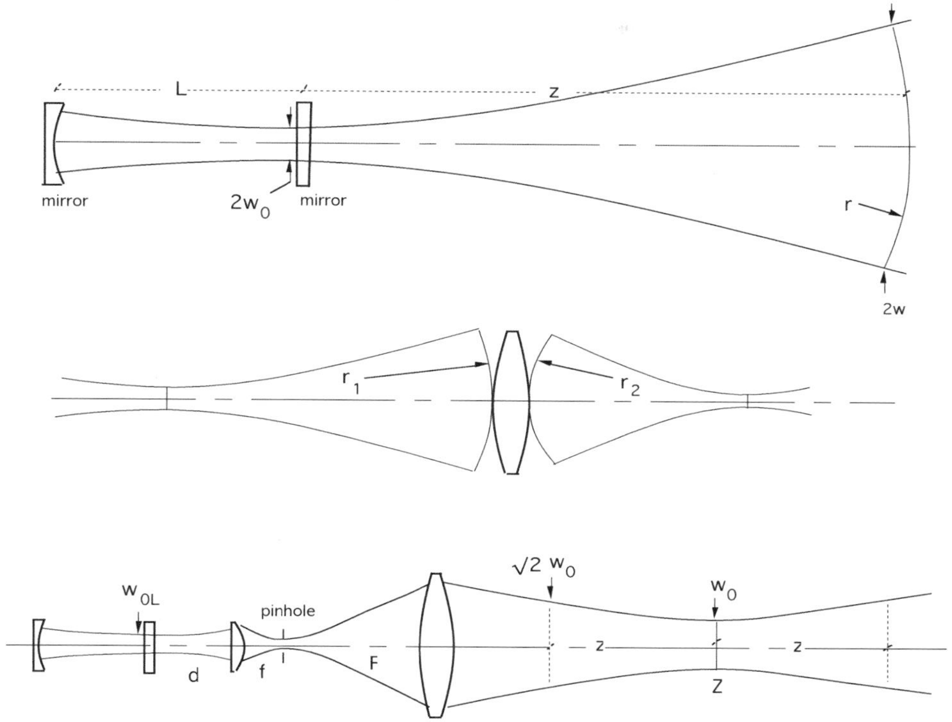

Fig. 2-1 Alignment using the Gaussian beam of a laser. Top: free propagation out of the laser; middle: conjugation through a lens; bottom: projection of the beam through a collimating telescope.

Now, we may look for the optimum waist w_0 that minimizes the spot size w on a given span of distance, let's say between –z and +z. To do that, we find the minimum of w^2 in Eq.2.1 by differentiating with respect to w_0^2 and equating to zero. We thus obtain:

$$\partial w^2/\partial w_0^2 = 1 - (\lambda z/\pi w_0^2)^2 = 0,$$

whence the optimum value of beam waist to be projected at a distance z from the laser is:

$$w_0 = \sqrt{(\lambda z/\pi)} \qquad (2.4)$$

By comparing Eq.2.4 and Eq. A1.5, it turns out that the beam waist minimizing the size of the beam on the desired distance –z...+z is the same as a laser oscillating on mirrors placed at a distance 2z apart.

Fig. 2-2 An early alignment instrument, based on a low-power He-Ne laser and collimating telescope, and (right) the application to installation of pipelines. This unit was introduced in 1978 by LaserLicht, Munchen.

In the span from -z to +z, the beam size is minimum at the midpoint, $w=w_0$, and, from Eq.2.1, it becomes $w=\sqrt{2}w_0$ at the ends of the span. To be able to project a beam waist w_0 at distance z from the laser, a collimating telescope is used (Fig.2.1 bottom). The telescope works better than a single lens because it can easily adjust the distance z by a small focusing of its ocular. The magnification $m=w_0/w_{0L}$ is easily calculated as $m=(f/d)(z/F)=Z/Md$ where M is the magnification of the telescope.

Example. For a laser with a plano-concave cavity and L=20 cm, K=5, we get from Eq. A1.5: w_{0L} =0.282 mm as the laser beam waist. The location of the waist is just at the output mirror. If we wish to cover a span of z=±20m, then Eq.2.4 gives w_0 =2 mm as the necessary beam waist. Then we need a magnification of $m=w_0/w_{0L}=7.1$. Taking z=20m and d=10 cm, the collimating telescope shall have a magnification $M=F/f= =z/md=20/7.1\times0.1=28$.

In Fig.2-1 bottom, the pinhole in the telescope filters out spots extraneous to the laser mode, for example, originated by the reflection at the second surface of the laser output mirror. This mirror flat is slightly wedged to prevent the second surface from contributing to reflection into the cavity.

An early example of application of laser alignment for the placement of pipelines is shown by Fig.2-2. Here, a HeNe laser was employed, and alignment was checked by sight. More recently, similar products employing a laser diode have been developed with nearly the same resolution of the HeNe, but with better position sensing thanks to the use of quadrant or PSD sensors.

Nearly the same instrument (see Fig.2-3) was introduced in the mid-1970s in Australia to establish a *marine channel light* helping boats in their approach to maneuver to the harbor and to avoid dangerous sand bars. In the application, the beam was switched periodically by a small angle in a horizontal plane through a prism inserted in the beam path, so as to provide an error signal that helps the boat keep the correct homing direction. Using a He-Ne laser emitting approximately 10 mW in the red (633 nm) and a 10-cm-diameter telescope, the beam can be seen by eye at a distance of several miles.

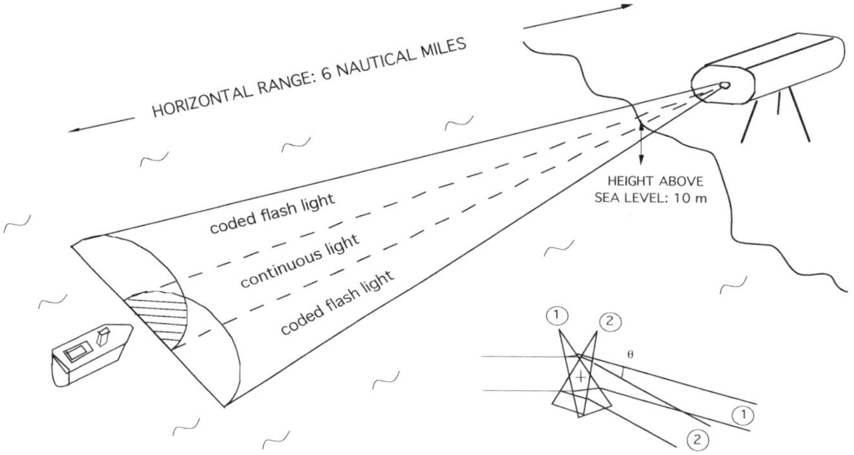

Fig. 2-3 Alignment instrument used as a marine channel light to help ships homing to the harbor. The beam is switched by an angle θ by a prism rotating at approximately 3 Hz and is transmitted by a telescope above the sea surface to the ship. In the region of superposition, light appears steady, whereas outside it flashes.

The actual range is limited primarily by weather conditions, introducing atmospheric attenuation and turbulence (see App.A3). A problem with this early marine channel aid based on direct viewing was laser safety (see Appendix A1.5). To ensure safe viewing, power density is designed to be less than the Maximum Permissible Exposure (MPE) from a minimum distance onward.

More recently, laser technologies for alignment and distance measurement (see telemeters in Ch. 3) have been demonstrated to be a solution competitive with the conventional radar and ultrasound technologies used at the time. In the harbor of Koper (Slovenia), a laser installation (Fig. 2-4) helps berthing and docking operations of large ships, minimizing the risk of collisions with crane, bow, and stern [4].

Fig. 2-4 Laser sensors (small white LDS) installed in the harbor of Koper supply data to the communication network (straight lines) for controlling the docking operation of large ships (from [4], ©Elsevier, reprinted by permission).

Another recent product designed for the consumer market is the Black & Decker BDL-120, an alignment tool for interior decor and home bricolage (Fig.2-5). It uses a simple DL (diode laser) with a collimating objective, and offers 1 mm accuracy on a 1 m span, with beam visibility up to 3 m in daylight. Intrinsically safe, the product is tagged at US$ 20.

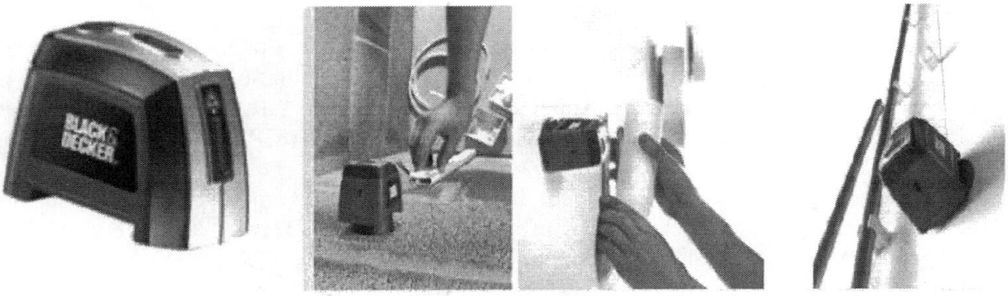

Fig. 2-5 The laser level BDL-120 (left) is an inexpensive tool for bricolage and alignment of objects along the floor and walls of interiors (center and right) [by courtesy of Black and Decker].

2.2 POINTING AND TRACKING

Centering of the laser spot can be performed just by sight, as in construction applications where the resolution we need is a fraction of the beam spot size, or roughly 1 mm.

For more exacting applications, we can use a photodetector to generate an error signal proportional to the alignment error. The photodetector may either be a special type designed for position sensing and pointing purposes, or may be a normal one acting in combination to a rotating reticle or mask establishing a spatial reference for the photodetector.

2.2.1 The Quadrant Photodiode

This device has the usual structure of a normal photodiode (see Ref.[1], Sect.5.2), but one of the access electrodes, for example, the anode in a *pn* or *pin* junction, is sectioned in four sectors (Fig.2-6, left). Light input is from the chip backside when sections are defined by metallization, and from the top when a p^+ layer with side-ring contact is used (Fig.2-6, right). Size of the quadrant photodiode ranges from 0.2 to 2 mm in diameter. The gap between sectors is made as small as possible and is usually 5 to 10 μm.

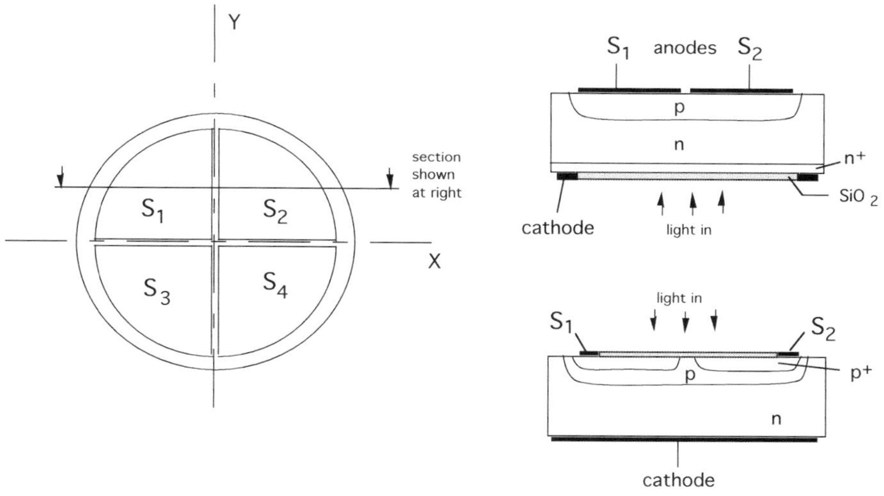

Fig.2-6 A quadrant photodiode for pointing applications. Left: the segmented electrodes structure; right: section of the *pn* photodiode structure with metal electrodes for backside light input (top) and section of a p^+pn structure for front light input (bottom).

Light impinging on a certain sector originates a photocurrent collected by the correspondent electrode. Thus, we have four signal outputs, S_1, S_2, S_3, and S_4, from the four-quadrant photodiode.

We can compute two coordinate signals S_X and S_Y as:

$$S_X = (S_2+S_4) - (S_1+S_3) \quad \text{and} \quad S_Y = (S_1+S_2) - (S_3+S_4) \tag{2.5}$$

A practical circuit that interfaces the four-quadrant photodiode and computes the S_X and S_Y signals is shown in Fig.2-7.

Based on operational amplifiers, the circuit terminates the four anodes of the photodiode on the virtual ground of transresistance preamplifiers ([1], Sect.5.3), that is, on a low-impedance value useful for enhancing the high-frequency performance.
The common cathode is fed by the positive voltage ($+V_{bb}$) to reverse-bias the photodiodes. Each transresistance stage converts the photodetected current I_{ph} in a voltage signal V_u=-RI_{ph}, where R is the feedback resistance of the op-amp. With the four voltages available, it is easy to carry out the sum and difference operation (Eq.2.5) with the op-amps arranged as shown in the second stage.

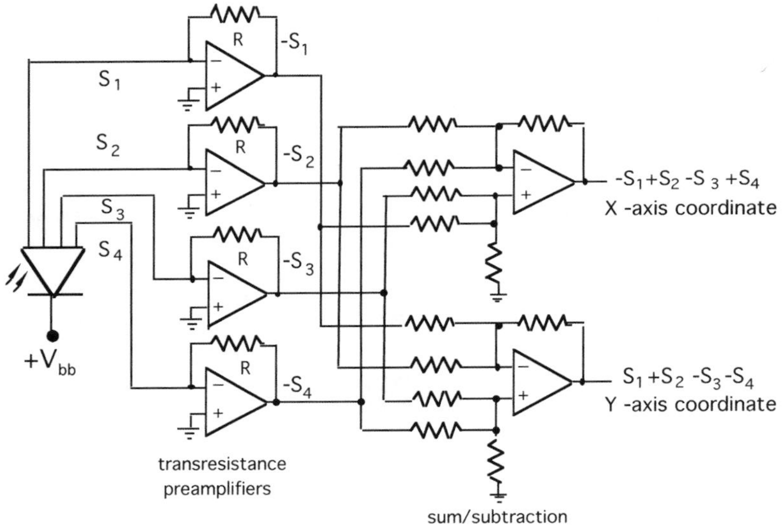

Fig.2-7 Op-amp circuit for computing by op-amps the coordinate signals S_X and S_Y from the outputs of the four-quadrant photodiode.

To analyze the dependence of signals S_X (or S_Y) from the coordinate X (or Y), let us suppose that a probe-spot scans the photodiode along a line parallel to the X axis. The spot will be smaller than the photodiode usually, but finite in size. Let X represent the coordinate of the spot center, and let R_{ph}, R_s be the photodiode and spot radii, while w_{dz} is the dead-zone width.

In Fig.2-8 we plot the coordinate signal S_X versus X. As can be seen, when X is outside the photodiode by more than the spot radius, the signal is zero; when X is well inside the

photodiode, the signal attains its maximum value, and when X approaches zero, the signal decreases and reverses its sign, as indicated by Eq.2.5.
Thus, the response around the X=0 is nearly linear (as shown in Fig.2-8), and the width of the linear regime, before S_X reaches saturation, is about $\pm R_s$.
In *tracking* applications, the outputs from the position sensor are used as the error signals to control the system motion. In this case, the dependence shown in Fig.2-8 is just the desired one to get a good control action, both in acquisition and regulation regions:

- in the *regulation region* (near X=0), we need a corrective action proportional to the error so that the equilibrium point near X=0 remains smooth (without oscillations).
- in the *acquisition region* (large X), the signal shall saturate so that the tracker is driven toward X=0 at the maximum speed.

The quadrant photodiode and other position-sensitive devices described later can be used as a *position sensor* that measures the X and Y coordinates of the impinging light spot.
In this type of sensing, we will mount the quadrant photodiode on a positioning stage with reference marks to set the X=0 and Y=0 position.

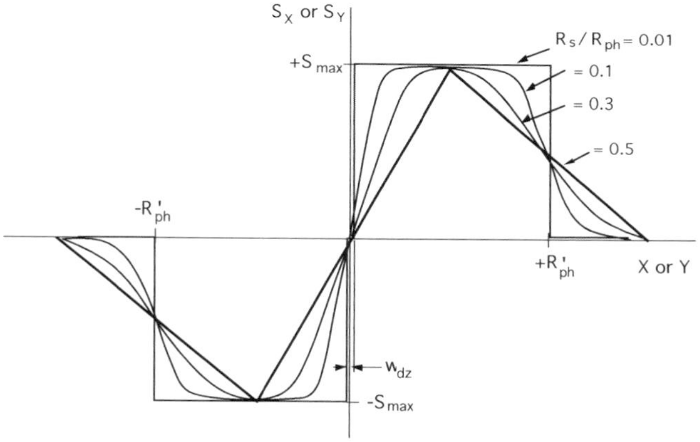

Fig.2-8 Dependence of the coordinate signal S_X (or S_Y) from the coordinate X (or Y) of a light spot falling on the device. When the spot size R_S is very small (R_S/R_{ph}=0.01 or less), the response is a square wave with a small dead band near X=0 (or Y=0). When R_S is not so small, the response is smoothed, and we find an almost linear response close to X=0 (or Y=0). The width of the linear region is $\pm R_S$. The dead-zone width w_{dz} only affects the response for a very small R_S/R_{ph} ratio.

The error (σ_X or σ_y) of the coordinate measurement depends on several parameters, such as optical power in the spot, spot shape and its fluctuations, and photodiode dimensions like R_{ph} and w_{dz}. In a well-designed device, typical values are in the range σ_X =0.01 to 0.1 R_{ph}.

The quadrant photodiode becomes an *angle sensor* when we place it in the focal plane of an objective lens as indicated in Fig.2-9. The sensor will then measure the angular coordinates θ_x and θ_y of the spot presented in its field of view. Because of the optical conjugation, the coordinates θ_x and θ_y are tied to the spatial coordinate by $X= \theta_x F$ and $Y= \theta_y F$, where F is the focal length of the objective lens. The error of the angle measurement follows as $\sigma_\theta= \sigma_X/F$.

2.2.2 The Position Sensing Detector

The Position Sensing Detector, or PSD, is a multi-electrode photodiode similar to the four-quadrant photodiode. Developed as an evolution of the four-quadrant concept, the PSD offers much better linearity over the entire active area and a response curve independent of the spot size and shape.

Actually, there are mono- and two-dimensional PSDs for sensing one or two coordinates. In the following, we describe the two-dimensional or array PSD only, because the mono-dimensional or linear PSD follows as a simpler case of the two-dimensional.

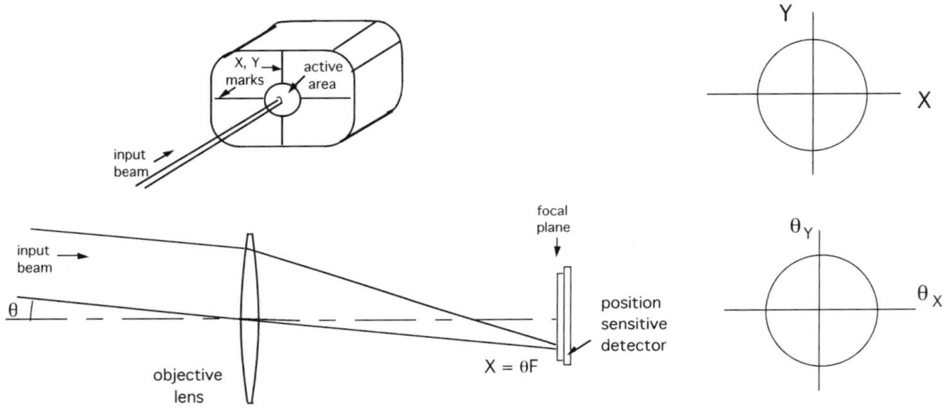

Fig.2-9 A position-sensing photodiode (either four-quadrant or reticle) can be used as an X-Y coordinate detector (top) or, when it is placed in the focal plane of an objective lens, as an angle-coordinate detector (bottom). In this case, the field of view is $\theta = R_{ph}/F$.

The basic structure of a PSD (see Fig.2-10) is that of a normal *pin* junction with thin *p* and *n* regions. The electrodes for the X and Y coordinates are metal stripes deposited on opposite faces of the chip. Electrically, the stripes are like the normal anode and cathode of a photodiode, yet they are placed at the edges of the p and n regions and sectioned. The doping level of p and n regions is kept much lower than in a normal photodiode to enhance the series resistance of the undepleted *p* and *n* regions facing the junction, contrary to what is commonly done. The typical resistance R between the stripe contacts (S_{x2} and S_{x1}) is about 10-100 kΩ.

Following the absorption of photons at point x,y (Fig.2-10), photo-generated electrons and holes are swept away quickly by the electric field at the junction.

Thus, the current I_{ph} to be collected reaches the undepleted p and n regions without appreciable error in transversal (x,y) position.

The hole current arrives at point x in the p region (Fig.2-10, top) and shall cross series resistances xR and (L-x)R before reaching the electrodes S_{x1} and S_{x2}. The electron current has a similar response with respect to the electrodes S_{y1} and S_{y2}, but has formally the opposite sign ($-I_{ph}$) as it conventionally enters the electrodes.

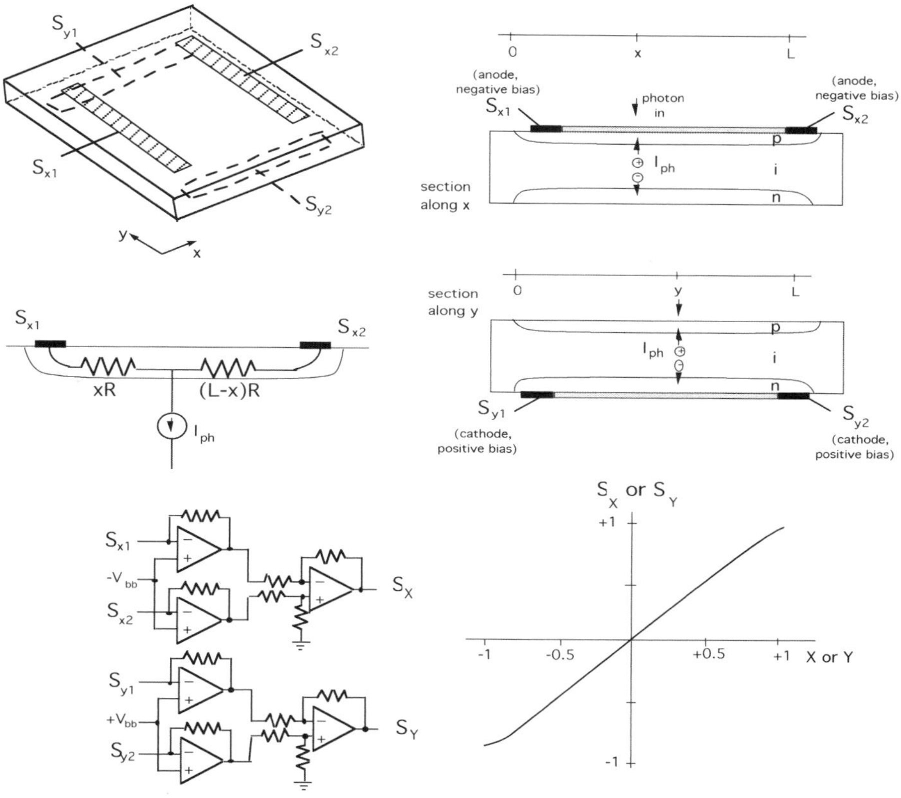

Fig.2-10 Structure of the PSD photodiode is a normal pin junction, with thin *p* and *n* regions, lightly doped or implanted. Light impinging at the point x,y generates a photocurrent that is shared between the X and Y electrodes. Coordinate signals are obtained as the difference between the currents exiting from the electrodes, S_{x2}-S_{x1} and S_{y2}-S_{y1}. To ensure good linearity and to have a negligible dependence from R, the currents are sunk by the virtual ground of a transresistance stage. The non-inverting inputs of the op-amp stage are used to set the bias, $+V_{bb}$ and $-V_{bb}$, of the X and Y stripes.

Because of the series resistance, the current I_{ph} is shared by the electrodes proportional to the divider ratio of the respective conductance $(xR)^{-1}$ and $(L-x)^{-1}R^{-1}$.
Therefore, the currents from S_{x1} and S_{x2} are:

$$I_{x1} = (xR)^{-1}/[(xR)^{-1}+(L-x)^{-1}R^{-1}]\, I_{ph} = (1-x/L)\, I_{ph}$$
$$I_{x2} = (L-x)^{-1}R^{-1}/[(xR)^{-1}+(L-x)^{-1}R^{-1}]\, I_{ph} = x/L\, I_{ph} \tag{2.6}$$

Similarly, the current $-I_{ph}$ arriving in the n region is shared by the S_{y1} and S_{y2} electrodes, and Eq.2.6 holds with y in place of x and $-I_{ph}$ in place of I_{ph}:

$$I_{y1} = (1-y/L)(-I_{ph}), \quad I_{y2} = y/L\,(-I_{ph}) \tag{2.6'}$$

To recover the currents without errors induced by a finite load resistance, the best approach is to terminate the photodiode outputs (S_{y1}, etc.) on the virtual ground of op-amps, as shown in Fig.2-10. With this scheme, we get a very low-resistance termination, and the current I_x is deviated in the feedback resistance where it develops an output voltage $-R'I_x$.

In addition, the use of the op-amp allows an easy biasing of the PSD electrodes. Because of the virtual ground, the inverting input voltage is dynamically locked to that of the non-inverting input, and here we will apply the appropriate bias, positive ($+V_{bb}$) at the cathode and negative ($-V_{bb}$) at the anode. The best for V_{bb} is half the value of the supply voltage ($+V_{BB}$ and $-V_{BB}$) feeding to the op-amps.

With op-amps, it is easy to compute the coordinate signals S_X and S_Y as the differences of the transresistance stage outputs (Fig.2-10, bottom left):

$$S_X = -R(I_{x2}-I_{x1}) = -(2x/L-1)RI_{ph}, \quad S_Y = R(I_{y2}-I_{y1}) = -(2y/L-1)RI_{ph} \tag{2.7}$$

Even better, if we add a voltage-controlled gain stage (not shown in Fig.2.10), we can divide outputs S_X and S_Y by I_{ph} and obtain coordinate signals independent of the impinging optical power $P= I_{ph}/\sigma$ (or, a PSD spectral sensitivity σ).

Often, instead of a fine spot, we may have an extended source as the optical input to the PSD. In this case, the coordinates supplied by the PSD are the center of mass of the extended source. This can be seen by considering that the power p(x) between x and x+dx gives a contribution xp(x), and total current is $\int xp(x)dx = \langle x \rangle$, equal to the mean value of p(x).

Commercially available PSDs have a square active area with a 0.5- to 5-mm side and are mostly in silicon for a spectral response from 400 to 1100 nm.

Linearity error is less than 0.5% within 80% of the active area, and cross-talk of the coordinate signal is usually less than 10^{-4}. Response time is usually limited by the large junction capacitance, $C_b \approx 50$-200 pF, and by the op-amp frequency response. With 100-MHz op-amps, the response time of coordinate signals may go down to ≈ 1–10 µs.

2.3 Laser Level

The laser level is an extension of the concept of point-like alignment described in Section 2.1. When we do not need a direction to follow, or two coordinates (spatial x,y or angular ρ,ψ) as exemplified by the pipeline alignment, but rather a single coordinate (the height z or the horizontal ϕ) with the others free, the laser level is the instrument we need.

The most common application of the laser level is the distribution of the horizontal plane for construction works (Fig.2-11) and for leveling (or slope removing) of cultivation terrain. Also, some products have been developed for the consumer market.

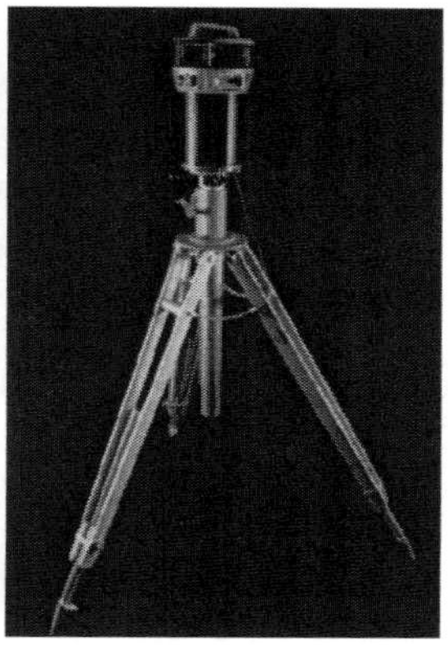

Fig. 2-11 Alignment with a laser level. Left: use of the level for height relief in construction works. A fan beam is distributed down a 20- to 50-m radius area, and its position is checked at sight with the graduated stick. Right: a typical tripod-mounted, laser-level instrument (by courtesy of Spektra).

As shown in Fig.2-11 and 2-12, a laser level can be implemented by a tripod-mounted laser carrying a collimating telescope to minimize the spot size on the range to be covered, as explained in Section 2.1.

The easy way to obtain a fan-shape for the output is to reflect the laser beam with a rotating mirror, oriented at 45 degrees to the vertical. To be a truly horizontal plane, the mirror shall be carefully oriented at 45 degrees, and the beam from the laser shall be oriented vertically. Errors from the nominal angles result in a tilt error of the fan-beam plane.

The mirror angle-error can be eliminated by using a pentaprism in place of the mirror. The pentaprism has a dihedral angle of 45 degrees and therefore (Section A2.2) steers the input beam of 90 degrees, irrespective of the angle of incidence.

Now, we have to adjust the laser beam to the vertical. For maximum simplicity and low cost, a spirit level may be used, incorporated in one tripod leg. However, doing so, the resolution is modest (perhaps ≈10 mrad or 30 arcmin). Yet, we may provide the instrument with a mechanical reference plane perpendicular to the laser beam, and the user will adjust the legs to verticality using a spirit level sliding on this reference plane. For good resolution, however, we may need a relatively bulky spirit level (50-cm side for ≈1mrad).

Alternatively, we may use an optical method to check the verticality of the laser beam. One method to do so is comparing the beam wave vector to the perpendicular of a liquid surface. As illustrated in Fig.2-12, the liquid (water) is kept in a hermetic box, and we use a small reflection (≈4%) of the main laser beam passing through it toward the output. The other glass surfaces of the bow are antireflection coated. Another small (≈4%) portion of the laser beam is picked at the beamsplitter (Fig.2-12 right), which is also antireflection coated at the second surface.

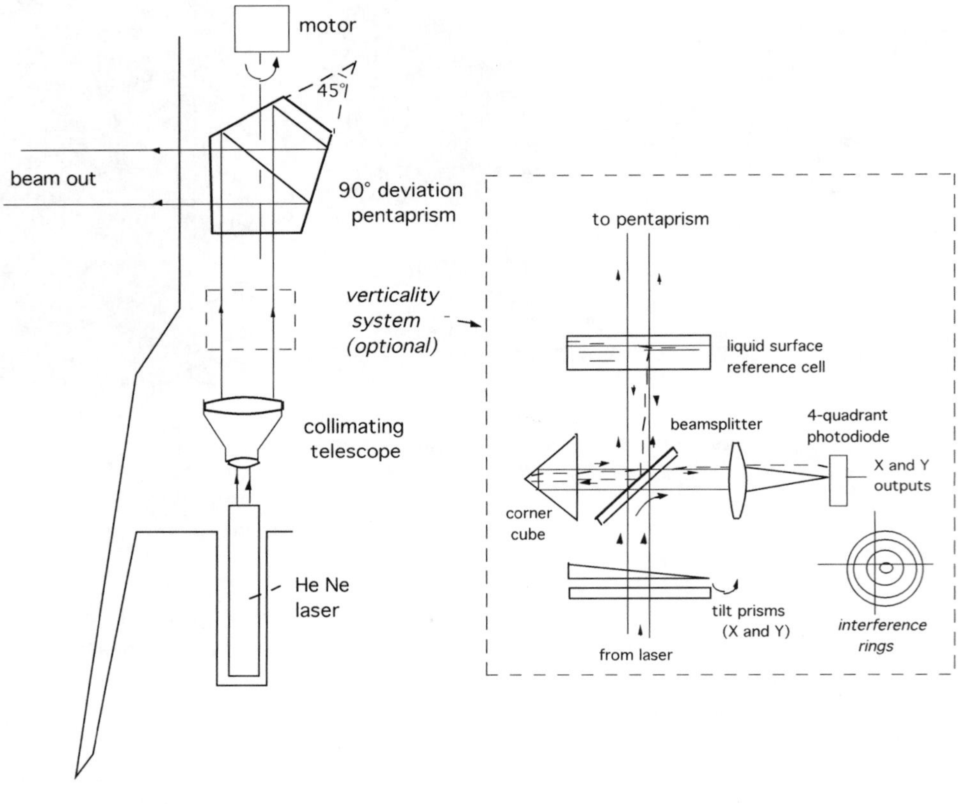

Fig. 2-12 A laser level uses a pentaprism (left, top) to deflect the incoming beam exactly by 90 degree, independent of the angle of incidence. To check verticality, a partial reflection of the laser beam from the surface of a liquid in a cell is superimposed on part of the beam itself and is reflected backward. Looking at the interference rings generated by the superposition with a four-quadrant detector brings out the error (right). Tilt prisms are used for providing the automatic compensation of the error.

The two beams are recombined by the beamsplitter and come to an objective lens acting as the input of an angle-sensing detector (Fig.2-9). An angular error of verticality produces a tilt in the wave vector reflected by the liquid surface, and then the two beams come to the focal plane of the objective with different directions. Thus, they will produce interference rings in an offset position with respect to the optical axis of the angle-sensing detector (Fig.2-12, bottom right). The detector supplies the coordinate signals θ_X and θ_Y of the center of the interference figure. These signals can be used directly, displaying them to the user, for a manual adjustment of the legs of the tripod, or they may be fed to actuators moving a pair of deflecting prisms to adjust the beam and dynamically lock its direction to the vertical.

In agriculture, a popular application of the laser levels is terrain leveling. Some crops, and rice in particular, require a large quantity of flooding. To save water, leveling of terrain within 1 cm on 1 acre (or 100 ton of saved water) is demanded, see Fig.2-13.

Fig. 2-13 Left: a laser level for terrain leveling and (right) its use in the field (by courtesy of Spektra).

2.4 Wire Diameter Sensor

In industrial manufacturing, instruments for measuring dimension are common for testing and online control purposes. In particular, two laser-based instruments for noncontact sensing are successfully employed: the wire diameter sensor and the particle size analyzer. Both instruments deal with relatively small-dimension D, in a range such that λ/D is favorable for exploiting diffraction effects.

Diameter instruments are used mostly as the sensor for process control in wire manufacturing. The wire may be a metal or a plastic wire extruded from a strainer or drawplate, or an optical fiber, or the feed for a 3D printer. The real-time measurement allows feedback on process parameters, and tolerance control of the fabricated wire or 3D printing process.

The wire to be measured is passed inside an aperture of the instrument (Fig.2-14), and illuminated by the beam waist of an expanded laser beam. For good beam quality, we use a He-Ne laser and a collimating telescope.

For the illuminating field, the wire is an obstruction or diffraction stop, and, in view of the Babinet principle, it is equivalent to a diffraction aperture. The diffracted field is collected by a lens and its distribution is converted into an electrical signal by a photodetector placed in the focal plane of the lens.

From this signal, the unknown diameter D can be calculated.

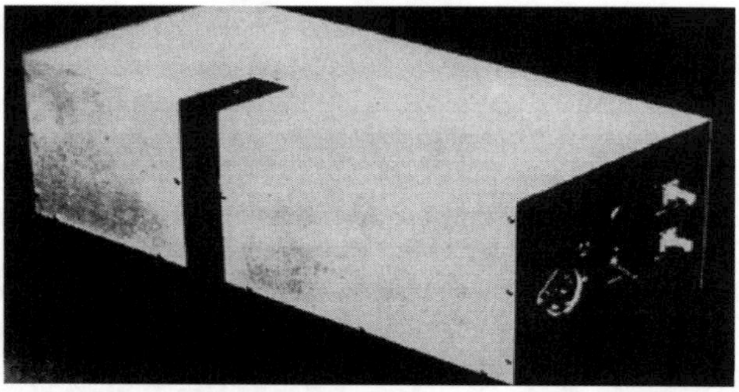

Fig. 2-14 An early noncontact sensor for diameter measurement. Top: the instrument envelope, with the U-slit for wire passage; bottom: internal details include the He-Ne laser (center left), the collimating telescope (bottom left), the receiver (bottom right), and the electronics box.

As it is well-known from diffraction theory (see Appendix A4.3), the distribution of diffraction from the wire is the Fourier transform of the wire aperture. Assuming a constant illuminating field E_0, the field and the radiant power density diffracted at the angle θ are given by:

$$E(\theta) = E_0 \quad \text{sinc}\ \theta D/\lambda \tag{2.8}$$

$$P(\theta) = E_0^2 \quad \text{sinc}^2\ \theta D/\lambda, \tag{2.8'}$$

In these expressions, sinc ξ =(sin $\pi\xi$)/$\pi\xi$. The first zero of the sinc distribution is at $\xi=\pm1$, which corresponds to a diffraction angle $\theta_{zero} = \pm\ \lambda/D$.

To work out D, we first manage to make available the $P(\theta)$ distribution and then look for the measurement best suited to extract D from the distribution.

We use a lens (Fig.2-15) to convert the angular distribution P(θ) in a focal plane distribution P(X) to be scanned by the photodetector. The scale factor is X=Fθ, with F being the focal length of the lens.

Usually, we need a long focal length F for a sizeable X, and yet we want to keep the overall length short (much less than F).

As an example, if D=200 μm and λ=0.5 μm, we need F=400 mm to have a zero at X=±1mm. A good choice is a telephoto lens [3] with two elements, a front positive and a rear negative (or Barlow's focal multiplier). By this, an F=400-mm lens may be only 50-mm long. On the source side (Fig.2-15), a collimating telescope is used to expand and collimate the laser beam, which makes a nearly plane wave front available in the measurement region.

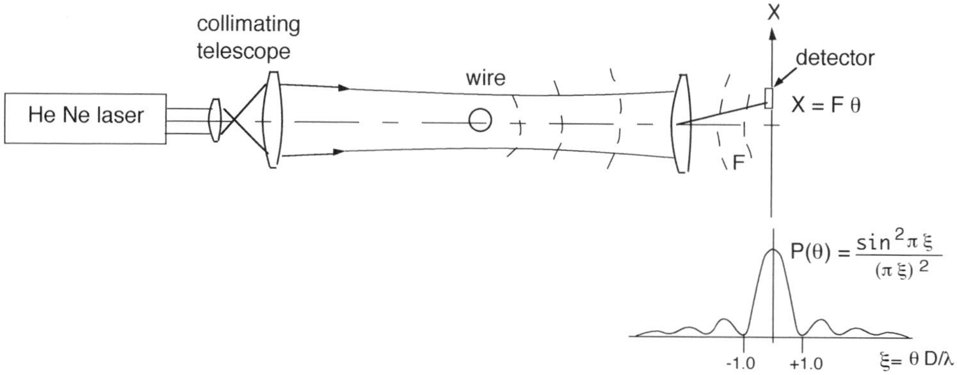

Fig. 2-15 In a wire-diameter sensor, the wire is illuminated with a laser beam, properly expanded and collimated. The far-field distribution of diffracted light is converted by a lens to a spatial distribution in the focal plane. By scanning the focal plane with a photodetector, we reveal the zeros of the distribution, located at X/F = θ = ±λ/D, hence the measurement of D.

The size of the expanded beam D_{beam} is usually kept several millimiters to a few centimeters, according to the user's specifications.

With a larger size, a larger transversal movement is tolerated, and the errors associated with the beam waist non-idealities are reduced. On the other hand, a larger size means less power on the wire and therefore a weaker signal and S/N ratio at the detector.

To measure D from the experimental distribution $P(X)= P_0 \text{ sinc}^2 XD/F\lambda$, we may in principle look at several different features in the waveform P(X), for example, the main pulse width at half maximum, the secondary-maximum positions, the flex points, etc.

However, we shall take into account two non-idealities in the P(X) detection: (i) the finite size of the detector W_{det}, by which the P(X) distribution is averaged on W_{det} and (ii) the relatively strong peak coming from the undiffracted beam and superposed to the useful signal near the origin X=0 (this contribution is not shown in Fig.2-15).

If D_{beam} is the beam diameter at the waist illuminating the wire, the undiffracted peak has a width of the order of $F\lambda/D_{beam}$ and an amplitude D_{beam}^2/D^2 times larger than the wire diffraction amplitude.

Then, a better feature to look at in the P(X) distribution is the location of the first zeroes near the peak. These are positioned at $\theta=\pm\lambda/D$, or $X=\pm F\lambda/D$. We may detect them conveniently with a linear-array Charge Coupled Device or CCD (see Ref. [1], Sect. 12.2) placed in the focal plane.

The CCD is a self-scanned detector that provides a signal $i(t)= i_0\ \mathrm{sinc}^2\ vtD/F\lambda$, where v is the scanning speed. By time differentiating the signal i(t), we get an i'(t) signal, which passes through the zero level at the minima of the sinc^2 function.

Further, we can arrange a digital processing of the measurement as follows: a threshold discrimination of the i'(t) signal picks up the zeroes at time $t=\pm F\lambda/vD$ and gates a counter to count clock transitions between two start-and-stop triggers. If the clock frequency is f_c, the number of counted pulses is $N_{count}=2f_cF\lambda/vD$, and we have a digital readout of the diameter D. However, we still require calculating a reciprocal.

In commercially available instruments, the range of measurable diameters extends from about 10 μm to 2 mm, and the corresponding accuracy is 1% and 5%, respectively, at the border of the measurement range.

Reasons for limits in the measurement are as follows. At large diameters, the diffraction angle becomes small, and the P(θ) is difficult to sort out, whereas errors due to illuminating wave front planarity and detector finite size become increasingly important. At the lower end of the diameter range, where diffraction angle is conveniently large, limits are due to the approximations in P(θ) and to vignetting effect in the collecting lenses. It is worth noting that the user's specifications set a requirement on small diameters at ≈50 μm because very small wires are uncommon in industrial manufacturing.

As a central design value for the instrument, the allowed transversal movement of the wire is typically 2 to 5 mm with no degradation of specifications. The overall size of the diameter-sensing instrument (Fig.2-14) is typically 50 by 20 by 10 cm^3.

Variants of the above-described progenitor of diameter-sensing instruments are (i) units using a diode laser as the source, leading to a more compact and cheaper unit at the expense of some performance sacrifice; (ii) units providing double-beam measurements along orthogonal axes to measure ellipticity of the wire; (iii) units for large diameter, based on measuring the shadow cast by illuminating the wire with collimated light by means of a linear CCD, to cover the range 0.1 to 30 mm with 2- to 5-μm resolution and error.

2.5 PARTICLE SIZING

Another instrument based on diffraction is the particle-size analyzer. With this instrument, we can measure the distribution of particle diameters in a fine-conglomerate mixture. Mixtures interesting for particle sizing range from manufacturing and industrial processes (ceramic, cement, and alloys powders) to particulate monitoring in air and water and sorting of living material for medical and biological purposes.

The first commercial particle-size analyzers date back to the 1970s [5]. They were developed originally to measure the size distribution of ceramic and alloy powders for control in manufacturing. Since then, the field has steadily matured, and the new products accepted in a variety of applications. The measurement range now extends from the sub-μm particulate to relatively large particles, up to a few millimiters.

The basic scheme of a particle-size analyzer is shown in Fig.2-16. This scheme is similar to the wire sensor, and with it we look at the small-angle diffraction or elastic scattering of light from the particles. Because of this, the scheme is classified as *Low-Angle Elastic Light Scattering* (LAELS) [6].

As a source, we may use a low power (typ. 5mW) He-Ne or diode laser. The collimating telescope expands the beam and projects a nearly plane wave front on the measurement cell that contains the particles to be measured.

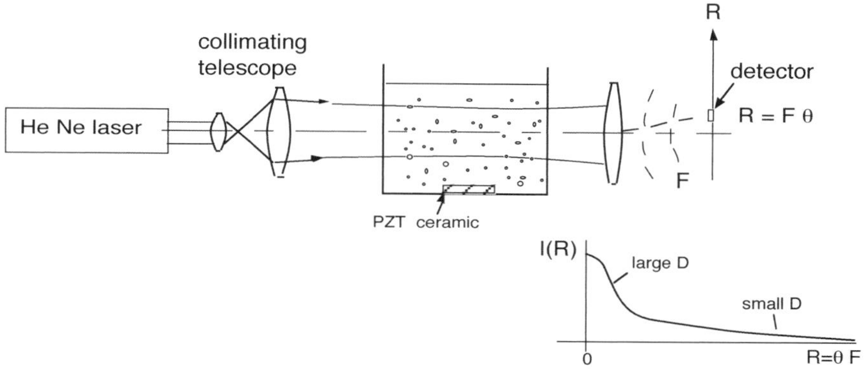

Fig. 2-16 A particle-size analyzer is based on the diffraction of light projected on a cell containing the particles in a liquid suspension. At the output of the cell, a lens collects light diffracted at angle θ and converts it in a focal plane distribution I(R), with R=θF. Then, I(R) is measured by the photodetector, and the desired diameter distribution p(D) is computed.

In the cell, the particles are kept in suspension by means of a mechanical stirrer or an ultrasonic shaker (the PZT ceramic indicated in Fig.2-16). In this way, we avoid flocculation, or, the sticking together of particles.

If the cell is relatively thin, the wave front can be assumed plane throughout the cell, and particles are illuminated by the same wave vector. Each particle contributes to the far field distribution collected at the angle θ according to Fourier transform of an aperture of diameter D, as given by Eqs.A4-16 and A4.16'. The contribution is A_0 somb $[(D/\lambda) \sin\theta]$ in field amplitude and I_0 $\text{somb}^2[(D/\lambda) \sin\theta]$ in power density, where somb $x = 2J_1(\pi x)/\pi x$ (see App. A4.3). The total field amplitude and power density are the integrals of somb and somb^2 terms, weighted by p(D), the distribution of diameters in the particle ensemble.

In summing up the contributions, we shall recall that particles are randomly distributed in the cell, and their positions fluctuate well in excess of a wavelength. Because of this, the interference terms originated by the superposition of field amplitudes are averaged to zero. Thus, in the far field of the cell, we shall add the intensities or the squares of the fields contributed by individual particles.

Using Eq.A4.16' and integrating on the diameter distribution, the far field intensity $I(\theta)$ at the angle θ is written as:

$$I(\theta) = I_0 \int_{0-\infty} \text{somb}^2[(D/\lambda)\sin\theta]\ p(D)\, dD \tag{2.9}$$

Now, we want to calculate from Eq.2.9 the unknown distribution p(D) of particle size, after $I(\theta)$ is determined through a measurement of the far field intensity. The kernel of Fredholm's integral, $\text{somb}^2[(D/\lambda)\sin\theta]$, is a given term because we can compute it for any pair of given θ and D values.

The problem of inversion is known as an ill-conditioned one, because p(D) is multiplied by the kernel, and the result is integrated on D so that details in p(D) are smoothed out in the result $I(\theta)$. For example, very large particles affect only a small interval in $I(\theta)$ near $\theta \approx 0$, whereas very small ones give only a little increase of $I(\theta)$ at large θ.

Another source of error is the constant scattering coefficient Q_{ext} tacitly assumed in writing Eq.2.9. This holds for large-particles, $D \gg \lambda$, whereas for intermediate and small particles the Q_{ext} varies appreciably (see Fig.A3-6 and Eq. A3.2).

From the previous considerations, the measurement range best suited for the LAELS will normally go from 1 to 5 μm to perhaps 1 to 3 mm in diameter. At small D, the limit comes from the Rayleigh approximation, whereas at large diameter it depends on the size of the photodiode, due to measure at $\theta \approx 0$ without disturbance from the undiffracted beam.

Because we are unable to solve Eq.2.9 exactly in all cases, we will try to maximize the information content picked out from the experiment through the measurements of $I(\theta)$. To do so, we will repeat the measurement at as many angles θ as the outcomes $I(\theta)$ are found appreciably different. With information provided by the $I(\theta)$ set, the kernel integral will limit resolution to a certain value, beyond which the solution becomes affected by a large oscillating error. We can then optimize the solution by increasing resolution (or number of diameters) until the computed p(D) starts being affected by oscillation errors.

Let us now briefly discuss the mathematical approaches that have been considered and applied to solve the inversion problem given by Eq.2.9.

Analytical Inversion. Although Eq.2.9 looks intractable, Chin et al. [7] have been able to solve it for p(D) using hypergeometric functions, and the result is:

$$p(D) = -[(4\pi/D)^2/\lambda] \int_{\theta=0-\infty} K(\pi D\sin\theta/\lambda)\ d[\theta^3\, I(\pi D\sin\theta/\lambda)]\ /I_0 \tag{2.10}$$

In this equation we let $K(x)= xJ_1(x)Y_1(x)$, J and Y are the usual Bessel functions, and $d[..] =(\partial[..]/\partial\theta)\, d\theta$ are the differential of the product $[..]=\theta^3 I(..)$, in the variable θ.

Despite being a remarkable mathematical result, Eq.2.10 is difficult to use because it requires integrating $I(\theta)$ on θ from 0 to ∞. Unfortunately, the Fraunhofer approximation and

the assumption of Rayleigh regime are valid only for $\theta \approx 0$. To avoid an abrupt truncation of $I(\theta)$ and the resulting oscillation error in p(D), the data $I(\theta)$ is apodized (i.e., smoothed) to zero. A typical smoothing profile is $P(\theta) = 1-(\theta/\theta_{max})^2$, and the maximum θ is chosen as $\theta_{max} \approx (2...5)\lambda/D$ to cover the range of interest in diffraction.
In the range of diameter 10 to 50 μm and with $\lambda = 0.633$ μm, the analytical inversion works nicely if measured data is very clean and accurate, that is, affected by an error of 1% or less.

Least Square Method. Using a discrete approximation, we may bring Eq.2.9 to a set of linear equations and then apply the Least-Square Method (LSM) to improve the stability of solution, i.e., reduce spurious oscillations.
To do so, we write the integral as a summation on the index k of the diameter variable D_k and repeat the equations for as many variables n of the angle of measurement θ_n. The matrix of known coefficients connecting θ_n and D_k is the discrete kernel S_{nk}:

$$S_{nk} = \mathrm{somb}^2[(D_k/\lambda)\sin\theta_n] = 4J_1^2[(\pi D_k/\lambda)\sin\theta_n] / [(\pi D_k/\lambda)\sin\theta_n]^2 \tag{2.11}$$

The range of variables n and k is n=1..N and k=1..K, and the associated variables are $I_n = I(\theta_n)$ and $p_k = p(D_k)$. With this, Eq.2.9 becomes:

$$I_n = \Sigma_{k=1..K} S_{nk}\, p_k \qquad (n=1..N) \tag{2.12}$$

In this set of equations, the number of equations N may be different from the number of unknown K. Usually, N is the number of angular measurements performed by the photodetector. If we use an array or a CCD to obtain a number of separate channels in angle θ, N is fixed and may be usually 50 to 100. The number of unknown diameters K is variable because we want to determine the calculated distribution p(D) on the largest number of separate diameters that are compatible with the absence of oscillating errors. Then usually we may have K=6...20, depending on the average D and the shape of p(D).

To solve for the unknown p_k, we need to bring the number of unknowns to coincide with the number of equations, whatever N and K are, and we do so by applying the LSM. With the LSM, we require that the solution p_k is such that the error ε^2, or mean square deviation of the computed values $\Sigma\, S_{nk}\, p_k$ and measured ones I_n, is a minimum:

$$\varepsilon^2 = \Sigma_{n=1..N} [\, I_n - \Sigma_{k=1..K} S_{nk}\, p_k]^2 = \min \tag{2.13}$$

The condition of minimum is obtained by setting to zero the partial derivative of the error ε^2, with respect to each value p_k of the unknown:

$$0 = \partial(\varepsilon^2)/\partial p_k = \Sigma_{n=1..N} [\, I_n - \Sigma_{k=1..K} S_{nk}\, p_k]^2$$
$$= \Sigma_{n=1..N}\, 2\, [\, I_n - \Sigma_{k=1..K} S_{nk}\, p_k]\,(-S_{nk}) \tag{2.13'}$$

By rearranging the terms in this expression, we get:

$$J_h = \Sigma_{k=1..K} Z_{hk}\, p_k \qquad (h=1..K) \qquad (2.14)$$

where we let:

$$J_h = \Sigma_{n=1..N} I_n S_{nh} \text{ and } Z_{nk} = \Sigma_{n=1..N} S_{nk}^2 \qquad (2.14')$$

Now, the number of equations in 2.14 is equal to the number of unknowns and we can solve for p_k with standard algebra. Using the LSM, the accuracy of solution improves, and the range of diameter distribution is larger with respect to the analytical inversion method. However, if data are affected by errors or the number of diameters is excessive, the error of the solution increases. Also, the error may be large if the waveform p(D) is bimodal or contains secondary peaks.

Thus, one has to choose a reasonable number of diameters to get the best result. A strategy is to increase the number K of diameters and repeat the calculation until oscillations or wrong features start to show up in the new result.

Usually, the range of diameters of interest may be large (for example, two decades from 2 to 200 μm), but the number of affordable diameters is modest (e.g., K=6). Then, we get more information from the cumulative distribution P(D) than from the density function p(D). The relationship between the two is:

$$P(D) = \int_{x=0-D} p(x)\, dx \quad \text{and} \quad P_k = \Sigma_{i=1..k} p_i \qquad (2.15)$$

Of course, using Eq.2.15 with Eq.2.14 allows us to solve for P(D) with the LSM.

Errors in measured data I_n can be reduced by working on integral quantities in place of the differential ones considered so far. In fact, the photodetector has a finite area, and we usually strive to shrink its active aperture, but with a loss of signal-to-noise ratio. Then, it is better to start the problem of solving Eq.2.9 by assuming the signal is collected in a finite angular aperture $\theta_1...\theta_2$.

The integrated signal is written as $Y(\theta_1,\theta_2) = \int_{\psi=\theta_1..\theta_2} I(\psi)d\psi$ if the photodetector is made of elements with the same area at all θ, or as $Y(\theta_1,\theta_2) = \int_{\psi=\theta_1..\theta_2} 2\pi\psi\, I(\psi)d\psi$ if the photodetector is made of annulus elements for which the area increases with radius. In any case, by inserting Eq.2.9 in these expressions, we can integrate the kernel of the Fredholm's integral, $\text{somb}^2[(D/\lambda)\sin\theta]$, with respect to θ, either numerically or analytically. The result, $X(\theta_1,\theta_2)$, becomes X_{nk} when we use the discrete approximation, and this value can be used to solve for p_k or P_k with the LSM expressed by Eq.2.14.

With this approach, monodispersed powders can be measured, in the range of diameter 20 to 50 μm, typically with a <2% error, in K≈5 intervals of the cumulative distribution. At the expense of reduced accuracy (≈5% error), the range can be extended to 5 to 200 μm and 7 to 10 intervals. A typical result of particle-size measurement is shown in Fig.2-17.

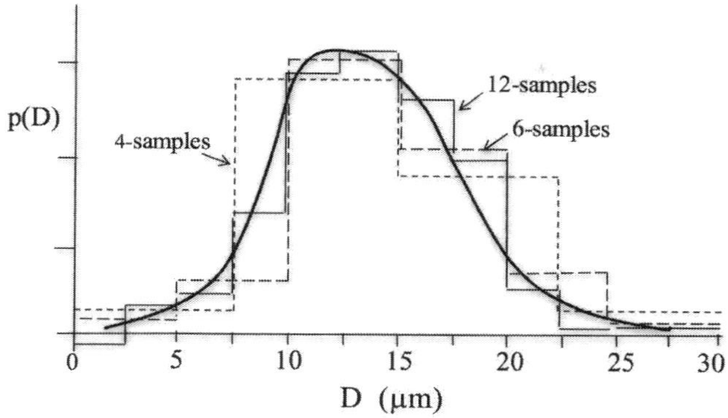

Fig. 2-17 Typical result of particle-size measurement obtained by LAELS. Thick line is the true distribution; step-wise curves are the reconstructed distributions at increasing number of resolved diameters, N= 4, 6 and 12. At N=12, one value of p(D) goes negative.

Other Linear Methods. The basic LSM we have outlined previously is the starting point for several extensions [8,9]. In all of them, we try to convey a physical constraint or information in a set of linear equations, just like the LSM concept leading to Eq.2.14.

A straight example is weighting of terms contributing to the total error ε^2 in Eq.2.13. To do that, we may add, under the summation term of Eq.2.13, a multiplying factor w_n that is taken =1 where the measured I_n is clean (or, has a good S/N ratio), and <1 where I_n is noisy. Of course, the choice of the exact weight is somehow arbitrary or left to our knowledge of the powder nature. Accordingly, the quality (or error) of the results will depend on the w_n values we have chosen.

Other examples of information that may be added through linear equations are (i) the nonnegative character of the distribution p(D), which leads to an inversion method known as Non-Negative LSM (NNLSM); (ii) the known shape of the p(D) waveform, for example, monodispersed or bimodal; (iii) the minimum-ripple or smoothness constraint.

With the addition of the linear constraint, both accuracy and dynamic range of the reconstruction are significantly improved. However, linear methods have the general disadvantage of a success dependent on our ability to tune the added information and to adjust free parameters describing it [7-9].

Iterative Methods. The prototype of these methods is Chahine's method [6,10] which is used to invert Fredholm's equation in a variety of applications involving ill-conditioned problems. We start letting K=N, that is, using a number of unknown variables p_k equal to the number of measured intensities I_n (K=N is also provided by using the LSM first).

The iterative method is based on the following reasoning. If the set of diameter distribution p_k is correct, it should give the measured distribution $I_{n.calc} = \Sigma_k C_{nk} p_k$. If the set of calculated values $I_{n.calc}$ differs from the experimental values, we may expect to approach the solution by multiplying p_k by $I_{k.meas}/I_{k.calc}$, where $I_{k.meas}$ is the measured value, or explicitly:

$$p_{k+1} = I_{k.meas}/I_{k.calc}\ p_k \qquad (2.16)$$

[a variant is to use for p_k the discrete Fourier transform p_k of the diameter distribution p_k]. By repeating the procedure an adequate number of times, p_k should converge to the correct solution. As a trial value of p_k, we can use either an estimated distribution or the result of LSM calculation. Starting with an all-positive trial distribution, the results in all the iterations are necessarily positive, and we get the inherent benefit of suppressing oscillations with negative swings. The results obtained by this method are definitely better than the normal LSM. Errors are greatly reduced (by a factor 3 to 5), and the range of diameters is significantly increased. On the other hand, the tendency exists to generate spurious spikes at small D, unfortunately, in place of the oscillations suppressed by the method.

Another problem with Chahine's and other iteration methods is that no clear sign exists when the optimal result is reached. Thus, we usually define a quality function, for example, the error ε^2 of calculated results to measured ones, and stop the iteration when a minimum of ε^2 is reached.

A refinement of the method has been recently introduced [6]. It consists of weighting the iteration by the normalized kernel, $S_{nk}/\ \Sigma_{n=1..N}S_{nk}$. With this position, Eq.2.16 reads:

$$p_{k+1} = (S_{nk}/\Sigma_{n=1..N}S_{nk})(I_{k.meas}/I_{k.calc})\ p_k \qquad (2.16')$$

At the generic iteration, each term p_k is weighted by S_{nk}, that is, proportionally to the efficiency of transfer of p_k to the output I_n. In consequence, spurious peaks found in the iteration method are suppressed, and resolution and dynamic range are improved further [6].

With regard to the optical layout of the particle-size analyzer, several *sources of error* can influence the results. The finite size of the photodetector has already been considered, and found that it can be accounted for by reformulating Eq.2.8 in terms of integrated intensity. Another error comes from the un-diffracted beam at $\theta=0$, which requires a stop on the focal plane to be blocked out (the stop is omitted in Fig.2-16 to avoid confusion). Even using the stop, some light will inevitably scatter or leak aside on the detector, which affects the $\theta\approx0$ measurements.

To get rid of it, we can use the filtering arrangement reported in Fig.2-18, known as reverse Fourier-transform illumination. Light projected through the cell by lens L1 converges at a focal plane position. The contribution diffracted at the angle θ reaches the focal plane of L1 at position $x=\theta F$ with rays parallel to the optical axis.

Instead of reading the intensity in the focal plane, we place a lens L2 there and perform a Fourier transform of the incoming angular distribution. All the θ-diffracted rays arriving parallel to the optical axis are focused on the axis of L2 (Fig.2-18), whereas un-diffracted light arriving at an angle will reach the focal plane out of the axis. Placing a pinhole at the focal plane of L2, we can allow the diffracted contribution through and block out the un-diffracted one. With another lens L3, we can then retransform the filtered distribution to one with rays parallel to the optical axis.

Because of the scattering suppression, the arrangement of Fig.2-18 is much better than a stop in the focal plane, and we can measure down to small angle θ (typ. <0.1mrad). Accordingly, the range of large diameters that can be reconstructed is expanded up to several hundred micrometers or even to the millimeters.

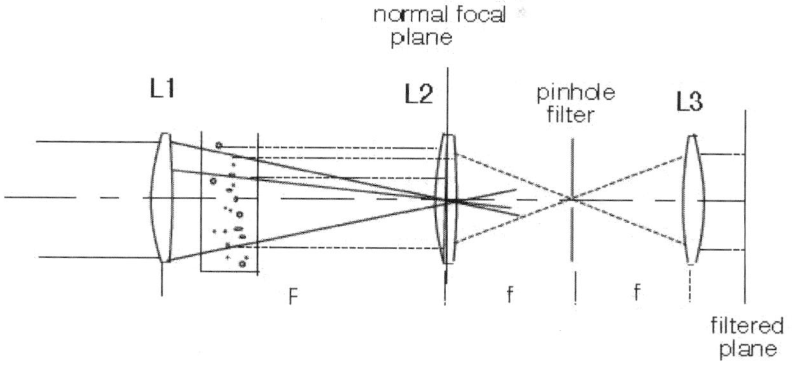

Fig. 2-18 Using a convergent beam to illuminate the cell, we are able to filter out the undiffracted beam more efficiently than with stops. Diffracted rays indicated by dotted lines are focused on axis and can pass through the pinhole, whereas undiffracted rays arrive out-of-axis and are blocked.

At the other end of the measurement range, that of large diffracted angle or small diameter, the errors are not only due to collection efficiency, vignetting, and the like, but also to the Fraunhofer approximation $D \gg \lambda$. Errors start becoming significant at, say, $D \approx 2...5\lambda$, so the practical range of measurement for particle sizing based only on the LAELS is typically 2 to 500 μm.

We can extend the range of measurement on small particles by taking advantage of the Mie ($D \approx \lambda$) and Rayleigh ($D \ll \lambda$) scattering regimes. These are described by the appropriate scattering function $f(\theta)$ and extinction factor Q_{ext} (Appendix A3.1), which are known and can be calculated from Mie's theory [11].

In the Rayleigh regime, the scattering is nearly isotropic in angle, and the extinction factor Q_{ext} varies as $(D/\lambda)^4$. When D increases up to about $D \approx \lambda$, the scattering function peaks forward, and the extinction factor increases up to ≈2-4 (see Fig.A3-6).

Now, we can supplement the LAELS measurement with a measurement of light scattered from the cell at a fixed angle, i.e., large enough to be out of the range of LAELS data.

For example, a choice is to look at the cell from $\theta \approx 45$ deg so that the detector is safely away the undiffracted beam leak. To collect information on the particle size distribution, we now scan the wavelength instead of the diffraction angle. By varying the ratio D/λ, the extinction factor $Q_{ext}(D,\lambda,n)$ varies and so does the scattered power too.

For a density distribution $p(D)$ of diameter, and given an extinction factor $Q_{ext}(D,\lambda,n)$ at the measurement wavelength λ, the power density collected within a solid angle $\Delta\Omega$ around $\theta=45$ deg is given by:

$$I(\lambda, 45\ \text{deg}) = f(45\ \text{deg})\ (\Delta\Omega/4\pi)\ I_0 \int_{0-\infty} Q_{ext}(D,\lambda,n)\ p(D)\ dD \quad (2.17)$$

where $f(\theta)$ is the scattering function and Q_{ext} is the extinction factor defined in App. A3.1. Eq.2.17 is clearly the counterpart of Eq.2.9 for the case at hand of extinction-related measurement. All the previously discussed methods of inversion of Fredholm's integral (Eq.2.17) can now be repeated [6] on the variables D_k and λ_n.

This method of particle sizing is called *Spectral Extinction Aerosol Sizing* (SEAS). With the computation of the distribution p(D) from Eq.2.17, it goes down to 0.02...0.1 μm as the minimum measurable size, whereas the maximum range overlaps the minimum of LAELS (≈ 2...5 μm).

About the source, we need one with a high radiance and a wide spectral emission because the minimum detectable signal depends on radiance, and the minimum measurable size depends on the shortest wavelength available. To cover the visible and UV range, a cheap choice is a halogen lamp with 3000 to 3300 K color temperature. The lamp output is collimated by a lens and filtered by a monochromator. The output beam can either be combined with the laser beam of the LAELS by means of a wavelength-selective beamsplitter or used as a stand-alone source. The detector to be used is usually a photomultiplier (see Ref.[1], Ch.4) which is well suited for the UV-visible spectral range, and provides the best sensitivity, even in the case of small particulate concentration.

An example of a modern particle-sizing instrument is shown in Fig.2-19. This instrument is the technical evolution of the first products released in 1971 and represents a big commercial success.

A last method of importance is the *Dynamical Scattering Size Analyzer* (DSSA), which is useful for measurements of very small (1...100 nm) particles [12,13].

When particles are very small, the scattered light is frequency shifted. The shift is due to the Doppler effect $f_D=(\underline{k}_o-\underline{k}_i)\cdot\underline{v}$, or equivalently by the interferometric phase $(\underline{k}_o-\underline{k}_i)\cdot\underline{s}$, of the particle displacement $\underline{s}$ observed from the direction $\underline{k}_o$ and illuminated from $\underline{k}_I$ (all underlined quantities being vectors). Velocity is provided by the thermal agitation of the small particles, given by $v=k_BT/m$, where k_B is the Boltzmann's constant, T the absolute temperature and m the particle mass. The shift is measured by frequency-spectrum analysis or by time-domain autocorrelation measurement of the detected signal, eventually with the Self-Mixing configuration (Ch.5) like that introduced by Ref. [13] (see also Sect.6.4).

The autocorrelation function $C(\tau) = (1/T)\int_{0\text{-}T} i(t)i(t+\tau)dt$ is found to depend on the diffusion constant δ of the particles, as $C(\tau) = C_0 \exp -\delta\,(k_o-k_i)^2\tau$. From the measurement of the exponential decay of $C(\tau)$, the diffusion constant is determined. Then, we use the relation $\delta = k_BT/3\pi\eta D$ to determine the diameter D after the viscosity η of the particle in the surrounding medium is known. As an example of the method, Ref. [13] reports the measurement of the diameter of polystyrene particles dispersed in water, in the range 20 to 200 nm. For these diameters, the Doppler frequency f_D distribution follows a Lorentzian curve and has a HWHM ranging from 2.5 kHz to 200 Hz, for small and large diameters, respectively [13].

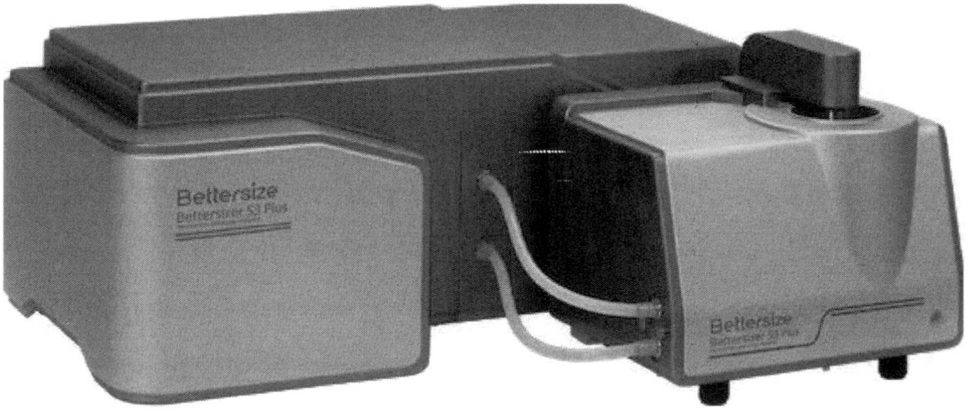

Fig. 2-19 A modern particle-size analyzer (Bettersizer S3) based on diffraction and extinction performs diameter measurements from 0.01 to 3500 μm (courtesy of Quantachrome, Odelzhausens, Germany).

REFERENCES

[1] S. Donati: "*Photodetectors*", 2nd ed., Wiley IEEE Press, Hoboken 2021.
[2] R.D. Hudson: "*Infrared System Engineering*", Wiley Interscience, New York, 1969.
[3] W.J. Smith: "*Modern Lens Design*" 2nd ed., McGraw Hill (USA) 2004.
[4] M. Perkovic, M. Gucma, B. Luin, L. Gucma, T. Brcko: "*Accommodating larger container vessels using an integrated laser system for approach and berthing*", Micropr. and Microsyst. vol.52 (2017), pp.106-116.
[5] J. Cornaillault: "*Particle Size Analyzer*", Appl. Opt. vol.11 (1972), pp.265-269.
[6] F. Ferri, G. Righini, and E. Paganini: "*Inversion of Low-Angle Elastic Light Scattering Data with a New Method Modifying Chanine Algorithm*", Appl. Opt. vol.36 (1997), pp.7539-50; see also: Appl. Opt. vol.34 (1995), pp.5829-5839.
[7] J.H. Chin, C.M. Sliepcevich, and M. Tribus: "*An Improved Least Square Method Routine to Compute Particle Distribution*", J. Phys. Chem. vol.56 (1955), pp.841-848.
[8] S. Twomey: "*Introduction to Mathematics of Inversion in Remote Sensing and Indirect Measurements*", Elsevier Science, Amsterdam 2013, Chapter 7.
[9] N. Wolfson et al.: "*Comparative Study of Inversion Techniques, part I and part II*"", J. Appl. Meteorology vol.18 (1979), pp.543-561.
[10] M.T. Chahine: "*Determination of the Temperature Profile in an Atmosphere from its Outgoing Radiance*", J. Opt. Soc. of Am. vol.58 (1968), pp.3074-3082.
[11] H.C. Van de Hulst: "*Light Scattering by Small Particles*", J. Wiley, New York, 1957.
[12] B.J. Berne and R. Pecora: "*Dynamic Light Scattering*", J. Wiley, New York, 1976.
[13] C. Zakian, M. Dickinson, T. King: "*Particle Sizing and Flow Measurement Using Self-Mixing Interferometry with a Laser Diode*", J. Optics A, vol.7 (2005), pp.S445-S452.

Problems and Questions

P2-1 *How small is the beam waist of a typical L=20 cm long He-Ne laser? Assume a plano-concave mirror cavity with* $R_1=\infty$ *and* $R_2=2L=40$ *cm.*

P2-2 *Using the He-Ne laser of P2-1 for alignment purposes, how small can the beam size be maintained down a total path length of 10 m? Which telescope is needed to project it?*

P2-3 *Instead of the telescope used in Prob.P2-2, could we use a single lens? What is the advantage or disadvantage of both choices?*

P2-4 *What about the beam size if we remove the telescope of P2-2 and let the beam propagate, using just the intrinsic collimation of the laser beam?*

P2-5 *How can I estimate the effect of air turbulence in a real experiment with alignment beams propagated through the atmosphere? Up to which distance L will the small spot size calculated previously still be obtained?*

P2-6 *In the 4-quadrant pn photodiode of Fig.2-6 (top) the two output anodes do not appear electrically isolated between them, because of the finite resistance of the p region. Does this introduce any errors?*

P2-7 *What is the dynamic range of position signal of the 4-quadrant photodiode? What is its minimum position change that can be detected? Can these two quantities be changed through design in a specific given device?*

Q2-8 *Could a photoconductor structure be employed in place of the photodiode structure to perform like a 4-quadrant detector?*

Q2-9 *Why shall we consider three types of position sensing devices, the 4-quadrant, the PSD and the reticle-assisted? Isn't one superior to the others?*

Q2-10 *What is arguably the fastest position sensing device among the three types of detectors?*

Q2-11 *In the laser level, why use a pentaprism to deflect the beam instead of a plain mirror inclined at 45 deg?*

Q2-12 *How collimated can the fan-beam at the output of the laser level be maintained at a distance z?*

Q2-13 *Can a semiconductor laser diode be used in place of the He-Ne to build a laser level?*

Q2-14 *Is the automatic verticality servo really superior to the manual setting version of the laser level?*

Q2-15 *In the wire diameter sensor, what about the shape of the photodetector that is needed at the focal plane to collect the scattered field with good efficiency?*

P2-16 *In the wire diameter sensor, calculate the amplitude of the signal collected by the photodetector.*

Q2-17 *In the wire diameter sensor, does the perpendicularity error affect the measurement?*

P2-18 *Is the measurement of first two zeroes width really the best way to extract the diameter information from the distribution of diffracted field?*

Q2-19 *Can any further information about the wire (such as absorption coefficient and index of refraction of the material) be inferred from the measurement of the diffracted field?*

Q2-20 *Can the spatial frequency (or the deniers) of a fabric be measured with the diffracted field setup of the diameter sensor?*

Q2-21 *In the particle size sensor, which errors may be caused by the scattering regime of the particles in the optical cell, and how can they be taken into account or be mitigated?*

Q2-22 *In particle sizing measurements, which is the most effective method for extracting the desired particle distribution* $p(D)$ *from the measured scattered intensity* $I(\theta)$*?*

P2-23 *Estimate the amplitude of the signal received at the focal plane in a particle sizing measurement.*

CHAPTER 3

Laser Telemeters

A telemeter is an instrument for measuring the distance to a remote target. Synonymous with telemeter is the term *rangefinder* and also *LiDAR* (light detection and ranging) especially when used in the automotive. Basically, the three main techniques to perform an optical measurement of distance are the following:

- *Triangulation*: the target is aimed at from two points separated by a known base D, placed perpendicular to the line of sight. By measuring the angle α formed by the line of sight, the distance is found as $L=D/\alpha$.
- *Time of flight*: a light beam from a high-radiance source is propagated to the target and back, and the time delay $T=2L/c$ is measured (c is the speed of light). The distance follows as $L=cT/2$.
- *Interferometry*: a coherent beam is used in the propagation to the target. The returned field is detected coherently by beating with a reference field on the photodetector, and a signal of the form cos 2kL (where $k=2\pi/\lambda$) is obtained. From cos 2kL, we can count the distance increments in units of $\lambda/2$ (in the simplest approach).

In view of the obtained performances, the three approaches are complementary.

Triangulation is the simplest technique to implement and may operate in daylight even without any source. Until a few decades ago, it survived in construction applications with the theodolite (to measure α) and the rulers (to set D). However, it has poor accuracy on long distances because, when L is much larger than the base D, the angle α to be measured becomes very small and is affected by a large relative error.

The time-of-flight technique requires a pulsed or a sine-wave modulated laser. In both cases, we measure the time of flight $T=2L/c$ of light to the target at distance L. This telemeter works with a constant uncertainty ΔT (or, in distance, $\Delta L=c\Delta T/2$), in principle. Thus, performance is excellent on medium and long distances. Because of this, it has superseded the theodolite on distances of 100 m to 1 km, and has proven to be a new powerful technique in a number of long-distance (> 1 km) applications.

Last, the interferometric technique is by far the most sensitive because it works with the phase of the optical field, but it requires a coherent laser source with a coherence length at least equal to the go-and-return distance to the target. The phase provides basically an *incremental* measurement of distance, not an *absolute* one, as the other two techniques. For this reason, interferometric techniques are treated separately in Chapter 4, whereas in this chapter we will concentrate mainly on time-of-flight telemeters.

From the point of view of the remote target features, we may have either (i) a cooperative target made of a retroreflector surface to maximize the returning signal (ii) a non-cooperative target, simply diffusing back the incoming radiation with a $\delta<1$ diffusion coefficient.

According to the measurement technique employed, time-of-flight telemeters are classified as one of the following:

- *Pulsed telemeters* when the delay $T=2L/c$ between transmitted and received (either short or long) pulses is measured directly in the time domain;
- *Sine-wave* modulated *telemeters*, when the source is modulated in power by a sine wave at a frequency f, and the delay T is measured from the phase $\Phi=2\pi fT$.

In the first case, using a *short pulse* with duration τ in the range of 0.5- to 5-ns, and relatively high peak power, we get the long-range telemeters useful for geodesy research and military applications, with operational ranges up to 100 km or more, whereas using a *long pulse* lasting $\tau=T$ (the go-and-return propagation time), signal processing is much simplified and we get distance up to hundred meters with centimeter-resolution for the application to the automotive, recently identified with the acronym LiDAR (Light Detection and Ranging).
In the second case, we get the topographic telemeters used on medium ranges (<1 km) for civil engineering and construction work applications.

The laser sources most suitable in the two cases are: (i) solid-state lasers (like Nd, YAG) operated in the Q-switching regime and semiconductor diode lasers, pulsed at 10- to 500-ns durations, and (ii) quasi continuous-wave semiconductor lasers like GaAlAs and GaInAsP modulated in amplitude up to hundreds of megahertz.

3.1 Triangulation

Let us consider the basic scheme for triangulation illustrated in Fig.3-1. Here, a distant object O is aimed from two observation points, A and B, along a base of width D.

The object is assumed self-luminous for the moment, which is a case referred to as passive triangulation. Active triangulation using a laser source to illuminate the target is considered later on.

The beamsplitter and rotatable mirror combination allows superimposing the images seen at A and B. The mirror M is rotated by an angle $\alpha/2$ from the initial position $\alpha=0$ parallel to the beamsplitter until the object images are brought to coincide. The distance L then follows as:

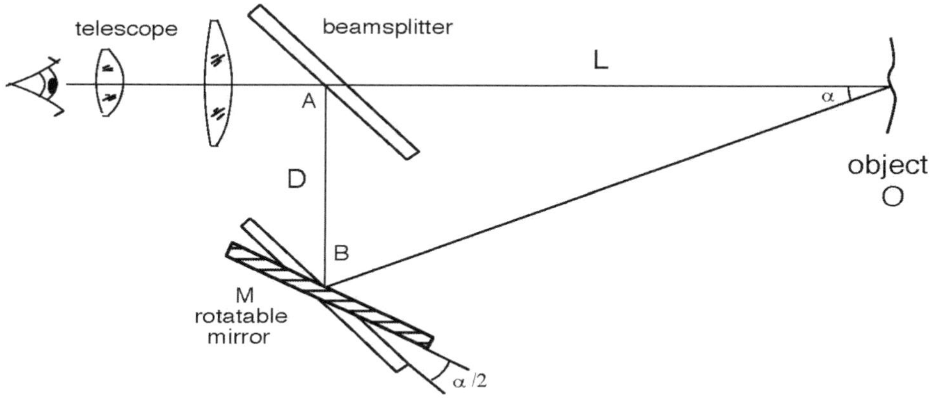

Fig. 3-1 Basic scheme of a triangulation measurement

$$L = D / \tan \alpha \approx D / \alpha \quad (3.1)$$

Of course, we need a good measurement of the small α to determine a long-distance L with a reasonably short base D. The absolute and relative errors, ΔL and $\Delta L/L$, due to an angle error $\Delta\alpha$, are found from Eq.3.1 as:

$$\Delta L = - (D/\alpha^2)\ \Delta\alpha = - (L^2/D)\ \Delta\alpha \quad (3.2)$$

$$\Delta L / L = - (L/D)\ \Delta\alpha \quad (3.2')$$

and they both increase with distance. For design purposes, the errors are plotted in Fig.3-2 as a function of L and with $\Delta\alpha$ as a parameter.

With regards to the error of the angle readout, a good micrometer screw with gear reduction and backlash recovery is representative of medium-accuracy performance and may resolve $\Delta\alpha$=10 arc-min (≈3 mrad). On the other hand, an angle encoder may provide a high-accuracy readout, going down to the limit of the (small) viewing telescope, typically $\Delta\alpha$=0.3 arc-min (≈0.1 mrad).

In these two representative cases, it is interesting to evaluate the performance of the optical telemeter based on triangulation. Let us exemplify the results for two hypothetical telemeters, one intended for short distance (L≈1m) and the other for medium distance (say 100 m).

At a distance of L=1 m, we may choose a D=10 cm base as a reasonable value for a compact instrument. From Fig.3-2, we find the intrinsic accuracy as $\Delta L/L$=3% and 0.1% respectively, for $\Delta\alpha$=3 and 0.1 mrad. Turning to the 100 m telemeter, we may expand the base within the limits allowed by the application and perhaps go to D=1 m. Then, from Fig.3-2, we find $\Delta L/L$=30% and 1% for $\Delta\alpha$=3 and 0.1 mrad.

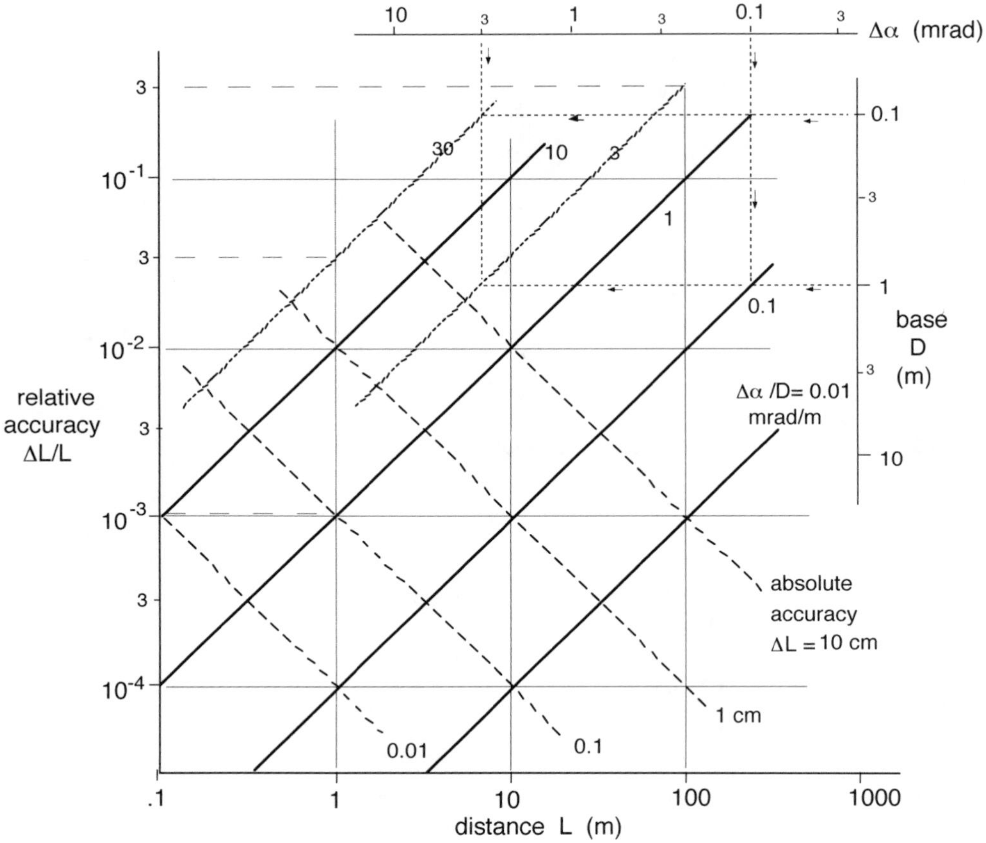

Fig. 3-2 Distance accuracy (relative $\Delta L/L$ and absolute ΔL) in triangulation measurements. Entering with $\Delta\alpha$ and D values in the upper right corner diagram (see small-dot line for 0.1 mrad and 1 m) gives the parameter $\Delta\alpha/D$ and hence the line of relative accuracy $\Delta L/L$ versus distance. Large-dot lines supply the absolute distance accuracy ΔL. Because the approximation $\tan\alpha \approx \alpha$ has been used, the diagram is valid for D<<L.

As is clear from these examples, triangulation can achieve a quite respectable performance, provided the application allows using a not-too-small base-to-distance ratio D/L.

A triangulation telemeter can be developed straight from the basic concept outlined in Fig.3-1. Such an instrument is classified as a passive optical telemeter because it does not require a source of illumination or a detector.

However, if we add a source to aim the target and a position-sensitive detector to sense the return, we can improve performance, eliminate the moving parts, and get a faster response. The best scheme for the active triangulation scheme can take very different configurations, depending on the requirements of the specific application (e.g., dynamic range, accuracy, size, and cost) [1].

To substantiate a design example, we report in Fig.3-3 the layout of an active triangulation telemeter intended for short distances (1...10 m) that uses a semiconductor laser and a CCD. The laser wavelength is chosen in the visible range for ease of target aiming, and the power is usually 1- to 5-mW emitted from an elliptical near-field spot of 1×3 μm size (typically). An anamorphic objective lens circularizes the beam to a radius w_l (typ. 5μm) and projects an image of it on the target.

On the target, the spot size radius is then $w_l\, L/F_{ill}$ (=5μm×1000/125=40 μm for L=1 m and F_{ill} =125 mm). As a viewing objective, we use a telephoto lens with focal length F_{rec} (typically 250 mm) and get an image of the target on the CCD.

The CCD (see [2], Sect.12.2) is a silicon device composed of a linear array of N individual photosensitive elements of width w_{CCD} (typically, we may have N=1024 and w_{CCD}= 10 μm). The size of the target-spot that is imaged on the CCD by the objective is easily computed as $w_l\, L/F_{ill}\, (F_{rec}/L)= w_l\, F_{rec}\, /F_{ill}$ =5μm×250/125=10 μm, which is equal to the pixel size w_{CCD}.

We may assume that the accuracy of localization in the focal plane is limited by the pixel size (see Ref.[3] for a refinement to take account of speckle errors). Then, the angular resolution is $\Delta\alpha= w_{CCD}/F_{rec}$ =10 μm/250 mm =0.04 mrad. By taking an axis separation D=50 mm, and from the data in Fig.3-2 we have ΔL/L=0.1% at L=1 m, and ΔL/L=1% at L=10 m. Converted in absolute errors, our telemeter would resolve 1 mm at 1 m and 10 cm at 10 m, and perhaps the results may still be improved.

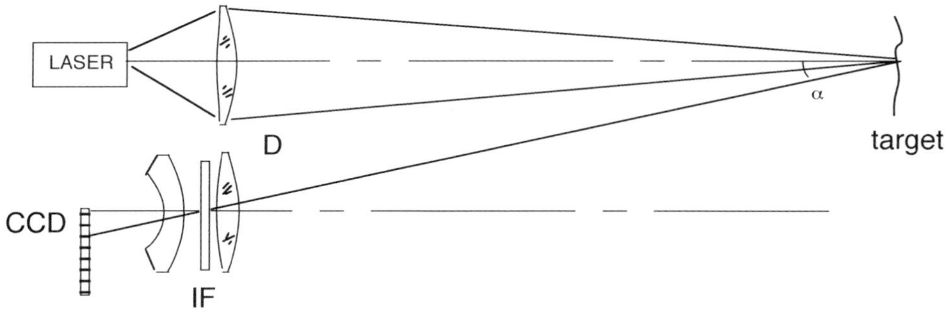

Fig. 3-3 Scheme of a triangulation telemeter with active illumination and a static angle readout. The optical axes of illuminating and viewing beams are parallel. The target-spot image is formed off-axis at a distance αF_{rec}, where F_{rec} is the focal length of the viewing objective. An Interference Filter (IF) is used for ambient light rejection.

The extension of the distance measurement by triangulation to a double multiplicity of points, so as to create a *3D* (tridimensional) map of the object, or have an *imaging* telemeter, is discussed in Section 3.4.

3.2 Time-of-Flight Telemeters

These telemeters are based on the measurement of the time-of-flight $T=2L/c$ of light to the target at distance L and back. The uncertainty ΔT of measurement reflects itself in a distance uncertainty $\Delta L=c\Delta T/2$. Accordingly, if we are using a pulsed light source, we require that the pulse duration τ be short enough for the desired resolution, or $\tau<\Delta T$.

Similarly, if we use a sine-wave modulated source, the frequency of modulation ω needs to be high enough, or $\omega>1/\Delta T$.

In the following sections, we first study the power budget of a generic time-of-flight telemeter, then evaluate the ultimate distance performance of pulsed and sine-wave modulated approaches, and conclude with the illustration of a few schemes of implementation [4].

3.2.1 Power Budget

The power budget of a time-of-flight telemeter can be analyzed with reference to Fig.3-4. The optical source emits a power P_s, and the objective lens projects it on the target with an angular divergence $\theta_s= d_s/F_s$, where d_s is the diameter of the source and F_s is the focal length of the objective lens.

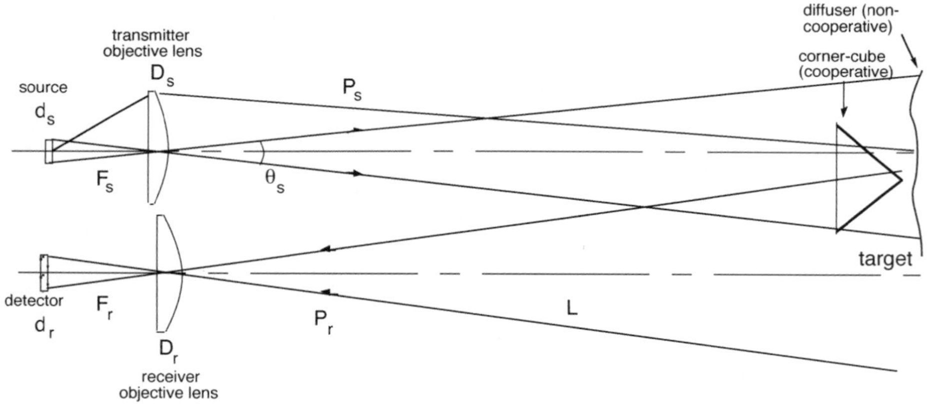

Fig. 3-4 General scheme for evaluating the received power P_r in a telemeter as a function of transmitted power P_t, distance L, and transmitter/receiver optics. The target may be either a corner cube or a diffuser. As distance L is much larger than other dimensions in the drawing, the field of view is actually superposed on the field of illumination.

A detector is aimed to the target through a receiving objective lens with diameter D_r.

We will consider the target either as *cooperative* (using a corner cube for self-alignment) *or non-cooperative* (to account for a normal diffusing surface, with diffusivity $\delta<1$).

In the first case, when the target is cooperative, the corner cube acts as a mirror, so the receiver sees the source as if it is at a distance 2L. Accordingly, the power fraction collected by the receiver is the ratio between the area of the collecting lens and the area of the transmitted spot at a distance 2L:

$$P_r/P_s = D_r^2/\theta_s^2\, 4L^2 \tag{3.3}$$

Eq.3.3 holds if the corner cube diameter D_{cc} is large enough and does not limit collection of radiation at the receiver. This requires that $D_{cc} \geq D_r/2$. If the reverse is true, we shall use D_{cc} in place of $D_r/2$ in Eq.3.3, so we get:

$$P_r/P_s = D_{cc}^2/\theta_s^2\, L^2 \tag{3.3A}$$

In the second case, when the target is non-cooperative, i.e., is a diffusing surface of area A_t, the power arriving on the target is P_s, and a fraction δ of it is re-diffused back to the receiver. Assuming a Lambertian diffuser, the radiance R of the target is $1/\pi$ times the power density $\delta P_s/A_t$, or $R= \delta P_s/\pi A_t$, and accordingly, the power collected by the receiver is $RA_t\Omega$, (see Ref.[2], Ch.2) where Ω is the solid angle of the receiver seen from the target, given by $\Omega=\pi D_r^2/4L^2$. Thus, we have:

$$P_r/P_s = \delta\, D_r^2/4L^2 \tag{3.4}$$

Also in this case, if the target has a diameter D_{tar} smaller than $\theta_s L$, the P_r/P_s ratio is reduced by a factor $(D_{tar}/\theta_s L)^2$, and

$$P_r/P_s = \delta\, D_r^2\, D_{tar}^2/4\theta_s^2\, L^4 \tag{3.4A}$$

and thus, the dependence on distance becomes of the type L^{-4}, the inverse fourth power of distance, which is commonly found in the case of a microwave radar. In the following, however, we restrict ourselves to the case $D_{tar}>\theta_s L$ because it is the most likely situation in distance measurements.

Eqs.3.3 and 3.4 describe the attenuation due to geometrical effects only. We also find an additional contribution from the transmittance T_{opt} of the transmitter/receiver lenses, and from the propagation through the atmosphere with an attenuation T_{atm}=exp-2αL, where α is the attenuation coefficient of the air (see App. A3.1). We can account for these terms by multiplying the second members of Eqs.3.3 and 3.4 by $T_{opt}T_{atm}$ so that we have in general:

$$P_r/P_s = T_{opt}T_{atm}\, D_r^2/\theta_s^2\, 4L^2 \tag{3.3'}$$

$$P_r/P_s = \delta\, T_{opt}T_{atm}\, D_r^2/4L^2 \tag{3.4'}$$

As a rule of thumb, we may take α=0.1 km^{-1} for an exceptionally clear atmosphere, α=0.33 km^{-1} for a limpid atmosphere, and α=0.5 km^{-1} for an incipient haze. The corresponding values of T_{atm} are reported in Fig.3-5 for medium/long ranges.
The previous values are an average in the range of visible wavelengths. However, because the telemeter will operate at a well-defined wavelength, it is important to have a closer look at the spectral attenuation $\alpha=\alpha(\lambda)$.

Information on $\alpha=\alpha(\lambda)$ is provided by Fig.A3-2 of Appendix A3. In addition, comparative data can be gathered from the spectrum of the solar irradiance, see Fig.A3-3. From here, we can see that there are some wavelengths to be avoided (e.g., 0.70, 0.76, 0.80, 0.855, 0.93 and 1.13 μm) because they coincide with absorption peaks of the atmosphere. On the other hand, wavelengths like 0.633 (He-Ne), 0.82...0.88 (GaAlAs), and 1.06 μm (Nd) are acceptable for long-range telemeters.

Now, we can generalize the power–budget equations by introducing an equivalent distance $L_{eq}=L/\sqrt{T_{atm}}$ and a gain factor G, so that:

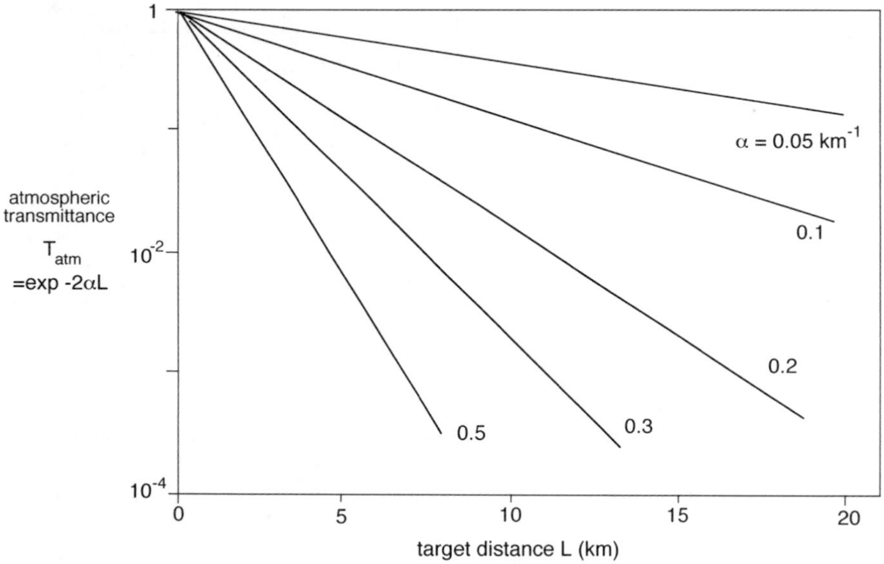

Fig. 3-5 Atmospheric transmittance as a function of target distance, with the attenuation coefficient as a parameter.

$$P_r/P_s = G\, D_r^{\,2}/4L_{eq}^{\,2} \quad (3.5)$$

By comparing with Eqs.3.3 and 3.4, G is given by:

$$G = G_{nc} = T_{opt}\,\delta \quad \text{(non-cooperative target)} \quad (3.6)$$

$$G = G_c = T_{opt}/\theta_s^{\,2} \quad \text{(cooperative target)} \quad (3.6')$$

With this position, Eq.3.5 is the *power-attenuation equation* of the telemeter, and tells us that the P_r/P_s ratio depends on the inverse square of target distance or, better, from the ratio $(D_r/L)^2$.

On the other hand, Eqs.3.6 and 3.6' are about the gain of the target, either cooperative or non-cooperative. The cooperative gain G_c is the counterpart of the antenna gain known in microwave, and is given by the inverse squared of the divergence angle, $1/\theta_s^2$. This factor may be very large indeed (for example, it is $G_c=10^6$ for $\theta_s=1$ mrad).

3.2.2 System Equation

Let us now consider the system performance of the telemeter. If P_r is the received power and the receiver circuit has a noise P_n, we shall require that the ratio of the two, i.e.:

$$S/N = P_r/P_n \qquad (3.7)$$

is at least, say 10, to get a good measurement. By combining Eqs.3.7 with Eq.3.5 we get:

$$GP_s = P_r\ 4L_{eq}^2/D_r^2 = (S/N)\ P_n\ 4L_{eq}^2/D_r^2 \qquad (3.8)$$

Eq.3.8 is the *system equation* of the telemeter because it relates the required S/N ratio to receiver noise P_n and to attenuation $4L_{eq}^2/D_r^2$. The diagram of the equivalent power GP_s versus distance L_{eq} and with the receiver noise P_n as a parameter is plotted in Fig.3-6.

Three central-design regions are indicated in Fig.3-6. One representative of a sine-wave modulated telemeter for topography, which may use a ≈1 mW transmitted power and has $GP_s \approx$ 1 W in virtue of the cooperative target (corner cube) gain. Because the measurement time T can be long (e.g., 10 ms to 1s), the bandwidth B=1/2T is small, and the receiver noise can go down to the nanowatt level.

The second case is representative of a pulsed telemeter intended for non-cooperative targets. The source is likely to be a Q-switched laser with high peak power (≈ 0.1–1 MW, resulting from ≈1 mJ energy per pulse and $\tau \approx 10$ ns pulse duration). Using a pin/APD photodiode (see [2], Ch.6.1) as the detector, and taking into account that the bandwidth $B \approx 1/\tau$ is now much larger, the receiver noise is in the microwatt range.

The third region is typical of the LiDAR or rangefinders for the automotive application, in which the source is a near-infrared laser emitting 1...10 mW of power and the target is non-cooperative.

Added to the useful received power we find noise coming from three contributions, namely:

- noise of the received signal P_s
- noise of the background light P_{bg} collected by the receiver
- noise of detector and front-end amplifier

To describe noise, we use the standard deviation p_n of the fluctuations referred to the detector input, a quantity called the Noise-Equivalent-Power (NEP) (see Sect.3.3.1 of Ref.[2]).

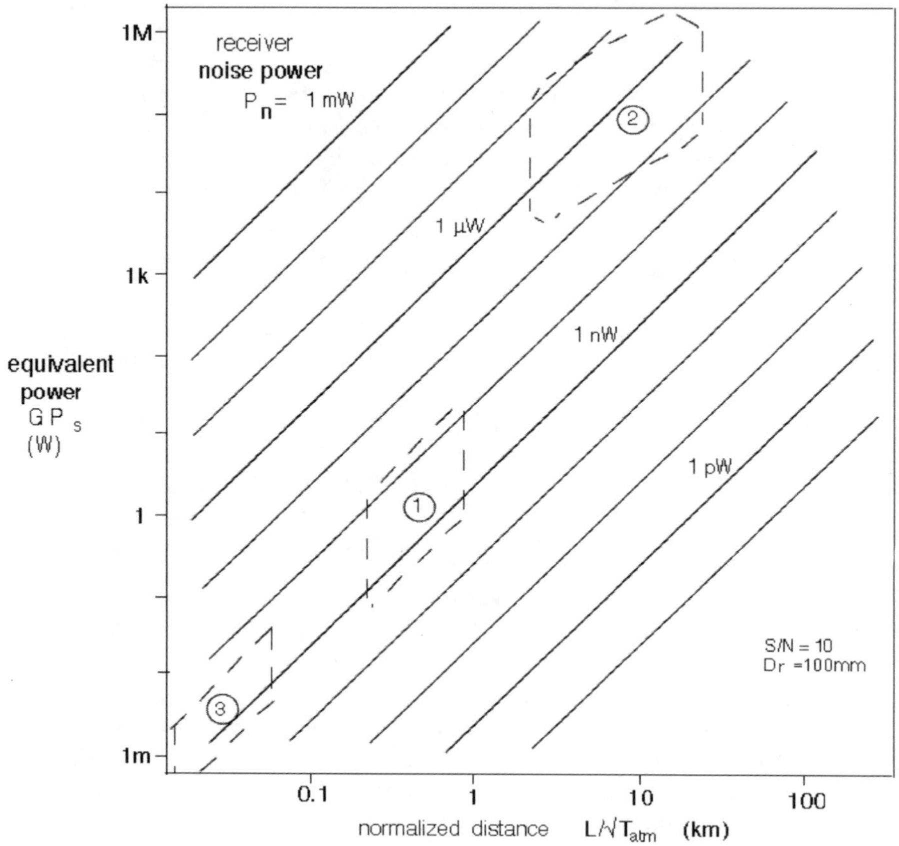

Fig. 3-6 Plot of the transmitted equivalent power versus normalized distance, with the receiver noise power as a parameter. It is assumed a S/N=10 and a receiver objective with D_r=100 mm. Zone 1 is representative of a sine-wave modulated topographic meter, zone 2 of a pulsed telemeter and zone 3 of an automotive LiDAR.

As a reasonable assumption, we assume that the photons of both signal and background light obey the Poisson statistics. Upon detection, each photon is converted into an electron with a success probability η, which is called a Bernoulli process. The photon flux F_{ph} is thus transformed in an electron flux $F_{el} = \eta F_{ph}$, where η is the quantum efficiency of the detector.

It can be shown that the cascade of a Poisson and a Bernoulli process is a Poisson process. Thus, the detected electrons follow the Poisson statistics, and the associated current has fluctuations described as shot noise with a white spectral density. Now, the average detected currents are written as $I_s = \sigma P_s$ and $I_{bg} = \sigma P_{bg}$, where $\sigma = \eta e/h\nu$ is the spectral sensitivity of the detector (Ref. [2], Ch.4). Superposed to the average currents, we find shot noise fluctuations i_{ns} and i_{nbg} whose variances are given by $i_{ns}^2=2eI_sB$ and $i_{nbg}^2=2eI_{bg}B$, with B being the bandwidth of observation.

The detector and front-end noise are summarized by an equivalent characteristic current I_{ph0} (see [2], Ch.3.2), defined as the current value for which the shot-noise is equal to the total noise of detector and front-end, that is $i_{nph0}^2=2eI_{ph0}B$.

Summing up the contributions (because they are mutually uncorrelated), the total variance of the detector current is $i_{rec}^2 = i_{nph0}^2+i_{ns}^2+i_{nbg}^2$, or:

$$i_{rec}^2 = 2eB\,(I_s+ I_{bg}+ I_{ph0}) \tag{3.9}$$

We can divide i_{rec}^2 by σ^2, the spectral sensitivity of the photodiode (coincident, in this case, with the responsivity), to get the (total) noise power variance at the input:

$$p_n^2 = 2h\nu/\eta\; B\,(P_s+ P_{bg}+ P_{ph0}) \tag{3.10}$$

In this equation, we let $P_{ph0}=I_{ph0}/\sigma$ for the power corresponding to I_{ph0}. An extra $1/\eta$ factor is left over in Eq.3.10 after the σ's ratios of powers to currents have been cleared and $2h\nu\sigma$ has been transformed in 2e.

This is the correct result, however, because a real detector with $\eta<1$ worsens the S/N ratio of detected current by η with respect to the S/N of incoming power.

The noise performances obtained by several combinations of detector and front-end preamplifiers are discussed in detail in Ref. [2], Sect.5.3. As a guideline for the reader, we summarize in Fig.3-7 the performance we may reasonably expect from a well-designed receiver for instrumentation applications.

Data are given in terms of the current noise spectral density $i_n/\sqrt{B}$, as a function of the maximum frequency of operation of the receiver f_2 (that is, of the cutoff frequency). To obtain p_n, the value read in Fig.3-7 shall be multiplied by the square root of the measurement bandwidth and divided by σ (Eq.3.10).

To evaluate the term P_{bg} in Eq.3.10, we shall consider the spectral irradiance E_s ($W/m^2\mu m$) of the scene on which the telemeter is aimed.

In daytime with direct sunlight illumination at different elevation angles, E_s is given by the diagram of Fig.A3-3. In other conditions (clouds, haze, etc.), the same diagram can be used as a first approximation, by properly rescaling the curve amplitude.

From the law of photography (Sect.2.1, Ref. [2]), we can write the power collected at the receiver objective lens as:

$$P_{bg} = \delta\, E_s\, \Delta\lambda\, NA^2\, (\pi\, d_r^2/4) \tag{3.11}$$

where:

δ = scene diffusivity

E_s = scene spectral irradiance ($W/m^2\mu m$)

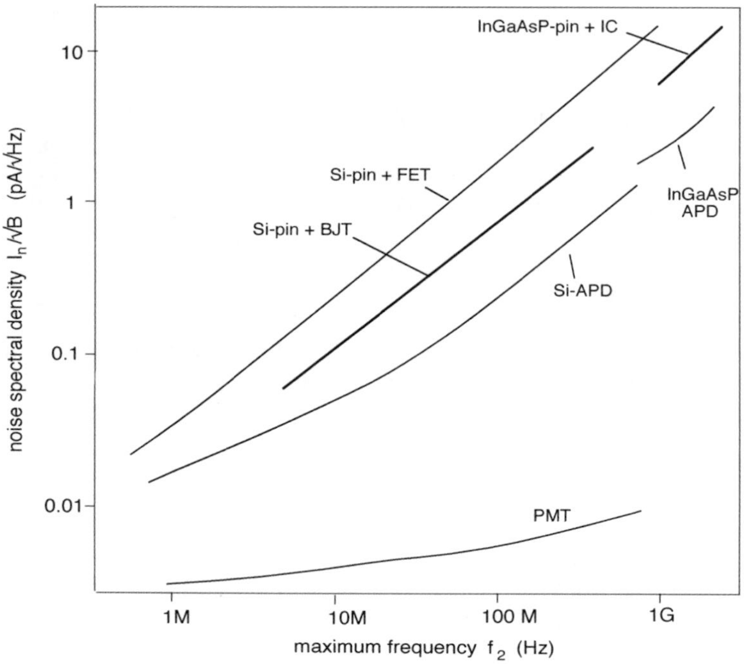

Fig. 3-7 Typical noise performance of receivers, as a function of the maximum frequency of operation f_2. Silicon-photodiodes with FET- and BJT-transistor input-stages (with eventual equalization correction) are compared to avalanche photodiodes (APD) and to photomultipliers (PMT). Si-photodiodes cover the range λ=400...1000 nm, whereas InGaAsP extends up to ≈1.6 μm and standard PMTs are centered in the range 300 to 900 nm. Data are for a detector with area A<0.5 mm^2 and capacitance C<0.5 pF, and front-ends with FET and BJT having a transition frequency f_T ≥2 GHz.

$\Delta\lambda$ = spectral width of the interference filter placed in the receiver lens
NA = arcsin $D_r/2F$ = numerical aperture of the receiver lens
d_r = diameter of the detector

To compare the relative importance of the noise terms, we can use the diagram of Fig.3-8, which is a plot of the quantity $p_n=\sqrt{[(2h\nu/\eta)\ P\ B]}$ for λ=1 μm and η=0.7.

Let us substantiate the previous considerations with the aid of numerical examples for a pulsed telemeter and a sine-wave telemeter.

For a Nd-laser pulsed telemeter operating at λ=1060 nm, we may have (Fig.A3-3) $E_s\approx$ 500 $W/m^2\mu m$ in direct sunlight at AM1.5 (sun elevation 42 deg). Other common design values we may assume are $\Delta\lambda$ =10 nm (for an 80% to 90% transmission of the filter), NA= 0.5, and d_r= 0.2 mm. Taking δ=1 as the worst case for background and inserting in Eq.3.11, we get P_{bg}= 40 nW.

The attenuation is $P_r/P_s = \delta(D_r/2L)^2$. Working at L=1 km with D_r=100 mm and δ=0.3 for the signal, we have $P_r/P_s = 0.075\ 10^{-8}$, that is, for a typical transmitted (peak) power of P_s =0.3 MW, we get a received power P_r=0.23 mW. In addition, the I_{ph0} noise for an APD receiver at 100 MHz is evaluated from Fig.3-7 as 2 pA/√Hz. Multiplying by (√B)/σ, we get $P_{ph0}\approx 4$ nW. Thus, we obtain the points labeled 'pulsed' in Fig.3-8, and we can see that the received signal shot-noise prevails.

For a sine-wave modulated telemeter operating at λ=820 nm, we may have $E_s\approx$ 900 W/m^{2}µm and P_{bg} approximately doubles at a value of 80 nW.

The attenuation is now $P_r/P_s = D_r^2/4\theta_s^2L^2$. Working at L=100 m with D_r=100 mm and θ_s=1 mrad, we have P_r/P_s= 0.25. That is, for a typical transmitted power of P_s =0.1 mW, we get a received power P_r=25 µW.

Again, using an APD receiver at 10 MHz, the noise spectral density is 0.2 pA/√Hz, and multiplying by (√B)/σ with a bandwidth of 100 Hz now, we get $P_{ph0}\approx$5 pW.

In both cases, the largest term is the shot-noise of the received photons. This corresponds to a good design result. On the contrary, if we had omitted the filter for the background or used too large a detector or noisy circuitry, the I_{bg} or I_{ph0} values could easily be increased by orders of magnitude and become the limiting factor of telemeter performance.

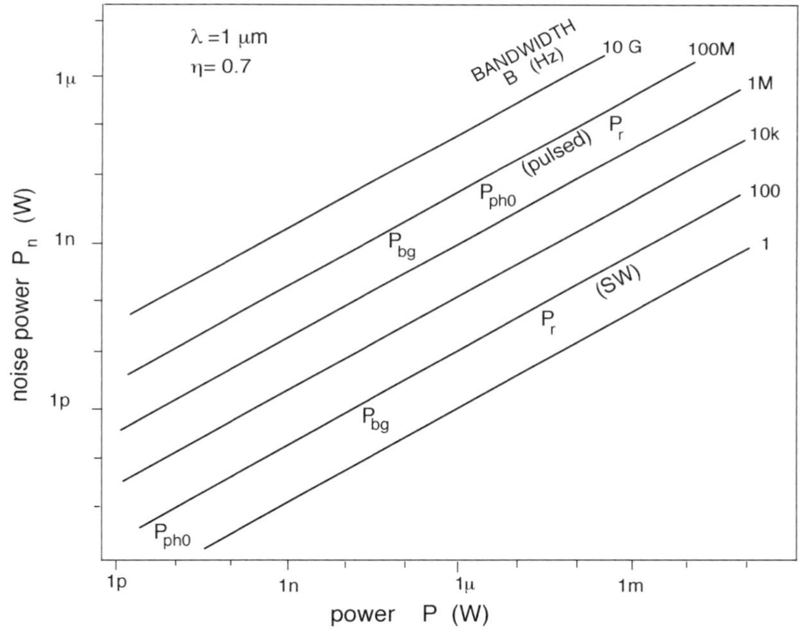

Fig. 3-8 Graph for evaluating the noise contributions associated with signal, background, and detector noise as a function of power. Two examples are reported for a hypothetical sine-wave telemeter on L=100 m (bottom, SW powers) and a pulsed telemeter for L=1-km (top, pulsed powers).

3.2.3 Accuracy of the Pulsed Telemeter

Let us now examine the accuracy of a short-pulse telemeter. In a short-pulse telemeter, we measure the time of flight T of an optical pulse going to the target and back. Distance is obtained as $L=cT/2$, and a timing error ΔT reflects itself in a distance error $\Delta L=c\Delta T/2$, or $\sigma_L=(c/2)\sigma_T$ in terms of the rms error. We shall therefore try to make ΔT as small as possible.

Generally, timing is performed on the electrical pulse supplied by the photodetector as a replica of the optical pulse. We may use a pair of timing circuits, one for the start and one for the stop pulse, or have a single circuit for both. The start is usually picked out as a fraction of the optical pulse leaving the transmitter and is then combined with the stop pulse arriving at the receiver from the distant target.
In any case, the time of flight is measured as $T= t_{st} - t_{sp}$, and the associated rms error is $\sigma_T^2= \sigma_{st}^2+ \sigma_{sp}^2$. However, because the start pulse can be large in amplitude, σ_{st} is usually negligible compared to σ_{sp}, and we may restrict ourselves to consider the stop timing error.

Several approaches and circuit solutions are available to implement time measurements on pulses. Most of them are based on threshold crossing, as illustrated in Fig.3-9.

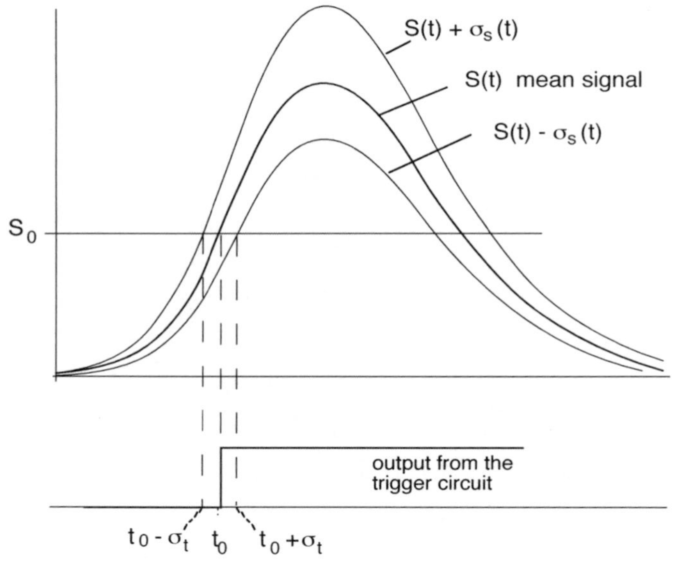

Fig. 3-9 Timing of the received pulse by means of an amplitude discriminator. When the signal crosses the threshold S_0 at time t_0, a timing signal is generated. Amplitude fluctuations $\pm\sigma_S$ produce a timing error $\pm\sigma_t$.

An amplitude discriminator with a threshold S_0 is used. Well-known circuit implementations for it are the Schmitt trigger, the tunnel-diode trigger, and other fast circuits.

When the signal crosses the threshold, the circuit switches and gives a step-wise waveform at the output. The circuit adds a negligible jitter in switching time (<10 ps in a well-designed circuit). Thus, the timing fluctuation σ_t around the average time t_0 is only due to pulse noise, specifically to the fluctuations σ_S on the mean amplitude waveform S(t). The time error σ_t can be related to the error in amplitude σ_S through a linear regression, as:

$$\sigma_t^2 = \sigma_S^2 / |\, dS/dt\,|^2 \qquad (3.12)$$

The linear regression is accurate when the amplitude fluctuation σ_S is small and the signal slope dS/dt is nearly constant in the crossing region, which is a reasonable hypothesis in most cases.

Now, to get a better insight, let us parameterize the problem, introducing normalized variables. The received pulse is written as a power:

$$S(t)= E_r\,(1/\tau)\; s(t/\tau) = P_s(t) \qquad (3.13)$$

where:

- τ is the time parameter (or scale) of the pulse, approximately equal to its duration;
- $s(t/\tau)$ is the dimensionless pulse waveform, with a peak value $s_{peak}\approx 1$ and an area normalized to unity, i.e., $\int_{0\text{-}\infty} (1/\tau)s(t/\tau)dt=1$. The frequency spectrum of $s(t/\tau)$ extends up to $B\approx\kappa/\tau$, with κ being a numerical factor usually not far away from $1/2\pi=0.16$;
- $E_r = h\nu N_{ph}$ is the energy contained in the pulse, as implied by the normalization of $s(t/\tau)$, with N_{ph} for the total number of photons in the pulse.

With these positions, the signal slope is $dP_s(t)/dt = E_r\,(1/\tau^2)\; s'(t/\tau)$. Using Eq.3.10 for the amplitude variance in Eq.3.12 and with Eq.3.13, we have:

$$\begin{aligned}\sigma_t^2 &= \{2h\nu/\eta\; (\kappa/\tau)\; [E_r\,(1/\tau)s(t/\tau)+P_{bg}+P_{ph0}]\}/[E_r(1/\tau^2)s'(t/\tau)]^2 \\ &= \tau^2\,\{2h\nu/\eta\;\kappa\;[s(t/\tau)/E_r + (P_{bg}+P_{ph0})\;\tau/E_r^2]/s'^2(t/\tau)\end{aligned} \qquad (3.14)$$

This expression tells us that the accuracy σ_t is primarily proportional to the pulse duration τ. In addition, we can write Eq.3.14 in the form:

$$\sigma_t = \tau\,(A\; h\nu/E_r + B/E_r^2)^{1/2} \qquad (3.15)$$

with A and B constants. The first term contributes to accuracy as $\sqrt{(h\nu/E_r)}=1/\sqrt{N_r}$, i.e. the inverse square root of the number $N_r = N_{ph}S_0$ of photons collected at the time of threshold crossing. In a well-designed receiver, the shot noise $h\nu/E_r$ term should dominate over the other noises, and the second term in parentheses should be negligible. In this case, we have:

$$\sigma_t \approx \tau/\sqrt{N_r} \qquad (3.16)$$

By introducing N_r in Eq.3.14, we obtain the general expression:

$$\sigma_t = \tau/\sqrt{N_r}\;\; [2\kappa/\eta\; s(t/\tau)/s'^2(t/\tau)]^{1/2}\,[1+\tau(P_{bg}+P_{ph0})/s(t/\tau)E_r]^{1/2} \qquad (3.16')$$

From this expression, we can see that, besides the main dependence on $\tau/\sqrt{N_r}$, the effects of quantum efficiency and pulse shape are summarized by the factor $\sqrt{2\kappa/\eta}$.

Also, the accuracy depends on the relative square slope $s'^2(t/\tau)/s(t/\tau)$, a dimensionless quantity less than, but not so far from unity, whose actual value is determined by the selected threshold $S_0=s(t/\tau)$. Of course, we will choose S_0 so that the threshold-dependent factor $s'^2(t/\tau)/s(t/\tau)$ is maximum.

Finally, if background and electronics noises are not negligible, their effect worsens the ideal accuracy by the factor in square brackets in Eq.3.16'.

This factor can also be written as $[1+N_{bg+ph0}/S_0N_r]^{1/2}$, where N_r is the total number of photons in the pulse, S_0N_r is the number of photons collected at the threshold crossing time, and $N_{bg+ph0}=\tau(P_{bg}+P_{ph0})/h\nu$ is the number of photons equivalent to background front-end powers P_{bg} and P_{ph0} collected in a pulse-duration time τ.
Explicitly, we may rewrite Eq.3.16' as:

$$\sigma_t = \tau/\sqrt{N_r}\ [2\kappa/\eta\ S_0/s'^2(t/\tau)]^{1/2}\,[1+N_{bg+ph0}/S_0N_r]^{1/2} \qquad (3.16'')$$

Example. To carry out a numerical evaluation, let us assume a Gaussian waveform for the telemeter pulse, $s(t/\tau) = [\sqrt{2\pi}]^{-1}\exp-(t/\tau)^2/2$, and a duration τ=5 ns, so that the full-width-half-maximum of the pulse is 2.36×5ns=12 ns. The Gaussian has a bandwidth factor equal to κ=0.13.
The optimum level for the threshold S_0 is found from the condition (see Eq.3.16') $s(t/\tau)/s'^2(t/\tau)$ = minimum, or by inserting the Gaussian function as $(t/\tau)^2[\sqrt{2\pi}]^{-1}\exp-(t/\tau)^2/2$ = maximum. Differentiating this expression with respect to t/τ and equating to zero, we find $(t/\tau)^2=2$ as the result. Accordingly, the optimum (fixed) threshold is $S_0=s(\sqrt{2}) = 1/[e\sqrt{2\pi}] = 0.147$ and the signal slope is $s'^2=2S_0^2=0.043$.
The threshold-dependent factor is therefore $s(t/\tau)/s'^2(t/\tau) = S_0/s'^2=0.147/0.043=3.42$. Assuming η=0.7 and being κ=0.13, the normalized time variance is calculated as $\sigma_t/(\tau/\sqrt{N_r}) = [2\kappa/\eta\times3.42]^{1/2}= [2\times0.13/0.7\times3.42]^{1/2} = 1.12$.

The timing accuracy we obtain is σ_t = 5.7 ns/$\sqrt{N_r}$, which is a value very close to that (5 ns/$\sqrt{N_r}$) given by the first factor in Eq.3.16.

From the data of Fig.3-8, we may expect $P_r\approx$1 mW as the typical received power of the pulsed telemeter, which corresponds to a number of photons $N_{ph}\approx P_r\tau/h\nu=10^{-3}\times5.10^{-9}/2.10^{-19}\approx2.5\ 10^7$. The accuracy of a single-shot measurement is then found as σ_t=5.7 ns/$\sqrt{(2.5\cdot10^7\ S_0)}$ = 2.8 ps, a very good theoretical limit of performance, even beyond the capabilities of electronic circuits (typ. in the range 10 to 50 ps).

As a more realistic pulse waveform, we may use an asymmetric Gaussian, with the leading edge τ_{le} faster that the trailing edge τ_{te}. Repeating the calculation with τ_{te}/τ_{le}=1.8 and τ_{le} =5 ns, we find that the result becomes σ_t = 3.2 ns/$\sqrt{N_r}$, i.e., changes only by about 10%.

Regarding the weight of the last factor in square brackets of Eq.3.16'', we need the number of photons at threshold S_0N_r to be larger than the corresponding number of background and circuit photons N_{bg+ph0} for the factor to be the minimum. Yet, even when it is $N_{bg+ph0}>S_0/N_r$, there is still a $1/\sqrt{N_r}$ dependence, which improves resolution at an increasing number of collected photons.

About repeated measurements, we know from statistics that summing up the outcomes $t_1,t_2,\ldots,t_N$ of N time measurements, we get an average value $\langle t\rangle= (t_1+t_2+\ldots+t_N)/N$ whose accuracy $\sigma_N= \sigma_1/\sqrt{N}$ is improved by $\sqrt{N}$ with respect to the single measurement σ_1. [By the way, Eq.3.16 could have been written directly, based on the consideration that a single photon has an uncertainty σ_t, so a pulse with N_r photons has uncertainty $\sigma_t/\sqrt{N_r}$]. Thus, for a measurement repeated N times, the total number of photons is NN_r and the accuracy $(\sigma_t/\sqrt{N_r})/\sqrt{N}$ is the same as a single measurement with NN_r photons. In practice, however, it is better to repeat several measurements within the permissible acquisition time and get the advantage of the $1/\sqrt{N}$ improvement rather than make a single measurement with a large total number NN_r whose accuracy can become smaller than the resolution time of the electronic circuit, which is then the dominant error. Also, the number of significant digits is increased by averaging because the discretization error $\Delta\tau_{ds}$ is reduced to $\Delta\tau_{ds}/\sqrt{N}$.

3.2.3.1 Effect of Non-Idealities

Let us consider the effect of non-idealities in the electronic circuits processing the pulse. A first non-ideality is the *discriminator noise*, that can be described as an amplitude jitter σ_{S0} superposed to the threshold S_0. Assuming σ_{S0} is homogeneous to S_0, that is, a dimensionless quantity, we may take it into account by adding $(E_r/\tau)^2\sigma_{S0}^2$ to σ_s^2 in Eq.3.12, the multiplying term being the scale factor. Propagating this quantity onward, we find that, at the right-hand side of Eq.3.16', there is an extra term, $\sigma_{S0}/s'(t/\tau)\,|_{S0}$. Simply stated, the amplitude jitter translates in timing jitter when divided by the signal slope.

A second non-ideality is the *finite resolution* of the time-sorting circuit, usually $\Delta\tau_{ds}$=5 to 20 ps for state-of-the-art circuits. As already noted, $\Delta\tau_{ds}$ is a discretization error and is reduced by $1/\sqrt{N}$ by making N measurements.

Third, the circuit receiving the signal from the photodetector before the amplitude discriminator may have a non-negligible *time response*, so that the signal waveform $s(t/\tau)$ becomes $s(t/\tau)*h(t/\theta)$, where $h(t/\theta)$ is the impulse response of the circuit, θ the scale factor and * is the convolution operation. With reasonable choices of $s(t/\tau)$ and $h(t/\theta)$, namely a Gaussian of standard deviation τ for $s(t/\tau)$ and an exponential with time constant θ for $h(t/\theta)$, the resulting time accuracy σ_T has been calculated in Ref.[5] and the result, standardized to $\tau/\sqrt{N_{ph}}$, where N_{ph} is the number of received photon per pulse, is plotted in Fig.3-10.

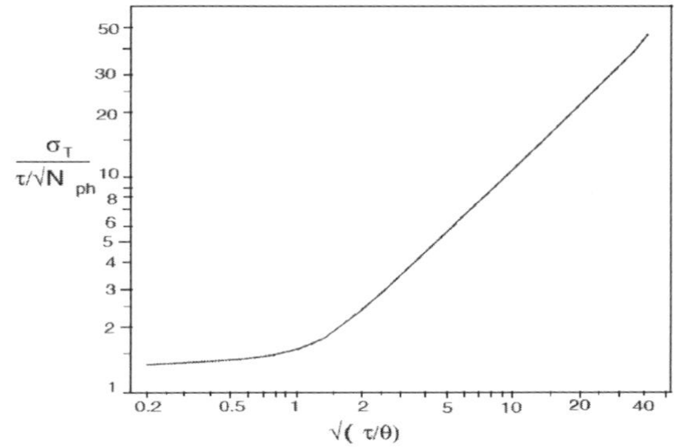

Fig. 3-10 Standardized accuracy $\sigma_T/\sqrt{N_{ph}}$ vs the ratio τ/θ of pulse standard deviation τ and circuit time constant θ (from [5], ©IEEE, reprinted by permission).

As we can see from Fig.3-10, the standardized accuracy approaches the value ≈1 of Eq.3.16 for $\sqrt{\tau}/\theta$ in the range 0.2 to 1.

Another result of interest is the dependence (Fig.3-11) of the fractional threshold S_0/S_{max} that yields the minimum error σ_T and of the corresponding mean crossing-time T_0, as a function of the time-constant ratio $\sqrt{\tau}/\theta$. The fractional threshold, also called constant fraction timing, means that the threshold operation is performed at a fraction h= S_0/S_{max} of the maximum amplitude S_{max} of the signal written as $P_s(t)=(E_r/\tau)S_{max}s(t/\tau)$, i.e., at a fraction $N_r(S_0/S_{max})$ of the number of detected photons. We can see from Fig.3-11 that the fractional threshold changes little (in the range 0.32 to 0.40) with the ratio $\sqrt{\tau}/\theta$, whereas the mean crossing time T_0 varies appreciably, from -0.6τ to -1.4τ as $\sqrt{\tau}/\theta$ swings from 0.2 to 5.

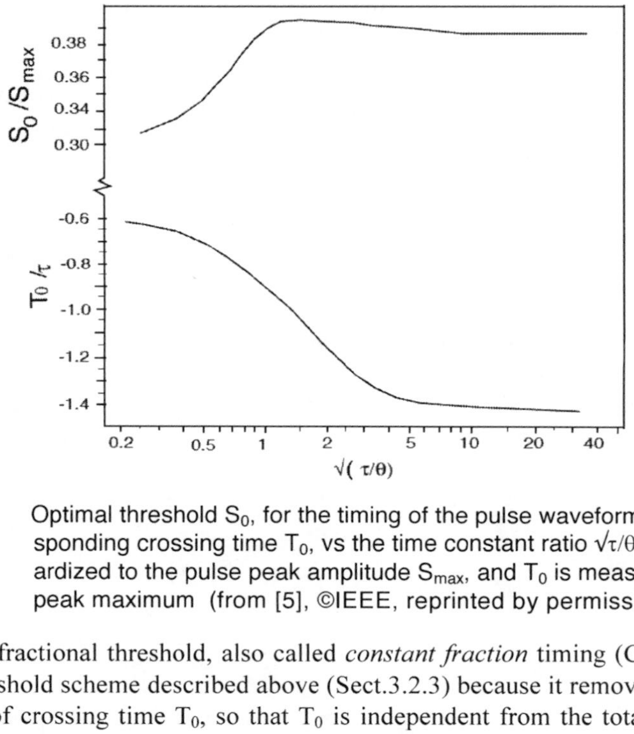

Fig. 3-11 Optimal threshold S_0, for the timing of the pulse waveform, and corresponding crossing time T_0, vs the time constant ratio $\sqrt{\tau}/\theta$. S_0 is standardized to the pulse peak amplitude S_{max}, and T_0 is measured from the peak maximum (from [5], ©IEEE, reprinted by permission).

Operation at fractional threshold, also called *constant fraction* timing (CFT) is superior to the fixed threshold scheme described above (Sect.3.2.3) because it removes the 'rigid shape' fluctuations of crossing time T_0, so that T_0 is independent from the total number of electrons N_{ph} contained in the signal, or has no walk-off error.

CFT is usually implemented by inserting a circuit making a negative integration summed to the delayed signal, so that the impulse response describing the operation is $h(t) = -\alpha\mathbf{1}(t) + \delta(t-T)$, where $\delta(t-T)$ is the delayed Dirac delta, and $\mathbf{1}(t)$ is the Heaviside step function weighted by α, the fractional area threshold value.

On the other hand, pulses coming from an experiment, like those after propagation to a target at different distances, are generally not the same *mean* value N_{ph}. Thus, if we use a constant, fixed value threshold I_0 irrespective of pulse amplitude N_{ph}, we will undergo an additional error, a *systematic* error σ_{T0}, also called the *walk-off* error, due to the change of the average crossing time T_0 with the change of mean amplitude N_{ph} (as depicted in Fig.3-9 when we imagine that the three curves are for three different values of the *mean* number of photons).

In Fig.3-12 we plot the error σ_{T0} standardized to τ, of the mean signal $(eN_{ph}/\tau)S(t)$ made by the above said Gaussian-exponential combination, using a fixed threshold S_0=3.86 S_{max} (optimal for N_{ph}=10), and plot it as a function of the average number N_{ph} of photons per pulse.

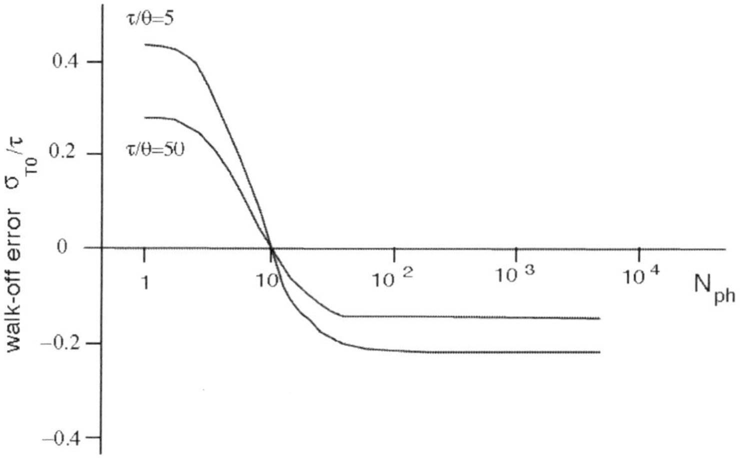

Fig. 3-12 The systematic or *walk-off* error σ_{T0}/τ, for a fixed threshold placed at S_0=3.86 S_{max} versus the mean value N_{ph} of photons per pulse. At increasing τ/θ, the absolute value of the walk-off decreases (from [5], ©IEEE, reprinted by permission).

Important to point out, the systematic error can be canceled out, other than with the CFT, in a fixed threshold timing if: (i), we know the dependence $\sigma_{T0} = \sigma_{T0}(N_{ph})$, [like the one plotted in Fig.3-12], (ii), we are able to measure N_{ph} with a separate arrangement (a beamsplitter at the input, and detector/integrator measuring the total pulse charge or N_{ph}), and (iii), calculate σ_{T0} and subtract it from the measured time T_0. This procedure has been successfully implemented in Ref.[6] to correct results of the time-transfer laser link.

3.2.3.2 Optimum Filter for Signal Timing

Threshold-crossing timing is a viable and widely used technique, but it is not necessarily the optimum one. Some timing information is lost because only the signal around the mean crossing time is used to produce the trigger, whereas all other signal portions are not. We may wonder if there is a better timing strategy, perhaps a variable-threshold or a pulse center of mass, or eventually something else.

To approach the problem, we look for the optimum filter [7,8] which, inserted between the detector and a threshold-crossing circuit, collects all the time-localization information contained in the pulse and makes it available at a threshold-crossing time (Fig.3-13).

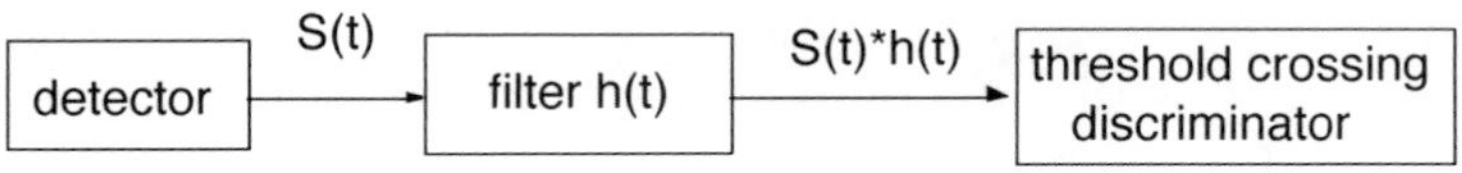

Fig. 3-13 Optimum timing of the pulse waveform. A filter with an impulse response h(t) to be determined is inserted between the detector and the threshold-crossing circuit.

If h(t) is the impulse (or Dirac-δ) response of a filter cascaded to the detector, the mean signal at the filter output is $S_{out}(t)=S(t)*h(t)$, where * stands for the convolution operation [explicitly, $a*b=\int_{0-\infty} a(\tau)b(t-\tau)d\tau$]. The slope signal is $S'(t)*h(t)$ and the amplitude variance is given by $\sigma_S^2(t)*h^2(t)$.
Thus, we can rewrite the time variance as:

$$\sigma_t^2 = \sigma_S^2(t)*h^2(t) \,/\, [S'(t)*h(t)]^2 \qquad (3.17)$$

The calculation of the minimum of σ_t^2 with respect to h(t) is carried out in Ref. [7], and the result is:

$$h(t) = S'(T_m-t) \,/\, S(T_m-t) \qquad (3.18)$$

As we can see from Eq.3.18, the optimum filter response is the time-reverse of the relative slope S'/S. The time of reversal T_m is also the measurement time, at which the mean signal is found to be zero (Fig.3-14).
Indeed, by writing the mean output $S_{out}(T_m)=S(t)*h(t)\,|_{Tm}$ at T_m, we get:

$$S_{out}(T_m) = \int_{0-\infty} S(\tau)h(T_m-\tau)d\tau = \int_{0-\infty} S(\tau)S'(\tau)/S(\tau)d\tau = \int_{0-\infty} S'(\tau)d\tau = 0 \qquad (3.19)$$

Thus, the threshold-crossing measurement with the optimum filter is a zero-crossing at time T_m. An interesting feature of the optimum filter is that the rigid-shape fluctuations superposed to the pulse are canceled out.

In fact, we may write the fluctuations as $\Delta S(t)=\xi S(t)+\zeta s_r(t)$, as the sum of a rigid-fluctuation $\xi S(t)$ proportional to the mean and a shape fluctuation $\zeta s(t)$, where ξ and ζ are random variables, and $s_r(t)$ is a random function orthogonal to S(t). Then, in view of Eq.3.19, just like the mean signal, the rigid fluctuation term gives a zero output at T_m. Only the shape fluctuations remain in the output after the optimum filter and contribute to the timing error.

Using Eqs.3.17 and 3.18, we can calculate the time variance of the signal supplied by the optimum filter as:

$$\sigma_{t\,(opt)}^2= \int_{0-\infty} S(\tau)[S'(\tau)/S(\tau)]^2\, d\tau \,/\, [\, \int_{0-\infty} S'(\tau) S'(\tau)/S(\tau)d\tau]^2 =$$

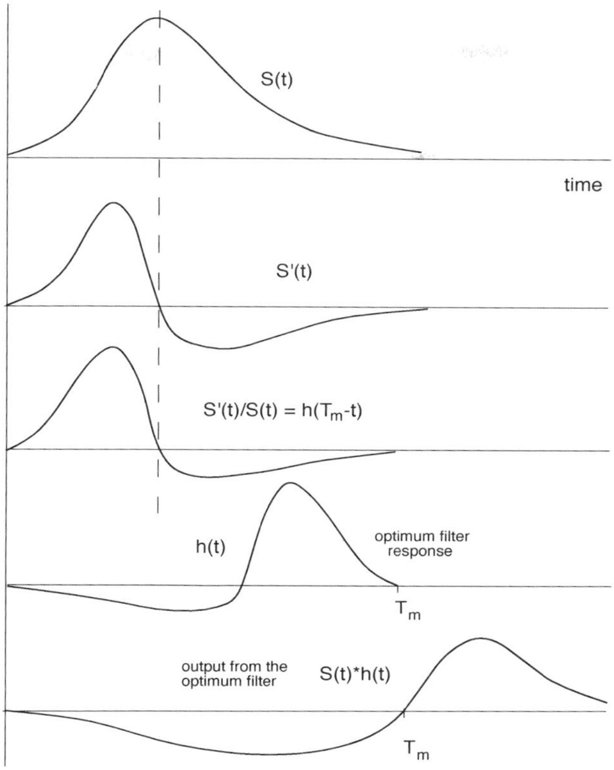

Fig. 3-14 Optimum timing waveforms (top to bottom): mean signal S(t), its derivative S'(t), the filter impulse response h(t), and the output signal S_{out} from the optimum filter. The output exhibits a zero-crossing at the measurement time T_m.

$$= 1 / [\int_{0-\infty} S'^2(\tau)/S(\tau)d\tau] \qquad (3.20)$$

Compared with Eq.3.16', which is valid for a fixed-threshold crossing, we see that the multiplying term S/S'^2 contained in it is now replaced by $[\int_{0-\infty} S'^2/S\, d\tau]^{-1}$, the time integral of the timing factor S'^2/S. In most cases, it is $[\int_{0-\infty} S'^2/S\, d\tau]^{-1} \approx 0.1\text{-}0.3\ S'^2/S$, or we may get an improvement of the timing variance by a factor 3 to 10.

Numerical example. For a Gaussian pulse waveform of the type: $s(t/\tau) = [\sqrt{2\pi}]^{-1}\exp\text{-}(t/\tau)^2/2$, we can evaluate the timing term $[\int_{0-\infty} S'^2/S\, d\tau]^{-1}$ as $\int_{0-\infty}\sqrt{(2\pi)}\ \tau^2\exp\text{–}(\tau^2/2)\, d\tau = 1$, or, the timing variance supplied by the optimum filter is $\sigma_{t\ (opt)}^2 = \tau /\sqrt{N_r(2\kappa/\eta)}$ to be compared with that of the best fixed threshold crossing previously calculated as

$$\sigma_t^2 = \tau /\sqrt{N_r}\,(2\kappa/\eta)\ s(t/\tau)/s'^2(t/\tau) = 3.42\times \tau /\sqrt{N_r}\,(2\kappa/\eta).$$

Thus, the optimum filter yields an improvement by a factor 3.42 in variance, or $\sqrt{3.42}=1.85$ in time rms error.

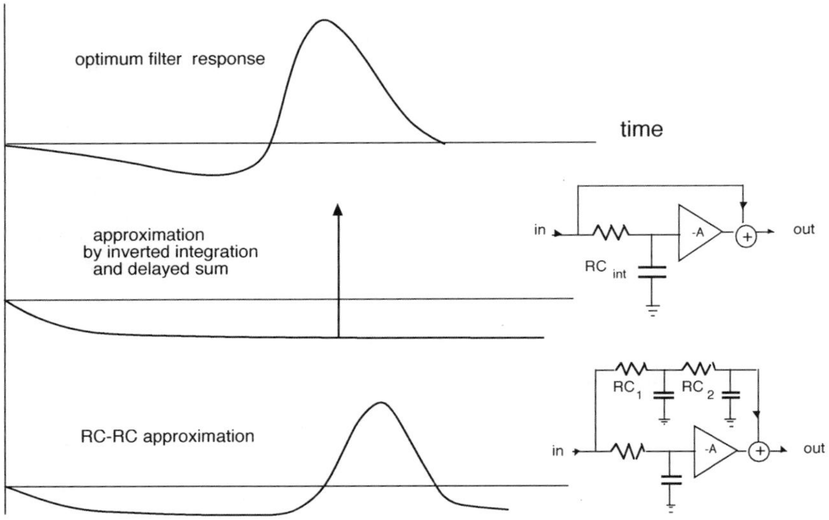

Fig. 3-15 Approximations to the optimum filter: (top) the ideal waveform, (center) the negative integration plus delayed pulse, and (bottom), the negative integration plus delayed approximate (RC-RC) integration.

A question is now appropriate: how sensitive is the improvement in time variance with respect to deviations from the theoretical optimum impulse waveform h(t)? Actually, h(t) may be hard to synthesize with conventional filter synthesis techniques because it has a slowly varying part at short times followed by a sudden jump later in time, which is the opposite of what is usually found.

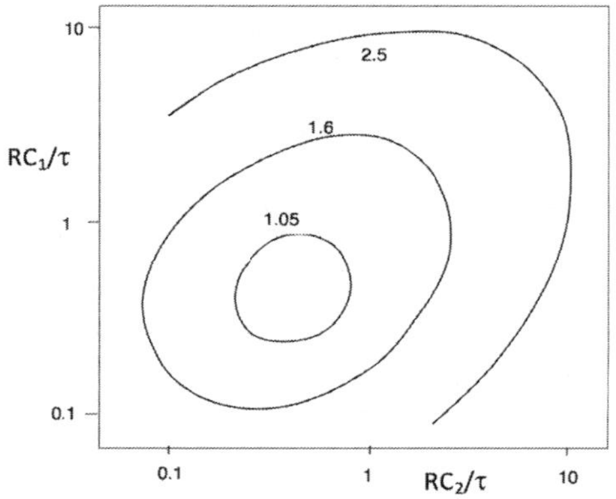

Fig. 3-16 The timing variance obtained by the RC-RC approximation in units of the optimum value, $\sigma_t^2 / \sigma_{t\,(opt)}^2$, as a function of some ratios of RC time constants to pulse duration τ.

An approximation to h(t) may be constructed as follows. We first integrate the signal S(t) and then sum on the integrated signal an inverted and delayed replica of S(t) (Fig.3-15).
A refinement is to smooth the response, by adding two approximate integrators with time constants comparable to the pulse duration.
Summing the pulse with inverted polarity on the integrated level is equivalent to generating a zero-crossing dependent on the pulse total area, a concept called Constant-Fraction-Timing (CFT). A calculation of the resulting timing variance $\sigma_t^2/\sigma_{t\,(opt)}^2$ is shown in Fig.3-16. As it can be seen, values are not far away from the optimum one given by Eq.3.20.

3.2.4 Accuracy of the Pulsed Long-Wave Telemeter

The discussion of previous sections is relevant to a *direct* ToF (time-of-flight) measurement, that is, the traditional readout employed for high-performance rangefinders based on the start and stop timing followed by a time-interval counter or TDC (time-to-digital converter), for use in a variety of applications like topography and geodesy. The direct ToF is capable of resolving centimeters up to distances in the kilometer range and meters up to hundreds of kilometers.

However, when the accuracy requirements are less demanding, like for a car anti-collision telemeter, usually called LiDAR, much simpler arrangements can be employed. Two of them have gained general acceptance, recently, the *indirect* ToF [5,9] and the FMCW [10]. The FMCW is a coherent technique and will be described in Ch.4, whereas the indirect ToF is considered below.

As we can see in Fig.3-17, a long rectangular pulse is used to illuminate the scene, with a duration $T_{max}=2L_{max}/c$ large enough to cover the maximum distance L_{max} of interest. The detector is kept open in the time interval 0–T_{max}, so that it will collect a constant rate of arriving photons in the time interval T–T_{max}, where T=2L/c for the distance to be measured.

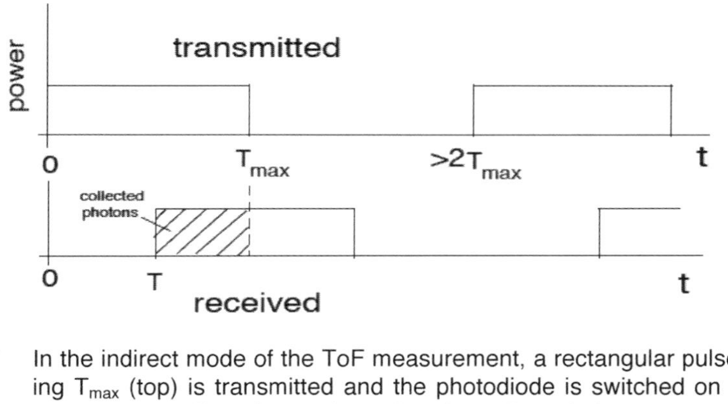

Fig. 3-17 In the indirect mode of the ToF measurement, a rectangular pulse lasting T_{max} (top) is transmitted and the photodiode is switched on in the time interval 0-T_{max} (bottom), so that the collected photons are a fraction $(T_{max}-T)/T_{max}$ of the total transmitted N_{ph} photons (dashed area) (from [9], ©IEEE, reprinted by permission).

For this scheme, the waveform of Eq.3.13 is given by $s(t/\tau)$ = rect $(0, T_{max})$, where rect $(0, T_{max}) = \mathbf{1}(t)-\mathbf{1}(t-T_{max})$ is the rectangle waveform of transmitted pulse. Letting N_{ph} be the average number of transmitted photons, the mean number $<n>$ of photons collected in the interval 0, T_{max} is expressed as:

$$<n> = (T_{max}-T) <P(t)>/h\nu = (N_{ph}/T_{max}) (T_{max}-T) \quad (3.21)$$

and is itself Poisson distributed, so it has a variance equal to the mean value, $<\Delta n^2>=<n>$. By taking the differential of both sides in Eq.3.21, and squaring and averaging, we get:

$$(N_{ph}/T_{max})^2 \sigma^2_T = (N_{ph}/T_{max}) (T_{max}-T)$$

whence, upon rearranging the terms, we find the rms timing error σ_T as:

$$\sigma_T = (T_{max}/\sqrt{N_{ph}}) (1-T/T_{max})^{1/2} = \kappa_i (T_{max}/\sqrt{N_{ph}}) \quad (3.22)$$

where $\kappa_i = (1-T/T_{max})^{1/2}$ is a factor not much different from unity [for example, if we limit the measurement range of T to, say, 0 to 0.75 T_{max}, we have κ_i=0.5 to 1.0].

In conclusion, also for the indirect ToF-telemeter, the main dependence of the timing error is from a characteristic time, the maximum-range time delay T_{max} in this case, divided by the square-root number of detected photons.

3.2.5 Accuracy of the Sine-Wave Telemeter

Sine-wave modulation of power is a good strategy to impress time-localization information on a quasi-continuous source. It is especially used with semiconductor lasers that are capable of supplying substantial dc power and being modulated at high frequency, but are not good for pulsed operation at a high peak power level.

Equally important, the use of modest power is also a system requirement in applications, like topography, that call for intrinsic compliance to laser-safety standards.

To analyze the accuracy of the sine-wave telemeter, let us write the transmitted power as:

$$P_s(t) = P_{s0} [1+m \cos 2\pi f_m t], \quad (3.23)$$

where P_{s0} is the mean transmitted power, f_m is the modulation frequency, and m is the modulation index. The power signal returning from the target at a distance L=2cT is:

$$P_r(t) = P_{r0} [1+m \cos 2\pi f_m(t-T)] \quad (3.24)$$

In view of the system equation (Eq.3.5), the received mean power P_{r0} is calculated from P_{s0} as $P_{r0}=P_{s0}GD_r^2/4L_{eq}^2$. The photodetected signal then follows as $I_r(t)=\sigma P_r(t)$, with σ being the spectral sensitivity of the photodetector, and is given by:

$$I_r(t) = I_{r0} [1+m \cos 2\pi f_m(t-T)] \quad (3.24')$$

The distance to be measured is contained in the phase-shift $\varphi=2\pi f_m T$ of the received signal relative to the transmitted one. An error $\Delta\varphi$ in phase reflects itself in a time error $\Delta T=\Delta\varphi/2\pi f_m$. By squaring and averaging, the relation between phase and time accuracy is:

$$\sigma^2_T = \sigma^2_\varphi \, (1/2\pi f_m)^2 \tag{3.25}$$

and, of course, the corresponding distance accuracy is $\sigma_L = (c/2)\,\sigma_T$.

Now, contrary to the pulsed telemeter, we have a long measurement time T_r available for averaging on the large number of sine-wave periods marking the time delay, not a single pulse, so that noise is reduced.

To implement the phase-shift measurement, we mix the received signal with a reference local oscillator I_{lo} at the same frequency and with an adjustable phase, that is, with a signal of the form $I_{lo}=I_0 \cos(2\pi f_m t+\varphi_0)$.

Basically, the mixer is a circuit made of a square-law element and a time integrator to average the result, and it produces an output $S_\varphi= \langle I_r\, I_{lo}\rangle$ of the two inputs I_r and I_{lo} applied to it. We may also regard the mixer as the circuit performing homodyne (electrical) detection of the signal $I_r(t)$, with the aid of a reference, the local oscillator I_{lo}. The result of the homodyning is a signal carrying the cosine of the phase difference.

Indeed, by inserting the expressions of I_r and I_{lo} in $S_\varphi= \langle I_r\, I_{lo}\rangle$ and developing the cosine product in sum and difference of the arguments we get after some easy algebra:

$$S_\varphi = \left\langle\, I_0 \cos(2\pi f_m t+\varphi_0)\, I_{r0}\,[1+m \cos 2\pi f_m(t-T)]\,\right\rangle =$$

$$= (I_0\, I_{r0}/2)\, m \cos[2\pi f_m T +\varphi_0] \tag{3.26}$$

The cos ϕ dependence in Eq.3.26 reveals that the mixer output is sensitive to the *in-phase* component of the received signal, a well-known feature of homodyne detection. For $\phi\approx 0$, the sensitivity of the mixer signal cos ϕ to small variations of ϕ is about zero. To have the maximum sensitivity, we shall work with signals in quadrature, and this can be done by adjusting the local oscillator phase to $\varphi_0=-2\pi f_m T+\pi/2$.

With this strategy, the mixer signal S_φ is kept dynamically to zero, and the time measurement is obtained as $T=(\pi/2-\varphi_0)/2\pi f_m$.

We can write the total phase in Eq.3.26 as $\phi=\pi/2+\varphi$ to indicate that the mixer works on a small phase signal φ around the $\pi/2$ quadrature condition. The mixer output is then:

$$S_\varphi = (I_0\, I_{r0}/2) m \cos(\pi/2+\varphi) = -(I_0\, I_{r0}/2) m \sin\varphi \approx -(I_0\, I_{r0}/2) m\varphi \text{ for } \varphi<<1.$$

To take account of the fluctuation ΔI_r superposed to the received signal, we insert $I_r+\Delta I_r$ in $S_\varphi= \langle I_r\, I_{lo}\rangle$ and change φ in $\varphi+\Delta\varphi$. Developing the product and averaging, we obtain $S_\varphi = (I_0\, I_{r0}/2)m \sin\varphi + \Delta I_r (I_0/2) + (I_0\, I_{r0}/2)m\Delta\varphi$.

From this expression, it is easy to find the relation between the phase variance $\sigma_\varphi^2 = \langle\Delta\varphi^2\rangle$, and received signal variance $\sigma^2_{Ir} = \langle\Delta I_r^{\,2}\rangle$ as:

$$\sigma_\varphi^2 = \sigma_{Ir}^2 / m^2 I_{r0}^2 \tag{3.27}$$

The corresponding relation, in terms of equivalent powers at the detector input (see Eqs.3.9 and 3.10) is:

$$\sigma_\varphi{}^2 = p_n{}^2 \,/\, m^2 P_{r0}{}^2 \qquad (3.27')$$

Now, combining Eqs.3.27' and 3.25 and using Eq.3.10, we obtain the time variance of the sine-wave modulated telemeter as:

$$\sigma_T{}^2 = (1/2\pi f_m)^2 [2h\nu/\eta\, B\, (P_{r0}+ P_{bg}+ P_{ph0})]/m^2 P_{r0}{}^2 \qquad (3.28)$$

As we can see from Eq.3.28, the primary dependence of accuracy in the sine-wave modulated telemeter is from $1/2\pi f_m$, the inverse of the angular frequency of modulation. By comparing with Eq.3.16', we see that $1/2\pi f_m$ is the counterpart of the pulse duration τ in a pulsed telemeter.

Further, in Eq.3.28 we can let $B=1/2T_r$ for a measurement lasting a time T_r, and can introduce the number of received photons $N_r= mP_{r0}T_r/h\nu$ contributing to the measurement. With these positions, we can rewrite Eq.3.28 as:

$$\sigma_T = (1/2\pi f_m\sqrt{N_r})\,(1/\eta m)^{1/2}\,[1+(P_{bg}+P_{ph0})/P_r]^{1/2} \qquad (3.29)$$

As for the pulsed telemeter, the dependence of timing accuracy is from the inverse square root of the total number N_r of signal photons collected in the measurement time T_r.

The counterpart of Eq.3.16'', expressing the accuracy in terms of the equivalent photon number N_{bg+ph0}, is:

$$\sigma_T = (1/2\pi f_m\sqrt{N_r})\,(1/\eta m)^{1/2}[1+(N_{bg+ph0})/N_r]^{1/2} \qquad (3.29')$$

By comparing the two equations, we can see that the performances are theoretically equivalent when we make $1/2\pi f_m$ equal to τ and $2\kappa S_0/s'^2$ equal to $1/m$.

Finally, let us consider the optimum filter treatment of the sine-wave modulated telemeter. If the signal waveform is given by Eq.3.24', from Eq.3.18 the impulse response of the optimum filter is found as:

$$h(T_m\text{-}t) = \sin 2\pi f_m t\,/[1+m \cos 2\pi f_m t] \qquad (3.30)$$

At the measurement time T_m, the optimum filter weights the signal as

$$s{*}h = \int_{0\text{-}T_m} m \cos 2\pi f_m t \times \sin 2\pi f_m t\,/[1+m \cos 2\pi f_m t]\, dt.$$

For $m<<1$, the weight is $\sin 2\pi f_m t$, exactly that of an in-quadrature homodyne detector.

3.2.6 The Ambiguity Problem

In the preceding sections of this chapter, we have studied the theoretical performance of time-of-flight telemeters and analyzed the dependence of accuracy from the physical parameters. This is the ultimate limit we can achieve in a well-designed telemeter and, as such, is the first step of design before developing the details.

Now, we shall review the specific problems, if any, that originate from the measurement approach (the time of flight) we are considering. After doing so, we will be ready for the instrument development. Incidentally, this procedure is generally valid as a good engineering approach to develop instrumentation.

A problem with time-of-flight telemeters is the measurement ambiguity arising because the signal is inherently periodic (Fig.3-18).

In the sine-wave modulated telemeter, ambiguity shows up when the period $T_m=1/f_m$ of the modulation waveform is shorter than the time of flight T=2L/c (and actually, we want a small T_m to get good accuracy). Thus, the phase-shift $\varphi=2\pi f_m T$ to be measured will be a multiple $n2\pi$ of the round angle, plus a residual φ', or $\varphi= n2\pi+\varphi'$. The phase meter measures φ' but cannot tell anything about n.

In the pulsed telemeter, we get an ambiguity when the repetition time T_r is shorter than the time of flight T=2L/c, and we send a second pulse before the first pulse has returned to the receiver (Fig.3-18). Thus, for a pulse transmitted at t=0, we receive a correct return at t=T, but also spurious returns at $t=T-T_r$, $T-2T_r$, etc., due to the previous pulses still in flight.

A trivial cure for ambiguity is, of course, to keep $T_{r,m} \geq T$ at all times. Using T=2L/c, the condition reads $L \leq cT_{r,m}/2$.

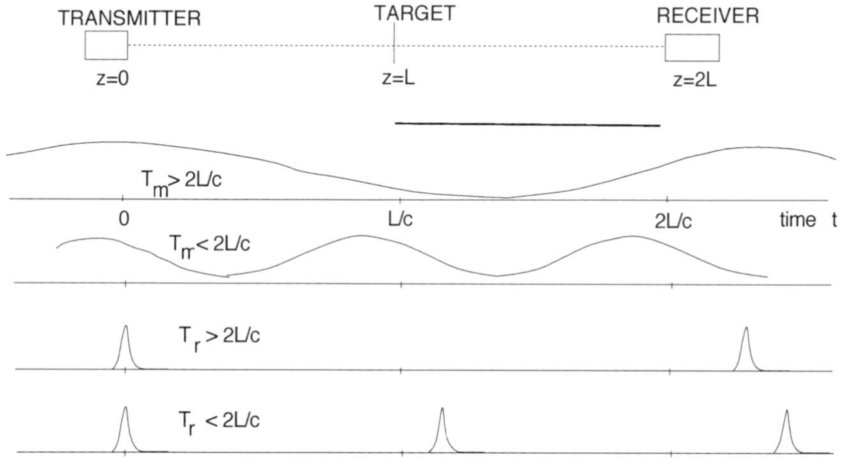

Fig. 3-18 In a time-of-flight telemeter, here drawn with the propagation path unfolded, the measurements present an ambiguity when more than a single period of modulation (T_m or T_r) is contained in the time-delay interval 2L/c. To avoid this, we shall have $T_{m,r}$ >2L/c or, equivalently, shall limit the useful range of distance to less than $cT_{m,r}/2=L_{NA}$, the non-ambiguity distance.

With the equal sign, the distance L becomes the maximum non-ambiguity range L_{NA} of the telemeter. This quantity is given by $L_{NA}=cT_{r,m}/2=c/2f_{r,m}$ at a given repetition (or modulation) period T_r (or T_m) or frequency f_m (or f_r).

Now, we can generalize the argument by writing the modulation waveform as a periodic function of time $\Gamma(t/T_{r,m})$, where Γ is a sinusoid for the sine-wave telemeter and is a pulse sequence for the pulsed telemeter. After a propagation on a distance 2L, the waveform is $\Gamma[(t-2L/c)/T_{m,r}]= \Gamma(t/T_{m,r}-L/L_{NA})$, where L_{NA} assumes the meaning of modulation wavelength. The maximum non-ambiguity range is then the wavelength of modulation L_{NA}.

In a pulsed telemeter, we usually start with a laser source capable of supplying a short, high peak-power pulse, like a Q-switch solid-state laser or a semiconductor laser diode-array. These sources work intrinsically at a low duty cycle, and the typical repetition rate may be f_r =10 Hz to 10 kHz. The corresponding non-ambiguity range is calculated as $L_{NA}=c/2f_r$= 15,000 km to15 km and poses no problem in most cases. Only at the highest repetition rate may we face ambiguity on very long-distance operation.

In this case, however, we may devise a strategy to manage the superposition of a moderate number N of pulses down the measurement distance. For example, we may use N time-sorters and feed each of them with the correct start-stop pair of pulses, selected by a multiplexer circuit. The multiplexer is implemented by a fully decoded base-N counter, whose outputs open linear gates connecting the start/stop pulses to the inputs of the time sorters. Operation is initialized at a very low repetition rate ($f_r<c/2L_{max}$, for non-ambiguity) with the measurements being distributed cyclically on the N time sorters. Then, the repetition rate is gradually increased up to N times the maximum frequency of non-ambiguity, or $f_r = Nc/2L$. This strategy is equivalent to expanding by a factor N the maximum non-ambiguity range.

In a sine-wave modulated telemeter, the ambiguity problem is more severe. Indeed, as we work with a quasi-CW power, we cannot trade duty cycle for non-ambiguity because received power and hence accuracy would be sacrificed.

We can only use a modulation frequency f_m low enough to avoid ambiguity on the distance range to be covered. The non-ambiguity range is L_{NA} = $c/2f_m$, and for L_{NA}= 0.15 km...1.5 km we need to keep the modulation-frequency f_m as low as f_m= 1 MHz...100 kHz.

In contrast, both a semiconductor laser diode and the phase-measuring circuit can easily handle signals with modulation frequency up to several hundred MHz (that is, 3 decades higher). Theoretically, the distance accuracy is $\sigma_L=c\sigma_T\approx c/2\pi f_m\sqrt{N_r}$ from Eq.3.27, and the relative accuracy is $\sigma_L/L_{NA}=1/\pi\sqrt{N_r}$.

Using $N_r \approx 10^7$ detected photons, accuracy would be theoretically limited to $\approx 10^{-4}$ of the maximum range L_{NA}, or we would get a measurement with no more than a four-decade dynamic range.

In practice, the phase-measuring circuit may introduce an additional limit. The phase resolution $\Delta\varphi_r$ is usually 10^{-3} or 10^{-2} of the round-angle 2π.

Because the phase 2π corresponds to L_{NA}, the phase $\Delta\varphi_r$ corresponds to one unit of the least-significant digit in our measurement.

This means that the dynamic range is limited to $\approx 1/\Delta\varphi_r$, or only two to three decades in this example. To overcome this limit, we may take advantage of the fact that phase resolution

does not worsen appreciably as the modulation frequency is increased up to the maximum that can be handled by the phase-measuring circuits.
Because distance resolution is $(c/2f_m)\Delta\varphi_r/2\pi$, we can then recover at least three decades of dynamic range by working at increased frequency.

Of course, to exploit this possibility, we need to combine a high frequency of modulation for resolution and accuracy to a low frequency as required by the non-ambiguity regime, which will be shown in the next sections.

3.2.7 Intrinsic Accuracy and Calibration

In a time-of-flight telemeter, we measure a time T and convert it to a distance L=cT/2. The intervening multiplication factor is the speed of light, c=299,793 km/s in vacuum [11] and c/n in a medium with index of refraction n.
For propagation through the atmosphere, the most common case for a telemeter, n, differs from unity by a modest amount, yet about $n-1=300 \cdot 10^{-6}$ or 300 ppm (part-per-million).

Thus, as the time-of-flight measurement of distance comes to resolve the fourth or fifth decimal digit, we have to care about the index of refraction of the air and apply the n-1 correction to calibrate the instrument. In this way, we can have a calibration up to the fifth decimal digit and eventually be able to reach the sixth.

The problem of air index of refraction correction is shared by interferometers, where we may eventually go to the 7th or 8th decimal digit.
Further information on the index of refraction correction can be found in Sect. 4.4.4.

3.2.8 Transmitter and Receiver Optics

With regard to the optical system of a time-of-flight telemeter, we may use either two separate objective lenses for the transmitter and receiver or a single one serving both transmitter and receiver (Fig.3-19).

The single-objective solution has the advantage of component saving and is the simplest to implement. The field of view (fov) of the receiver and field of illumination (foi) of the transmitter coincide, and there is no dead zone in front of the instrument. A disadvantage of the optical axes of transmitter and receiver being the same is the particularly strong back-scattered power from the atmosphere collected by the receiver. This may limit the performance of weak-echo detection, especially if the target is non-cooperative. In addition, an insertion loss is incurred because we insert the beamsplitter to deviate the returning beam onto the photodetector (Fig.3-15). In a double-objective system the back-scattered disturbance is much reduced, but field of view and field of illumination become superposed only after a distance L_{dz}, similar to the triangulation distance.

Thus, unless provisions are taken to move the lens axes or the detector position at short distance, the measurement has a dead zone. A dead zone is, of course, a minor problem with long-distance telemeters for geodesy, whereas it may be of importance in short-range telemeters intended for topography.

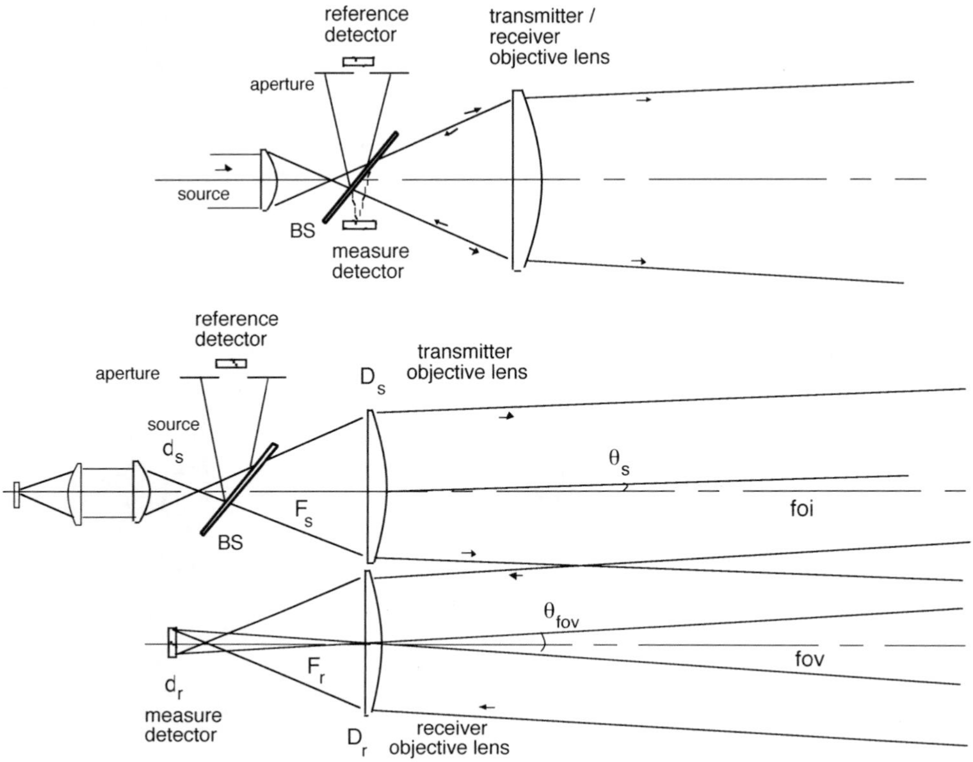

Fig. 3-19 In a time-of-flight telemeter, we may use either a single-objective lens for both transmitter and receiver (top) or a double-objective system (bottom). The beamsplitter allows collecting a fraction of the outgoing optical signal for the reference photodetector to be used in the reference channel.

In both cases, a beamsplitter (BS in Fig.3-19) is inserted in the transmitter lens to pick up a minute fraction of the outgoing optical signal and to detect it by a reference-channel photodetector. If the same type of photodetector is used in the reference and measurement channels, errors due to residual phase-shifts or delays of them are canceled out.

The beamsplitter reflectivity at the surface hit by the outgoing beam is kept at 1% or less, whereas the other surface is either antireflection coated (double-objective case) or coated to 50% reflection (single-objective case). In the double passage across the beamsplitter of the outgoing and returned beams, the loss is TR=0.25 (or 6 dB) in the single-objective optical system, as opposed to $\approx$ 0 dB in the double-objective system. If we take advantage that the laser output is usually linearly polarized and add polarization-control elements to combine the beams, the loss can be reduced to 3 dB. Last, we may go down to $\approx$0 dB if we replace the beamsplitter with optical circulator, but the cost of this component is usually too high to be afforded. The zero-distance edge of the telemeter yardstick is determined by measurement and reference-channel photodetectors. Considering their virtual position along the transmitter path, as produced by the beamsplitter, it is easy to see that the zero-distance location is the midpoint of the segment connecting these two positions.

Let us now consider the design of receiver and transmitter lenses. The constraint to start with is the maximum allowable size of the objective lenses D_r and, of course, we want to be able to use the smallest detector size d_r for the best noise and speed performance. To minimize d_r, we need to keep the spot size w_t on the target as small as possible.

For a Gaussian beam, the minimum obtainable size as set by the diffraction limit is given by $w_t = \sqrt{(\lambda L/\pi)}$, where L is the target distance (see Sect. 2.1). Correspondingly, the transmitter objective lens shall have a diameter $D_t = \sqrt{2} w_t = \sqrt{(2\lambda L/\pi)}$ to project the spot w_t at distance L. For L=100 m...10 km and at λ=1 μm, we get D_t = 0.8...8 cm, quite reasonable values.

At the distance L, the spot w_t is seen under the angle $\theta_{fov} = w_t/L$ (=0.1...0.01 mrad in the previous example). Working with a cooperative target (a corner cube), the retro-reflected beam has a size $2w_t$ at the receiver, and this value is the lens-size D_r we require.

When the target is non-cooperative (a diffuser), the received signal is proportional to D_r^2 (Eq.3.4) and we will use the largest D_r that is practically allowed by the application.

To collect from a θ_{fov} field of view, we need a detector size $d_r > \theta_{fov} F_r$, where F_r is the focal length of the receiver objective. Letting F_r=100 mm, we get d_r >10 μm for θ_{fov} = 0.1 mrad.

With respect to the transmitting objective, if the source collimated beam has a waist w_l, we need a magnification $M = w_t/w_l$.

Then, we set $M = F_t/f$, the ratio of the focal lengths of the two lenses constituting the collimating telescope indicated in Fig.3-15, top. If the source provides a near-field spot as an output (for example, is the output facet of a diode laser), we will use a separate collimating lens to enter the telescope.

This lens will include an anamorphic corrector to circularize the beam.

3.3 Instrumental Developments of Telemeters

In the following, we describe some conceptual approaches that have been used to develop pulsed and sine-wave modulated telemeters. As it is well known in instrumentation, the optimum design approach strongly depends on the available components and may become much different as the performance of components undergoes even minor changes. A telemeter is no exception, and the examples reported in the following are illustrative of the design approaches and ingenuity of researchers in the last two decades.

3.3.1 Pulsed Telemeter

A very simple pulsed telemeter can be developed starting from the block scheme shown in Fig.3-20. Two objective lenses are used for the transmitter and receiver, and the start pulse is derived with a beamsplitter from the outgoing optical beam. The stop pulse is detected with the second objective lens aimed at the target. If the light pulse is well correlated to the electrical drive of the laser, that is, the laser response to pulse drive has a low time jitter, we can also use the signal of the pulser circuit itself as the start pulse.

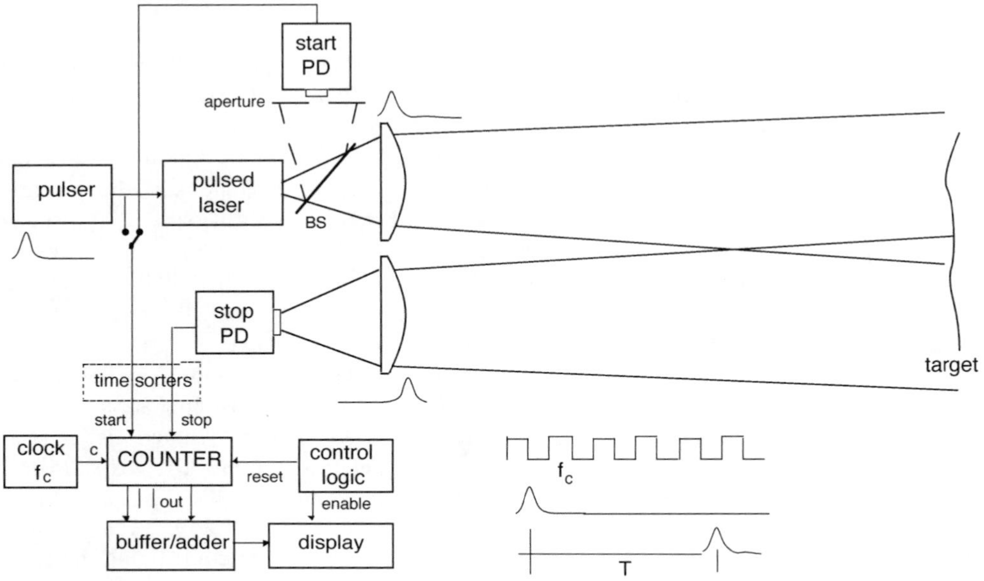

Fig. 3-20 Scheme of a pulsed telemeter using separate objective lenses for transmitter and receiver. Start and stop pulses are used to gate on and off the counter, for which the input is a clock at frequency f_C. After a time 2L/c, the counter content is then N_{count}= f_C 2L/c, as a result of a single measurement. With the aid of the buffer/adder, we may sum the results of N_m successive measurement, getting a total N_m N_{count} as the output of the display. If the laser pulse response has a low time jitter, we may even use the driving signal of the pulser as a start pulse.

This variant allows us to save a few parts (BS, PD, and preamp in Fig.3-20), and it can be used with semiconductor-diode sources, whereas crystal Q-switched lasers have excessive jitter.

Start and stop pulses enter time sorters to improve time localization and to standardize the waveform for the subsequent input to the counter circuit. The counter is fed by a clock at frequency f_c, and the periods of f_c are counted in the time interval T=2L/c defined by start and stop pulses. Thus, in a single measurement, the content of the counter we get is:

$$N_{count} = f_c\,T = f_c\,2L/c = L/(c/2f_c) \tag{3.31}$$

[here, we implicitly assume to take the integer part of the quantities at the right-hand side of the equation]. Eq.3.31 tells us that the quantity $c/2f_c = L_1$ is the scale factor representing the length of a single unit of N_{count}, so that the mean value of distance L is:

$$L = N_{count}\, L_l \tag{3.32}$$

The quantity L_l also represents the unit U of rounding off in the measurement of L. Because the rounding-off error has a variance $U^2/12$ and we make two truncation errors with the start and stop pulses sampling the clock, the total of the rounding-off error is twice as much, or $U^2/6$. In terms of distance, this amounts to a variance $\sigma^2_{L(ro)} = L_l^2/6$.
Adding the timing error σ^2_t (Sect.3.2.3) multiplied by $(c/2)^2$, we get the total distance variance of our telemeter as:

$$\sigma^2_{L(tot,\,1)} = (c^2\sigma^2_t)/4 + L_l^2/6 \tag{3.33}$$

We may repeat the measurement N_m times to improve resolution and accuracy, provided the application at hand allows for an increased total measurement time, now given by $N_m T_r$ where T_r is the pulse repetition period. Repeating the measurement N_m times and summing up the results of each measurement, we get as an average of the accumulated counter content:

$$N_{count} = N_m\, f_c\, T = N_m\, L/(c/2f_c) \tag{3.34}$$

Also, the distance variance decreases by a factor N_m, or:

$$\sigma^2_{L(tot,\,Nm)} = [c^2\sigma^2_t/4 + L_l^2/6]\,/N_m \tag{3.35}$$

Now, we choose the clock frequency f_c according to two criteria: (i) it shall have numerals such that N_{count} represents the target distance L in metric units, and (ii) it shall be high enough to attain the desired resolution.
Because of requirement (i), we need $L_l = N_m\, c/2f_c$ to be an exact decimal number, for example 0.1, 1 or 10 m.

As a first approximation, it is $c = 3\cdot 10^8$ m/s, and hence the frequency should be $f_c = 1.5\cdot 10^9$, $1.5\cdot 10^8$, and $1.5\cdot 10^7$ Hz in the previous example for $N_m = 1$.
More precisely, if we take n=1.0003 and c_{vac}= 299,793 km/s (see Sect.3.2.7), we have c= c_{vac} /n=299,6731, and the numerals of f_c are 1.4983655 (in place of 1.5). This value requires a further correction against wavelength and temperature/pressure as we go to resolve the 5th-6th decimal digit (see Sect.3.2.7).
Regarding the clock, we will use a good quartz-oscillator circuit, for which its accuracy in frequency may easily reach the 5th-6th decimal numeral, and even the 6th-7th in special units.

Because of requirement (ii), the order of magnitude of f_c shall be compatible with the optical pulse duration τ, and its value determines the required speed of the counter circuits.

As a rule of thumb, we should choose f_c not much exceeding the inverse of the pulse duration $1/\tau$, so that the two terms in Eq.3.35 contributing to the distance variance have approximately equal weights.
Indeed, if we just take $f_c = 1/\tau$, we have one count per pulse duration. This loosens the speed requirement of circuits, but probably wastes some of the accuracy available in the pulse waveform. On the other hand, taking $f_c = 10/\tau$, we need faster circuits and better time sorters, but are able to optimize the time accuracy down to the limits discussed in Sect.3.2.3.

In general, when writing $f_c = M/\tau$ where M=1...10 in the previous examples, the single count of the telemeter represents a length-equivalent increment $c/2f_c = c\tau/2M$.

Examples: with a pulse duration τ=1, 10, 100 ns we have $c\tau/2$= 0.15, 1.5, 15 m, and $L_l = c\tau/2M$ is a submultiple of these values, for example, 0.05, 0.5, 5 m for M=3. Using $f_c = 3/\tau$, the required clock frequency is f_c = 3 GHz, 300 MHz, and 30 MHz, respectively.

The case τ=30 ns, M=3, and f_c =100 MHz is representative of a pulsed-telemeter based on a semiconductor laser source. Counters and sorters can work easily at this medium speed, and the circuits can be implemented at a relatively low cost. The least significant digit in a single measurement of this hypothetical telemeter is 1.5 m.

By repeating the measurement N_m (say 15...150) times, the resolution is increased by a factor N_m and reaches 10... 1-cm, while accuracy improves by $\sqrt{N_m}$ (=3.9...12). At a typical pulse repetition rate f_r=3 kHz, the total measurement lasts N_m/f_r =5...50 ms.

The case τ =5 ns, M=7.5, and f_c =1.5 GHz is representative of a pulsed telemeter based on a Q-switched solid-state laser. The least significant digit is now 0.1 m in a single measurement, a figure already adequate for applications requiring good accuracy and a single-shot measurement time.

Typical applications calling for the previously mentioned specifications are (i) altimeters for helicopters; (ii) terrain profilers, like the MOLA telemeter mission (Fig.1-5, see [12]); (iii) satellite ranging observatories (see [13] as an example); (iv) military telemeters and (v) long-distance geodesy [14].

In all these applications, as we go to long distances, we have to increase the peak power of the pulse and the collecting aperture of the receiver by using a large-aperture telescope.

Then, it is advisable to use just one optical element for both transmitter and receiver, as shown in Fig.3-19 (top).

In this configuration, a beamsplitter inserted in the outgoing optical path collects a fraction of the optical pulse for use as the start signal.

Several different approaches for pulsed telemeters have been proposed in the scientific literature, and an interesting selection of papers on the subject is provided by Ref.[1]. More results were presented recently at topical conferences [15].

3.3.1.1 Improvement to the Basic Pulsed Setup

By slightly modifying the time interval measurement described in the previous section, we can get rid of the start/stop rounding-off error and improve the distance variance down to the ultimate limit of pulse-timing variance σ^2_t.

The idea consists of combining the coarse measurement of time interval, performed by clock counting, with a fine measurement of the fraction-of-period excess ΔT_{ex} in both start and stop pulses.

As illustrated in Fig.3-21 for the start pulse, we can recover this fraction ΔT_{ex} by using the discriminator output to trigger the start of a Time-to-Amplitude Converter (TAC).

The stop to the TAC is provided by the first clock pulse arriving after the discriminator switch-on.

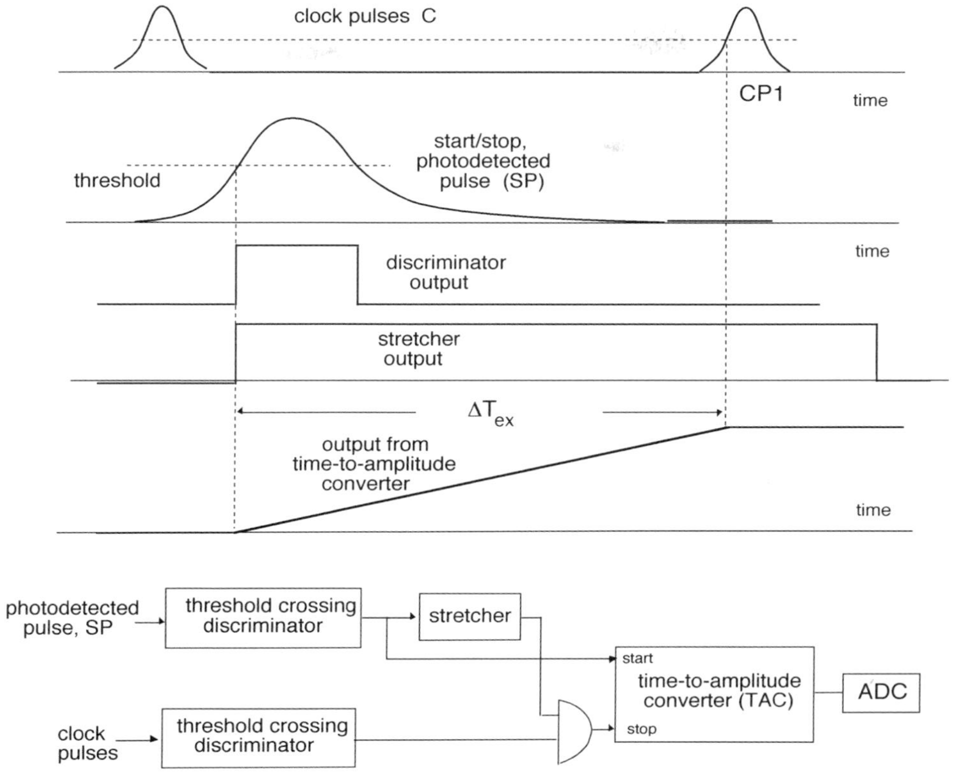

Fig. 3-21 The fraction-of-period time interval ΔT_{ex} between start pulse SP and the first counted clock pulse (CP1) that can be recovered by this circuit to improve accuracy of the telemeter. The scheme is duplicated for the stop pulse.

Using a stretcher to increase the duration of the discriminator output, in combination with an AND gate receiving the clock pulse, we can sort out this first clock pulse (CP1 in Fig.3-21).

Then, the amplitude of the output from the TAC is proportional to the required time ΔT_{ex}. By converting the TAC signal amplitude to a digital output with the aid of an Analog-to-Digital Converter (ADC), we get ΔT_{ex} (and the corresponding L_1) digitized in, say, one or two decades. Of course, the same procedure will be used for the stop pulse to recover the fraction of period in excess of it and to correct the result accordingly.

Note that the two more discriminations in Fig.3-21, performed on the clock pulses, do not appreciably degrade the accuracy. In fact, like the start (photodetected) pulse, clock pulses are large in amplitude, and their time variance σ^2_t is negligible [from Eq.3.16] as compared with the σ^2_t of the stop (photodetected) pulse.

3.3.1.2 Enhancing resolution with slow pulses

Two techniques have been demonstrated to improve resolution down to nanoseconds when the pulse available from the laser source is slow, for example, μs-long as for CO_2 laser sources. They are the *vernier* technique and the *pulse-compression* technique described below, and used in high-performance instrumentation for military applications.

Vernier Technique. This is a technique well known in nuclear electronics [16] to increase time resolution in measurements with slow waveforms.

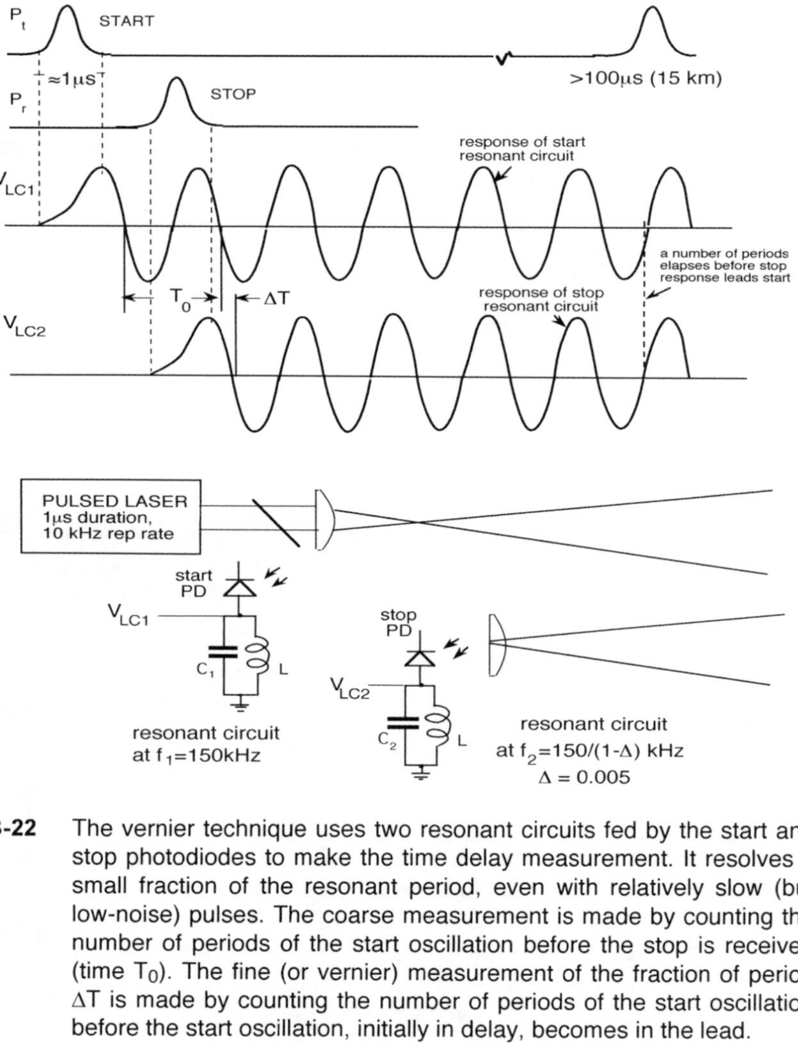

Fig. 3-22 The vernier technique uses two resonant circuits fed by the start and stop photodiodes to make the time delay measurement. It resolves a small fraction of the resonant period, even with relatively slow (but low-noise) pulses. The coarse measurement is made by counting the number of periods of the start oscillation before the stop is received (time T_0). The fine (or vernier) measurement of the fraction of period ΔT is made by counting the number of periods of the start oscillation before the start oscillation, initially in delay, becomes in the lead.

As shown in Fig.3-22, the optical start and stop pulses are detected by a photodiode. The currents excite two resonant, parallel LC-circuits. The LC circuits are designed to have a high Q-factor, and the voltage V_{LC} across them is read through a high input impedance. With a reasonably high Q (typ. 50–100), the voltage V_{LC} is a nearly sinusoidal oscillation, and it damps out slowly after excitation, making several periods of oscillation ($N \approx \pi Q$) available for the measurement. Timing is performed by looking at the zero-crossing of the V_{LC} waveform. Indeed, V_{LC} starts oscillating after the pulse ends and its charge has been completely collected by the capacitor. It can be shown that V_{LC} carries the timing information associated with the center of mass (or time-centroid) of the pulse, provided the pulse duration is short as compared with the oscillation period.

Now, let us compare the start and stop waveforms V_{LC1} and V_{LC2} (Fig.3-22). The time difference between the start and stop centroids (or, between the negative-going zero-crossings of V_{LC1} and V_{LC2}) can be written as $T=T_0+\Delta T$. Here, T_0 is the coarse delay between the V_{LC1} and V_{LC2} oscillations and is given by an integer multiple of the resonant period $1/f_0$, whereas ΔT is a fine delay, given by a fraction of $1/f_0$ (Fig.3-22).

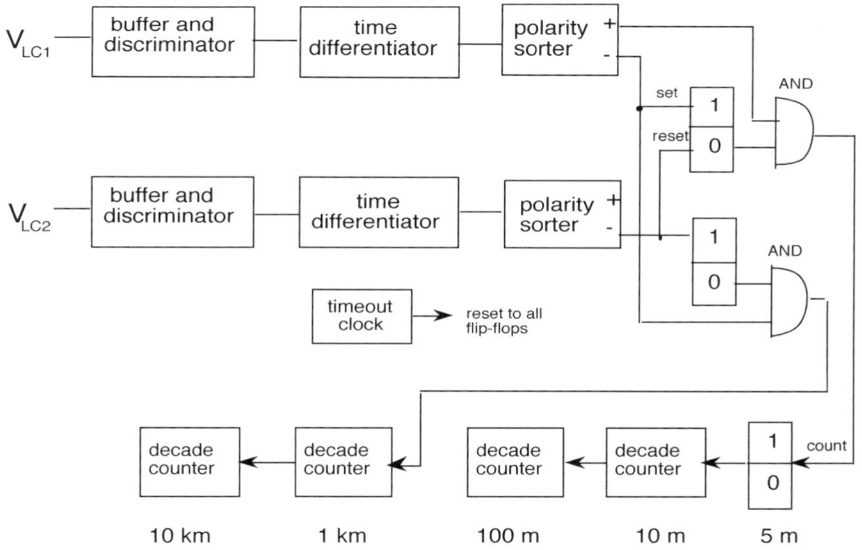

Fig. 3-23 Processing of signals from the vernier telemeter of Fig.3-22 can be made with a variety of approaches. Here, resonant signals are buffered and passed through a discriminator. By time differentiation, we get positive and negative pulses marking the periods. A logic circuit (middle flip-flop and AND gate) counts the negative transitions of V_{LC1} until V_{LC2} makes the first negative-going oscillation. These counts go to the 1-km and 10-km decades. The vernier measurement is made by the top flip-flop and AND gate. Here, we count the positive transitions of V_{LC1} until the negative-going V_{LC2} leads the negative-going V_{LC1}. This gives the two-and-a-half decades with resolution down to 5 m.

The coarse measurement is readily obtained by counting the number of periods N_c, which are contained in V_{LC1} after its first zero-crossing and before V_{LC2} makes its own first zero-crossing. For the fine measurement, the resonant frequencies of the start and stop circuits are made slightly different, let us say f_0 (start) and $f_0/(1-\Delta)$ (stop).

Then, signal V_{LC1} initially leads V_{LC2}, but suffers a small delay, with respect to V_{LC2} at each period. The small delay is $1/f_0 -(1-\Delta)/f_0 = \Delta/f_0$. After a number N_f of periods has elapsed, the stop waveform V_{LC2} finally leads the start waveform V_{LC1}.
Counting the number N_f of periods we find in V_{LC1} after V_{LC2} has done its first zero-crossing, until V_{LC2} leads V_{LC1}, we get the measurement of ΔT in units of Δ/f_0. The block scheme of Fig.3-23 illustrates the signal handling.

Example. With a τ=1-µs pulse (for which the resolution is $c\tau/2$=150 m), we may choose f_0=150 kHz as the resonant frequency of the LC circuits. The period $1/f_0$ corresponds to a distance increment $c/2f_0$ =1000 m and is the least significant digit of the coarse measurement. In addition, by taking Δ=0.005 as the resonance offset, we resolve $\Delta c/2f_0$ =5 m. This is the least significant digit of the fine measurement. The maximum number of periods $1/f_0$ we have to count in the vernier operation is $1/\Delta$ =200 and, therefore, we need a Q factor of about 200 to keep damping of V_{LC} waveforms reasonably low. Of course, the $\tau\Delta$= 5 ns resolution will be actually obtained if the received pulse carries an adequately large number of photons (Eq.3.16).

The centroid-timing corresponds to weighting the pulse waveform by $t-t_0$. This is actually the operation performed by the optimum filter when the received pulse has a Gaussian waveform (see Sect.3.2.3.2). Because the Gaussian is a very popular approximation, we may conclude that the LC timing and vernier technique are near-to-optimal choices in most practical cases.

Pulse-Compression Technique. This technique has been demonstrated with a CO_2 laser since the early times of laser telemeters [17] as an adaptation to optical frequency of a well-known method used in the microwave radar. As shown in Fig.3-24, we may start with a Continuous Working (CW) source, such as one like the CO_2 laser that may be difficult to pulse, but that makes a large power available, e.g., a few Watts.

A reason to use a CO_2 laser source is that the Far-Infrared (FIR) wavelength of emission, λ=10.6 µm, is favorable for a much larger range of operation in turbid media, such as fog and smokes (see App.A3.1 and Fig.A3-7). The relatively large CW power is readily provided even by compact CO_2 sources and is desirable to compensate for the decreased detector sensitivity and increased propagation attenuation.

Using an acousto-optics modulator [18], we are able to impress a modulation on the average power in the form of sine-wave bursts, as indicated in Fig.3-24.

The repetition period T_b of the burst is dictated by the desired ambiguity range (Sect.3.2.6), and a typical choice for an airborne telemeter is T_b=30 µs (L_a=4.5 km). A burst duration τ_b which is relatively short for timing purposes, but long enough to accommodate the chirp modulation, is the value τ_b= 4 µs shown in Fig.3.24.

With a CO_2 laser, these times are short as compared with the active-level lifetime ($\tau_{21} \approx 400$ µs), and therefore the average power is the same as the CW value P_{CW}. Then, the peak power of the burst is $P_p=P_{CW}/\eta$, or, it is increased by the inverse of the duty cycle $\eta=\tau_b/T_b$ (a factor of 7.5).

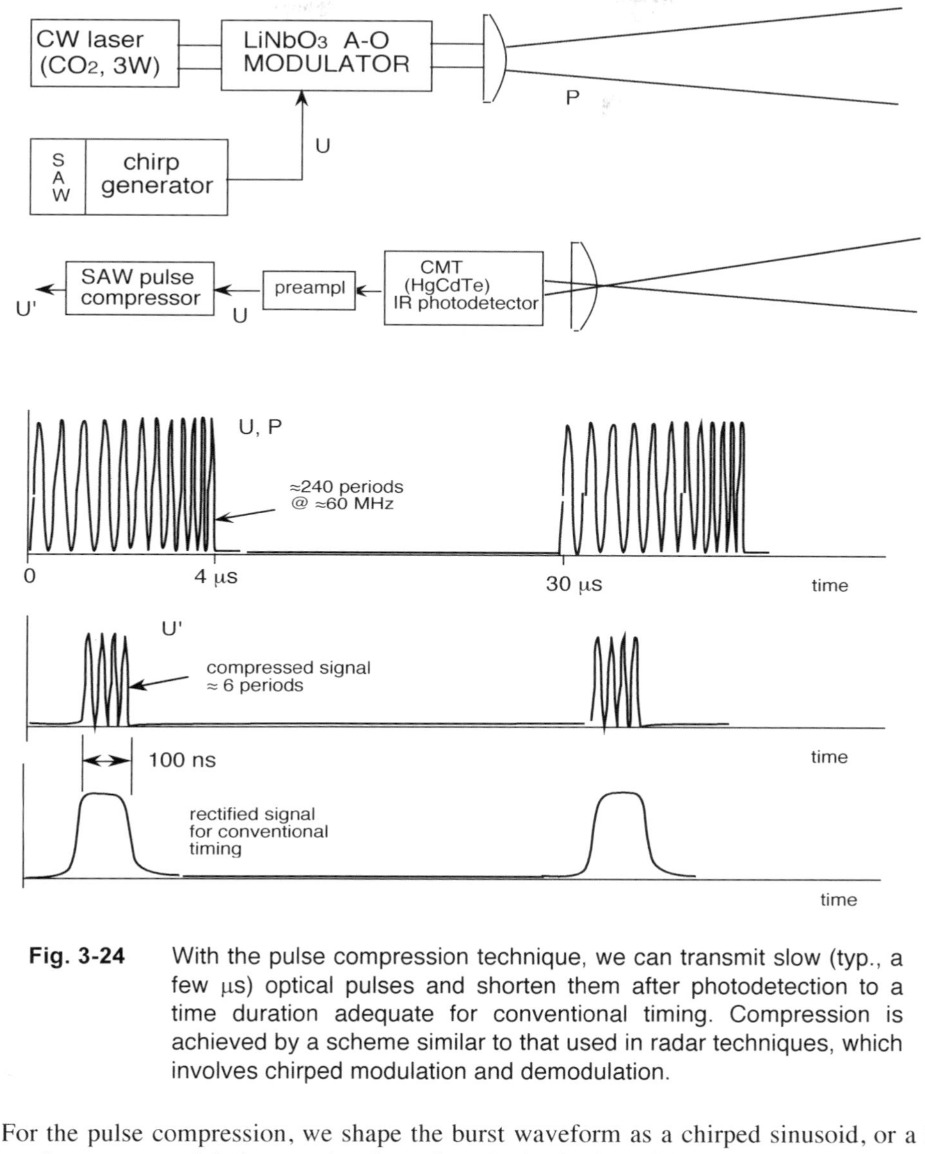

Fig. 3-24 With the pulse compression technique, we can transmit slow (typ., a few μs) optical pulses and shorten them after photodetection to a time duration adequate for conventional timing. Compression is achieved by a scheme similar to that used in radar techniques, which involves chirped modulation and demodulation.

For the pulse compression, we shape the burst waveform as a chirped sinusoid, or a linear-sweep frequency modulation ranging from f_0 at the beginning of the burst to $f_0+\Delta f$ at the end of the burst. Such a chirped sinusoid can be synthesized by a number of circuits, yet the Surface Acoustic Wave (SAW) is the neatest and simplest to implement [18,19]. Typical values compatible with the performance of the lithium-niobate modulator are f_0=60 MHz and Δf =10 MHz. Thus, we have ≈60 MHz×4 μs=240 periods of sinusoidal oscillation in the burst, with an instantaneous frequency increasing from 60 to 70 MHz. The chirp is a sort of (frequency) dispersion of the impulse response of the circuit generating the modulating wave.

We may cancel it by cascading, after detection, a second chirp-generating circuit with the same dispersion, but opposite sign. The circuit shall have a resonant response lasting the same time τ_b, but with $f_0+\Delta f$ at the beginning and f_0 at the end of the burst. With a SAW device, this is particularly easy because we need only to interchange input and output of the device.

Theoretically, the compression ratio is found as $\tau_b\,\Delta f$ [18]. With the previously chosen values, it is a factor 10 MHz×4 μs =40, and then the compressed duration is 4 μs/40=100 ns, and rise/fall times may be ≈1/60 MHz≈15 ns.

In the compression, we gain an increase of (equivalent) peak power of $\tau_b\,\Delta f$. In addition, a factor $1/\eta=T_b/\tau_b$ comes from the repetition rate. Thus, the total increase with respect to the CW value is factor $(T_b/\tau_b)(\tau_b\Delta f)=T_b\,\Delta f=300$, a remarkably good value indeed.

After rectification of the compressed envelope, we may proceed in the timing operation by conventional processing of signals.

3.3.1.3 Slow Pulse Telemeters for the Automotive (LiDAR)

The two methods illustrated in the last section are employed for achieving short-time (typ. nanoseconds) resolution with long pulses (typ. microseconds) in high-performance telemeters used especially in military applications.

Recently, the automotive segment has prompted the use of long pulses for application on medium distance (typ. up to one hundred meters) and with moderate accuracy (typ. several centimeters), and the concurrent mandatory requirements are: (i) very low cost, as the telemeter is just a part of a much more complex autonomous driving system, and (ii) intrinsic laser safety.

According to the international Laser Safety Standards (see also Appendix A.1.5), the laser should emit no more than 1 mW at λ=900 nm, the wavelength operation of GaAs laser diodes, or no more than 10 mW at 1550 nm, the so-called eye-safe wavelength of quaternary GaAlAsP diodes.

As power is so low, to obtain the desired resolution, which from Eq. 3-22 is seen to depend on the square number of received photons N_{ph}, we shall accumulate the largest possible number of photons by using the longest possible pulse duration, which is $T_{max}= 2L_{max}/c$, where L_{max} is the maximum distance to be covered.

So, the source shall emit a rectangular light pulse as already considered in Fig.3-17. In this way, the accuracy starts from a large value, i.e., the maximum distance L_{max} (about one hundred meters) at $N_{ph}=1$, but thanks to the $1/\sqrt{N_{ph}}$ dependence, it attains the desired centimeter-level accuracy at sufficiently high N_{ph}.

Thus, the class of accuracy of the automotive telemeter, σ_L/L_{max}, is about 1 part in 10^4 instead of the 1 part in $10^6...10^7$ of the other high-performance telemeters described in this chapter, but processing of the photodetected signal is particularly simple and low cost.

As depicted in Fig.3-25, the processing starts with an astable multivibrator biasing the laser diode with a square waveform. Half period of the wave is used for illuminating the scene up to L_{max}, and this requires a time $2L_{max}/c$, while in the other semiperiod the laser is switched off to allow resetting of the circuit. Thus, the square wave period is $4L_{max}/c$ and its frequency is $c/4L_{max}=1/4T_{max}$.

The signal received by the photodiode is amplified and passed to a zero discriminator to remove its amplitude dependence on distance and obtain a standardized signal (Fig.3-25, center right). The output S_{zd} is complemented and sent together with the drive rectangular signal V_{sw} to the input of a NAND gate, which supplies as an output a rectangle of duration proportional to distance L (Fig.3-25, bottom right). This output is integrated (for example by a voltage-to-current converter feeding a capacitor) and sampled at the end of the drive square wave, giving the distance L in analogue format.

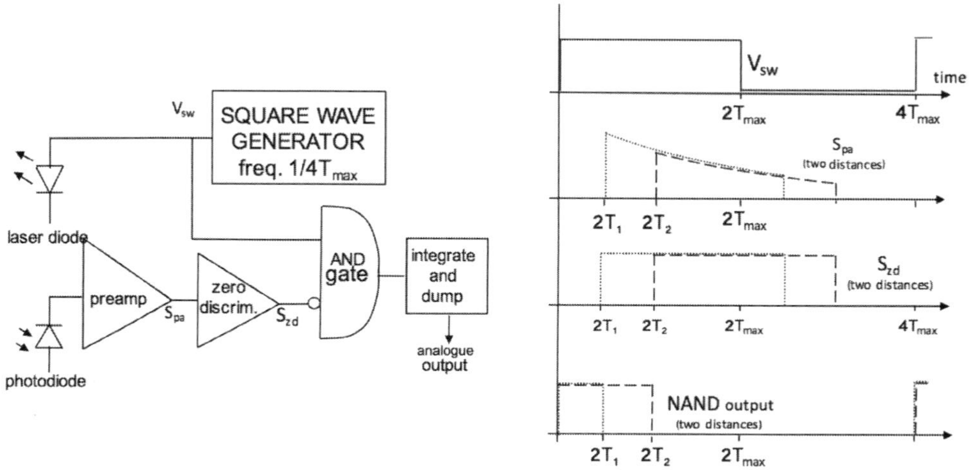

Fig. 3-25 Left: the slow pulse of an automotive LiDAR telemeter is processed by simple circuitry to sort out distance with centimeter resolution on a range of one hundred meters; right: waveforms and timing of signals in the circuit.

3.3.2 Sine-Wave Telemeter

The sine wave telemeter allows using a low-power laser, provided it can be modulated at reasonably high frequency. Indeed, the theoretical accuracy of the sine-wave telemeter equals that of the pulsed telemeter when the inverse angular frequency $1/2\pi f_m$ of modulation is equal to the pulse duration τ (Sect.3.2.5), and we use the same number of collected photons.

Low power is desirable and may even be a mandatory requisite in view of laser safety issues (App.A1.5). This is especially true for outdoor applications in the presence of the public, for example, the topographic telemeter replacing the theodolite, the laser level and alignment equipment. In such applications, either we use barriers to prevent exposure of humans to a hazardous level, or work with the maximum power allowed by a Class-1 laser.

For a semiconductor laser diode operating at a wavelength λ=700-900 nm, this power is P_t = 0.2-0.6 mW, a reasonable value readily provided by ternary GaAlAs lasers. These lasers are cheap, reliable, and compact, and can be easily modulated through the injected current up to several hundred megahertz.

The problem of ambiguity (Sect.3.2.6) has been approached by a number of methods [1,15], and in the following we describe two of them, perhaps the most practicable because they have been incorporated in products.
They are the *frequency-sweep method* and the *multi-frequency method.*

Frequency-sweep method. We use a sinusoidal waveform for modulating the laser power and vary the modulation frequency f_m in time. We start with a frequency f_{m0} low enough to ensure the desired non-ambiguity distance L_{NA} and sweep it up to a large value f_{M0} (see Fig.3-26) that allows us to reach the desired distance resolution ΔL. Specifically, we will use (Sect.3.2.6) $f_{m0} \leq c/2L_{NA}$ to obtain the non-ambiguity range L_{NA}, and $f_{M0} \geq c/2N\Delta L$ to obtain the resolution ΔL in a phase measurement with a 1/N resolution of the round angle. For example, if our telemeter measures up to 15 km, we choose $f_{m0} = 3 \cdot 10^8/2 \cdot 15 \cdot 10^3 = 10$ kHz. Also, if we want a 1-cm resolution out of a phase sorter with a 1/100 resolution of 2π, we use $f_{M0} = 3 \cdot 10^8/2 \cdot 100 \cdot 0.01 = 150$ MHz.

During the frequency sweep, the instantaneous phase signal that we obtain by mixing the transmitted and received waveforms is written, from Eq.3.28 with $\varphi_0=0$, as:

$$S_\varphi(t) = (I_0 I_{r0}/2)\, m \cos\,[2\pi f_m(t) T] \tag{3.36}$$

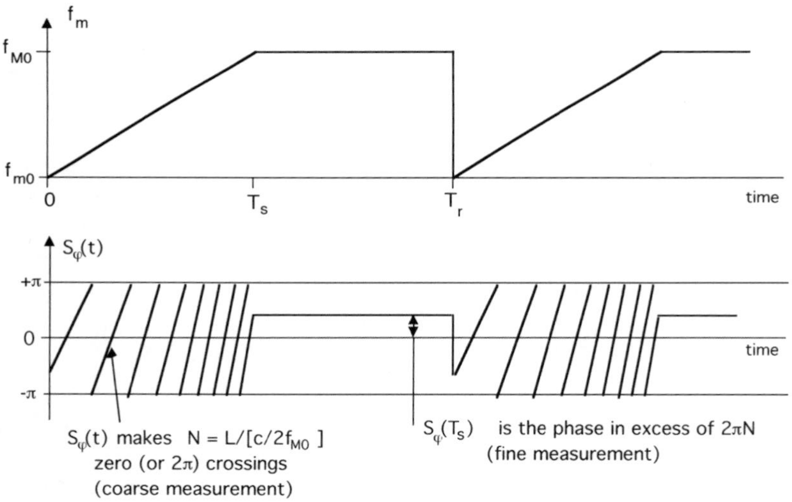

Fig. 3-26 In a sine-wave modulated telemeter, we want to avoid the distance ambiguity and yet be able to use a high-modulation frequency. To this end, frequency is made to sweep from a very low value f_{m0} (typically 15 kHz, avoiding ambiguity up to 15 km) to a fairly high f_{M0} (typically 150 MHz, adequate for a 1-cm resolution). The coarse measurement is recovered from the number of zero-crossings contained in the phase signal S_φ, while frequency goes from f_{m0} to f_{M0}.

At the beginning of the sweep, we start from a phase $2\pi f_{m0} T$ less than 2π because of our choice $f_{m0} \leq c/2L_{NA}$.Then, as frequency increases, we reach a value for which $2\pi f_m T = 2\pi$. At this value, the actual distance L contains exactly one full period of the modulation waveform, or $2\pi f_m(2L/c) = 2\pi$, and the phase signal S_φ is zero. As time elapses and f_m increases further, we find new zero-crossings of the phase S_φ. This happens each time the actual distance L becomes an integer multiple M of the instantaneous modulation period $c/2f_m$, or $L=M(c/2f_m)$. This can also be written as $2\pi f_m(2L/c) = 2\pi f_m T=2\pi M$, so (Eq.3.36) $S_\varphi=0$. When the sweep finally reaches its maximum frequency f_{M0}, the number of zero-crossings that the phase signal S_φ has done is exactly the number of times M_0 the distance $c/2f_{M0}$ is contained in the distance L to be measured. We may count the zero-crossings of S_φ by using the (positive) slope of the sweep signal as a gate input to a 4-decade counter (Fig.3-27). In this way, we obtain the coarse measurement of distance in units of $c/2f_{M0}$ ($=3\cdot 10^8/2\cdot 150$ MHz = 1 m). In the schematic of Fig.3-27, this processing is done by mixing the signal of the modulator and the signal received from the photodetector. Actually, as the signal from the photodetector comes with some noise, to filter it, a Phase-Locked-Loop (PLL) block is used.

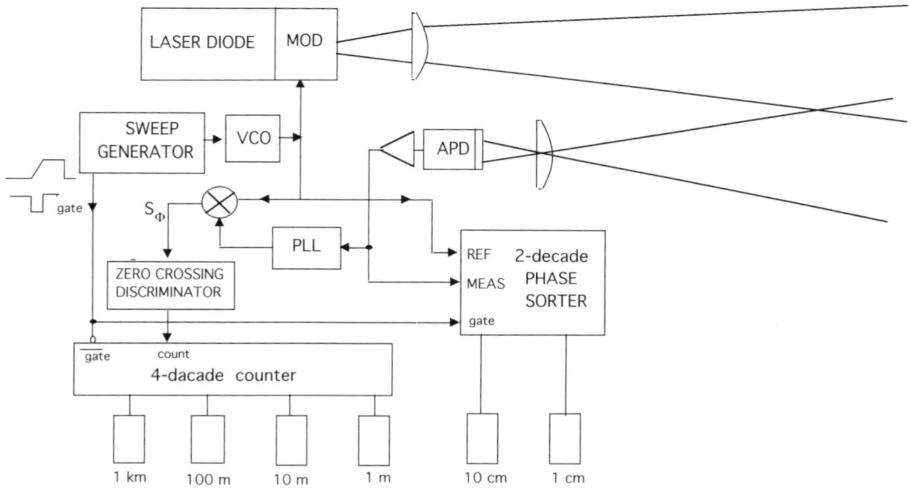

Fig. 3-27 The sweep generator feeds a modulator (or directly) the laser diode current) with the frequency sweep of Fig.3-26, and also gates the coarse- and fine-measurements decade counters. The phase signal is generated by mixing the sweep waveform and photodetected signal, properly amplified and filtered by the Phased-Locked-Loop (PLL).

The PLL is equivalent to a narrow pass-band filter with the central frequency dynamically tracked to the signal frequency. A narrow bandwidth is desirable to effectively filter out noise from the phase signal, to avoid counting spurious zero-crossings. The smallest bandwidth we can use in the PLL filter is determined by the sweep duration T_s as $B\approx 1/T_s$.

The fine measurement is performed on the phase-signal S_φ during the period from T_s to T_r (Fig.3-26) when the modulation-frequency is kept constant ($f_m= f_{M0}$). A conventional 2-decade digital phase meter is adequate to sort out the measurement (Fig.3-27).

As detailed in Fig.3-28, a typical design uses a conversion of the signal (MEAS) to an intermediate frequency, usually f_{IF}=1 kHz. The periods of a clock running at $f_{ck} = Nf_{IF}$ (100 kHz for N=100) are counted using REF and MEAS signals as start and stop.

Fig.1-4 is an illustration of a sine-wave modulated telemeter developed on the frequency-sweep concept for the application in topography. The resolution of such a telemeter may be 1 cm (for a 10-s measurement time). Distance of operation can go up to 100 to 300 m when using a cooperative target. Usually, the target is a corner-cube array with ≈50 to 200 cm^2 area. The optical source is typically a 0.1 mW GaAlAs laser, and the entire instrument, including the electronic processing unit, fits in a briefcase for use in the field.

Multifrequency method. Another approach to the sine-wave telemeter is using a frequency f_{m1} low enough to avoid distance ambiguity, together with its multiples, 10 f_{m1}, 100 f_{m1}, etc., to get the desired resolution. The ingenuity is in properly combining the measurements

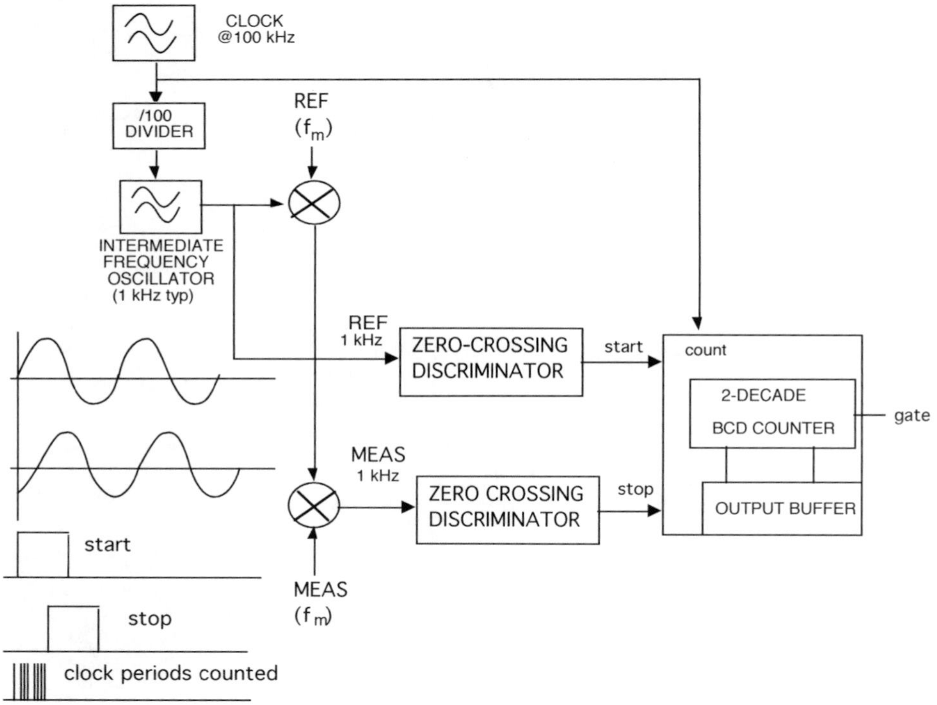

Fig. 3-28 Simplified block scheme of a typical 2-decade phase meter. The measurement is performed at an intermediate frequency (1 kHz), that is tied through a :100 divider to a main oscillator (100 kHz), which is also used as a clock of the counter. The MEAS signal is down-converted to 1 kHz by mixing with the REF signal shifted by 1 kHz. Start and stop of the counter are obtained by zero-crossing discrimination of the MEAS and REF signals.

performed by the different frequencies. For example, if we choose three frequencies that are a multiple of a factor 100 and use a 1/100 round-angle phase measurement, each frequency supplies two decades, and as a whole we cover 3×2= 6 decades in distance. As shown in Fig.3-29, the three measurements can be made in sequence by multiplexing the modulation waveforms to the laser and distributing the results to the appropriate counters.

To substantiate the concept, let us take f_{m1}=15 MHz as the highest frequency. The corresponding distance is $L_{m1} = c/2f_{m1}$ = 10 m, and the 10 cm and 1 m digits come out from the $2\pi/100$ phase measurement. The second frequency is f_{m2} = 150 kHz to supply the 10 m and the 100 m digits, and the third frequency f_{m3} =1.5 kHz, gives the 1 km and the 10 km digits.

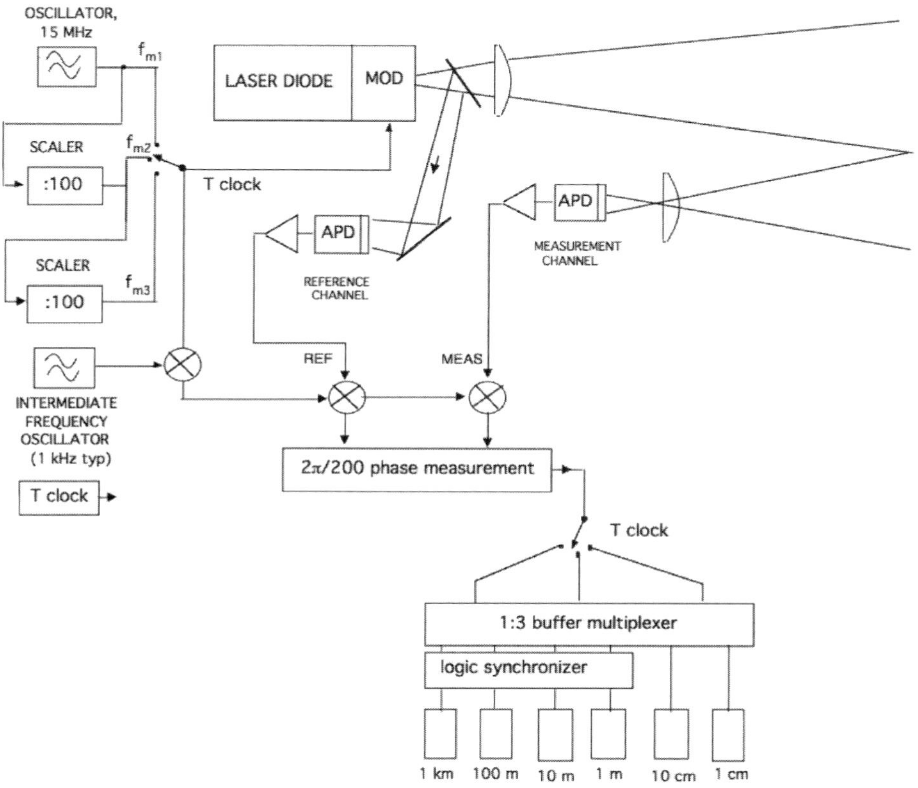

Fig. 3-29 The multi-frequency method of the sine-wave modulated telemeter uses three frequencies, multiples of 100, to obtain a large non-ambiguity range with the smallest of them ($f_{m3} \approx$1.5 kHz), yet has a good resolution with the largest available ($f_{m1}\approx$15MHz). The 1-kHz local oscillator shifts all three frequencies so that, after mixing with REF and MEAS, the phase measurement is always performed at the same intermediate frequency. A logic circuit provides synchronization of the three pairs of digits.

Theoretically, the measurement associated with each frequency is the pair of digits in L remaining from the next lower frequency. In fact, from Eq.3.26, the phase argument of the cosine, the quantity to be measured in $2\pi/100$ intervals, can be written as:

$$\varphi_{m1} = 2\pi f_{m1} 2L/c = 2\pi\ L/L_{m1}$$

where L_{m1} is the scale factor associated with f_{m1}. As the phase is modulo 2π, we measure the remainder of L/L_{m1} from the integer part, or the first two decimals.

Considering that $f_{m2}=f_{m1}/10^2$ and $f_{m3}=f_{m1}/10^4$, we have $\varphi_{m2} = 2\pi 10^2\ L/L_{m1}$, and $\varphi_{m3} = 2\pi 10^4\ L/L_{m1}$. Thus, the measurements with f_{m2} and f_{m3} are the remainders of $10^2\ L/L_{m1}$ and $10^4\ L/L_{m1}$ from the integer part, or the second two decimals and the third two decimals, respectively, of L/L_{m1}.

A subtle problem arises, however. We may even tie the three frequencies very precisely, i.e., deriving them from a single oscillator through dividers, but unavoidably, measurement errors will prevent the results from being exact multiples of each other. Thus, when the 10 cm and the 1 m digits pass from 99 to 00 (or vice versa), the digit of the 10-m does not increment (or decrements) the content by one. Instead, the 10 m digit may switch erratically, when the 10 cm and the 1 m digits are somewhere, e.g., between 90 and 10 (as they pass through 99 and 00).

This problem is clearly a very basic one, common to all readouts from de-multiplied scales and calls for a technique of digital synchronization.

To synchronize the digits, we add a redundancy to the coding scheme. Instead of the $2\pi/100$ resolution for the $\times 100$ frequency ratio, we double the resolution to $2\pi/200$ in f_{m2} and f_{m3}.

In this way, we can make available a logic variable, say $X_{1/2}$, which indicates the state (00 to 49 or 50 to 99) of f_{m2}. Nominally, $X_{1/2}$ replicates the most significant bit of the f_{m1} readout, let us say Y_{50}. The bit Y_{50} changes from 0 to 1 and from 1 to 0 when the digits of f_{m1} go from 00 to 49 and from 50 to 99, respectively. Because of measurement errors, $X_{1/2}$ deviates from Y_{50}, however. By comparing the two variables, it is easy to determine if the f_{m2} (and f_{m3}) bit shall be incremented by one or not. Thus, synchronization is obtained at the expense of a modest loss in resolution (a factor of 2).

The multi-frequency technique is incorporated in commercially available telemeters, and it yields performance equivalent to those of frequency-sweep, sine-wave modulated telemeters.

3.4 Imaging and 3D Telemeters

A number of applications, ranging from military to biomedical and to industrial, call for a spatially resolved measurement of distance, that is, for *imaging telemeters*. Also called imaging laser radar and three-dimensional (3-D) imaging, they enjoy the benefit that distance information adds to the image morphology. This greatly helps solve the problem of Automatic Target Recognition (ATR), both in the military and robotic segments of application, as well as for automatic driving [15]. Of course, an imaging telemeter can be developed by combining a normal single-point telemeter and an *optical scanner* device that aims, in a raster or TV-like format, to each pixel of the 2-D scene in an ordered sequence.

We need also an optical switch in the path outward from the laser, because we shall turn off the beam when the scanner reaches the end of a line and the beam shall fly back to the beginning of the next line. As a scanner, we may use a pair of X and Y mirrors, mounted on galvanometer actuators.

The *scanner* solution enables us to form an image with typically N=100×100 pixels, a 5 deg×5 deg angular field of view, and a reasonable frame repetition rate (10 to 30 frames per second). The switch could either be an electro-optical cell (KDP or $LiNbO_3$ crystals), or an acousto-optical switch ($LiNbO_3$ or CdTe). Both switches are fast enough (settling time of microseconds) to match the speed required by galvanometer-scan systems. Wavelengths of operation of the imaging telemeter can go, depending on the application, from the visible, using diode lasers, to the near and far infrared. The benefit of far infrared wavelengths is the mitigation of scattering attenuation and turbulence effects (Ref. [19] and App.A-3), both in a normal atmosphere and in light fogs and hazes. A few imaging laser radars operating at several wavelengths, from 1.0 to 10 μm, have been evaluated in [19] in good and low visibility conditions.

Because the measurement time is subdivided between N pixels, the theoretical accuracy of the imaging telemeter is worse than that of a still staring telemeter by a factor $\sqrt{N}$ (≈100 in the previous example). However, this is not a severe hindrance from the practical point of view. It turns out that the extra information gathered from the frame-size image largely alleviates the resolution or accuracy requirement with respect to the single-point measurement.

About the measurement configuration, scanner imaging telemeters were first demonstrated with time-of-flight telemeters operating in connection with XY deflection mirrors scanning the scene. Both pulsed laser sources for long distance operation (L>100 m) and sine-wave configuration for medium/small distance (L=5...50 m) have been reported [9,19, 20].

A recent improvement has been the replacement of the galvo scanner with a *DMD* (Digital Micromirror Device), a chip of 5- to 10-mm size based on the MEMS technology and made up of an array of micromirrors (each typ. 10 to 50 μm on a side) steerable up to a ±12 deg deflection in one or two angular directions [21,22].

Anyway, as mechanical scanning of mirrors is rather bulky and slow, two directions have been pursued for scanner-less operation of 3-D telemeters: *parallel structure* and *beam steering*. In the parallel structure, the receiver photodetector is sectioned in n×m individual elements, as many as the pixels in the scene, and the laser source uses a beam of large aperture angle to illuminate all the n×m pixels of the scene, while an objective lens conjugates the scene pixels to the photodetector array pixels.

Both pulsed and sine-wave approaches can be implemented in the parallel structure, of course after a redesign of the processing circuits aimed at minimizing the footprint, that is, the area of the semiconductor they occupy around the photodetector sensitive area of the "smart" pixel.

The ideal photodetector for the pulsed approach is the SPAD (Single Photon Avalanche Detector) [see Ref.[2], Ch.6.2], fast enough (typ. 50 ps response time) for operation on even short distances (e.g., 1 m or 3.3 ns delay). The size of the typical elementary cell (see Fig.3-30) is 50 μm on a side, and the SPAD is 8 μm in diameter, resulting in a fill-factor FF =

$(\pi/4)8^2/50^2 = 0.02$, a low value that would seriously impact the quantum efficiency of the detector, but that is mitigated by the use of a microlens array placed in front of the 3-D array.

The microlenses can recover the FF up to a factor of 20...30 [23]. The 3-D chip based on the cell shown in Fig.3-30 contains 32×32 smart pixels fabricated by a 130-nm CMOS standard process, and it occupies a 10×13 mm^2 area of Si.

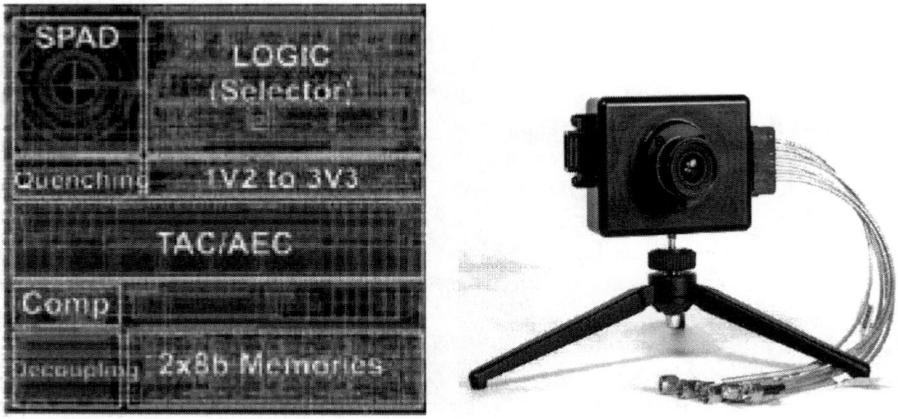

Fig. 3-30 Left: the structure of the pixel integrating a pulsed telemeter in a 3-D array of 32x32 pixels, each of 50 μm on a side and incorporating a SPAD detector and ancillary circuits, fabricated by in130-nm standard CMOS technology (by courtesy of the Megaframe project [24]); right: a turnkey module of SPAD-based pulsed telemeter array, including a 64x32-pixel chip (by courtesy of Micro Photon Devices, Bolzano).

With an illuminator made of an array of 50 diode lasers working in the mode-locking regime so as to supply short (15-ps) pulses and an average power of 100 mW, the SPAD pulsed telemeter array has a resolution time of <100 ps in pixel, and can cover a measurement range up to 20 m, typically.

The sine-wave modulated circuitry can also be integrated in a smart pixel of a 3-D telemeter scanner-less array [25]. As the signal processing is relatively simpler than the pulsed case, we obtain a smaller chip (2.5×2.5 mm^2) for a larger 128×128-pixel array, see Fig.3-31.

In the application, a wide-angle beam from a 900-mW LED-array source illuminates the scene [25]. The LED-array is amplitude-modulated at 20 MHz, and the measurement is limited to the non-ambiguity distance L_{NA}=7.5 m. A CCD array (see Ref. [2], Sect.12.2) is used as a detector, receiving light diffused by each pixel of the scene through an objective lens. In the CCD, the electrode commanding the storage of the photogenerated charge is driven by the modulating waveform at 20 MHz. The efficiency of charge collection depends on the voltage on the storage electrode, so the useful CCD output current is the product of the received radiant power and of the modulating waveform, or the phase ϕ=2ks. The phase is read with <10^{-3} resolution resulting in a distance error of a few centimeters up to 7.5 m.

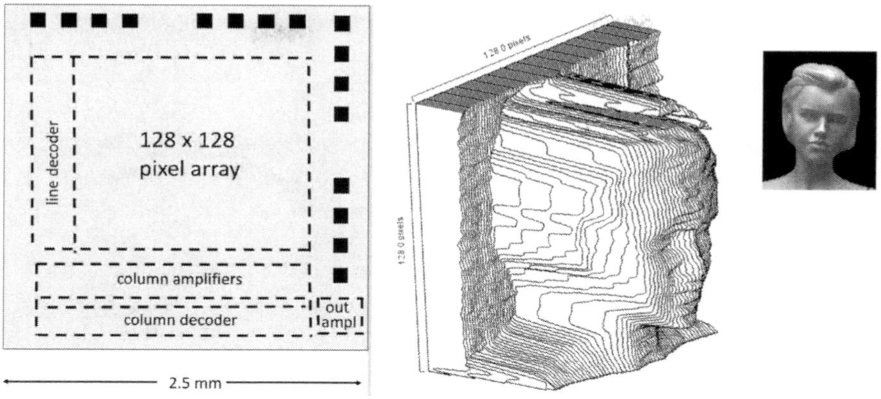

Fig. 3-31 Left: layout of a 2.5×2.5-mm^2 chip realized by 180-nm CMOS standard process and incorporating an array of 128×128 smart pixels, each a full sine-wave telemeter, plus the ancillary circuitry. Right: example of 3-D detection obtained with the system described in the text, for facial recognition (by courtesy of the Megaframe project [25]).

Last, the truly motionless *beam steering* is of great interest and can indeed be implemented when the source used for the distance measurements is a coherent one, like for example in the FMCW technique for the automotive (see Sect.4.2.3). Then, we can take advantage of the properties of the laser beam that can be deflected by diffraction when impinging on a grating-like structure.

We may recall (see App. A4) that the distribution of the optical field $E(\theta,\psi)$ at angles $\theta=x/z$ and $\psi=y/z$ in the far field is the Fourier Transform of the source field $A(x,y)$, or explicitly $E(x/z, y/z) = FT\,A(x,y)$. Thereafter, we can use two different schemes to deflect the beam:

(i) by coding the deflection angle (either θ or ψ) in wavelength through a *chirp* impressed to the laser λ, so that by impinging on the dispersive element (typically a grating-like structure) the beam deflection is made proportional to the impressed wavelength shift [26]. This method has, however, the limitation of a single-angle operation;

(ii) by constructing a *phased array,* either 1-D or 2-D, to diffract the impinging beam. The phased array shall contain fully addressable individual pixels, so as to allow synthesizing the phase distribution $\Phi_A(x,y)$ that corresponds to the desired deflection at angles θ and ψ. The phase distribution will be multiplied by $A(x,y)$ as a phase term $e^{i\Phi}$. The concept has been demonstrated in the Si-photonics technology [27], yet the fully addressable and reconfigurable array has still to be developed.

3.5 LIDAR AND LADAR

Nowadays, the acronym LiDAR (Light Detection and Ranging) is widely used to identify a telemeter for the automotive application, one typically operating on a hundred meters range and a 3- to 10-cm accuracy and employing the pulsed long-wave technique (Sect.3.3.1.3).

Also used, but the less frequently, is the name LaDAR or LADAR (for Laser Detection and Ranging) for this application.

Actually, the LIDAR (acronym for Laser Identification, Detection and Ranging) was originally developed in the 1970s to indicate an instrument capable of performing the remote sensing of natural media (mainly the atmosphere, but also water masses) based on the analysis of the light returning to the receiver from the backscatter of the medium rather than from a remote target.

So, while not properly a telemeter, but similar to it because it uses a short-pulsed source, the LIDAR has become the instrument to probe pollutants of the atmosphere and to obtain distance-resolved maps of them for ambient protection (Fig.1-6). The LIDAR has also been successfully used in a number of scientific surveys, including ozone and vertical temperature profile in Antarctica, plankton and chlorophyll on the sea surface, etc. [28-31].

In the LIDAR, we use a pulsed source with high peak power, usually a Q-switched laser, and choose the wavelength in correspondence to an absorption peak of the atmospheric component to be measured. The wavelength range of interest spans from the near UV (300 nm) to the far IR (10 μm) complementing the range of millimeter to centimeter wavelengths employed by the microwave radar to probe the atmosphere. Plenty of useful lines are found at optical wavelengths [28], for virtually any kind of gaseous species we can find in the atmosphere. These include, for example: nitrogen, oxygen, and carbon oxides; water vapor, ozone, and dangerous species like sulfur oxides, hydrocarbons, fuel gases, chlorofluorocarbons (CFCs), etc. From one side, the abundance of lines means that hardly a species will be found that cannot be measured in the optical range by a LIDAR technique. On the other hand, the crowding of many lines pertaining to different species will make it hard to distinguish between different constituents. Thus, the LIDAR will make measurements and actually work with unparalleled sensitivity on constituents known to be present in the medium under test, whereas the discrimination between similar species is not pursued, in general. With regard to the sensitivity, the actual figures will depend on parameters like species, measurement technique, power source, and wavelength range. The typical sensitivity, expressed as the minimum detectable concentration per unit volume, may range from 10 ppb×m (part per billion, or 10^{-9}, per meter of propagation path) up to 10 ppm×m (part per million per meter).

To tune the wavelength on a wide spectral range and cover most species of interest, the laser may be an Optical Parametric Oscillator (OPO) [31]. The OPO is pumped by a Q-switched laser at fixed λ, like Nd:YAG at 1.06 μm or at 0.53 μm with second harmonic generation (SHG), and can be tuned from about 0.6 to 3.5 μm. Other options for tunable sources in the visible and near-infrared are the alexandrite laser (0.8-0.9 μm) and several dye lasers (0.4 to 1.0 μm) [32]. For the infrared, CO_2 and CO lasers have been widely used, as well as lead-salt semiconductor diodes.

The optical beam is collimated by the transmitting telescope and sent through the atmosphere along the line of sight. There is no need to place a distant target, except for safety issues. The line of sight of successive pulses is arranged in a raster sequence so that a map of measurement can be constructed. For each pulse, power back-scattered by the atmosphere is collected back by the receiving telescope and is converted into an electrical signal by the photodetector. The time dependence of the photodetector signal is easily translated into a distance dependence of the back-scattering strength.

To calculate the signal received by the LIDAR, we start assuming that the transmitted pulse is very short and write it as $P(t)=E\ \delta(t)$, where δ is the Dirac-impulse function and E is the energy contained in the pulse. Upon propagation to a distance $z=ct$, the optical signal is delayed and attenuated, and it can be written as:

$$P(t) = E\ \delta(t\text{-}z/c)\ \exp -\int \alpha\ dz$$

Here, z/c is the time delay, and $\alpha=\alpha(z)$ is the attenuation coefficient of the atmosphere. The back-scattered power going toward the source in the interval z—z+dz is then:

$$dP(t) = E\ \delta(t\text{-}z/c)\ [\ \exp -\int \alpha dz\ ]\ f(\theta)\ \alpha\ dz\ d\Omega$$

where θ and $d\Omega$ are the scattering angle and the solid angle of collection, respectively, and $f(\theta)$ is the scattering function (see Appendix A3-1 and Fig.A3-5).

By integrating $d\Omega$ around θ=180-deg, the direction along which we collect the back-scattered radiation, we get $\Delta\Omega=\pi D^2/4z^2$ as the total solid angle subtended by the receiving lens of diameter D. Letting $\sigma_{bs}(z)= f(180\text{-deg})\alpha$ for the back-scattering coefficient, and also considering the delay and the attenuation experienced in the backward propagation, the back-scattered power collected at the receiver is obtained as:

$$dP_{bs}(t) = E\ \sigma_{bs}(z)\ (\pi D^2/4z^2)\ \delta(t\text{-}2z/c)\ [\exp\text{-}2 \int \alpha dz]\ dz \qquad (3.37)$$

From Eq.3.37 we can see that, by launching a short δ-like pulse at time t=0, we receive at time $t=2z/c$ an elemental power contribution $dP_{bs}(t)$ in an elemental time $dt=dz/c$. The intensity $dP_{bs}(t)/dt$ is proportional to the back-scattering of the atmosphere $\sigma_{bs}(z)$, to the exponential attenuation $\exp -2 \int \alpha dz$, and to a distance-dependent term $(D/z)^2$.

As an illustration, Fig.3-33 reports the typical $P_{bs}(t)$ waveform received from a LIDAR instrument. Here, a relatively slow pulse (duration $\approx$150 ns) is used, and a remote beam stop at $L\approx570$ m cuts off the returning signal for $t > 4\ \mu s$. The negative-exponential trend of the back-scattering signal (Eq.3.37) is clearly seen in Fig.3-33.

Now, we may take account of the finite response time T of the detector and associated circuitry. For simplicity, let us suppose that the detector integrates the signal for duration T, so that its output is:

$$P_{out}(t) = \int_{0\text{-}T} dP_{bs}(t) \ = \ E \int_{0\text{-}cT/2} \sigma_{bs}(z)\ (\pi D^2/4z^2)\ [\exp\text{-}2 \int \alpha dz]\ dz \qquad (3.38)$$

where $z=ct/2$. Eq.3.38 tells us that the finite response time has the consequence of averaging the distance-dependent factors on an interval $cT/2$. Equivalently, we can say that the distance resolution in the determination of the distance-dependent waveform is limited to $cT/2$ or $c/4B$ because of the response time T or of the bandwidth B of the detector.

The same argument applies if we take account of the finite duration of the transmitted pulse.

In general, we may take the pulse waveform as $\tau^{-1}F(t/\tau)$, where τ is the time duration and $F(t/\tau)$ is the dimensionless pulse-waveform normalized at unit area.

Then, it is easy to see that we shall multiply each elemental contribution given by Eq.3.37 by $\tau^{-1}F(t/\tau)$ and integrate it on the time variable dt=dz/c from 0 to ∞ to obtain the received power as

$$P_{bs}(t) = E\,\tau^{-1} \int_{0\text{-}\infty} \sigma_{bs}(z)\,(\pi D^2/4z^2)\,F(t/\tau)\,\exp\text{-}2 \int \alpha dz\,dz \Big|_{z=ct/2}$$
$$\approx E\,\tau^{-1} \int_{0\text{-}c\tau/2} \sigma_{bs}(z)\,(\pi D^2/4z^2)\,\exp\text{-}2 \int \alpha dz\,dz \Big|_{z=ct/2} \qquad (3.39)$$

In the second integral of Eq.3.39, we have used the finite duration of $F(t/\tau)$ to reduce the integration limits to 0—$c\tau/2$, the pulse-duration length. Thus, also in this case, the resolution is limited by the value $c\tau/2$ associated with the pulse duration.

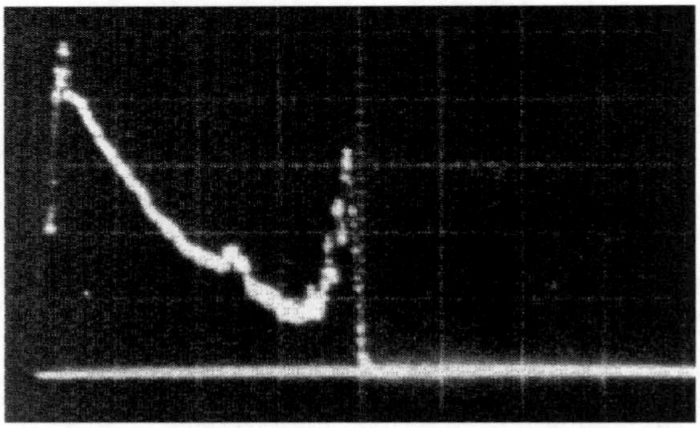

Fig. 3-32 Typical back-scattering waveform $P_{bs}(t)$ received by a LIDAR. The horizontal scale is 1 μs/div, which corresponds to a distance z=150 m/div. The signal is close to a negative-exponential in the first three divisions, and then it is clipped (at about 3.8 div or L≈570 m) because of a remote beam stop. The beam stop gives a strong back-scattering peak followed by a sudden return-to-zero of the signal.

About processing of the received signal, we can assume $\alpha \approx$const. and write the signal as

$$S(t) \approx E\,\sigma_{bs}(z)\,(\pi D^2/4z^2)\,\exp\text{-}2\alpha z$$

(in which z=ct/2). To extract the physical parameters contained in S(t), we can first remove the inverse square distance dependence by multiplying electronically the received signal by a term $(t/t_0)^2$, which increases with the square of time, so that $S(t)(t/t_0)^2 = S(t)(z/ct_0)^2$. Then, we take the log of the result and get a signal L(t) given by:

$$L(t) = \ln [E\,(\pi D^2/4c^2 t_0^2)] + \ln \sigma_{bs}(z) - 2\,\alpha\,z$$
$$= \text{const.} + \ln \sigma_{bs}(ct/2) - \alpha ct \qquad (3.40)$$

In this equation, the main dependence of L(t) is in the last term $-\alpha ct$, a negative-going ramp for constant attenuation α along z. The second term $\ln[\sigma_{bs}(ct/2)]$ is significant at those locations where the scattering σ_{bs} is unusually high and is revealed by a (positive) spike at time t= 2z/c (see Fig.3-32). The spike is followed by a negative-going step in L(t) (Δ in Fig.3-31) because, where the local attenuation $\alpha(z)$ is higher than the normal average value, the ramp $-\alpha ct$ has a larger slope.

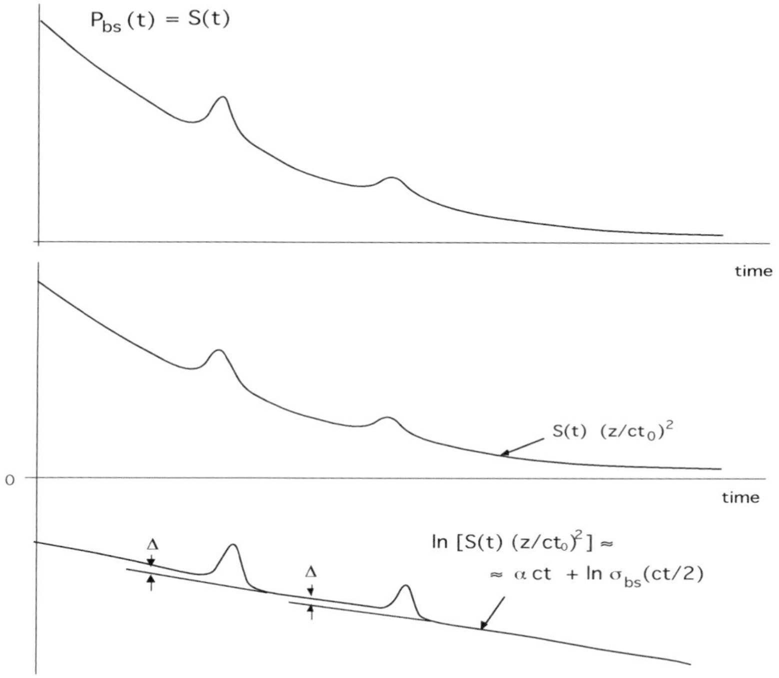

Fig. 3-33 By multiplying the back-scattering power $P_{bs}(t)$ (top) by the square-inverse distance dependence (middle), and taking the logarithm, a signal is generated (bottom) that unveils (i) the attenuation $\alpha(z)$ of the atmosphere (given by the negative slope of the diagram) and (ii) the local back-scattering $\sigma_{bs}(z)$ coefficient of the medium along the propagation path (positive-going spikes).

A sample of LIDAR waveforms is provided in Fig.3-33 to illustrate the above considerations. The received power $P_{bs}(t)$ waveform is smoothed by the distance correction (Fig.3-33, middle). After taking the logarithm, the final result (bottom in Fig.3-33) is a nearly linear trend with a few superposed peaks, which is indicative of local high-scatter centers.

A very useful way to represent distance-dependent scattering maps is that of using the data of Fig.3-34 (bottom) to intensify the trace (z-axis input) of an oscilloscope with a signal proportional to the scattering at the given distance z. Maps of scattering intensity versus distance are reported in Fig.3-34. These maps reveal the state of pollution, or of the scattering concentration, as a function of distance or of the elevation angle.

The LIDAR can be operated in a number of ways [27-30]. The one considered so far, with the optical axis of the illuminating beam parallel (or coincident) to the optical axis of the viewing telescope, is classified as *monostatic* (or *coaxial monostatic*). When the two axes are separated by a significant angle, the LIDAR is called *bistatic*.

Regarding the back-scattering signal we shall look at, we may use a single wavelength for the transmitter and receiver. In this case, we have an *elastic-scattering* LIDAR, which supplies information on the distance-resolved concentration of the atmospheric scattering, but basically is unable to sort out the nature of scatterers.

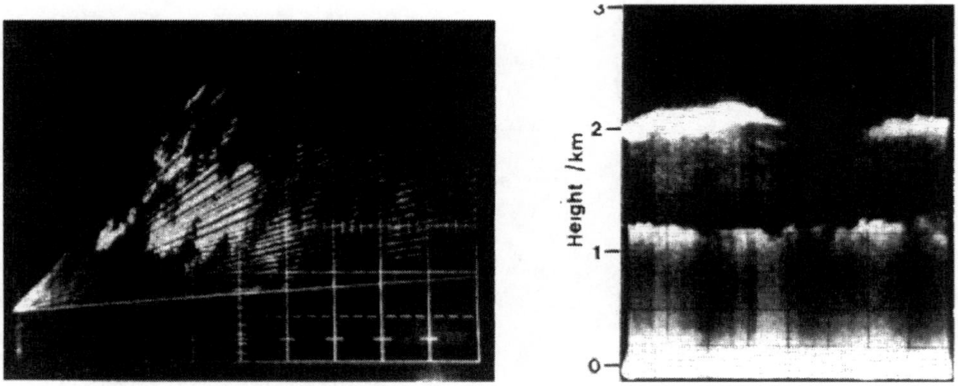

Fig. 3-34 Maps of back-scattering signals detected by the LIDAR. Left: signal as a function of the elevation angle at which the beam is aimed (10 to 60 deg; horizontal scale: 200 meter/div). Right: signal as a function of height above ground when the beam is aimed vertically and time elapses (x-scale: 10 minute/div).

With the advent of tunable laser sources, another mode of operation has become feasible and is frequently implemented, i.e., the *Differential-Absorption-Lidar* (DIAL). In the DIAL, the scattering measurement is carried out at two wavelengths: one centered at the absorption peak of the species or pollutants we want to monitor and the other just off the peak. Off the peak, absorption is contributed only by the normal atmosphere.

Therefore, subtracting the two measurements gives the differential (or extra) absorption caused by the species, a quantity proportional to the concentration (see Eq. A3-3) of it.

The wavelength most suited for DIAL monitoring of the atmosphere depends, of course, on the considered species, their targeted concentration, and coexisting lines of normal atmospheric constituents. Normal constituents of the atmosphere strongly attenuate the probe beam outside the transmittance windows, so an obvious choice is to operate inside or at the boarder of one of these windows. For example, to measure methane, we can operate in the near–IR window around 1.6 μm as shown in Fig.3-35. With an OPO pulsed source pumped by a Q-switched Nd:YAG laser, a sensitivity of 50 ppm×m has been reported on a 50- to 300-m range of operation [33]. The unit is mounted on a medium-size van for outdoor operation and can also sense NO, NO_2, O_3, and SO_2 with a 0.25-1 ppm×m resolution on a range of 15-250 m.

Another mode of operation is that of using fluorescence. In the targeted molecules, fluorescence can be induced by an intense beam illuminating the atmosphere with a suitable wavelength, usually in the blue or UV range, typically λ_{ex}<400 nm. This source can be obtained by second- or third-harmonic generation (SHG or THG) of a Q-switched solid-state laser. Following the excitation, the molecules decay back to the ground state and emit a fluorescence photon, at a specific wavelength $\lambda_{fl} > \lambda_{ex}$, usually in the red to near-infrared range.

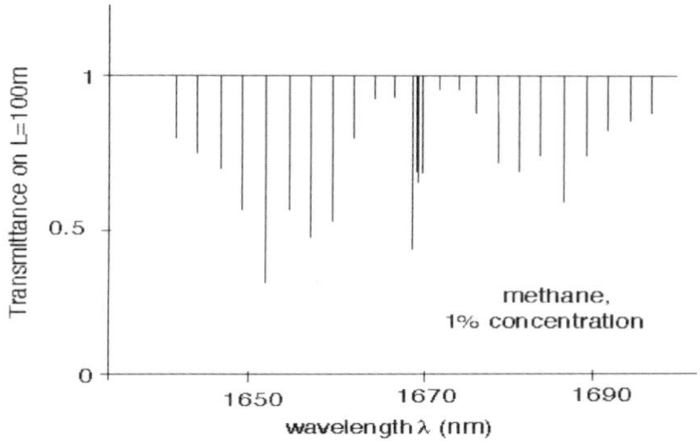

Fig. 3-35 Transmittance of the atmosphere with diluted methane (C=1% in volume) for a 100-m path. Several lines of absorption are shown, for which amplitudes are proportional to concentration C. Other pollutants (hydrocarbons, nitrogen oxides, etc.) have similar lines in the near-infrared.

It is easy to see that Eq.3.39 can be modified to describe the *fluorescence-LIDAR* by changing $\sigma_{bs}(z)$ in $\sigma_{fl}(z)$, the fluorescence cross-section, and $\exp\text{-}2\int\alpha dz$ in $\exp\text{-}\int(\alpha_{fl}+\alpha_{ex})\, dz$ to account for the different wavelengths λ_{fl}, λ_{ex} of the go-and-return propagation. Thus, a signal of the form L(t) of Eq.3-40 is still obtained, which gives the concentration of the targeted species. Because the fluorescence signal is at a wavelength different from the illumination, the contribution of a given species or pollutant is much better distinguished as compared to a scattering signal where a lot of species do contribute to absorption. A variant of the two-λ LIDAR is the *Raman LIDAR*, where we look at the Raman, λ-shifted, backscattering signal. In water, the Raman signal is strong and nearly constant with wavelength so that its variations can be used to monitor pollutants (especially hydrocarbons at the surface) and particulate. In the atmosphere, the Raman signal is weak, but we can take advantage of its dependence from temperature to measure a height-resolved air temperature.

REFERENCES

[1] M.-C. Amann, T. Bosch, M. Lescure, R. Myllyla, M. Rioux: "*Laser ranging: a critical review of usual techniques for distance measurement*", Opt. Eng., vol.40 (2001), pp.10-19.
[2] S. Donati, "*Photodetectors*", 2nd ed., Wiley IEEE Press, Hoboken 2021.

[3] R.G. Dorsch, G. Hausler, J.M. Hermann: "*Laser Triangulation: Fundamental Uncertainty in Distance Measurements*", Appl. Opt., vol.33 (1994), pp.1306-1314.
[4] S. Donati, A. Gilardini: "*Advanced Techniques of Laser Telemetry*", Selenia Technical Review, vol.8 (1982), pp.1-12.
[5] S. Donati, G. Martini, W.-H. Cheng and Z. Pei: "*Analysis of Timing Errors in Time-of-Flight LiDAR using APDs and SPAD Receivers*", IEEE J. Quant. Electr., vol. QE-57 (2021).
[6] E. Samain, R. Dalla, I. Prochazka: "*Time walk compensation of an Avalanche Photodiode with a linear Photo-detection*", Proc. 13th Int. Workshop on Laser Ranging Instr., 2002.
[7] E. Gatti, S. Donati: "*Optimum Signal Processing for Distance Measurements with Lasers*", Appl. Opt., vol.10 (1971), pp.2446-2451.
[8] S. Donati, E. Gatti, V. Svelto: "*Statistical Behaviour of the Scintillation Counter*", in Adv. in Electr. and Electr. Phys., ed. by L. Marton, vol.26 (1969), pp.251-307.
[9] R. Lange, P. Seitz: "*Solid-State Time-of-Flight Range Camera*", IEEE J. Quant. Electr., vol.37 (2001), pp.390-397; see also: A. Spickermann, et al: "*CMOS 3D image sensor based on pulse modulated TOF principle and intrinsic LDPD pixels*", Proc. of 2011 Eur. Solid-State Circ. Conf., pp.111-114.
[10] M. Okano, C. Chong: "*Swept Source Lidar: simultaneous FMCW ranging and nonmechanical beam steering with a wideband swept source*", Opt. Expr., vol.28 (2020), pp.23898-23915.
[11] P. E. Ciddor: "*Refractive index of air: new equations for the visible and near infrared*", Appl. Opt. vol.35 (1996), pp.1566-1573, and: J. Stone, J. Zimmerman: "*Index of Refraction of Air*", NIST 2001, [online], http://emtoolbox.nist.gov/Wavelength/Documentation.asp
[12] D.E. Smith and M. Zuber: "*The Mars Observer Laser Altimeter Investigation*", J. Geophys. Res., vol.97 (1992), pp.7781-7797; see also: Proc. ODIMAP III, ed. by S. Donati, Pavia, 20-22 Sept.2001, pp.1-4; and: Physics Today (Nov.1999), pp.33-35.
[13] G. Bianco and M.D. Selden: "*The Matera Ranging Observatory*", in: Proc. ODIMAP II, ed. by S. Donati, Pavia, 20-22 May 1999, pp.253-260; see also: "*The Matera Ranging Observatory: Observational Results*", in: Proc. ODIMAP III, ed. by S. Donati, Pavia, 20-22 Sept.2001, pp.90-96.
[14] B. Querzola: "*High Accuracy Distance Measurement by Two-Wavelength Pulsed-Laser Source*", Appl. Opt., vol.18 (1979), pp.3035-3047.
[15] S. Donati (editor): "*Proc. ODIMAP II, 2nd Int. Conf. on Distance Measurements and Applications*", Pavia, 20-22 May 1999, LEOS Italian Chapter, 1999; see also: "*Proc. ODIMAP III, 3rd Int. Conf. on Distance Meas. and Appl.*", Pavia, 20-22 Sept. 2001, LEOS Italy Chapter.
[16] K. Kowalski: "*Nuclear Electronics*", J. Wiley and Sons: Chichester, 1975.
[17] C.J. Oliver: "*Pulse Compression in Optical Radar*", IEEE Trans. on Aerosp. Syst., vol. AES-15 (1979), pp.306-324.
[18] F.K. Hulme, B.S. Collins, and G.D. Constant: "*A CO_2 Rangefinder using Heterodyne Detection and Chirp Pulse Compression*", Opt. Quant. Electr., vol.13 (1981), pp.35-45.
[19] G.R. Osche and D.S. Young: "*Imaging Laser Radar in the Near and Far Infrared*", Proc. of the IEEE, vol.84 (1996), pp.103-125.
[20] J. Shan, C.K. Thot (Ed.): "*Topographic Laser Ranging and Scanning*", 2nd ed., CRC Press, Boca Raton (FL) 2018.
[21] N. Druml, I. Maksymova, T. Thomas, D. Lierop, M. Hennecke, A. Foroutan: "*1D MEMS micro-scanning LiDAR*", Proc. 9th Int. Conf. Sensor Device Techn. and Applic., Venice, 2018.
[22] Texas Instrument model DLP A200PFP.

[23] S. Donati, G. Martini, E. Randone: "*Improving Photodetector Performance by means of Microoptics Concentrator*", IEEE J. Lightw. Techn., vol.29, (2011), pp.661-665.
[24] see Ref.[2], page 211, the EU Megaframe project, http://www.megaframe.eu/, and also: D. Stoppa, F. Borghetti, J. Richardson, R. Walker, R.K. Henderson, M. Gersbach, E. Charbon: "*A 32x32-pixel Array with In-pixel Photon Counting and Arrival Time Measurement in the Analog Domain*", Proc. ESSCIRC, *2009*, pp. 204-207, see also Ref. [2], Sect.6.4.
[25] Results of the EU Megaframe project, http://www.megaframe.eu/, see also: G.-F. Dalla Betta, S. Donati, Q. D. Hossain, G. Martini, D. Saguatti, D. Stoppa, G. Verzellesi: "*Design and Characterization of Current Assisted Photonic Demodulators in 0.18-μm CMOS Technology*", IEEE Trans. Electr. Dev., vol.58, (2011), pp.1702-1709.
[26] M. Okano, C. Chong: "*Swept Source Lidar: simultaneous FMCW ranging and nonmechanical beam steering with a swept source*", Opt. Expr. vol.28 (2020), pp.23898-23915.
[27] C.V. Poulton, A. Yaacobi, D.B. Cole, M.J. Bird, M. Raval, D. Vermeulen, M.R. Watts: "*Coherent solid-state LIDAR with silicon photonic optical phased arrays*", Opt. Lett., vol.42 (2017), pp.4091-4094.
[28] W.B. Grant: "*LIDAR for Atmospheric and Hydrospheric Studies*", in Tunable Laser Applications, ed. by F.J. Duarte, M. Dekker, Inc.: New York, 1995, pp.213-306.
[29] R.M. Meausers: "*Laser Remote Sensing*", J. Wiley: New York, 1984.
[30] L. Pantani, G. Cecchi: "*Fluorescence LIDAR in Environmental Remote Sensing*", in Optoelectronics for Environmental Science, ed. by A.N. Chester and S. Martellucci, vol.131, Plenum Press: New York 1990.
[31] E. Galletti, R. Barbini, A. Ferrario et al.: "*Development of a CO_2 Pulsed Laser for Spaceborn Coherent Doppler LIDAR*", 5th Conf. on Coherent Laser Radar, ed. by C. Werner and J.W. Bilbo, SPIE Proc. vol.1181 (1989), pp.113-124.
[32] R.L. Sutherland: "*Handbook of Nonlinear Optics*", M. Dekker, Inc.: New York, 1996, see Chapter 3: Optical Parametric Generation, Amplification and Oscillation.
[33] R. Barbini: "*Measurement of Physical Parameters of the Atmosphere with a LIDAR*", Proc. Elettroottica'90, Milan (Italy), 16-18 Oct. 1990, pp.335-342.

Problems and Questions

P3-1 *Calculate the accuracy of a triangulation telemeter operating at a distance of L=1 meter, using the following: a laser source with near-field spot size w_l =20 μm, a parallax base D=100 mm, a multi-element detector with pixel size (and period) p=10 μm, and an objective lens with focal length F=200 mm. Repeat the calculation for L= 0.3, 3 and 10 meters.*
P3-2 *Evaluate the length of the detector used in Prob.P3-1 needed to cover the distance range 1 to 10 m.*
P3-3 *What do we need for $\Delta\alpha$ in order to keep the relative accuracy $\Delta L/L$ a constant, independent from the distance L?*
P3-4 *How many speckles are contained in the spot imaged back on the receiving photodetector of a triangulation telemeter? Do they affect the accuracy of the instrument?*
Q3-5 *In a direct-sight triangulation telemeter (as the one depicted in Fig.3-1), how can we incorporate an angle sensor capable of resolving less than the arc-minute of parallax angle α?*
P3-6 *Evaluate the background noise P_{bg} in a typical pulsed and sinewave- modulated telemeter.*
P3-7 *Calculate the accuracy of the threshold-crossing timing in a pulsed telemeter operating with a Gaussian waveform of standard deviation τ=5ns and with a peak received power of 1 mW.*
P3-8 *Repeat the calculation of Prob.P3-7 using a Gaussian-like waveform but with the leading edge faster than the trailing edge. Assume a realistic ratio 1.8 of the two.*
Q3-9 *What about the contribution of background noise to the timing accuracy?*

P3-10 *Evaluate the improvement that can be obtained by optimum filter processing of the photodetected signal, in the case of a Gaussian waveform like in Prob.P3-7.*
Q3-11 *How important is it, in a sine-wave modulated telemeter, to use a modulation index m close to unity?*
Q3-12 *Can a waveform different from the sinusoid be used in the sine-wave modulated telemeter, or a penalty is incurred when distortion of the sine-wave is appreciable?*
Q3-13 *In a pulsed telemeter, couldn't we overcome the ambiguity problem by slowly increasing the repetition frequency at the switch-on of the instrument, so that we can determine if 1, 2 or more pulses are contained in the distance span of the measurement?*
Q3-14 *Discuss the advantage/disadvantage of using a single- versus double-objective optical system in the transceiver (transmitter-receiver) of a time-of-flight telemeter.*
P3-15 *Design a simple pulsed telemeter operating with the technique of start-stop clock counting, using a semiconductor laser source and intended for a medium-range application (e.g., an altimeter for a 0 to 3000 m span).*
P3-16 *With the data of the pulsed telemeter of Prob.P3-15, calculate the accuracy of the timing measurement performed by threshold crossing on a 30-ns pulse with Gaussian waveform, at the distances 100, 300, 1000 and 3000 m. Assume a diffuser with* $\delta=1$ *and an objective lens with diameter* $D_r = 100$ *mm. Assume a quantum-limited performance set by the noise of the received power.*
P3-17 *Using the data of the pulsed telemeter of Prob.P3-15 and P3-16, check whether the quantum noise contribution associated with the received photons is larger than the background and electronic noises, so that the latter can be actually neglected.*
P3-18 *What would be changed in the results of Prob.P3-17 if the peak power of the laser is decreased by ten times?*
P3-19 *How would the use of a better (or worse) photodetector help (or penalize) the accuracy calculated in Prob.P3-16 and P3-17?*
P3-20 *How does atmosphere attenuation affect the received signal in the pulsed telemeter of Prob.P3-15 through P3-17?*
P3-21 *What happens to the resolution of the pulsed telemeter of Prob. P3-15 through P3-17 using a solid-state Q-switched laser providing 200 kW of peak power, all the other parameters being the same?*
P3-22 *Design a sine wave modulated telemeter operating with the frequency-sweep method, using a semiconductor source for a medium-range application (e.g., an altimeter working on a 0-3000 m span with a 10-cm resolution). Evaluate the quantum-limited accuracy at 300 m and 3000 m for a measurement time of 100 ms. Then, evaluate the actual accuracy when the background and electronic noises are taken into account.*
P3-23 *Recalculate the performance of the sine wave modulated telemeter of Prob.P3-22 when the target is a corner cube.*
Q3-24 *Explain why the PLL is so important in filtering the received signal during the frequency sweep (from the minimum* f_m *to the maximum* f_M*) of the sine-modulated telemeter.*
Q3-25 *Are the calculations of power, number of collected photons and accuracy developed in Prob.P3-22 and P3-23 applicable also to a multi-frequency telemeter?*
P3-26 *A LIDAR employs a Q-switched laser providing a 10-ns pulse width and the preamplifier and log-converter circuits have a passband of 10 MHz. What is the distance resolution we can achieve?*
P3-27 *In a LIDAR, the Q-switched laser provides a pulse of* $E=1$ *mJ energy and* $\tau=10$*-ns duration. The collecting objective has a* $D=30$*-cm diameter. Calculate the power collected from the atmospheric backscatter with* $\alpha=1\ km^{-1}$.
P3-28 *Can the power level found in Prob.P3-27 be enough to perform a distance resolved measurement of backscattering, with a resolution of, say, 5 meters?*
P3-29 *What is the minimum variation of received power and hence of attenuation coefficient you can resolve with the S/N ratio found in the LIDAR of Prob.P3-27? What is the typical concentration we can resolve accordingly?*

CHAPTER 4

Laser Interferometry

Interferometry is one of the most powerful tools in measurement science, because it provides very high sensitivity, unequalled by any other techniques, and is widely used in virtually every segment of engineering and physics. Interference of light was one of the phenomena marking the birth of optics and challenging seventeenth-century scientists to explain the nature of light, notably Newton and Huygens [1]. In the nineteenth century, though a good understanding of interference was reached, interferometry had few notable applications because of the limited coherence length of the sources then available. With the advent of lasers in the 1960s, the full capability of this powerful technique was finally unleashed. Many impressive sensors and measurement devices were developed: sub-micrometer displacement interferometers, laser Doppler velocimeters, optical coherence tomography, and gyroscopes, just to name the big engineering achievements of photonics science.

In this chapter, we adopt a bottom-up presentation, first describing the operation of the instrument with a minimum of basics. Then, we expand the coverage by considering the many facets of modern interferometry. Among these, we will treat optical configurations, fundamental limits of sensitivity and accuracy, different schemes of superposition, and all the topics of general relevance. Other chapters in this book will deal specifically with applications of interferometry to velocimetry, gyroscopes, and fiberoptic sensors.

As a primer, let us now briefly outline the paradigm of interferometry.
A suitable source, usually a laser, is used to direct an optical beam into the experiment or the physical ambient to be sensed (Fig.4-1).

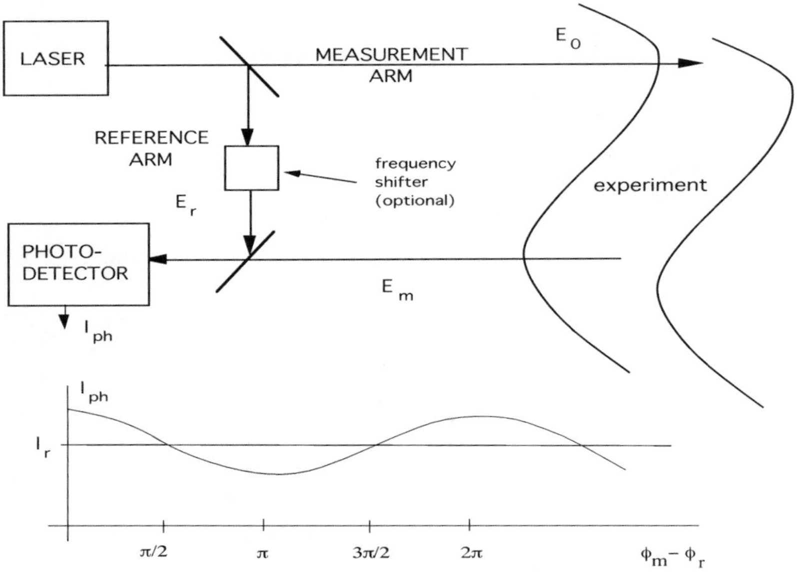

Fig.4-1 Top: the conceptual scheme of an interferometer, with a measurement arm going into the experiment and a reference arm used to superpose on the photodetector a fraction of the outgoing field to the returning field; bottom: the signal generated by the photodetector is a cosine-type waveform periodical in the optical phase shift difference of the two arms, ϕm-ϕr.

A fraction of the beam is kept in the instrument to be used as a reference. Upon returning from the experiment, the measurement beam is superposed to the reference beam onto a photodetector. If E_m and E_r are the optical fields of the two beams, the photogenerated current I_{ph} obtained as a signal out from the photodetector is:

$$I_{ph} = \sigma\ |E_m+E_r|^2 = \sigma\ |\ E_m \exp i\phi_m + E_r \exp i\phi_r|^2 \qquad (4.1)$$

where the fields $E_{m,r} = E_{m,r} \exp i\phi_{m,r}$ have been represented as rotating vectors with amplitude $E_{m,r}$ and phase $\phi_{m,r}$, the bars (|..|) denote modulus and σ is a conversion factor between current and squared field (or power). Developing Eq.4.1 yields:

$$I_{ph} = \sigma\, [\, E_m^2 + E_r^2 + 2E_m E_r\, \mathrm{Re}\{\exp i(\phi_m - \phi_r)\}]$$

$$= I_m + I_r + 2(\sqrt{I_m I_r}) \cos(\phi_m - \phi_r) \tag{4.1'}$$

Here, $I_{m,r} = \sigma E_{m,r}^2$ are the currents that the measurement and reference fields would provide individually, and the last term is the interference signal. We can also write Eq.4.1', letting $I_0 = I_m + I_r$ for the constant term on which the interferometric signal is superposed, as:

$$I_{ph} = I_0\, [1 + V \cos(\phi_m - \phi_r)] \tag{4.2}$$

where

$$V = [2(\sqrt{I_m I_r})]/(I_m + I_r) \tag{4.2'}$$

is called the *fringe visibility* factor or *fringe modulation depth* as it represents the peak amplitude $2(\sqrt{I_m I_r})$ of the interferometric signal as compared to the dc component $(I_m + I_r)$.
In the ideal case of equal strengths $I_m = I_r$ of measurement and reference intensities, the fringe visibility reaches the maximum value V=1, whereas when the signal is weak with respect to the reference, $I_m << I_r$, the factor decreases to $V = 2\sqrt{(I_r/I_m)} << 1$.

In addition to an unbalance of intensities, other factors discussed later in the book may contribute to reduced fringe visibility, namely spatial and temporal coherence (see Sect.4.4.2), polarization matching, and stray light (see also App.A2).
For example, if ϕ_m is affected by a random added term ϕ_{rm} due to the fluctuations of the laser frequency, so that $\phi_m = <\phi_m> + \phi_{rm}$, upon developing Eq.4.1' we get again Eq.4.2, but now the visibility factor becomes $V = \mu\, [2(\sqrt{I_m I_r})]/(I_m + I_r)$ where $\mu = <\cos\phi_{rm}>$ is the *temporal coherence factor*, ranging from 0 (phase uncorrelation) to 1 (complete correlation). Similarly, spatial coherence contributes with a factor μ_s, polarization with μ_p, etc.

The interference signal is proportional to the cosine of the optical phase shift $\phi_m - \phi_r$ between measurement and reference beams.
If we keep the reference beam fixed and unaffected by any disturbance (ϕ_r =const.), while the measurement beam carries the phase shift collected in the propagation, the signal from the photodetector is a quantity measuring the optical phase ϕ_m.
Usually, the beam propagates to a distant reflector and back, totaling a length 2s which amounts to a phase shift $2\pi(2s/\lambda) = 2ks$, where λ is the wavelength, 2π is the number of radians per wavelength and $k = 2\pi/\lambda$ is the wavenumber.

Apart from a constant term I_0, Eq.4.2 tells us that the interferometric signal is of the type cos 2ks (the cosine-signal), and its amplitude is proportional to $\sqrt{(I_m I_r)}$, the geometric mean between reference and measurement powers.
This amplitude is $2\sqrt{(I_m I_r)}/I_m$ times larger that of the signal I_m being detected alone. Thus, in the interferometer, there is a sort of internal gain of photodetection, $G = 2\sqrt{(I_r/I_m)}$. This is a consequence of the superposition of coherent beams, resulting in a detection known as a coherent scheme (see Ref.[2], Ch.10), one also used in optical communications and character-

ized by the welcome property of working at the *quantum-noise limit* of the received signal. Because of that, the sensitivity of the interferometer is generally very good, a feature that is detailed later in this chapter.

4.1 Overview of Interferometry Applications

Fig.4-2 presents a pictorial overview of the big tree of interferometry. Applications may be divided into two main branches, industrial/avionics and scientific measurements.
In the first, we find the well-established applications to the following:

- Mechanical metrology (positioning of tool machines, mechanical workshop calibrations, and measurements)
- Fluid anemometry (also called LDV laser-Doppler velocimetry)

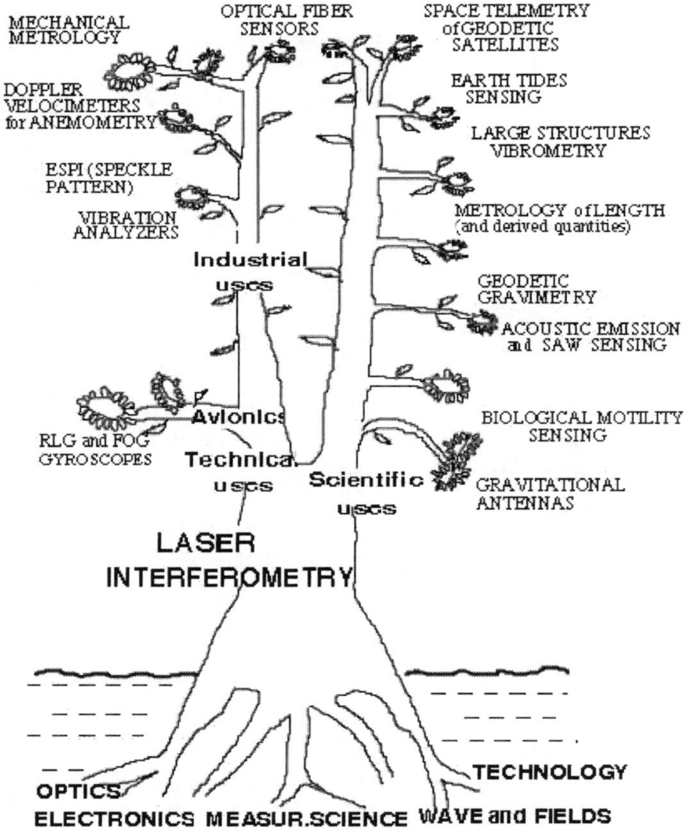

Fig.4-2 Pictorial overview of interferometry applications.

- Vibrometry (rotating machinery diagnostics and control)
- Optical Coherence Tomography (for biomedicine)

In avionics, a wide segment of application is covered by the gyroscope, discussed in Chapter 8. This is an interferometer and senses the Sagnac phase shift induced by rotation. The gyroscope is the heart of inertial sensors like Inertial Navigation Units (INU), Heading Reference Systems (HRS), and attitude and spin control systems.

About scientific uses, a number of amazing examples of application have been successfully reported. Classifying them as in Fig.4-2 according to the distance or the baseline on which the measurement is performed, we find the following:

- Terrestrial tide monitoring (for geodesy)
- Satellite-to-subsatellite displacement measurement (to unveil mass concentrations)
- Long-distance vibrometry (for testing the structural integrity of towers, bridge and dams) and the metrology of length and derived quantity (gravimeter, thermal expansion of materials)
- Pickup of biological signals, motility (respiration sounds and cells motility for fertility) and tissue tomography or OCT
- Pickup of surface acoustic waves (study, design, and testing of SAW devices and AOM modulators)

Most recent on the long baseline end, we find the biggest interferometers ever built (and operating successfully), the Virgo and LIGO (Fig.1-13). These experiments are gravitational antennas based on an interferometer. They are able to detect extremely minute ($\approx 10^{-20}$m) bumps on a target mass acting as an antenna, coming from very remote (Megaparsec away) sources of gravitational waves, like giant-stars or black-hole collapses (Fig.1-14).

4.2 The Basic Laser Interferometers

The term *laser interferometer* is commonly used to indicate a photonic instrument capable of performing the measurement of displacement of a corner cube target, with a fraction-of-wavelength resolution and a 10^{-6} accuracy and precision. The target is usually mounted on the carriage of a tool machine or the like, and the distance range covered by the measurement is up to the scale of meters.

Soon developed after the invention of the He-Ne laser in 1961, the laser interferometer has become a well-established instrument for measurements and especially *calibration* of tool machines. The precision is exceptionally good because the scale factor is directly connected to the wavelength of the laser. This quantity can be easily stabilized to a relative accuracy better than $\Delta\lambda/\lambda \approx 10^{-8}$ in commercial, relatively cheap He-Ne frequency units (see Appendix A1).

Diode lasers are interesting alternatives in this application, but to obtain at least an accuracy of $\Delta\lambda/\lambda \approx 10^{-6}$, we should have either a temperature-stabilized DBR laser diode, or an external-grating stabilized diode, both of which are rather expensive and still lag in performance with respect to He-Ne's.

Therefore, in the following, we refer to interferometers based only on the He-Ne laser. In the next sections, we describe the two main approaches considered the best for the development of practical laser interferometers: the *dual-beam* and the *two-frequency*. Both approaches are used in commercial products, and also give rise to variants with respect to frequency stabilization of the laser and/or electronic processing of the interferometric signals.

4.2.1 The Two-Beam Laser Interferometer

It may be instructive to follow the evolution of the Michelson interferometer, see Fig.4-3a, up to the modern two-beam laser interferometer of Fig.4-3c.

As soon as the laser was available for use as the source in a Michelson interferometer, displacement measurements on a mirror target at a distance s_m were readily performed with $\lambda/2$-resolution. This was done simply by counting the transitions (e.g., the up-going semi-periods) of the cosine signal out of the photodetector (Fig.4-3a). However, to avoid losing counts while moving the mirror target in a displacement measurement, angular alignment had to be ensured, which is a critical point that makes the setup impractical. The solution was to adopt the Twyman-Green interferometer with corner cubes in place of mirrors (Fig.4-3b). In a glass cube corner with dihedral angles of 90°, an incoming beam is returned parallel to the incidence, irrespective of the incidence angle. The lateral displacement of the returning beam is beneficial because the returning beam does not fall into the laser like with mirrors. In fact, back-injection disturbs the laser oscillation, and may spoil the coherence length, generally. With the Twyman-Green interferometer, both back injection and alignment criticality are eliminated. It suffices that the measurement corner cube is moved with a small transversal error, less than the beam size, for a proper superposition with the reference to take place on the photodetector.

Also, a glass-cube beamsplitter is better than a thin beamsplitter because it is easier to mount in a stable holder, and the reference corner cube can be cemented to one of its faces for increased stability of the reference-arm length s_r. The cube beamsplitter has the semi-transparent surface coated by multilayer dielectric film, giving the desired splitting ratio at the wavelength of operation and has antireflection coating on the entrance/exit surfaces.

Last, to also work with a large arm mismatch s_m-s_r (that is, at sizeable distance) without losing the beating signal, we shall use a long-coherence length source, such as a frequency-stabilized laser, which also brings about the benefit of wavelength calibration.
Thus, we come to the configuration of Fig.4-3b, which is actually working and satisfactory in some applications. In it, the source is normally a small-power ($P_{laser} \approx 1\,mW$), compact He-Ne unit, which is frequency-stabilized by means of one of the several possible schemes (see App.A1.2).

Entering the optical interferometer, the beam is divided by the cube beamsplitter BS into two equal parts with powers P_r and P_m, one reflected and one transmitted toward the reference and the measurement corner cubes. The corner cubes reflect the beams back to the beamsplitter, where the beams are recombined after propagation on a distance $2s_r$ and $2s_m$, respectively. After recombination, we get two equal beams, one of which is collected by the photodiode PD. The signal out from the photodiode is given by:

$$I_{ph} = {}^1/_2\, I_m + {}^1/_2\, I_r + \sqrt{I_m I_r}\cos 2k(s_m - s_r) = I_m\,[1 + \cos 2k(s_m - s_r)] \tag{4.3}$$

where $I_m = \sigma P_m$ and $I_r = \sigma P_r$ are the currents separately given by measurement and reference beams, taken equal in power, $P_m = P_r = P_{laser}/2$.

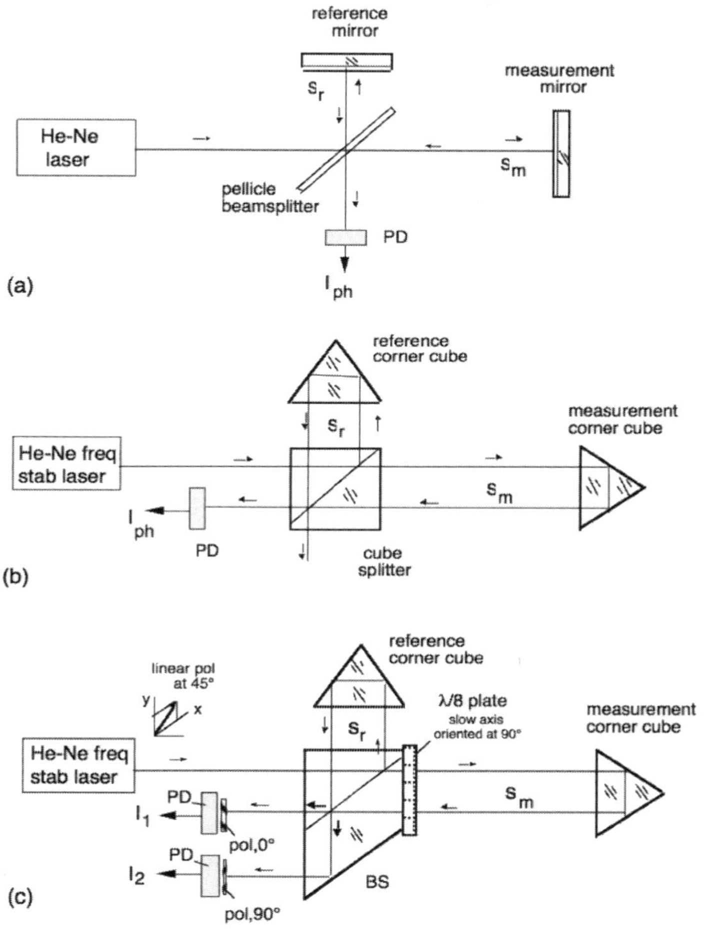

Fig. 4-3 Evolution from the Michelson interferometer (a) to the Twyman-Green interferometer (b) using angular alignment-tolerant corner cubes, and to (c), the modern two-beam laser interferometer measuring the displacement with its sign.

The interferometric signal $\cos 2k(s_m - s_r)$ is superposed to a dc term (of amplitude unity) and can be processed in several ways, analogue and digital. For example, if we want to measure small displacements with subwavelength amplitudes, we can try stabilizing the quiescent working point of the interferometer at the *half-fringe* point, that is midway between the maximum and minimum of the cosine function.

To do so, assume $2ks_m << 1$ (or, $s_m << \lambda/4\pi$) and take $2ks_r = \pi/2$. Then, Eq.4.3 becomes:

$$I_{ph}/I_m = 1+\cos(2ks_m - \pi/2) = 1 + \sin 2ks_m \approx 1+2ks_m \quad (4.4)$$

or, we get a linear replica of the displacement s_m waveform directly from the photodetector current. This is the analogue *vibrometer*-regime of operation, discussed later in this chapter.

On the other hand, if we want to measure fairly large displacements, we can count the interferometric signal transitions. Looking at the signal $\cos 2k(s_m - s_r)$, which can be written as $\cos(2ks_m - \phi)$ with ϕ=const., we get a positive-going transition at each period of $2ks_m$ or for each variation $\Delta 2ks_m = 2\pi$, that is, for $\Delta s_m = \lambda/2$ being $k = 2\pi/\lambda$. In this way, the displacement s_m is measured in steps of $\lambda/2$ =316 nm (He-Ne laser), which is a digital readout without limit in the dynamic range other than accuracy of the wavelength yardstick.

However, a serious drawback of the configuration in Fig.4-3b is that it cannot distinguish increasing from decreasing displacement because the cosine function is even. The other unused output from the beamsplitter [see Fig.4-3b] cannot help either because the signal it supplies is $1 - \cos 2k(s_m - \pi/2)$, i.e., has the same ambiguity. Therefore, the single-channel scheme of Fig.4-3b only works correctly with monotonic displacements $s_m(t)$.

To recover the sign of displacement, we may double the measurement channel and take advantage of polarization diversity, getting the *two-beam interferometer* scheme of Fig.4-3c. The linear-polarization of the laser mode is adjusted to enter the beamsplitter at 45° with respect to the incidence plane. Thus, two independent components, linearly polarized at 0° and 90°, are provided. Both components share the same physical path, down the corner cubes and back.

By inserting a $\lambda/8$-retardance plate at the beamsplitter output, as in Fig.4-3c, we add an extra path in one component (e.g., the one at 90°, parallel to plate slow axis) with respect to the other component (at 0°, the plate fast axis). In the go-and-return path, the total path length is $\lambda/4$, or $\pi/2$ in phase. Then, after the (polarization-independent) recombination at the beamsplitter, we have two beams feeding each photodetector PD1 and PD2. The beatings of 0° and 90° components are obtained by polarizers which are oriented at 0° and 90°.

Taking into account the 1/2-attenuation introduced by the polarizer, the photodetected currents are written as:

$$I_{ph1} = {}^1/_4 I_m + {}^1/_4 I_r + {}^1/_2 \sqrt{I_m I_r} \cos 2k(s_m - s_r) = {}^1/_2 I_m [1+\cos 2k(s_m - s_r)]$$

$$I_{ph2} = {}^1/_4 I_m + {}^1/_4 I_r + {}^1/_2 \sqrt{I_m I_r} \cos 2k(s_m - s_r + \lambda/8) = {}^1/_2 I_m [1-\sin 2k(s_m - s_r)] \quad (4.5)$$

Now, with the pair of *cosine* and *sine* interferometric signals, the argument $s_m - s_r$ of the trigonometric functions can be recovered without ambiguity. One possible processing strategy is illustrated in Fig.4-4.

Each signal out from the photodetector is passed through a comparator, with the threshold placed at the average ($\frac{1}{2} I_m$) photocurrent so that *amplitude* logic signals A_C and A_S are generated (A=1 for high amplitude, A=0 for low amplitude).

Next, *slope* logical signals S_C and S_S are generated for each channel, according to the sign of the signal slope (S=1 for positive slope, S=0 for negative).

Count pulses are generated for the switching of both sine and cosine signals, for example, by differentiating the discriminator's outputs (either up-going or down-going) and by rectifying the obtained pulses so that they are all positive. Because the sine and cosine signals provide four switchings per $\lambda/2$-period, one count represents a $\lambda/8$ (=0.079 μm for a 633-nm He-Ne laser) increment of displacement s_m.

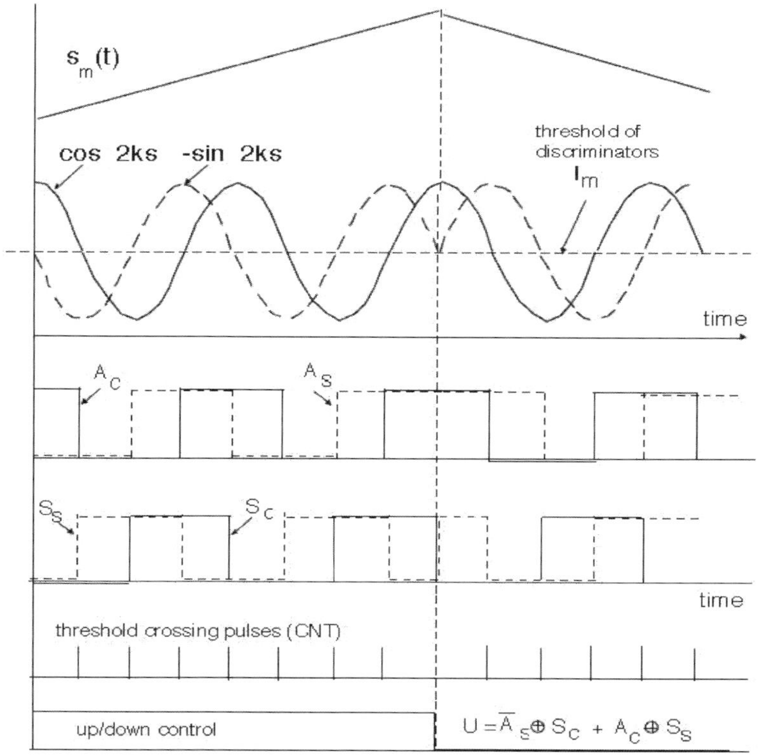

Fig. 4-4 Signal handling in a two-beam interferometer. Cosine and sine signals are passed through discriminators, yielding four transitions per period. Combining the logic signals of amplitude (A_C and A_S) and slope (S_C and S_S) by means of an appropriate logic circuit (Fig.4-5), we obtain the command for the up/down counter input, with four pulses per period, and have a resolution of $1/4(2\pi) = \lambda/8$.

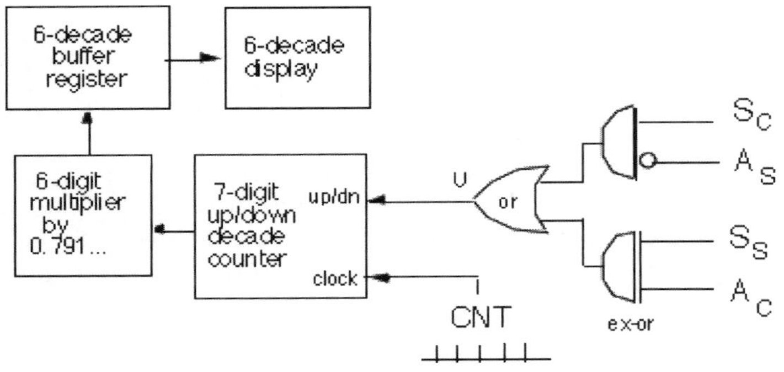

Fig. 4-5 The up/down 7-decade decimal counter stores the displacement under measurement in λ/8 units. A 6-digit multiplier brings the reading to metric units for the buffer register and the display.

Count pulses are sent to an up/down counter as clock (or count) pulses, and a logic combination of amplitude and slope enables the up/down input in the correct direction. With reference to Fig.4-4, let us suppose that the displacement first increases and then decreases. At the first switching, it is A_S=0 and S_C=0, at the second A_C=0 and S_S=1, at the third A_S=1 and S_C=1, and at the fourth A_C=1 and S_S=0, whereas in the decreasing portion the reverse is true. The appropriate logic for the up-count command is therefore written as:

$$U = S_C{}^*A_S{}^*+A_C{}^*S_S+S_CA_S+A_CS_S{}^* = S_C{}^*\oplus A_S+A_C\oplus S_S \tag{4.6}$$

where * stands for the negation (or complement) and ⊕ is the exclusive-or logic operation. The calculation of U is easily implemented by using a few logic gates, as shown in Fig.4-5. The counter is normally a 7-decade counter, so that it can nominally accommodate readings of displacements from 0.079 μm up to 0.79 m.

Now, after the counter, we shall bring the counter content to decimal. This requires a 7-digit multiplier, with the other number entering the multiplier being the length corresponding to one count, e.g., 6328162/8=791020 in air, usually settable in a register for calibration. The result is stored in a 6-digit buffer register, which has 0.1 μm as the least-significant digit and 0.1 m as the most-significant digit. Therefore, a displacement up to ±0.999999 m can be measured. The buffer is connected to the numeric display. This allows the display to be refreshed periodically (e.g., every 0.1 s) and to avoid the last digits from changing fast and becoming unreadable during count buildup. It is not advisable to go beyond 6 digits because of several errors intervening at the ppm (or 10^{-6}) resolution level.

The processing scheme described previously works well with well-behaved displacement waveforms, but has a weak point. The slope logical signals $S_{S,C}$ are obtained necessarily by a high-pass circuit that rejects quasi-dc components of $I_{ph1,2}$. We can indeed design it with a very low frequency cutoff covering the practical range of $s_m(t)$ expected frequency content, but it can never reach zero. Therefore, a possible loss of counts for very slow waveforms $s_m(t)$ is unavoidable.

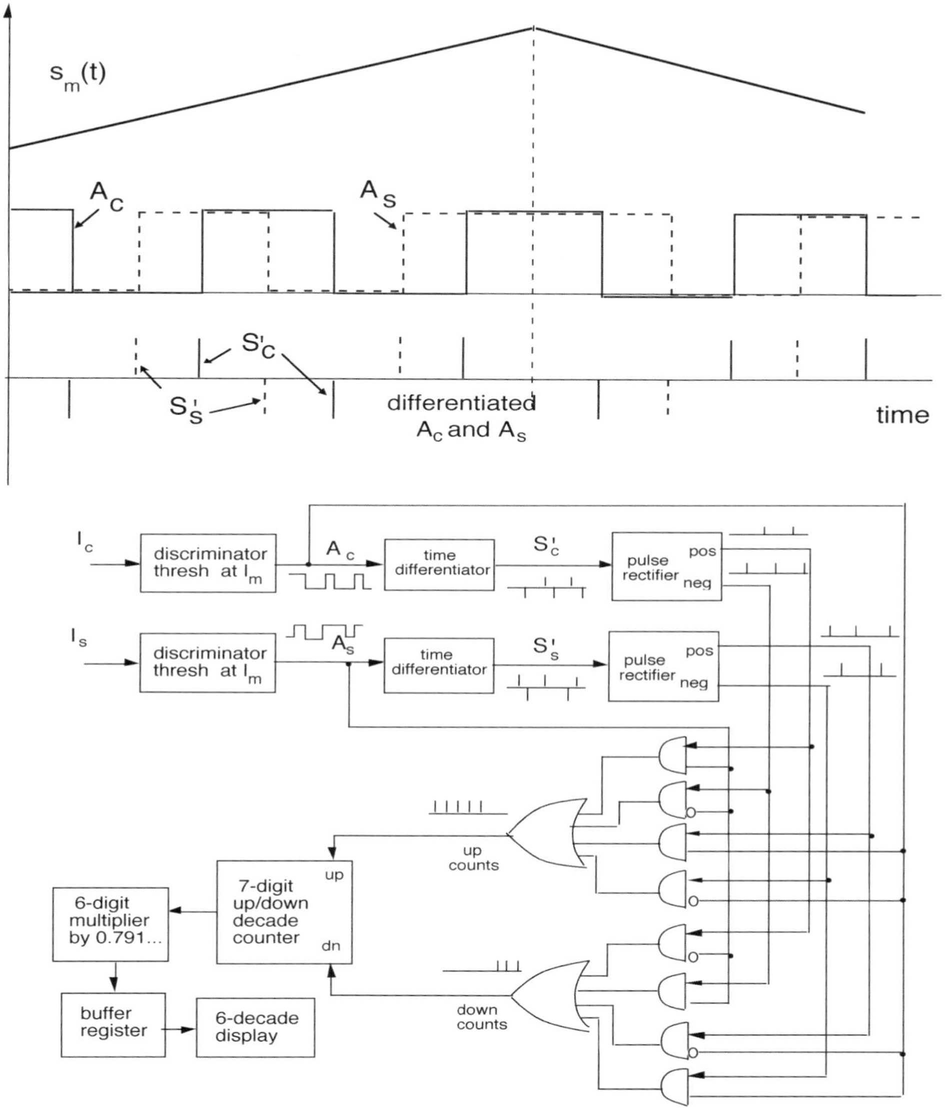

Fig. 4-6 Another more robust strategy for pulse counting in a two-beam interferometer. Pulses S_C' and S_S' from the switching edges of the discriminator outputs are processed directly by the AND/OR gates, and the results are counted in a 7-digit up/down counter.

The drawback can be overcome with the more robust processing scheme of Fig.4-6. Here, we perform the up/down logic operation directly on the switching signal edges, derived from the amplitude comparator outputs, which make the same logic function as described by Eq.4.6, but without the need for a slope logic signal. Accordingly, the scheme of Fig.4-6 removes the low-frequency constraints on $s_m(t)$.

The position of the discriminator threshold shall be set at $^1/_2\, I_m$ in the electronic processing, and this is a critical factor in the performance of the two-beam interferometer. As the photodetected signals I_{ph1} and I_{ph2} may go well below the average value simultaneously (see Fig.4-4), the strategy to derive the threshold is not trivial, unless an assumption on the expected behavior of $s_m(t)$ is made. Perhaps the less stringent assumption is that, because of ambient-related vibrations, $s_m(t)$ undergoes several λ-cycles in a medium-term period. Then, a good practical choice is to take the threshold as the semi-sum $(I_{ph1}+I_{ph2})/2$ of the interferometric signal amplitudes, integrated on, say, a 1- to 10-s time interval.

Regarding the high-frequency cutoff of the measurement, if all of circuits have at least (a specified) bandwidth B, the interferometer can correctly count $\lambda/8$ steps of displacement at a rate of one per period 1/B. The maximum velocity allowed to target displacement is then: $v_{target}= (\lambda/8)$ B. For B=10 MHz and $\lambda/8=0.079$ μm, we get the quite satisfactory value of covered maximum velocity $v_{target}= 0.79$ m/s.

When going to the field, the two-beam interferometer is recognized as satisfactory by users. However, it reveals the following drawbacks that need to be eventually corrected:

- if the optical beam is interrupted, counts are lost and the measurement is incorrect;
- high-frequency (EMI) disturbances sometimes lead to counting errors;
- ambient-induced vibrations can occasionally lead to incorrect counts.

For the first point, a strategy is to monitor the amplitudes of signal I_{ph1} and I_{ph2} and give a warning (with a panel-mounted LED) when both fall below, say, 5% of the time-averaged threshold. The second and third points call for a good shielding from electromagnetic, as well as mechanical disturbances, but are difficult to be eliminated intrinsically. The basic reason is that the two-beam interferometer works on threshold crossing by *baseband signals*. Thus, all the disturbances falling in a spectral range 0 to B are indistinguishable from signal and even if small in amplitude, may lead to incorrect switching.

4.2.2 The Two-Frequency Laser Interferometer

In this approach, we take advantage of the frequency difference between the two-orthogonal polarization modes emitted by a Zeeman-stabilized He-Ne laser (see App.A1.1.3) to make the interferometric signal available on a *carrier* frequency instead of the baseband. The basic schematic of the two-frequency interferometer is reported in Fig.4-7. The Zeeman laser is one with an axial magnetic field and generates two counter-rotating, circularly polarized modes, that are spaced in frequency by f_1-f_2 = 5 MHz approximately. The circular polarizations are transformed into linearly polarized waves by a quarter-wave retarder inserted at the laser output. Before entering the optical interferometer, we first get a reference signal I_{phR} at the frequency difference f_1-f_2. To do so, a small fraction η of the beam power (for example, $\eta \approx 5\%$) is deviated to the reference photodiode PD_R, in front of which a polarizer oriented at 45° is placed. The photocurrent is therefore written as (cf. with Eq.4-1'):

$$I_{phR} = {}^1/_2\, \eta \,\{I_1+I_2+2\sqrt{I_1 I_2} \cos [2\pi(f_1-f_2)t +\varphi]\} = \eta\, I_{av} \,\{1+\cos [2\pi(f_1-f_2)t + \varphi]\} \qquad (4.7)$$

where $\varphi = \varphi_1 - \varphi_2$ is the relative phase of the two modes, and we let $I_1=I_2=I_{av}$ for simplicity.

In the optical interferometer, we use the usual normal cubes and a Glan-cube beamsplitter (or polarization splitter), one with the property of dividing linear polarizations. The Glan is usually made of calcite, a well-known birefringent crystal.

The two halves of the Glan are cut along different orientations of the crystal, to have a large difference of index of refraction for the two polarizations. In this way, at the separation surface, incidence is beyond total reflection angle for one polarization (the perpendicular to incidence plane), which is accordingly reflected, while it is below the total reflection for the other polarization (the parallel to incidence plane), which is transmitted.

Thus, by virtue of the Glan-cube splitter, we are able to direct one polarization (the perpendicular, at frequency f_1) along the reference path, and the other (parallel, f_2) to the measurement path. On the return path, the two polarizations recombine in a beam directed on the photodetector (Fig.4-7).

In the propagation to the corner cubes and back, the two waves have cumulated a phase shift $2ks_m$ and $2ks_r$ and, accordingly, the signal I_{phM} from the photodetector PD_M is:

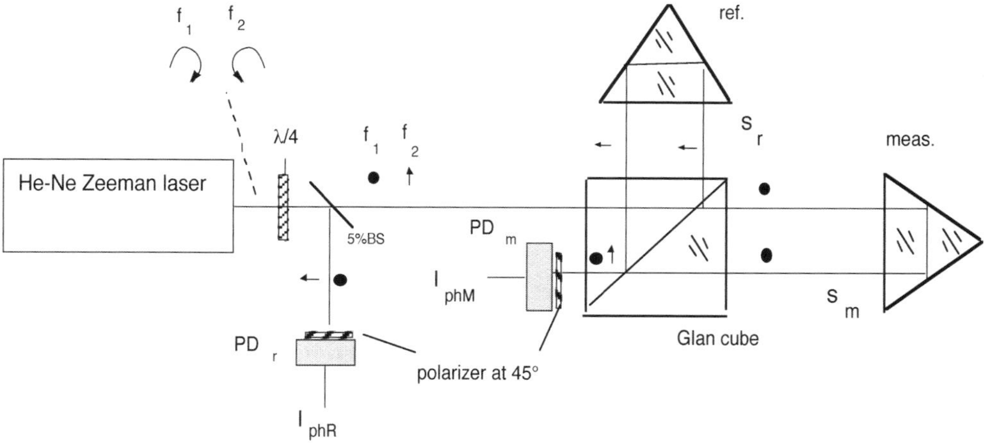

Fig. 4-7 The two-frequency interferometer uses the two modes of a Zeeman laser, separated by f_1-$f_2 \approx 5$ MHz and with orthogonal polarizations. The polarization Glan beamsplitter sends one mode to the reference path and the other mode to the measure path. Upon recombination, the phase difference $2k(s_m-s_r)$ is found on the carrier frequency f_1-f_2.

$$I_{phM} = {}^1/_2\,(1\text{-}\eta)\ \{I_1+I_2+2\sqrt{I_1 I_2}\cos[2\pi(f_1\text{-}f_2)t+\varphi+2k(s_m\text{-}s_r)]\ )$$
$$\cong {}^1/_2\,I_{av}\ \{1+\cos\,[2\pi(f_1\text{-}f_2)t+\varphi_t+2ks_m]\} \qquad (4.8)$$

where we have taken $I_1= I_2=I_{av}$, neglected the small η, and let $\varphi_t= \varphi+2ks_r$ for the phase term, which is a constant if $2ks_r$ is kept constant. Eq.4.8 shows that the quantity to be measured, $2ks_m$, is now a phase superposed on a carrier at frequency $f_1\text{-}f_2$. To recover it, we have several possible choices.

For example, we may want to go back to analog interferometric signals. Then, we can electrically mix the photodetected signals I_{phM} and I_{phR} and generate sum-and-difference frequencies. The difference frequency term is $\cos 2k(s_m\text{-}s_r)$ and is recovered with a low-pass filtering of the mixer output. Also, by phase-shifting the reference I_{phR} by $\pi/2$ and mixing it again to I_{phM}, we obtain $\sin 2k(s_m\text{-}s_r)$. This approach works, but does not fully exploit the advantages of the two-frequency arrangement, which are made clear later.

A better signal processing is illustrated in Fig.4-8. The photodetected signals, I_{phM} and I_{phR}, are squared by comparators and, from the transitions (e.g., positive going), one pulse per period is obtained. We use two counters, one for the measurement and the other for the reference channel. The counters are allowed to count freely, and their content is transferred to buffer registers by a gate pulse G1. This pulse has a period T, which is the renewal time of measurement (typically 0.1 s). The reason for the buffer registers is that pulses may occur simultaneously in the two channels, both running close at $f_1\text{-}f_2\approx 5$ MHz, and they cannot be handled directly by an up/down counter (unless we use a complicated logic circuit).

To find the content of the counters at the end of period T, we use Eqs.4.7 and 4.8 with $(d/dt)\varphi_t=0$, i.e., assume a still reference-arm. Also, we recall that frequency is the time derivative of the phase, $f=(1/2\pi)d\varphi/dt$, and that counters perform an integration operation, which yields the integer part of the quantity $C= \int_{0\text{-}T} f\,dt$. Thus, we obtain:

$$C_R = \int_{0\text{-}T} (1/2\pi)(d\varphi/dt)\ dt\ =\ \int_{0\text{-}T} (f_1\text{-}f_2)\ dt = (f_1\text{-}f_2)\ T,$$

$$C_M = \int_{0\text{-}T} [(f_1\text{-}f_2)+(1/2\pi)\ d/dt(2ks_m)]\ dt\ = \int_{0\text{-}T} [(f_1\text{-}f_2)+2v_m/\lambda)\ dt$$
$$= (f_1\text{-}f_2)\ T + 2\Delta s_m/\lambda \qquad (4.9)$$

In this equation, $v_m=(d/dt)s_m$ is the velocity of the measurement-arm target. The term v_m/λ in the second line of Eq.4.9 can be recognized as the change in frequency induced by the Doppler effect, usually written as $\Delta f=(v/c)f$. The term Δs_m in the last line of Eq.4.9 is the displacement that occurred in the period T, in units of $\lambda/2$.

The subtractor (Fig.4-8) makes the difference of the buffer contents, and the result is:

$$S = C_M - C_R = 2\Delta s_m/\lambda \qquad (4.10)$$

Last, the content of the subtractor is transferred to the main output register by the gate pulse G2 (delayed with respect to G1, see Fig.4-8, to allow for subtraction time). In the output register, the $2\Delta s/\lambda$ counts from the subtractor are added (with sign) to the content already present there. Thus, the content at time t represents the counts in $\lambda/2$ units of $s=\int_{0-\tau} \Delta s$, which is the total displacement from time t=0 (the general reset) to current time t.

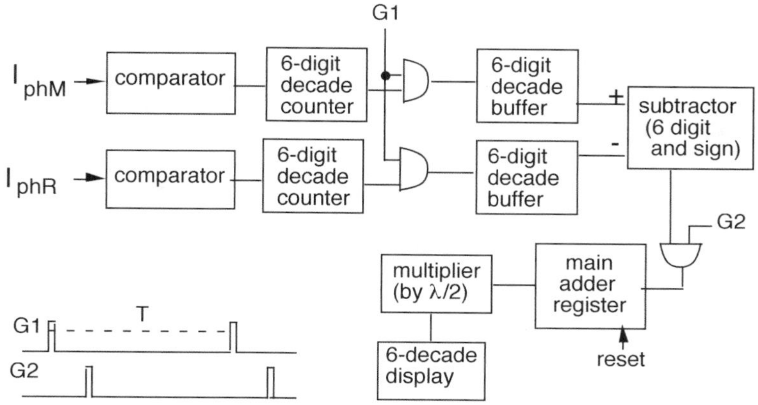

Fig. 4-8 Signal processing for the two-frequency interferometer. Measurement and reference signals are counted for a period T, and then are transferred to buffers. The results are subtracted to yield 2kΔs. A main adder gives 2ks, and a multiplier converts the counts in decimal units for the display.

The measurement resolution is now $\lambda/2$, and we can bring it readily to $\lambda/4$ by doubling the pulses obtained from the comparators (that is, by adding the negative-going transitions in both channels, properly rectified).

In a two-frequency interferometer, the maximum speed of the corner cube we can follow with correct counts is again $v=(\lambda/4)B$, where B is the bandwidth of processing circuits and, also, of the signal I_{phM} and I_{phR} filtering centered on the carrier f_1-f_2=5 MHz. Allowing for a reasonable fraction of the carrier frequency, i.e., B=1 MHz, we get a speed limit typically about $v=0.15\ \mu m \cdot 10^6\ s^{-1}=0.15$ m/s.

The advantages of the two-frequency method over the two-beam method can be summarized as follows:

- the threshold of discrimination can be placed at zero because the dc components of I_{phM} and I_{phR} can be removed without loss of information.
- the rejection of electromagnetic disturbances is much better at the 5±1 MHz frequency, as compared to the baseband.
- any beam loss or interruption is readily detected, looking at the I_{phM} signal amplitude.

Based on the two-frequency approach, a commercial laser interferometer was developed as early as 1967 by Hewlett-Packard. The popular HP5525 *Laser Interferometer* (Fig.4-9) is nowadays considered one of the greatest commercial successes of photonic instrumentation. The concept was later taken over by Renishaw who offers a large family of *laser interferometers* based on model XL-80, single and multi-axis (see Sect.4.2.5.1) and with the option of roll, pitch and yaw angle measurement (6-axis interferometer).

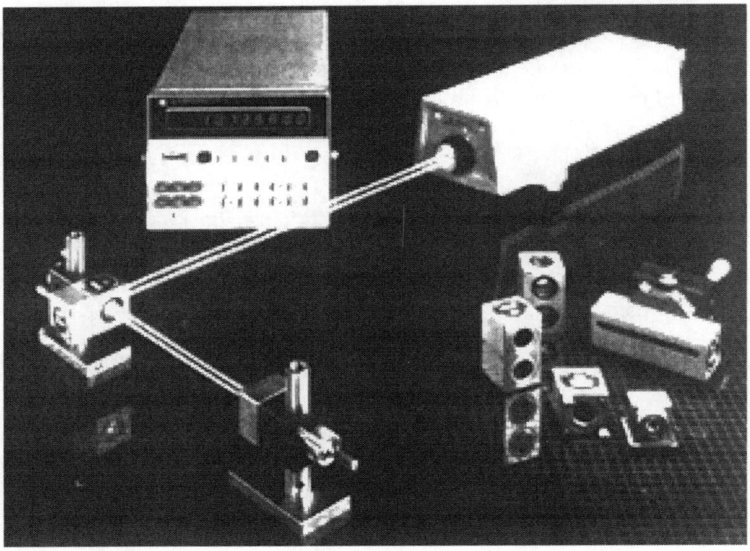

Fig. 4-9 The first laser-interferometer instrument, commercialized in 1967: upper right, the frequency-stabilized He-Ne laser; upper left, the 6-digit readout instrument; bottom, beam bender, corner cube and accessories (from Ref. [3], by courtesy of Elsevier, reprinted by permission).

4.2.2.1 Extending the Digital Displacement Measurements to Nanometers

We can go well beyond the $\lambda/4$ resolution and down to the ultimate limits of sensitivity by means of an electrical phase-shift measurement performed directly on the I_{phM} and I_{phR} waveforms, as illustrated in Fig.4-10.

To do so, it is customary to electrically mix both signals with a local oscillator at a frequency $f_{LO} = (f_1\text{-}f_2)\text{-}f_{IM}$, where f_{IM} is the intermediate frequency (usually $f_{IM} \approx 10...100$ kHz) at which we will perform the electrical phase-shift measurement. A feature to be noted is that the phase shift 2ks is transferred exactly from the $f_1\text{-}f_2$ carrier to a lower intermediate frequency f_{IM}, i.e., with no error.

Then, by measuring the phase shift with a 100-interval subdivision of the 2π angle, as provided by a counter at 100 times the intermediate frequency, we obtain a $\lambda/200$ resolution from the interpolator.

The output of the interpolator is in digital format and is suitable for supplying the user with two additional decades of counts that represent, say, 0.01 and 0.001 μm. Alternatively, we may also amplify and low-pass filter the ANALOG OUT signal of Fig.4-10 to make available an analog-format signal that supplements the digital readout for small displacements.

It is interesting to remark that sub-μm resolution is not usually required in displacement measurement for industrial (e.g., machine-tool) applications and, if we actually want to attain it, operation of the instrument on an anti-vibration table is mandatory, as well as strict control of ambient-induced microphonic disturbances.

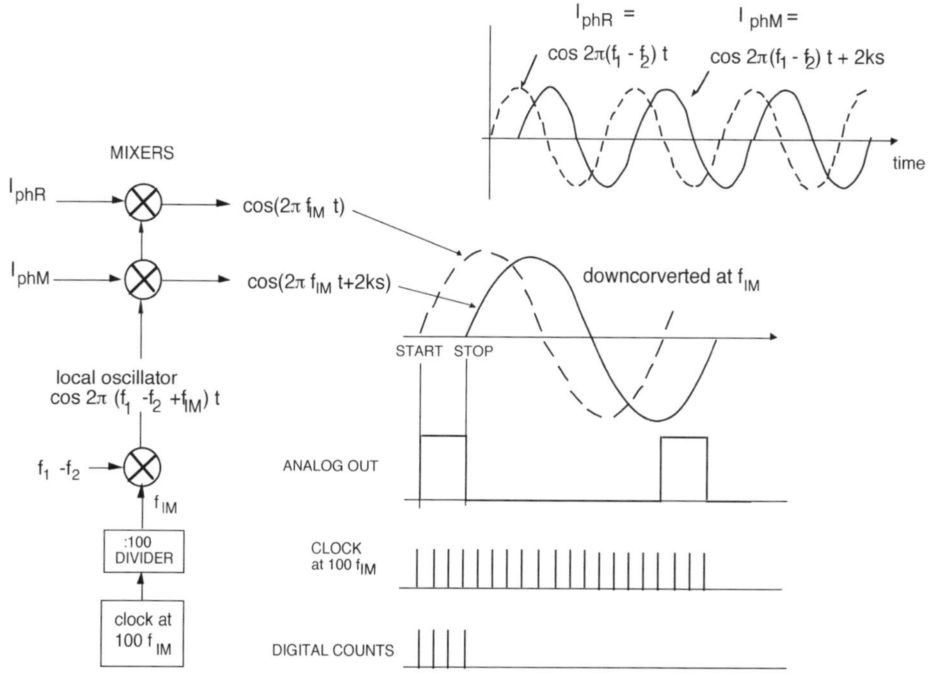

Fig. 4-10 The concept of interpolation to extend the resolution, here to λ/200, in a two-frequency interferometer.

On the other hand, measurement of *vibrations* (that is, periodic phenomena) of nano- and pico-meter amplitudes really makes sense and has been actually developed for a number of applications (see also next sections). The oscillating character of the phenomenon is useful for a much simpler approach based on the *analogue processing* of signals.

However, when we try to extend operation of the two-beam interferometer by resolving a sub-micrometer displacement Δs from the two signals given by Eq.4.5, we incur an ambiguity. The reason is that, while cos2ks and sin2ks uniquely define the argument 2ks, the signals C=1+cos2ks and S=1+sin2ks do not, as they lead to a second-degree equation in 2ks with two solutions for 2ks. Indeed, as a function of 2ks the two signals C and S describe a circle tangent to the axes, and given a point of coordinates C, S, there are two possible circles passing through the point (Fig.4-11, top left).

To remove the ambiguity, we may triplicate the measurement channels of the interferometer as illustrated in Fig.4-11, top right. It consists of segmenting the propagating beam aperture in three sectors, each detected by a separate photodetector. On the ongoing propagation path, a segmented retardance plate is inserted, with 0°, 120°, and 240° (that is, 0, π/3 and 2π/3) of retardance. Thus, the signals are:

$$S_1 = I_{ph1}/I_{ph0} = A[1+V\cos 2ks],$$

$$S_2 = I_{ph2}/I_{ph0} = A[1+V\cos(2ks+\pi/3)],$$

$$S_3 = I_{ph3}/I_{ph0} = A[1+V\cos(2ks+2\pi/3)].$$

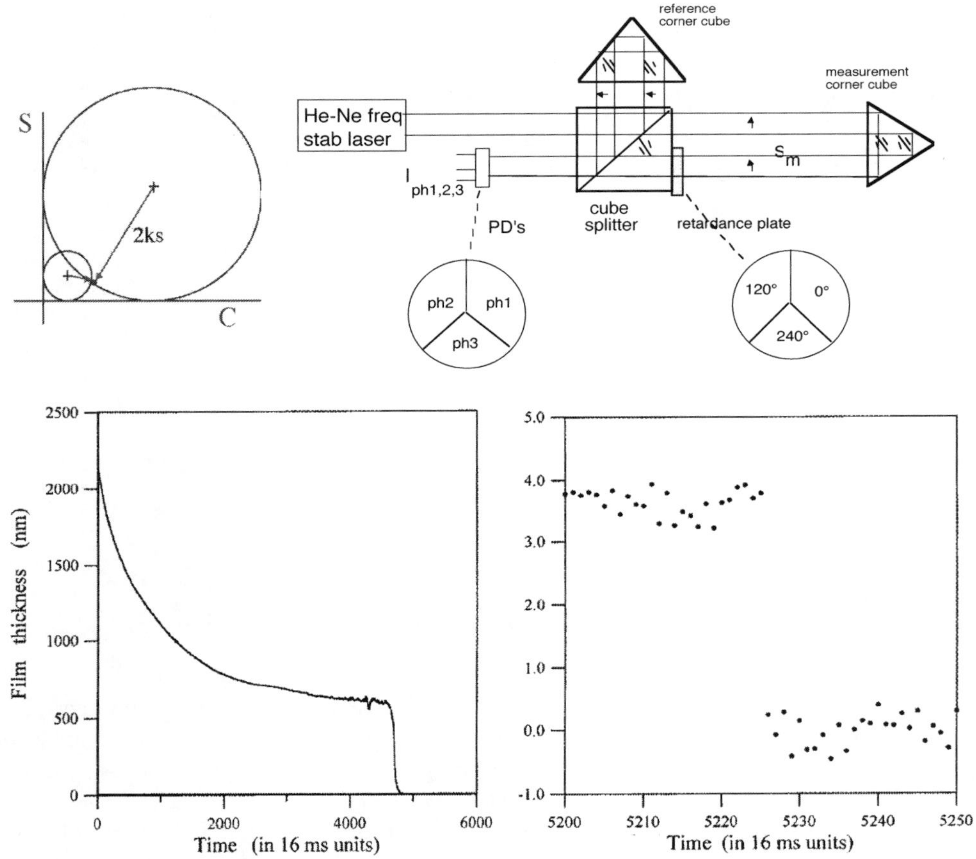

Fig. 4-11 Top left: the ambiguity of signals C and S is explained by their circle being tangent to axes; (top right) extension of the interferometer to nm-resolution by means of triple-sectored waveplate and photodiode combination. Bottom: the thickness of an evaporating droplet is measured while decreasing from 2000 nm to about 500 nm, and (bottom right) it sudden disappears around 4 nm (from [4], ©Institute of Physics, reprinted by permission).

where A is the amplitude of detected signals, assumed equal for the three channels, and V allows for non-unity fringe visibility. To solve for phase 2ks, we start computing:

$$X = S_2 + S_3 - 2S_1 = (2+\sqrt{3})\ AV \cos 2ks, \quad \text{and} \quad Y = (S_2 - S_3) = AV \sin 2ks$$

Then we have:

$$2ks = \text{atan}\ (2+\sqrt{3})\ Y/X \tag{4.11}$$

and this solution is independent from amplitude A and fringe visibility V.

In laboratory experiments [4], the approach has demonstrated a 0.5- to 2-nm resolution and a dynamic range of $\pm\lambda/4$ (displacements larger than $\lambda/4$ require unfolding because of the tan dependence of Eq.4.11). The accuracy of the measurement, $\approx$1 nm, has been evaluated by watching at the Lissajous figure of C and S. Panels in Fig.4-11, bottom left and right, show the result of the measurement of a droplet thickness while the droplet is evaporating, and reveals the sudden, collective vanishing at the final 4 nm [4].

4.2.2.2 Integrated Optics Interferometers

Over the years, there have been considerable efforts to miniaturize the laser interferometer and integrate the optical parts into an integrated optics (IO) chip, so as to develop a com-

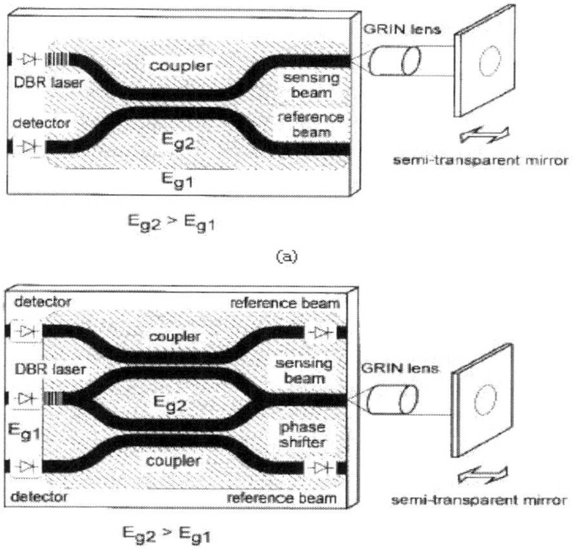

Fig. 4-12 An Integrated Optics chip in GaAlAs includes all the functions to build a laser interferometer: the laser (DFB), the photodiode and coupler/beamsplitters. Only an external transmitting objective is necessary. Top: single-channel version, bottom: double-channel to get *cos* and *sin* signals (from [5], ©IEEE, reprinted by permission).

pact sensor, easier to use than the bulky counterpart and cheaper thanks to integration, so that potentially new fields of application could be reached and exploited.

The integration technologies used for the IO chip are: (i) silica-on-silicon, (ii) Silicon photonics, and (iii) GaAlAs semiconductor [5-7].

The first two are *passive*, because they can only integrate beamsplitters and the phase shift controller of the interferometer, whereas the semiconductor platform (iii) is *active* because the laser and photodiode(s) are integrated in the chip; all that is needed is a GRIN-fiber lens or a microlens to transmit the beam onto the target and collect the returning light.

In addition to the lens, the passive chips (i) and (ii) require pigtailing to the laser and to one or two photodiodes. As the assembly operations of these parts are time consuming and/or require specially developed manufacturing equipment, the advantage of integration is in this case not fully exploited.

The best active IO chip reported so far for interferometric measurements is shown in Fig.4-12. Introduced by Hofstetter et al. [5] it uses GaAlAs laser and photodiodes fabricated with low energy gap E_{g1}, and waveguides fabricated in higher energy gap E_{g2}, for optical confinement and no absorption in the guides.

To obtain the two signals cos/sin necessary for up/down fringe counting, the interferometer is doubled (Fig.4-12 bottom) and the reference arm lengths are made different of $\lambda/8$ so as to introduce a $\pi/2$ phase shift on the go-and-return path.

4.2.3 The FMCW Interferometer for Distance Measurement

In all of the previously described implementations, we have seen that the laser interferometer measures the *incremental displacements* of a target (the retroreflector), not really the distance of the target, because the phase Φ=2ks is a modulus-2π variable, and we can only measure phase differences $\Delta\Phi$ from the interferometric signal. To work on phase differences, we apply increments Δs to the target distance, so as to develop counts and recover the number of periods 2π of phase (or $\lambda/2$ in distance) contained in the path to the target.

But, as Φ also depends on k, we can as well modulate the wavevector k (or, the laser frequency ν, being $\nu = \omega/2\pi = k/2\pi c$) while keeping distance constant, so that counts are developed to obtain a true *distance measurement*. This is the principle of the FMCW (Frequency Modulated Continuous Working) interferometer [8-10], in which the returning field is delayed by T=2s/c with respect to the outgoing field, and is therefore shifted by a frequency difference $\Delta\nu = \chi T$, where χ indicates the rate of change of frequency applied to the laser (Fig.4-13, top).

By letting the returning field beat and at the photodetector with (a portion of) the laser field, we obtain the frequency difference $\Delta\nu$ at the output signal of the photodetector and upon measuring $\Delta\nu$, we can trace back the distance s as:

$$s = (\Delta\nu/\chi)\, c/2 \qquad (4.12)$$

Denoting the optical field leaving the laser as $E_l = E_{l0}\exp\{2\pi i[\nu_0+\chi t]t\}$ and the received field back from the target as $E_2 = E_{20}\exp\{2\pi i[\nu_0+\chi(t-T](t-T)\}$, where T=2s/c is the time delay of propagation to the target and back, the generated current (Eq.4.1') is:

$$I_{ph} = \sigma\ |E_1+E_2|^2 = \sigma\ |\ E_{10}\exp i\phi(t) + E_{20}\exp i\phi(t-T)\ |^2$$
$$= \sigma\ [\ I_{10} + I_{20} + 2\surd(I_{10} + I_{20})\cos[\phi(t) - \phi(t-T)] \qquad (4.13)$$

and, developing the phase difference $\phi(t) - \phi(t-T)$, which the instantaneous frequency ν is $\nu_0+\chi t$ for the outgoing beam, and $\nu_0+\chi(t-T)$ for the received beam, we can write:

$$\phi(t) - \phi(t-T) = 2\pi[\nu_0+\chi t]\ t - 2\pi[\nu_0+\chi(t-T]\ t]$$
$$= 2\pi\ (\nu_0 T + \chi\ T\ t) \qquad (4.14)$$

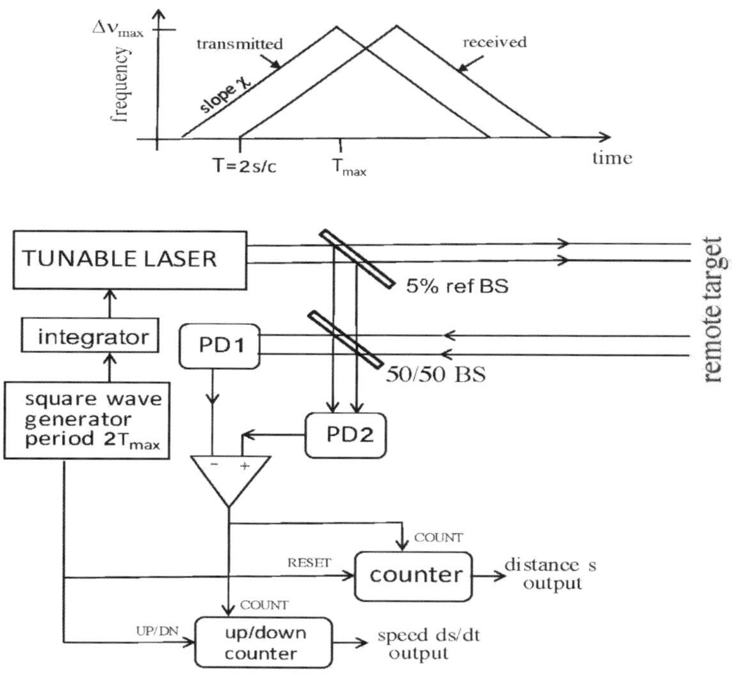

Fig. 4-13 Top: in the FMCW interferometer, the laser frequency is modulated linearly from t=0 to $t=T_{max}$ in a triangular waveform, so that the beam returning from the target is shifted in frequency by an amount $\Delta\nu=\chi T$, where T=2s/c, and this frequency appears at the output signal from the photodetector; bottom: circuit for processing the photodetector signal to obtain distance: a balanced coherent scheme of detection (PD1 and PD2) is followed by a gated counter. Also, the difference of semi-period counts is computed by means of an up/down counter fed by the square wave semi-periods, and the speed $\Delta s/\Delta t$ is also obtained.

In Eq.4-14, the term $\nu_0 T$ is a constant phase shift and can therefore be dropped. Then, using Eqs. 4.2 and 4.2', Eq.4.13 becomes:

$$I_{ph} = \sigma(I_{10}+I_{20}) [1 + V \cos 2\pi (\chi T) t] \qquad (4.15)$$

that is, the photodetected signal carries an oscillation at a frequency $\Delta\nu=\chi T$. From $\Delta\nu$ we can calculate distance s, as already written with Eq.4-12.

Processing of signals is particularly simple in an FMCW interferometer, as illustrated in the block scheme of Fig.4-13. Here, the field returning from the target is detected by the homodyne balanced detector (see also Ref.[2], Sect. 10.2) made up of photodiodes PD1 and PD2 which receive a small fraction of the outgoing beam to act as the local oscillator.

A single photodiode could suffice to perform the coherent detection, but the balanced arrangement with two photodiodes takes advantage of the phase-shift introduced by the beam-splitter, and avoids its 3-dB loss, thus doubling the signal amplitude, and also cancels the common-mode disturbances (Ref.[2], Sect. 10.2).

An important difference of FMCW with respect to the time-of-flight telemeter based on the direct (i.e., incoherent) detection that we have considered in Ch.3, is that the coherent instrument always works in the *quantum limited* regime of detection, thanks to the coherent gain that makes negligible the background and preamplifier (or electronics) noise discussed in Sect.3.2.2. So, the FMCW instrument can operate in full daylight illumination with little or no wavelength filtering, and in general, thanks to the better noise performance, the FMCW may reach a maximum range typically 2.5 to 4 times larger than intensity-based telemeters described in Ch.3.

After photodetection, the frequency-carrying signal is sent to a counter and, upon an integration time T_{max}, the counter will accumulate a content N given by:

$$N = \chi(2s/c)T_{max} \qquad (4.16)$$

The measurement is readily made decimal by appropriately choosing the frequency swing $\Delta\nu_{ma}$ (see Fig.4-13) of slope $\chi=\Delta\nu_{max}/ T_{max}$, because then we have, substituting in Eq.4.16:

$$N = \Delta\nu_{max}(2s/c) \qquad (4.16')$$

so $\Delta\nu_{ma}$ shall be the same mantissa of c/2, that is, 299792.../2 for s to be exactly decimal. Of course, to improve accuracy of the distance s measurement we can repeat the measurement of N on several periods, say N_{ave}. In this way, we have a total number of counts NN_{ave} and the accuracy is $\sigma_s = \sigma_{s0}/\sqrt{N_{ave}}$ with respect to the single measurement result σ_{s0}.

In addition to absolute distance, velocity is also readily measured by the FMCW interferometer: we simply subtract counts N of successive semi-periods (of up- and down-going $\Delta\nu$) of the frequency sweep. Letting them be N(t) and N(t+T), speed is then calculated as $v= [N(t+T)-N(t)]/T_{max}$.

The good performance of the FMCW distance-measuring interferometer have been recognized as particularly attractive for the automotive LiDAR application [11], because the instrument can be implemented around a compact diode laser, easily tuned in wavelength by the bias current, and exploiting the benefits of coherent detection (low detection noise and easy rejection of ambient light, and eventually using a motionless scanning of the scene (Sect.3.4 and Ref.[26-27] of Ch.3).

About the choice of $\Delta\nu_{ma}$ and T_{max}, the frequency difference generated by the 2s/c delay, that is $\chi(2s/c)$, should be large enough to minimize the truncation error of ±1 count at the beginning and end of the time counting interval, but also be the minimum possible for easy operation of electronic counters; the scanning time shall be short enough to satisfy the system requirement of fast updating of the measurement, but also the largest possible to minimize noise (proportional to the square root of measurement bandwidth, or $1/\sqrt{T_{max}}$).

Example. Assuming a diode laser (DL) wavelength of λ=800 nm, we can start with $\Delta\lambda$= 1 nm, a wavelength interval readily achieved by a few milliampere change of bias current, and have $\Delta\nu_{max} = \nu\,(\Delta\lambda/\lambda)$ = 470 GHz for the corresponding frequency swing; let's also assume T_{max}=1 ms for fast, real-time acquisition. Then, targets at distance s= 1, 10, 100 m (with 2s/c= 6.6, 66, 660 ns) will develop a frequency beating, in the detected signal, of $\chi(2s/c) = (\Delta\nu_{max}/T_{max})(2s/c)$=3.1, 31, 310 MHz, a frequency large enough (for a low error) at the minimum distance and still manageable (with medium speed counters) at the maximum distance. Of course, these numbers shall be interpreted as central design values, and can be modified according to the user's specification and/or to device characteristics.

About the DL, units with Fabry-Perot structure and active waveguide in ternary compounds offer a large $\Delta\lambda$ swing by bias current modulation (see Fig. A1-13) but with dangerous mode hops that shall be carefully avoided by mounting the chip on a TEC (App. A1.3.1) to stabilize chip temperature and by employing a well-regulated power supply to work in a linear $\lambda=\lambda(I_{bias})$ region, in the middle of two mode hops. In any case, the bias current modulation applied to the DL to change frequency also produces a power change, that shall be corrected by an appropriate circuit.

With VCSEL DL we can use an external mirror, a MEMS, to tune the wavelength [11], and we have no feedthrough to power.

One issue about DLs for FMCW is the demanding linewidth $\Delta\nu_{lw}$ requirement. For the returning beam to superpose coherently to the reference beam and develop the frequency signal, the delay time 2s/c shall be smaller than coherence time $T_c=1/2\pi\Delta\nu_{lw}$, or $\Delta\nu_{lw}=1/2\pi)(2s/c)$. For the above ranges of distance, s= 1, 10 100 m, we need $\Delta\nu_{lw}$= 24, 2.4, 0,24 MHz, and especially the last value may be difficult to obtain at least in both VCSEL and FP DL of traditional design.

Let's consider now the resolution of the FMCW distance measurement. Writing Eq.4.16' in terms of the wavelength swing, i.e., with $\Delta\nu_{max}=\nu\,(\Delta\lambda_{max}/\lambda)$ gives:

$$N = \nu\,(\Delta\lambda_{max}/\lambda)\,(2s/c) = s/[\lambda^2/2\Delta\lambda_{max}] = s/L_0 \qquad (4.17)$$

Eq.4.17 tells us that the unit of measurement of the distance measurement performed by the FMCW interferometer is:

$$L_0 = \lambda^2/2\Delta\lambda_{max} \qquad (4.17')$$

inserting the values of the example above, we get $800^2/2{\cdot}1$=320 μm, a value that is much larger than the $\lambda/2$=0.4 μm resolution of a plain interferometer, yet one good enough for civil engineering surveying, and overabundant for automotive Lidar requiring resolution of 3- to 10-cm. For this last application, we could reduce the $\Delta\lambda_{max}$ swing by a factor of one hundred still obtaining frequencies 31k, 310k, 3.1M Hz, adequate to keep the round-off error small.

4.2.4 Comb Frequency Interferometry

An important new approach to absolute distance measurement using a coherent technique is provided by the mode-locked laser supplying a sequence or comb of very short pulses (sub-picosecond or femtosecond pulse duration).

This source was developed by Hänsch and co-workers [12] who were able to lock all the individual modes of the comb sequence directly to a radio frequency reference – without the need of clumsy, high-order harmonic multipliers – achieving an unprecedented 5.10^{-16} relative frequency stability, exploited in distance measurements but also as in spectroscopy and clock distribution [13].

Typically, the fs-mode-locked laser is built around a Ti:sapphire laser, pumped by a few Watts from a green diode laser, and exploiting the Kerr-lens nonlinear focusing that leads to spontaneous mode-locking. The emission is a frequency comb centered around 820 nm, with a spectral width of 25 nm and the pulses are as short as 50 fs with a repetition frequency of 0.5 to 1 GHz (for a 30- to 15-cm cavity length).

The output beam is passed through a (typ.) 35-cm long PBF (photonic bandgap fiber), a micro-structured fiber with a small core and many small holes in the clad, so that light is spectrally broadened to more than an octave (a factor 2) in wavelength, for example from 600 to 1200 nm, thanks to the high-power density in the core which enhances nonlinear self-phase modulation and four-wave mixing. The many voids of the clad holes contribute to shift down the zero-dispersion or GVD in the visible.

Now, the frequency f of the comb modes can be written as

$$f = n\, f_r + f_0 \tag{4.18}$$

where n is the order of the mode (a large number that can go up to 10^5), f_r=c/2L is the longitudinal mode spacing for a cavity length L, and f_0 is the frequency offset (or remainder) due to the oscillation frequency not being an exact multiple of mode spacing.

To obtain a precise frequency, we measure both f_r and f_0 and lock them to an electrical reference like a quartz oscillator or a quartz-based synthesizer, and apply the error signal to the piezo actuator on which we will mount the high-reflectivity mirror of the cavity.
The measurement of f_r is straightforward, because the photodetector receiving the laser beam will provide the beating of modes in the comb and therefore their spacing f_r.

On the contrary, the measurement of f_0, which is not directly accessible, requires a more sophisticated strategy. Usually, the *f:2f frequency chain* [12] is adopted as shown in Fig.4-14. Here, a low-order n_1 mode is frequency doubled by means of a SHG crystal (typ. a 3-mm long KTP), so that we get a frequency $f_{double}=2\,(n_1 f_r + f_0)$, and this signal is made to beat on a photodetector together with a high-order n_2 mode of frequency $f_2 = n_2 f_r + f_0$. The photodetector then supplies the frequency difference of the two inputs, or $f_{double} - f_2 = 2\,(n_1 f_r + f_0) - (n_2 f_r + f_0) = f_0 + (2n_1 - n_2)\, f_r$. If we let $n_2 = 2n_1$, the difference $f_{double} - f_2$ is just equal to the offset frequency f_0 and we can stabilize it against another frequency reference. In [12], typical values were f_r = 625 MHz and f_0 = 64 MHz.

Actually, it's not even necessary to select the n_1 and n_2 modes from the two combs because, thanks to the beating process, only the mode n_2 whose frequency is approximately double of n_1 will provide a low-frequency beating, all the others being at $>f_r$ and easily filtered out.

After stabilizing f_r and f_0, all the lines of the comb turn out to be stabilized in frequency, and the typical (2-sample) Allan variance depicted in Fig. 4-14 (bottom) fits nicely with the $1/\tau$ dependence from the inverse integration time up to $\tau = 100$ s, and has an error σ_f/f of a few10^{-15}. Such a high-stability source is of interest as a potential optical clock and for high-resolution spectroscopy in addition to the long-distance measurement.

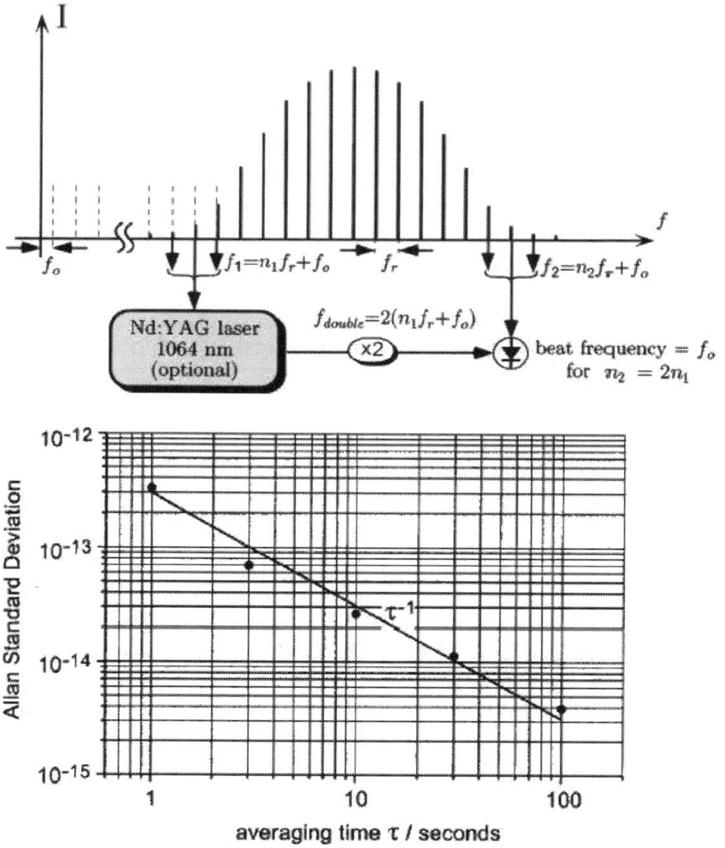

Fig. 4-14 Top: the Hänsch [12] *f:2f* strategy for measuring the offset frequency f_0 of the wavelength-broadened comb of the mode-locked fs-pulse laser: a high-order mode n_2 is made to beat on a photodetector with the second harmonic of a low-order mode, so that the difference is f_0. Bottom: the Allan standard deviation of two samples of frequency chain (from [12], ©APS, reprinted by permission).

In distance measurements, the low value of σ_f reflects itself in a long coherence time $\tau_{coh} = 1/2\pi\sigma_f$, or also, in a very long coherence length $L_{coh} = c\tau_{coh} = c/2\pi\sigma_f$, opening the way to very-long-arm interferometers (e.g., orbiting gravitation observatory) of the near future.

Starting from a stabilized comb frequency laser similar to that described above, several researchers have developed absolute distance measurements [14-15].

Minoshima et al. [14] have used the frequency comb as an amplitude-modulated signal to probe the phase shift of the propagation path to a distant reflector with respect to the reference beam (Fig.4-15). In the photodiode output, they select a high-frequency beating component, in the GHz range (e.g., f_{PD} =10.61 GHz in Fig.4-15), that corresponds to a large number of mode-order difference $\Delta n = n_2 - n_1$, (for f_r=48 MHz it is Δn =221). In this way, the corresponding wavenumber difference Δk and also the phase $\Delta\phi = 2\ \Delta k\ L$ are crucially increased by a factor Δn. Both measure and reference signals at photodetector outputs are mixed with a local oscillator at a frequency $f_{PD} + \Delta f$, differing from $f_{PD} = \Delta n\ f_r$ of a convenient intermediate frequency Δf (about 10 MHz) at which the phase measurement of $\Delta\phi$ is then carried out. The reported resolution at a standoff distance L=240 m is 50 μm (p-p) for a swing of distance (ΔL=12 mm, Fig.4-15) within the ambiguity range of $c/2f_{PD}$ = 30 mm, and the contribution of cyclic error (see below Sect.4.2.4) is reduced with respect to other configurations, thanks to the minimum number of components along the optical path.

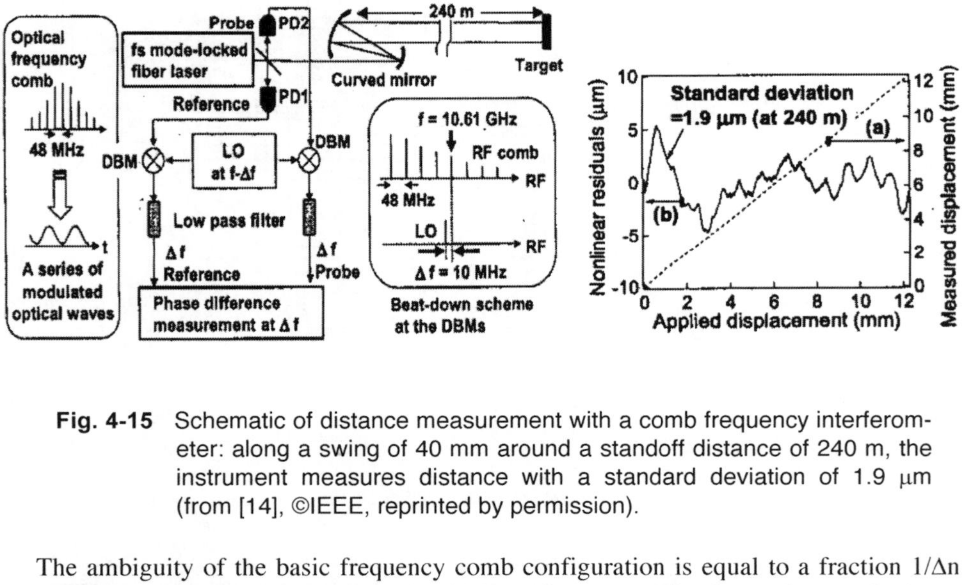

Fig. 4-15 Schematic of distance measurement with a comb frequency interferometer: along a swing of 40 mm around a standoff distance of 240 m, the instrument measures distance with a standard deviation of 1.9 μm (from [14], ©IEEE, reprinted by permission).

The ambiguity of the basic frequency comb configuration is equal to a fraction $1/\Delta n$ of the cavity length. This is not a limitation in long-distance applications when we look at minute variation ΔL of the standoff distance L, whereas it shall be corrected if we need a true distance measurement. By either an amplitude modulation [16] or a frequency sweeping [17] of the comb pattern of the laser, or by extracting a single pulse for a time-of-flight auxiliary measurement, the ambiguity range could be increased to distances of $10^2 ... 10^3$ m.

Finally, it has to be noted that, because of the more complicated setup and signal processing, the femtosecond comb interferometer is not likely to become a contender of traditional laser interferometers for industrial applications.

4.2.5 Measuring with the Laser Interferometer

The target distance is dynamically measured by counts and requires that regular counts are developed and that no *beam interruption* occurs at any time during target movement. Interruption of the beam while the target is moving results in a loss of counts, and thus a wrong measurement that cannot be corrected anymore. In this case, we shall go back to the initial reset and restart operation.

For the same reason, immunity of the instrument to *electro-magnetic interference* (EMI) is of the utmost importance, because we have no way to discriminate spurious EMI pulses from true displacement pulses. Interference may result in a wrong measurement with no warning to the user.

Another source of spurious counts comes from *microphonics*. This term means sensitivity to ambient-related mechanical disturbances that induce vibrations in the measurement path. Vibrations generate up/down counts that have nominally zero average and are harmless, but may introduce an error when the content is sampled instantaneously.

In interferometers with fraction-of-λ resolution, vibration-induced spurious counts are kept low or negligible by mounting the measuring setup on a suitable *antivibration table*. On the other hand, if we deal with instruments reaching the nanometer resolution, mechanical isolation is even more demanding and may require a special design or arrangement of the experimental layout.

Another specific requirement of the laser interferometer is the need for a corner cube reflector. The corner cube is mounted on the moving carrier of the tool-machine under measurement to serve as the mechanical reference. The device is generally compact (typically, 1-2 cm in diameter), but insertion in the experiment means that the measurement is invasive. Additionally, we need to keep it reasonably clean in the surrounding environment.

In the practical operation of the instrument, several systematic errors may occur.

One is the *cosine* error (also called Abbe's error), arising because the beam wave vector $\underline{k}$ and the motion vector $\underline{s}$ are not exactly parallel, but form an angle α_{ks}. As the target is moved, the displacement s is measured as if it was s cos α_{ks}. To adjust the parallelism of $\underline{k}$ and $\underline{s}$, despite vector $\underline{k}$ being immaterial, we must check alignment of an optical component (e.g., the corner cube) at the beginning and at the end of a displacement stroke [3].

Another systematic error comes from minute spurious reflections of the reference and/or the measurement beams on a parasitic optical path. If ε is the fraction of optical power leaking to the unwanted path, either reference or measurement, a cyclic error $\sigma_{cyc}= (\lambda/16)\sqrt{\varepsilon}$ is generated. This *cyclic error* is a ripple error, affecting the true value of measurement with an amplitude σ_{cyc} and a periodicity $\lambda/2$ versus displacement [3,18].

With good engineering practice, the above errors can be minimized and the laser interferometer can be used in a number of circumstances, as detailed in the following sections. Frequently, the ultimate limits of performance discussed in Sect.4.4 are closely approached.

4.2.5.1 Multiaxis Extension

Tool machines with numerical control usually require *three-axis* positioning. We do not need to triplicate the interferometer to perform a three-axis measurement, however.
To save parts, we may share the laser, taking advantage of the power being still adequate when reduced to 1/3 per leg, and use for each axis the two-frequency scheme, with the same reference for all axes, as illustrated in Fig.4-16 (left). A universal tool machine equipped with the three-axis laser interferometer is illustrated in Fig.4-16 (right).

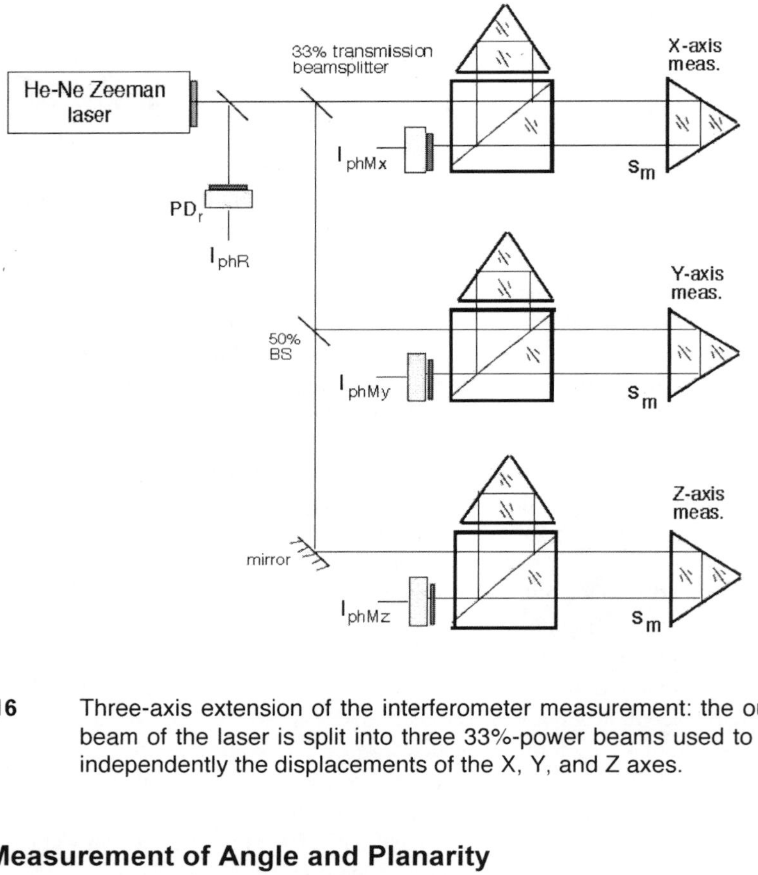

Fig. 4-16 Three-axis extension of the interferometer measurement: the output beam of the laser is split into three 33%-power beams used to read independently the displacements of the X, Y, and Z axes.

4.2.5.2 Measurement of Angle and Planarity

Another measurement commonly performed in mechanical workshops is the *angle* and from here, of the surface *planarity*. We may use the scheme shown in Fig.4-17 where, by aid of a beamsplitter and mirror combination, we direct two beams to a couple of corner cubes mounted on a square.

The square with the corner cubes is moved along the plane to be tested, whereas the reference mirror and beamsplitter assembly is kept fixed. The measurement photodiode receives the recombined beams returned from the corner cubes, and it supplies a signal containing the interferometric phase difference $2k(s_{up}\text{-}s_{low})$, where s_{up} and s_{low} are the path lengths of the upper and lower beams.

When the corner cube square is moved and it remains parallel to itself, s_{up} and s_{low} increase by the same amount, and no count is developed. However, if a tilt α is suffered, the path length difference $2\alpha L$ is counted in units of $\lambda/4$, with L being the distance between the corner cubes (Fig.4-17). Thus, the angle α is measured with an angular resolution cor-

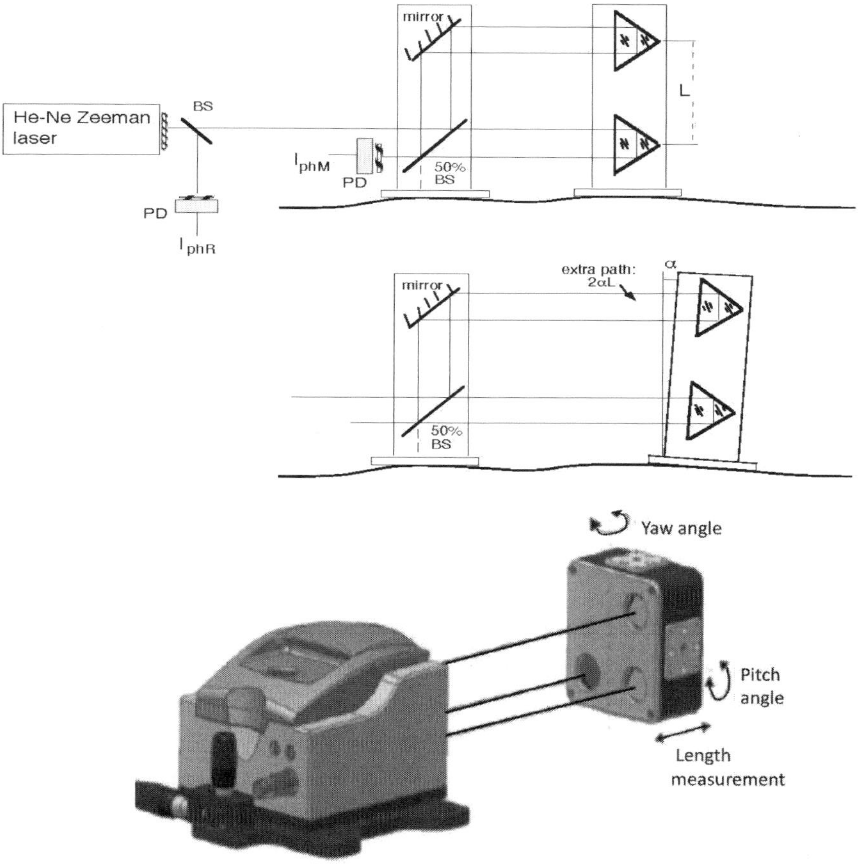

Fig. 4-17 Top: scheme of angle/planarity measurement by the laser interferometer; bottom: measurement of displacement and pitch and yaw angles of the surface under test by employing the three-beam scheme of Fig.4-16 (left) and using two beams for angles (courtesy of SIOS, Messtechnik).

responding to the single count, or $\alpha_{1c} = \lambda/8L$. Taking L=100 mm and being λ=0.633 μm, we get α_{1c} =0.8 10^{-6} rad or 0.16 arc-sec, a very good resolution indeed.

Now, considering the surface under test divided in individual cells, typically of the size of the corner cubes square basement, we can measure the profile z(x,y) of the surface by adding step by step the counts L_c of vertical displacement $\Delta z = \alpha L_B = N_c L_B\, \lambda/8L$ from one cell to the next. The vertical resolution we obtain in the planarity deviation of the surface under test is $\Delta z = N_c\lambda/8$ (for L_B =L), or, $\lambda/8$=0.08 μm using a He-Ne laser interferometer.

Still another possibility offered by the scheme of triple beams is to simultaneously measure displacement and the two-angle components of the surface normal, that is, pitch and yaw, as shown in Fig.4-17, bottom. This is achieved simply by duplicating for each angle the scheme of Fig.4-17, top.

4.2.5.3 Rectangularity Measurement

Another variant to the planarity scheme is obtained by adding a 90° deviation of the beams by means of a pentaprism, as indicated in Fig.4-18. The pentaprism has a dihedral angle of γ=45°, and it is easily seen to deviate the incident beam by 2γ = 90° irrespective of the incidence angle. Compared to using a mirror oriented at 45°, the pentaprism eases the alignment and introduces no error associated with the incidence angle.

The counts developed in the arrangement of Fig.4-18 are related to the planarity errors, plus the deviation error from rectangularity of the surface under test on which the mobile corner cube square sits. Again, the resolution in rectangularity is the same as for planarity, $\alpha_{1c} = \lambda/8L$, while the accuracy is of course affected by the pentaprism dihedral angle error $\varepsilon = \gamma$-45° (typically <1 arcsec in best units).

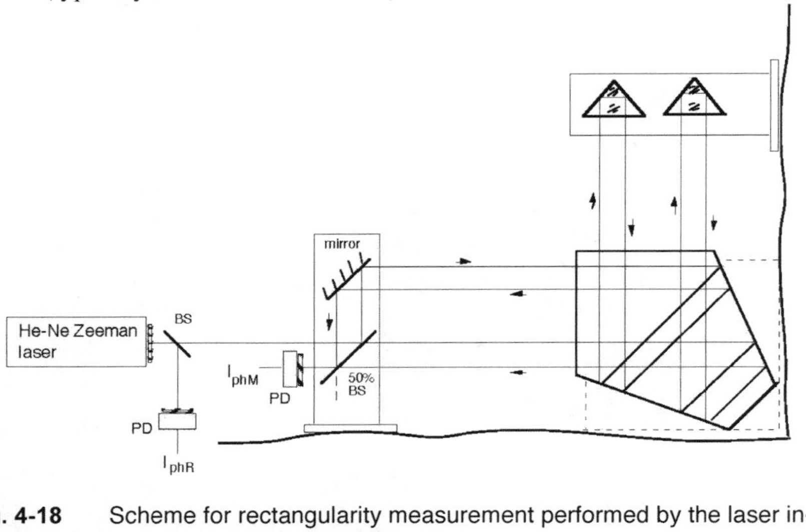

Fig. 4-18 Scheme for rectangularity measurement performed by the laser interferometer and a pentaprism.

4.2.5.4 Extending the Measurement on Diffusing Targets

An important extension of the basic interferometer, either dual-beam or two-frequency, is operation on a *diffusing surface* target. Indeed, from the user's point of view it is preferable to have the instrument capable of working on the native surface of the target, be it a tool-carrying turret or guide, rather than having to deal with a corner cube which, even when small, is still an invasive item that requires mounting space and surface cleaning.

When a diffusing target is used in place of the corner cube, we may consider using a focusing lens in front of the target, as shown in Fig.4-19, so as to improve collection of light returning to the photodetector.

On the target, the spot focused by the lens ideally has a (radial) dimension $w_t = \lambda/\pi NA$ (see Ref. [2]), where NA is the numerical aperture of the lens used by the incoming beam, given by $NA = w_l/F$, where F is the lens focal length and w_l is the input beam spot size.

Assuming the diffuser is ideal, light radiated back to the target follows Lambert's law (see [2], Ch.2) and the amount of it superposed to the reference path spot of area πw_l^2 on the photodetector PD_m is readily found as $P_m = (1/\pi)\, P_0\, (\pi w_l^2)/F^2 = P_0\, (w_l/F)^2$ where P_0 is the power from the laser leaving the beamsplitter.

Thus, the power returning to the readout section of the interferometer is attenuated by a factor $(w_l/F)^2$, usually <<1. However, as the interferometer is a coherent detection scheme, when the target moves, a signal of the form $I_{phM} = I_m + I_r + 2\sqrt{I_m I_r} \cos 2ks$ is developed, that can be processed like in a conventional interferometer, even though its amplitude $2\sqrt{I_m I_r} = 2\sqrt{[\sigma P_0\, \sigma P_0 (w_l/F)^2]} = 2\, \sigma P_0 w_l/F$ is attenuated by w_l/F with respect to the signal from a corner cube target.

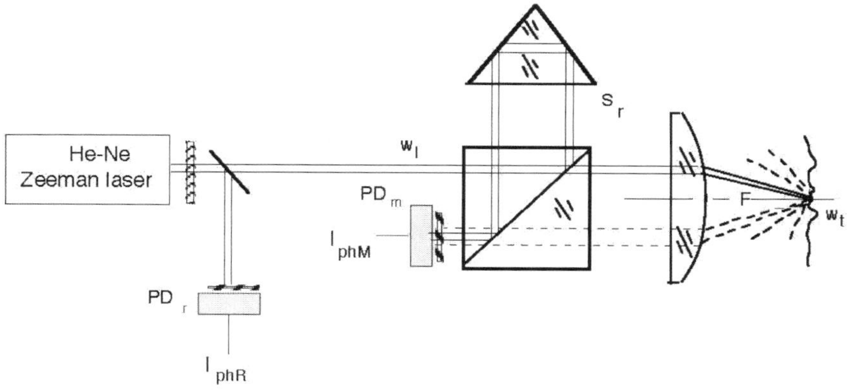

Fig. 4-19 Modification of the basic optical setup for operation on a diffusing target by means of an objective lens focusing a small spot w_t on the target. A fraction of light diffused back (dotted lines) is superposed to the reference beam on the detector PD_m, to detect small-amplitude vibrations.

Of more concern is the range of displacement allowed for the measurement before a large speckle error is suffered. Indeed, as the diffuser moves along the z-axis appreciably, it runs out of focus and the sample of random elemental areas contributing to the returned field will change, adding a random phase error ϕ_{sp} to the expected phase shift 2ks.

This error is called the *speckle-pattern error*, and will be analyzed in detail in Sect. 5.

For the moment, let us just mention that ϕ_{sp} becomes comparable with 2π as the displacement Δz brings the target surface out of focus of the lens. This happens when $\Delta z = w_t/NA$ where $NA= w_l/F$ is the numerical aperture used by the beam of spot size w_l. Using $w_t=\lambda/\pi NA$ in this expression, we get $\Delta z =\lambda/\pi NA^2$ as the dynamic range of maximum displacement for a λ-error. Even at a small NA, the resulting range is clearly much less than the tens of centimeters to meters range that we require in applications of the interferometer to mechanical metrology. On the contrary, the dynamic range allowed by the speckle statistics is usually adequate for vibration measurements, as in this case the amplitude of the periodic displacement is much smaller.

4.3 Operation Mode and Performance Parameters

About the measurement of the optical phase shift ϕ=2ks and its variation accumulated along the path to the target at distance s(t), we can distinguish two modes of operation and two possible signal processing, as illustrated in Fig.4-20, top:

(a) *displacement* measurement, when s(t) is *aperiodic* and can cover a very large number of wavelengths (for example $10^2...10^6$), or a span ranging from micrometers to meters.
This is the mode of operation of the laser interferometer for machine tool control described in Sects.4.2.1 and 4.2.2. As the signal spans many decades, *digital processing* of the interferometric signal is ideally suited and most frequently used;

(b) *vibration* measurement, when s(t) is *periodic* and has a small amplitude (typ. micrometers, but also may go down to nanometers or picometers). Then, this small signal is best handled by *analogue processing*. This is the case of instruments for detecting minute vibrations in a number of application fields (e.g., biology, geodesy and material science) as well as to perform non-destructive-testing of structures including the measurement of mechanical transfer function and of hysteresis cycle.

About performance, interferometers are characterized by a number of parameters, the most important being the following:

(i) the minimum detectable amplitude of displacement, or *noise equivalent displacement* (NED) when limited by noise, even though frequently the limit is set by incidental factors like circuit noise, microphonics and ac hum, and electromagnetic disturbances;

(ii) the maximum amplitude of displacement or *dynamic range* accommodated by the system, usually traced back to the dynamic range of circuits processing the signal, especially the preamplifier of the photodetector;

(iii) the maximum and minimum *frequency* of detectable displacement, due to the frequency cutoffs of the circuits handling the interferometric signal;

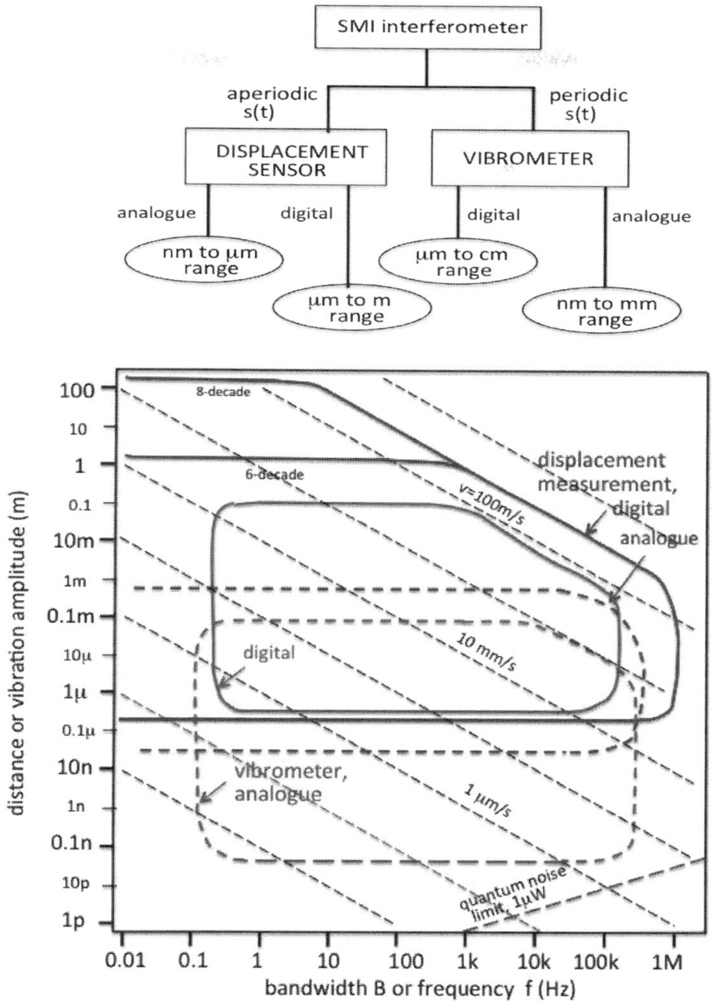

Fig. 4-20 Top: classes of interferometric measurements of target distance s(t): *displacement sensors* are for aperiodic, large swing signals s(t) and *vibrometer* for small, micrometer and sub-micrometer amplitude, periodic signals s(t). In both cases, processing can be either *analogue* or *digital*, with corresponding different working range of amplitudes; bottom: diagram of interferometer performance, with amplitude of displacement plotted versus bandwidth or frequency of vibration. Operation of state-of-the-art design is inside the thick-line perimeters of the curves. Typical performances are indicated, for both analogue (dotted lines) and digital (full lines) processing. At the bottom right corner, the quantum noise NED limit is reported.

(iv) last, we shall consider the speed or *Doppler limit*, because even a slow displacement can develop a high-frequency interferometric signal. Indeed, if the target moves at a speed v, the developed signal is $\phi=\cos 2kvt$, or its frequency content is $f=(d\phi/dt)/2\pi= =2kv/2\pi =2v/\lambda$. This quantity shall be less than the circuit bandwidth B, indicating that large and fast displacements have a $v< \lambda B/2$ limitation in performance.

We can represent these four parameters in a diagram of performances (Fig.4-20, bottom), also called a Wegel (or lemon) diagram because of its shape, by plotting the detectable signal amplitude (distance covered by displacement or vibration amplitude) as a function of bandwidth or frequency of vibration; the Doppler limits are lines running at -45°.

In the diagram, each square is a 10 by 10 factor of covered performance, thus, the more the squares are covered in the plot, the better the instrument is performing.

Several other parameters are necessary for a complete description of the interferometer performance, like for example: (v) resolution, and accuracy of the readings (depending on the λ calibration and stability), (vi) dependence on temperature and atmospheric pressure and composition, (vii) immunity to quasi-static magnetic fields and to EMI disturbances.

4.4 ULTIMATE LIMITS OF PERFORMANCE

We have seen in the preceding sections that laser interferometers easily reach several digits (e.g., six) of dynamic range and sub-micrometer resolution. To ascertain the fundamental limits of operation posed by physical laws, we analyze in this section several factors that affect the instrument performance.

4.4.1 Quantum Noise Limit

The ultimate limitation to the sensitivity of a laser interferometer is the quantum limit associated with the detection of the signal returning from the optical interferometer.

Let us consider it in detail, starting with the photocurrent written in the generalized form as:

$$I_{ph} = I_0 [1+V\cos Rk(s_m-s_r)] \qquad (4.18)$$

where R is the interferometer responsivity and V the fringe visibility. For simplicity, the reader may start taking V=1 and R=2, as in previous sections. We assume that we can adjust the reference path so that the interferometer works in a quiescent point of maximum sensitivity, that is at $Rk(s_m-s_r)=\phi-\pi/2$ (or, at half fringe). In this case we have from Eq.4.18:

$$I_{ph}/I_0 = 1+V\sin\phi \approx 1+V\phi \quad \text{(for small } \phi\text{)}$$

Explicitly, the phase ϕ will then be given by the small displacement to be measured, $\phi=Rk\Delta s$. For a small deviation around the quiescent point, the photocurrent deviation ΔI_{ph} is:

$$\Delta I_{ph}/I_0 = V\,\phi \qquad (4.19)$$

Superposed to the useful signal I_{ph}, we find a fluctuation ΔI_{ph} associated with the shot (or quantum) noise (see [2], App.A3) of the quiescent point current I_0. The rms value i_n of such noise is proportional to the square root of the average current I_0 and of the observation bandwidth B:

$$i_n = (2eI_0B)^{1/2} \tag{4.20}$$

The associated signal-to-noise ratio of the current amplitude measurement around I_0 is S/N= $i_n/I_0 = (2eB/I_0)^{1/2}$. Equating $\phi=\phi_n$ and $\Delta I_{ph}= i_n$ in Eqs.4.19 and 4.20, we get:

$$\phi_n = i_n/VI_0 = (2eB/I_0)^{1/2}/V = [(S/N)V]^{-1} \tag{4.21}$$

Now, we may introduce the *Noise-Equivalent-Displacement* (NED), which is defined as the value of displacement Δs giving the same effect as the intrinsic noise ϕ_n. By definition and Eq.4.18, it is $Rk\Delta s = \phi_n$ or:

$$NED = \phi_n/kR = (\lambda/2\pi)/(S/N)\ RV\ = (\lambda/2\pi)\ (2eB/I_0)^{1/2}/RV \tag{4.22}$$

Eq.4.22 is plotted in Fig.4.21 for the case of λ= 633 nm, η = 0.9, and R=V= 1.

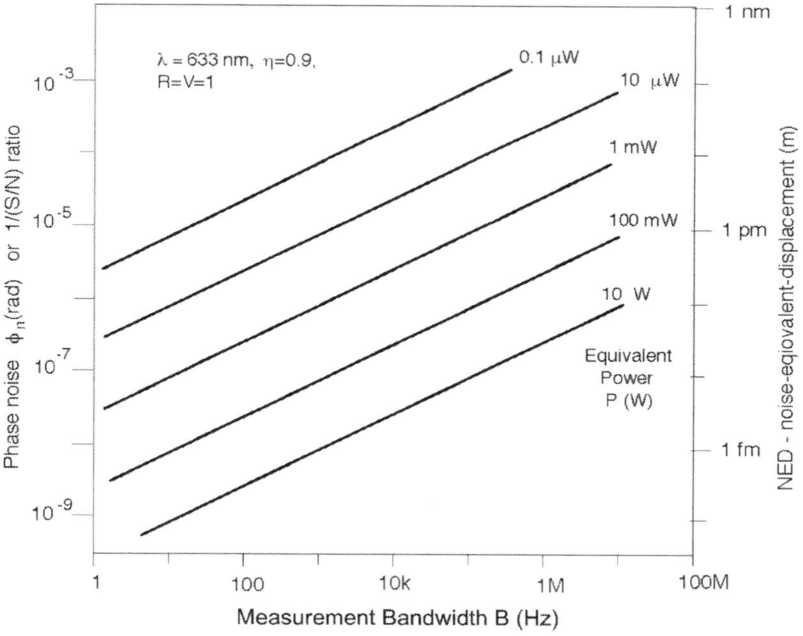

Fig. 4-21 The phase noise (left scale) and the NED (right scale) versus measurement bandwidth B, and with the detected power P as a parameter, at the quantum noise limit of operation in an interferometer.

As we can see from the figure, the quantum limit is indeed very low and theoretically allows us to reach very high sensitivities (e.g., $\phi_n \approx 10^{-6}...10^{-8}$ rad and NED $\approx$ 0.1pm...10 fm). For example, a displacement sensor working with an instrumental bandwidth of, say, B=1 Hz, and with 1 mW of detected power, we can go down to a resolution of just a few femtometers (10^{-15} m). For a vibrometer aimed at detecting oscillations up to a few hundred kHz with 1 mW power, the quantum noise limit is around a few picometers (10^{-12} m).

These very challenging figures are indeed obtained in practice, for example, in vibrometers and gyroscopes. However, we shall stress that generally they require a very careful design to first get rid of several perturbing effects, usually much larger than the quantum limit. For RV≠1, the values of ϕ_n read in Fig.4.21 shall be divided by V, the fringe visibility, and those of NED by RV.

4.4.2 Temporal Coherence

When recombining two beams propagated on different lengths along the measurement s_m and reference s_r paths, we actually superpose two field contributions delayed by a time $T=(s_m-s_r)/c$. If the optical frequency ν is not truly constant, but fluctuates in time, a decrease from the ideal unity visibility of fringes is incurred, and additionally, a phase error in the measurement of $k(s_m-s_r)$ is generated.

An easy way to describe the decrease of fringe visibility is to consider the finite line width Δν of the source. If frequency ν_0 is defined not better than Δν in the time-dependent term $\exp i2\pi\nu t+i\varphi = \exp i2\pi(\nu_0+\Delta\nu)t+i\varphi$, after a time $T_{coh}= 1/\Delta\nu$ the initial phase φ has changed by 2π, and coherence is lost (see also Appendix A1.1.2).

Time T_{coh} is therefore called the coherence time of the source. Alternatively, we talk of coherence length $L_{coh}=cT_{coh}$ as the distance traveled by light in the coherence time T_{coh}. The quantity L_{coh} represents the maximum value allowed to the path length difference s_m-s_r of the interferometer if a beating is to be obtained.

Usually, if the shape of the line is Lorentzian, $p(\nu)=[1+4(\nu-\nu_0)^2/\Delta\nu^2]^{-1}$, the decrease of fringe visibility is a negative exponential: $A=\exp -T/T_{coh}$.

To study the phase error due to incomplete temporal coherence, let us assume that the frequency is a constant ν_0 and ascribe all the fluctuations to the phase term φ(t). Writing the superposed fields as:

$$E_m = E_0 \exp i2\pi\nu_0 t + i\varphi(t) - iks_m, \quad \text{and} \quad E_r = E_0 \exp i2\pi\nu_0 t + i\varphi(t+T) - iks_r$$

The beating on the photodetector becomes:

$$I_{ph}/I_0 = 1+ \cos[k(s_m-s_r) + \varphi(t) - \varphi(t+T)] \tag{4.23}$$

Now, let us express the phase term $\varphi(t) = \varphi_0+\varphi_n(t)$ as the sum of a mean value φ_0 and a random fluctuation $\varphi_n(t)$ with zero mean value, or $\langle\varphi_n(t)\rangle=0$. Then, the phase difference in Eq.4.23 becomes $\varphi(t)-\varphi(t+T)=\varphi_n(t)-\varphi_n(t+T)= \phi_n$ and we get by inserting in Eq.4.23:

$$I_{ph}/I_0 = 1+ \cos[k(s_m-s_r)+\phi_n] = 1+ \cos k(s_m-s_r) \cos \phi_n + \sin k(s_m-s_r) \sin \phi_n \tag{4.24}$$

The mean photodetected current is obtained by taking the average at both sides of Eq.4.24. Because $\langle\phi_n\rangle=0$, it is also $\langle\sin\phi_n\rangle=0$ because $\sin\phi_n$ is an odd function of a zero-mean argument, and therefore we have:

$$\langle I_{ph}\rangle/I_0 = 1+\mu_{tc}\ \cos k(s_m-s_r)$$

where $\mu_{tc}=\langle\cos\phi_n\rangle$ has the meaning of coherence factor (see also [2], Sect.10.1.2). Developing the cosine in Taylor's series, $\cos\phi_n=1-\phi_n^2/2+\ldots$, reveals that $\mu_{tc}\approx 1-\langle\phi_n^2\rangle/2$ is related to the variance $\sigma_\phi^2=\langle\phi_n^2\rangle$ of the random-phase difference $\varphi_n(t)-\varphi_n(t+T)$ at times t and t+T. If T is short, we may expect that ϕ_n remains much less than unity and $\mu_{tc}\approx 1$. The fringe visibility then decreases as $V=\mu_{tc}$ in Eq.4.2.

Moreover, the fluctuation i_n around the mean value $\langle I_{ph}\rangle$ adds a phase error. To evaluate it, let us note first that the $\cos\phi_n$ term in Eq.4.24 is close to unity, in the case of practical interest of not so bad temporal coherence ($\mu_{tc}\approx 1$), and therefore it can be assumed as a constant not contributing to fluctuations. By taking the differentials at both sides of Eq.4.24, and then squaring and averaging, we get for the variance $\langle i_n^2\rangle$:

$$\langle i_n^2\rangle/I_0^2 = \sin^2 k(s_m-s_r)\ \langle\sin^2\phi_n\rangle$$

Again, assuming the case of $\mu_{tc}\approx 1$, the term $\langle\sin^2\phi_n\rangle$ can be approximated to $\langle\phi_n^2\rangle=\sigma_\phi^2$. Now, we recall that the phase error is related to the amplitude error by Eq.4.21, and assume that the maximum sensitivity condition $k(s_m-s_r)=-\pi/2$ to obtain the result:

$$\phi_n^2=\sigma_\phi^2 \tag{4.25}$$

The phase error σ_ϕ^2 can be traced to the two-sample *Allan's variance* $\sigma_V^2(2,T,\tau)$ describing the frequency stability of a generic oscillator. This quantity is defined as:

$$\sigma_V^2(2,T,\tau) = (1/\tau)\langle \textstyle\int_{0\text{-}\tau} [\nu(t)-\nu(t+T)]^2 dt\rangle \tag{4.26}$$

or, it is the mean square deviation of two frequency samples separated by a delay T and averaged on a time interval τ. The phase error is found to be related to the frequency variance by $\sigma_\phi=(2\pi\tau)\,\sigma_V(2,T,\tau)$, and in turn, σ_V is related to a two-consecutive sample variance by $\sigma_V(2,\tau,\tau)=(\tau/T)\,\sigma_V(2,T,\tau)$ for $T<<\tau$. Using Eqs.4.22, 4.25 and 4.26, we then get the NED_{tc} due to temporal coherence effects as:

$$NED_{tc} = (\lambda/2\pi)\phi_n = (\lambda/2\pi)\,(2\pi T)\,\sigma_V(2,\tau,\tau) = cT\,\sigma_V(2,\tau,\tau)/\nu$$

Recalling that $T=(s_m-s_r)/c$, we may write the result in the expressive form:

$$NED_{tc}=(s_m-s_r)\,(\sigma_V/\nu) \tag{4.27}$$

This result tells us that incomplete temporal coherence, beyond reducing the fringe visibility, produces a noise-equivalent random error equal to a fraction of the arm mismatch (s_m-s_r). The fraction is given by the relative frequency stability σ_V/ν of the source and is evaluated in the integration time τ (where $\tau=1/2B$ in terms of observation bandwidth).

With practical frequency stabilized lasers, frequency stability of 10^{-9} to 10^{-11} are obtained, so a 1-count or $\lambda/4=0.15$ μm error is generated for an arm mismatch $s_m-s_r= 0.15$ μm $/(10^{-9}..10^{-11}) \approx 0.15..15$ km. Using a comb frequency-locked source (Sect.4-2.4), well stabilized at 10^{-17}, we can achieve $s_m-s_r=15$ million km, the figure necessary for orbiting LIGA interferometers.

4.4.3 Spatial Coherence and Polarization State

In the superposition at the photodetector, the reference E_r and measurement E_m fields shall have the same spatial mode distribution. If not, their interference term, and accordingly the fringe visibility, will be reduced from the full value to a factor:

$$\mu_{sp} = \int_A E_m(x,y)\, E_r^*(x,y)dxdy \,/ \left[\int_A |E_m(x,y)|^2 dxdy \int_A |E_r(x,y)|^2 dxdy\right]^{1/2} \qquad (4.28)$$

Eq.4.28 explains why interferometers shall use single-mode beam spatial distributions. If the beam contains a mixture of N modes sharing the total power, because the integral product of different modes is zero, only homologous modes will contribute to μ_{sp}. The net result is that μ_{sp} cannot be larger than 1/N. More commonly, it will be $\mu_{sp} << 1/N$ because, in addition, the different modes may have a slightly different propagation constant and smear out the interferometric phase difference.

Similarly, we shall use the same State of Polarization (SOP) for both modes, either linear or circular, or generically elliptic. If not, the signal will be reduced by a factor:

$$\mu_{pol} = \mathbf{E_m} \cdot \mathbf{E_r} / |\mathbf{E_m}|\,|\mathbf{E_r}| \qquad (4.29)$$

where $\mathbf{E_m} \cdot \mathbf{E_r}$ is the product of Jones matrices of signal $\mathbf{E_m}$ and local oscillator $\mathbf{E_r}$, and $|..|$ indicates the modulus of vectors.

Thus, the combined effect of spatial, temporal, and polarization matching is to reduce the fringe visibility to:

$$V = \mu_{tc}\, \mu_{sp}\, \mu_{pol} \qquad (4.30)$$

No extra noise is generated by the loss of visibility due to spatial and polarization effects, however, because of the deterministic nature.

4.4.4 Dispersion of the Medium

A nice feature of interferometric measurements is the very precise yardstick on which they are inherently calibrated: the wavelength. However, we usually propagate the beams in air, and therefore the unit of measure is λ/n_{air}. Though not too different from unity, the index of refraction n_{air} can indeed introduce a calibration error in measurements with several significant digits.

In standard conditions (T=15°C and p=760 mbar) and with its standard composition, air has an index of refraction that is well approximated by the following expression, known as Edlen's equation (see Ref.[11] of Ch.3):

$$(n_{air}-1)|_{st} = 272.6 + 4.608/\lambda_{(\mu m)} + 0.061/\lambda_{(\mu m)}^{2} \text{ (ppm)} \tag{4.31}$$

We can see from this expression that in standard conditions the interferometric measurement requires a correction of about 280 ppm (ppm = parts per million) at λ=633 nm. In addition, the correction is dependent on the wavelength of operation, and the amount is a few parts per million along the visible to near-infrared regions.

At temperatures and pressures other than those of the standard conditions, the excess to 1 of the index of refraction is easily computed by noting that this quantity is proportional to the mole number per unit volume n/V, and therefore is equal to P/RT for the law of perfect gases. Therefore, we can write:

$$n_{air}-1 = (n_{air}-1)|_{st}\,(P/760)\,(288/T) \tag{4.32}$$

From Eq.4.26 we can calculate the temperature and pressure coefficients of the wavelength and of the correction to the corresponding interferometric measurement:

$$\begin{aligned} d(n_{air}-1)/dT &= -\,(n_{air}-1)|_{st}\,(288/T^{2}) \approx -1 \text{ ppm/°C} \\ d(n_{air}-1)/dP &= -\,(n_{air}-1)|_{st}\,(1/760) \approx +0.36 \text{ ppm/mbar} \end{aligned} \tag{4.33}$$

If we take reasonable values for ΔT and ΔP, for example 10°C and 10 mbar, we can see that temperature and pressure variations affect the 5th and 6th decimal digits of the displacement measurement. As already noted in Sect.4.2.1, the correction is performed by two sensors with outputs that change the multiplication factor (as shown in Figs.4-6 and 4-8) used to convert the fraction-of-λ counts to a metric scale.

Ideally, the sensors should average T and P on the actual optical path, but this is clearly difficult to perform in a practical operation.

With point sensors, located within the instrument, or close to the measurement optical path covered by the instrument, the correction is effective to the 6th decimal digit in normal operation in laboratory and outdoors conditions.

4.4.5 Thermodynamic Phase Noise

In optical fiber sensors with interferometric readouts (see Fig.9-28 and App.A2), the thermodynamic fluctuation of index of refraction of the fiber introduces a phase noise ϕ_{th} that may become comparable, or even larger than quantum noise. By an analysis of the phenomenon [19], the NED_{th} is found as:

$$NED_{th} = \phi_{th}/k = 0.37\ 10^{-5}\ [k_B T^{2} LB/\kappa]^{1/2} \tag{4.34}$$

where k_B is the Boltzmann constant, T the absolute temperature, L the fiber length, and κ the thermal conductivity. Because of the interference of counter-rotating waves sharing the same medium, the Sagnac configuration (App.A2) provides a partial cancellation of ϕ_{th} and improves immunity to thermodynamic phase noise [18].

4.4.6 Brownian Motion

When aiming at small-mass targets, Brownian motion may add a fluctuation that has to be taken into account in interferometric measurements of vibrations or displacements, despite not being a fault of the instrument.

On the line of sight along which we are performing the measurement, we find a thermodynamic degree of freedom and hence an energy (1/2)kT. This energy shall be equated to kinematic energy $(1/2)m\langle v^2\rangle$, thus obtaining:

$$\langle v^2\rangle = kT/m \qquad (4.35)$$

In the same way, for a rotating target for which we want to measure the angular speed of rotation Ω, we have a kinematic energy $(1/2)I\langle\Omega^2\rangle$, where I is the inertia momentum of the target. Equating to (1/2)kT gives a variance of the angular speed:

$$\langle\Omega^2\rangle = kT/I \qquad (4.36)$$

Letting numbers in Eqs.4.35 and 4.36, reveal that even for not so small masses (e.g., 1 to 10 g), the Brownian-induced speed or displacement may become comparable to the quantum noise NED.

Example. If we let T=300 K and being $k = 1.38 \cdot 10^{-23}$ J/K the Boltzmann constant, we get from Eq.4.35 for a 1-mg mass: $\langle v^2\rangle = 4 \cdot 10^{-21}/10^{-3} = 2 \cdot 10^{-18}$, or $\sqrt{\langle v^2\rangle} = 1.4 \cdot 10^{-9}$ m/s, which is a value well in the reach of a typical interferometer (it corresponds to s=1 nm and f=1 Hz in the diagram of Fig.4-19).
Also, for a disk of mass 0.01 g and radius 1 mm, the inertia momentum is $I = (1/2)mr^2 = 5 \cdot 10^{-3} \cdot 10^{-6} = 5 \cdot 10^{-9}$, and we get from Eq.4.36: $\sqrt{\langle\Omega^2\rangle} = \sqrt{(4 \cdot 10^{-21}\ 0.2 \cdot 10^{9})} = 0.9 \cdot 10^{-6}$ r/s = 0.18 deg/h, again a value within the readout sensitivity of a gyroscope (see Ch.7).

4.4.7 Speckle-Related Errors

In virtually all the applications of interferometry, the *corner cube* used as the remote target is considered *invasive*, that is, it may somehow disturb the device or system we are trying to measure.

To be *non-invasive*, we should work directly on the native surface of the device to be measured, that is on a diffusing surface. A *diffuser* emits according to Lambert's law (see Ch.2 of Ref.[2]) that is, with a cosine radiation pattern. Because of that, only a minute fraction of power impinging on the target is re-diffused into the solid angle subtended by the receiver to the diffuser, so that a sizeable attenuation is introduced with respect to operation with the corner cube.

Luckily, despite the much smaller amplitudes of the returning signal, the interferometer can work anyway thanks to the high sensitivity of coherent detection helped by the reference beam superposition to the returning beam.

The problem is that the diffuser surface adds a phase error. Indeed, the individual small areas making up the surface have random heights, differing from one another for more than λ, and thus the contributions re-radiated back to the detector are randomized in phase.

This regime of phase errors generated by the diffusing target is known as *speckle-pattern* and will be treated in detail later in Chapter 5, where we will study the effect of the combined random phase contributions on the accuracy of the interferometric measurement, in the three possible schemes of operation (Fig.4-20): longitudinal displacement (along the line of sight) Δs_l, transversal displacement (perpendicular to the line of sight) Δs_t, and projection (change of the diffuser portion illuminated by the beam) Δs_w.

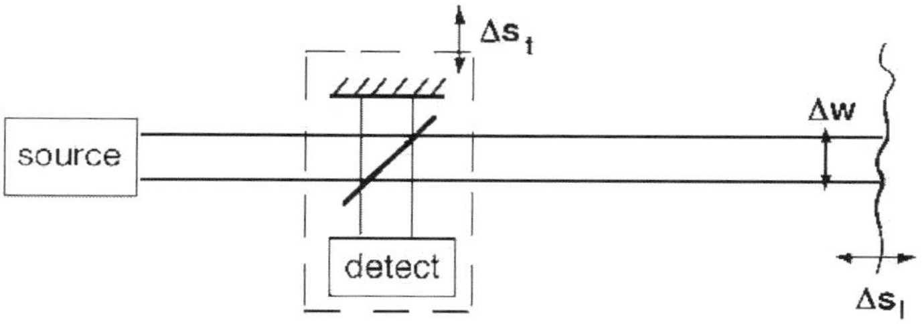

Fig. 4-22 Conceptual scheme to evaluate the NED of an interferometric measurement performed on a diffusing surface.

In all cases, as we will develop in detail in Ch.5, the NED (or noise equivalent displacement) introduced by the speckle pattern is given in general by NED = $\lambda\ \Delta s/S$, where S is a characteristic dimension. This means that, for displacements Δs small enough, the NED due to the speckle is a small fraction of wavelength, or non-invasive operation will be in general possible, without severely affecting the accuracy performance, of course, under appropriate conditions.

4.5 VIBRATION SENSING

We describe here the development of vibration sensing instruments based on the Michelson interferometer, and applied to the measurement of sub-micrometer amplitude of vibrations s(t) superposed to a large ($>10^4\ \lambda$) standoff distance s_0. The problem with the resulting interferometric signal $I=I_0 \cos 2k[s_0+s(t)]$ is once again linked to the cosine function, that around a value $s_0 = N(\lambda/2)$ reaches a maximum or a minimum and thus washes out the small s(t).

Solutions that have been demonstrated for many decades, and successfully incorporated in instruments based on Michelson interferometers, are the half-fringe feedback loop and the frequency shift reference described below.

4.5.1 Short Stand-off Vibrometry

Early experiments on short stand-off distance vibrometers were developed in the 1970s around a Michelson configuration modified by adding lenses focusing the reference and measurement beams and by adding dynamical tracking of the reference path to compensate for ambient-induced disturbances (Fig.4-23).

This is accomplished by the piezo actuator in the reference path, which has either a diffuser (as shown in Fig.4-23) or a reflective surface. Light scattered by the diffusing target in the measurement arm is collected back by the lens and recombined on the photodetector. Signal I_{ph} is preamplified by the front-end and becomes an output $V_{ph}= RI_0[1+\cos 2k(s_m-s_r)]$, then is low-pass filtered with corner frequency f_1. A constant voltage V_{ref} is then subtracted from V_{ph}, and the result, amplified by A, constitutes the error-signal supplied to the piezo.

Because of the feedback loop, the system adjusts itself at a quiescent point where the error signal is nearly zero, or $V_{ph}-V_{ref} \approx 0$. Setting $V_{ref}=RI_0$, we get the condition $0= V_{ph}-RI_0 = RI_0 \cos 2k(s_m-s_r)$. For s_m very small, the quiescent point is $2ks_r= \pi/2$, and we get:

$$V_{out} \approx RI_0\, 2ks_m = V_{ref}\, 2ks_m \tag{4.37}$$

The best choice for the low-pass frequency f_l depends on the actual level and frequency content of the disturbance from the ambient. In most cases, the disturbance spectrum is found to damp off quickly between, say f_{dist}= 100-1000 Hz. Choosing $f_l \approx f_{dist}$, the low-frequency components ($f< f_1$) are suppressed because of the feedback effect. Instead, signals $s_m(t)$ with frequency content $f>f_l$ find the loop ineffective (interrupted by the filter) and are passed on to the output. From a well-known result of feedback theory, the actual suppression factor is given by the loop-gain G.

By inspection of Fig.4-23, the dc loop gain is easily found as $G_0=R\, I_0 A\kappa$, where $\kappa =\Delta s/V$ (μm/Volt) is the conversion factor of the piezo. We can readily make $G_0=10^3$ and nearly eliminate the disturbance.

In frequency, the loop gain is $G(f)= G_0/[1+(f/f_1)^2]^{n/2}$, where n is the order of the filter (n=1 for a simple RC). Signals with $f>f_l$ find a small G(f) and are not reduced, so we can take them either from the transimpedance output or from the filter high-pass output as indicated in Fig.4-23.

Of course, also the signals with frequency content $f<f_l$ are suppressed, like the disturbances. To minimize loss of useful spectrum, we will make the filter frequency f_l adjustable and trim it to optimize the outcome of the measurement by inspecting the result.

The speckle-pattern regime of operation is of little problem for this vibrometer. Should the reference or measurement fields eventually fall on a dark (weak amplitude) speckle, it will suffice to move the target surface transversally just a little bit (by a few mirometers) or finely trim the position of the lenses, to find another, bright high-amplitude speckle.

Regarding the phase error generated by the speckle-pattern statistics, the re-radiated field received through the lens is in the far-field of the target and generates a negligible error, whereas changes in the spot projected on the target due to longitudinal or transversal displacements may be of concern.

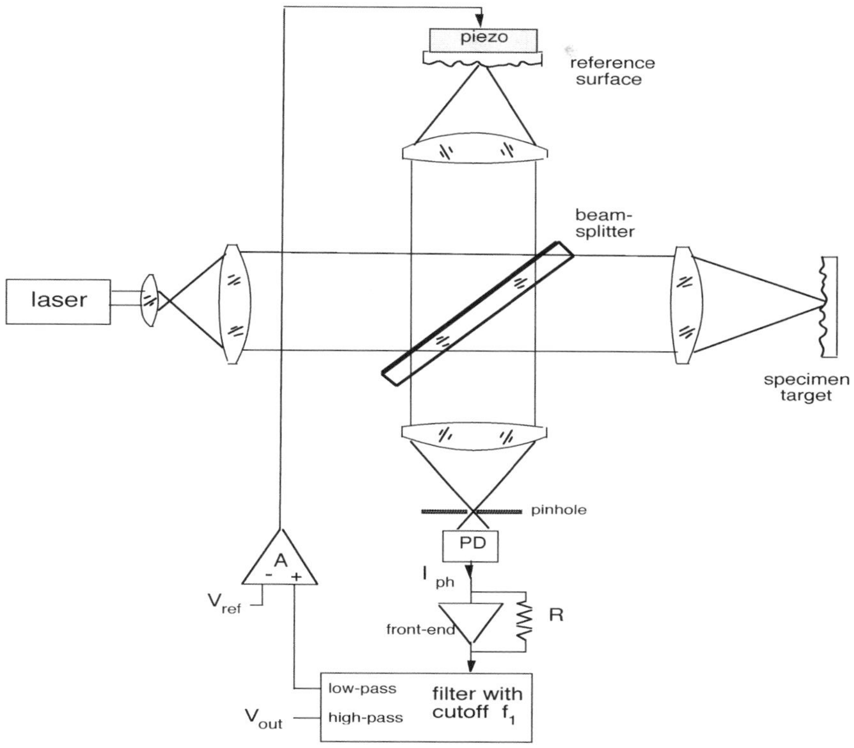

Fig. 4-23 A short-standoff vibrometer for measurements on a small specimen. The reference beam is focused on the diffuser surface driven by a piezo actuator. The interferometer signal is low-pass filtered, amplified, and sent to the piezo. Because of the feedback loop, the working point is locked to $I_{ph}=V_{ref}/R$, the half-fringe point of the interferometric signal $I_{ph}=I_0[1+\cos 2k(s_m-s_r)]$. The vibration signal, at $f>f_1$, is taken from the high-pass output V_{out} of the filter as $RI_0\, 2ks_m$.

To summarize the results developed later in Sect. 6.1 and 6.2, longitudinal displacement introduces a negligible error as far as the amplitude of vibration (signal plus disturbance) is small as compared to the depth of focus of the lens Δf. This is given by $\Delta f = \lambda/\pi NA^2$, where NA is the lens numerical aperture and typically is $\Delta f \approx$ 0.1-1 mm. On the other hand, transversal displacements of the target are to be more tightly controlled, as they are small with respect to the focused spot size $\Delta w = \lambda/\pi NA$, which may have a typical value of ≈5 to 30 μm. So, it may be important to align the specimen under test to get the vibration to be parallel to the impinging beam axis.

In addition to the mechanical segment, a short-distance vibrometer is useful for noncontact measurements of otherwise undetectable minute vibrations in a variety of devices, including electronic and piezoelectric components. As an example, we report in Fig.4-24 the measurement of sub-nanometer amplitude waves in a Surface Acoustic-Wave (SAW) filter, obtained with a setup similar to that reported in Fig.4-23. This surprising result has pointed out a subtle difference between the Stoneley regime and the normal SAW regime [20] of surface waves, a useful diagnostic for the design of filters based on acoustic surface waves.

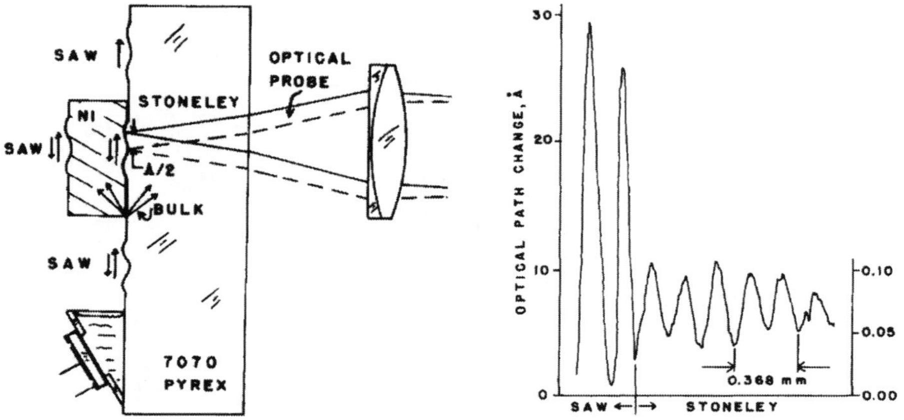

Fig. 4-24 Detection of sub-nanometer (1Å=0.1 nm) vibrations in a SAW device. Bulk Stoneley waves and ordinary surface waves are discriminated by their amplitude. Central frequency of the SAW is 8 MHz (from [20], ©American Institute of Physics, reprinted by permission).

To evaluate the ultimate sensitivity (or NED, noise-equivalent-displacement) of the vibrometer, we may start considering that the optical power P_0 supplied by the laser is sent half to the measurement and half to the reference path. All this power reaches the target, but, in the retro-diffusion process, only a fraction η of it is sent back into a single spatial mode for coherent superposition at the photodetector.

If w_l is the spot size of the Gaussian beam and $\theta=\lambda/\pi w_l$ is the associated diffraction angle, the power fraction is $\eta=(\lambda/\pi w_l)^2$. With typical values for a He-Ne laser, P_0=1 mW, λ=0.633 μm and w_l =0.5 mm, we have $\eta=1.6\ 10^{-7}$. The useful power is then $\eta P_0/2=0.8\ 10^{-10}$ W, and, from the diagram of Fig.4-21, we can find the minimum measurable amplitude of vibration as NED = $(0.6\ 10^{-6}\text{m})\ \sqrt{B}/P= 6.7\ \text{pm}/\sqrt{B_{(Hz)}}$, where B is the bandwidth of the measurement in Hz.

Concerning the spatial resolution of the vibration measurement, this is given by the spot size Δw projected by the lens on the target in the focal plane.

In general, we have (see Ref. [2]): $\Delta w=\lambda/\pi NA$, where NA=arcsin D/2F is the numerical aperture of the lens. However, if the lens diameter D is not filled completely by the beam, NA is determined by the spot size radius R as NA= arcsin R/F. Using λ=0.633 μm, w_l =0.5 mm, and F=20 mm, we obtain a spot size Δw=16 μm, representing the elemental pixel of spatial resolution.

Last, the bandwidth of response of the vibrometer is primarily determined by the high-frequency cutoff of the photodiode transimpedance amplifier and associated electronic processing circuits. With available optical power in the $\eta P_0 \approx 100$ pW range (or, $I_{ph} \approx 100$ pA), we may use a 10- to 100-kΩ feedback resistance and get, from state-of-the-art design [2], a bandwidth of 1 to 10 MHz.

In conclusion, the *half-fringe* stabilization approach described above is a good technical choice to develop a short-distance vibrometer with nanometer-sensitivity, ≈20 μm sampling pixel, and MHz bandwidth.

4.5.2 Long Stand-off Vibrometry

Early experiments in long-distance vibrometry were aimed at remote, noncontact diagnostics of large structures, like towers, bridges, and other construction hardware of industrial, civil, or historical interest [21,22].

The approach selected has been that of *heterodyne* detection of the remote target echo, quite similar to the two-frequency interferometer (Sect.4.2.2).

Of course, as a source for long distance operation, we need a frequency-stabilized laser with a coherence length in excess of the distance 2s to be covered. For s≈ 300 m, this requires a laser linewidth $\Delta\nu = c/4\pi s = 90$ kHz or better. Also, the power should be adequate to allow for an increased attenuation, which calls for units with power typically in the range 20 to 50 mW (He-Ne) or ≈1 W (CO_2).

The basic configuration to start with is illustrated in Fig.4-25. The source is a single-mode frequency-stabilized laser (He-Ne or a CO_2). The beam is sent to a beamsplitter (BS1) giving two output beams. One is directed to a telescope aimed at the target, while the other is used as the local oscillator for heterodyning. To get a frequency-shifted reference, an acousto-optical modulator (or Bragg cell) is employed and provides an output shifted by $\Delta\nu$ =10-50 MHz typically. After propagation to the target and back, the optical signal collected by the telescope is superposed to the local oscillator with the aid of beamsplitter BS2, and ends on a balanced detector (see Ref.[2], Ch.10.2).

With the beating of fields, we obtain as photodetected signals $I_{ph1}=(I_0/2)[1+\cos(\Delta\omega t+2ks)]$ and $I_{ph2}=(I_0/2)[1-\cos(\Delta\omega t+2ks)]$ and, after subtraction, we get $S= I_0 \cos(\Delta\omega t+2ks)$. By mixing it with the original frequency difference $\Delta\omega$ or a fraction of it, the vibration-signal $2ks=A_0\cos\omega_0 t$ is brought to the baseband (or to an intermediate frequency) for further processing.

Several improvements can be introduced in the basic scheme of Fig.4-25. First, the beamsplitter BS1 splits the optical signal returning from the telescope and directs half of it back into the laser. We can eliminate this 3-dB loss by using a Glan-polarizing beamsplitter for BS1 and by adding a polarizer before entering BS1. Second, superposed to the useful signal coming back from the target, we find a back-scattering contribution, generated by the atmosphere along the propagation path. This contribution can even be larger than the useful echo when distance increases. To get rid of it, we may take advantage of the polarization properties of scattering, by which a propagating circularly polarized (CP) beam (say right-handed, RH) scatters back as a left-handed (LH) contribution. In contrast, the target depolarizes the echo for any impinging state of polarization, and the useful returning signal contains as much RH as LH power. We may use a CP beam in transmission and cut off the back-scattered power by adding a polarizer and a quarter-wave plate at the telescope entrance. If the laser beam has a linear polarization and we align it to the polarizer, no loss in transmission will be experienced, while the back-scatter power will find the polarizer in the crossed state and be eliminated. The outgoing CP signal is then combined at the balanced detector, again using a Glan beamsplitter for BS2, with the reference beam either in a CP or in a linear, 45° oriented state.

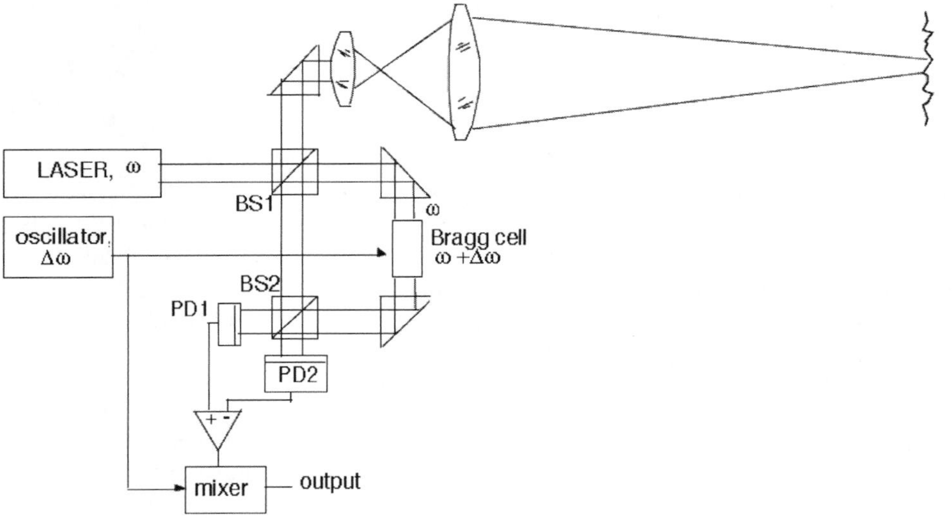

Fig.4-25 Scheme of a long distance vibrometer. The frequency-stabilized laser is frequency-shifted by a Bragg acoustooptic cell to generate a reference local oscillator. Beamsplitter BS2 recombines beams on the balanced detector. The mixer brings the signal to the base band.

To protect the frequency-stabilizer laser source from even minute, unwanted reflections eventually from optical devices within the instrument, we may add an Optical Isolator (OI) at the laser beam output. The OI is a well-known device based on magneto-optical nonreciprocal rotation and is widely used for protecting lasers in fiber-optics communications. It has ideally a T=1 transmission in one direction, and T=0 in the reverse direction.

Still another magnetooptical device may be useful, the Optical Circulator (OC). The circulator has four access ports for the beam. A beam entering in the i-th input will exit from the (i+1)-th port with T=1 transmission, while it is T=0 for all other ports. Placed at the entrance of the telescope, the OC provides us with an input and output port spatially (or physically) separated without introducing any loss, different from normal beamsplitters.

Typical sensitivity (or, NED noise-equivalent-displacement) of the vibrometer built around the heterodyne concept are reported in Fig.4-26. Sensitivity is dominated at low frequency (<100 Hz) by an excess noise. This is due to turbulence effects, overwhelming the fundamental, flat-spectrum shot noise and electronics noise, which is finally reached at higher frequency (≈1kHz).

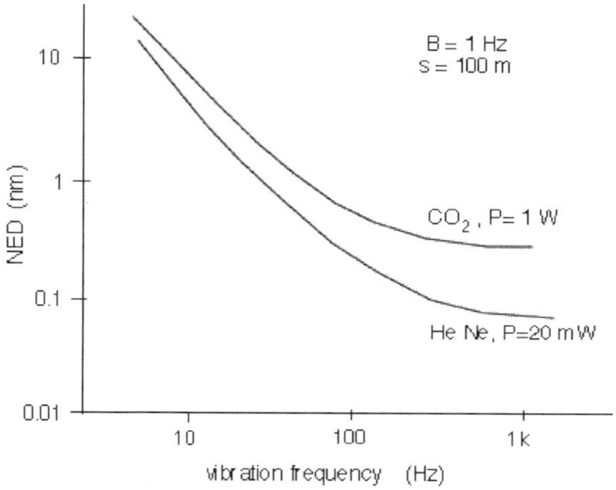

Fig.4-26 Typical NED of a long-distance vibrometer, as a function of frequency, for the detection of small remote vibrations on a diffusing target with He-Ne and CO_2 lasers. Operation is at s=100 m distance, B=1 Hz, in a clear atmosphere.

In the propagation, turbulence (App. A3) may introduce a random fluctuation of the index of refraction and hence of the optical path length summed to the useful vibration signal. In addition, we find a beam wandering around the aiming direction and distortion of the wave front in the form of scintillation or bright spots. Both effects combine with the speckle regime of the remote-diffusing surface to add an error.

Thus, operation of the vibrometer on substantial distance (say, in excess of a few tens of meters) requires taking care of propagation condition to keep turbulence effects low.
We can try to mitigate or almost compensate for these effects of turbulence by electronic and optical, as well as measurement strategies.

Electronically, we may subtract the spectrum of the fluctuations, in a measurement with the exciter (vibrodyne) off, from the useful spectrum with the vibrodyne on. This helps reduce the noise floor at low frequency, perhaps by a factor of 10, but not much better because turbulence is not strictly stationary in successive measurement sessions. By optical means, we can attempt to compensate for turbulence with adaptive optics, but practical schemes are rather involved and have not yet demonstrated a valuable improvement, so far. Last, we can take advantage of the variability of turbulence from day to day and in daily hours. By waiting for a calm day and the best hours to carry out the measurement, turbulence effects are greatly reduced. This strategy is indeed acceptable when evaluating the structural integrity of buildings, which requires an outcome be determined within, say, a few days rather than instantly.

In performing the measurement, we should prefer operating on the shortest distance that is practicable. As a rule of thumb, if the article we are going to test has an out-of-ground height H, we will position our instrument at a distance $d \approx H$ to be able to test the bottom and top parts of it in a single session (Fig.4-27).

To generate a test vibration, a vibrodyne is used, fastened securely atop the structure. The vibrodyne is an electromechanical device consisting of an electrical motor (typ. ≈50 W in power) with an eccentric-mounted mass (typ. ≈1 kg) rotating at a frequency f_{vib} (typ $f_{vib} \approx 0.1$ to 10 Hz). Because of the mechanical torque generated by the vibrodyne, even a modest mass like 1 kg generates amplitude of displacements 10 to 100 nm in the tower top surface. Of course, this amplitude is well within safety limits, and corresponds to the natural stress imparted by a moderate wind (say, with a speed 0.1 to 1 m/s).

The test session consists of measuring the vibration amplitude on a number of points selected along the structure and repeating the measurements for a set of frequencies.

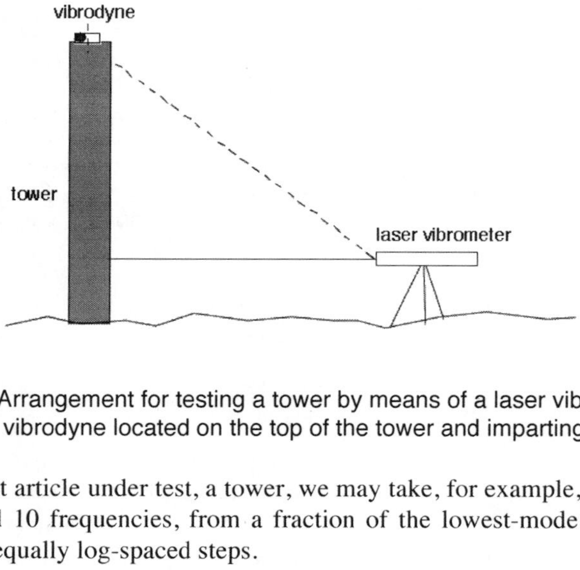

Fig. 4-27 Arrangement for testing a tower by means of a laser vibrometer and a vibrodyne located on the top of the tower and imparting oscillations.

For the simplest article under test, a tower, we may take, for example, 10 sampling points in H/10 steps, and 10 frequencies, from a fraction of the lowest-mode frequency f_0 to 10 times as much, in equally log-spaced steps.

In this way, we are able to obtain the frequency response $F(h,\omega)$ to a mechanical excitation, a response which is resolved spatially. Now, the measured frequency response can be compared with what is expected from a solid structure, as calculated by standard methods of structural engineering. For example, tower or chimney has bending modes at frequencies $f_n = n\,\eta\sqrt{(k/M)}$, where n = 1,3,5… is the order of the mode, k is the elastic constant (N/m), M is the mass, and η is a dimensionless factor depending on height, diameter, wall thickness, etc.

The lowest mode has a node of vibration at h=0 and an antinode (or maximum swing) of vibration at h=H, or, the standing wave of vibration has half-period in the height H. Higher order modes have n/2 periods in H (Fig.4-28). Additional modes can be found in a structure lacking radial symmetry or having finer structural details, but here we ignore them for simplicity.

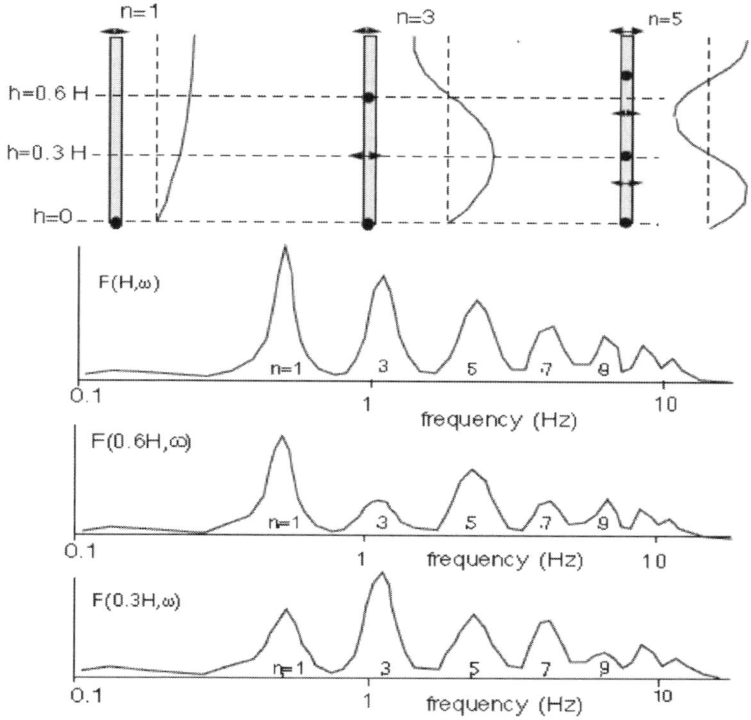

Fig. 4-28 Top: a tower or chimney has bending vibration modes with nodes (indicated by points) and antinodes (arrows) at integer fractions of the height H. The modes are revealed by the frequency response (center and bottom) taken at different heights h.

The frequency of the n=1 mode is found at f_1=0.02 to 1 Hz, typically (Fig.4-29), in most practical cases of interest (i.e., for 10 to 50 m towers), and higher modes have resonant frequencies that are odd multiples of the fundamental one.

As a function of height, the amplitude of vibration changes with the mode order (Fig. 4-28), and we can therefore recognize them by looking at the relative obtained at different heights. The sharpness of the resonance decreases with the order of resonance and, more important, with the deviations from the structural integrity of the article. A solid structure has a well-defined resonance spectrum, whereas a near-to-collapse article usually exhibits a broad spectrum of frequency response F(h,ω), with a barely recognizable resonance pattern.

The raw interferometric signal of the vibrometer has a frequency spectrum with a strong background due to turbulence (Fig.4-29). After background subtraction through a separate measurement with the vibrodyne off, the resonance peaks of a solid structure are clearly highlighted. For a weak structure, the spectrum is rather flat without the expected resonances, and the measurement of the frequency response F(h,ω) at various heights adds further evidence.

At a closer look, in the background spectrum due to wind and turbulence we find, for a solid structure, modest peaks at resonances (Fig.4-29, top). This can be explained because the wind is a sort of white-noise excitation of the structure, and the mechanical response is enhanced at resonance. Thus, under favorable circumstances, like for example a wind strong enough to impart a mechanical excitation to the structure, but not so strong to overwhelm the useful signal by turbulence, we can dispense with using the vibrodyne and need not climb the article to mount it.

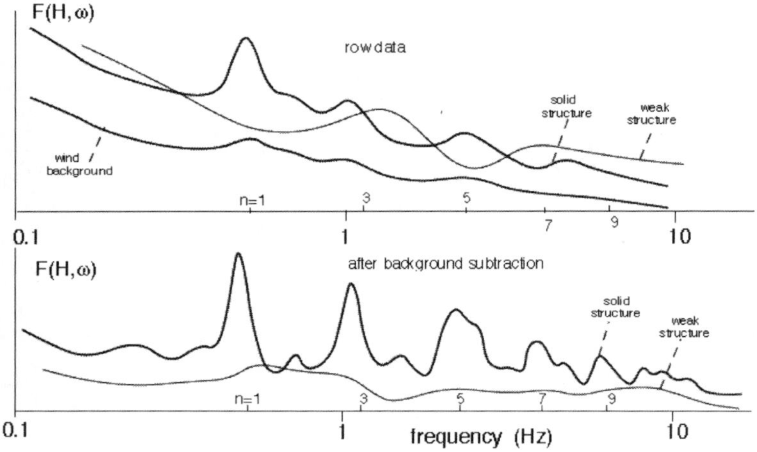

Fig. 4-29 Vibration measurement (top) and the result after background subtraction (bottom), revealing the state of conservation by means of the resonance spectrum.

Subtraction of the background cannot be performed as it was previously, in this case. However, from the set of measurements at different heights, and knowing the relative amplitudes of the expected resonances, we can process the data by subtracting the common background with a least-mean-square error routine.

In conclusion, despite the limitation of turbulence, long-distance vibrometry offers a powerful diagnostic tool for noncontact testing of buildings.

This is exemplified by examination of masterpieces like the historical Pisa tower [56] and similar articles, as well as in the diagnostics of structural integrity of industrial articles, notably cooling towers and chimneys of cement factories, electrical generating plants, etc.

About the choice of the wavelength best suited for long-distance operation of the vibrometer, let us consider the factors involved, by comparing He-Ne (λ=0.63 μm) and CO_2 (λ=10.6 μm) lasers. The turbulence-induced Δn fluctuation and the phase noise (App.A3.2) is proportional to $\sigma_\chi \propto \lambda^{-7/12}$, thus noise decreases of a factor $(\lambda_{He\text{-}Ne}/\lambda_{CO2})^{7/12}$ =0.24 going to the infrared.

The NED is proportional to wavelength and to the $(S/N)^{-1}$ ratio, on its turn proportional to λ: so, it increases by $\sqrt{(\lambda_{CO2}/\lambda_{He\text{-}Ne})}$ = 4.1 in the IR.

Optical device transmission and detector quantum efficiency η are worse in IR as compared to visible, and we may take a factor ≈3 of increase in the IR.

Last, power available from a small CO_2 laser is readily ≈1W, instead of 50 mW of a He-Ne laser, and this gives a factor $\sqrt{0.05/1}$=0.22 of NED decrease in IR.

Collecting the contribution, the NED we should expect in IR is $0.24 \cdot 0.4 \cdot 16.8 \cdot 3 \cdot 0.22 = 1.06$ times that of the He-Ne, or, substantially the same.

However, the effects of turbulence are smaller in the IR, as indicated by Fig.4-26, where the corner frequency of noise for the CO_2 wavelength is smaller.

4.6 Read-Out Configurations of Interferometry

The laser interferometer we have considered so far can be classified as an *external configuration*, and in applications it is by far the most commonly used configuration. In it, the laser feeds an optical interferometer external to the source, and from the recombination of the propagated beams, a signal $I_0 \cos\Phi$ is provided, that is, an intensity signal carrying the *phase* information Φ=2ks (see Fig.4-30, top).

This is not the only possibility we have available to make an interferometric readout, however. We may also think of using the mirrors of the laser itself as the optical interferometer, as shown in Fig.4-30, middle.

This is the *internal configuration*, which generates an output signal of the form $I_0 \cos \Omega t$, i.e., a *frequency* signal Ω proportional to the optical phase shift, $\Omega=\chi ks$, χ being a suitable constant. The internal configuration is put to advantage in the Ring-Laser-Gyro (RLG) version of the gyroscope (Ch.8).

A third configuration is the *injection-modulation*, or self-mixing, interferometer, (Fig.4-30, bottom). Here, the optical interferometer external to the source is absent, and we rely just on the interaction of the returned field from the remote target into the laser cavity field to produce a modulation of the emitted field, both in amplitude (AM) and frequency (FM), related to the phaseshift Φ= 2ks of the go-and-return path to the target.

The external configuration is the most developed and we have described the developments of displacement measuring instruments in the previous section of this chapter. The internal configuration is more recent and we will describe some basic features of it in the next section, as an introduction to the detailed development of the electro-optical gyroscope of chapter 8.

The injection-modulation, or self-mixing configuration, will be the topic of the next chapter and here we will just introduce some of its basic features.

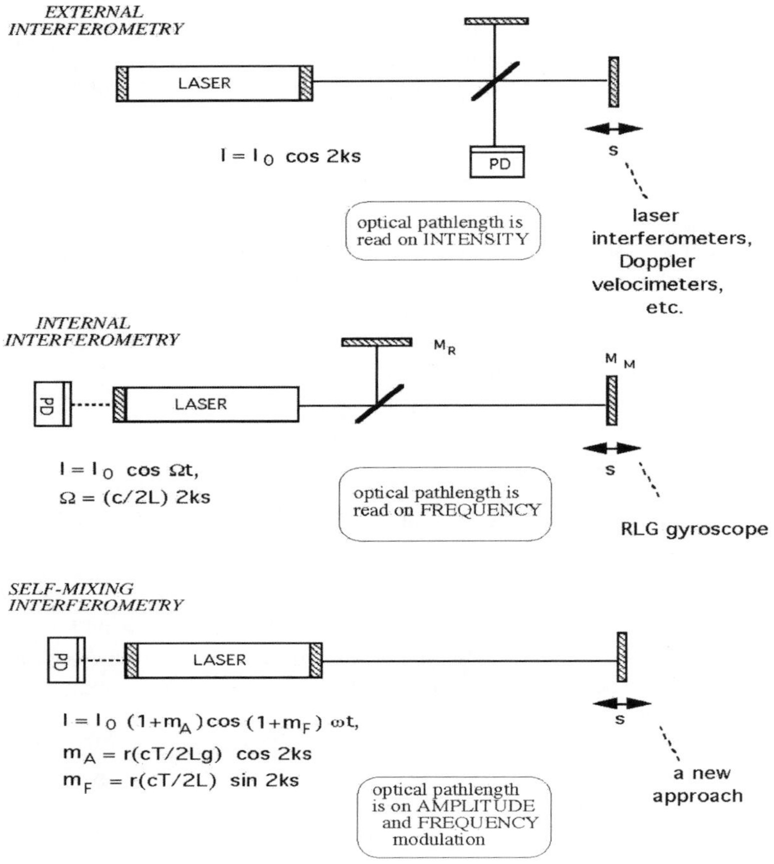

Fig. 4-30 Configurations of interferometry: external (top), internal (middle), and injection or self-mixing (bottom).

4.6.1 Internal Configuration

Let us consider a laser oscillating in a single spatial and longitudinal mode, for example, a 20-cm He-Ne tube with external mirrors (see App.A1-1). Moving one mirror by a small displacement Δs along the cavity axis, the oscillation frequency will change by:

$$\Delta f = (c/2L)\, \Delta s/(\lambda/2) \qquad (4.38)$$

This result is explained because the spacing of longitudinal modes is c/2L, where L is the mirror distance (see Fig.A1-1) and, for each $\lambda/2$-increase of the laser cavity length L, the frequency changes by c/2L (and the order of the mode increases by 1). Therefore, the ratio R_f of frequency variation Δf to displacement Δs, called responsivity R_f, is given by:

$$R_f = \Delta f/\Delta s = (c/\lambda L) \tag{4.39}$$

By letting λ=0.633 µm and L=20 cm so that c/2L=750 MHz, we get from the internal configuration the remarkable responsivity to mirror displacement R_f = 750 MHz/316 nm = 2.4 MHz/nm, which is a very large value.
However, we cannot measure optical frequency directly and, to exploit this result, we need a second mode to be used as a reference frequency so that the frequency shift Δf can be converted down to an electrical frequency.

Thus, we have the arrangement of Fig.4-31, where a beamsplitter is used to allow the laser to sustain the oscillation of two modes, defined by the mirrors M_M and M_R.
When M_M is moved, the frequency difference f_2-f_1 carries the interferometric signal and has a frequency variation proportional to Δs as given by Eq.4.38, a value that we will measure from the photodetector output.

The maximum range of displacement is $\Delta s=\lambda/2$=316 nm, and the corresponding frequency is $\Delta f= f_2-f_1=c/2L$=750 MHz. As we now have a Michelson interferometer brought inside the active cavity, the name internal configuration is justified.

In an actual experiment, however, the linear response expected from the configuration is obtained only when the frequency difference f_2-f_1 is not too small (Fig.4-31), whereas when f_2 approaches f_1, first a decrease in response is found, and then, for $f_2 \approx f_1$, the signal suddenly disappears [23].

The dead band is due to frequency *locking* of the two oscillations. Locking is a very general phenomenon in coupled oscillators, which comes from even very minute coupling of power from one oscillation to the other.
A residual very small interchange of power is unavoidable (for example, because of mirror scattering and gain-medium coupling), and therefore the dead band will never be zero (see also Ch.8). However, we first try to reduce it as much as possible, and an improved scheme is shown in Fig.4-31. Here, because we use a Glan beamsplitter, two modes oscillate with linear orthogonal polarizations.

At the rear mirror output, we can insert a polarizer in front of the photodetector, oriented at 45° for the modes to beat on the photodetector.
Because of the orthogonal polarizations, the gain coupling in the active medium is nearly suppressed. In addition, we get rid of the 50% loss of a normal beamsplitter used in Fig.4-31, which is too high a loss in a normal He-Ne laser.

Using low-scatter mirrors and a high-quality Glan cube to realize a He-Ne internal interferometer, we may go down to a few MHz of locking range as compared to several hundred megahertz of the basic scheme.

Then, we can bias the interferometer to operate far away from f_2-f_1=0, e.g., at f_{bias}=100 MHz, so we can also detect the sign of Δs.

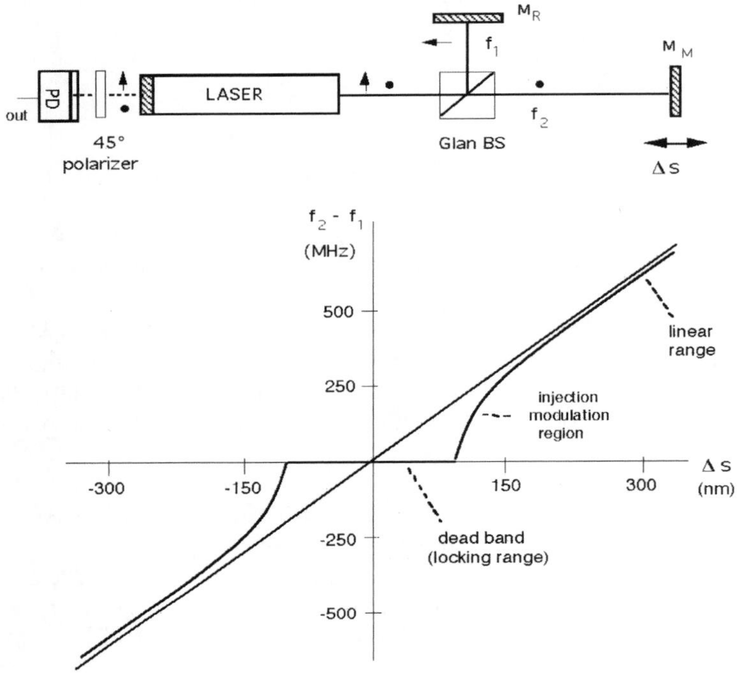

Fig. 4-31 Internal interferometry. Top: scheme with polarization-split modes. Bottom: the theoretical response of frequency difference f_1-f_2 versus displacement Δs is linear, with a range of c/2L=750 MHz for a $\Delta s = \lambda/2$ =316 nm displacement (values for a 30-cm He-Ne). In practice, a dead band around $f_1 - f_2 \approx 0$ is found because of locking.

This can be done by finely adjusting the position of mirror M_R. A further refinement is to mount M_R on a piezo actuator and make a servo loop on the reference mirror position to stabilize the interferometer at $f_2 - f_1 = f_{bias}$=const. The Δs signal is then obtained from the error signal of the feedback loop.

The internal configuration is critical to operate because mirror M_M shall be kept aligned to the cavity during motion, and the tolerable error is less than the diffraction limit, <1 arcsecond in practice. Though the displacement we are aiming at might be very small, alignment is still very critical.

On the other hand, the internal configuration concept is vital in Sagnac interferometers like the gyroscope (Ch.8). In this case, optical path length variations are induced in a balanced cavity from outside, and the very high responsivity of the readout is put to great advantage.

4.6.2 Injection (or Self-Mixing) Configuration

Even without using any optical interferometer external to the laser source, we can make a laser interferometer by using the interaction, produced in the laser cavity, by a small fraction of the field emitted from a laser and returned after propagation from the remote target.

First reported in 1978 [24], the resulting scheme of interferometry is variously referred to in the literature as *induced-modulation*, *injection*, *self-mixing (SMI)*, or *feedback* interferometry, terms that we will use as synonymous in this book.

A straightforward explanation of the phenomenon comes from considering the field oscillating in the laser cavity as a rotating vector (Fig.4-32).

The cavity field E rotates in the phase plane at optical frequency ω. If a fraction αE of it is allowed out for propagation to a remote target at distance s, the optical phase shift accumulated in the go-and-return path is $\Phi=2ks$. Thus, upon re-entering in the laser cavity through the mirror with transmittance t, we find a rotating vector of complex amplitude $aE \exp i\Phi$, where $a=t\alpha\eta$, and η is an eventual propagation loss. This contribution adds as a vector to the existing field E to give the instantaneous new cavity field. The addition of rotating vectors is well known in communication theory, and the result can be stated as follows. The in-phase component $aE \cos \Phi$ produces *amplitude modulation* (AM) of the pre-existing field E, and its modulation index (or depth) is $a \times \cos \Phi$. The in-quadrature component $aE \sin \Phi$ produces *frequency modulation* (FM) of the pre-existing field E, and its modulation index (or depth) is $a \times \sin \Phi$.

Thus, the laser cavity field acts as the optical carrier of AM and FM modulations induced by the perturbation returning from the remote target. These modulation indexes are exactly the $\sin \Phi$ and $\cos \Phi$ signals we were attempting to procure in conventional configurations based on an optical interferometer. Unfortunately, while the AM is easily detected as impressed on the laser power, the FM term requires some addition to the basic setup, or a modification of the laser source to make it a dual-mode oscillator. Yet, we can have proper working of the interferometer with just the AM channel, and this is the key advantage of the configuration, a minimal part-count one with performance equaling the traditional external configuration.

It is worth noting that in other applications of lasers, injection phenomena are undesired, and we want to get rid of them. For example, in fiber-optics communications, it is common practice to protect, with an optical isolator, the narrow-line laser transmitter, from the unwanted back-scattering that spoils the laser line and adds amplitude noise, just the AM and FM modulations described above but random-like because uncontrolled.

In the injection interferometer, on the other hand, injection is a well-controlled way to measure the returning field. We only need a narrow-line, single-mode laser to make clean modulation waveforms available.

It is interesting to observe that the injection phenomena are quite general, and the discussion here reported for a laser source actually applies with minor changes to virtually all kinds of oscillators. For example, echo detection by feedback modulation has been reported in microwaves as well as in ultrasonic measurements.

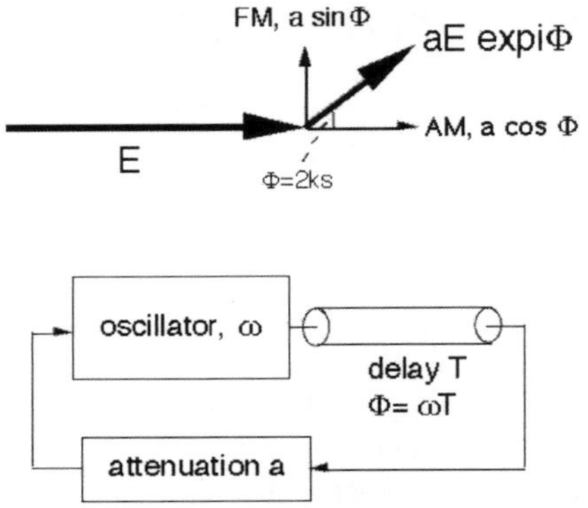

Fig. 4-32 Rotating vector description of injection modulation (top), and circuit analogy of the injection phenomenon (bottom).

In addition, if we make an Op-Amps version of the circuit model of Fig.4-32, we will readily obtain for voltage signals the same induced-modulation waveforms.

It is also interesting to note that, while we focus here on the interferometric measurements application, the injection configuration scheme of Fig.4-25 is conceptually a coherent detection scheme [2] that allows applications based on the measurement of phase as well as amplitude of the returning weak signal [24].

4.7 White Light Interferometry and OCT

In recent years, a totally new approach to interferometry has been demonstrated, gaining general acceptance and providing several successful examples of application in different fields, from engineering to manufacturing and to biological areas.

It's *white light interferometry*, which looks at first sight like an oxymoron (because interferometry is usually developed around narrow-line source, not at all a wideband one) and is also referred to with the synonym of OCT (Optical Coherence Tomography).

The principle is surprisingly simple and opposite to "normal" interferometry: it overturns completely the concept that "we need coherence" because instead of requiring a long coherence length L_{coh} to the source, so as to obtain measurable returns from long distances $s \approx L_{coh}$, now we use a very short L_{coh}, down to a few micrometers, so as to have a coherent detection and signal only for $s < L_{coh}$. In other words, we detect details with a spatial resolution dictated by the (small) coherence length L_{coh}. We use a low-coherence length source, like an LED, a super-luminescent LED (SLED), or a filtered incandescent lamp with a line width so large ($\Delta\lambda \approx 20$-100 nm around $\lambda \approx 1$ μm) that the coherence length is limited to a very small value like $l_c = \lambda^2/\Delta\lambda \approx 10$ to 50 μm.

Now, we extend the strategy of operation by looking at both the incoherent and coherent regimes of superposition. For $2(s_m - s_r) > l_c$, the superposition is incoherent and the photode-

tector yields a dc current which is just the sum of the intensities provided by the reference and measurement beams. On the other hand, when the arms are nearly balanced, $2(s_m-s_r)<l_c$, the photodetector yields the interferometric beat signal added to the previous dc level.
Thus, by waiting for a spike to appear in the photodetector output signal while we scan somehow the reference arm length s_r, we are able to distinguish the balance condition $s_m=s_r$ with a $l_c \approx 10$ to 50 μm resolution.

Recalling Eq.4.18, we may write the photodetected signal from a Fabry-Perot interferometer in white light (Fig.4-33) as:

$$I_{ph} = I_0 [1+V\cos 2k(s_m-s_r)] \tag{4.40}$$

Following the result of Sect.4.4.2 and 4.4.3 and assuming a Lorentzian line, the visibility V can be expressed as

$$V= \mu_{sp}\, \mu_{pol} \exp-2(s_m-s_r)/l_c \tag{4.41}$$

To illustrate the practical implementation of the concept, let us refer to Fig.4-33.

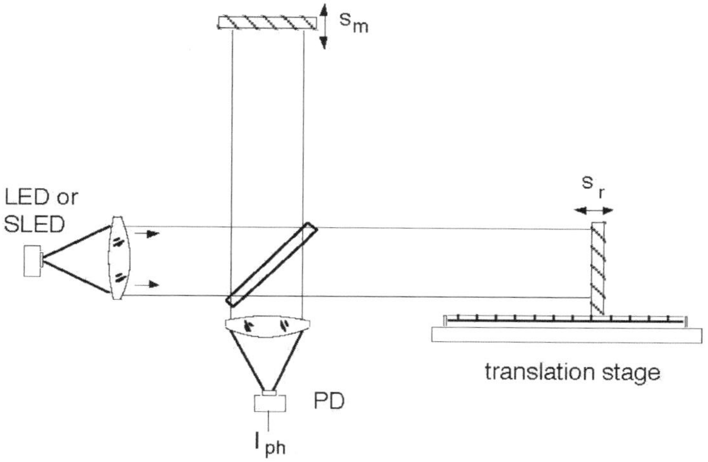

Fig. 4-33 In a white-light interferometer, the interference signal only appears when arms are very close to balance ($s_m=s_r \pm l_c$) because the coherence length l_c is very short (≈10 to 50 μm). The measurement condition is searched by scanning the reference arm length s_r, with the aid of a motorized translation stage and is revealed by a fringe-like signal in photocurrent.

We may use either an LED or a SLED as the source of the interferometer. The basic SLED is a diode with a laser structure, but with the output mirror missing (the output facet is antireflection coated). Thus, laser oscillation is prevented and emission from the device is in form of spontaneous emission generated in the bulk and amplified down the path to reach the output facet, hence the name of super-luminescent light.

The SLED has an emitted power (a few milliwatts) and spectral width (typically $\Delta\lambda \approx 20$ to 50 nm) comparable to a normal LED, but emission is from a small spot size, nearly single-mode area, like that of a laser. Thus, the SLED brilliance is high, which is a very useful feature. In view of Eq.4.41, the SLED is better because it has $\mu_{sp} \approx 1$, whereas it may be $\mu_{sp} \approx 10^{-3}$ for an LED. For the same reason, though in principle useable, an incandescent lamp source would provide a very low $\mu_{sp} \approx 1/N$ and useful signal, because of the large number of modes emitted ($N \approx A\Omega/\lambda^2$, see [2], Ch.2).

The other factor in Eq.4.41 is usually $\mu_{pol} \approx 0.5$ because either the source is randomly polarized or polarization is not preserved in back diffusion from a rough target.

Considering a Fabry-Perot interferometer (Fig.4-33), one mirror is secured to the target being measured, and the other mirror is mounted on a precision translation stage. In this way, the reference arm can be scanned to search for the $s_r = s_m$ condition while keeping the interferometer aligned. Of course, we can replace the mirrors with corner cubes to relax the angular alignment criticality.

Now, as we pass through the $s_r = s_m$ condition, oscillations corresponding to fringes show up in the photocurrent signal I_{ph}, as illustrated in Fig.4.34, in accordance with Eq.4.40. If the translation stage is moved with a constant speed, the waveform of I_{ph} versus time is a replica of the dependence versus s_r.

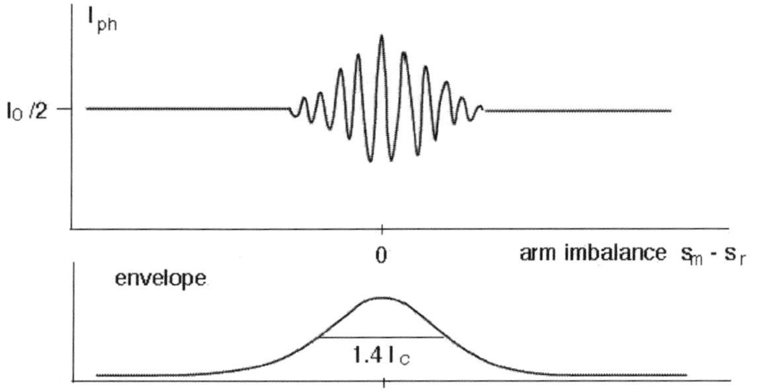

Fig. 4-34 The output signal of a white-light interferometer is an oscillation with a pulse envelope, peaking in correspondence to the balance condition $s_r = s_m$. The envelope width is 1.4 times the coherence length l_c and determines the resolution of the measurement.

To accurately localize the spike position, we may implement the following electronic processing. The a.c. photocurrent signal I_{ph} is first rectified, thus obtaining its envelope. By time differentiation of the envelope, we get a zero crossing in correspondence to its peak.

Finally, a zero-crossing comparator provides the logic signal to sample the s_m value, and is then read by an encoder mounted on the translation stage.

About the measurement accuracy, several systematic, as well as random contributions, are to be considered. By analyzing them, it is found that the accuracy σ_m is a fraction of the response Full-Width at Half Maximum (FWHM), equal to $2 \ln 2\ l_c = 1.4\ l_c$ for a Lorentzian line (see more in Sect.4.7.2). In practice, we can achieve an accuracy of 10 to 50 μm, not as good as in a conventional interferometer, yet a respectable value that can be adequate in a number of applications.

As a comment, the main drawback of the white-light technique is that we need to move the reference arm inside the instrument, covering a distance D equal to the range 0-D to be measured. The practical range of measurement is accordingly limited to, say, a few tens of centimeters, and this makes the technique unsuitable for long-distance measurements.

Of course, we can add a fixed delay in the reference arm to extend the range. For example, a coil of single-mode fiber with objective lenses to focus the input beam and to collimate the output beam readily provides a compact device, which increases the path length by nL and the distance range to nL+D.

This feature coupled with the inherent single-mode propagation in fibers, makes white-light interferometry very attractive [28] for optical fiber sensors (Ch. 10). Also, when the distance range to be covered is rather small (say <1 cm), the technique is easily implemented and can be extended to the image format, as described in the next section.

4.7.1 Profilometry for Industrial Applications

When incorporated in a microscope, the white-light interferometer provides a measurement of the surface height of the specimen, an application widely used in industry for the inspection of manufactured samples.

Spatial resolution is determined by the size of the illuminating spot. By moving the spot across the specimen in an X-Y raster arrangement, we can obtain a map of surface height or a profilometry of the surface [25-27].

Specifically, the surface under test can be either a polished or a rough surface. A working setup is illustrated in Fig.4-35. The light source can be either a lamp filtered in wavelength by a colored filter (for the desired λ_0 and $\Delta\lambda$) or an LED. Light from the source is also spatially filtered by a pinhole to work with a few-mode spatial distribution. With a collimating lens, the beam is sent along the axis of the microscope with the aid of a beamsplitter (BS1). After passing through the objective lens focused on the specimen surface, the illuminating beam encounters a pair of optical flats acting as an interferometer (called Mirau interferometer).

The first flat F has antireflection coatings on both surfaces and a small central obstruction, which is black (absorbing) on the upper side and metallized (reflecting) on the bottom side. The second flat is partially reflecting and acts as a beamsplitter (BS2) on one of its surfaces. At BS2, a part of the beam is directed to the specimen and constitutes the measurement beam, whereas the other part is reflected toward F to be used as the reference beam.

The portion of the reference beam falling on the central metallization of F is reflected back to BS2 where it recombines with the measurement beam returning from the specimen. The two superposed beams are collected by the objective lens in front of the photodetector PD.

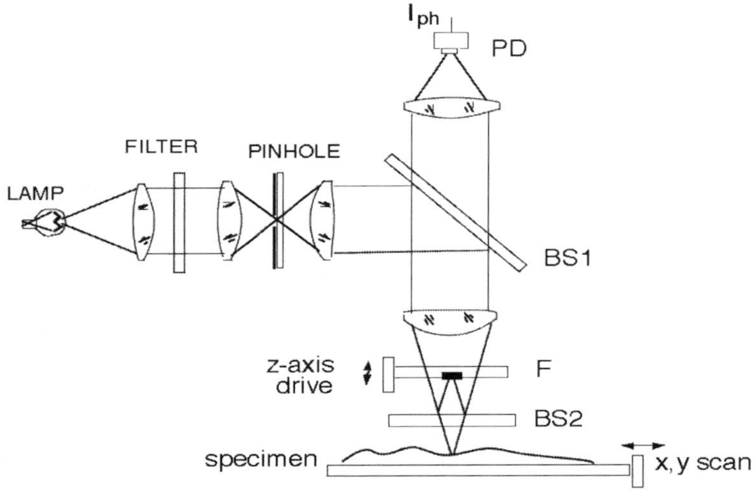

Fig. 4-35 Setup of a profilometer based on white-light interferometry. A first beamsplitter BS1 combines the beam from the source (a filtered lamp or an LED) and the returning beam. Flat F and beamsplitter BS2 act as a Mirau interferometer (F has a small central portion made reflective). The path balance is obtained when the distances of BS2 to F and to the specimen are equal. This condition is searched by moving the z-axis drive until the I_{ph} exhibits the spike shown in Fig.4-34. The x,y scan performs a sequential pixel-by-pixel readout to build up the image.

The interferometer arms are balanced when the distances of the beamsplitter BS2 to the specimen and to the flat F are equal, or differ by less than l_c. This condition is checked by the signal I_{ph} of the photodetector (Fig.4-35). To obtain the condition of balance, the flat F is moved by a micrometer motorized stage. With a phase detector, we are able to lock the stage position at the I_{ph} maximum, with an accuracy of a few hundredths of l_c. The corresponding resolution in surface height is thus well below the micrometer.

About the stage drive, we do not actually need a precise z-axis movement, but rather a precise measurement of it, as can be supplied by a sensor secured on it. The sensor may be one of the several, well-known linear-motion transducers (e.g., a differential transformer, a linear potentiometer, or even a low-cost interferometer), capable of micrometer accuracy or better [29].

As the profile variations from point to point are usually small, and the moving part is of little weight, the time response of the system is adequately short, and acquisition of a single point may require $t_r \approx$1-10 ms.

The spot size is determined by the numerical aperture NA of the objective lens as s=λ/πNA (see [2], Ch.2). This is the sampling dimension (or pixel minimum size) of the x-y profile measurement, and as a typical value, for λ=0.5 μm and NA=0.5, we get s=0.3 μm.

To obtain a full-image profile measurement, we move the specimen by a micromotorized (or piezo) stage in an x-y raster fashion. Again, precision encoders will provide the x-y position readout to which the z-axis measurement is assigned. The total scan time of a 100×100 pixels image may be 10^4×(1-10 ms) =10-100 s, a fairly long but acceptable time.

Also, we may go back to the Michelson interferometer of Fig.4-33 and use a CCD array as the photodetector to cover the full specimen image, and simply scan the z-axis. When an individual detail in the specimen reaches the balance condition, the corresponding pixel will exhibit a current spike, and the z-measurement is assigned to the pixel coordinates.

A recent refinement to white-light profilometry has been that of improving the vertical resolution down to nanometers [30].

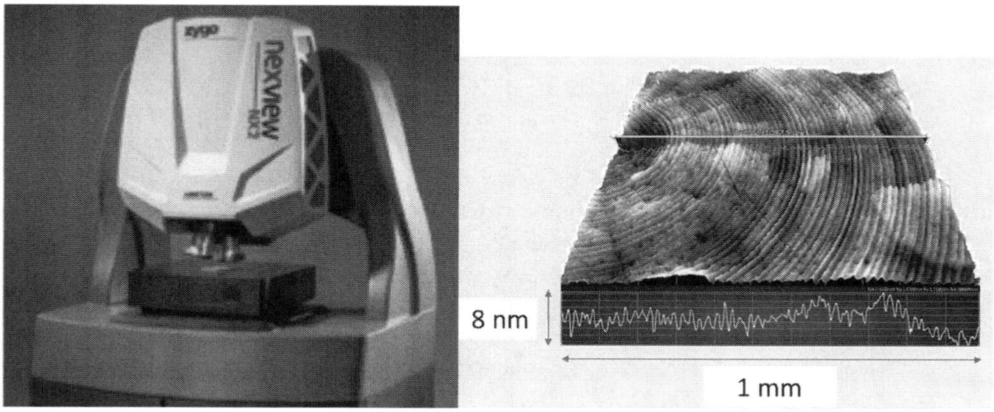

Fig. 4-36 A typical profilometer based on white-light interferometry (left), and (right) an example of diamond turned plate taken with nanometer height resolution (courtesy of Zygo, Inc.).

This is achieved by adaptation of the phase-shifting technique discussed in Sect.4.2.2.1 to the white-light interferometer scheme, when the fringes show up inside the coherence balance condition.

For points in the x-y plane close to the point under test, and deviating vertically by less than $\Delta z < l_c$, a fringe system is generated in the detector plane. Using a CCD, we can sample this fringe pattern and process the data in a computer.

By successive acquisitions on adjacent pixels covered by the X-Y scan (Fig.4-35), the image of the entire specimen is built up. A commercial profilometer is shown in Fig.4-36, along with typical examples illustrating the results of x-y profile measurements.

4.7.2 OCT for Biomedical Applications

Optical Coherence Tomography (OCT) was first proposed by Huang et al. [31] as early as 1991, and since then this method has gained increasing consensus as a powerful tool for the analysis of 3D images of biologic tissues in fields like ophthalmology, dermatology, oncology, and angiography [32], reaching in a few years revenue comparable to the that of the laser interferometer described in Sect.4.2.2.

The reason for the success is that OCT is a revolutionary new imaging technology, different from previously known approaches because it takes advantage of low-coherence interferometry in combination with coherent detection to supply 3D images reaching a depth down to a few millimiters in tissues and to 10...20 mm in the eye, depending on the absorption of tissues at the wavelength of operation.

As biological tissues have a minimum of absorption in the near-infrared [33], preferred sources are the SLED (super luminescent LED) at 850 nm (with a $\Delta\lambda \approx 20$ nm) and the Ti:sapphire tunable laser emitting in a wide $\Delta\lambda$ range, from 700 to 880 nm [34].

The wavelength width $\Delta\lambda$ is important because it determines the coherence length L_c and hence the spatial resolution, as shown by the response curve (Fig.4-34) of the interferometer.

About configurations for the OCT, several variants have been developed. The schematic of Fig.4-35 has been the first to be used and is identified as *time-domain OCT* (or *TD-OCT*) because the photodetected signals are processed in the time domain to sort out the depth of details, much in the same way as a time-of-flight telemeter, albeit using coherence as the timing variable. The TD-OCT requires scanning on three axes and doesn't use light efficiently because only at the coherence balance do we get a useful signal (the constant signal $I_0/2$ in Fig.4-34 is wasted power). In another approach, called *Fourier domain OCT* (or *FD-OCT*), we process the signal in the spatial frequency domain, using a spectrometer and a line-array of photodetectors, and then make the Fourier transform to get back to the time domain. In this way, called *spectral domain OCT* (SD-OCT), we don't need the z-scan anymore, and the power is all used to generate the useful signal, with improved SNR. Another important variant of the Fourier domain approach is to use a frequency-swept laser source, that avoids using the spectrometer and can improve SNR and frame rate, an approach called *swept source OCT* (SS-OCT).

Time-domain OCT (TD-OCT). To analyze the signal of the time-domain OCT, let $E_{m,r} = E_{0m,r}(\nu)$ exp $i(ks_{m,r}-\omega t)$ indicate the fields returning to the photodetector after propagation along the measure (m) and reference (r) arms (Fig.4-33), with the spectral content of fields being represented by $E_{0m,r}(\nu)$.

The photogenerated current I_{ph} is like in Eq.4.1, that is, the square modulus of the total field $E_m + E_r$ times σ, but now we shall also integrate these terms on frequency ν. Omitting the d.c. constant terms $E_{0m,r}^2$ and assuming σ is independent from ν, we can write I_{ph} as:

$$I_{ph} = 2\,\sigma\ E_m\, E_r^*$$
$$= 2\,\sigma \int_{0-\infty} d\nu\ E_{0m}(\nu)\ E_{0r}(\nu)\ \exp i2\pi\nu(s_m/c\ -t) \cdot \exp -i2\pi\nu(s_r/c\ -t)$$
$$= 2\,\sigma \int_{0-\infty} d\nu\ I_0(\nu)\ \exp i2\pi\nu(s_m - s_r)/c \qquad (4.40)$$

where, in the third line of Eq.4.40, we let $E_{0m}(\nu)E_{0r}(\nu)=I_0(\nu)$ for the intensity (or power) dependence from ν. The last term is recognized as the kernel of the Fourier transform of intensity $I_0(\nu)$ on the conjugate variable, the time difference $(s_m-s_r)/c$.

Thus, the profile $I_{ph} = I_{ph}(s_m-s_r)$ depends on the frequency distribution $I_0(\nu)$, and determines the waveform of the peak seen in Fig.4-34 around the zero, or $s_m-s_r \approx 0$.

Let's now discuss the shape of $I_0(\nu)$. The shape depends on the dominant mechanism of line broadening [35]. We may have a Lorentzian line, with a negative exponential as the corresponding transform in the space s_m-s_r domain:

$$I_0(\nu) = (\Delta\nu_L/\pi)\ /[(\nu-\nu_0)^2 + \Delta\nu_L^2]\ \leftarrow FT \rightarrow\ I_{ph} = \exp - \Delta\nu_L(s_m-s_r)/c \qquad (4.41)$$

or a Gaussian linewidth and a Gaussian in the space s_m-s_r domain:

$$I_0(\nu) = [\sqrt{(2\pi)}\Delta\nu_G)\ \exp -(\nu-\nu_0)^2/2\ \Delta\nu_G^2\ \leftarrow FT \rightarrow\ I_{ph} = \exp -[(s_m-s_r)/c]^2 \Delta\nu_L^2/2 \qquad (4.42)$$

The line is Lorentzian when broadening is due to collision or is radiative, and Gaussian when the Doppler broadening is dominant [35]. A third case, the inhomogeneous (or Voigt) broadening [35], is not considered here.

Anyway, when two mechanisms intervene independently, the resulting $I_0(\nu)$ profile is the weighted sum of the two components and (for the linearity of the *FT* operation) the I_{ph} is a weighted sum as well. If one mechanism is cascaded in part to the other, the relative proportions are multiplied in the frequency domain $I_{0a}(\nu)\ I_{0b}(\nu)$, and go in convolution in the space domain as $I_{pha} * I_{phb}$.

The envelope of the interference distribution around $s_m-s_r \approx 0$ of Fig.4.34 is given by the right-hand expressions in Eqs.4.41 and 4.42. These quantities are the visibility factor V of the fringes, as expressed by Eq.4.2.

Now, we can calculate the width of the photodetector envelope I_{ph} that represents the localization resolution of the measurement arm (Fig.4-43).

For a Lorentzian line, Eq.4.41 tells us that the envelope I_{ph} decreases exponentially with a spatial constant $c/\Delta\nu_L = l_c$. Then, the half-width at half-maximum (HWHM) is found by letting $\exp -(s_m-s_r)/l_c = 1/2$, whence $(s_m-s_r) = (\ln 2) l_c = 0.7\ l_c$, and therefore the full-width is FWHM $= (2 \ln 2)\ l_c = 1.4\ l_c$.

For a Gaussian line shape, the rms value of displacement is l_c, and we may similarly require $\exp -(s_m-s_r)^2/2\ l_c^2 = 1/2$ and obtain $(s_m-s_r) = (2\sqrt{\ln 2})\ l_c = 1.66\ l_c$, and have a FWHM$= (4\sqrt{\ln 2})\ l_c = 3.33\ l_c$ as the width of the envelope of photodetector response.

The reader is warned that, in literature, resolution may be found expressed by the rms value, or by HWHM, or by FWHM. So, from the above examples, the multiplicative factor of l_c may range from 0.7 to 3.3. With this caveat, and being aware of the multiplicative factor, we simply assume in the following that resolution is given by $(s_m-s_r) = l_c$, that is, by the coherence length.

About the coherence length $l_c = c/\Delta\nu$, we can use the relation $\Delta\nu/\nu=\Delta\lambda/\lambda$ and $\lambda\nu=c$ and obtain the dependence from the wavelength interval $\Delta\lambda$ as:

$$l_c = \lambda^2/\Delta\lambda \tag{4.43}$$

this quantity, apart from the multiplicative factor discussed above, is the *axial resolution* of the OCT instrument, that is the resolution w_z along the depth or z-axis coordinate.

Another resolution of importance is $w_{x,y}$ along the x and y coordinates, that is, the *transversal resolution,* also called *en-face* resolution. This is totally independent from l_c or w_z, and is determined by the spot size w_0 of the projection objective (Fig.4-35). At the diffraction limit, the spot size is $w_0 = n_m\,\lambda/\pi NA$ (see Ref. [2], Ch.2.4), NA being the numerical aperture of the objective lens, and n_m the number of spatial modes of the source. So, resolutions are:

$$w_z = \lambda^2/\Delta\lambda, \qquad w_{x,y}= n_m\,\lambda/(\pi NA) \tag{4.44}$$

Connected to the objective lens performance there is a third quantity of importance: the *depth of focus* z_{df}. This is the axial length within which the beam is kept focused and will give the same amplitude signal back to the photodetector. In general, if the ideal focus is at depth z_0 with a spot size w_0, then we have a spot size less than $\sqrt{2}\,w_0$ in the range $z_0-z_{df}< z<z_0+z_{df}$. Applying the acceptance invariance, we find the depth of focus as (compare with Eq.6.1):

$$z_{df} = n_m\,\lambda/(\pi\,NA^2) \tag{4.45}$$

To maintain axial and transverse resolutions along all the depth Z of the scan, while focusing at Z/2, the middle of it, we need that $z_{df} \geq Z/2$. But, as lateral resolution and depth of focus require opposite NA, i.e., large for the former and small for the latter, we need to make a trade-off to ensure satisfactory performance.

For example, at λ=800 nm and NA=0.5, with $\Delta\lambda$=180 nm, n_m =3 for the Ti:sapphire laser, the lateral resolution is $w_{x,y}$=3 0.8/3.14 0.5= 1.5 μm, and the depth of focus is w_{df} = 3 0.8/3.14 0.5^2= 3 μm, too low a value, and we may prefer to use NA=0.1, which gives $w_{x,y}$= 7.5 μm, and w_{df}= 75 μm.

The working scheme of the TD-OCT for bio-applications is the same as in Fig.4-35 described in last Section, with the only difference that the target is not reflective as in industrial profilometry, but rather transmissive. In profilometry, there is no requirement of depth-of-focus (other than the maximum height swing ΔZ, usually <1 μm) and therefore we don't have to sacrifice NA, whereas in TD-OCT we can trade the smaller NA with increased source power, replacing the SLED with the Ti:sapphire laser.

Both applications pose the requisite of three mechanical scans: one in depth (z-axis) and two laterals (along x and y axes) to form the 3D image of the target.

In literature, the depth (or z-axis) scan is called *A-scan*, the scan in x or y is called *B-scan*, and the raster or zigzag scan to totally cover the *en-face* x-y surface is called *C-scan*. To acquire a full 3D image on a size XY, we need to perform $XY/w_{x,y}^2$ A-scans of duration τ_A each. Assuming, say, X=Y=1 mm and $w_{x,y}$= 7.5 μm to have an in-depth coverage of ΔZ=2×75 μm= 150 μm, and a duration of the A-scan τ_A =10 ms, we get a total frame time $XY/w_{x,y}^2\ \tau_A = (1/0.0075)^2$ 10 ms=180 s.

It's worth noting that this scanning of the TD-OCT is rather time consuming, or, it requires trading the total pixel number of the image, N= $XY/w_{x,y}$, for frame rate or acquisition time. In addition, the TD-OCT uses only a small fraction of the power sent to the specimen, because only during the balance condition $s_m \approx s_r$ does power contribute to the useful signal.

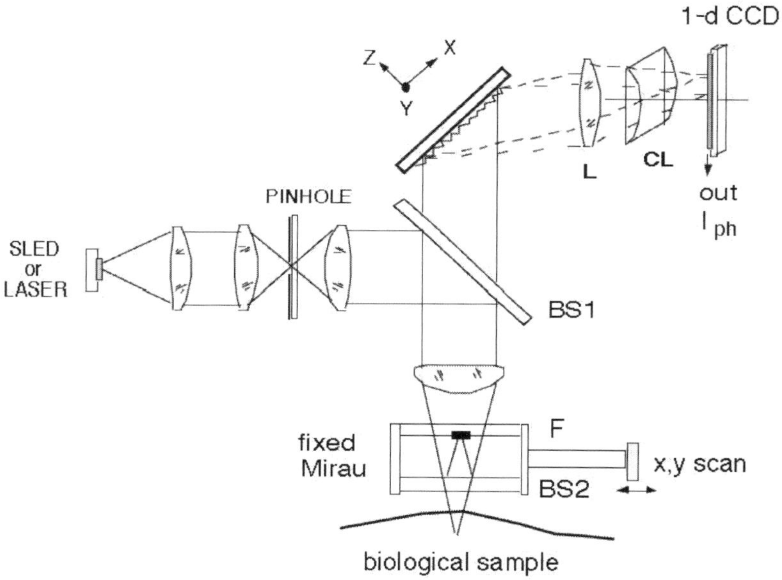

Fig. 4-37 The spectral domain OCT (SD-OCT) uses a grating placed before the photodetector to obtain the wavelength-resolved interference signal I(λ). A Fourier transform then yields the depth-resolved signal I(z). There is no need to scan the Mirau interferometer and the optical power is entirely used for the signal, with an improvement of sensitivity of typically 10 dB.

Spectral domain OCT. The setup of the SD-OCT is that of a low-coherence interferometer, and is built around a Michelson or a Mirau interferometer (the one shown in Fig.4-37), but with a spectrometer added before the photodetector. Another difference is, the mirror of the reference arm is fixed, i.e., need not be scanned. To understand the frequency domain processing, let's write the power P_{out} reaching the grating (before the detector) as the integral over the target distance s_m, of the fields from the measure and reference arms, that is:

$$P_{out} = \int | [E_m(s_m)R(s_m)+E_r(s_r)] |^2 ds_m \tag{4.46}$$

where $R(s_m)$ indicates the reflected (or diffused) signal backscattered from distance s_m. Developing the square modulus in Eq.4.46 and ignoring the constant terms, we have:

$$P_{out} = \int 2 \,\text{Re}\, [E_m(s_m)R(s_m)E_r^*(s_r)]\, ds_m \tag{4.46'}$$

where * indicates the complex conjugate. We can express the fields E_m and E_r (compare with Eq. 4.40) as:

$$E_m(s_m) = E_{0m}(\lambda) \exp i2\pi(s_m/\lambda - \nu t) \quad \text{and} \quad E_r(s_r) = E_{0r}(\lambda) \exp i2\pi(s_r/\lambda - \nu t)$$

and, upon inserting in Eq.4.46' we get:

$$P_{out} = \int 2\, E_{0m}(\lambda)\, E_{0r}(\lambda)\, R(s_m) \exp i2\pi(s_m - s_r)/\lambda \;\; ds_m \tag{4.47}$$

Now, let assume the simplifying hypothesis that $E_{0m,r}(\lambda) \approx$const. on λ in the interval λ_1 to λ_2, drop the constant term s_r, and rename s_m=z to indicate the depth, ranging from 0 to Z. In this way Eq.4.47 becomes:

$$P_{out}(1/\lambda) = 2\, E_0^2(\lambda) \int_{0\text{-}Z} R(z) \exp i2\pi z/\lambda \;\; dz$$
$$= F\, [R(z)] \tag{4.48}$$

and we can see that the power P_{out} is the Fourier transform of the object reflectivity R, as a function of depth z.

Note that the same power P_{out} is found at the photodetector input also in the basic scheme of Fig.4-35. Now, as in Fig.4-37, we enter the grating with the spectral distribution $P_{out}(1/\lambda)$ ranging from $1/\lambda_1$ to $1/\lambda_2$, the wavelength range $\Delta\lambda$ of the source.
At the output of the grating, diffracted rays sent in different directions θ, a focusing lens converts direction into a focal plane coordinate z=F θ, and finally the CCD image detector gives out a signal $I_{CCD}=\sigma P_{out}$ (1/λ). Technically, if the grating grooves lie along the X-direction, the arriving beam is dispersed along the Y-direction, and we have to sum all the charge of all the X-column pixels to build up the output signal I_{CCD}.

After the acquisition of signal and its conversion to the digital format through an ADC converter, we will numerically compute the inverse Fourier transform and get the A-scan of the object, namely F^{-1} [I_{CCD} (1/λ)) = R(z).

We still need the X and Y scans to build up an image, but the advantage of the SD-OCT is that available power is fully used, not wasted like in a TD-OCT that scans most of the time out of the arm balance condition. Thus, sensitivity and SNR of SD-OCT is found improved by about 10 dB [36] with respect to TD-OCT, obtaining an increased depth of the A-scan, as we can tolerate more propagation attenuation.

About the CCD detector (see Ref.[2], Ch.12.2), we may use a 1-D (or linear array) device. Indeed, if the grating grooves lie parallel to the Y-direction (Fig.4-37), the beam is diffracted along the X coordinate and every line along the Y-direction carries the same current

$(1/n_Y)\ I_{CCD}(1/\lambda)$, where n_Y is the number of Y-lines. We can sum up all the currents of the Y-lines using a cylindrical lens (fg.4-37) that focuses the Y-line in a single point (a CCD pixel). The output signal is obtained by scanning the CCD, and we need a time $T_f = n_X T_c$ to complete the transfer, where n_X is the number of pixels and T_c the clock period of the pixel shift (see Ref.[2], Ch.12.2). In commercial CCDs, the clock time T_c is around 200 ns, so the acquisition of a 250-pixel spectrum takes T_f=5 μs, typically.

For the SD-OCT, time T_f is the acquisition time of a single A-scan, that shall be repeated $N_X N_Y$ times to complete the C-scan of a 3D tomography containing $N_X N_Y$ pixels, thus the total acquisition time of a 3D image is $N_X N_Y T_f$ (=0.45 s for a 300x300 pixel C-scan resolution).

It is instructive to see how the signal propagates through the SD-OCT chain. The signal arriving at the grating is $P_{out}(1/\lambda)$, a distribution of power vs wavelength, or better vs spatial frequency $h=1/\lambda$, as this is the variable conjugated through the Fourier transform to the depth z. Let $h_1=1/\lambda_1$ and $h_2=1/\lambda_2$ be the spatial frequencies corresponding to the $\Delta\lambda$ of the laser emission. Upon arriving at the grating, the spectrum components are deflected angularly according to the grating equation,

$$\sin\theta_{in} + \sin\theta_{out} = 2p/\lambda \qquad (4.49)$$

where $\theta_{in,out}$ are the input and output angles of the beam and p is the period of the grating. As $\sin\theta_{in}$ =const. = S, we get from Eq.4.49: $\sin\theta_{out} = 2p\ (1/\lambda)\ -S$ and, for small θ_{out}, the corresponding x on the focal plane of the lens is $x=F\ \theta_{out} = 2Fp\ (1/\lambda)\ -FS$. This equation tells us that, on the focal plane of the CCD lens, power P(x) is distributed along coordinate x proportional to the spatial frequency of the spectral content in the arriving signal, and we may write it as $P(1/\lambda)$.

So, the max and min frequency components are deposited at $x_{min}=2Fp\ (1/\lambda_{max})$-FS and x_{max} = =2Fp $(1/\lambda_{min})$ - FS with a linear swing of the $1/\lambda$ spectrum between them.

On the signal from the CCD output, we perform the inverse Fourier transform of $P(1/\lambda)$ and obtain as a result (see Eq.4.48) $E_0^2\ R(z)$, that is, the desired A-scan map of reflectivity.

From the basic properties of Fourier transform [37], it follows that:

(i) the spatial frequency range (inverse wavelength span) Δh, coincides with the coherence length, as $\Delta h = 1/\lambda_{min} - 1/\lambda_{max} = \lambda_{min}\lambda_{max}/(\lambda_{min}-\lambda_{max}) = l_c$ and is a rectangular window in λ that determines the point spread transfer function of the measuring system, given by

$$F(z) = \sin(z\Delta h/2)/(z/2) \qquad (4.50)$$

F(z) has a FWHM=$2.44/\Delta h$ [2,37] and hence the spatial resolution is

$$\Delta z = 2.44/(1/\lambda_{min} - 1/\lambda_{max}) = 2.44\ \lambda_{min}\lambda_{max}/\Delta\lambda \approx 2.44\ \lambda^2/\Delta\lambda \qquad (4.51)$$

As noted, this value is comparable to the coherence length l_c given by Eq.4.44 when the multiplicative factors discussed therewith are taken into account;

(ii) Although not very obvious, there is a sampling process of the spectrum $P_{out}(1/\lambda)$, that affects dynamic range and aliasing. Indeed, the CCD (like any other array photodetector) has, say, N elements (or pixels) along a line, and each pixel separately takes a sample of the distribution $P_{out}(1/\lambda)$.

Thus, the sampling frequency is $N/\Delta h$, the Nyquist frequency is half as much, $N/2\Delta h$, and the maximum depth z_{ma} that can be reconstructed without aliasing (Ref.[2], Sect.12.3.3) is given by $z_{max}= 2\pi\ N/2\Delta h = \pi\ N/\Delta h$. This limit is important because all the frequency components at $1/\lambda > N/2\Delta h$ will be folded back in the baseband and introduce spurious signals known as aliasing interference (Ref.[2], Sect.12.3.3).

Swept Source OCT (SS-OCT). With the swept source OCT (Fig.4-38) we can dispense with the grating (and use a photodiode in place of the CCD) because its function is carried out by scanning the wavelength of the laser source. Eqs.4.47 and 4.48 still hold and tell us that the signal reaching the photodetector is the Fourier transform of reflectivity R(z). Of course, we need again a large wavelength tuning range $\Delta\lambda$ to ensure a wide dynamic range $\pi N/\Delta h$, and a fast sweeping mechanism because the period of sweep repetition is the A-scan acquisition time.

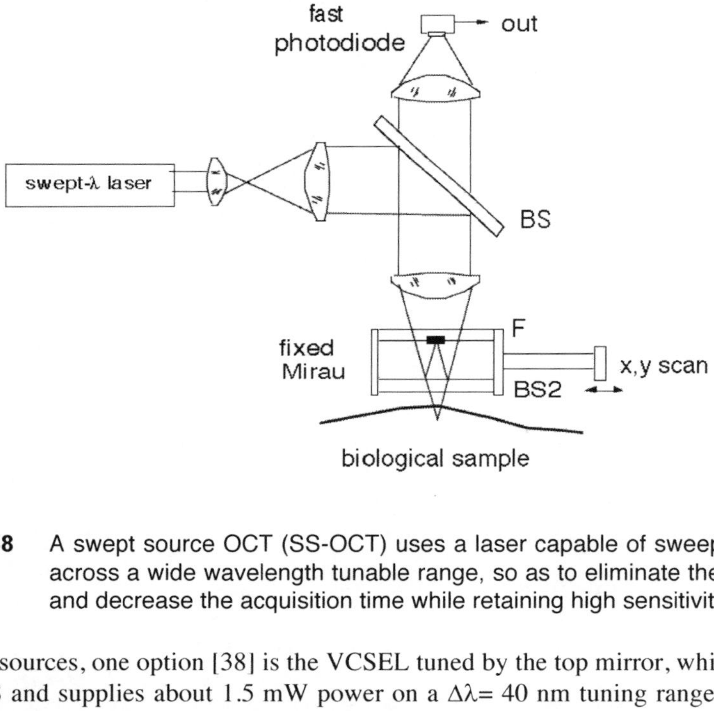

Fig. 4-38 A swept source OCT (SS-OCT) uses a laser capable of sweeping fast across a wide wavelength tunable range, so as to eliminate the grating and decrease the acquisition time while retaining high sensitivity.

About laser sources, one option [38] is the VCSEL tuned by the top mirror, which is actuated by a MEMS and supplies about 1.5 mW power on a $\Delta\lambda$= 40 nm tuning range at 1.15 MHz repetition rate. A second solution [39] is a ring laser including a pigtailed SOA and a Fabry-Perot tunable etalon in a fiber loop of optical pathlength 2.5 km; power is 1.4 mW on $\Delta\lambda$= 63 nm at 236 kHz repetition rate.

An example of OCT images detected by a commercial instrument [40] is shown in Fig.4-39.

Finally, an interesting account of the historical developments of the OCT technology, along with a rich bibliography is provided by Ref. [41].

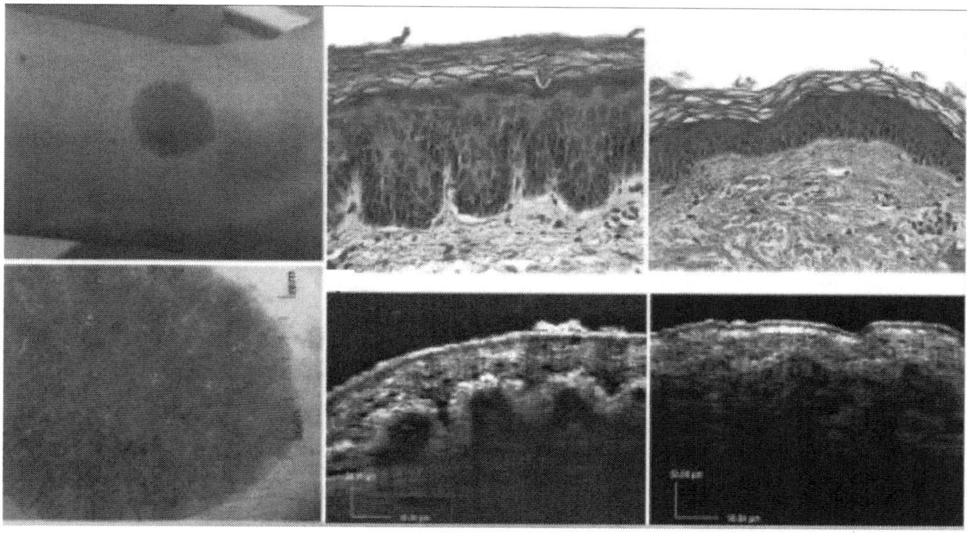

Fig. 4-39 Images from an OCT, based on a Ti:sapphire crystal fiber laser and a Mirau interferometer. Resolution of the instrument: lateral 1.5 μm, axial 1.3 μm, on a scanning depth of 400 μm and a field-of-view of 0.5x 0.4 mm^2. Left, the skin images; top center and right: the hematoxylin and eosin stained sample; bottom center: a large acanthoma and bottom right: normal skin (by courtesy of Apollo Medical Optics, Taipei).

REFERENCES

[1] I. Newton: "*Lectiones Opticae*", 1678 manuscript published in Cambridge, 1727; C. Huygens "*Traité de la Lumiere*", Cambridge 1690; A. Schuster and J.W. Nicholson "*Theory of Optics*", E. Arnold: London, 1904.
[2] S. Donati: "*Photodetectors*", 2nd ed., Wiley IEEE Press, Hoboken 2021.
[3] L.J. Wuertz, R.C. Quenelle: "*Laser interferometer system for metrology and machine tool applications*" Precision Engineering, vol.3 (1983) pp.111-114
[4] V. Greco, G. Molesini: "*Monitoring the thickness of soap films by polarization homodyne interferometry*", Meas. Sci. Techn., vol.7, (1996) pp.96-101, see also: G. Molesini et al., "*Multiphase Homodyne Interferometry*", Proc. Elettroottica'94, edited by S. Donati, pp.53-57, AEI: Milan, Italy, 1994.
[5] D. Hofstetter, H.P. Zappe, R. Dandliker: "*Optical Displacement Measurement with GaAs/AlGaAs-Based Monolithically Integrated Michelson Interferometers*", J. Lightw. Techn., vol.15 (1997), pp.663-670.

[6] O.G. Helleso, P. Benech, R. Rimet: "*Interferometric Displacement Sensor made by Integrated Optics on Glass*", Sensors and Actuat. A vol.46-47 (1995), pp.478-481.
[7] R. Baetz et al.: "*Detecting carotid stenosis from skin vibrations using Laser Doppler Vibrometry – An in vitro proof-of-concept*", Plos One, June 20, 2019, https://doi.org/10.1371/ journal.pone.0218317.
[8] T. Harimaya, P.A.M. Sandborn, M. Watanabe, M.C. Wu: "*High-accuracy Range Sensing System based on FMCW using low-cost VCSEL*", Opt. Expr., vol.26 (2018), pp.9285-9297.
[9] E. Baumann, F.R. Giorgetta, J.-D. Deschênes, W.C. Swann, I. Coddington, and N.R. Newbury: "*Comb-calibrated laser ranging for three- dimensional surface profiling with micrometer-level precision at a distance*", Opt. Expr., vol.22 (2014), pp.24914-24928.
[10] D.J. Lum, S.H. Knarr, J.C. Howell: "*Frequency Modulated Continuous Wave LiDAR Compressive Depth-Mapping*", Opt. Expr., vol.26 (2018), pp.15420-15432.
[11] N. Druml, I. Maksymova, T. Thomas, D. Lierop, M. Hennecke, and A. Foroutan: "*1D MEMS micro-scanning LiDAR*" Proc. 9th Int. Conf. Sensor Device Techn. and Appl., Venice, Italy, 2018.
[12] R. Holtzwarth, T. Udem, T.W. Hansch: "*Optical Frequency Synthesizer for Precision Spectroscopy*", Phys. Rev. Lett., vol.85 (2000), pp.2264-2267.
[13] J. Ye, S.T. Cundiff (Editors): "*Femtosecond Optical Frequency Comb: Principle, Operation, and Applications*", IX+359 pages, (2005) Kluwer Academic Publ., Norwell.
[14] K. Minoshima, T. Tomita, H. Matsumoto: "*Femtosecond-comb distance meter: ultra-high-resolution distance measurement using a mode-locked laser*" Proc. CLEO Pacific Rim 2003, vol. 2, p.394, paper TH1B-(10)-1.
[15] J. Ye: "*Absolute measurement of a long, arbitrary distance to less than an optical fringe*", Opt. Lett. vol. 29 (2004), pp.1153-1155.
[16] G. Li, Y. Fang, H. Zhang, J. Sun: "*High-precision long-distance measurement with an intensity-modulated frequency comb*", Appl. Opt. vol.59 (2020), pp.7292-7298.
[17] X. Xu, Z. Zhang, H. Zhao, W. Xia, M. He, J. Li, J. Zhai, H. Wu: "*Long distance measurement by dynamic optical frequency comb*", Opt. Expr. vol.28 (2020), pp.4398-4411.
[18] P.G. Halverson and R.E. Spero: "*Signal processing and Testing of Displacement Metrology Gauges with pm-Scale Cyclic Nonlinearity*", J. of Opt. A, vol.4 (2002), pp. S304-10.
[19] V. Annovazzi Lodi, S. Donati, and S. Merlo: "*Thermodynamic Phase Noise in Fiber Interferometers*", J. Opt. Quant. Electr., vol.28 (1996), pp.43-49.
[20] R.O. Claus and C.H. Palmer: "*Direct Measurement of Ultrasonic Stoneley Waves*", Appl. Phys. Lett., vol.31 (1977), pp.547-548.
[21] M. Corti, F. Parmigiani, and S. Botcherby: "*Description of a Coherent Light Technique to Detect Vibration of an Arch Dam*", J. Sound and Vibrat., vol.84 (1981), pp.35-45.
[22] G. Macchi and A. Pavese: "*Diagnosis of Masonry Towers Using Interferometric Transducers*", in Proc. ODIMAP III, Pavia 20-22 Sept.2002, pp.51-56; see also A. Miks and J. Novak: "*Non-contact Measurement of Deformation in Civil Engineering*", ibidem, pp.57-62.
[23] F. Aronowitz: "*The Laser Gyro*", in Laser Applications, edited by M. Ross, Academic Press: New York, 1971, pp.134-199.
[24] S. Donati: "*Laser Interferometry by Induced Modulation of the Cavity Field*", J. Appl. Physics, vol.49 (1978), pp.495-497.
[25] M. Davidson, K. Kaufman, I. Mazor, and F. Cohen: "*Application of Interference Microscopy to IC Inspection and Metrology*", SPIE Proceedings, vol.775 (1987), pp.233-247.
[26] T. Dresel, G. Hausler, and H. Venzke: "*3-D Sensing of Rough Surfaces by Coherence Radar*", Appl. Opt., vol.31 (1992), pp.919-925.

[27] J.C. Wyant and K. Creath: *"Advances in Interferometric Optical Profiling"*, Intl. J. Mach. Tools Manufact., vol.32 (1992), pp.5-10.
[28] D.A. Jackson et al.: *"An Interferometric Fibre Optic Sensor Using a Short Coherence Length Source"*, Electr. Lett., vol.23 (1987), pp.1110-1112.
[29] P.J. Caber: *"An Interferometric Profiler for Rough Surfaces"*, Appl. Opt., vol.32 (1993), pp.3438-3441.
[30] A.G. Podoleanu et al.: *"OCT En-face Images from the Retina with Adjustable Depth Resolution in Real Time"*, IEEE Sel. Top. Quant. El., vol.5 (1999), pp.1176-1184.
[31] D. Huang, E.A. Swanson, C.P. Lin, J.S. Schumann, W. G. Stinton, W. Chang, M.R. Hee, T. Flotte, K. Gregory, C.A. Puliafito, J.G. Fujimoto: *"Optical Coherence Tomography"* Science, vol.254 (1991), pp.1178–1181.
[32] W. Drexel, J.G. Fujimoto: *"Optical Coherence Tomography - Technology and Applications"*, Springer 2015, Berlin.
[33] G. Kaiser: *"Biophotonics: Concepts to Applications"*, 2nd ed., Springer, Berlin 2022.
[34] T.T. Yang, Y.-I. Yang, R. Soundararajan, P.S. Yeh, C.Y. Kuo, S.-L. Huang, S. Donati: *"Widely Tunable, 25-mW Power, Ti:sapphire Crystal-Fiber Laser"*, IEEE Phot. Techn. Lett., vol.31 (2019), DOI 10.1109/LPT.2019.2950020.
[35] P.W. Milonni, J.H. Eberly: *"Laser Physics"*, 2010, J. Wiley & Sons Inc., Hoboken (NJ).
[36] R. Leitgeb, C.K. Hitzenbergen, A.F. Fercher: *"Performance of Fourier Domain vs Time Domain OCT"*, Opt. Expr., vol.11 (2003), pp. 889-894.
[37] J.D. Gatskill: *"Linear Systems, Fourier Transforms and Optics"*, J. Wiley, N.Y., 2010.
[38] K. Li, C. Chase, P. Qiao, C.J. Chang-Hasnain: *"Widely Tunable 1060-nm VCSEL with High Contrast Grating Mirror"*, Opt. Expr., vol. 25 (2017), pp.11844-11856.
[39] R. Huber: *"Fourier Domain Mode Locking: a Laser for Ultrahigh-Speed Retinal OCT imaging at 236 kHz Line Rate"* Proc. CLEO 2007, paper CThAA5.
[40] Y-J. Wang, Y.-K. Huanga, J.-Y. Wanga, Y.-H. Wu: *"In Vivo Characterization of Large Cell Acanthoma by Cellular Resolution Optical Coherent Tomography"*, Esevier Photodiagn. Photodyn. Ther., vol.26 (2019), pp.199-202.
[41] C.K. Hitzenberger: *"Optical coherence tomography in Optics Express"*, Opt. Expr., vol.26 (2018), pp. 24240-24259.

Problems and Questions

Q4-1 *Does the conversion of field to current really correspond to performing a square of the electrical field, as express by Eq.4.1, and then, why there is no second harmonic generation?*
Q4-2 *In the schematics of Fig.4-3, isn't there a phase-shift between transmitted and reflected components introduced by the beamsplitter, that should be taken into account as a π-difference added to the $2k(s_m\text{-}s_r)$ arguments in Eq.4.3?*
Q4-3 *What about the reflections from the surfaces crossed by the beams in the schematics of Fig.4-3, can they add to the useful beam? Don't they alter the operation with respect to the case considered?*
P4-4 *How do you implement the multiplier of Fig.4-5, which is necessary to bring the counted fringes to decimal?*
P4-5 *Isn't there another approach to implement the multiplication using a reasonably small number of SS-ICs, eventually at the expense of speed of output updating?*

P4-6 *Calculate how parallel the stroke of the measurement corner cube should be with respect to axis defined by the propagation vector of the beam, to keep the cosine error below 1 count.*
Q4-7 *Is the cosine-error a positive or negative deviation from the true value?*
Q4-8 *In Fig.4-11 of the text, we have considered a 3-sector phase shifter to provide 3 signals dephased by 120 and 240 degrees. Why can't the same thing be done with just two signals?*
P4-9 *A two-frequency laser interferometer shall be measuring a carriage with a corner cube moving at a speed up to 30 cm/s. Determine the required bandwidth of the interferometer signal.*
P4-10 *In Fig.4-16 we see the 3-axis extension of the interferometric measurement. Is it really useful for part saving? How many more axes can realistically be fed by a single laser source?*
P4-11 *The working principle of the planarity-measurement setup shown in Fig.4-17 is very clear if we look at the angle α as the output, whereas it is not so straightforward if we are looking to the height distribution $z(x,y)$. How are the data of $\alpha(x,y)$ used to compute $z(x,y)$?*
Q4-12 *Can we measure the angles (e.g., roll, pitch, and yaw) of a turret carrying the tools of the machine-tool as in Fig.4-16 so that we control in it six axes?*
P4-13 *An interferometer operates on a diffuser target like that indicated in Fig.4-19. The laser power is P_l=1 mW and we use a lens with F=100 mm and D =30 mm, that focalizes the laser beam (with w_l= 0.5 mm) on the target in a focal spot of w_{tar}=100 μm. What is the collected power P_{ph}, on the 1-mm dia. photodiode? Does it depend on w_{tar} ? What is the dynamic range of the measurement? What is the loss of sensitivity performance with respect to operation on a corner cube?*
P4-14 *How does the quantum efficiency η of the detector impact the shot-noise-limited NED? How important is it, compared to visibility and responsivity?*
P4-15 *What is the frequency stability $\Delta\nu$ of the laser that we need in an interferometer to be able to resolve a NED of, say: i) 1-nm on a Δs=1-m distance (arm) imbalance, ii) 1 μm on a Δs =10-m distance imbalance.*
P4-16 *Are the spatial-coherence and polarization-state effects the only effects related to beam parameters, or can other possible errors arise, for example in wave-front radius mismatch?*
P4-17 *If the incoming wavefront is spherical and the detector surface is plane, isn't there an error because of the varying phase of the field along the photodetector surface?*
P4-18 *How large is the error of the interferometric measurement which is caused by an error in wavelength and the associated variation of the index of refraction of the air? Evaluate it for a single mode-hop.*
P4-19 *Evaluate the thermodynamic fluctuation arising in a fiber of length L=1000 m with B=10 kHz.*
P4-20 *Evaluate the Brownian rms velocity fluctuation affecting a 1-milligram mass.*
P4-21 *Evaluate the Brownian rms angular velocity fluctuation affecting a disk, like a MEMS gyroscope, of 10-milligram mass and a 1-mm radius.*
P4-22 *Calculate the minimum displacement that can realistically be measured with the internal configuration of interferometry working with a L=20-cm He-Ne tube.*
P4-23 *Derive the loop gain and the transfer function of the short-distance vibrometer shown in Fig.4-34. Find the filter response to low-frequency (<100 Hz) components (representative of microphonics).*
P4-24 *Suppose you remove the low-pass filter in the vibrometer of Fig.4-34. What happens to the output signal voltage? How can we recover the interferometric signal? What is its dynamic range?*
P4-25 *Calculate the axial and the transversal resolution of a time-domain OCT at 800 nm, assuming a Ti:sapphire laser with $\Delta\lambda$=180 nm as a source with n_m=3 modes, and an objective lens with NA=0.5. If the depth to be measured by the OCT is down to Z=200 μm, what is the required NA?*
P4-26 *How large are the axial and the transversal resolutions of the time-domain OCT change of we use an objective with NA=0.5, and a SLED with $\Delta\lambda$=20 nm, and n_m=30?*
P4-27 *An SD-OTC uses a Ti-sapphire laser emitting from λ_1=620 nm to λ_2=880 nm, and uses a CCD with 64 pixels. Calculate: the resolution limit due to the $\Delta\lambda$ swing, the Fourier transform resolution limit and the depth z_{max} of operation to avoid undersampling; finally, the necessary NA of the objective.*

CHAPTER 5

Self-Mixing Interferometry

To describe the self-mixing phenomena, we have to distinguish three levels of injection according to the fraction of power returned to the cavity, or feedback strength, and to the ratio of target distance s to cavity length L. The dependence is summarized by the *feedback parameter C*, introduced by Acket et al. [1], defined as:

$$C = [a\ s\ \sqrt{(1+\alpha_{en}^2)}]\ /\ n_l L \qquad (5.1)$$

where a is the field-amplitude feedback parameter (or the square root of returning power fraction), α_{en} is the linewidth enhancement factor [2], introduced by Henry [3] ($\alpha_{en} \approx 1$ in He-Ne and $\approx 2...6$ in diode lasers), L is the cavity length, and n_l is the refractive index of the active medium. Then, we have the following cases [4-6]:

- *Weak feedback*, for $C \ll 1$. If the output mirror has a high reflectivity ($R \approx 0.99$ and cavity is $n_l L = 10\text{-}30$ cm, as in a He-Ne laser) and distance is $s < 10$ m, we are probably in this case. The simple picture of rotating vectors is applicable, the laser has nearly the same properties as in the unperturbed state, and the interferometric waveforms are sinusoidal.
- *Moderate feedback*, for $C \approx 1$. This is easily found in semiconductor laser diodes ($R \approx 0.05...0.3$, $n_l L \approx 300$ µm). The interferometric signal becomes a distorted-sinusoid wave-

form, and switching between levels in each $2ks=2\pi$ period may occur. Both spectral line and coherence exhibit significant deviations from the unperturbed state [5-7].

- *Strong feedback,* for $C>>1$. The interferometric waveform exhibits multiple switching in each $2ks=2\pi$ period, coherence length and line width are strongly affected and the laser enters in the chaos regime [7-8].

5.1 INJECTION AT WEAK-FEEDBACK LEVEL

In weak feedback conditions, we can apply the description based on the well-known Lamb's equations [6-8] used in the original derivation [9] to analyze injection in He-Ne lasers. In the standard treatment, we start writing the laser cavity field **E** and the returned signal field $\mathbf{E}_s$ as rotating vectors, i.e.:

$$\mathbf{E} = E \exp i\varphi \quad \text{and} \quad \mathbf{E}_s = a\, E \exp i(2ks+\varphi),$$

where $\varphi = \Omega t+\psi=2\pi\nu t+\psi$.

Amplitude E and phase φ of the oscillating field are assumed to be slowly varying quantities, and a balance is written of their time derivatives dE/dt and dφ/dt. Taking into account the external field re-entering the cavity after propagation to a remote target at distance s with *field* attenuation a, the Lamb's equations are written as:

$$dE/dt = [(\alpha-\beta E^2)c - \Gamma]\, E + (c/2L)\, aE \cos(\varphi+2ks)$$

$$d\varphi/dt = \Omega_c + \zeta(\alpha-\beta E^2)c + (c/2L)\, a \sin(\varphi+2ks) \qquad (5.2)$$

where:

$a = At_1^2\eta$ is the total *field* loss suffered by the reinjected field because of: (i) attenuation A in the go-and-return propagation to the target, (ii) double passage through the output mirror with a field transmission t_1, and (iii) mismatch η of the mode (cavity vs. returning) spatial distributions;

$t_1 = \sqrt{(1-r_1^2)}$ is the *field* transmission of the input mirror of field reflectance r_1 (explicitly, the *power* transmittance and reflection are $T= t_1^2$ and $R=1-T= r_1^2$);

$\alpha = \lambda^2(n_2-n_1)/8\pi\tau_{21}\Delta\nu_{at}$ is the active medium gain rate per unit length (of the field)

n_2-n_1 is the population inversion concentration (cm^{-3});

$\Delta\nu_{at}$ is the (atomic) gain line width and τ_{21} is the active level lifetime;

β is the gain saturation coefficient;

$\Gamma= \Omega_c/2Q$ is the cavity *field* loss-rate (per unit time), including scattering in the medium and mirror transmission loss, explicitly $\Gamma=t_1t_2c/2L+\Gamma_{sc}c$;

Ω_c is the cavity resonant frequency;

L is the cavity length, and c/2L is the longitudinal mode spacing;

$\zeta = (\nu_0-\nu_{at})/\Delta\nu_{at}$ is the frequency detuning with respect to the gain line center ν_{at}.

Letting (d/dt)E=0 in Eqs.5.2, we get the steady-state solution. For the solitary laser (a=0), the quiescent values E_0, Ω_0 are readily found as:

$$E=E_0=\sqrt{[(\alpha-\Gamma/c)/\beta]} \quad \text{and} \quad \Omega=\Omega_0=\Omega_c+\zeta\Gamma/c\,.$$

For $a\neq0$, we look for a small-perturbation solution of Eq.5.2, by letting $E=E_0+\Delta E$ and considering $\Delta E<<E_0$ so that only first-order terms in ΔE are retained. By neglecting the pulling term $\zeta<<1$ and dropping the unessential constant phase φ_0, after some algebra, we obtain the solution as:

$$E = E_0\,[1+ (a/2L\gamma_0) \cos 2ks] = E_0\,[1+m_A \cos 2ks] \quad (5.3a)$$

$$d\varphi/dt = \Omega_0 + a(c/2L) \sin 2ks = \Omega_0 + \Delta \quad (5.3b)$$

where $\gamma_0=\alpha-\Gamma/c$ is the net gain per unit length available in the medium when oscillation is not yet started (it equals βE^2 because, in the permanent regime of oscillation, $\alpha+\beta E^2-\Gamma/c=0$).

Thus, at the first order of perturbation, the cavity field has an AM with modulation term $m_A \cos 2ks$ proportional to the cosine of the external pathlength, and an FM with frequency deviation $\Delta/2\pi=a(c/4\pi L)\sin 2ks$ proportional to the sine of the external pathlength.

Numerical example. The proportionality coefficient of AM is primarily the total attenuation $a=A\eta t_1^2$, while the term $2L\gamma_0$ at the denominator has an order of magnitude not far away from unity in practical cases. Working on a diffusive target with a Gaussian beam with waist w_0, we find $A=\lambda/\pi w_0$ in the near field and $A=w_0/s$ in the far field. For a retroreflector, it is $A=1$.

The mode superposition efficiency is $\eta=1$ for the diffuser and $\eta^2=2/[(w/w_0)^2+(w_0/w)^2]$ for the retroreflector, where $w^2=w_0^2+(\lambda 2s/\pi w_0)^2$ is the spot size at distance 2s. In the far field, the product $A\eta$ is in both cases $A\eta\propto 1/s$, for which we get an inverse-distance dependence of the signal level, which is typical of a coherent detection scheme and well confirmed by experiments. Last, the output mirror (power) transmission is usually $t_1^2\approx0.01$ in a He-Ne laser. About the AM modulation index, $\alpha-\Gamma/c$ can be estimated from the amount of extra gain allowed for the startup of oscillation. In a He-Ne laser, usually $\alpha\approx0.0005$ cm^{-1} and with $L=20$ cm, we have $2L\alpha\approx0.02$; with $t_1^2\approx0.01$, it would be $\Gamma/c=t_1^2/2L\approx0.00025$ cm^{-1} and $\gamma_0=0.00025$ cm^{-1}.

To conclude our example, let us take a 20-cm He-Ne laser with a 0.2-mm waist w_0 and a 1-mrad divergence. Working at s=500 mm distance on a diffuser target (far-field condition), we have: $m_A=A\eta t_1^2/2L\gamma_0\approx$ (0.2/500)0.01 /0.00025 · 40=4 · 10^{-4}, while, on a retroreflector, it would be $m_A\approx$0.2·0.01/0.01 =0.2. Correspondingly, the frequency deviation $\Delta/2\pi$ in the two cases is 750 MHz (0.2/500)0.01/6.28=500 Hz and 750 MHz· 0.2· 0.01/6.28 =250 kHz, respectively.

Eqs.5.3a and 5.3b are an adequate description of the weak-injection regime ($a<<1$) in media with a population inversion n_2-n_1 only dependent on the pumping rate ($\alpha,\beta\approx$ constant), a condition identifying a Class A laser, according to Arecchi's categorization [10]. These equations are well suited to He-Ne lasers and other media with small gain per unit length, for which they provide a good fit of theory to experimental data.

5.1.1 Bandwidth and noise of the SMI

The frequency response of the interferometric signal (E, $d\varphi/dt$) to a variation of the distance s is composed of two contributions: (i) the finite time taken by the laser medium to reach a new equilibrium condition established by the external distance, and (ii) the delay associated with the propagation time of the field reflected by the target.

About the laser medium response time, we may note that the cold laser cavity has a decay time $\tau_d=1/\Gamma$; in addition, the laser oscillator has a loop-gain $G=1/2L\gamma_0$. Thus, the resulting time constant of response is then $\tau=\tau_d/G=2L\gamma_0/\Gamma$, and bandwidth is $B= 1/2\pi\tau=\Gamma/4\pi L\gamma_0$.
With the usual values of parameters, B is in the range of some hundreds of MHz in He-Ne lasers and up to gigahertz (GHz) in semiconductor lasers.

To evaluate the noise, let us recall that the signal returned to the cavity is aE. If no amplification mechanism were present, the detected power would be $a^2E^2=a^2P$. Because the detection is coherent, the quantum noise limit is attained.

Thus, we may conclude that the minimum detectable displacement (or NED) of the injection interferometer is given by the diagram of Fig.4-21, in which we shall use a^2P as the value of power, P being the power output from the unperturbed laser.
This result can be used as a default estimate of the injection interferometer noise. To improve the evaluation, however, we may account for the amplification undergone by the injected field.

From Eqs.5.3, we can see that the laser supplies an AM signal $1/2L\gamma_0$ larger than the cosine component. In power, we have thus a gain $G=1+(1/2L\gamma_0)^2$.
Using a result from optical amplification theory (see Ref. [11], Sect.11.1), we have that the noise figure is given by $NF=1+2n_{sp}(G-1)/G$, where $n_{sp}=n_2/(n_2-n_1)$ is the inversion factor (≈ 1 for $n_2>>n_1$). Therefore, we may conclude that, at most, noise is increased by a modest 3 dB (or a factor 2) with respect to the a^2P level when $G>>1$ and $n_{sp}\approx 1$.

5.1.2 The He-Ne SMI

In the practical implementation, the first experiment was carried out with a frequency-stabilized He-Ne laser (Appendix A1.2) built around a Zeeman-split two-frequency source. Splitting of the two modes was obtained by applying a magnetic field to the laser capillary, and the control of the cavity length simply made by thermal expansion through a resistive wire wound on the He-Ne capillary tube [12].

The two frequency split modes oscillate with linear, orthogonal polarized states. With a polarizer P placed in front of the main output mirror (Fig.5-1), one mode (say f_1) is selected for propagation to the remote target. The other mode (at f_2) is prevented from exiting the laser and is kept in cavity to act as a reference local oscillator for detection.
The propagated mode undergoes the AM and FM injection modulations and carries the optical path length 2ks information.
When both modes are superposed on the photodetector placed on the rear mirror, they will beat and down-convert the AM and FM modulations to an electrical carrier at f_1-f_2.
The experimental waveforms of the photodetector current are shown in Fig.5-2.
These are obtained with a 20-cm long tube, 0.5-mW commercial He-Ne laser aimed at a diffuser-surface target (a loudspeaker driven by a triangular waveform at audio frequency, with a piece of white paper glued on the cone) placed at a s=40-cm distance.

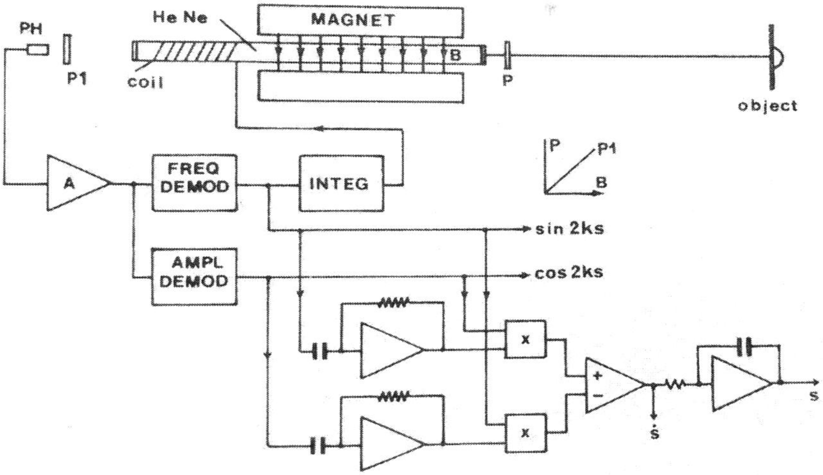

Fig. 5-1 Self-mixing interferometer with a He-Ne laser: two modes at slightly different frequencies (f_1-f_2≈100kHz) are generated by Zeeman-splitting due to a transversal magnetic field applied to the capillary tube. The two modes have linear orthogonal polarizations. One mode is passed through the polarizer P and the other is kept in the cavity. Polarizer P1 in front of the photodetector PH is oriented at 45° to allow beating on the photodetector of the propagated mode with the unperturbed mode kept in cavity. After FM and AM demodulation, signals sin 2ks and cos 2ks are obtained. The frequency signal is also used for frequency stabilization through the thermal actuator. Through differentiation and cross-multiplying of sin 2ks and cos 2ks, the v and s are recovered (from Ref.[12], ©AIS, reprinted by permission).

After AM and FM demodulation of the photodiode electrical signal, the waveforms of the interferometric signals S=sin 2ks and C=cos 2ks are obtained, as shown in Fig.5-3. Starting from these signals, several schemes can be devised to recover the displacement s.
For example, we can acquire S and C through an A/D interface and use a small dedicated computer to calculate s. For small displacements $\Delta s<\lambda/2$, Δs is simply computed as $\Delta s = (1/2k)$ atan S/C. For large displacements, because atan is a multivalued function, we shall compute s looking at the evolution of Δs.

With digital processing, we may add to (1/2k) atanS/C an increment $\pm\lambda/4$ for each zero-crossing of C, with the sign (+) for U=1 and (–) for U=0, where U is given by Eq.4.6.
Alternatively, we may prefer to use a straightforward analogue processing performed by a few electronic analogue circuit blocks.
Taking the time derivatives of S and C, we get:

$$dS/dt = 2kv \cos 2ks, \qquad dC/dt = -2kv \sin 2ks \tag{5.4}$$

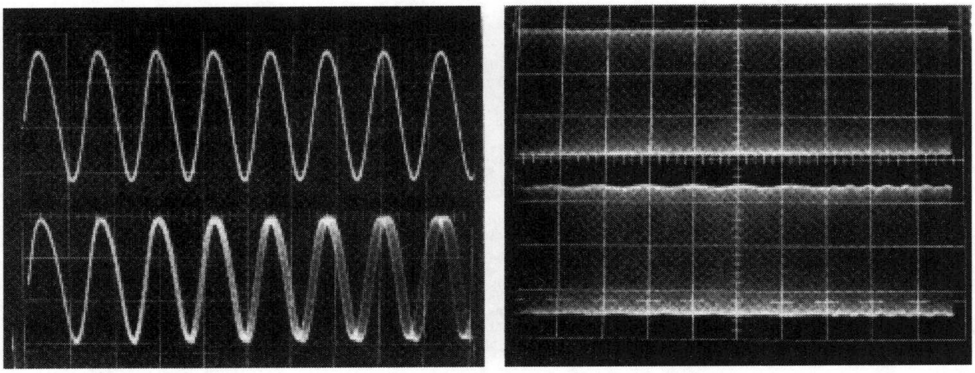

Fig. 5-2 Waveforms of the signal out from the photodiode when the target is a diffuser (white paper on a loudspeaker driven at audio frequency). Bottom traces are for injection while top traces are for the beam blocked off. Left: on a 20 μs/div scale, the waveform reveals the FM signal in the form of a time jitter; right: on a 1 ms/div scale, the AM signal shows up in the form of a ripple (from Ref.[12], ©AIS, reprinted by permission).

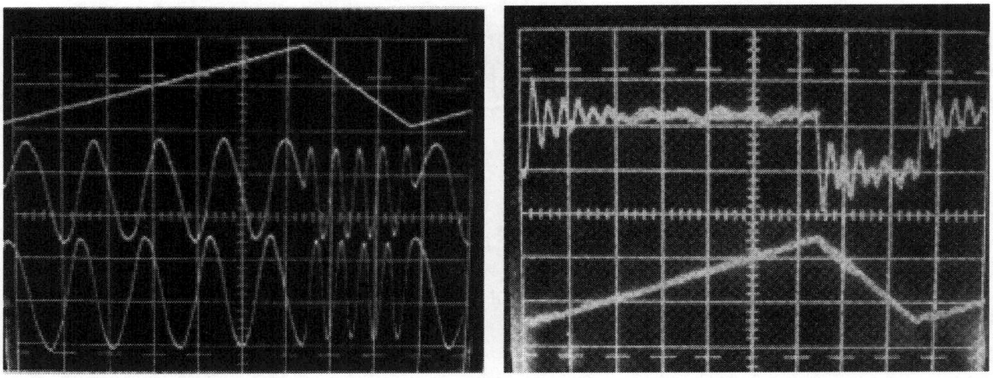

Fig. 5-3 (left): waveforms of target driving signal and of the interferometric signals (S=sin 2ks and C=cos 2ks) demodulated from the photodiode output; right: the reconstructed velocity signal v=ds/dt and its integral, the displacement s. Time scale: 1ms/div (from Ref.[12], ©AIS, reprinted by permission).

where v=ds/dt is the velocity. By cross-multiplying the two derivatives to C and S and subtracting, we have:

$$C\, dS/dt - S\, dC/dt = 2kv \cos^2 2ks + 2kv \sin^2 2ks = 2kv \tag{5.5}$$

and, with a time integration, we obtain the displacement $s = \int v\, dt$, see Fig.5-3 bottom.

The circuit performing this analog processing is shown in Fig.5-1. With this circuit, the reconstructed velocity and displacement for a triangular waveform excitation of the target are those reported in Fig.5-3.

Note that the s(t) and v(t) waveforms deviate somehow from those expected from the drive waveform. However, this is not due to the circuits, it is the loudspeaker real response and is actually a measurement of the response itself.

As the target is a piece of white paper, the back-scattered field received by the laser obeys the speckle-pattern statistics. The dynamic range is then limited by the longitudinal speckle dimension $S_l = \lambda s^2/w^2$, where w is the spot size on the target and s is the target-laser distance (see Chapter 6). A peculiar feature of the injection interferometer is the ability to self-filter the returned field spatially.

The transversal speckle size is $S_t = \lambda s/w$, and the far-field spot size is $w=(\lambda/w_0)s$. By substitution, we get $S_t=w_0$, that is, the returning speckle (or coherence) size S_t exactly matches the laser mode size w_0, and no waste of power occurs.

Besides displacements, another field of application is the pickup of biological motility signals, like respiratory sounds and blood pulsation [13]. In Fig.5-4, we can see the experimental trace of the blood pulsation detected on a finger, revealing the cardiac pulsation, detected remotely without contact with the tissues under measurement.

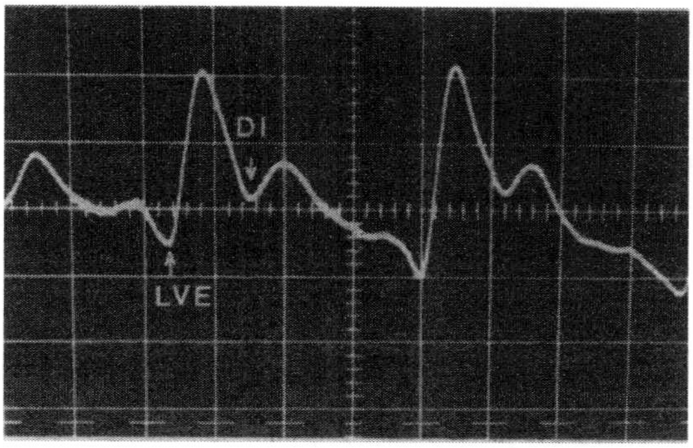

Fig. 5-4 Pickup of biological motility with the He-Ne laser injection interferometer. The blood pulsation waveform is measured on an (immobilized) finger and reveals details of clinical importance like the left ventricular ejection (LVE) as well as the dicrotic incisure (DI). The vertical scale is 0.1 μm/div, and the horizontal scale is 0.2 s/div.

Of course, the injection interferometer of Fig.5-1 can operate on a corner cube target as well [14]. In this case, the returned signal is much larger than with the diffusing target, and an optical attenuator is inserted on the beam to avoid high-level injection effects. When counts are drawn out of signals C and S, the performance is much the same as the two-beam interferometer (Sect.4.2.1).

As a final comment, the He-Ne laser is well suited for weak-level injection interferometry, however, but its layout is rather bulky and lifetime is questionable. So, we may think of a semiconductor laser. In diode lasers, unfortunately, it is difficult to obtain a Zeeman dual-frequency oscillation, and just the AM signal is available as a modulation of emitted power. We may recover the FM at the expense of adding additional optical components (Sect. 5.3.4), but this approach is not the preferred one.

Assuming that only the AM channel is available, we have the problem of *ambiguity* of the cos 2ks = cos Φ function, because we cannot distinguish $\Phi = 2k\Delta s$ from $\Phi = 2k(-\Delta s)$, [and also $\pi+2k\Delta s$ from $\pi+2k(-\Delta s)$] that is, we cannot tell the sign of incremental displacement Δs when phase Φ of the cosine function is passing through 0 or π. Luckily enough, at the medium feedback level (C >1) a switching occurs in the waveform and, as shown in the following, we will be able recovering the sign of Δs and remove the ambiguity and apply the interferometer to displacement measurements with arbitrary waveforms. On the other hand, for small-vibration measurements of amplitude Δs less than or comparable to $\lambda/2$, the single interferometric AM signal cos 2ks is sufficient (as shown in Sect.5.4.2).

5.2 Analysis of Injection at Medium-Feedback Level

Experimentally, we have the setup of Fig. 5-5, in which a semiconductor laser is aimed through a collimating objective lens to a mirror target. A variable attenuator is used for adjusting the level of power returned into the laser cavity.

Detection can be performed by a photodiode placed everywhere on the beam (because the AM signal is carried by the beam itself), a distinctive feature compared to conventional interferometers. Most conveniently, however, we may use the monitor PD usually incorporated in the diode-laser package at the rear mirror of the laser.

Thus, a signal $I=\sigma P$ is obtained, which is proportional to the power P emitted by the laser and that carries the AM modulation waveform, that is, cos2ks at a low-injection level.

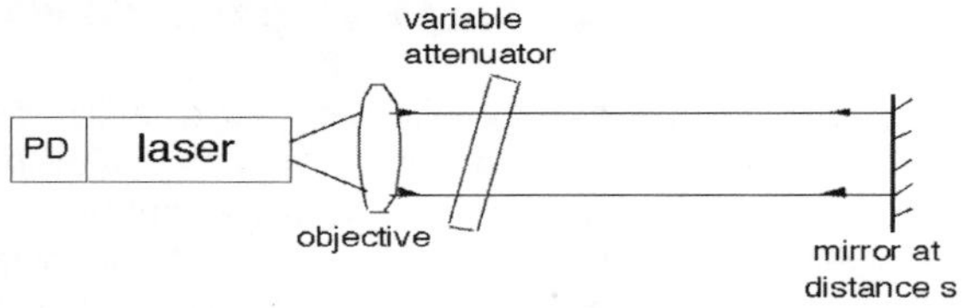

Fig. 5-5 Basic arrangement for experiments with an injection interferometer: using a variable optical attenuator, we adjust the level of injection and look at the AM waveform using the rear-mirror photodiode PD.

We enter the moderate or medium level of feedback when, at increasing back-reflected power, the interferometric signal cos 2ks becomes appreciably distorted (Fig.5-6). First, the waveform exhibits a leading semiperiod slower than the trailing semiperiod.

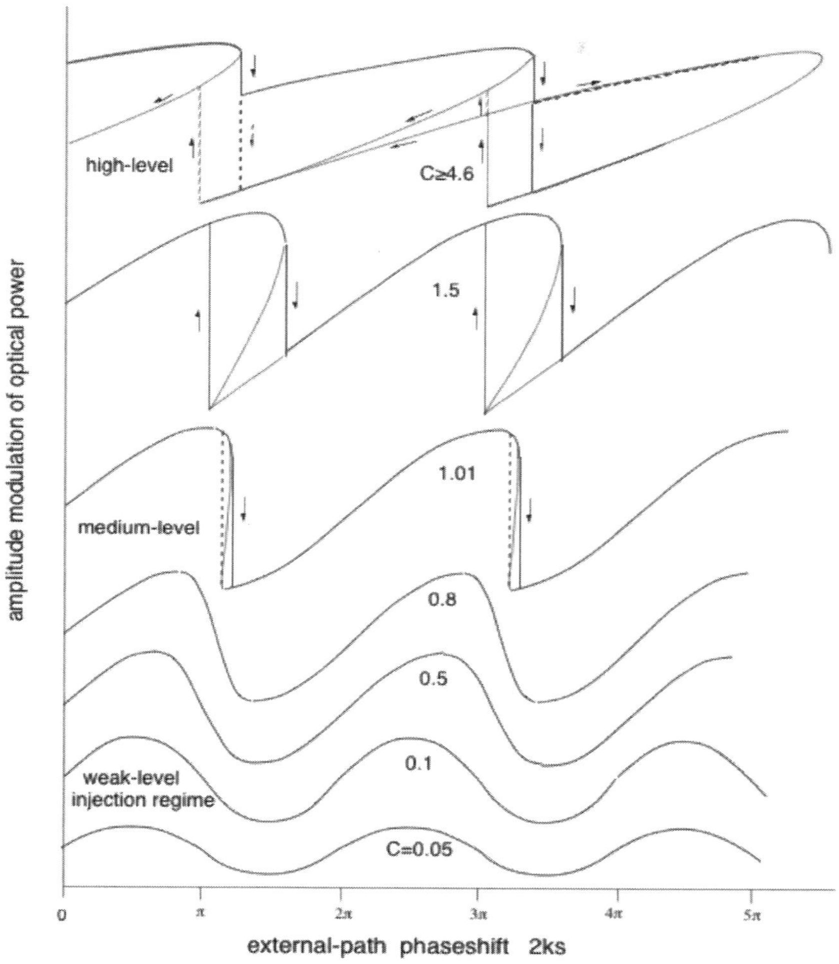

Fig. 5-6 Waveforms of optical power in an injection interferometer. At weak levels of injection (C<<1), the dependence cos2ks from the external phase shift of a normal interferometer is found. Increasing the injection, the waveform becomes progressively distorted and asymmetric. As C reaches unity, a switching occurs with period 2π. If the target reverses its motion, the phase shift decreases, and the waveform becomes time-inverted, except at the switching point where now it follows the up arrow. Beyond $C \geq 4.6$, more than one switching per period occurs, and operation of the interferometer becomes affected by errors. At still larger values of C (e.g., 10...20), the laser enters regimes of multiperiodicity and chaos oscillations.

Then, when injection is increased further and C (Eq.5.1) reaches unity, the waveform becomes sawtooth-like, with a sharp switching in amplitude. We find one switching per period of the optical path length 2ks, that is, one for each $\Delta s=\lambda/2$ increment of displacement and, interesting to note, the switching is negative-going when 2ks is increasing (Fig.5-6); when motion of the target is reversed, the diagram still applies, but with 2ks runs from right to left. The waveform has now a leading semiperiod faster than the trailing semiperiod and, most important, the switching is now positive-going (and occurs at a slightly changed value of 2ks because of hysteresis). Interesting to note, the period is independent of the waveform with periodicity 2π whatever the C.

Then, a strategy to make a displacement measurement without ambiguity easily follows. By operating the injection interferometer at $1<C<4.6$, we count each switching corresponding to $\lambda/2$. Counts will be taken with the positive sign if the transition is up-going and with the negative sign if it is down-going.

Operation is limited to the range $1<C<4.6$ to ensure that there is just one switching per period (Fig.5-6) [15]. The experiment is developed in more detail in the next section. In the basic setup of Fig.5-5, the FM signal cannot be detected. However, it is found theoretically that the sin2ks signal, similar to the cos2ks, is also asymmetrical and has one switching per period at $1<C<4.6$ [15].

5.2.1 Analysis by the Three-Mirror Model

Let us consider the injection interferometer as a three-mirror cavity laser, as shown in Fig.5-7. Analyzing the round-trip loop, we can write the field returning to the initial point as the sum of two contributions, from the paths internal and external to the laser cavity, as:

$$E\, r_1\, r_2 \exp 2\alpha^* L \exp i2kL + E\, a \exp i2ks \tag{5.6}$$

where r_1 and r_2 are the mirrors' field reflectivity, $\alpha^*=\alpha-\beta E^2-\Gamma/c$ is the net gain (or, gain less loss) per unit length, L is the laser cavity length, and $a=A(1-r_1^{\,2})\eta$ is total field loss in the external propagation to the third mirror. Thus, the loop gain is:

$$G_{loop} = r_1\, r_2 \exp 2\alpha^* L \exp i2kL + a \exp i2ks = 1 \tag{5.7}$$

Applying the well-known Barkhausen criterion of oscillators [i.e., G_{loop} shall be exactly unity in the permanent regime of oscillation, and phase ϕ_{loop} exactly zero] we get:

$$G_{loop} = |G_{loop}| \exp i\phi_{loop} = 1, \quad \text{or also: } |G_{loop}|=1,\ \phi_{loop}=0 \tag{5.8}$$

When a=0, Eq.5.8 requires $|G_{loop}| = r_1 r_2 \exp 2\alpha^* L=1$ (net gain equal to mirror losses) and $\phi_{loop}=2kL=0_{[mod 2\pi]}$ (cavity resonance condition), or that 2ks is an integer multiple of 2π.

Writing the condition as $2(2\pi n_l \nu_0/c)L=2\pi N$, the unperturbed frequency is found as $\nu_0 = Nc/2n_lL$, where N is the order of the mode and n_l the index of refraction of the active medium. Near to resonance, if the actual frequency ν deviates from ν_0, the part of 2kL in excess of 2π can be expressed as $2kL = 4\pi n_l L(\nu-\nu_0)/c$ [as $4\pi n_l(\nu_0/c)L=2\pi N$].

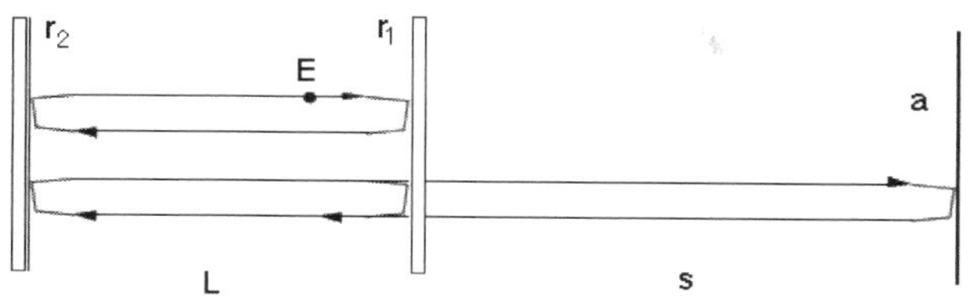

Fig. 5-7 Schematization of the injection interferometer as a three-mirror laser.

In presence of feedback ($a \neq 0$), the phase condition is ϕ_{loop}= atan $[ImG_{loop}/ReG_{loop}]$ =0 or, more simply, $Im\{G_{loop}\}=0$, and explicitly reads:

$$r_1\, r_2 \exp 2\alpha^* L \sin [4\pi L n_l(\nu-\nu_0)/c] + a \sin 2ks = 0 \qquad (5.9)$$

Using $r_1 r_2 \exp 2\alpha^* L=1$ in Eq.5.9, assuming in the first term that $\nu-\nu_0$ is small (so that $\sin x \approx x$) and noting that $2ks=4\pi s\nu/c \approx 4\pi s\nu_0/c$, we obtain:

$$\nu-\nu_0 + (c/4\pi L n_l)\, a \sin 4\pi s\nu_0/c = 0 \qquad (5.10)$$

This equation tells us that, in presence of injection, the actual frequency ν deviates from the frequency ν_0 of the solitary laser by an amount $(c/4\pi L n_l)$ a sin2ks, which is easily recognized to coincide with the quantity derived by Lamb's equations (Eq.5.3b) for small a.

Now, if we plot ν versus ν_0 as given by Eq.5.10, we get a sinusoid curve superposed to a straight line (Fig.5-8). When the distance s is changed by $\lambda/2$, the argument of the sine changes by 2π, and ν covers the full swing of the sine curve. We can draw a horizontal line (the dotted line in Fig.5-8) to represent the working point determined by Δs, which moves vertically along a full period of the sinusoid as Δs varies of $\lambda/2$.

As long as the amplitude of the sinusoid is small, only one intersection is found, and we are in the case of weak-feedback injection. However, at increased injection, we may have a curve with three intersections of a horizontal line.

The middle intersection is found to be unstable [7] and thus the frequency jumps from the left to the right intersection. In this condition, frequency ν doesn't depend anymore merely on the cavity length L, but also on target distance s, and it switches according to the excess of s from multiples of λ. Accordingly, we say that the laser is oscillating on the *external cavity modes* (or ECM) [7]. The boundary from the weak- to the moderate-feedback regime is that of the curve with a horizontal flex. Mathematically, the condition of a horizontal flex in the y-x diagram, for a function $y=x+A \sin Bx$, is $AB=1$. Comparing with Eq.5.10, we find the condition $(c/4\pi L n_l)\, a 4\pi s/c=1$, or also: $as/Ln_l=1$.

We may now define a factor describing the strength of injection as $C= as/Ln_l$ and state that, for $C>1$, we enter in the regime of frequency switching. This factor coincides with that already quoted in Eq.5.1, except for the term $\sqrt{(1+\alpha_{en}^2)}$ specific to the semiconductor lasers.

About the AM amplitude signal, applying the amplitude Barkhausen criterion (loop gain shall be unity at permanent oscillations) we write $|G_{loop}|=1$, or $Re^2(G_{loop})+Im^2(G_{loop})=1$.

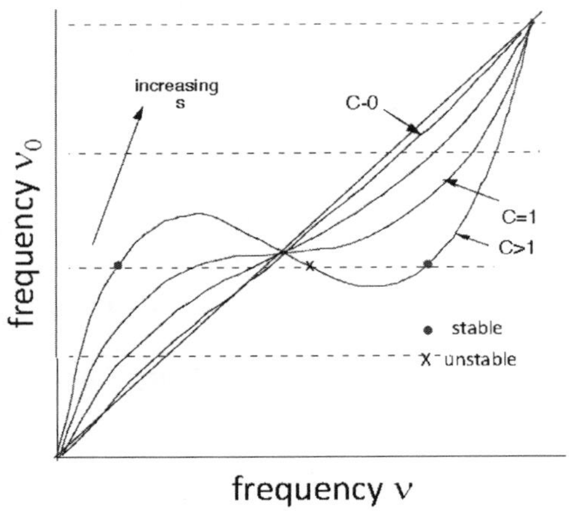

Fig. 5-8 Frequency of the injection-perturbed oscillation ν as a function of the solitary laser (unperturbed) frequency ν_0. At increasing feedback level (C values), the curve has three intersects (two stable and one unstable) and exhibits switching.

Letting $E=E_0+\Delta E$ in Eq.5.6, developing the modulus $|G_{loop}|$, and considering $\Delta E<<E_0$ in it, after some lengthy algebra we find:

$$\Delta E = E_0\,(a \cos 4\pi s\nu/c)/[2L\gamma_0(1+a\cos 4\pi s\nu/c)] \tag{5.11}$$

This expression is coincident with the weak-feedback result (Eq.5.3a) for a<<1. Moreover, when the factor a increases, the waveform becomes distorted because of the cosine dependence at the denominator. At C=1, when frequency switches, ΔE and the AM signal will too, because of the ν-dependence of the cosine term at the numerator. By numerically solving Eq.5.10 (or using the diagram of Fig.5-8), we can find a function $\nu= F(4\pi s\nu_0/c)$ that replaces ν in $\sin 4\pi s\nu/c$ at the moderate feedback level, and using $\nu=F(s)$ in Eq.5.11, the waveforms of Fig.5-6 are obtained.

As a conclusion of the previous analysis, by applying Barkhausen's criteria (loop gain equal to one) to the three-mirror cavity model, we have derived the basic results describing the moderate-level feedback regime [6,7,15-17]. In particular, switching of frequency and amplitude signals are found to occur, starting from the critical value C =1.

5.2.2 Analysis by the Lang and Kobayashi Equations

A rigorous treatment of moderate-level feedback, including the effects specific to a semiconductor laser, is provided by the Lang and Kobayashi equations [4]. These equations rephrase Lamb's equations for amplitude E and phase φ in a more detailed way, and also add a third equation describing the active-level population and its dependence on pumping rate and cavity field. In presence of feedback, the Lang and Kobayashi equations read:

$$dE(t)/dt = [g_N(N-N_{tr}) - 1/\tau_p]E(t) + (ac/2n_lL)\, E(t-\tau) \cos[\omega_0\tau+\varphi(t)-\varphi(t-\tau)] \quad (5.12a)$$

$$d\varphi/dt = \alpha_{en} g_N(N-N_T) - (ac/2n_lL)\, E(t-\tau)/E(t) \sin[\omega_0\tau+\varphi(t)-\varphi(t-\tau)] \quad (5.12b)$$

$$(d/dt)N = J/ed - N/\tau_s - g_N (N-N_{tr}) E^2(t) \quad (5.12c)$$

where: g_N is the gain coefficient [typically $8\cdot10^{-7}cm^3s^{-1}$]; in terms of the Lamb's coefficient α, (Eq.5.2) it is $g_N = \alpha c/(n_2-n_1) = \lambda^2 c/8\pi\tau_{21}\Delta\nu_{at}$.
$N=N(t)$ is the carrier (electron-hole pairs) density (cm^{-3}) in the active region.
N_{tr} and N_T are the carrier density (cm^{-3}) at transparency and at threshold of the solitary laser [typically a few $10^{18}cm^{-3}$] and $g_N(N_T-N_{tr}) = 1/\tau_p$ for $N= N_T$ at threshold
τ_p is the photon lifetime [typically a few ps] related to absorption and mirror loss.
$a = At_l^2\eta$ is the total *field* loss suffered by the reinjected field in the roundtrip path.
L is the laser cavity length, and n_l is the active-medium index of refraction.
τ =2s/c is the external roundtrip time.
ω_0 is the (angular) frequency of the unperturbed laser.
α_{en} is the linewidth enhancement factor [3] [typically 3 to 6 in diode lasers].
J is the pump current density (A/cm^{-2}), d =active layer thickness.
τ_s is the charge-carrier lifetime [typically a few ns].

The analysis of the moderate-level injection [4-6] starts with letting dE(t)/dt=0 and (d/dt)N =0 in Eqs. 5.12a and 5.12c, and solving for the stationary values E_{0f} and N_{0f} in presence of feedback. From Eq. 5.12a we get:

$$N_{0f} = N_T - (ac/2n_lLg_N) \cos \omega_{0f}\tau \quad (5.13)$$

The term on the right-hand side shows that feedback induces a modulation of the carrier density N. As a consequence, a variation of voltage developed across the laser-diode junction is induced [6,17]. This is a signal that can indeed be recovered and used as the interferometer output in place of the normal photodiode output, as usually done when self-mixing is performed with quantum cascade lasers at THz frequencies [18].
Now, we write Eq.5.12b two times, one in perturbed conditions by inserting Eq.5.13 and letting $d\varphi/dt=\omega_{0f}$, $N=N_{0f}$ and $a\neq0$, and one in unperturbed conditions letting $d\varphi/dt=\omega_0$ and a=0. Subtracting term by term the equations, the perturbed oscillation frequency ω_{0f} is:

$$\omega_{0f} = \omega_0 - (ac/2n_lL)\, [\alpha_{en} \cos \omega_{0f}\tau + \sin \omega_{0f}\tau] \quad (5.14)$$

This equation can be brought to coincide with the Barkhausen phase-condition. Now, we can use the feedback parameter C defined by Eq.5,1 and repeated here for convenience:

$$C = a\, s\, (\sqrt{1+\alpha_{en}^2}) / n_l\, L$$

By multiplying both members of Eq.5.14 by τ=2s/c, it becomes:

$$\omega_{0f}\tau = \omega_0\tau - C\,(1+\alpha_{en}^2)^{-1/2}\,[\alpha_{en}\cos\omega_{0f}\,\tau + \sin\omega_{0f}\,\tau\,] \quad (5.15)$$

or also, using identities $\alpha_{en}\,(1+\alpha_{en}^2)^{-1/2}$= sin atan α_{en} and $(1+\alpha_{en}^2)^{-1/2}$ =cos atan α_{en} we get:

$$\omega_{0f}\tau = \omega_0\tau - C\sin(\omega_{0f}\,\tau + \mathrm{atan}\,\alpha_{en}) \quad (5.16)$$

By deriving Eq.5.16 with respect to τ in both sides, we obtain

$$\omega_{0f} = \omega_0 - C\,\omega_{0f}\cos(\omega_{0f}\,\tau + \mathrm{atan}\,\alpha_{en}) \quad (5.16a)$$

and thereafter we can divide both sides by 2π, rearrange and get:

$$\nu_0 = \nu_{0f}\,[1 + C\,\nu_{0f}\cos(\omega_{0f}\,\tau + \mathrm{atan}\,\alpha_{en})] \quad (5.16b)$$

Eq.5.16 has just one solution for C<1, and three solutions for 1<C<4.6 (moderate-level injection) at C>1, the laser starts oscillating on external cavity modes.
Then, at C>4.6 there are multiple switchings in a single 2π-period of $\omega_{0f}\tau$ and the laser is said to operate in the strong-level injection regime.
At even larger C, we get 5, 7... n, and 2 C/π modes, asymptotically. Finally, at C=20...50 the laser enters a high-level dynamics regime, characterized by spontaneous oscillations (called period-1 and multi-periodic oscillations) and finally breaks into chaotic oscillations [7,8] as it is destabilized by feedback as a class B laser [10]. In measurement applications, these regimes shall be avoided by properly limiting the C factor to values of a few units, for example by means of a variable attenuator as indicated in Fig.5-5. The variable attenuator may be a slab with a gradually increasing absorbing (black) along its length, or more conveniently, a linear polarizer whose rotation of an angle θ introduces a (double pass) attenuation $\cos^2\theta$.
From Eq.5.12c, letting (d/dt)N=0, the electric field with feedback E_{0f}^2 is:

$$E_{0f}^2 = [J/ed - N/\tau_S]/g_N(N_{0f} - N_{tr}) \quad (5.17)$$

whereas, from Eqs.5.12a and 5.12c, with a=0, the quiescent point is $E_0^2 = \tau_p g_N(N_0 - N_{tr}) = \tau_p(J/ed - N_0/\tau_S)$, and $P_0 \propto E_0^2$. With this result and N_{0f} as given by Eq. 5.13, we obtain, after some algebra, the optical power variation $\Delta P \propto E_{0f}^2 - E_0^2$ induced by feedback as:

$$\Delta P = P_0\,(ac\tau_p/n_l L)\cos\omega_{0f}\tau\,/[1 + (ac\tau_p/n_l L)\cos\omega_{0f}\tau] \quad (5.18)$$

which becomes the familiar $\Delta P_0 m_A \cos 2ks$ expression for a<<1.

A last result worth mentioning is the reconstruction of the displacement waveform s(t) from the AM interferometric signal given by Eq.5.18. By combining Eqs.5.16 and 5.17, we can obtain the following expression relating $\omega_{0f}\tau$=2ks to the instantaneous $\Delta P/\Delta P_0$=F(t):

$$2ks(t) = \omega_{0f}\tau = \pm \arccos F(t) - C\{-\alpha_{en}F(t) \pm \sqrt{[1-F(t)^2]}\}/\sqrt{(1+\alpha_{en}^2)} + 2m\pi \quad (5.19)$$

where the sign shall be taken '+' for dF(t)/dt×(ds/dt) <0, and '-' for dF(t)/dt×(ds/dt) >0, and m shall be increased by 1 every two zero-crossings of F(t) [7,19].

5.3 THE LASER-DIODE SMI

Starting from the basic setup of Fig.5-5, a number of researchers [34-41] have reported on the development of laser-diode injection (or Self-Mixing, SMI) interferometers for measurements of displacements, velocity and for the detection of remote targets. The typical waveforms of the current detected under moderate feedback conditions with single-mode laser diodes emitting in the near-infrared (λ=800 nm) are shown in Fig.5-9. In Fig.5-10 we can see the corresponding theoretical waveforms, calculated by Eq. 5.18.

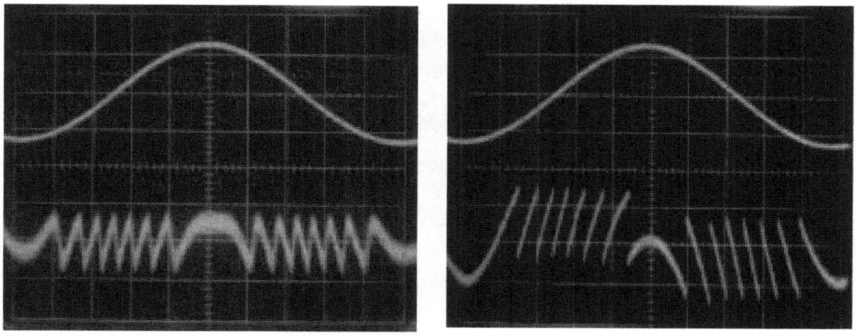

Fig. 5-9 Injection interferometry experimental signals, in response to sinusoidal drive s(t) signal (top trace in both pictures). Left: at C=0.6, the cos 2ks waveform is slightly distorted, with falling semi-periods faster than the rising ones; right: at C=2.2, a switching occurs in the fall semi-period (λ=850 nm, tme scale: 2 ms/div) (from Ref.[15], ©IEEE, reprinted by permission).

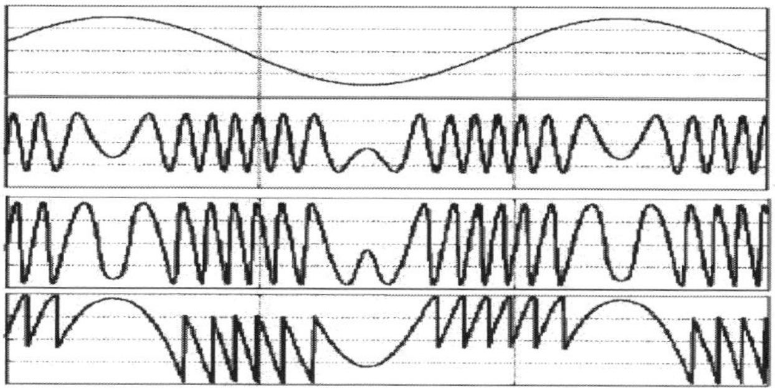

Fig. 5-10 Drive s(t) (top) and theoretical $\Delta P/P$ waveforms corresponding to the experimental data of Fig.5-9, for some values of the feedback parameter C, ranging from weak to moderate level of injection (C=0.01, 0.7, and 3.3 from top to bottom). The relative vertical scale of $\Delta P/P$ signal is 0.5, 3, and 10 (top to bottom). Horizontal scale: 1 ms/div.

The block scheme of a displacement interferometer working on $\lambda/2$-step counting is shown in Fig.5-11. The target is aimed through a collimating objective lens (typically a 5-mm focal length) or by a telescope (with typical $\times 20$ magnification). An optical attenuator is inserted in the target path to adjust the feedback parameter to be in the range $1<C<4.6$, the desired moderate-feedback regime of a single switching per period.

The signal out from the photodiode (in most practical schemes, the monitor photodiode of the laser chip is adequate for use) is passed to a transimpedance preamplifier and then to a differentiator with a short time constant $\tau_d = RC$. Out of the differentiator, the up/down switchings are a sequence of positive/negative short ($\approx\tau_d$) pulses. We can easily separate the pulses in two channels, rectify the negative ones, and send one channel to the up-count input of a counter, and the other to the down-count input. Last, multiplying by the scale factor $\lambda/2$ brings the readout of the display to decimal metric units.

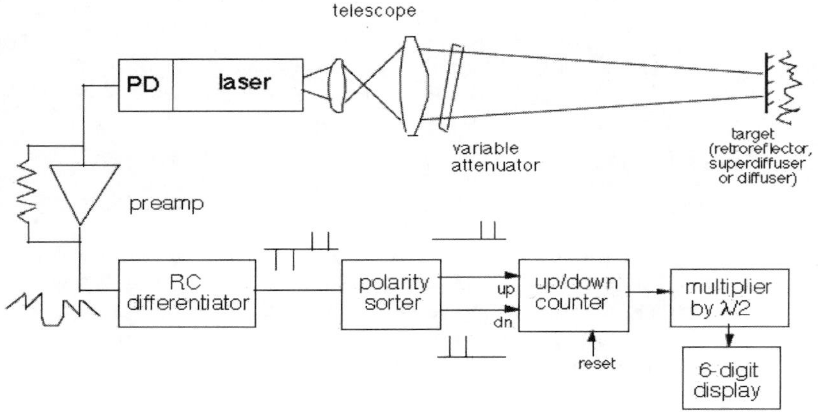

Fig. 5-11 Block scheme of the laser-diode injection interferometer working on the waveform switching to count displacement in steps of $\lambda/2$.

Thus, with a minimum part count, we are able to build an interferometer measuring target displacement in steps of $\lambda/2$, or 425 nm, per least-significant digit (at λ= 850 nm).

As in all interferometers, the maximum distance that can be covered in the measurement is limited by the coherence length $L_{coh} = c/2\pi\Delta\nu$ of the laser diode, which can be up to several meters in selected Fabry-Perot lasers.

The target for the measurement may be a corner cube, to allow for a sizeable attenuation, or a Scotchlite™ tape or varnish. This last provides a gain of $\approx$40 with respect to the normal Lambertian diffuser. Also, we may use a simple plain diffuser, that is, white paper or the like. This requires a careful design of the objective optics when using the plain diffuser because the signal from it is barely sufficient. With the corner cube or the superdiffuser, the signal is even too large, and we shall adjust it by the variable attenuator, eventually changing its setting with the working distance. It may then be advisable to add a servo-loop control of the attenuation to get a self-adjusting operation.

In all cases, we shall keep the signal in the single switching ($1<C<4.6$) regime, eventually acting on the attenuator as shown in Fig.5-11.

Note. A strategy to stabilize the signal amplitude in the $1<C<4.6$ range is the following. Superposed to the dc laser diode current, we add a small modulation signal i(t) at audio frequency, typically a sin wave. The emitted wavelength then varies as $\Delta\lambda(t)=\alpha_I\ \Delta i(t)$, where α_I is the current coefficient (see App.A1.3.1). We choose $2\Delta ks=4\pi s\Delta\lambda/\lambda^2 \geq 2\pi$ so that at least one switching occurs in a period of modulation, even for a target at rest. Then, we look at the differentiated signal and, if it is missing, we decrease the optical attenuation until the signal shows up and allow for a moderate (e.g., x2) overdrive. As a simple and cheap controlled-voltage attenuator, we can use a Liquid Crystal (LC) sandwiched between two polarizers. The birefringence of the LC is controlled by the applied voltage, resulting in an adjustable attenuation. The up and down switching of the modulation has a zero-mean value and does not interfere with the displacement measurement.

About precision, the laser-diode interferometer is different from the He-Ne laser because it does not possess an inherent calibration of wavelength. In a Fabry-Perot semiconductor laser, wavelength is markedly dependent on drive current and temperature, mode-hopping can occur, and the actual $\lambda=\lambda(I,T)$ curve is also dependent on the previous story of I,T changes.

In the most common and inexpensive Fabry-Perot cavity lasers, the typical temperature and current coefficients are $d\lambda/\lambda dT=2\text{-}5\ 10^{-6}(°C^{-1})$ and $d\lambda/\lambda dI=1\text{-}3\ 10^{-6}(mA^{-1})$ (see also App.A1.3.1). Thus, we may expect an error in the displacement measurement of several ppm for 1 °C or a 1 mA variation of chip temperature or diode drive current.

The situation is much better in DFB or DBR lasers, where the incorporation of a grating reflector alleviates the temperature dependence and nearly eliminates the current dependence. However, these lasers are not easily available for wavelengths other than the fiber third-window ($\lambda\approx1500$ nm) and are more expensive than Fabry-Perot types.

Altogether, the previous points call for a thermal stabilization of the laser chip (with a TEC or Peltier cell module) and of the drive current (by a regulated and low-ripple supply). With Fabry-Perot lasers, using a stabilization of 0.01 °C in temperature and ±10 μA in current, the instrumental accuracy of the injection interferometer has been evaluated to be ±1 count (≈0.4 μm) at 50 cm distance, or ±1 ppm [34]. In addition, DFB lasers with a long-term stability of $\Delta\lambda/\lambda=0.1$ ppm have been reported in the laboratory [20].

To extend the operation of the interferometer to sub-λ resolution, the photodetected signals can be acquired by a computer, and the calculation of 2ks is then performed with the aid of Eq.5.19. This scheme of analogue processing has been shown [19] to be able to reconstruct the displacement waveform s(t) with ±5-nm accuracy, with no constraint imposed on the signal s(t) waveform.

In conclusion, the advantages and disadvantages of the laser-diode injection interferometer, as compared to conventional types, can be summarized as follows:

- it has minimum optical part-count and relatively simple electronics;
- the optical head (laser, lens, attenuator) is very compact with a distinct advantage for ease of mounting in experiments short of space;
- no critical alignment on the target is required;
- no spatial, wavelength or stray-light filtering are necessary; the laser cavity itself provides these functions;

- operation on a normal diffusing target surface is readily achieved, without modifications of the setup;
- the interferometric signal can be detected everywhere on the beam, and also from the target side;
- $\lambda/2$ resolution with digital processing and sub-λ sensitivity with analogue are easily achieved up to potentially several hundred kilohertz or megahertz bandwidth;
- performances are comparable to those of conventional interferometers.

Yet, there are some drawbacks, at least in the basic configuration with Fabry-Perot laser diodes, like:

- wavelength accuracy is in the range of 100 to 1000 ppm compared to the 10^{-7} typical of He-Ne frequency stabilized lasers;
- temperature drift may be of concern;
- no reference is available for differential measurements (at least in the basic setup);
- flexibility of reconfiguration is modest.

5.3.1 Design of an SMI Displacement Instrument: A Case Study

Following the schematic of Fig.5-11, we can design and develop a displacement measuring instrument based on self-mixing interferometry.
Blocks like: photodiode preamplifier, amplifier and RC-differentiator, polarity sorter and up/down counter in Fig.5-11 are realized by commonplace electronics design and the resulting circuits can be fitted in a PCB of a few cm^2 area.

Typical performances [21] are: a self-mixing signal bandwidth of 80 kHz, upon using a Fabry-Perot diode laser ML-2701 (manufactured by Mitsubishi) emitting 6.15 mW at 853 nm and biased at 40 mA (2.2 times the threshold current), with a stability of $\Delta I_{bias}=\pm 0.01$ mA. The B=80 kHz bandwidth limits the maximum speed of the target to $B(\lambda/2)=34$ mm/s, a number that could be easily improved but is sufficient for the measurements to follow.
A TEC control of laser diode temperature was employed, limiting the temperature error to ±0.01 °C, compared to the ±5 °C of the unregulated unit: with a laser thermal dependence $\alpha_\lambda=d\lambda/dT=$ 2-5 10^{-6} °C^{-1}, this corresponds to a wavelength thermal stability of $\Delta\lambda/\lambda=$0.02-0.05 ppm/°C. The contribution of bias current drift was $\Delta I_{bias}\ d\lambda/dI_{bias}=\pm 0.01$ mA (1-3 ppm/mA) = ±0.01-0.03 pm, corresponding to a negligible relative error, typically <0.08 ppm, whereas the plain laser (without TEC and regulated bias control) would have a calibration error up to 3.5 ppm. Control is also necessary in Fabry-Perot diode lasers to avoid mode hopping (App. A1.3.1) that causes the wavelength jump of 0.2-0.4 nm causing loss of counts and a calibration error as large as a few parts per thousand. Mode hopping is missing in DBR diode laser (but not the temperature and bias dependence of λ).

The monitor photodiode of the DL was used to detect the SMI signal, and supplied about 1 mA of dc photodetected current to an AC-coupled trans-impedance amplifier with $R_f=100$ kΩ feedback resistance, yielding a noise equivalent current of 75 pA at B=50 kHz bandwidth of operation. Given a spectral sensitivity of the photodiode σ=0.5 A/W, the associated noise equivalent power is 150 pW and corresponds to a maximum distance covered (with SNR=1) of s= 1 m for the return from a diffusive target, and an SNR=20 from a corner cube target.

The accuracy of the instrument was tested by carrying out a sample of 65 measurements on the corner cube target, moved along a displacement of 66 cm by a translational stage (repositioning error was estimated to be ±1.5 μm).

A corner cube was employed to eliminate, from the measurement of accuracy, the speckle pattern error of a diffusive-surface target.

Yet, the diffuser is the preferred condition for a self-mixing interferometer because operation is then *non-invasive*. The amount of speckle-generated error, and the scheme for reducing it are treated later in Sect. 6.2.

A first batch of measurements on an s=66-cm displacement revealed a strong drift with temperature (Fig.5-12 left), but was soon recognized as simply due to the aluminum antivibration table on which the experiment was sitting. From the slope of the s=s(T) curve (open dots) a thermal expansion coefficient α_T=22 10^{-6} °C^{-1} was calculated. [By the way, this measurement tells us that we can measure thermal expansion coefficients by interferometry].

Thereafter, the displacement was corrected against this error, obtaining the points indicated with full circles in Fig.5-12, left. Also, at constant temperature of the laboratory room (±0.5 °C) the spread of successive measurement was 2.2 ppm (rms), as shown in Fig.5-12 right.

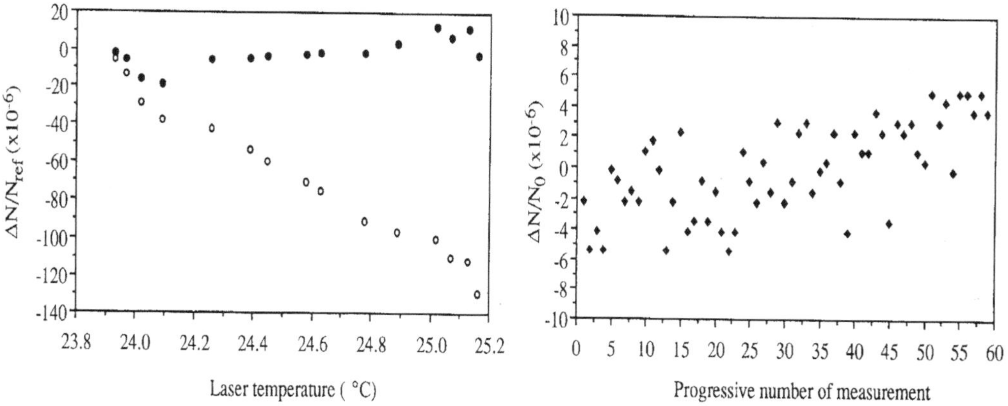

Fig. 5-12 Precision of a laser-diode injection interferometer based on a single-channel waveform λ/2 switching, operating on a corner cube at distance s=66 cm. Left: thermal drift of the Al-table (open circles) is corrected of the 22 10^{-6} temperature coefficient (full dots); right: the spread of successive measurements is limited to a few ppm (from Ref.[21], ©IEEE, reprinted by permission).

One concern about the development of the self-mixing interferometer is the burden of circuits necessary to bring the measurement output to decimal, by multiplication of measured counts by λ/2, a six-digit number to attain the ppm precision involved. If we think of using small-scale of integration ICs, the multiplication can occupy half-European card PCB. On the other hand, PC can perform the calculation, but would spoil the intrinsic simplicity and low cost of the self-mixing instrument. The ideal solution is to employ a microcontroller platform, like the Arduino chip, to acquire the pulses to be classified, up/down counted and brought to decimal metric.

5.3.2 Recovering the FM Channel in an SMI

Two approaches have been reported in the literature to recover the FM signal of the self-mixing interferometer in diode lasers:

(i) applying a frequency shift thanks to an AOM (acoustic-optical modulator) [22] or a modulation by a phase modulator (PM) [23], and

(ii) using a sharp wavelength filter [24,25].

In both cases, the new device has to be inserted in front of the laser, down the path to the target under measurement. The addition of a device raises an issue of increased size of the instrument as well as an increase of cost. Therefore, it has to be carefully checked that the solutions for unambiguous processing by these methods offering two signals, cos2ks and sin2ks, is really advantageous with respect to the single-signal processing at medium level 1 of feedback described in the previous section.

With the *AOM modulator,* the frequency of the beam crossing the device is shifted by an amount Ω, and therefore the beam going to the target and back is shifted by 2Ω. So, the signal detected by the receiving photodiode is $\cos(2\Omega t+2ks)$.

Therefore, the phaseshift 2ks is superposed on the carrier frequency 2Ω, and we can apply the same processing described in Sect.4.2.2 for the signal of the two-frequency interferometer to measure s in $\lambda/2$ steps.

A variant of the above approach is to employ a *phase modulator* in place of the AOM. In this case we add a phase term $\Phi_m \cos \omega_m t$ to the usual phase signal 2ks, so that the detected signal becomes $S= S_0 \cos [2ks +\Phi_m \cos \omega_m t]$.

Thereafter, by demodulating S with the aid of a lock-in amplifier on two separate references, $\cos \omega_m t$ and $\cos 2\omega_m t$, we obtain the signals cos 2ks and sin 2ks, much in the same way as in the processing used for a fiber gyroscope (see also Eqs.8.26 and 8.27, and Ref.[23]), and then proceed by counting $\lambda/8$ steps (Sect.4.2.1)

About the *sharp-response wavelength* filter, the idea is that of converting a frequency deviation (or FM) into an amplitude deviation (or AM), taking advantage of the very sharp response that can be realized in optical filters like, e.g., the Fabry-Perot (see App.A2.1 and Fig.A2-2) or the Mach Zehnder optical interferometers.

Letting $F(\lambda)$ for the wavelength response of the filter and $S_F=dF(\lambda)/F(\lambda)$ for the slope of its (leading or trailing) edge, an FM signal will be converted into a AM signal of amplitude (assuming the case $C<<1$):

$$\Delta P_{CFM} = P_{out} S_F \Delta \nu_p \tag{5.20}$$

where P_{out} is the power and $\Delta\nu_p$ the frequency deviation of the FM-carrying signal.

Recalling that $\Delta\nu_p=-(C/2\pi\tau_{ext}) \sin (2ks+\text{atan}\ \alpha)$, where $\tau_{ext}=2s/c$ and C is given by Eq.5.1, and inserting in Eq.5.20 we get:

$$\Delta P_{CFM} = P_{out} S_F (C/2\pi\tau_{ext}) \sin(2ks+\text{atan}\,\alpha) \tag{5.20a}$$

Compared with the normal AM signal, which is written as [24]

$$\Delta P_{AM} = 2P_0 C (1+\alpha^2)^{-1/2} \tau_p/\tau_{ext} \cos(2ks+\text{atan}\,\alpha) \tag{5.21}$$

we get a gain $G_{FM\text{-}AM}$ of the FM-to-AM peak-amplitude ratio of the sin and cos signals:

$$G_{FM\text{-}AM} = [S_F (1+\alpha^2)^{1/2}/4\pi\tau_p] (P_{out}/P_0) \tag{5.22}$$

With a typical power ratio $P_{out}/P_0 = 0.08$ of the FM and AM measuring arms, and a slope value $S_F = 19\ \text{GHz}^{-1}$ of a Mach-Zehnder filter [24,25], we get a gain $G_{FM\text{-}AM}=91$ in satisfactory agreement with the experimentally observed value of 70. Also, it can be shown [24] that, surprisingly, the SNR of the converted FM signal is better than the AM signal. Fig.5-13 illustrates the typical results obtained using a Mach-Zehnder filter.

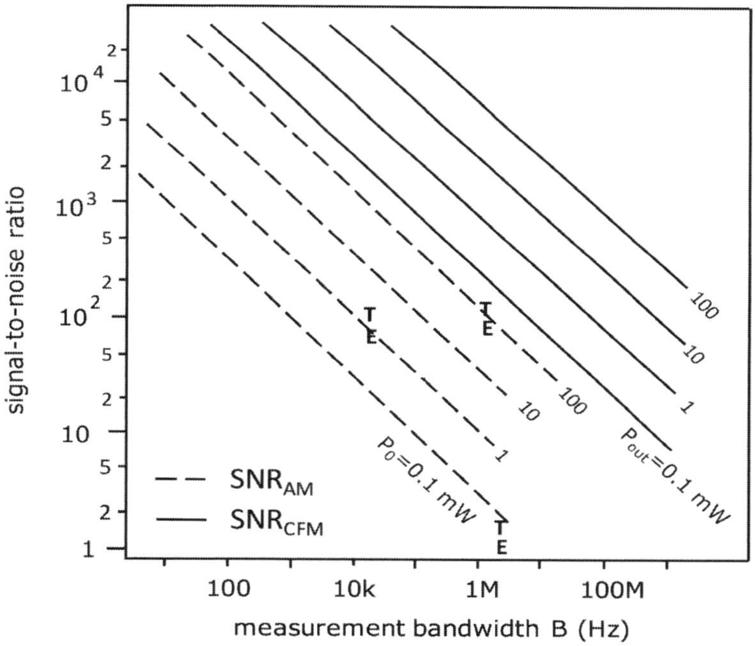

Fig. 5-13 Signal-to-noise ratio of AM (dashed line) and converted-FM (full line) signals plotted versus measurement bandwidth B and with powers P0 and Pout as parameters. Points E and T are experimental and theoretical pairs. Assumed values are: C=2, α=6, τ_p = 4 ps, τ_{in}= 9 ps, and τ_{ext} =3.67 ns (from Ref.[24], ©IEEE, reprinted by permission).

5.4 Self-Mixing Vibrometers

A laser vibrometer is an interferometer especially designed for use as an instrument capable of detecting and measuring small, usually periodical, vibrations in a variety of applications related to mechanical [26], automotive, avionics, and construction engineering (Sect.4.5.1), as well as in biological structures [27, 28].

Vibrometry is also an important diagnostic tool commonly used to measure the frequency response of a structure under test using a periodic excitation applied to it. This allows collecting important information about the structure for design, verification, and testing purposes. With respect to other measurement techniques based on electronic sensors, the laser vibrometer has the key advantage of noncontact operation and does not disturb the structure under test. Usually, vibration means a periodic displacement waveform, with typical amplitudes of interest ranging from less than 1 nm to tens of micrometers, and frequency ranging from less than 0.1 Hz to perhaps 100 kHz, as indicated by the performance chart of the interferometric measurements in Fig.4-20.

Basically, we may also use a displacement interferometer (Sect.4.2.5.4) as a vibrometer, but three specific features shall be considered for design optimization.

First, superposed to the small displacement, we commonly find a relatively large (up to one hundred micometers) ambient-induced disturbance. This can be strongly reduced if we are permitted to work on an antivibration laboratory table, but cannot be avoided if we aim at in-the-field operation of our instrument. Second, we are forbidden to use anything that disturbs the target and its vibration state, like a corner cube, and the vibrometer is to work directly on the natural target surface, normally a gray (δ<1) diffuser following Lambert's law (see Ref. [11], Ch.2). Thus, the returned field is much attenuated and subjected to the speckle-pattern errors. Last, the range of displacement to be measured is very small and it may permit relaxing the coherence length needed when operation is on short distance, or when we can balance the interferometer arm path lengths.

In the following we report some development of self-mixing vibrometers.

5.4.1 An SMI Vibrometer Locked at Half Fringe

Let us consider a small signal s(t) (say $<<\lambda$) superposed to a large stand-off distance s_0. If we are able to dynamically adjust the distance s_0 so that it exceeds an integer number of half wavelengths exactly by a quarter wavelength, that is $s_0 = N(\lambda/2)+\lambda/4$, then the signal developed by the interferometer will be

$$I(t) = I_0 \{1+\cos 2k[s_0+s(t)]\} = I_0 \{1+\cos[\pi/2+2ks(t)]\} = I_0 [1-\sin 2ks(t)]$$

or, the amplitude I(t) is a linear replica of phase 2ks(t) (for small vibrations $s(t)<< \lambda$).
It is even better if the interferometer works at C>1, because we start with a large linear region of response (see Fig.5.14 left) to accommodate the phase signal Φ=2ks(t) and convert it in an amplitude signal readily available with no additional processing.

Doing so, we are locking the interferometer at *half-fringe*, and perform an *analogue* processing of the vibration signal, which has good linearity provided s(t) is less than the $\lambda/2$ excursion of the self-mixing signal, as a starting result.

Yet, to ensure that the quiescent point of operation remains at half fringe against large fluctuations of the distance s_0, we shall introduce an active control of s_0.
This function is accomplished by the feedback loop shown in Fig.5-14, right (dotted box), developed from the idea that we can correct eventual drift of the quiescent point phase of the interferometer by feeding back an opposite-sign phase signal.

To implement the phase correction, we take advantage of the diode laser wavelength dependence on bias current, $\alpha_I=d\lambda/(\lambda dI_{bias})$, and force a wavelength variation $\Delta\lambda$ by impressing a bias current change ΔI_{bias}. Upon a change $\Delta\lambda=\alpha_\lambda\ \Delta I_{bias}$ of wavelength, the phase read by the interferometer will change by the quantity:

$$\Delta\Phi = 2\Delta k\ s= -(4\pi s/\lambda^2)\ \Delta\lambda \qquad (5.23)$$

being $\Delta k/k =-\Delta\lambda/\lambda$, whence $\Delta k=-(4\pi/\lambda^2)\Delta\lambda$. This phase $\Delta\Phi$ sums up to the phase variation $\Delta\Phi_s$ induced by the displacement $2k\Delta s$. The resulting phase $\Delta\Phi_s$-$\Delta\Phi$ is detected by the photodiode as a signal $RI_0\cos(\Delta\Phi_s-\Delta\Phi)$, amplified by the difference amplifier A that has the V_{ref} voltage set at the half-fringe level to the negative input, then is converted to a current through the block G_m, and finally fed to the diode laser to produce a wavelength change.

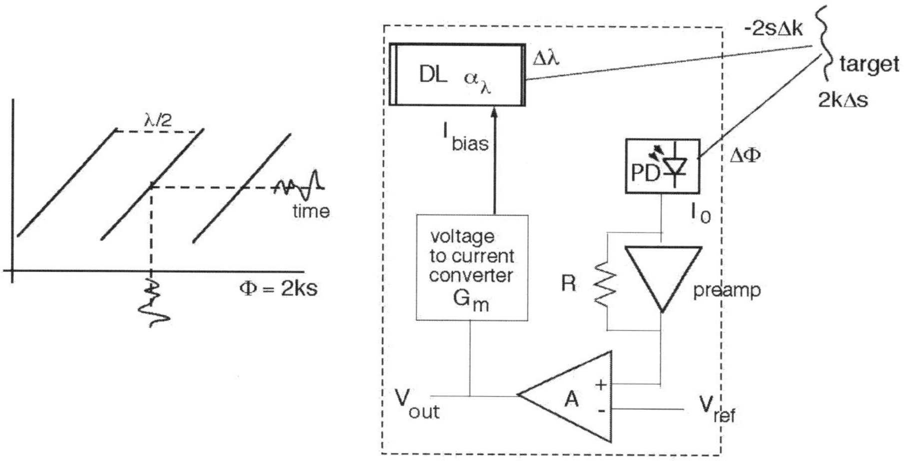

Fig. 5-14 The SMI vibrometer (simplified schematic) uses analogue processing for locking the working point of the interferometer at half fringe (left), thanks to a feedback loop that takes the signal detected by the photodiode PD, compares it to a threshold voltage V_{ref}, converts it to current, and feeds the bias of the diode laser. By virtue of feedback, with a high loop gain G_{loop} (typ. =1000), nonlinearity errors are reduced by 1+ G_{loop} and dynamic range of response is expanded of the same quantity, up to about 1 millimeter.

Now, starting from the target generating phase $\Delta\Phi_{in}$ =2kΔs and going along the feedback loop till the output of the DL, we can write the self-mixing phase signal reaching the output of the feedback chain as:

$$\Delta\Phi_{out} = \Delta\Phi_{in}\, I_0 RAG_m \alpha_I\, (-4\pi s/\lambda^2) \tag{5.24}$$

where factor $(-4\pi s/\lambda^2)$ comes from Eq.5.23 relating $\Delta\lambda$ to $\Delta\Phi$.
Thus, the loop gain is given by:

$$G_{loop} = \Delta\Phi_{out}/\Delta\Phi_{int} = -\, I_0 RAG_m \alpha_I\, (4\pi s/\lambda^2) \tag{5.25}$$

The factor G_{loop} may be made as large as 500 or 1000 in practice, and therefore the control loop behaves like an operational amplifier that, thanks to the feedback effect, keeps the difference of signals applied to its differential inputs dynamically equal to zero.

In particular, at the input of the amplifier A, the signal arriving from the preamplifier will be pinned at the value V_{ref} of the half fringe we have selected, and at the virtual node of the target (Fig.5-14) the two phases will be made zero sum, $\Delta\Phi + \Delta\Phi_s = 0$, so that $2\Delta ks = -2k\Delta s$. From one side, this equality means that $\Delta s/s = \Delta k/k = -\Delta\lambda/\lambda$ or, that only a small wavelength variation $\Delta\lambda$ will be required to measure a small Δs, making the λ-control of the DL easily feasible.

Much more important, the loop G_{loop} gain will impact all the linearity errors, included the one introduced by the $\pm\lambda/2$ limit of response of the fringe (Fig.5-14 left), and will reduce them by a factor given by the loop gain plus one, as well known by feedback theory. Therefore, the dynamic range of linear response of the half-fringe stabilized, analogue vibrometer is no longer $\pm\lambda/2$, but $1+G_{loop}$ time larger, and may go up to about 1 millimeter with state-of-the art design.

Intuitively, this important result is easily explained: when the phase signal increases and tends to exit out of the $\lambda/2$ fringe boundary, the feedback loop pulls it back to the middle of the fringe, so a very large ($>>\lambda/2$) amplitude is necessary to finally overcome the $\lambda/2$ limit, or, stated in other terms, the dynamic range is much increased.

Another feature of the vibrometer, the output signal can be taken at the output V_{out} of the main amplifier (A in Fig.5-14) because, thanks to the virtual ground effect of feedback, the two phases at the target are equal and opposite, $2k\Delta s = -2s\Delta k$ or $\Delta s/s = -\Delta\lambda/\lambda$, and going back in the loop and scaling $\Delta\lambda$= of α_λ and of G_m, we get just $V_{out} = -(\Delta s\ \lambda/s)/\alpha_I\, G_m$. Or, solving for Δs, we may write

$$\Delta s = -V_{out}\, \alpha_I\, G_m\, (s/\lambda) \tag{5.26}$$

Eq.5.26 reveals that the instrument shall be calibrated once for all against the current coefficient α_λ and in the actual use against the distance s. Experimentally, the scale factor matches well the 1/s dependence (see Fig.5-15) and is well reproducible (with a <1% error).
In practice, s can be measured in the same setup by applying the method described later in Sect.5.5.

Note that Δs in Eq.5.23 doesn't depend on C, that is, it is independent from the strength of feedback (unless C is high enough to drive the laser into chaos), and also, very important, is independent of speckle-pattern fading of the self-mixing signal.

Indeed, fading is a cause of signal loss that is once again compensated for by the feedback loop gain (at least as far as the fading attenuation $1/a_f$ is less than loop gain G_{loop}).

Numerical Example. With the typical value of λ-coefficient α_I= 0.004 nm/mA (Sect.A1.3.1), and taking reasonable values for the quantities in Eq.5.25, that is, λ=800 nm, s_0=50 cm, G_m=10 mA/V, R=10 kΩ, I_0= 0.1 mA and A=1, we obtain a loop gain of G_{loop}=400. With the same values, Eq. 5.26 gives for the scale factor of the measurement $V_{out}/\Delta s$ = 0.04 V/μm, in accordance with the diagram of Fig.5-15.

In the practical implementation of the injection vibrometer, we may add a PID controller (a network including a variable gain, a low-pass, and a high-pass filter) to optimize the frequency response of the regulated loop. Also, an automatic reference circuit, made by a long time-constant integrator and a sample-and-hold can be used to avoid readjusting the reference V_{ref} when optical power changes.

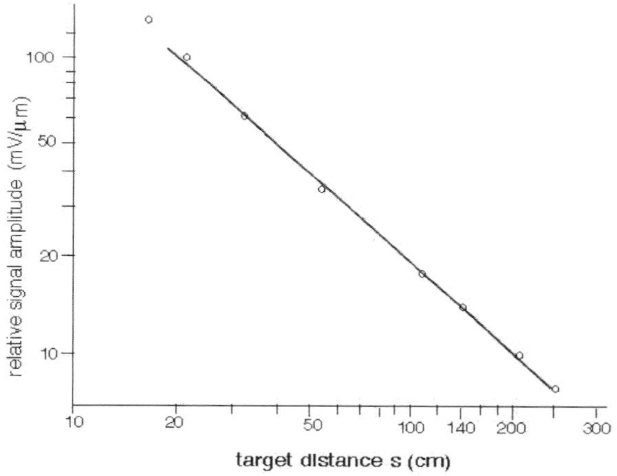

Fig. 5-15 The scale factor of the output signal from the injection vibrometer. Line: theoretical 1/s dependence; points: experimental values.

The performance of an instrument developed [29] with the half-fringe stabilization concept and operating in the injection regime described above, are shown in the diagrams of Fig.5-16. At a stand-off distance of 80 cm, the limit sensitivity is down to ≈100 pm/√Hz, and the maximum amplitude is up to several 100 μm (in total, the instrument covers six decades in amplitude dynamic range). The frequency response is up to several hundred kilohertz.

The linearity of the measurement is better than 1%. Also, the instrument operates on untreated target surfaces with diffusivity as low as 0.3. The minimum detectable signal is down to 20 nm (peak) and dynamic range is up to 100 μm, see Fig.5-16 and Ref. [29].

An exemplary measurement of mechanical response performed by the SMI vibrometer is reported in Fig.5-17, revealing structural details through the several resonances appearing in the frequency spectrum of the SMI signal.

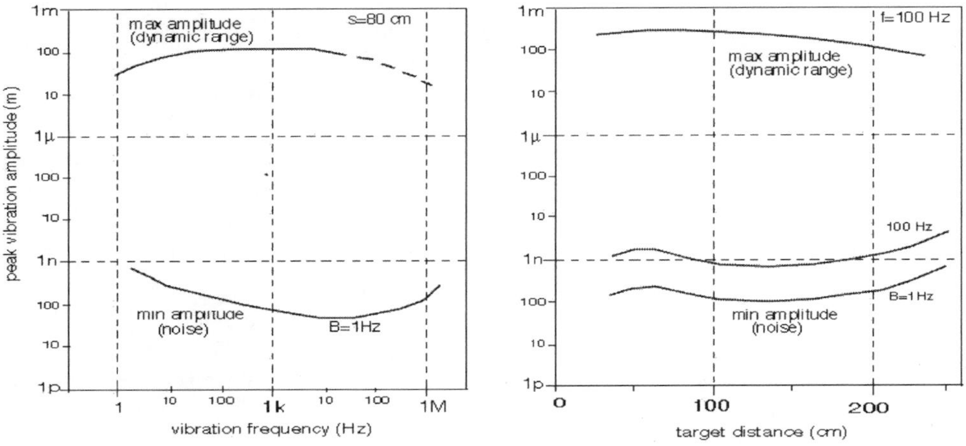

Fig. 5-16 Performance of the half-fringe stabilized SMI vibrometer, working on a diffusing target surface [29]. The laser source is a 25-mW, λ=780-nm laser diode. Left: at s= 80 cm vibration is measured with 100 pm (or better) resolution and up to 100 μm as limited by the tracking circuit. Right: at a representative 100-Hz signal frequency, resolution is 0.1 to 1 nm, from s=30 to 250 cm, with up to 100 μm max amplitude.

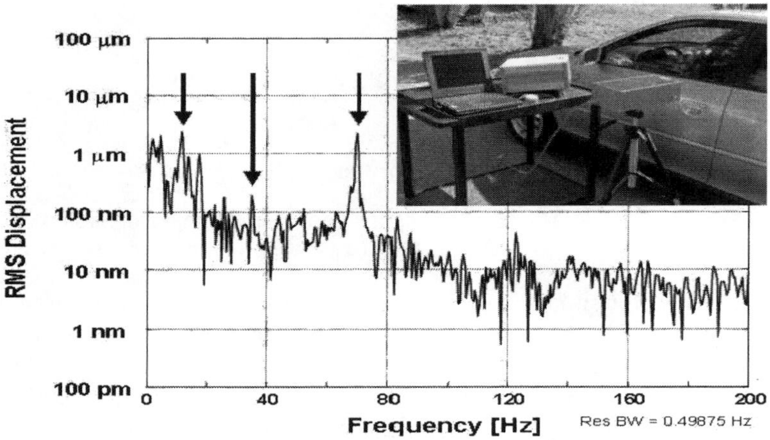

Fig. 5-17 The mechanical response of the car structure taken by aiming at a door of the car, with the motor rotation peak at 35 Hz and two other major peaks; the sparkle events at 70 Hz and the suspension resonance at 13 Hz. Many other peaks, above the 100-pm noise floor, belong to structure but are not identified (from Ref.[29], ©IoP, reprinted by permission).

5.4.2 Differential Vibrometer for Measuring Mechanical Hysteresis

The self-mixing interferometer has just one leg, that is, no reference leg is available like in the external configuration (Fig.4-23). Thus, it seems unrealizable to obtain the differential signal of optical phases $S=\cos(\phi_m-\phi_r)$, the one useful to subtract a common mode optical phase ϕ_r from the measurand phase ϕ_m. Actually, in the optical phase differential measurement, we work at half fringe (Sect.4.5.1) setting $\phi_r=\phi_{r0}-\pi/2$, and the signal becomes $S=\sin(\phi_m-\phi_{r0}) \approx \phi_m-\phi_{r0}$ for small phases. While the difference of phases was made at the optical level, the conversion into electrical signals brings along the difference as $I_0\,\phi_m - I_0\,\phi_{r0}$.
In a self-mixing interferometer locked at half fringe (Sect.5.4.1) we get directly the phase at the electrical level, $I_0\,\phi_m$, so if we procure a reference $I_0\,\phi_{r0}$ by a second identical self-mixing interferometer, we can make the difference at the electrical level, an operation equivalent to make the difference $\phi_m-\phi_{r0}$ at the optical level, provided the two self-mixing channels are linear and identical.

In the practical implementation [30] shown in Fig.5-18, we use two nominally identical self-mixing vibrometers stabilized at half fringe (Sect.5.4.1) ending in a difference amplifier that allows a fine trimming of the channels balance. The main parameters of the two channels (responsivity, max and min signal levels, as in Fig.5-16) were tested to match within ±5%, and the final trim adjusted levels within ±0.5%.

The instrument was then used to measure the differential vibration exhibited on a small item, a damper bead, when it is excited by a shaker moving the basement on which the bead is standing, while a pressure for keeping the bead firm onto the base is applied (Fig.5-19).

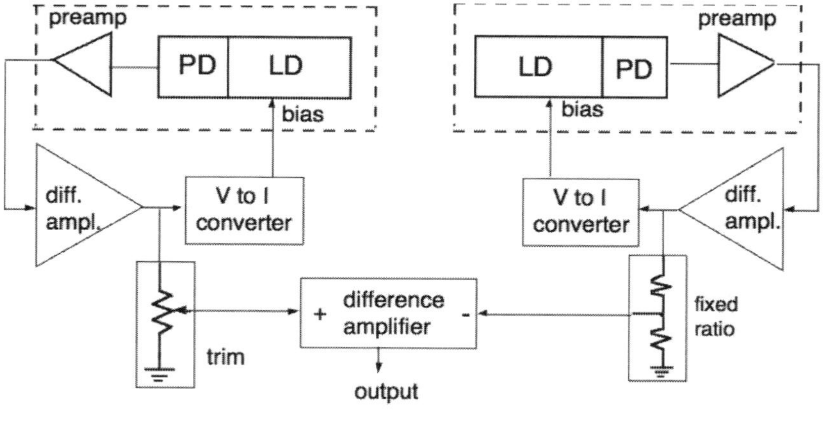

Fig. 5-18 The differential configuration of the self-mixing vibrometer, based on electronic subtraction of the phase signal out of two channels. To obtain a good balance, a final trimming allows adjustment of the responsivity of channels within ±0.5%.

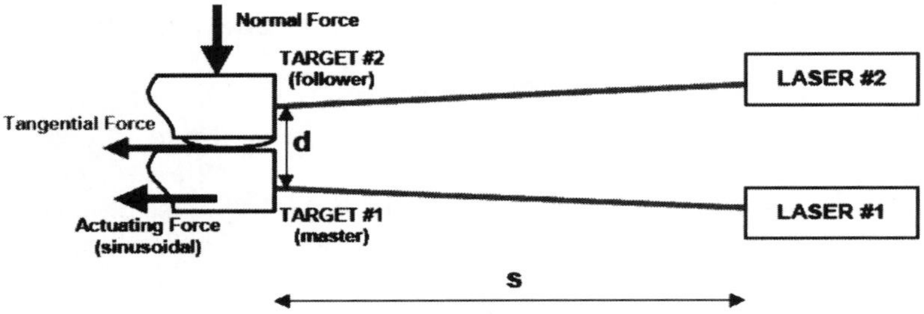

Fig. 5-19 Setup of differential measurement: target #2 is the damper bead under test, and target #1 is the basement of the shaker. The instrument makes the difference of the two vibrations, from a stand-off distance s = 50 cm typ.

With the differential vibrometer, we have tested damper beads mounted in a shaker machine at increasing amplitude of stress, and measured the total and differential displacements reported in Fig.5-20 (right).

As the total displacement is proportional to the stress T imparted to the bead, and the differential displacement is proportional to the strain S of the bead, we can plot one quantity against the other and obtain the *mechanical hysteresis cycle* of the bead under test shown in Fig.5-21.

The hysteresis cycle is of interest for the design and testing of damping elements, because its area $\int TdS$ is equal to that of force times the displacement (per unit volume), that is $\int Fds/V$, and thus it represents the energy dissipated per unit volume in a cycle [30].
In the application as a damper, we need the largest possible area but also wish to stay safely away from the mechanical breakdown, which occurs at a value of drive force a little bit larger than that of the widest cycle shown in Fig.5-21 (specifically, at ≈18).

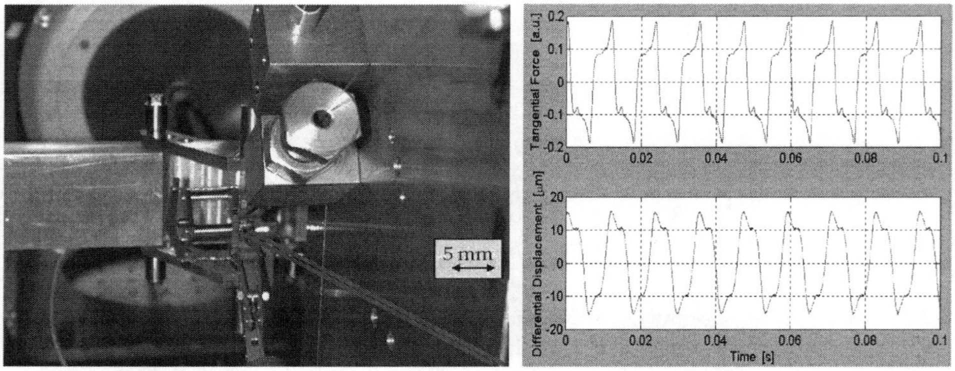

Fig. 5-20 Left: detail of the machine for the measurement of differential vibration [red lines indicate the measuring beams] and right: a sample of shaker-induced vibration (top) and differential displacement (bottom).

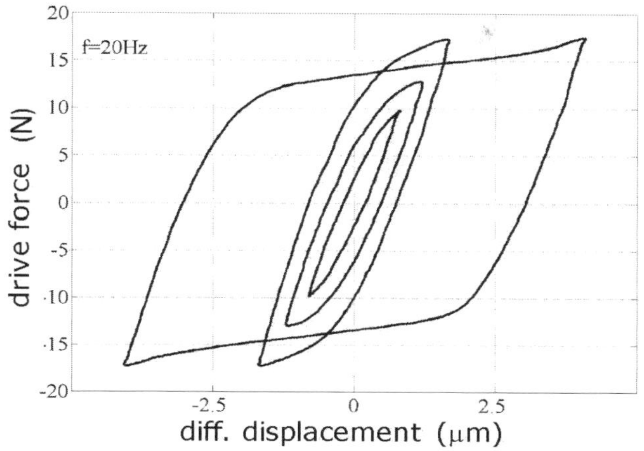

Fig. 5-21 Mechanical hysteresis cycle of a bead sample measured by the differential self-mixing vibrometer. Up to a force F=±5 N the regime is Newtonian, then the sample enters the plastic regime and hysteresis area (and dissipated power) increases, and at about F=±18 N the sample breaks down.

5.4.3 A Plain Vibrometer for Micro Targets

In some circumstances, we don't really need to develop a sophisticated scheme or processing to make useful measurements with the self-mixing interferometer, also because we may wish to trade performance for low cost and layout simplicity while retaining the basic advantages of the injection concept (like self-alignment, self-filtering, etc.).
Then, we can use a minimum-count injection vibrometer made by an optical section including a diode laser, an objective lens, and an attenuator, and an electronic section made by the front-end preamplified connected to the laser monitor photodiode.

With this arrangement, we get the plain interferometric signal $I_{ph}=I_0\,(1+\cos 2ks)$, where $s=s_m(t)+s_0(t)$, where $s_m(t)$ stands for the useful vibration signal and $s_0(t)$ for the ambient-induced disturbances (or microphonics). As previously noted, we can separate these two addends if they lie in different frequency bands. Also, the cosine ambiguity is no problem if we look to a multi-λ amplitude vibration $s_m(t)$, of which we can easily count (looking at the oscilloscope trace) periods of half-wavelength variations $\Delta s(t)$.

An example of application is the self-mixing instrument intended for MEMS testing [31] shown in Fig.5-22. To diagnose the mechanical properties of the structure, the beam is focused on the 500 μm by side MEMS Si-chip, which incorporates a comb-like spring structure and a mass. The mass is aimed at by the focused laser spot (of typ. 100 μm size) at a stand-off distance of 25 cm, and, even if some nonmoving parts are illuminated, they do not disturb the detection of the useful signal thanks to the injection process.

In the measurement setup (Fig.5-22), the MEMS is mounted inside a vacuum chamber to test it at different ambient pressures. The window of the glass chamber need not be high optical quality, because the wave front distortion is filtered out by the laser diode. The out-of-plane vibration of the MEM mass is viewed at an angle Φ ($\approx 20°$ in Fig.5-22), and the appropriate $\cos\Phi$ correction on s(t) is applied to the measurement.

The signal out from the preamplifier is acquired by a digital oscilloscope, and the desired s(t) waveform may be computed from the $I_{ph}=I_0[1+\cos(2ks_p\cos\omega_0 t)]$, using a routine that minimizes the rms errors of the desired quantities, amplitude s_p and frequency ω_0. But, even easier, we can just estimate s_p by simply counting the periods of the self-mixing signal (we find 5.5 in Fig.5-23). Each period corresponds to $\lambda/2=680\text{nm}/2=340$ nm, so we obtain the peak amplitude of the square wave exciting the MEMS as 5.5×340 nm= 1.87 μm.

Now, by repeating the measurement of amplitude of s(t) for a number of frequencies of the drive signal, we can measure the *electromechanical frequency response* of the MEMS in

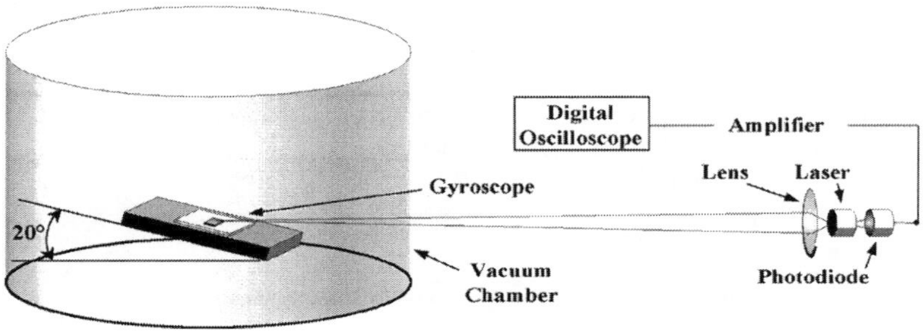

Fig. 5-22 Setup for measuring the mechanical properties of a Si-MEMS by means of an injection vibrometer. Light from the laser is focused on the vibrating mass of the MEMS chip through the glass wall of the vacuum chamber (from Ref.[31], ©IEEE, reprinted by permission).

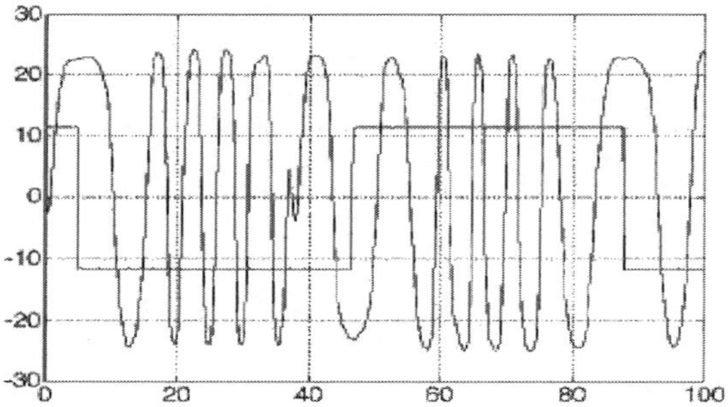

Fig.5-23 The MEMS drive signal (square wave) and the interferometric signal output, as obtained from the vibrometer of Fig.5-22 (from Ref.[31], ©IEEE, reprinted by permission).

several conditions of interest and test the dependence on drive voltage and ambient pressure. Experimental results in Fig.5-24 show the effect of increasing drive voltage (and vibration amplitude).

Another application of the above technique has been the testing of MEMS micromirror arrays to measure their frequency response. In this case, the laser diode has been pigtailed with a single mode fiber (with the near end surface angled to avoid backreflection), whose far end was spliced and fused into a spherical lens of 15-μm radius, so as to focus a small spot (≈5 μm) on the device under test [32].

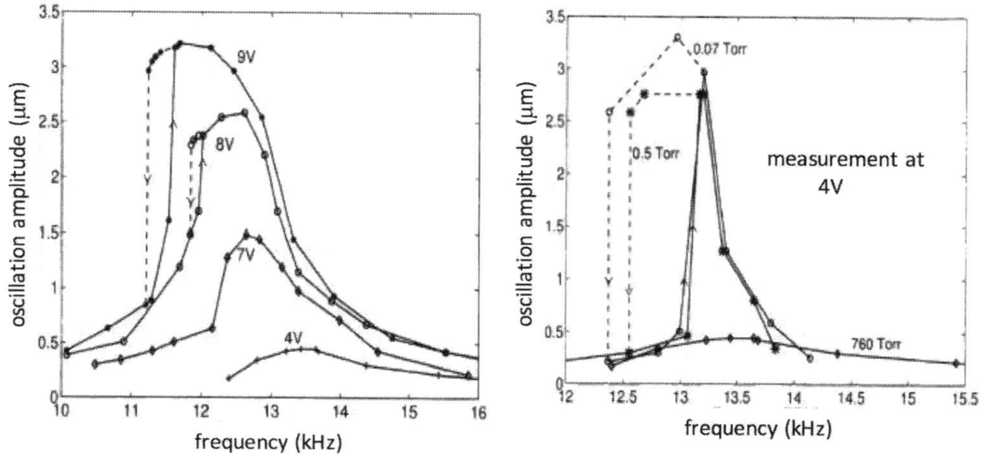

Fig.5-24 The MEMS frequency response is measured by the self-mix vibrometer (horizontal scale in kHz, vertical scale in relative units). Left: at increasing drive voltage we observe incipient hysteresis; right: at increasing pressure we observe the damping of Q-factor resulting in decreased selectivity (from Ref.[31], ©IEEE, reprinted by permission).

5.5 Absolute Distance Measurement by SMI

All laser interferometers, including the self-mixing version, are actually *displacement* measuring instruments, and require moving the remote target to develop a phase change and count periods. This is different from telemeters (Ch.3) that can operate on still targets and are truly absolute *distance* meters. In other words, the measurement performed by the interferometer is incremental, not absolute, because it is based on the optical phase.

To circumvent the basic limitation and realize an absolute distance measurement, let us consider the phase signal cos 2ks. Usually, we change distance by Δs keeping wavelength λ (or $k = 2\pi/\lambda$) constant, so as to develop 2π-periods of phase $2k\Delta s$ and count them, each period corresponding to a $\lambda/2$ increment of distance: this is the displacement measurement.

Yet, we can keep distance s constant and apply a sweep of Δk, so that the phase will change by $2s\Delta k$ and a self-mixing signal is developed. The number of 2π-periods of the self-mixing signal, viz. $2s\Delta k/2\pi$, is a number proportional to distance s.

So, by sweeping λ (or k), the number of wavelengths $N=2s/\lambda$ contained in the go-and-return path to the target will change. Each time one more wavelength fits in 2s, the self-mixing signal undergoes a 2π phase shift, and a switch in the C>1 injection regime. By detecting it, we are able to trace back the absolute distance s [33-38].

To develop the idea, we can sweep the wavelength of the source, a feature readily available in diode lasers through modulation of the drive current (App.A1.3).

As the waveform of modulation, let us now consider a linear sweep on a swing $\Delta\lambda$, from λ_0 to $\lambda_0+\Delta\lambda$. This modulation is readily obtained by superposing a linear sweep on the bias current I_{dc}, with an amplitude $\Delta I=\Delta\lambda/\alpha_I$ (App.A1.3). The interferometric signal $I_{ph}/I_0=1+\cos 2ks$ at the start and at the end of the sweep period is, respectively:

$$I_{ph}/I_0 = 1+\cos 4\pi s/\lambda_0, \quad \text{and}$$

$$I_{ph}/I_0 = 1+\cos 4\pi s/(\lambda_0+\Delta\lambda) \approx 1+\cos 4\pi s(1-\Delta\lambda/\lambda_0)/\lambda_0$$

Thus, the phase variation $\Delta\Phi$ produced by the sweep is:

$$\Delta\Phi = 4\pi s(1-\Delta\lambda/\lambda_0)/\lambda_0 - 4\pi s/\lambda_0 = -4\pi s\,\Delta\lambda/\lambda_0^2$$

By dividing $\Delta\Phi$ by 2π, we get the number of interferometric waveform periods found in the photodetected current as $N= 2\Delta\lambda/\lambda_0^2 s$. For C>1, N is the number of switchings to be counted. From N, the distance is obtained as:

$$s = N\,\lambda_0^2/2\Delta\lambda \tag{5.27}$$

Each count (N=1 in Eq.5.27) corresponds to a distance

$$L_{res} = \lambda_0^2/2\Delta\lambda, \tag{5.28}$$

and write the distance as

$$s = N\ L_{res} \tag{5.27'}$$

thus, L_{res} represents the *resolution* (or the unit of discretization) of the measurement.

In a Fabry-Perot diode laser, the practical limit to the wavelength swing we may impose with a reasonable excursion of the drive current is $\Delta\lambda \approx 4$ nm (Fig.A1-13), and thus the theoretically attainable resolution at $\lambda=1000$ nm is $L_{res} = (1000)^2/2\cdot 4 = 125\ \mu m$, an attractive value indeed. Unfortunately, Fabry-Perot laser diodes are affected by mode hopping (Fig.A1-13) at large $\Delta\lambda$ swings. As mode hopping is detrimental to signal processing, to avoid it, reported experiments [34,35] limit the useable wavelength swing to about $\Delta\lambda \approx 0.2$ nm, the interval of a step between two successive mode hops. Therefore, the resolution actually attained is $L_{res} \approx (1000)^2/(2\cdot 0.2) \approx 2.5$ mm, up to a distance of a few meters.

An experimental arrangement is shown in Fig.5-25. A triangular modulation of a few mA peak-to-peak is applied to the laser drive current. This produces the wavelength modulation and, in addition, also a power amplitude modulation of the same waveform.

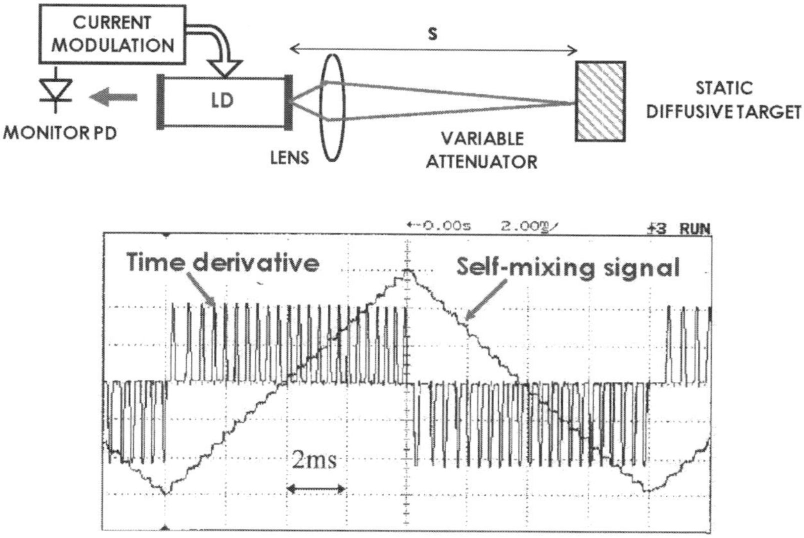

Fig.5-25 Absolute distance meter based on self-mixing (top): wavelength is swept with a triangular waveform, using current modulation superposed to the laser diode bias; (bottom): the self-mixing signal detected by the photodiode signal exhibits a small periodic ripple corresponding to $\Delta\Phi=2\pi$ variations. After a time-derivative, each period shows up as a pulse. Counting the pulses, the distance is found as $s_{res}=N\ \lambda^2/2\Delta\lambda$ (from Ref.[33], ©IoP, reprinted by permission).

Superposed to the triangular power waveform we get the self-mixing signal in the form of a small ripple, see Fig.5-25 (the phase $2s\Delta k$ periodic of 2π). With a time-differentiation, we obtain a sequence of pulses, one for each $L_{res}=\lambda_0^2/2\Delta\lambda$ contained in the distance s to be measured, and distance follows from Eq.5.27' as $s=NL_{res}$.
As a refinement, we may count, say, N_+ positive pulses in the increasing semiperiod of the triangular wave, and N_- negative pulses in the decreasing semiperiod. Then we can compute the average distance using the semisum $N_{av}=(N_++N_-)/2$ in Eq.5.27', so that $s= N_{av}\ L_{res}$. In addition, we can compute the difference $N_\Delta= N_+-N_-$ that represents the change of distance occurred in a semiperiod T, and obtain the velocity of the target as $v=ds/dt= N_\Delta\ L_{res}/T$.

Eq.5-27' gives the distance in units of L_{res} for a single measurement, and is affected by the discretization error, because we incur a ±1 count as a round-off error at the beginning and end of the sweep. Recalling that the variance of a uniform distribution is 1/12, we have a random rms error $\sigma_{dis}=\sqrt{(1+1)/12}=1/\sqrt{6}$ counts. This error is reduced by a factor $1/\sqrt{N_T}$ if we repeat the measurement N_T times. Thus, the resolution L_{res}= 2.5 mm cited above becomes an error of 2.5/√6 = 1 mm for a single measurement, and eventually may go down to 0.1 mm upon averaging on 100 measurements. However, while the accidental error can be reduced by averaging, systematic errors are not and will set the limit of precision of the measurement. Fig.5-26 shows the typical error that can be obtained by the basic scheme of self-mixing distance measurement (Fig.5-25) using Fabry-Perot diode lasers [42].

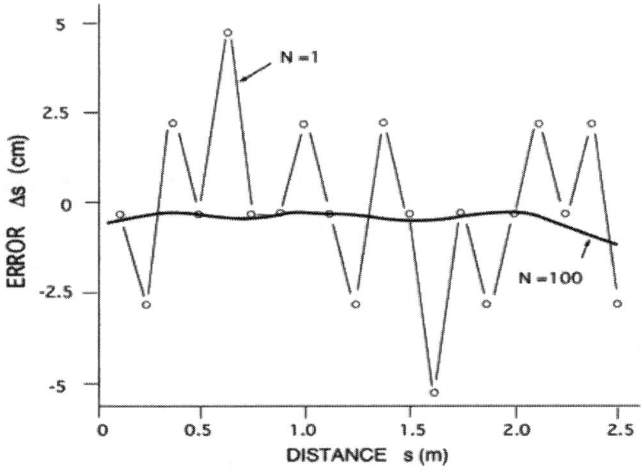

Fig.5-26 Typical uncertainty of a plain absolute distance meter based on self-mixing, for N=1 and averaged on N=100 measurements. Laser is a Fabry-Perot diode with a $\Delta\lambda \approx$ 0.2-nm swing, and L_{res}= 2.5 mm.

Analyzing the source of systematic errors [45], it is found that the coefficient laser diode response α_l has a nonlinearity with respect to current and a tilt in the frequency response. Using a digital signal processor (DSP), both errors can be corrected [45] and the resolution is improved, as shown in Fig.5-27.

Other possible improvements to the basic scheme of Fig.5-25 are:

(i) removing the dependence of the self-mixing signal from the triangular modulation of bias, also affecting the emitted power, by subtracting the triangular waveform from the self-mixing signal, so as to better reveal the cos2sΔk ripple, especially at short distance where the number N of periods contained in the sweep are few, and frequency discrimination of them from the triangular waveform is marginal;

(ii) better calibration of $L_{res}=\lambda_0^2/2\Delta\lambda$ which, by separate measurements of $\Delta\lambda$ and λ_0 is unlikely to be known better than a few percent, by introducing a separate reference channel ending on a target of known distance L_{REF}. Making two measurements on distance s and L_{REF}, (for example aided by an optical commuting element) that will supply counts $N=s/L_{res}$ and $N_{REF}= L_{REF}/L_{res}$, the distance follows as:

$$s = (N/ N_{REF})\, L_{REF}, \qquad (5.29)$$

independent from λ_0 and $\Delta\lambda$.

(iii) using the vibrometer configuration of Fig.5-14 to accommodate by $\Delta\lambda$ just a single fringe (N=1 in Eq.5.27) in the distance to be measured, thus removing discretization error, provided $\alpha_I = \Delta\lambda/\Delta I_{bias}$ is calibrated vs [38] so that we get

$$s = \lambda_0^2/2\ \alpha_I \Delta I_{bias} \qquad (5.30)$$

(iv) improving resolution by using widely tunable laser sources, like external grating tuned (see A1.3.2) or VCSEL cavity tuned by MEMS. With achievable $\Delta\lambda$ of several tens of nanometers, we can reach resolution L_{res} down to 2...5 μm, and ppm accuracy on a few meters distance. Of course, this requires a larger detection bandwidth up to $(s/L_{res})/T$, where T is the semi-period of the triangular wave, as well as a narrow line $\Delta\nu$ of the laser (Sect.4.4.2) to ensure the necessary coherence length.

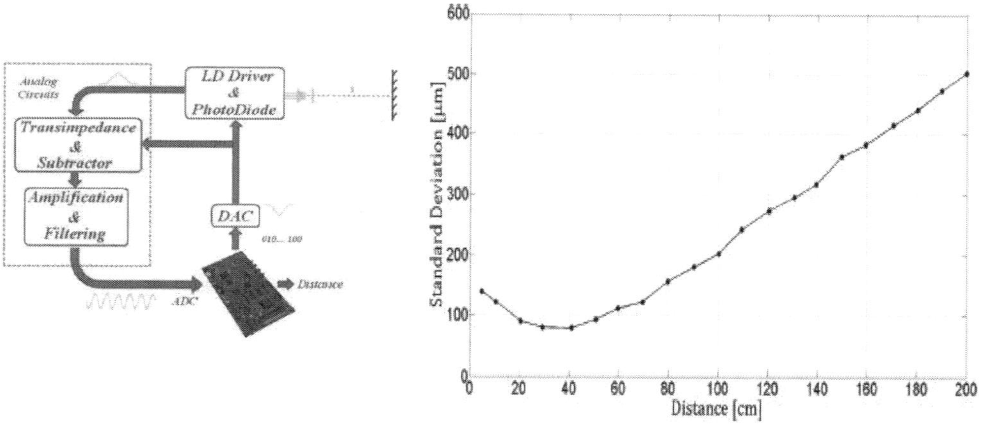

Fig.5-27 Left: scheme of the α_I-corrected absolute distance meter. Right: the accuracy averaged on N=100 measurements (from Ref. [37], ©AIP, reprinted by permission).

5.6 Alignment and Angle Measurements

Alignment of mirrors along the line of sight defined by a laser-beam wave vector is one of the early applications reported [39] for injection interferometry. Fig.5-28 shows the straight arrangement for such an autocollimator measurement. The laser beam is expanded by a telescope and sent to the mirror target. On a substantial distance (say, >30 cm) the ambient-induced vibrations (or microphonics) readily provide an interferometric signal.

The amplitude of the signal reaches a maximum when the alignment is best. Specifically, the maximum is attained when the angle error α is below the diffraction limit of the transmitted beam, with a typical resolution $<\approx 3$ arcsec (≈ 15 μrad) as indicated in Fig.5-28.

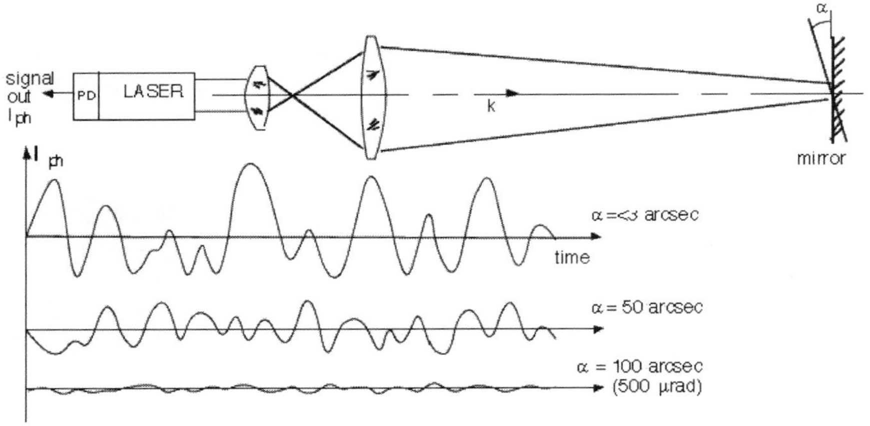

Fig. 5-28 Alignment of a remote mirror to the propagation wave vector k of the laser: ambient vibrations provide an interferometric signal, whose maximum is found when the error angle α is at a minimum (typ. < 3 arcsec).

Like when working with a normal optical autocollimator, the alignment of several reflecting surfaces is achieved by separately aligning each surface to the k vector, which acts as an angular yardstick. The injection-interferometer autocollimator has been found useful in alignment with IR lasers, originally, with a 3.39 µm He-Ne laser [39]. However, it does not provide a true angle measurement, but just a sensitive null detection. By adding a line-of-sight modulation, we can work out a true angle measurement. The beam is steered by an actuator, e.g., a PZT ceramic drive moving a x-y stage carrying the laser chip or the first lens of the collimating telescope [61].

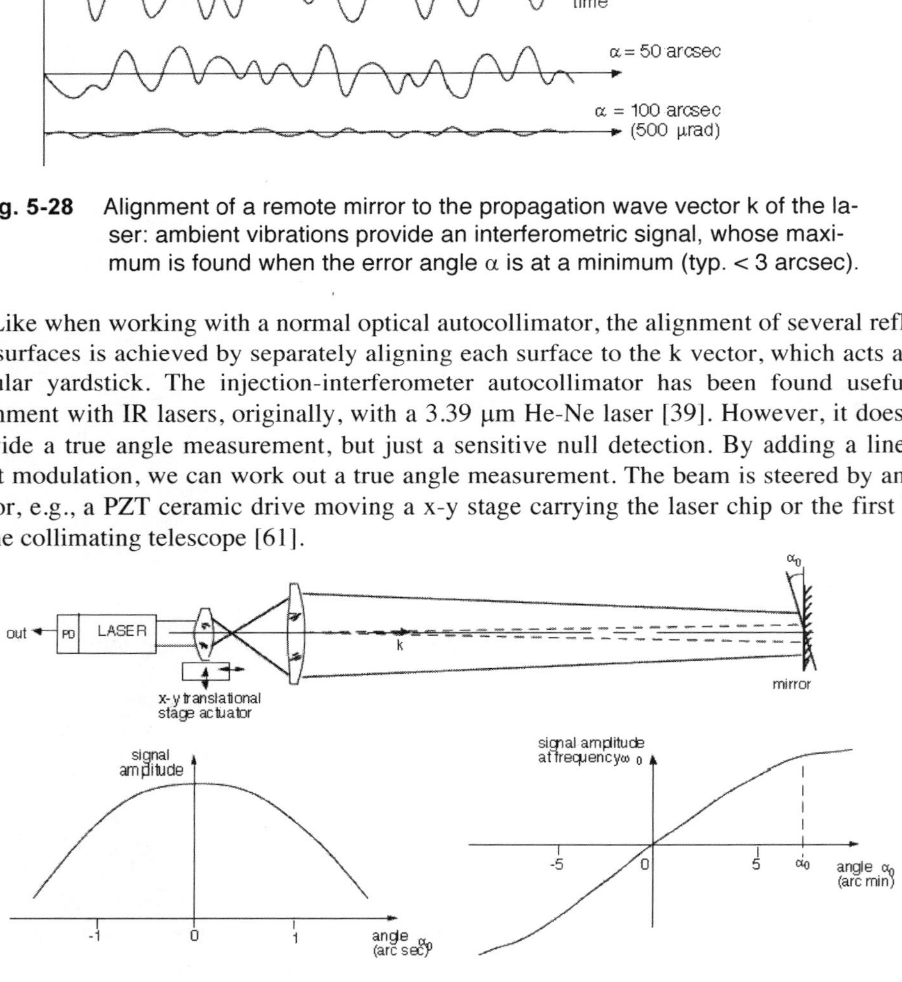

Fig. 5-29 Angle measurement by injection interferometry: the aiming direction is modulated so that the parabolic amplitude response is transformed into a linear dependence on the angle α_0 to be measured.

To substantiate the design, let us assume a deflection $\Delta x = \Delta x_0 \cos\omega_0 t$ of the first lens. Then, the angle of wave vector k is modulated with a deflection $\alpha = \Delta\alpha_0 \cos\omega_0 t$, where $\Delta\alpha_0 = \Delta x_0/F$ and F is the focal length of the first lens. As illustrated in Fig.5-29, with no modulation the response versus α is parabolic, whereas adding the modulation and using phase detection, the response is linear up to the amplitude $\Delta\alpha_0$ of the angle swing.

Reported performance of the angle meter implemented with a laser diode in the injection-interferometry configuration [40] is a noise-limited resolution of ≈ 0.2 arcsec (≈1 μrad) on a dynamic range $\alpha_0(\max) \approx \pm 5$ arcmin (≈1.5 mrad).

Simultaneous displacement and measurement of two angles. An extension of the angle measurement is the demonstration that a single channel of a self-mixing interferometer can carry the signals of the usual displacement s(t) together and of two angles, *tilt* and *yaw* of the target surface, as indicated in Fig.5-30 [41]. These quantities are of interest for the application to mechanical engineering and the control of tool machines.

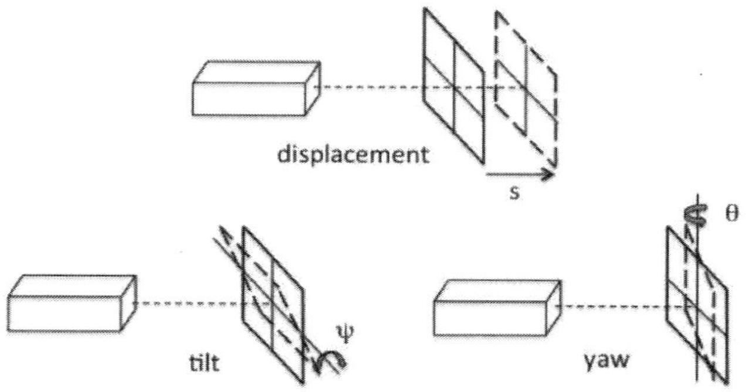

Fig. 5-30 Three quantities can be measured with a single self-mixing channel: displacement s(t) and the target surface angles tilt ψ and yaw θ.

Different from the above methods [39, 40] that look at the amplitude of ambient-induced spurious vibrations to find angle α, we now use *incoherent feedback*, which is the dependence $P(\alpha)$ of average emitted power from angle α of the target surface. Indeed, if the target surface is reflective, the return in cavity acts as a change of exit mirror reflectivity so that the power emitted is found as [41]:

$$P(\alpha) = P_0(\alpha, R_3)\,(1+m\cos 2ks) \tag{5.31}$$

where R_3 is the target reflectivity (at the target, a mirror is needed in this measurement) and α may be either ψ or θ. The reason that angles affect P_0 and not m or cos 2ks is because the angular misalignment is a source of attenuation, not a phase change, as explained in more detail in Ref. [41]. The dependence of power from target reflectivity is reported in Fig.5-31 for a VCSEL emitting a 1 mW at 850 nm. Also shown in Fig.5-31 is the dependence from angle α, a nearly Gaussian shape with a full-width half maximum of about $FWHW_\alpha$=20 mrad, or standard deviation σ_α = FWHW$_\alpha$ /2.36 ≈ 8.5 mrad. Dropping the R_3 dependence, we can rewrite $P_0(...)$ in Eq.4.64 as

$$P_0(\alpha) = P_{00} \exp -\alpha^2/2\sigma_\alpha^2 \tag{5.31a}$$

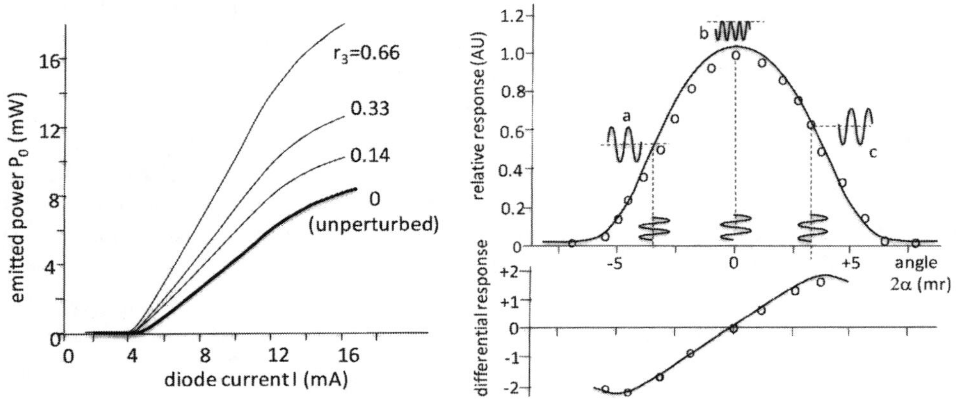

Fig. 5-31 Left: the power emitted by a VCSEL laser (Optowell model PH85-F1P1S2-KC) when subjected to the return from an external mirror of reflectivity R_3 placed at s=60 mm distance; right: the $P(\alpha)$ dependence of power on the angle α, (either tilt ψ or yaw θ) as measured by the photodetected current I_{ph} at the monitor photodiode. Note how the modulation of α allows to linearization of the response [from Refs. [41, 42], ©IEEE, reprinted by permission).

Now, we can measure a single tilt angle α (e.g., pitch or yaw) of the target by mounting the objective lens on a piezo actuator (typically a small 1x1x2 mm^3 PZT ceramic) and driving with a sinusoidal modulation $s_0 \cos \omega_m t$, so as to add to α a deflection angle $\alpha_m \cos \omega_m t$ where $\alpha_m = s_0/F$ and F is the focal of the objective. As a consequence, the feedback power received by the detecting photodiode is:

$$P_0(\alpha) = P_{00} \exp -[\alpha+\alpha_m \cos \omega_m t\,]^2/2\sigma_\alpha^2 \tag{5.32}$$

developing Eq.5.32 at the first order of the exponential and rearranging terms, we get:

$$\begin{aligned} P_0(\alpha)/P_{00} - 1 &\approx -[\alpha+\alpha_m \cos \omega_m t\,]^2/2\sigma_\alpha^2 \\ &\approx -(\alpha^2+\alpha_m^2 \cos^2 \omega_m t + 2\alpha\alpha_m \cos \omega_m t\,)/2\sigma_\alpha^2 \end{aligned} \tag{5.33}$$

The last term in parentheses of Eq.5.33 has an amplitude proportional to α and is impressed on the carrier $\cos \omega_m t$; thus, it can be demodulated by the same waveform driving the piezo, and this operation yields the result shown in Fig.5.32.

In addition, we can improve the linearity and extend the measurement range or dynamics by computing the relative derivative of received power with respect to angle. The result of this operation is shown in Fig.5-32, demonstrating the improvement in linearity and dynamic range obtained with respect to the plain processing of the Gaussian.

Last, we can measure simultaneously two angles ψ and θ by modulating the objective lens tilt along two orthogonal axes, as shown in Fig.5.33. Here, each axis is actuated by a pair of piezo-ceramic (PZT), driven in push-pull, that is, by applying opposite sign drive signals $+V_x$ and $-V_x$, so that the lens is not stressed by the PZT action. To be able separate the X and Y signals summed up in the photodiode detected power, one axis is driven with a cosine term, $V_x= V_{x0} \cos \omega_m t$ and the other with a sine term, $V_y= V_{y0} \sin \omega_m t$. Thanks to the orthogonality of the sin and cos functions, by demodulating the photodiode signal in two channels with $\cos \omega_m t$ and $\sin \omega_m t$, we can sort out the two angle signals separately.

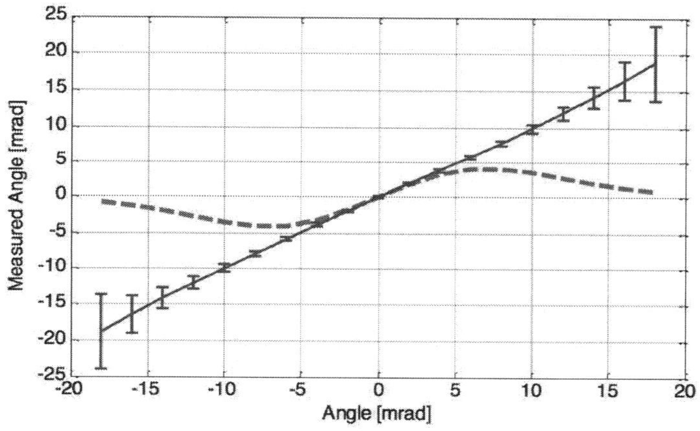

Fig. 5-32 Dotted line: the angle measured by the Gaussian dependence sensed by a small amplitude angle modulation; full line: by computing the relative received power $dP(\alpha)/P(\alpha)d\alpha$, dynamic range and linearity are improved. Bars are the uncertainty of experimental points (from Ref. [41], ©IEEE, reprinted by permission).

Finally, if the target undergoes a displacement Δs, the usual self-mixing signal $\cos 2k\Delta s$ will be generated in form of fringes (Fig.5-33), summed up to the angle contributions. Generally, we will be able to separate it from the angle signal by taking advantage of the different bandwidth occupation of the two (angle signal can be suppressed by a narrow-band notch filter at the modulation frequency ω_m).

About the performance of the double-angle measurement, the reported error of both angles is less than 12 μrad around the zero angle, and the dynamic range is ± 6 mrad [41].
Still another angle, the *roll angle* (not shown in Fig.5-30), that is, the rotation in the plane of the target surface, is also of interest for mechanical engineering applications, but it is difficult to measure because the roll movement doesn't develop any path length change along the line of sight. A preliminary demonstration of roll measurement has been reported using a double beam interferometer with μrad resolution, to read the phase shift introduced upon rotation of a quarter wave birefringence plate [43, 44].

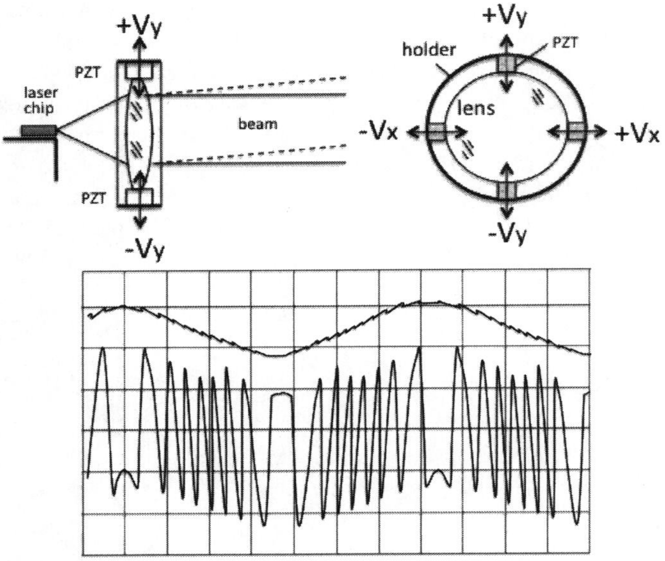

Fig. 5-33 Top: arrangement for two-angle scanning with push-pull actuating piezo mounted along two axes; bottom: the angle signal and the composite angle plus displacement signal (from Ref.[41], ©IEEE, reprinted by permission).

5.6.1 Radius of Curvature Measurement

In semiconductor manufacturing, it's interesting to measure the curvature of a substrate after an epitaxial layer has been grown on it, because mismatch of thermal expansion coefficient leads to bending of the substrate and originate internal stress, affecting the optoelectronic properties (band alignment, carrier mobility, etc.) of the layer, just the case of GaN grown on sapphire. A relationship exists [45] between stress S and radius of curvature R, written for a thin slab as:

$$S = 0.166\ [E/(1-\nu)]\ (t_{sub}^2/t_f)\ /R \qquad (5.34)$$

where E is the substrate Young's modulus, and t_{sub}, t_f are the substrate and film thickness, respectively. Thus, measuring the wafer radius of curvature R we can trace back S.

The most commonly employed method to measure R is triangulation. A He-Ne laser beam is collimated and projected on the surface of the substrate, and the reflection is observed using a position sensitive detector (PSD), resolving minute deviations of the spot off the position of the plane surface reflection [46]. Deflections of $\Delta\alpha \approx 10$ µrad with a maximum measurable deflection up to 2 mrad were reported in Ref. [46], corresponding to curvature radius of 15 m to 3 km, and to an internal stress ranging from 0.2 to 10.6 10^7 Nm^{-2}. However, a long baseline (L=450 mm) between wafer and PSD was necessary to get a measurable deviation on the PSD), and the integration time was relatively long, about 1 s.

Using a self-mixing interferometer as the sensor of the deflection angle generated by the reflection on the surface under test [42], we are able to measure small deflections, down to about 0.5 μrad with short response time (about 1 ms), without any additional components and with a very compact setup.

The experimental arrangement is shown in Fig.5-34. A VCSEL laser emits 8 mW at λ=850 nm, and the beam is focused on the wafer with a spot of about 1 mm. The beam is reflected by the wafer under an angle 2α, where α is the inclination of the wafer edge (see Fig. 5-34). As well-known from geometry, it is $dz(x)/dx = \tan \alpha(x) \approx \alpha(x)$ (for small α), so by taking the second derivative, we get $d^2z(x)/dx^2 = d\alpha(x)/dx = 1/R$. Therefore, we get:

$$R = [d\alpha(x)/dx]^{-1} \tag{5.35}$$

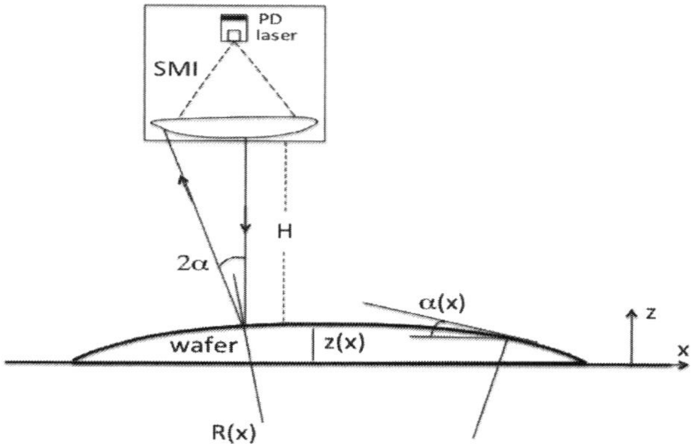

Fig. 5-34 Wafer curvature radius is measured by the self-mixing interferometer SMI by looking at the deflection angle 2α. As the wafer is moved along the x-axis by the wafer carrier, we get an R=R(x) measurement spatially resolved (from Ref. [42], ©IEEE, reprinted by permission).

by taking the second derivative, we get $d^2z(x)/dx^2 = d\alpha(x)/dx = 1/R$. Therefore, we get:

$$R = [d\alpha(x)/dx]^{-1} \tag{5.35}$$

Thus, after measuring the angle 2α while the wafer is moved under the measuring station, we divide it by 2, make the x-derivative, and make the inverse, so obtaining the radius of curvature R(x) resolved in position x along the wafer size W=50 mm. The angle modulation was 0.5 mrad as shown in Fig.5-31 (inset at far right). Noise was about α_n = 5 μrad, corresponding to a maximum measurable radius R_{max}= $[\alpha_n/W]^{-1}$= 50 mm/5 μr=10 km. Yet, quantum noise limit is much smaller, ≈0.05 μrad; thus, potentially, radii up to 1000 km are in the reach of the SMI measurement. Finally, the dynamic range is limited by the response curve width to a maximum angle of about 3.5 mrad, that is, a minimum measurable radius of $R_{min} = [\alpha/W]^{-1} = 16$ m. An example of a typical radius measurement is reported in Fig.5-35.

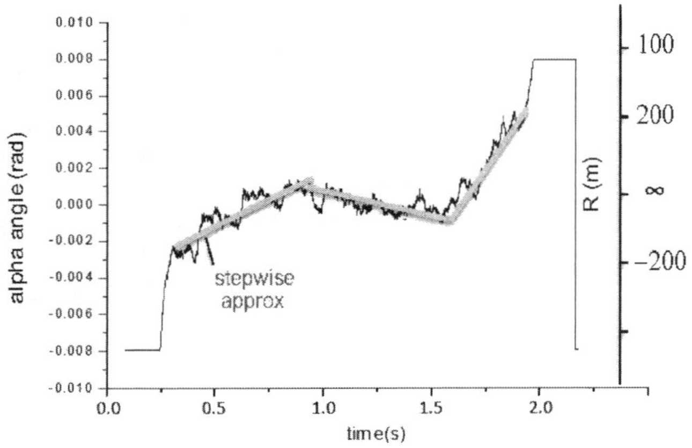

Fig. 5-35 Typical result of a wafer curvature measurement by the SMI interferometer.

5.7 Detection of Weak Optical Echoes

The interferometric signal (Eq.4.2) contains not only the phase difference $\varphi_m - \varphi_r$ –the commonly used quantity exploited in interferometry– but also the amplitude of the field E_m received at the photodetector. Actually, any interferometer can be regarded as a coherent detection scheme (see Ref. [2], Ch.10) in which E_m is the signal and $E_r >> E_m$ is the local oscillator. By beating the fields E_r and E_m at the photodetector, an internal gain is generated, and the quantum limit of detection is attained. Thus, as an amplitude detector, the interferometer has the advantage of a very good sensitivity to small signal and echo returned after a strong attenuation. Coherent schemes are classified as homodyne or heterodyne according to whether $f_m = f_r$ or $f_m \neq f_r$ and as an injection scheme if the returning signal is originated by the source itself. Classical homodyne or heterodyne detection is used in fiber optics communication, whereas injection detection is attractive in instrumentation.

Using Eq.5.18, we can write the photocurrent signal $i_{ph} = \sigma \Delta P$ of the self-mixing configuration in the weak-injection regime as:

$$i_{ph} = I_0 \kappa \, a \cos 2ks / (1 + \kappa \, a \cos 2ks) \quad (5.18')$$

Here, $I_0 = \sigma P_0$ is the dc photocurrent, $\kappa = c\tau_p / n_l L$ is a constant of the order of unity, and a is the total (field) loss in the path to the target at distance s and back, explicitly $a = \sqrt{A}$ in terms of the total power attenuation A.

Eq.5.18' tells us that, at weak injection levels ($a << 1$), the signal has an amplitude proportional to the square root of power attenuation, $a = \sqrt{A}$, as expected from a coherent detection process. When $\kappa a \approx 1$ or $C>1$ (moderate feedback), the amplitude does not increase anymore and we get a saturation of the signal versus attenuation (Fig.5-36).

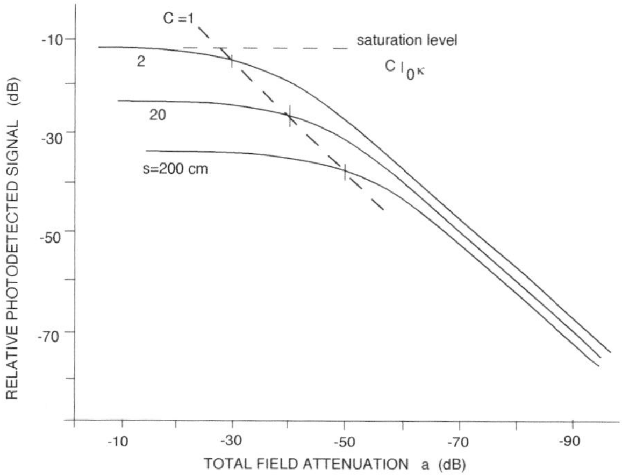

Fig. 5-36 Theoretical dependence of signal amplitude from external attenuation a in an injection interferometer, for some values of distance s.

To measure the amplitude $I_0\kappa a$ without being disturbed by the phase term cos2ks, we need either the distance to be a constant or vary it in a known way. When operating on a substantial distance (say, s >50 cm), ambient-related microphonics usually contributes with a random jitter $s_j(t) >> \lambda$ added to the mean $\langle s \rangle$, and therefore cos2ks is a random waveform with zero mean value and an rms value $1/\sqrt{2}$. In this case, we will measure the rms i_{ph} value of the signal and get the attenuation as $a=\sqrt{2} i_{ph(rms)}/I_0\kappa$.

If distance is short or we want to move the signal off the dc, we may add, in the optical path, a phase modulation $\Phi=\Phi_0 \cos\omega_m t$, with a deviation $\Phi_0>2\pi$ large enough to have a phase term $\cos(2ks+\Phi)$ swinging from -1 to $+1$. In this way, the signal is modulated on a carrier frequency ω_m, at which we will perform the measurement of amplitude.

Typical examples of echo attenuation measurement [47-49] are shown in Fig.5-37. The first setup [47, 48] is for the measurement of the return loss from a fiber device (DUT, device under test).

The optical path length is modulated with the aid of a piezo ceramic (PZT) phase modulator driven at frequency ω_m, and the output signal at frequency ω_m is proportional to the square root of the ratio $P_{back}/P_0=A_{RL}$.
In this scheme, we may also add a second photodetector PD2 to measure the ratio P_{tr}/P_0, of the DUT, that is, its insertion loss.

In the second example, aimed to test the isolation factor of an optical isolator mounted in front of the laser chip [47], we shall use a path-length modulation external to the device. This can be a mirror aligned to the transmitted beam and mounted on a loudspeaker, driven at the desired frequency ω_m.

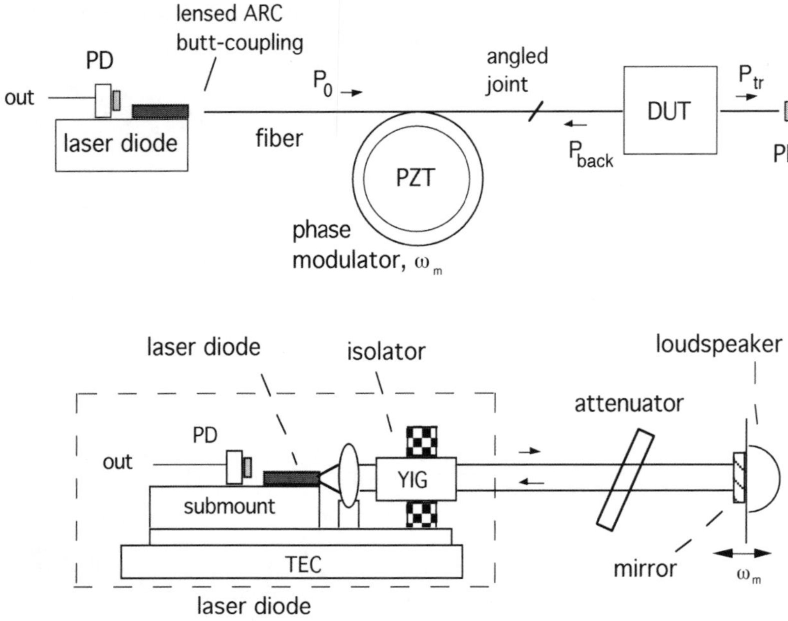

Fig. 5-37 Typical arrangements for the measurement of weak echoes by self-mixing interferometry. Top: an in-line PZT phase modulator is used for measuring the return loss of an all-fiber device (DUT); bottom: to test the isolation factor of an optical isolator mounted in the laser package, a remote vibrating mirror mounted on a loudspeaker supplies the path length modulation.

In this case, the signal is proportional to the square root of the isolation factor P_{back}/P_0. [To make the point clear, 10 dB (or a decade) of attenuation corresponds to half a decade of variation in the current or voltage signal].

Typical performances of the echo detector based on the injection interferometer are illustrated in Fig.5-38. Several experiments have been performed with single-mode laser diodes operating at different wavelengths, either plain Fabry-Perot or DFB (grating reflector) structure. For all of them, the range of measurable attenuation spans from –25 dB to –80 dB [48].

By adding an attenuator in the optical path as in Fig.4-58 (bottom), we can extend the upper limit to about 0 dB.

A systematic study of optical attenuation that can be measured by the self-mixing interferometer, either looking at an external photodiode or measuring the diode anode-cathode voltage, is provided by Ref. [50]. At the quantum limit, attenuations down to -80 dB are measurable with a 10 μW received power and on a bandwidth of 1 kHz, whereas using the voltage across the laser diode terminals, we can theoretically reach the same level, but in practice the performance is penalized up to 40 dB by the thermal (or Johnson) noise of the junction, as observed experimentally at THz waves [51].

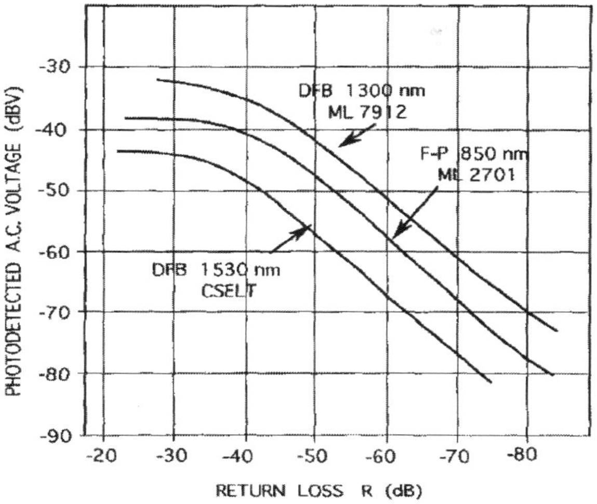

Fig. 5-38 Typical range of optical attenuation or isolation factor that can be measured by a diode laser in the injection interferometer regime of echo detection. Limits are found at -80 dB because of the SNR ratio, and at ≈ -30 dB because of saturation (from Ref. [47], ©IEEE, reprinted by permission).

5.7.1 Consumer Applications of Self-Mixing

Other interesting applications of the echo detection measured by the injection-interferometer have been reported in the fields of optical testing [52], compact-disk readout [53], and scroll sensors [54]. These are applications belonging to the consumer market, which is becoming increasingly important for the self-mixing technology.

In *CD readout*, optical pits correspond to the bits of recorded information, and the conventional design to read the bits employs a laser and photodiode combination, a beamsplitter to divide input and output beam paths, and a conjugating lens to focus on the spot.

Using injection detection, we can dispense with the beamsplitter and the lens by placing the laser diode close to the disk (typically at 10-20 μm off the surface). The rear photodiode will supply the readout signal in the form of spikes corresponding to the bits superposed to the dc quiescent current [53].

Another reported example of an application of the readout of echoes from a remote surface is the *scroll sensor* [54, 55]. The scroll sensor is used in mobile telephone sets and is one of the two axes of the computer mouse. It supplies a signal indicating the up/down direction of a finger or another diffusive surface presented at the viewing window of the sensor, and operates with no physical contact or a particularly clean window.

To be able to determine the direction of the scroll movement, the sensor is interferometric that is, it looks at the phase developed by the moving finger, and has a wavelength modulation to discriminate the sign of the phase change.

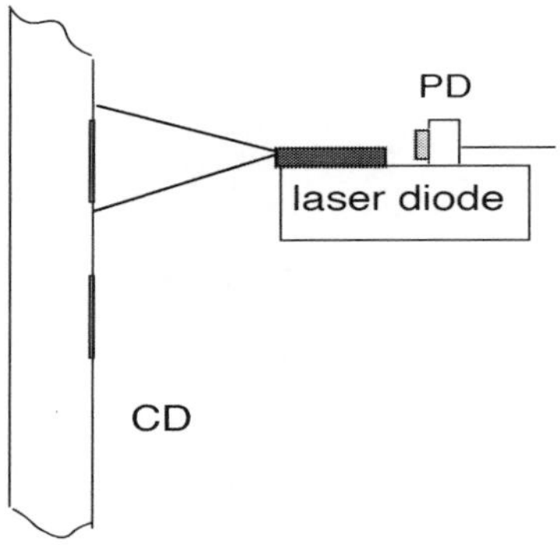

Fig. 5-39 Readout of optical bits in a CD by self-mixing technology: unwritten portions of the CD surface reflect light and give a large signal, whereas pits diffuse light and give a small injection signal at the rear photodiode.

The scroll sensor is designed [55] around low-cost VCSEL lasers in a package including a photodiode sitting at the bottom mirror of the laser (Fig.5-40). The wavelength of operation is typically 620 to 680 nm, easily compliant with safety issues (App.A1.5) because the laser power is modest (typ. ≈ 0.2 mW or less). The incidence angle of the beam is ≈ 45° to develop a large $\Phi=2\,\underline{k}\cdot\underline{s}$ signal [equal to 2ks cos 45° = $2ks/\sqrt{2}$].

The direction of movement is detected by applying a triangular-wave sweep $\Delta\lambda$ to the wavelength, much in the same way as the absolute distance meter described in Sect. 5.5. As distance s is nearly constant, the number of periods $N= s/(\lambda^2/2\Delta\lambda)$ and the frequency of fringes $f_{fr}=N/\Delta T$ (with ΔT = period of the sweep) will also be constant.

Yet, phase Φ changes also because the target (or finger) moves, and $\underline{s}=\underline{s_0}+\underline{\Delta s}$ contributes to the phase difference as

$$\Delta\Phi = \Delta(2\,\underline{k}\cdot\underline{s}) = \Delta(2\,\underline{k}\cdot s_0 + 2\,\underline{k}\cdot\Delta s) = \Delta(2ks_0 + 2k_{\sqrt{2}}\Delta s)$$
$$= 2\,s_0\,\Delta k + 2k_{\sqrt{2}}\Delta s \qquad (5.36)$$

where $k_{\sqrt{2}}=k/\sqrt{2}$ and $\Delta k=-k\Delta\lambda/\lambda$ is the sweep given to wavenumber k. To obtain the self-mixing signal frequency, we divide $\Delta\Phi$ by $2\pi\,\Delta T$ and get:

$$f_{SM} = \Delta\Phi/2\pi\Delta T = 2s_0\,(\Delta k/2\pi\Delta T) + 2(k_{\sqrt{2}}/2\pi)(\Delta s/\Delta T) \qquad (5.36a)$$

Now, $\Delta s/\Delta T=v$ is the speed of the target, $2(k_{\sqrt{2}}/2\pi)=1/\lambda\sqrt{2}$ and $2(\Delta k/2\pi\Delta T)s_0=-2(\Delta\lambda/\lambda^2)\,s_0/\Delta T= f_{fr}$ is the frequency of fringes generated by the sweep. So, Eq.5.36a becomes:

$$f_{SM} = f_{fr} + v/(\lambda\sqrt{2}) \qquad (5.36b)$$

The increasing semiperiod of the λ-modulation has $\Delta\lambda>0$ and so $f_{fr}=-2(\Delta\lambda/\lambda^2)s_0/\Delta T$ is negative, whereas in the decreasing semiperiod $\Delta\lambda<0$ and f_{fr} is positive.
The frequencies of the semiperiods are formally $f_{incr}= -f_{fr}+v/(\lambda\sqrt{2})$ and $f_{decr} = + f_{fr}+v/(\lambda\sqrt{2})$.

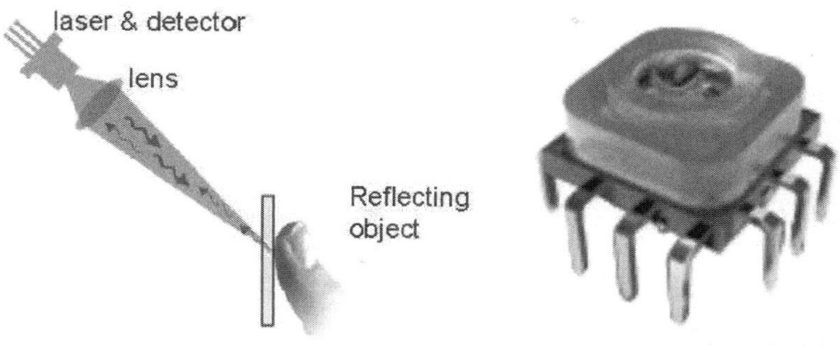

Fig. 5-40 The TwinEye of Philips detects the movement of a target using a self-mixing sensor aimed at the target at an angle (typically 45°) and using a triangle modulation like in Fig.5-25 to detect the direction of movement (from Ref. [55], ©IEEE, reprinted by permission).

As we make $f_{fr}>v/(\lambda\sqrt{2})$ in the experiment, and frequency is always read by the cosine signal as positive, we have $f_{incr}=f_{fr}-v/(\lambda\sqrt{2})$ as the actual frequency read by the interferometer. Finally, we just make the difference of increasing/decreasing semi-periods and obtain:

$$f_{decr} - f_{incr} = \sqrt{2}\ v/\lambda \qquad (5.36c)$$

that is, the velocity v of the target. For the consumer application, Philips (Holland) has developed the Twin-Eye™ Sensor, which incorporates a VCSEL and a preamp/counter circuit to detect the self-mixing signal at up 2...4 m on a diffuser target.

In particular, three chips in hybrid integrated technology have been released, that is, devices optimized for use as: (i) a scroll device, or human input XY sensor, (ii) translational measurement (paper movement) and (iii) car speed meter (automotive speed) [55].

5.8 SMI Measurements of Physical Quantities

Looking at details of the self-mixing waveform, we can measure important parameters of the laser diode we are using, like the linewidth $\Delta\nu$, or maximum distance on which we can make interferometry, and the linewidth enhancement factor α_{en} that tells the excess noise contribution due to frequency fluctuations introduced in Lang-Kobayashi equations (see Eq.5-12).

Then, we will show that the algorithm for α_{en} of the laser also supplies the C factor of the experiment. Last, we will describe the simultaneous measurement of thickness and index of refraction of a substrate.

5.8.1 Linewidth Measurement by SMI

With a plain self-mixing interferometer that receives the return from a loudspeaker placed at a fixed distance on an anti-vibration table, we detect the usual fringe signal shown in Fig. 5-41. Switching occurs for C>1 in the waveform, and to observe it we ensure a return strong enough by using a mirror (or a corner cube) glued on the loudspeaker, and an optical attenuator to trim the amplitude of attenuation (Fig. 5-41, top).

The switching exhibits some jitter, due to the linewidth of the laser [56]. Frequency fluctuations $\Delta\nu$ due to a finite linewidth reflect themselves in a wavenumber fluctuation Δk, being $\Delta k = (2\pi/c)\Delta\nu$. From Δk we can get a phase fluctuation $\Phi = 2s_0\Delta k$, responsible for the time-jitter of the switching we observe (Fig. 5-41, bottom).
Indeed, the phase read by the self-mixing interferometer is written as:

$$\Phi = 2s_0\Delta k + 2k\Delta s \qquad (5.37)$$

We apply a small displacement Δs to the loudspeaker, so as to develop the usual detected signal $I = I_0(1+\cos\Phi)$, see Fig.5-42 bottom right, and acquire the phase deviation $\Delta\Phi$.
Using $\Delta k=(2\pi/c)\Delta\nu s$ in Eq.5.37, and making the quadratic average of terms, we get:

$$\langle\Phi^2\rangle = (2s_0 2\pi/c)^2\langle\Delta\nu^2\rangle + [2k\Delta s]^2 \qquad (5.37a)$$

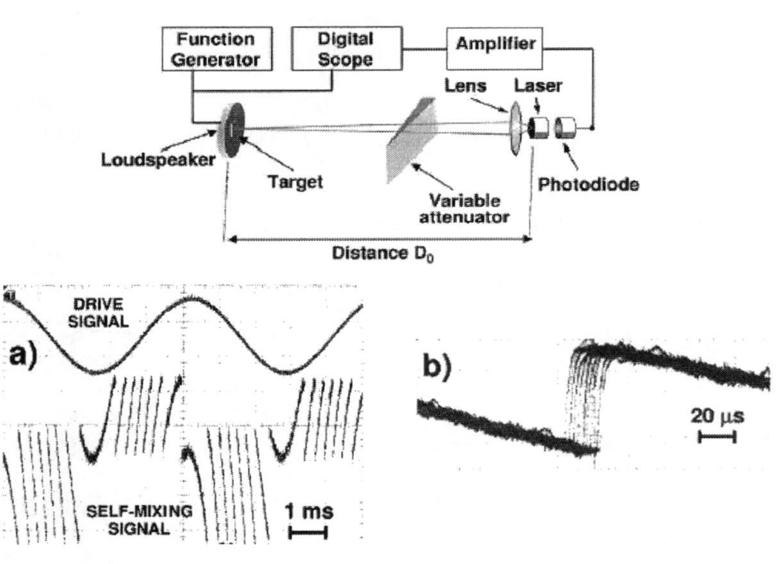

Fig. 5-41 Top: the measurement setup; bottom left: the usual self-mixing waveform, showing switching at C>1, on an exemplary distance s_0=2m; bottom right: the switching exhibits a time jitter due to non-zero linewidth of the laser (from Ref. [56], ©IEEE, reprinted by permission).

Thus, the measured rms phase is the square root of two quadratic contributions: the laser linewidth weighted by distance s_0, and the small displacement phase $k\Delta s$. Thus, the diagram of rms phase $\sqrt{<\Phi^2>}$ against distance is that of a quadratic sum, and becomes a linear dependence when term $[2k\Delta s]^2$ becomes negligible, when we can find $\sqrt{<\Delta\nu^2>}$ as the slope of the $\sqrt{<\Phi^2>}$ versus s_0 diagram. The result of this processing is shown in Fig.5-42 for two diode lasers at different bias currents [56].

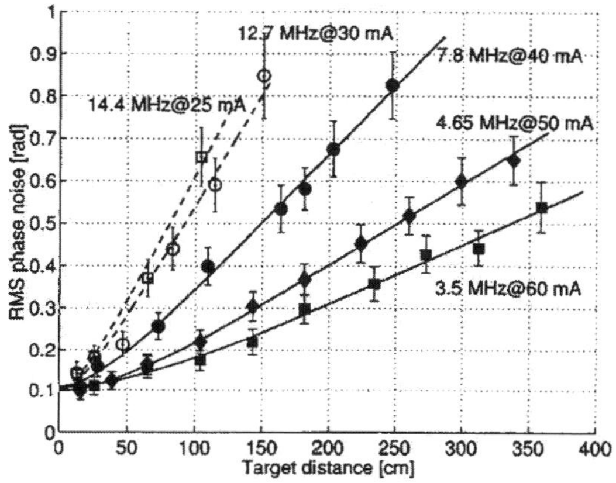

Fig. 5-42 Linewidth is calculated from the rms phase Φ vs distance measurement. Open dots are for the ML2701 diode laser, and full dots for the SDL540 (from Ref. [56], ©IEEE, reprinted by permission).

About the disturbance introduced by optical feedback to the linewidth measurement, it has been found [57, 58] that the perturbed linewidth $\Delta\nu_{SM}$ is related to the unperturbed one $\Delta\nu$ and obeys the following equation (cfr. Eq.5.16b):

$$\Delta\nu_{SM} = \Delta\nu / [1 + C \cos(2ks + \text{atan}\, \alpha_{en})] \tag{5.38}$$

Thus, the measured $\Delta\nu_{SM}$ may differ from the true $\Delta\nu$ by a factor up to $1/(1\pm C)$. However, as the switching at which we measure the phase jitter occurs at $2ks=\pi$ and α_{en} is usually large (e.g., ≈ 6) for diode lasers, so that $\cos(2ks+\text{atan}\, \alpha_{en}) \approx \cos(2ks+\pi/2) = \sin(2ks) \approx 0$, the method usually doesn't require a correction [that can, however, be introduced by applying Eq.5.38]. Last, it is worth noting that the above method based on self-mixing is much simpler and requires much less propagation space (max s_0=3.5m) than traditional methods based on delayed heterodyning.

5.8.2 SMI Measurement of Alpha and C Factors

The self-mixing waveform $F(\Phi)$ at $C>1$ deviates from the sinusoid in a characteristic way influenced by the alpha factor as well as by the C factor. There are several possible ways to describe the distortion of $F(\Phi)$. As a first choice, we take the time intervals t_{13} and t_{24} depicted in Fig. 5-43, i.e., the time from a positive (negative) going zero-crossing to the maximum (minimum) of the waveform, in correspondence with a negative (positive) going switching. Time intervals t_{13} and t_{24}, divided by the periods T_1 and T_2 of the upgoing (downgoing) periods of the waveform, give the two phases as a fraction of 2π, that is:

$$t_{13}/T_1 = \phi_{13}/2\pi = X_{13} \quad \text{and} \quad t_{24}/T_2 = \phi_{24}/2\pi = X_{24} \tag{5.39}$$

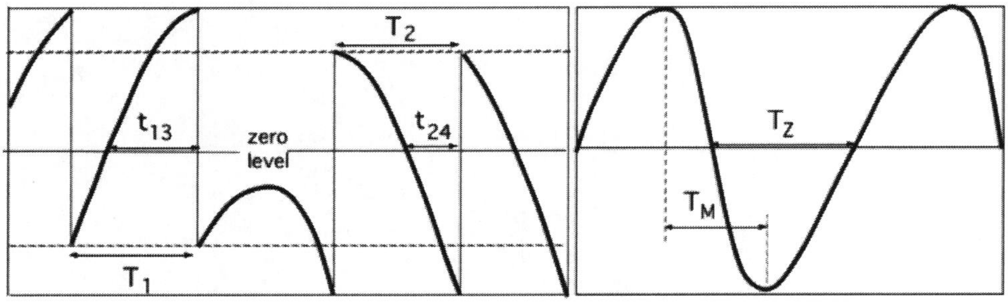

Fig. 5-43 Time intervals selected on the self-mixing waveform for the evaluation of factors α_{en} and C: at left, for C>1; at right, for C<1.

and now, for phases ϕ_{13} and ϕ_{24} an analytical expression can be found after some lengthy algebra, using Eq.5.16. These expressions read [58]:

$$\phi_{13} = (C^2 - 1)^{1/2} + C\,(1+\alpha_{en}^2)^{-1/2} + \arccos(-1/C) - \operatorname{atan}\alpha_{en} + \pi/2 \tag{5.40a}$$

$$\phi_{24} = (C^2 - 1)^{1/2} - C\,(1+\alpha_{en}^2)^{-1/2} + \arccos(-1/C) + \operatorname{atan}\alpha_{en} - \pi/2 \tag{5.40b}$$

Inverting Eqs.5.40 to find α_{en} and C as a function of ϕ_{13} and ϕ_{24} is not an easy matter, but we can use two approaches for the solution: (i) numerical calculation based on successive approximations until a specified error is reached; (ii) graphical solution based on plotting Eqs.5.40 in an X_{13}-X_{24} diagram with α_{en} and C as parameters.

In Fig.5-44 we show the result of the second method: entering the diagram with a pair of $X_{13}=t_{13}/T_1$ and $X_{24}= t_{24}/T_2$ values we identify a point and get a pair of α_{en} and C.
This method has been applied to a number of laser diodes with α_{en} ranging from 2.2 to 4.9 and C from 1.2 to 2.2 [59] and the resulting accuracy was ±5% for both α_{en} and C.

This is more than enough for use in the reconstruction equations of the displacement waveform s(t) measured by self-mixing (see Eq.5-19). The above method requires that C>1 to read the time intervals around the waveform switching. When C<1 we can use the semiperiods T_1 and T_2 of Fig.5-43 and solve Eq.5-16 as (see Ref. [19]):

$$T_1/T_2 = [\pi\,(1+\alpha_{en}^2)^{1/2} - 2\alpha_{en}C]\,/\,[\pi\,(1+\alpha_{en}^2)^{1/2} + 2\alpha_{en}C] \tag{5.41}$$

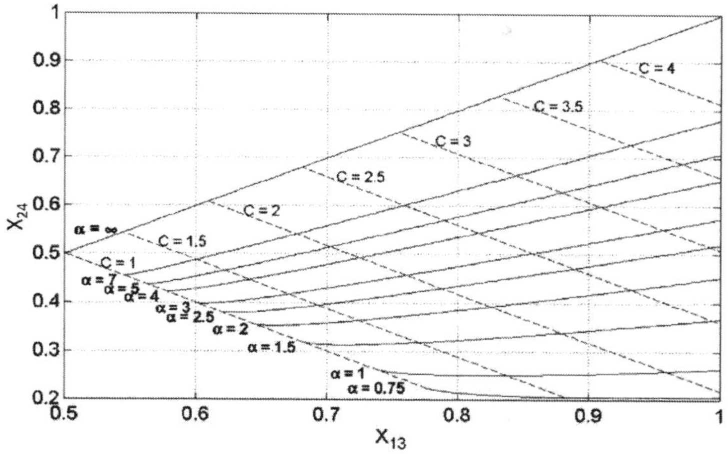

Fig. 5-44 Graph for the inversion of Eqs.5.40, giving α_{en} and C as a function of intervals X_{12} and X_{34} measured on the self-mixing waveform of Fig.5-43 (from Ref. [59], ©IEEE, reprinted by permission).

in which $T_1<T_2$, otherwise the equation holds for T_2/T_1. Assuming now that α_{en} >>1, Eq.5.41 is brought to the form:

$$C = [(1- T_1/T_2)/(1+ T_1/T_2)](\pi/2)[1+1/2\alpha_{en}^2] \quad (5.42)$$

or, the C factor calculated by this method has only a minor sensitivity on α_{en}.

In addition, always at C<1, introducing the time intervals T_M and T_Z defined as the positive peak-to-negative peak time, and the zero-crossing interval of the waveform, (Fig.5-43), it has been found [60] that the linewidth enhancement factor is given by:

$$\alpha_{en} = (T_M/T\text{-}0.5)/(T_Z/T\text{-}0.5) \quad (5.43)$$

where T is the total period of the waveform. Similarly, we get

$$C= \pi\, [(T_Z -0.5)^2+ (T_M -0.5)^2]^{1/2} \quad (5.43')$$

Eqs. 5.42 and 5.43 have an accuracy of a few percent in both C and α_{en}.

5.8.3 SMI Measurement of Thickness and Index of Refraction

Upon measuring the optical phase ϕ suffered by a beam crossing a slab of thickness L in a material of index of refraction n, whatever the method used, we get as a result ϕ = nL, that is, the optical pathlength given by the product of thickness and index of refraction. It is not possible in a single measurement to sort out either n or L. The schematic of Fig.5-45, based on the measurement made by a self-mixing interferometer superposed to a shearing interferometer, allows us to get n and L separately [61,62].

In the setup, a laser diode with back-mirror photodiode PD1 sends a beam to cross a slab of material, making a double pass thanks to the reflection at the bottom surface of the slab (dotted lines in Fig.5-45) and reaches photodiode PD2, where phase ϕ_{SH} is detected as the output of a shearing interferometer. In addition, the beam is reflected back from the entrance window of the photodiode (PD2), traces back the incoming rays and is collected by the self-mixing interferometer (combination of Laser Diode and PD1) where it gives rise to a phase signal ϕ_{SMI} [61]. Analyzing the optical path in Fig.5-45, after some lengthy calculation, the two phases are written as:

$$\phi_{SH} = 2\,k\,L\,n\cos\theta \tag{5.44a}$$

$$\phi_{SMI} = 2\,k\,L\,(n\cos\theta - \cos\alpha) \tag{5.44b}$$

where L is the slab thickness, α is the tilt angle of the slab, and θ is the refraction angle internal to the slab (Fig.5-45). By subtraction of the phase terms of Eq.5.44 we get:

$$\phi_{SH} - \phi_{SMI} = 2\,k\,L\cos\alpha \tag{5.45}$$

a result independent from n. Now, scanning the slab tilt α with a motorized stage, we get L as the multiplying coefficient of the cosine function shown in Fig.5-46. The curve depends on thickness only and is the same for any material, as it has been checked by measuring glass and silicon [61,62]. With the method described above, we can measure L in the range 20 to 1000 μm with a relative accuracy of ±1%.

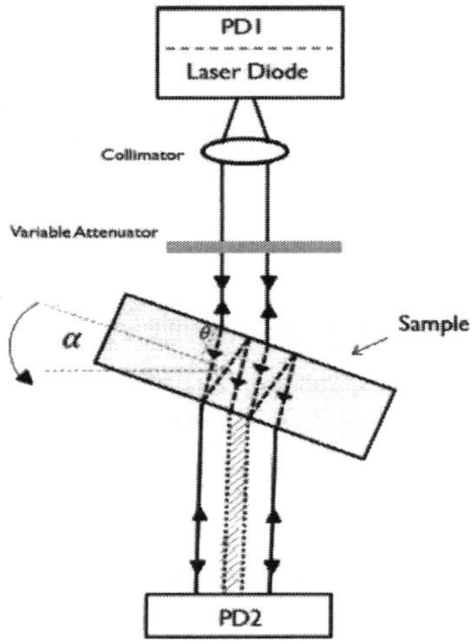

Fig. 5-45 Schematic of the combined self-mixing plus shearing interferometer for measuring thickness and index of refraction (from Ref.[61], ©Optica Publ., reprinted by permission).

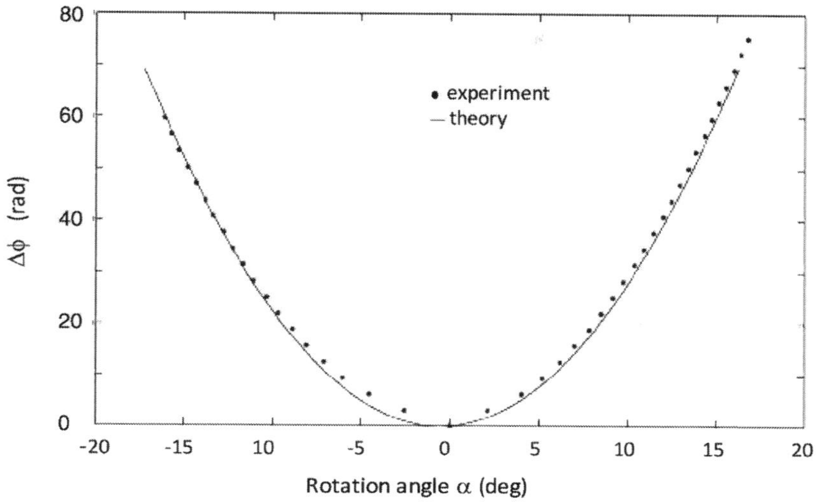

Fig. 5-46 Plot of Eq.5.45 as a function of tilt angle α scanned with a motorized rotating stage. Experimental points are for a 150-μm thickness slab of glass (from Ref.[61], © Optica Publ., reprinted by permission).

Once thickness has been measured, we can go back to Eq.5-44a, insert in it the measured value of L and calculate the index of refraction, too, obtaining the n-value with a typical accuracy of ±0.04. Attenuation of the beam in the slab and in setup doesn't affect the measurement, in principle. Accordingly, semiconductor material can be tested also below the photoelectric threshold λ_t (Ref. [11], Sect.1.2) where they absorb. For example, at λ<1100 nm for Silicon, the beam attenuation across the slab can be tolerated up to 20 dB.

5.9 SMI Measurements for Medicine and Biology

Applications to medicine and biology of interferometry and self-mixing technology is a new segment promising important developments [27, 28, 63]. The OCT (see Sect. 4.7) has already gained acceptance and is routinely used in clinical practice, and other instruments may follow in the near future. Here we present some applications still in their infancy but scientifically and technically interesting.

We have already described an early application of self-mixing to the pickup of arterial pulsation (Fig.5-4). More recently, blood pulsation on the finger was measured by a diode laser [64] and in several body areas [65] showing similarity of the vibration waveforms called VCG (vibrocardiograph) to those of the ECG (electrocardiograph), see Fig.5-47.
While the actual shape of the VCG waveforms and its details are quite peculiar and different from the EGG waveform in the area close to the heart (Fig.5-47), on the finger as we can see in Fig.5-4, and in peripheral areas the waveforms closely resemble the ECG.

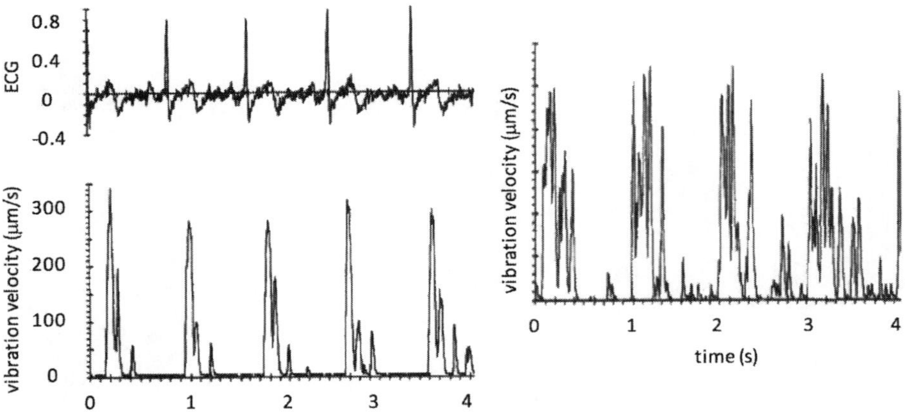

Fig. 5-47 Top left: the normal ECG; bottom left: the VCG at the radial artery; right: the VCG at the tricuspid valve region (from Ref. [65], ©IEEE, reprinted by permission).

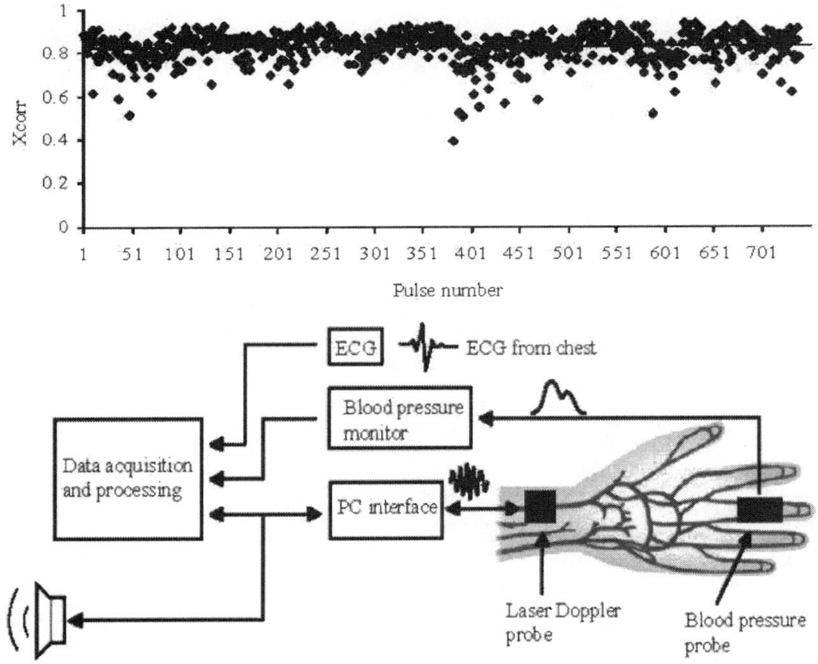

Fig. 5-48 Top: correlation of blood pressure derivative and VCG. Bottom: measurement setup (from Ref. [66], by courtesy of Turpion, Moscow).

Indeed, in a study carried out by Myllyla et al. [66] the VCG waveform has been compared to the pressure wave on the wrist, finding a high (>0.84) degree of correlation (Fig.5-48). So, while the optical pickup VCG has not yet reached the level of a routine diagnostic tool, when an exhaustive clinical method is developed it may likely supplement or even replace the traditional ECG testing.

Another measurement with the early He-Ne SMI was about respiratory sounds aiming at the upper back thorax of a patient [13]. Inspiration and expiration noise were neatly detected and, compared to the corresponding pickup by a stethoscope, they show no artifacts like in the acoustic counterpart, affected by noise of skin friction under the stethoscope (Fig.5-49).

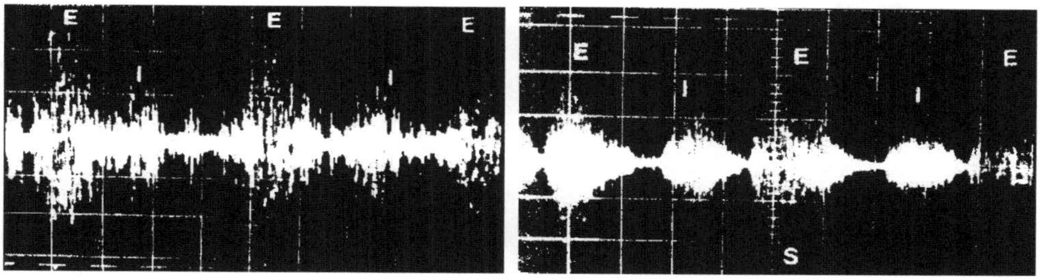

Fig. 5-49 Left: respiratory sounds measured by an acoustical stethoscope on the patient's back [E = expiration, I = inspiration] showing many artifact spikes due to skin friction; right: the same signal measured by a self-mixing VCG (vertical scale 200 nm/div, time scale 0.75 s/div) (from Ref. [27], ©IEEE, reprinted by permission).

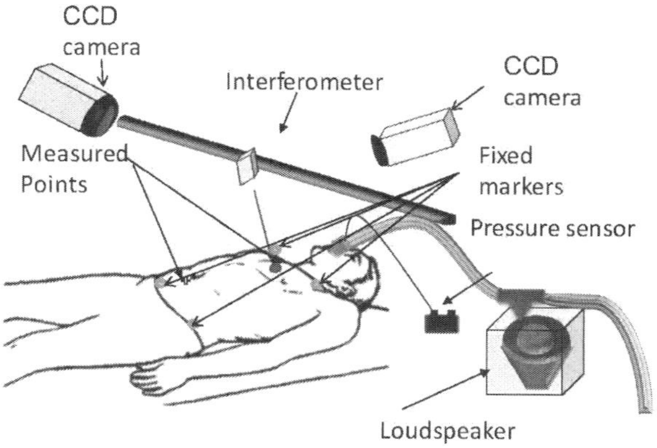

Fig. 5-50 Setup for measuring the acoustic impedance of the thorax: the self-mixing interferometer detects the frequency of fringes f_{SM} whence velocity v is calculated; a loudspeaker applies the driving pressure, read by a sensor at the mouth opening (from Ref. [27], by courtesy of the IEEE).

Acoustical impedance of the chest $Z_{tr} = \Delta P/v$ has been measured by a forced oscillation technique [67]. Impedance Z_{tr} provides information on tissues and airways, useful in checking the mechanical properties of lungs and associated disorders.

In the experiment, a variation of breathing pressure ΔP is applied to the mouth of the patient by a loudspeaker (Fig.5-50) and the VCG signal on the thorax is measured.

Velocity v was measured by making the frequency spectrum of the self-mixing signal and sorting the frequency f_{sm} of the self-mixing signal, and then computing $v = (\lambda/2)\, f_{SM}$; pressure ΔP was taken by a conventional sensor [67].

Blood flow measurement is another important application of self-mixing technology. But, as flow F is the product of velocity v and cross-section area A of the tube carrying the fluid, that is F=vA, we will treat this instrument as a by-product of a velocimeter in Ch.7.

REFERENCES

[1] G.A. Acket, D. Lenstra, A.J. DenBoef, B.H. Verbek: "*The influence of feedback intensity on longitudinal mode properties and optical noise in index-guided semiconductor lasers*", IEEE J. Quant. Electr. QE-20 (1984), pp.1163-1169.

[2] P.W. Milonni, J.H. Eberly: "*Laser Physics*", 2010, J. Wiley & Sons Inc., Hoboken (NJ).

[3] C.H. Henry: "*Theory of Spontaneous Noise in Open Resonators and its Application to Laser and Optical Amplifiers*", IEEE J. Lightw. Techn., vol. LT-4 (1986), pp.288-1297.

[4] R. Lang and K. Kobayashi: "*External Optical Feedback Effects on Semiconductor Injection Laser Properties*", IEEE J. Quant. Electr., vol. QE-16 (1980), pp.347-355.

[5] T. Bosch, N. Servagent, and S. Donati: "*Optical Feedback Interferometry for Sensing Applications*", J. of Opt. Engin., vol.40 (2001), pp.20-27.

[6] S. Donati: "*Developing self-mixing interferometry for instrumentation and measurements*", Laser Phot. Rev., vol. 6 (2012), pp.393–417.

[7] J. Ohtsubo: "*Semiconductor Lasers, Stability, Instability and Chaos*" 2nd ed., 2008, Springer, Berlin.

[8] S. Donati, S.-K. Hwang: "*Chaos and High-Level Dynamics in Coupled Lasers and their Applications*", Progr. Quant. Electr. vol.36, (2012), pp.293–341.

[9] M.B. Spencer, W.E. Lamb: "*Laser with a Transmitting Window*", Phys. Rev. vol.A5 (1972), pp.884-891.

[10] F.T. Arecchi et al.: "*Deterministic Chaos in Laser with Injected Signal*", Opt. Comm. vol.51 (1984), pp.308-314.

[11] S. Donati: "*Photodetectors*", 2nd ed., Wiley IEEE Press, Hoboken 2021.

[12] S. Donati: "*Laser Interferometry by Induced Modulation of the Cavity Field*", J. Appl. Physics, vol.49 (1978), pp.495-497.

[13] S. Donati and V. Speziali: "*Laser Interferometry for Sensing of Respiratory Sounds*", Proc. CLEA Conf. on Laser and Electrooptics Appl., Washington, June 1977, Digest in IEEE J. Quant. Electr., vol. QE-13 (1977), pp.798-87D; see also "*Interferometric Sensing of Respiratory Sounds*", Laser & Elektro-Optik, Munchen, vol.12 (1980), pp.34-35.

[14] S. Donati: "*A Novel Laser Interferometer for Distance Measurements*", Proc. Conf. on Precision Electromagn. Meas. Ottawa (1978), pp.75-77.

[15] S. Donati, G. Giuliani, and S. Merlo: "*Laser Diode Feedback Interferometer for the Measurements of Displacements without Ambiguity*", IEEE J. Quant. Electr., vol. QE-31 (1995), pp.113-119.
[16] N. Servagent, F. Gouaux, and T. Bosch: "*Measurement of Displacement using the Self-Mixing Interference in a Laser Diode*", J. Optics, vol.29 (1998), pp.168-173.
[17] W.M. Wang, K.T.W. Grattan, A.W. Palmer, and W.J. Boyle, "*Self-Mixing Interference inside a Single-Mode Laser Diode for Optical Sensing Applications*", IEEE J. Lightw. Techn. vol.LT-12 (1994) pp.1577-1587.
[18] P. Dean, Y.L. Lim, A. Valavanis, R. Kleise, M. Nikolic, S.P. Khanna, M. Lachab, D. Indjin, Z. Ikonic, P. Harrison, A.D. Rakic, E.H. Lindfield, G. Davies: "*Terahertz Imaging through Self-Mixing in a Quantum Cascade Laser*", Opt. Lett. vol.36 (2011), pp.2587-2589.
[19] S. Merlo and S. Donati: "*Reconstruction of Displacement with a Single-Channel Laser-Diode Feedback Interferometer*", IEEE J. Quant. El., vol. QE-33 (1997), pp. 527-531.
[20] R.S. Vodhanel, M. Krain, and R.E. Wagner: "*Long-Term Wavelength Drift of 0.01nm/y for 15 Free-Running DFB Lasers*", Proc. OFC-94, 20-25 Feb.1994, paper WG5, pp.103-104.
[21] S. Donati, L. Falzoni, and S. Merlo: "*A PC-Interfaced, Compact Laser-Diode Feedback Interferometer for Displacement Measurements*", IEEE Trans. Instr. Meas., vol.45 (1996), pp.942-947.
[22] X. Wan, D. Li, S. Zhang: "*Quasi-common-path Laser Feedback Interferometer based on Frequency Shifting and Multiplexing*", Opt. Lett. vol.32 (2007), pp.367-369.
[23] M. Laroche, C. Bartolacci, G. Lesueuer, H. Gills, S. Girard: "*Serrodyne Optical Frequency Shifting for Heterodyne Self-Mixing in a Distributed-Feedback Fiber Laser*", Opt. Lett. vol.33 (2008), pp.2746-2748.
[24] S. Donati, M. Norgia: "*Self-Mixing Interferometer with a Laser Diode: unveiling the FM channel and its advantages respect to the AM channel*", IEEE J. Quant. Electr. vol.53 (2017).
[25] M. Norgia, D. Melchionni, S. Donati: "*Exploiting the FM-signal in a laser-diode SMI by means of a Mach-Zehnder filter*", IEEE Phot. Techn. Lett., vol.29, (2017), pp.1552 -1555, DOI: 10.1109/LPT. 2017.2735899.
[26] S. Donati, M. Norgia: "*Overview of self-mixing interferometer applications to mechanical engineering*", Opt. Engin., vol.57 (2018), doi:10.1117/1.OE.57.5.051506.
[27] S. Donati, M. Norgia: "*Self-mixing Interferometry for Biomedical Signals Sensing*", (invited paper), IEEE J. Select. Topics Quant. El. 20 (2014), DOI 10.1109/JSTQE.2013. 2270279
[28] G. Kaiser: "*Biophotonics: Concepts to Applications*", 2nd ed., Springer, Berlin 2022.
[29] G. Giuliani, S. Bozzi-Pietra, and S. Donati: "*Self-mixing Laser Diode Vibrometer*", Meas. Sci. Technol., vol.14 (2003), pp.24-32.
[30] S. Donati, M. Norgia, and G. Giuliani: "*Self-mixing differential vibrometer based on electronic channel subtraction*", Appl. Opt., vol. 45 (2006), pp. 7264-7268.
[31] S. Donati, V. Annovazzi Lodi, S. Merlo, and M. Norgia: "*Measurements of MEMS Mechanical parameters by Injection Interferometry*", Proc. IEEE-LEOS Conf. Opt. MEMS, Kawai, HI, 21-24 Aug.2000, pp.89-90,
[32] V. Annovazzi-Lodi, M. Benedetti, S. Merlo, M. Norgia: "*Spot Optical Measurements on Micromachined Mirrors for Photonic Switching*", IEEE J. Sel. Top. Quant. Electr., vol.10 (2004), pp.536-544.
[33] G. Giuliani, M. Norgia, S. Donati, T. Bosch: "*Laser Diode Self-Mixing Technique for Sensing Applications*", IoP J. Optics A, vol.4, (2002), pp. S283-S294.

[34] F. Gouaux, N. Sarvagent, and T. Bosch: *"Absolute Distance Measurement with an Optical Feedback Interferometer"*, Appl. Opt., vol.37 (1998), pp. 6684-6689.
[35] P.J. de Groot, G. Gallatin, and S.H. Macomber: *"Ranging and Velocimetry Signal in a Backscattering Modulated Laser Diode"*, Appl. Opt. vol.27 (1988) pp.4475-4480; see also: Optics Lett., vol.14 (1989) pp.165-167.
[36] T. Bosch, S. Pavageau, D. d'Alessandro, N. Servagent, V. Annovazzi-Lodi and S. Donati: *"Low-cost Optical Feedback Laser Range-Finder with Chirp Control"*, IEEE Instr. and Measur. Techn. Conf., Budapest 2001.
[37] M. Norgia, A. Magnani, A. Pesatori: *"High resolution self-mixing laser rangefinder"*, Rev. Sci. Instr. vol 83 (2012).
[38] M. Norgia, G. Giuliani, S. Donati: *"Absolute Distance Measurement with Improved Accuracy Using Laser Diode Self-Mixing Interferometry in a Closed Loop"*, IEEE Trans. Instr. Measur. vol.IM-56 (2007), pp.1894-1900.
[39] H. Matsumoto: *"Alignment of Length-Measuring IR Laser Interferometer using Laser Feedback"*, Appl. Opt., vol.19 (1980), pp.1-2.
[40] G. Giuliani, S. Donati, M. Passerini and T. Bosch: *"Angle Measurement by Injection Detection in a Laser Diode"*, Opt.l Engin., vol.40 (2001), pp.95-99.
[41] S. Donati, D. Rossi, M. Norgia: *"Single Channel Self-Mixing Interferometer Measures Simultaneously Displacement and Tilt and Yaw Angles of a Reflective Target"*, IEEE J. Quant. Electr., vol. QE-51(2015).
[42] T. Tambosso, R.-H. Horng, S. Donati: *"Curvature of Substrates is Measured by means of a Self-Mixing Scheme"*, IEEE Phot. Techn. Lett., vol. 26 (2014), pp. 2170-2174.
[43] Q. Jingya, W. Zhao, J.H. Huang, B. Yu, G. Jianmin, S. Donati: *"Enhancing the Sensitivity of Roll-Angle Measurement with a Novel Interferometric Configuration based on Waveplates and Folding Mirror"*, Rev. Sci. Instr. vol.87 (2016).
[44] J. Qi, Wang Zhao, J. Huang, J. Gao: *"Resolution-enhanced heterodyne laser interferometer with differential configuration for roll angle measurement"*, Opt. Expr. vol. 26, (2018), pp.9634-9644.
[45] D.S. Campbell: *"Mechanical properties of thin films"*, in: Handbook of Thin Film Technologies, ed. by L.I. Maissel and R. Glang, (1970), pp.12-13, McGraw Hill, New York.
[46] S.N. Sahu, J. Scarminio, F. Decker: *"A Laser Beam Deflection System for Measuring Stress Variations in Thin Film Electrodes"*, J. Electr. Soc., vol.137 (1990), pp.1150-1154.
[47] S. Donati and M. Sorel: *"A Phase-Modulated Feedback Method for Testing Isolators Assembled into the Laser Package"*, Phot. Techn. Lett., vol.PTL-8 (1996), pp.405-407.
[48] S. Donati and M. Sorel: *"High-Sensitivity Measurement of Return Loss by Self-Heterodyning in a Laser Diode"*, OFC'97, Proc. Opt. Fiber Conf., Dallas, 1997, p.161.
[49] S. Donati and G. Giuliani: *"Return Loss Measurement by Feedback Interferometry"*, Proc. WFOPC'98, Workshop on Fiber Opt. Comp. Pavia, 1998, pp.103-106.
[50] S. Donati: *"Responsivity and Noise of Self-Mixing Photodetection Schemes"*, IEEE J. Quant. El., vol.47 (2011), pp.1428-1433.
[51] S. Lui, T. Taimre, K. Bertling, Y.L. Lim, P. Dean, S.P. Khanna, M. Lachab, A. Valavanis, D. Indjin, E.H. Lindfield, G. Davies, A.D. Rakic: *"Terahertz Inverse Synthetic Aperture Radar Imaging using Self-Mixing Interferometry with a Quantum Cascade Laser"*, Opt. Lett. vol.39, (2014), pp.2629-2632.
[52] P. de Groot, G. Gallatin, and G. Gardopee: *"Optical Testing using Laser Feedback Metrology"*, Conf. on Laser Interf., San Diego, SPIE vol.1162, (1989), pp.435-442.

[53] H. Ukita, Y. Uenishi, and Y. Katagiri: "*Application of an Extremely Short Strong-Feedback External-Cavity Laser Diode System Fabricated with GaAs-based Integration Technology*", Appl. Opt. vol.33 (1994), pp.5557-5563; see also: S. Shinohara, H. Naito, H. Yoshida, H. Ikeda, M. Sumi: "*Compact and Versatile Self-Mixing Type Semiconductor Laser Doppler Velocimeter with Direction Discrimination Circuit*", IEEE Trans. Instr. Measur. vol.38 (1989), pp.574-577; see also IEEE Trans. Instr. and Measur., vol.41 (1992), pp.40-44.
[54] J. Hewett: "*Optical Interface to Give Mobile Phone a New Look*", Opt. Laser Eur., vol. 97 (Jul./Aug.2002), p.9; see also J. of Meas. Sci. and Techn. vol.13 (2002), pp. 2001-2006.
[55] A. van der Lee, M. Carpaij, H. Monch, M. Schemmann, A. Pruijmboom: "*A miniaturized VCSEL based sensor platform for velocity measurement*", IEEE Conf. I^{2}MTC, Victoria, (Canada), May 12-15 2008, DOI: 10.1109/IMTC.2008.4547019
[56] G. Giuliani, M. Norgia, S. Donati: "*Laser Diode Linewidth Measurement by means of Self-Mixing Interferometry*", Proc. LEOS Annual Meet. San Francisco, Nov. 8-11, 1999, pp. 726-727.
[57] K. Petermann: "*Laser Diode Modulation and Noise*", Kluwer: Dordrecht, 1988.
[58] N. Schunk and K. Petermann: "*Stability Analysis of Laser Diodes with Short External Cavity*", IEEE Phot. Techn. Lett., vol. LT-1 (1989), pp.49-51.
[59] Y. Yu, G. Giuliani, and S. Donati: "*Measurement of the linewidth enhancement factor of semiconductor lasers based on the optical feedback self-mixing effect,*" IEEE Phot. Techn. Lett. vol.16 (2004), pp.990–992.
[60] Y. Yu, J. Xi, E. Li, J. F. Chicharo: "*Measuring the Linewidth Enhancement Factor of Semiconductor Lasers with Weak Optical Feedback*", Proc. SPIE vol.5628, (2005) pp.34-39.
[61] M. Fathi, S. Donati: "*Thickness Measurement of Transparent Plates by a Self-Mix Interferometer*", Opt. Lett., vol.35, (2010), pp.1844-46.
[62] M. Fathi, S. Donati: "*Simultaneous Measurement of Thickness and Refractive Index by a Single-Channel Self-Mixing Interferometer*", Proc. IET, Optoelectr. vol.6 (2012) pp.7–12.
[63] S. Donati, M. Norgia: "*Self-mixing Interferometry for Biomedical Signals Sensing*", IEEE J. Sel. Topics Quant. El. vol.20 (2014), DOI 10.1109/JSTQE.2013 .2270279
[64] A. Arasanz, F.J. Azcona, S. Royo, A.J. Pladellorens: "*A new method for the acquisition of arterial pulse wave using self-mixing interferometry*", Optics & Laser Techn., vol.63 (2014), pp. 98-104.
[65] H.D. Hong and M.D. Fox: "*No Touch Pulse Measurement by Optical Interferometry*", IEEE Trans. Biomed. Engin., vol.4 (1994) pp.1096-98.
[66] J. Hast, R. Myllylä, H. Sorvoja, J. Miettinen: "*Arterial pulse shape measurement using self-mixing effect in a diode laser*" Quant. Electr. vol.32 (2002) pp.975-982.
[67] I. Milesi, M. Norgia, P. P. Pompilio, C. Svelto, R. Dellacà: "*Measurement of Local Chest Wall Displacement by a Compact Self-Mixing Laser Interferometer*", IEEE Trans. Instr. Meas. vol. IM-60 (2011), pp. 2894-2901.

Problems and Questions

P5-1 *A self-mixing interferometer uses a 20-cm He-Ne laser with output mirror reflectivity of 99% and has a round-trip gain* $2\gamma_0 L=0.02$. *Evaluate the AM index and the FM frequency deviation. Assume a spot size* $w=0.2$ *mm and a target to be either diffusive or reflective, placed at a distance* $s=50$ *cm.*

P5-2 *Repeat the calculation of Prob.P5-1 for the AM modulation of a semiconductor laser, with output mirror reflectivity of 30% and a round-trip gain $2\gamma_0 L=3$. Assume that an output objective lens makes the beam circular, into a spot size w=0.25 mm. Allow for a target, either diffusive or reflective, placed at a distance s=50 cm, and calculate the factor C.*

P5-3 *Calculate the speed of response of the AM/FM modulations for an interferometer that uses the He-Ne laser with the data of Prob.P5-1, or the diode laser with the data of Prob.P5-2. Assume s=50 cm.*

P5-4 *How can we calculate the minimum-displacement or NED performance of a single AM channel self-mixing interferometer considered in problems P5-1 and P5-2?*

P5-5 *Evaluate accuracy and precision of self-mixing interferometers that use as a source: (i) a frequency-stabilized He-Ne laser; (ii) a monomode He-Ne laser; (iii) a thermal stabilized DFB diode laser; (iv) a Fabry-Perot thermal stabilized laser diode; (v) a normal Fabry-Perot laser diode.*

P5-6 *Evaluate the NED of a short-distance vibrometer in: (i) the external configuration, and (ii) the self-mixing configuration of interferometry. Assume 1-mW power and a 10-kHz bandwidth.*

P5-7 *The half-fringe servo loop is incorporated in a self-mixing interferometer operating at s=10 cm from a diffuser target. Calculate the necessary current swing that shall be impressed to the diode drive.*

P5-8 *Can the feedback loop based on the current control be implemented in a normal Fabry-Perot laser, without any detrimental effects from mode hopping? What happens at mode hops?*

P5-9 *How can you tell the amplitude of the square-wave driving signal exciting the MEMS in Fig.5-23 and 5-24 from the waveform of the interferometric signal?*

P5-10 *An absolute distance self-mixing interferometer is aimed at a target at s=1 m. Consider a current sweep of $\Delta I=0.85$ mA as in Prob.P5-7 and of 10 times as much. How large is the yardstick s_{yst} of the measurement? How many periods N_c are developed when aiming at s=1 m?*

P5-11 *What is the frequency we have to measure in a self-mixing absolute distance measurement working with a wavelength scan $\Delta\lambda=1$-nm around $\lambda=1$ μm, on a maximum distance of 100 m. Calculate also the necessary linewidth of the laser source.*

P5-12 *How can resolution and error of the absolute distance self-mixing interferometer be improved?*

P5-13 *In the angle-measurement scheme of Fig.5-28, how can we change (for example, increase) the dynamic range of the measurement?*

P5-14 *Calculate the maximum value of measurable attenuation a when the weak-echo injection-detector is used (Sect.5.7).*

P5-15 *For the half-fringe locked vibrometer (Fig.5-14) calculate the responsivity $V_{out}/\Delta s$ and the loop gain for operation at $s_0=200$ mm, assuming: $\lambda=1$ μm, $G_m=20$ mA/V, R=100 kΩ, $\alpha_I=4$ pA/mA and A=10; the detected current is $I_0=10$ μA.*

P5-16 *Derive equations 5.43 for the alpha factor and C measurements at C<1 through the time interval of zeroes and peaks. Hint: use Eq.5.16 for the perturbed frequency and recall that the self-mixing signal is cos $\omega_F\tau$, while $\omega_0\tau$ is the phaseshift 2ks=Φ we are looking for at zeros and peaks.*

P5-17 *With the same approach and hints of Prob.P5-16, derive Eq.5.41.*

P5-18 *Derive Eq.5.44a of the shearing interferometer, which looks at first sight odd, because upon tilting a slab we expect the optical pathlength change as $L/\cos\theta$ and not as $L\cos\theta$ like indicated by Eq.5.44a. Hint: consider the three segments of the beam reflecting twice internally on the slab surface, and the path of beam passing through without internal reflections.*

P5-19 *Derive Eq.5.44b of the shearing interferometer. Hint: draw the tilted and untilted slab superposed, and consider the segments of the beam refracted internally as well as those (upper and lower) external to the slab.*

CHAPTER 6

Speckle-Pattern and Applications

When light from a laser source is shed on a plain surface like a sheet of paper or a white wall, the spot appears granular (Fig.6-1). This appearance is termed *speckle pattern* and came as a surprise to the first researchers experimenting with lasers. Indeed, spatial coherence was a predicted feature of the laser beam, and common sense was that a laser spot should look smooth. Actually, the spatial distribution of a mode is smooth, and as such is preserved when the beam impinges on a smooth surface. However, a smooth surface is one with roughness much smaller than wavelength, whereas only a mirror satisfies this condition.

In a normal diffuser or diffusing surface, one with a brightness that looks the same from any line of sight, the vertical scale of roughness $\Delta z=f(x,y)$, see Fig.6-2, is large compared to wavelength. Each elemental area of the diffuser is then a source of secondary wave that leaves the surface with an added random phase $\phi=2k\Delta z$, replica of the random profile Δz.
The random phase is responsible for destroying the spatial coherence of the near field distribution at the diffuser and for the grainy appearance of the far field.

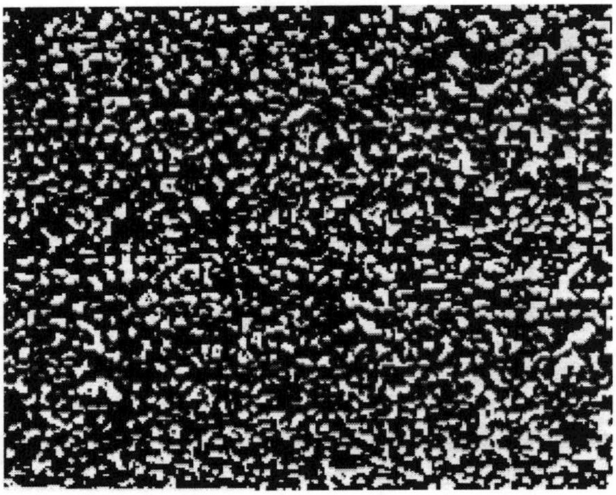

Fig.6-1 Coherent light projected on a diffusing surface appears granular, and this phenomenon is termed speckle pattern.

In this chapter, we start by considering the statistical properties of the speckle pattern field, first described on a qualitative base and then through a more rigorous analytical approach. We proceed by analyzing the errors of speckle in measurements with interferometers and will conclude with the description of an interferometer imaging technique known as Electronic Speckle Pattern Interferometry (ESPI).

6.1 SPECKLE PROPERTIES

In treating speckle properties, it is customary to assume [1] an ideal diffuser to schematize the statistical properties of surface roughness. The ideal diffuser is the model of total randomness and gives rise to the so-called *fully developed* speckle statistics [1]. A real diffusing surface may approach the ideal diffuser rather well, or have a limited randomness of surface roughness that translates itself in a reduced variance. Thus, the ideal diffuser case is the limit case of statistics.

6.1.1 Basic Description

In an ideal diffuser, the height profile $\Delta z=f(x,y)$ is described by a random function f(x,y). The function is invariant to translations (i.e., properties are the same for all points of the diffuser), has a zero-mean deviation, $\langle\Delta z\rangle=0$, and a quadratic mean value or variance $\sigma_z^2=\langle\Delta z^2\rangle$ much larger than λ^2. As a consequence, the phase shift ϕ introduced by the profile corrugation on diffusion at any point x,y is larger than 2π, or $\phi=k\sigma_z>>2\pi$.

In an ideal diffuser, the profile has no spatial correlation on short-scale, or f(x,y) is uncorrelated to f(x',y') as soon as |x'-x| and |y'-y| >λ.
This condition can be expressed in terms of the profile correlation function, expressed as:

$$C_z(\Delta x, \Delta y) = \int\int \Delta z(x,y)\, \Delta z(x+\Delta x, y+\Delta y)\, dxdy.$$

We have spatial uncorrelation when $C_z(x-x',y-y') = \sigma_z^2\, \delta(x-x',y-y')$, where δ is the Dirac impulse function and σ^2_z is the variance defined previously.

Now, let us consider the electrical field $\underline{E}(x,y)$ leaving the diffuser at z≈0, after illumination from the coherent beam $\underline{E}_1$. If we move just a little away from the diffuser, the amplitude E(x,y) is substantially equal to the illuminating one, that is E_1. The total phase of E(x,y) is the sum of a deterministic phase contribution $\phi_1(x,y)=kx\sin\zeta$, due to the incidence angle ζ of the illuminating beam (Fig.6-2) and a random phase term φ(x,y) due to the diffuser roughness.

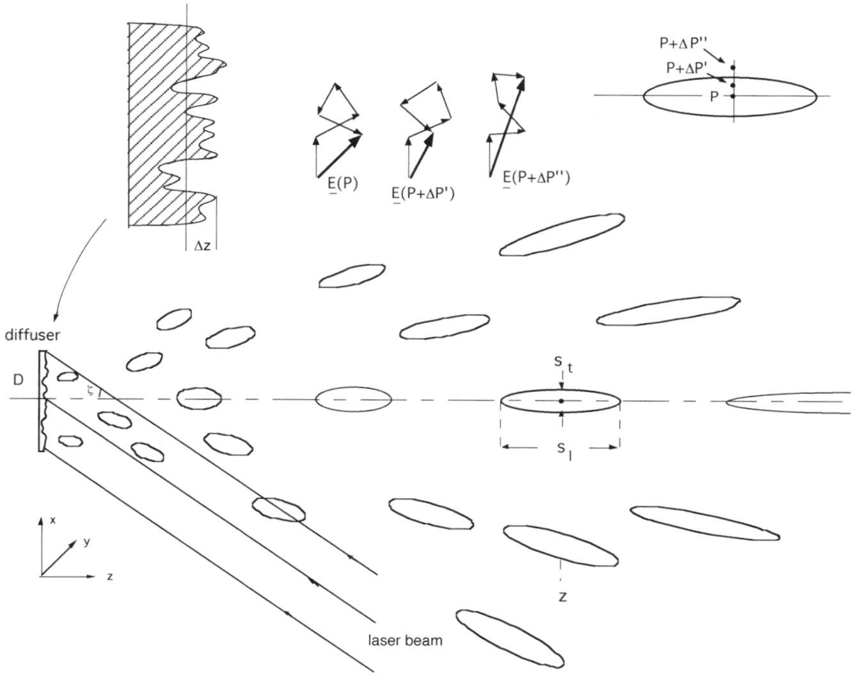

Fig.6-2 Speckle pattern is the field radiated from a diffuser, that is, a surface with roughness larger than wavelength, Δz>>λ (top left). The field at a generic point P is the summation of many individual vectors with a random phase (top center). Moving away from P to P+ΔP' and P+ΔP'', the field gradually loses correlation. A speckle grain is a region within which correlation is C≥0.5. Grains are cigar shaped and point out from the diffuser center.

Because the random phase shift is large, $k\sigma_z >> 2\pi$, and has no spatial correlation, the deterministic term is negligible. Therefore, field correlation $C_E(x-x',y-y')$ duplicates the statistics of $C_z(x-x',y-y')$ or, it is δ-like.
Near the diffuser ($z \approx 0$), coherence areas are delta-like regions, and the spatial coherence of the original laser beam is lost, destroyed by the diffuser's roughness.

Because the δ-like condition holds irrespective of the actual amplitude and phase dependence of the illuminating field E_1 from x,y, we may take for simplicity E_1=const. in the following.

The field $\underline{E}(x,y,z)$ at any point P in space (Fig.6-2) is evaluated as the vector summation of contributions emitted by all dxdy elemental areas in the source. Each contribution has constant amplitude, but random phase. Such a vector sum is a statistical process known as random walk. The process is the same in all points P of the space, and only the descriptive parameters change with P, not the distributions.

Now, in a generic point P, we write the field as $\underline{E} = E_R + iE_I$ and recall the results known from statistics about the properties of the *random-walk* process [1-3]. Because of the law of large numbers relative to the addition of many individual contributions, both the real E_R and the imaginary E_I component of the field are normal distributions with zero mean value, $\langle E_R \rangle = \langle E_I \rangle = 0$. Moreover, the intensity $I = E^2$ distribution is found to be a negative exponential. Explicitly, the probability density of intensity is written as $p(I)=[1/\langle I \rangle] \exp -I/\langle I \rangle$, where $\langle I \rangle$ is the average intensity of the speckle field and coincides with the mean square value $\langle E^2 \rangle$ of the field components.

When we move at a substantial distance from the diffuser, the statistical properties remain unchanged, but the δ-like correlation is smeared to a finite-width distribution. In fact, let us go back to point P and consider the field resulting from the addition of elemental vectors with a random phase. Let $\underline{E}(P)$ be the vector sum, as indicated in Fig.6-2. Now we move to point $P'=P+\Delta P'$. If $\Delta P'$ is small, it is reasonable to expect that the phase of the individual contributions is not so much changed or that $\underline{E}(P+\Delta P')$ is correlated to $\underline{E}(P)$. This shall certainly happen, because *ab absurdo*, if $\Delta P' \to 0$, $\underline{E}(P+\Delta P')$ cannot but converge to E(P).

Second, if we move substantially away from P to $P''= P+\Delta P''$ and let the deviation $\Delta P''$ increase, we shall ultimately collect the field contributions with new phase terms, different from the initial ones, whence the resulting vector addition $\underline{E}''= E(P+\Delta P'')$ becomes different from the initial result E(P), or, is not correlated to it anymore.

Now, the substance is about how large $\Delta P'$ shall be before correlation is lost (or, quantitatively, it becomes less than a specified value, e.g., $C<0.5$). This quantity defines the so-called speckle size. The speckles then represent the coherence areas we find at a distance z from an ideal diffuser illuminated with a spatially coherent beam.

In the next section, we will calculate the speckle size, both *longitudinal* (s_l along z) and *transversal* (s_t along x or y). For a circular diffuser of diameter D, uniformly illuminated, the results are:

$$s_t = \lambda\, z/D, \quad \text{and} \quad s_l = \lambda\, (2z/D)^2 \tag{6.1}$$

As can be seen in Fig.6-2 and from Eq.6-1, speckles are cigar shaped, with the longitudinal size larger than the transversal, and become more elongated as we move away from the diffuser. Speckles point out from the diffuser center, and the projection of off-axis speckles is equal to the on-axis speckles.

Subjective and Objective Speckles. The previous discussion holds for the free propagation of the diffused field depicted in Fig.6-2. This case is referred to as one generating *objective speckles*, in the sense that speckles are only dependent on the diffusing object or target.
Another case of interest is when an optical element, for example a lens as in Fig.6-3, is interposed between the conjugated image and source planes. This is the case of *subjective speckles*, so called because then the speckle depends on the observer, too.

We may analyze the optical conjugation of the diffuser (with diameter D) to an image plane by a lens (with diameter D_L and focal F_L) as drawn in Fig.6-3. For ease of notation, let us take $z>>F_L$ so that the distance p from the lens is $p=F_L$.
Along its diameter, the lens accommodates a number of speckles $N= D_L/s_t$ given by the ratio of lens diameter to speckle transversal size. Explicitly, here we consider N as a ratio of linear size, not of surface.

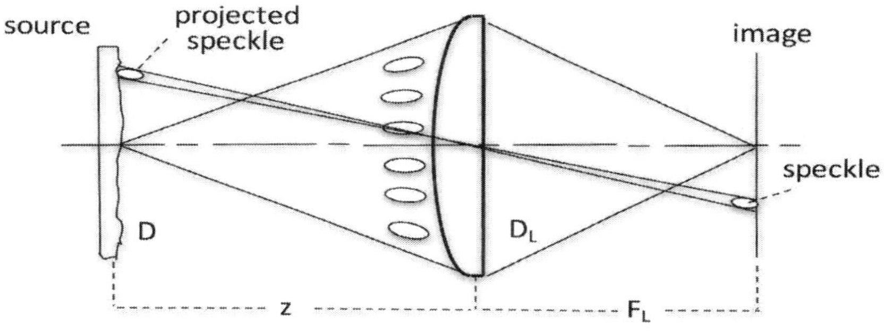

Fig.6-3 A lens looking at the source collects N speckles and, because of mode invariance, forms N (smaller) speckles inside the image. As the image is conjugated to the source, N virtual speckles appear projected back on the source. This is the *subjective* speckle pattern.

Because speckles point to the diffuser center, they are each oriented at a different angle and thus come to a different position at the focal (or image) plane, filling there the image of the diffuser. The image has a diameter $D(F_L/z)$ and contains $N=D_L/s_t$ speckles. Then, the diameter of speckles (transversal size $s_{t(fp)}$) found in the image at the focal plane is:

$$s_{t(fp)} = D(F_L/z)\,(s_t/D_L) = \lambda\, F_L/D_L \qquad (6.2)$$

This speckle dimension is the same that we get for objective speckle with a diffuser of diameter D_L at distance F_L.

Now, from the focal plane we can see a virtual image of the source, equal to the one in the focal plane multiplied by z/F_L. On the diffuser, the apparent dimension (transversal size $s_{t(pr)}$) we see projected is given by:

$$s_{t(pr)} = (\lambda\, F_L/D_L)\,(z/F_L) = \lambda z\,/D_L \qquad (6.3)$$

Eq.6.3 explains why this speckle regime is called subjective: the diameter of speckles seen on the source is dependent on the lens diameter D_L. Moreover, the speckle size is that generated by the lens aperture D_L, not the target aperture. In addition, the longitudinal size of virtual speckles appearing on the diffuser is $s_{l(pr)} = \lambda(2z/D_L)^2$.

The effect of subjective speckle can be visualized easily, looking at a laser spot projected on a wall through a pinhole and varying the diameter of it. For a pinhole, you may use your own index finger, closing it fully folded leaving a small aperture in the middle of the skin fold that you can finely adjust.

6.1.2 Statistical Analysis

We can analyze the statistical properties of the speckle-pattern field starting from the Fresnel approximation to the diffraction equation (see Appendix A4.2).
Let the diffuser be illuminated by a uniform coherent beam in the plane ξ,η. The complex amplitude of the source can be written as $E_l(\xi,\eta) = E_l \exp i\phi(\xi,\eta)\ \mathrm{rect}\,(\xi,\eta,D/2)$, where:
- E_l is the constant field amplitude of the illuminating beam,
- $\phi(\xi,\eta)= k\ \Delta z(\xi,\eta)$ is the random phase function, associated with Δz,
- $\Delta z(\xi,\eta)$ is the surface roughness, and
- rect (x) is the step function defining the limits of integration, defined as:
- rect (x)=1 for $x=\xi^2+\eta^2 \le (D/2)^2$, and rect (x)=0 for $x=\xi^2+\eta^2>(D/2)^2$.

The function $\phi(\xi,\eta)$ has a delta-like $\delta(\xi,\eta)$ spatial correlation and a variance σ_z^2 such that $\sigma_z >> \lambda$. Each elemental area $d\xi d\eta$ contributes with a spherical wave delayed by kr_{12} to the field collected at point (x,y) in the image.

In the Fresnel approximation, we may assume $r_{12} \approx z+(x^2+y^2)/2z+(\xi^2+\eta^2)/2z-(x\xi+y\eta)/z$ and $r_{12}/z \approx 1$. By summing up all the elemental contributions in $d\xi\, d\eta$, we have in the image plane x,y at a distance z (Fig.A4-1):

$$E(x,y,z) = (E_l/\lambda z)\ \exp ikz\ \ \exp ik[(x^2+y^2)/2z]\ \int\int_{-\infty,+\infty} d\xi d\eta\ \ \mathrm{rect}(\xi,\eta,D/2) \times$$
$$\times\ \exp ik[(\xi^2+\eta^2)/2z]\ \ \exp\text{-}ik[(x\xi+y\eta)/z]\ \ \exp i\phi(\xi,\eta) \qquad (6.4)$$

At the right-hand side of this equation, after the amplitude $E_l/\lambda z$, the first term is the phase shift of propagation down the distance z, and the second is the field curvature term. Then we find the integral with the boundary truncation term (rect). The fourth term is again a field curvature term, the fifth is the mixed-coordinate term originating the Fourier transform kernel, and the last term is the diffuser random-phase term.

From Eq.6.4, we can deduce that the field E is a complex quantity with a real E_R and an imaginary E_I part. Both of them are affected by the random part of the phase $\phi(\xi,\eta)$, as it can be seen by developing the exponential under the integration $\exp i\phi(\xi,\eta) = \cos\phi(\xi,\eta) + i \sin\phi(\xi,\eta)$. As $\phi = k\Delta z >> 2\pi$, we have $\langle\cos\phi\rangle = \langle\sin\phi\rangle = 0$ and, as a consequence:

$$\langle E_R\rangle = \langle E_I\rangle = 0 \qquad (6.5)$$

Thus, the speckle field has *zero-mean* real and imaginary parts at all points in space.

Second, in the Δz profile, the number of scatterers is very large and, therefore, in view of the central-limit theorem [2,3], Δz obeys a normal (or Gauss) distribution. In the same way, $\phi = k\Delta z$ and its sine and cosine functions are normal distributions, too.

It is then not surprising that, after linear operations as those performed through the integration in Eq.6.4, the real and imaginary parts of the field, $E = E_R + i\, E_I$ are both normal distributed and uncorrelated. Accordingly, their probability distributions can be written as:

$$p(E_R) = (2\pi\sigma_E^2)^{-1/2} \exp -E_R^2/2\sigma_E^2, \quad p(E_I) = (2\pi\sigma_E^2)^{-1/2} \exp -E_I^2/2\sigma_E^2 \qquad (6.6)$$

In Eq.6.6, we have used Eq.6.5 and indicated with σ_E^2 the variance of the field components that shall be the same for both E_R and E_I for symmetry reasons. Writing the σ_E^2 variance as the mean of squares reveals that it is coincident to the mean intensity:

$$\sigma_E^2 = \langle E_R^2\rangle = \langle E_I^2\rangle, \quad 2\sigma_E^2 = \langle E_R^2\rangle + \langle E_I^2\rangle = \langle E_R^2 + E_I^2\rangle = \langle I\rangle \qquad (6.7)$$

Another quantity of interest is the field amplitude $|E| = \sqrt{(E_R^2 + E_I^2)}$. By transformation of the variables it is easily found that $|E|$ is *Rayleigh distributed*, or it can be written as:

$$p(|E|) = (2\pi\sigma_E^2)^{-1/2} |E| \exp -(|E|^2/2\sigma_E^2) \qquad (6.6')$$

About the intensity $I = |E|^2 = E_R^2 + E_I^2$, knowing that both E_R and E_I are normal distributions and using the rules of variable transformation [3], the probability density p(I) easily follows (see below) as a *negative exponential* distribution:

$$p(I) = \langle I\rangle^{-1} \exp -I / \langle I\rangle \qquad (6.8)$$

Because of the negative exponential distribution, weak (or dark) speckles are more probable than intense (or bright) ones. For example, 10% of speckles have ≤10% the average intensity, 1% have ≤1% the average intensity, and so on.

This is just *amplitude fading*, a feature hampering measurement that we shall appropriately tackle in all systems working on a diffusing surface rather than a mirror.

Last, the phase of the electric field at a given point P is a random distribution with no information on the initial distance-related phase kz. Computing the phase as $\varphi = \operatorname{atan} E_I/E_R$ gives, as a result, a *uniform distribution* of phase on 0-2π as:

$$p(\varphi) = 1/(2\pi) \qquad (6.8')$$

Recall on the change of variables. We want to find the probability distribution p(E) of a random variable E=f(I) related to another variable I, whose probability distribution p(I) is given. We equate the differential probability dp around I and E by writing dp= p(E)dE= p(I)dI. The desired p(E) is then found as: p(E)=p(I)dI/dE= p(I)/[df(I)/dI]. For example, let $|E|=\sqrt{I}$ and $p(I)= I_0^{-1}\exp\text{-}I/I_0$. As df(I) /dI $=1/2\sqrt{I}=1/2|E|$, then $p(|E|)= I_0^{-1}\, 2|E| \exp -|E|^2$, which is the Rayleigh distribution following from the exponential. For joint distributions, $p(I,\theta)=f(A_1,A_2)$, the generalization of the method is $p(I,\theta)=p(A_1,A_2)|J|$, where $|J|$ is the Jacobian determinant of the variable transformation, from A_1 and A_2 to I and θ [2].

A summary of the statistical properties of the speckle pattern is reported in Fig.6-4.

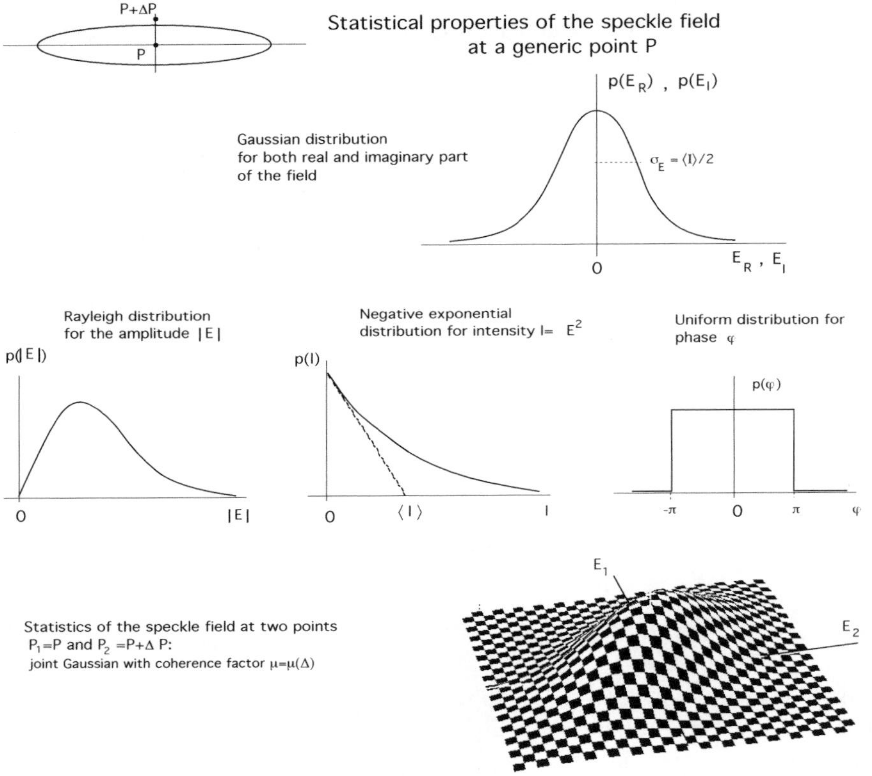

Fig.6-4 Summary of the speckle pattern statistics: field components E_R and E_I, amplitude |E|, intensity I and phase φ, and joint distribution E_1, E_2.

The quantities described so far are the first-order statistical properties of the speckle pattern, because they are related to the field in a specific point P, but they don't tell anything about the correlation of the speckle field. To get such information, we need to consider the

bivariate or joint probability function, a two-variable Gaussian which relates the statistical properties of the speckle in two points in space, for example, P and P+ΔP in Fig.6-4.
The starting point for analyzing the second-order statistical properties of the speckle is the Gaussian joint probability of intensity and phase in two points P_1 and P_2.
This is written as [1,3]:

$$p(I_1,I_2,\varphi_1,\varphi_2) = [16\pi^2\sigma_E^4(1-\mu^2)]^{-1}\times$$
$$\times \exp-\{[I_1+I_2-2(I_1I_2)^{1/2}\mu\cos(\varphi_1-\varphi_2-\psi)]/2\sigma_E^2(1-\mu^2)\} \quad (6.9)$$

In this equation, $\mu \exp i\psi = \mu_c$ is the *complex coherence factor* for the field at P_1 and P_2. The complex coherence factor is defined as the ratio of the mutual intensity normalized to the product of rms values of fields at points P_1 and P_2, or:

$$\mu_c = \langle E(P_1)E^*(P_2)\rangle / [\langle |E(P_1)|^2\rangle \langle |E^*(P_2)|^2\rangle]^{1/2} \quad (6.10)$$

where $\langle .. \rangle$ indicates the operation of ensemble average on the speckle set, and * is the complex conjugate. The coherence factor μ_c has a modulus μ that may go from 0 (complete uncorrelation) to 1 (full correlation). It is complex because, simply, $E(P_2)$ may be delayed with respect to $E(P_1)$ and then $\psi = ks(P_2-P_1)$, where s(..) is the distance between P_2 and P_1.
It's worth noting that, Eq.6.9 represents, in general, Gaussian correlated noise, including electrical noise of devices and circuits of interest for communications [3].
The numerator in Eq.6.10 is the *correlation* of the field, which is also called *mutual intensity* [1]:

$$C_\mu = \langle E(P_1)E^*(P_2)\rangle \quad (6.11)$$

The correlation has the important physical meaning of a beating signal between the fields at points P_1 and P_2. It is maximum for full correlation, or P_1 approaches P_2, and decreases to zero when fields are uncorrelated, or P_1 moves far away from P_2. The correlation length is defined as the distance $s(P_2-P_1)$ at which the correlation C_μ has decreased to a specified value (usually 0.5 of the maximum) or has reached the first zero (if C_μ oscillates).

6.1.3 Speckle Size from Acceptance

Transversal and longitudinal sizes of the speckle can be calculated as the correlation width of the field. As correlation is related to coherence, s_l and s_t can be interpreted as the coherence lengths (along longitudinal and transversal directions) and the speckle grain as a coherence region, or a mode of the field.

In this section, we show that speckle size can be traced to the size of the spatial mode found at a distance z from a diffuser. A theorem on the invariants of radiometry (Ref. [5], Ch.2) states that the acceptance $\mathfrak{a}$ of an aperture receiving power under a solid angle Ω and on an area A is equal to *N* times λ^2, or:

$$a = A\,\Omega = N\lambda^2, \qquad (6.12)$$

where N assumes the meaning of number of modes, or degree of freedom associated with the area-solid angle aperture.

Let us apply this theorem to speckle propagation, with reference to Fig.6-5. We let N=1 for a single mode, and note that radiation is received at distance z under the solid angle $\Omega = \pi(D/2z)^2$. The receiving area associated with a (transversal) spatial mode of width s_t is $A = \pi(s_t/2)^2$. Inserting in Eq.6.12, we get $\lambda^2 = \pi(s_t/2)^2\pi(D/2z)^2$ whence $s_t = (4/\pi)\lambda z/D$.

Second, the bundle of rays in the solid angle Ω keeps limited transversally to an extent $\leq s_t$ (Fig.6-5) for a longitudinal width s_t/θ, where $\theta = D/2z$ is the angular aperture of the bundle. Combining these expressions gives $s_l = (2/\pi)\lambda(2z/D)^2$ as the longitudinal size. A comparison with Eq.6.1 shows that these values are the same already found, except for a minor multiplying factor of the order of unity.

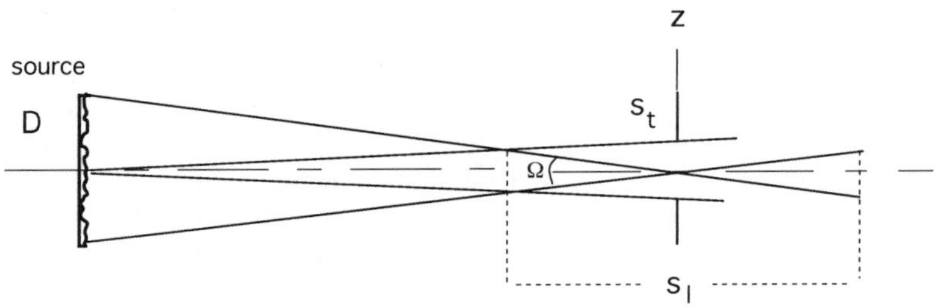

Fig.6-5 Geometry for the calculation of speckle size as a spatial mode.

6.1.4 Joint Distributions of Speckle Statistics

The joint statistics describes the properties of the speckle field in two points P_1=P and P_2=P+ΔP in space. When ΔP is comparable to the speckle size, the joint statistics gives relative-amplitude and phase-difference information important to displacement and vibration measurements made with interferometers on diffuser like target surfaces.

The starting point is the joint probability density of intensities and phases $p(I_1,I_2,\varphi_1,\varphi_2)$ given by Eq.6.9. This probability is a direct consequence of the Fresnel diffraction integral and of the diffuser surface randomness.

We can get the joint probability density of intensities $p(I_1,I_2)$ by a double integration of Eq.6.9 on φ_1 and φ_2 from 0 to 2π. In this way, we obtain [1,3,6]:

$$p(I_1,I_2) = [4\sigma^4(1-\mu^2)]^{-1} \exp-[(I_1+I_2)/2\sigma^2(1-\mu^2)]\, \Im_0[\mu(I_1I_2)^{1/2}/\sigma^2(1-\mu^2)] \qquad (6.13)$$

In this equation, $\Im_0$ is the modified Bessel's function of the first kind, zero order, μ is the modulus of the coherence factor (Eq.6.10), and σ^2 is the variance of the (real and imaginary) field components (Eq.6.7). About this variance, we may recall that it is also $2\sigma^2 = \langle I_1\rangle = \langle I_2\rangle$.

Similarly, the joint probability density of phases $p(\varphi_1,\varphi_2)$ is obtained by a double integration of Eq.6.9 on I_1 and I_2 from 0 to ∞, and the result is [3,6]:

$$p(\varphi_1,\varphi_2) = [(1-\mu^2)/4\pi^2](1-\beta^2)^{-3/2}\ [\surd(1-\beta^2)+\beta\ \arcsin\beta+(\pi/2)\beta] \qquad (6.13')$$

Here, $\beta=\mu\cos(\varphi_1-\varphi_2-\psi)$, and ψ is the argument of the complex coherence factor $\mu_c=\mu\exp\ i\psi$. The probability $p(\varphi_1,\varphi_2)$ is brought to the probability density of the phase difference $\varphi_1-\varphi_2$ by changing the variables φ_1,φ_2 in their sum and difference and integrating on the sum. This gives as a result $p(\varphi_1-\varphi_2)=2\pi\ p(\varphi_1,\varphi_2)$.

The joint probability $p(\varphi_1-\varphi_2)$ is plotted in Fig.6-6 as a function of $\varphi=\varphi_1-\varphi_2-\psi$ and with the coherence factor μ as a parameter. In the diagram, we start at μ=0 with the uniform distribution $p=1/2\pi$. As μ increases, the probability density becomes more and more peaked around φ =0, up to the limit case μ=1 when it becomes a Dirac delta, $p(\varphi_1-\varphi_2)=\delta(\varphi_1-\varphi_2)$.

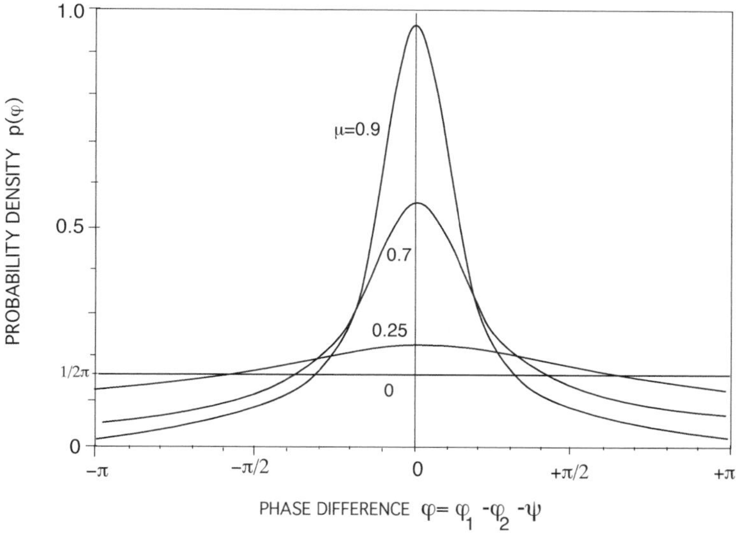

Fig.6-6 Probability density $p(\varphi_1-\varphi_2)$ of the phase difference plotted for some values of μ, the modulus of the coherence factor.

From the probability density, we can calculate the variance σ_φ^2 of the phase difference $\varphi_1-\varphi_2$, that is, the error in a phase measurement performed in the speckle regime between two points with a coherence factor μ.

As it is $\langle\varphi_1-\varphi_2\rangle=0$, the variance is given by $\sigma_\varphi^2=\int_{0-2\pi}\varphi^2 p(\varphi)d\varphi$, where $\varphi=\varphi_1-\varphi_2$. Collecting the terms and using Eq.6.13', one is left with the evaluation of an integral that appears intractable at first sight, but is finally solved with the result [6]:

$$\sigma_\varphi^2 = \pi^2/3 - \pi\ \arcsin\mu + \arcsin^2\mu - (1/2)\sum_{n=1,\infty}\mu^{2n}/n^2 \qquad (6.14)$$

Let us now deal with conditional probabilities. They are useful to consider when we can add any knowledge of one of the variables $(I_1, I_2, \varphi_1, \varphi_2)$, or combination of them. For example, let us assume that the intensity I_1 in point P_1 is known, and we want to compute the probability of intensity at point P_2. The conditional probability $p(I_2 | I_1)$ of having I_2, once I_1 is known, is given by Bayes' theorem [2] as the ratio of the joint probability to the probability density of the conditioning variable:

$$p(I_2 | I_1) = p(I_2, I_1)/p(I_1) \tag{6.15}$$

Combining Eqs.6.15 and 6.13 we obtain [6]:

$$p(I | I_1) = [2\sigma^4(1-\mu^2)]^{-1} \exp-[(I_2+\mu^2 I_1)/2\sigma^2(1-\mu^2)]\, \Im_0[\mu(I_1 I_2)^{1/2}/\sigma^2(1-\mu^2)] \tag{6.16}$$

We can now calculate the first and second moments of this distribution, that is:

$$\langle I_2\rangle |_{I1} = \int_{0-\infty} I_2\, p(I_2 | I_1) dI_2, \quad \text{and} \quad \sigma_{I2}^{\ 2} |_{I1} = \int_{0-\infty} I_2^{\ 2}\, p(I_2 | I_1) dI_2, \tag{6.17}$$

and obtain the conditioned mean and variance of intensity, that is:

$$\langle I_2\rangle |_{I1} = 2\sigma^2(1-\mu^2)+\mu^2 I_1, \text{ and } \sigma_{I2}^{\ 2} |_{I1} = [2\sigma^2(1-\mu^2)]^2+4\sigma^2\mu^2(1-\mu^2) I_1 \tag{6.18}$$

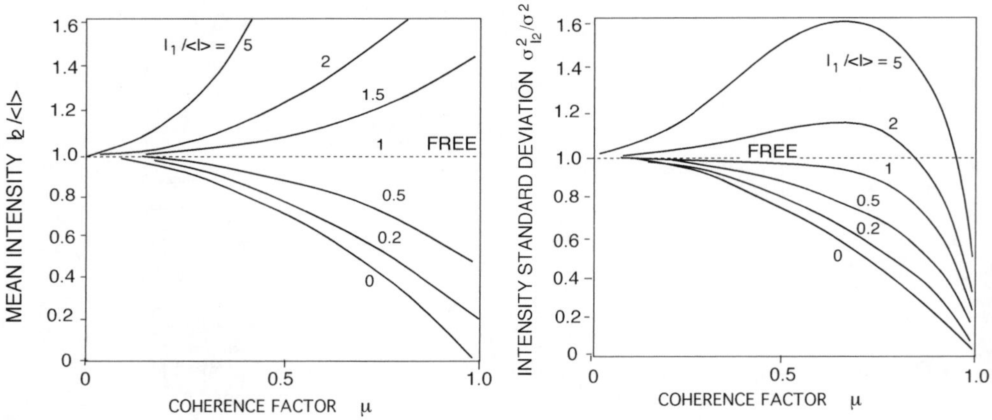

Fig.6-7 Mean value (left) and variance (right) of intensity I_2 conditioned to I_1, as a function of μ, and for some values of the intensity I_1 as a parameter. Values of mean and variance of the unconditioned distribution are also indicated (from Ref.[6], ©Optics Publ., reprinted by permission).

In Fig.6-7 we report the diagrams of mean and variance of intensity I_1 conditioned to intensity I_2 versus the modulus of the coherence factor. The mean value is standardized to $\langle I_1\rangle = \langle I_2\rangle = 2\sigma^2$ and the variance to σ^2, the 'free' value. From Fig.6-7 we can see that when the conditioning intensity I_1 is larger (smaller) than the mean, I_2 also is, on average.

The variance for large conditioning intensity I_1 is larger than the 'free' value, but not so much larger. For example, $I_1=5\langle I\rangle$ gives only $\sigma_{I2}^2\approx1.6\sigma^2$ (at $\mu\approx0.6$), or a relative standard deviation $\sigma_{I2}/I_1\approx0.25$. Of course, if we move to the high-correlation region $\mu\approx1$, this relative standard deviation is even smaller.

In summary, a bright speckle has comparatively less intensity noise.

Another conditioned probability is that of intensity I_2, when we know both intensity I_1 and phase φ_1 in the other point. This probability can be computed starting from Eq.6.9, and the result is given in [6]. Using the definitions (Eq.6.17), mean and variance are found as:

$$\langle I_2\rangle|_{I1,\varphi1}=\sigma^2(1-\mu^2)[3-D(\delta)+2\delta^2],\quad \sigma_{I2}^2|_{I1,\varphi1}=\sigma^4(1-\mu^2)[6-(1-2\delta^2)D(\delta)-D^2(\delta)+8\delta^2] \quad (6.19)$$

In Eq.6.19, erf is the standard error function [2] and we let $D(\delta)=[1+\delta\sqrt{\pi}\,(1+\text{erf}\,\delta)\exp\delta^2]^{-1}$ and $\delta=\mu\cos(\varphi_1-\varphi_2-\psi)\sqrt{[I_1/\sigma^2(1-\mu^2)]}$. Mean and variance of the intensity are plotted in Fig.6-8 for some values of the conditioning parameter $I_1\cos^2\varphi$, the projection of intensity I_1 on I_2, being $\varphi=\varphi_2-\varphi_1$ as usual. The abscissa variable is the modulus of the coherence factor, and is extended to negative values (by the multiplication to the sign of cos φ) to account for anti-correlation when $\varphi\approx\pi$. As we can see from Fig.6-8, similar to intensity, the phase also becomes more regular in correspondence to bright speckles.

Last, we want to compute the conditional probability density of phase φ_2, with the intensities I_1 and I_2 and the phase φ_1 as conditioning variables. Before proceeding, let us remark that, if phase φ_1 is known, then the phase difference $\varphi=\varphi_2-\varphi_1$ is the only statistical variable of significance.

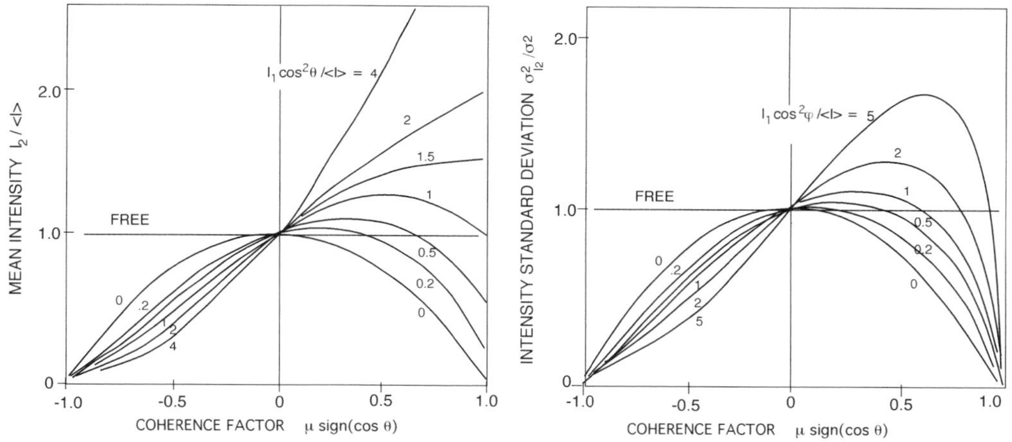

Fig.6-8 Mean value (left) and variance (right) of intensity I_2 conditioned to I_1 and to the phase difference φ, as a function of the coherence factor and for some values of the projected intensity $I_1\cos^2\varphi$ as a parameter (from Ref.[6], ©Optica Publ., reprinted by permission).

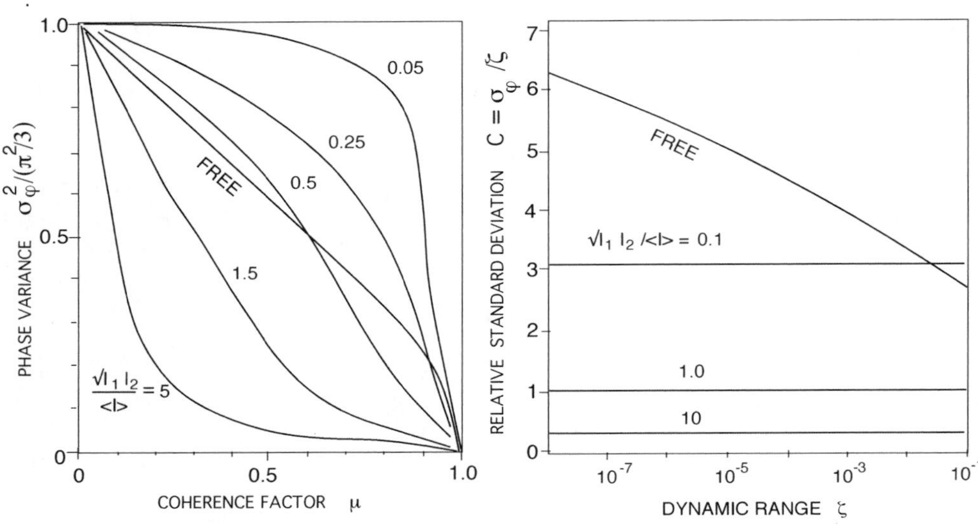

Fig.6-9 Variance of the phase difference σ_φ^2 normalized to $\pi^2/3$, as a function of the coherence factor μ (left), and of the dynamic range factor $\mu=1-\zeta^2$ (right), with the relative intensity $\sqrt{(I_1I_2)}/\langle I\rangle$ as a parameter. The 'free' variance given by Eq.6.14 is also plotted for comparison. The right-hand diagram is an expansion near $\mu\approx1$, the region of high coherence, and is plotted for $\mu=1-\zeta^2$ (from Ref.[7], ©IEEE, reprinted by permission).

Incidentally, the phase difference is just the quantity we are interested in for a measurement of optical path length. Thus, we consider the conditional probability $p(\varphi\,|\,I_2,I_1)$. As before, by integration of Eq.6.9, we find [6]:

$$p(\varphi\,|\,I_2,I_1)= (4\pi^2)^{-1}\exp[(\mu\cos\varphi)\cdot(I_1I_2)^{1/2}/\sigma^2(1-\mu^2)]\ \{\Im_0[\mu(I_1I_2)^{1/2}/\sigma^2(1-\mu^2)]\}^{-1} \qquad (6.20)$$

Because of the same symmetry of Eq.6.13', the mean value of phase $\langle\varphi\rangle|_{I2,I1}$ is still given by the 'free' value of Eq.6.14, for which the diagram is plotted in the left-hand part of Fig.6-9. The variance $\sigma_\varphi^2|_{I2,I1}$ follows from the definition (Eq.6.17) and is found as:

$$\sigma_\varphi^2|_{I2,I1} = \pi^2/3 + [4/\Im_0(z)]\ \Sigma_{n=1,\infty}(-1)^n\Im_n(z)/n^2 \qquad (6.21)$$

In Eq.6.21, we let $z=\mu(I_1I_2)^{1/2}/\sigma^2(1-\mu^2)$, and $\Im_n$ is the n-th order modified Bessel function of the first kind (usually denoted with I_n).

Fig.6-9 (left) shows the probability density of phase conditioned to intensity against the coherence factor μ. Also plotted is the 'free' value of phase variance, given by Eq.6.14, and relative to the full set of speckle realization. We can see from the diagram that the free value decreases from $\pi^2/3$ to zero as μ varies from 0 to 1. The conditioned variance is always smaller than the free variance for $\sqrt{(I_1 I_2)}\,/\langle I\rangle > 1$, that is, where the speckle is locally more intense than the average. The decrease of variance (and of measurement error) is particularly sizeable near $\mu \approx 1$ where the speckle field are well correlated.

Eqs. 6.14 and 6.21 become indeterminate forms when $\mu \rightarrow 1$, and cannot be used directly in the region of high coherence. Then, we let $\mu = 1 - \zeta^2$ and obtain the following asymptotic behavior for small ζ:

$$\sigma_\varphi^{\,2}\,|_{\,\mathrm{free}} = \zeta^2\,(3 - 2\,\ln\zeta^2) \tag{6.14'}$$

$$\sigma_\varphi^{\,2}\,|_{\,\mathrm{I2,I1}} = \zeta^2\,[\sqrt{(I_1 I_2)}/\langle I\rangle]^{-1} \tag{6.21'}$$

The main dependence of both free and conditioned phase variances is on ζ^2. The multiplying factor steadily increases for $\zeta \rightarrow 0$ for the former and is given by the inverse of the relative speckle intensity $\sqrt{I_1 I_2}/\langle I\rangle$ for the latter. In Fig.6-9 (right), we plot the ratio of the rms phase error σ_φ and dynamic range factor C:

$$C = \sigma_\varphi\,|_{\,\mathrm{I2,I1}}\,/\,\zeta \tag{6.22}$$

6.1.5 Speckle Phase Errors

Let us now evaluate, from the results of speckle pattern statistics found in the previous section, the noise-equivalent-displacement (NED) associated with the interferometric measurement. Let the phase measured on a displacement z_2-z_1 be $\varphi = k(z_2 - z_1)$, where $k = 2\pi/\lambda$, and the error superposed be σ_φ. Then, we may recall from Sections 4.4.1 and 4.4.7 that the NED can be written as:

$$\mathrm{NED} = \sigma_\varphi/k = C\,\zeta/k \tag{6.23}$$

where the second expression follows from the definition of Eq.6.22.

The factor ζ is found from the coherence factor as $\mu = 1 - \zeta^2$. For the two cases considered in Section 6.1.2, of longitudinal and transversal displacements from an initial point P to a final point P+Δ, the coherence factor has the expression (see Eqs.6.16 and 6.19):

$$|\langle E_{st}(P)\,E_{st}{}^*(P+\Delta)\rangle| = \mathrm{somb}\;X \tag{6.24}$$

In the two cases, it was $X = \Delta D/\lambda z$ and $X = \Delta D^2/\lambda z^2$. Recalling Eqs.6.17 and 6.20, we may write the factor as $X = \Delta/\lambda s_t$ and $X = \Delta/s_l$, s_t and s_l being the transversal and longitudinal size of the speckle. In a generalized form, we can write:

$$X = \Delta/s_{\mathrm{spckl}} \tag{6.24'}$$

and s_{spckl} is the current size for the experiment at hand. Now, we may develop at the third order of X the somb function and obtain μ as:

$$\mu = \text{somb } X = 2J_1(\pi X)/\pi X = 2\,[\pi X/2 - (\pi X)^3/16]/\pi X = 1-(\pi X)^2/8 \quad (6.25)$$

By comparing with $\mu=1-\zeta^2$ we get:

$$\zeta = \pi X / 2\sqrt{2} \quad (6.25')$$

Finally, we go back and insert Eqs.6.24', 6.25' into Eq.6.23. The equivalent displacement error due to the speckle turns out to be given by the expressive form:

$$\text{NED} = C\,\zeta/k = (C/4\sqrt{2})\,\lambda\,\Delta/s_{spckl} \quad (6.26)$$

Eq.6.26 is interpreted as follows. The speckle error, in units of wavelength (NED/λ), is given at first instance by the ratio of displacement Δ to speckle size s_{spckl}, either longitudinal or transversal, and more precisely by $C/4\sqrt{2}$ times as much. This factor is not too far from unity and depends on the modality of measurement (that is, free speckle statistics versus a conditional statistics), as indicated in Fig.6-9 (right).

Intra-Speckle Error. As Eqs.6.14' and 6.21' have been derived under the asymptotic assumption $\zeta<<1$ or $\Delta<<s_{spckl}$, the NED of Eq.6.26 applies to the case of *intra-speckle displacement*, the one mostly found in the operation of a vibrometer. For a longitudinal displacement, recalling Eq.6.1, we may rewrite Eq.6.26 in the form:

$$\text{NED} = C\,\zeta/k = (C/4\sqrt{2})\,\Delta\,(D/2z)^2 \quad (6.27)$$

where D is the spot size at the target, and $\Delta = z_2 - z_1$ is the longitudinal swing covered around distance z.

The NED of Eq. 6.27 is plotted in Fig.6-10 versus the normalized distance z/(D/2) and with the displacement $\Delta = z_2 - z_1$ as a parameter.

As we can see, the intra-speckle phase error may become even smaller than the quantum noise limit (for the representative case of 1 μA detected current and B=1 Hz), and of course decreases with increasing distance to the spot size ratio. For example, for a vibrometer reading a swing Δ=10 μm at z=0.1 m distance on a spot size D/2= 1 mm, the NED can go down to 0.1 pm.

Two experimental points in Fig.6-10 were obtained [6] with a vibrometer measuring a target at z=50 mm distance, with an amplitude Δ=10 μm and a spot size D/2 = 1 mm (dot) and 3 mm (cross). Agreement is good, although the points lie a little bit lower than the corresponding theoretical curve (of a factor $\approx$1.8); this is explained by the diffuser non-ideality, that is, the incomplete randomness, a condition known as *non-fully developed* speckle pattern [9] rather common in a real diffuser.

Inter-Speckle Error. For large displacements $\Delta > s_l$, we cross two or more speckles and Eq.6.27 doesn't hold anymore. Large Δ are commonly found in the operation of a displacement interferometric measurement on a diffuser.

In this case, we first try to work on the *last speckle*, at a distance z such that $s_l > \Delta$. This requires $z < \lambda(2z/D)^2$ or $z > \lambda\ (D/2)^2/ = z_F$, the Fresnel distance associated with aperture D, a far-field condition usually difficult to satisfy, in practice.

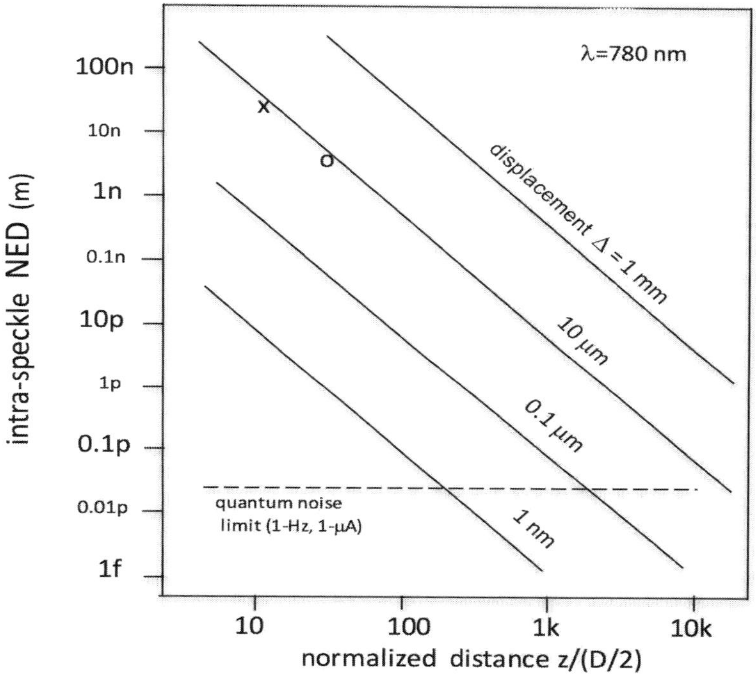

Fig.6-10 Noise-Equivalent-Displacement (NED) of the intra-speckle phase error plotted vs distance z normalized to spot size D/2 and with the displacement $\Delta = z_2 - z_1$ as a parameter. Two experimental points (cross and dot) are shown, in agreement with theory (from Ref.[7], ©IEEE, reprinted by permission).

To evaluate the *inter-speckle* phase error, consider that the density per unit length of quadratic error contribution is $\pi^2/3\ dz/s_l$ (the variance of a uniform distribution 0-2π times the differential elemental path dz/s_l), and by integration of this quantity on $\Delta = z_2 - z_1$ we get:

$$\sigma_\phi^2 = (\pi^2/3) \int_\Delta dz\ (D/2z)^2/\lambda\ = (\pi/24)\ k\ D^2\ \Delta/z^2 \qquad (6.28)$$

(for $\Delta \ll z$) and taking the square root and dividing by k we get the NED as:

$\mathrm{NED} = (D/z)\,(\lambda\ \Delta/48)^{1/2}$ or also

$$NED/\lambda = (\Delta/12\, s_l)^{1/2} \tag{6.29}$$

In addition to the random phase error, the propagation on a substantial distance gives rise to a systematic error, due to the field curvature term exp ik[$(x^2+y^2)/2z$ in Eq.6.4. Integrating this error from z_1 to z_2 (see Ref. [7]) yields a *systematic phase error* (SPE):

$$SPE \doteq -k(D^2/16z^2)\,\Delta \quad \text{or also} \quad SPE = (\pi/2)\,\Delta/s_l \tag{6.30}$$

Note the different trend of the random and systematic errors with respect to the displacement Δ and to distance z: the NED increases as the square root of Δ, whereas SPE is proportional to Δ, and they decrease as $1/z$ or $1/z^2$, respectively.

The diagram showing random phase error NED and systematic phase error SPE is shown in Fig.6-11. As we can see, even for large displacements such as Δ =1 m, both errors can be made smaller than say, 1 μm, by working with a sufficiently large z/(D/2) ratio.

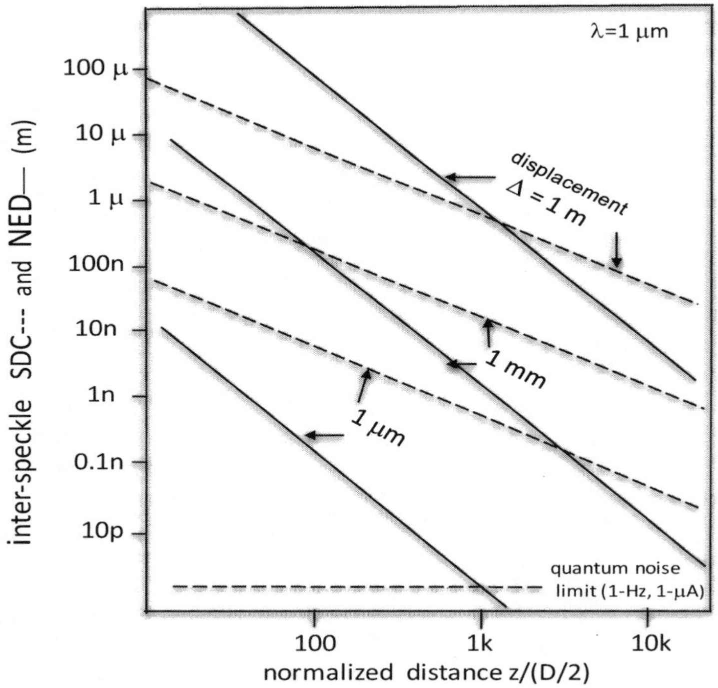

Fig. 6-11 The inter-speckle error contributions: the random error NED (full lines) and the systematic error SPE (dotted lines) plotted versus normalized distance z/(D/2), for a displacement interval going from $z_1=0.1z$ to $z_2=z$, one representative of a common measuring situation (from Ref.[7], ©IEEE, reprinted by permission).

6.1.6 Additional Errors Related to Speckle

In the previous section we derived the errors incurred in a displacement measurement performed on a diffusive target because of the speckle pattern statistics, both for small (intra-speckle) and large (intra-speckle) displacements (Eqs. 6.27 and 6.29) and also shown that an accompanying systematic error due to field curvature is found in the measurement (Eq. 6.30). These results are ideal in the sense that they tacitly assume a point-like detector for the field measurement, a still target, and a still illuminating beam. Now we shall consider what happens when these assumptions are removed.

Still observer and a moving target case. The target movement may be either transversal or longitudinal, as shown in Fig.6-12. and we assume the illumination beam doesn't move. So, the target illumination is unaltered in a longitudinal movement, whereas it will change in a transversal movement.

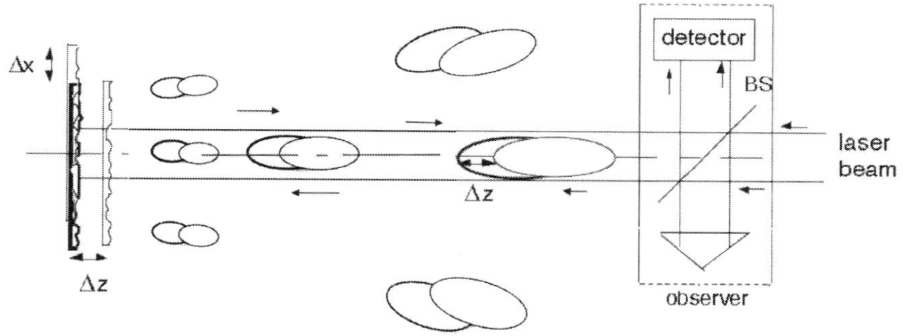

Fig.6-12 A moving target drags the speckle pattern field projected on the measuring system (for longitudinal displacement as shown here, the field is dragged along Δz). The correlation lengths for transversal Δx and Δz displacements are coincident to the speckle sizes s_t and s_l.

We can calculate the correlation function $C_\mu = \langle E(P)E^*(P+\Delta)\rangle$ with the Fresnel approximation (Eq.6.4). By repeating the calculations (Probl.6.9), we obtain in either case the same result, which reads $\langle E(P)E^*(P+\Delta)\rangle = \text{somb}(\Delta/s_{spckl})$, in which s_{spckl} is the longitudinal (or transversal) size of the field projected by the target.

Thus, it is like the target drags along its speckle field when it moves with respect to the laser beam of illumination as illustrated in Fig.6-9 for a longitudinal movement. Therefore, the above results apply also to interferometric measurement on a moving target and give the error caused by the speckle regime.

Movement of the illuminating beam on the target. While all other parts are still, a movement may be imparted to the beam by a scanning for a 3D pickup [8], or by wandering and turbulence effects (App.A3.2) especially when operation is on a sizeable path length (say L>50 m).

This case is treated by calculating the correlation function $C_\mu = \langle E(P)E^*(P+\Delta)\rangle$ when the diffuser coordinate ξ is changed to $\xi+\Delta$.

Again, we obtain the result $\langle E(P)E^*(P+\Delta)\rangle = \mathrm{somb}(\Delta/s_t)$, where s_t is the transversal speckle size of the field projected by the target. This result is intuitive and confirms that moving the target is equivalent to moving the illuminating spot on it.

Speckle Errors with a Focusing Lens. We may use an objective lens to focus the illuminating beam onto the target, as illustrated in Fig.4-19 and generally used in laser vibrometers (Sects.4.5 and 5.4). To treat this case, we shall just recall the results of Sect.6.1.1 about subjective speckle. At the focal plane of the objective lens, the transversal and longitudinal size of the speckles are given by (Eqs.6.2 and 6.3): $s_{t\,(fp)} = \lambda F_L/D_L$ and $s_{l\,(fp)} = \lambda(2F_L/D_L)^2$, in which F_L and D_L are the focal length and diameter of the lens. At the object plane, because of the optical conjugation provided by the objective lens, the speckles are seen stamped on the target with apparent dimensions: $s_{t\,(trg)} = \lambda z/D_L$ and $s_{l\,(trg)} = \lambda(2z/D_L)^2$, and thus when a lens is used, the aperture governing the speckle dimension is the lens diameter, not the target size, but all the other results still hold.

Speckle Errors Due to Detector Size. Let us consider the effect of finite detector size on the random phase contribution, and on the field curvature error. In the image plane of the focusing lens (Fig.6-3), if the detector diameter D_{det} is smaller than the speckle size $D_{det} \le s_{t(fp)} = \lambda F_L/D_L$, we can assume that the speckle distribution is point-wise sampled, and all the results of preceding sections directly apply.

A small detector delivers a small signal, however. We might increase the detector size so that it collects $N_{sp} = (D_{det}/s_{t(fp)})^2$ speckles. In this case, what about the statistics?

Integration of the received field on several speckles can be treated as the sum of independent samples of the same statistical ensemble. We can consider two cases:

- incoherent addition of speckles intensity when we perform a (normal) direct detection,
- coherent detection of the speckle field when we add a local oscillator or reference beam on the detector (see also Sect.10.1 of Ref. [5]) to perform a phase measurement.

In the case of incoherent addition, the sum of N_{sp} speckles gives N_{sp} times the intensity I of the single speckle, as an average. The probability of the sum of N terms $p_N(I)$ is the convolution of the single p(I), repeated N times, or $p(I)^* \, p(I)^* \ldots {}^*p(I)$. For the exponential distribution $p(I) = \langle I\rangle^{-1} \exp{-I/\langle I\rangle}$, we get [2,4] the result $p_N(I) = (I/\langle I\rangle)^N (N!)^1 \exp{-I/\langle I\rangle}$. Thus, small-amplitude speckles are no more the most probable, and for large N we go to the well known result of the central-limit theorem [2]. This is a Gaussian (or normal) distribution of intensity with a relative standard deviation equal to $1/\sqrt{N}$.

In the case of coherent detection, we have to deal with the sum of N independent field components E, each with a normal distribution p(E) (Eq.6.6) having zero mean value and $\sigma^2 = \langle I\rangle$ variance. The sum of such contributions is again the convolution of the single p(E), repeated N times, or $p_N(E) = p(E)^*p(E)^* \ldots {}^*p(E)$. The resulting distribution $p_N(E)$ is still normal, with a mean value and a variance both equal to N times the mean and variance of the single p(E). Thus, the mean value of the sum remains zero, while the variance is increased to $N\sigma^2$.

Considering the readout configuration of interferometers (Sect.4.6), the external configuration can use a large area photodetector, and the previous arguments are applicable, whereas the self-mixing configuration has an acceptance determined by the small laser cavity aperture of acceptance is λ^2. In other words, the equivalent area of the detector is that of the single mode. Therefore, to use a self-mixing configuration in the speckle regime, the speckle transversal size shall be matched to the spatial mode of the laser.

About the field curvature error, this is caused by the term exp $ik(x^2+y^2)/2z$ in Eq.6.4. As the detector has a radial extension from $r=\sqrt{(x^2+y^2)}=0$ to $D_{det}/2$, we integrate to field exp $ikr^2/2z$ on $2\pi r$ dr and average on the area $\pi D_{det}^2/4$ and find the field collected by the detector as:

$$E_{det} = 16 \exp i[k(D_{det}/4)^2/z] \; \mathrm{sinc}\,(D_{det}/\lambda z) \tag{6.31}$$

The phase is contained in the exponential term of Eq.6.31 as: $\varphi=k(D_{det}/4)^2/z$. The sinc function is close to unity in the reasonable assumption of a small detector, $D_{det}<<\lambda z$, and can be dropped. Phase φ is the SPE of the last section and, comparing it to Eq.6.30, we see that the result is the same but with D replaced by D_{det}.

6.2 Speckle in Single-Point Interferometers

Here, we treat the effect of speckle statistics in *vibration* and *displacement* measurements made with interferometers. Vibrations are the small-amplitude periodic displacements (roughly from a few μm to millimeter), and displacements are the aperiodic, large-amplitudes displacement (say, larger than a few μm) with arbitrary waveform (see also Fig.4-20). Our aim is to keep the speckle effects under control, so that a measurement on a diffusing target surface is performed with little error. Speckle statistics has two consequences: *phase noise* already considered in Sect. 6.1.5, and, even more serious, *amplitude fading*. In fact, the intensity of the speckle pattern field is distributed as the negative exponential given by Eq.6.8.

6.2.1 Speckle Regime in Vibration Measurements

In a vibrometer, it will be likely that we may fall on a weak or dark speckle, and find an intensity I substantially smaller than average value $\langle I \rangle$. Indeed, with the negative exponential distribution, the probability of getting $I<0.1\langle I \rangle$ is $p=0.1$, and that of $I<0.01\langle I \rangle$ is $p=0.01$. This is called amplitude fading because in a small, but not negligible, fraction η of cases of operation, the signal returning from the target will be smaller than $\eta\langle I \rangle$. No matter how sensitive or low threshold the signal-processing circuits are, we cannot avoid getting a signal below the minimum working condition.

There are several ways to alleviate amplitude fading.

(i) first, we may incorporate in the instrument a circuit sorting the intensity level and *warning* the operator that speckle intensity is too small, so the operator can move the beam a little bit and project it on the target to another spot. Then, if we fall on a bright speckle with $I> \langle I \rangle$, we get improved conditional statistics, as per Eq.6.21'.

(ii) second, we can use *Automatic Gain Control* (AGC) to raise signal amplitude to a level adequate for circuit processing. So, if M is the dynamic range improvement of the AGC, the threshold of signal loss is moved to $\eta\langle I\rangle/M$ and made less probable, but not eliminated.

(iii) a third strategy is exploiting sensor duplication, a technique known in radio engineering as *space diversity*. Two receivers, slightly apart in space, sample different realizations of the speckle statistics. The probability of fading on both signals below $\eta\langle I\rangle$ is therefore, is η^2, a much smaller value than a single channel, but not yet zero.

From the point of view of applications, the first strategy is usually quite acceptable for a vibration-detecting instrument. The second and the third can also conveniently be incorporated and reduce the probability of fading down to quite acceptable levels.

A possibility common to all the above strategies is to use a *superdiffuser* as the target surface. A superdiffuser is a piece of back-reflecting tape or varnish of the type used for enhancing the visibility of traffic sign at night. Also known as Scotchlite™ in a commercial product, the superdiffuser provides a gain of a factor 20 to 50 (typically) in the back-reflected signal reaching the interferometer. Again, this option improves the applicability but does not eliminate fading totally. A final option used in displacement interferometers and described in Sect.6.2.2 is the *dynamical tracking* of a bright speckle, one that at the same time cures fading and phase error.

6.2.2 Speckle Regime in Displacement Measurements

In a displacement measurement, fading and phase error problems are more serious than in vibration, because it spans a large dynamic range and cannot make sure that $s_{l(trg}>>\Delta_{max}$. This is the case of application to tool-machine numerical control, calling for displacement up to meters [10]. Tool-machines are usually equipped with an optical rule (see Ref. [5], Sect.12.3.4) which proves to be the least expensive solution and accurate to 1/100 mm. Occasionally they employ a multi-axis interferometer equipped with corner cubes (Fig.4-16) for increased performance. The use of a plain diffuser has the advantage of eliminating the problem of keeping the corner cube clean and the surface under test clear of invasive elements.

If we are to measure a large displacement, we will use a digital readout, by counting $\lambda/4$- or $\lambda/8$-transitions, to develop the several-digit figure. Then, our measurement is incremental and cannot tolerate a count loss. Nor can we move away from a dark speckle as in a vibrometer. Count loss due to fading is a concern because, when it happens, we are forced to reset the instrument to its mechanical zero and restart the measurement. Thus, we strive to eliminate fading.

We can use a combination of the techniques discussed in the previous section, including speckle size tailoring, ACG, detector diversity, and super-diffusing target.

As a first issue, we have to choose the *beam size*. A small beam size gives a large speckle that keeps the phase error small, but the signal collected is small and is affected by quantum noise. A large NA objective lens helps collect a substantial return signal so that the quantum noise is low, but increases the speckle error.

Example of evaluation. We may start requiring a large speckle size. With the external configuration, we shall collimate the laser beam on the full dynamic range Δ_{max} to be covered. Using Eq.2.4, we get a spot size $w=\sqrt{(\lambda\Delta_{max}/\pi)}$. Then, the longitudinal speckle size is $s_l=\lambda(2\Delta_{max}/w)^2$ and, inserting the previous expression, becomes $s_l=4\pi\Delta_{max}$, we are inside a single speckle (as $s_l>\Delta_{max}$) on the full dynamic range. The signal back from the last speckle is weak, however: If P_L is the laser power, and we use half in the reference path, then $P_L/2\pi$ is the intensity at the target, and $(P/2\pi)\pi(w/\Delta_{max})^2$ is the power collected by the receiver. The power attenuation is accordingly $(w/\Delta_{max})^2=\lambda/\Delta_{max}$. Using $\lambda=1$ µm and $\Delta_{max}=1$ m, we shall be prepared for a –60 dB loss. On the other side, if we take a sizeable diameter of the objective lens, let's say D_L =30 mm, in place of the small w (typ.$=\sqrt{\lambda\Delta_{max}}=1$ mm) of the laser, we limit the loss to -30 dB, but will have a small speckle dimension i.e., $s_l= \lambda(2\Delta_{max}/D_L)^2=4$ mm only. In a 1-m=Δ_{max} dynamic range, we would find 250 speckle passages and have a $\lambda\sqrt{250}$ error.

On another hand, the practical system shall tolerate residual walk-off movement of the target or of the illuminating beam (see last section). To illustrate a reasonable compromise, we take a dynamic range $\Delta_{max}=1$ m, and let n=10 speckle passages in it, so that the error is just $\sqrt{10}\lambda\approx3.1$ µm and the average speckle size is $s_l=100$ mm. At $\lambda=1$ µm, this means that we can use a lens diameter $D_L=6$ mm in $s_l=\lambda(2\Delta_{max}/D_L)^2$ to give room for a reasonable lateral walk off of the target on the dynamic range.

An efficient method to cure fading is adding a provision for *bright speckle tracking* (BST) [10,11]. The basic idea consists of moving the beam projected on the target to maximize the signal amplitude, and keep it locked to a bright speckle while the target eventually undergoes its movement. Locking to a bright speckle has two advantages: (i) the probability of fading is greatly reduced, and (ii) the phase statistics improves because we work consistently at a high $I/\langle I\rangle$ ratio (Fig.6-9). The setup employed to perform bright speckle tracking with a self-mixing interferometer configuration is shown in Fig.6-13.

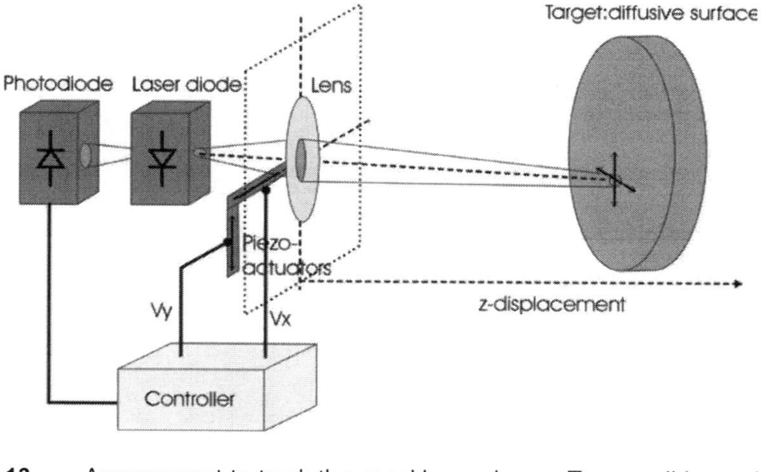

Fig.6-13 Arrangement to track the speckle maximum. Two small bars of PZT ceramic are mounted on the lens edge and move the lens along the X and Y axes, changing the spot position on the target enough (just a few micrometers) to track a bright speckle grain (from Ref.[11], ©IEEE, adapted by permission).

We use a pair of microactuators, two small bars mounted in the objective lens holder along the X and Y direction. Normally the bars are made of lead-zircon-titanate Pb(Zr,Ti)O_3 (PZT) ceramics. By actuating the PZT elements, a (small) lateral displacement Δ of the lens (along X or Y axis) is generated, and the incoming beam is deflected by an angle Δ/F, where F is the focal length of the objective lens. On the target located at a distance z, the beam deflection is zΔ/F. In the practical implementation, it suffices to deflect the beam of just ≈5-20 μm on the target to follow a local maximum intensity.

To detect the direction of signal increase and actuate the deflection accordingly, the well-known method of small signal sensing and phase detection is used (Fig.6-14).

For example, to track the maximum along X, we feed the X-axis PZT ceramic with a small-amplitude square wave at audio frequency. The photodiode on the laser rear facet detects a square wave in signal amplitude, and the square wave will be in-phase or antiphase (shifted by π) with respect to the PZT drive. When we find it in-phase/antiphase, we increase/decrease the actuator dc drive. When we reach the maximum, the photodiode signal is at twice the drive frequency. Thus, demodulation of the photodiode signal with the PZT drive square wave, followed by a low-pass filtering (Fig.6-14) provides the required tracking of the maximum, along one axis for the moment.

To derive the two X and Y error signals from a single photodiode output, we take advantage that phase and quadrature signals are orthogonal, a concept called *dither of phase*. Indeed, we use the same square wave, with 0° and 90° phase shift, as the drive of the X and Y actuators. Upon demodulating the photodiode signal by the corresponding X and Y waveforms, the contribution from the other channel is not seen because it is out-of-phase of 90° (that is, orthogonal).

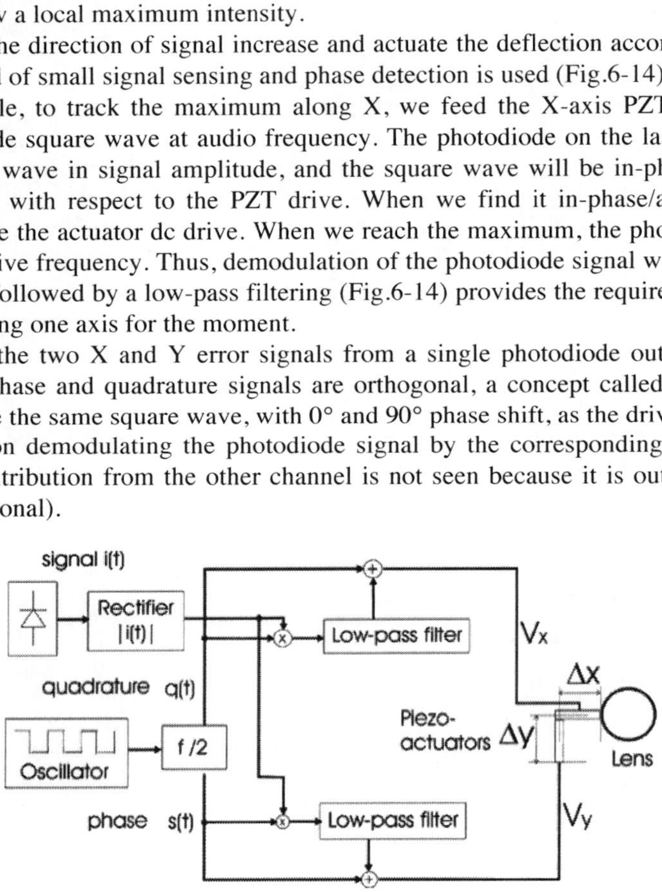

Fig.6-14 Block scheme of the bright speckle-tracking circuit. The signal from the photodiode is rectified peak-to-peak and demodulated with respect to the dither frequency, in phase and quadrature. The results are the X and Y error signals that, after a low-pass filtering, are sent to the piezo X and Y actuators to track the maximum amplitude or stay locked on the bright speckle (from Ref.[11], ©IEEE, reprinted by permission).

With the dither, the optical axis of the laser beam projected out the objective lens moves along a square path in the X-Y plane and the displacement adjusted to be a few microimeters, i.e., much less than spot size but enough to track the speckle maximum. In Fig.6-14, a circuit rectifies the interferometric signal of the photodiode then multiplies it with the two square waves driving the X and Y PZT-actuators. After a low-pass filter, the signals are sent to the PZT actuators. As we get a beam movement in the direction of increasing signal for both axes, a final state of equilibrium is reached when both axes are on a local maximum, or a bright speckle. The feedback loop speed is limited by the response time of the PZT actuators [typ. τ_{PZT} =0.1-0.3 ms for small (≈2×2 mm) bars]. This response time is adequate to follow the target movement up to a speed s_l/τ_{PZT}. For a speckle size of s_l= 100 mm, the target speed we can track without errors is s_l/τ_{PZT} =0.3-1 m/s.

To illustrate the results of the bright tracking method, the distribution of the field amplitude |E| (Eq.6.6') is the Rayleigh-distributed probability, one of the curves plotted in Fig.6-15 (left, thin line). Also plotted in Fig.6-15 (left, bars) are the experimental results measured on a white paper target. To improve the match, a small non-ideality of the real diffuser is accounted for, by assuming a 2% reflection from the target Fig.6-15 (left, thick line).

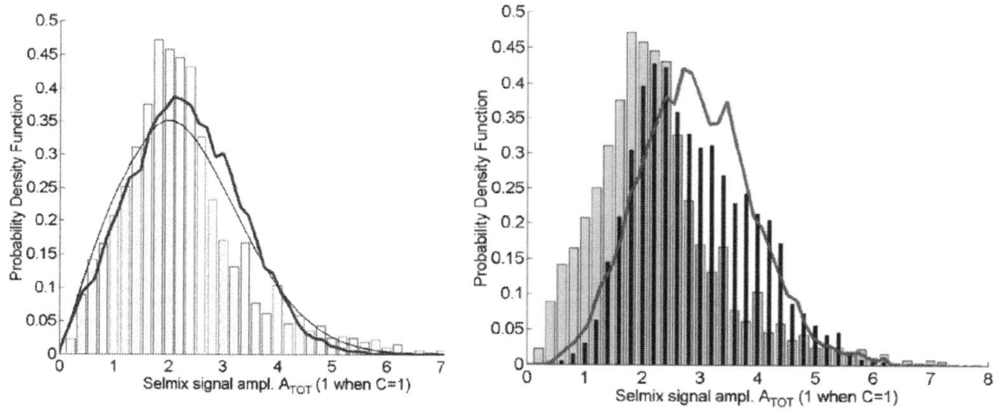

Fig. 6-15 Probability density function p(|E|) of the field amplitude |E|, with the amplitude normalized to 1 for a value C=1 of the self-mixing parameter. Left: thin line is the Rayleigh distribution for an ideal diffuser, and thick line is the result of a numerical simulation for a real surface, with 2% reflection and 98% diffusion, and fully developed speckle statistics. Bars are experimental data for plain white paper (z=50 cm, w_{las}=2 mm). Right: with the bright-speckle tracking on, the experimental distribution moves to significantly larger |E| (dark bars) as predicted by simulations, with respect to the circuit-off distribution (gray bars) (from Ref.[36], ©IEEE, reprinted by permission).

The field |E| is standardized to the value that makes C=1 in the self-mixing interferometer configuration. From Eq.5.1, we have $C=(E/sE_0)n_1L\sqrt{(1+\alpha_{en}^2)}$, where E_0 is the field leaving the laser. Thus, the field for C=1 is $E=sE_0/[n_1L\sqrt{(1+\alpha_{en}^2)}]$. As the circuit tracking the bright speckle maximum is switched on, the p(|E|) distribution becomes that of the right-hand diagram of Fig.6-15. The improvement of tracking is evident looking at the small-amplitude part of the diagrams. Occurrence of small amplitudes is greatly reduced.

In particular, with the speckle tracking circuit, we can immediately work with the self-mixing configuration of the interferometer, one demanding a given minimum signal amplitude (the condition $C\geq1$) to stay in the up/down switching regime. (Sect.5.3).

Data of Fig.6-15 are relative to a working distance z=50 cm and a laser spot w_{las}=2 mm on the diffusing target. In this condition, the probability of C<1 (loss of counts) is ≈10% in normal conditions, but drops to <0.5% with the tracking-circuit on.

Thus, we can operate on a substantial span of distance without amplitude fading, even with the self-mixing interferometer, and can expect that the phase error is kept low.

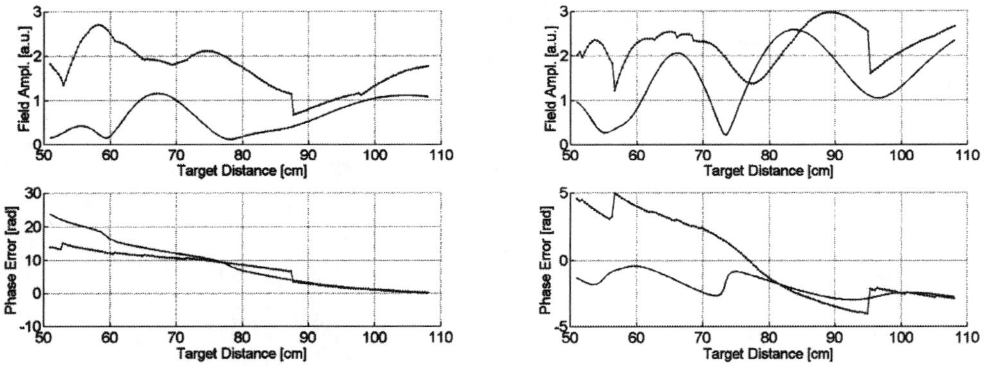

Fig.6-16 Simulations of amplitude (top curves) and phase (bottom curves) of returning field. The diffusing target moves from 108 to 51 cm. In the amplitude diagrams, top curves are with the tracking circuit on, and the bottom with the circuit off. In phase diagrams, the curves vary less when the tracking circuit is on. Abrupt jumps (near z=87 cm on the left and 95 cm on the right) are correct switches decided by the tracking system, which skips from one speckle becoming too weak to the next adjacent speckle, being brighter (from Ref.[11], ©IEEE, reprinted by permission).

In Fig 6-16, we plot the result of a simulation of speckle-regime amplitude and phase of the returning signal in a generic interferometer. In the calculation, the target moves about 50 cm. The phase error includes both speckle-related error and field curvature error. Note the jumps at z=87 cm on the left and 95 cm on the right diagrams. These are correct results, and are due to the dither that discovers a better speckle adjacent to the current one, and jumps on it. As the simulation goes from 108 to 51 cm, the jump is actually going upward.

Another result is about the total phase error. In a 1000-sample set of computed displacement (each similar to Fig.6-16), the unconditioned ensemble average phase error is 9.7 rad, and the standard deviation is 3.9 rad. In the same situation, when the bright-speckle tracking is on, we get an average error of 6.16 rad and standard deviation of 3.6 rad. Most important, amplitude fading is not a problem anymore. Translating the previous phase error to a displacement, we are left with a $\approx\lambda$ error on a displacement from 50 to 100 cm. This means a $\approx 10^{-6}$ relative accuracy, which is a value that is very good in several applications.

The self-mixing interferometer has been calibrated against an optical ruler, while in operation on a diffuser with the speckle tracking provision. Fig.6.17 shows an example of the signal waveform versus displacement, acquired while the target is moved from 70 to 80 cm by a motor-driven mechanism. Without speckle control, the signal amplitude has a strong fading at $\approx$76 cm, whereas fading is eliminated when the bright speckle control is on. In the same run, the jump at $\approx$73 cm is simply due to the tracking system that decides to change the speckle being tracked and to move to a brighter adjacent speckle, thus avoiding the count-loss error near z=76 cm of the measurement without bright-speckle tracking. In Fig.6-17, the signal with speckle control is noisier because of the dither of the spot position. Of course, this does not imply any loss of accuracy of the interferometer.

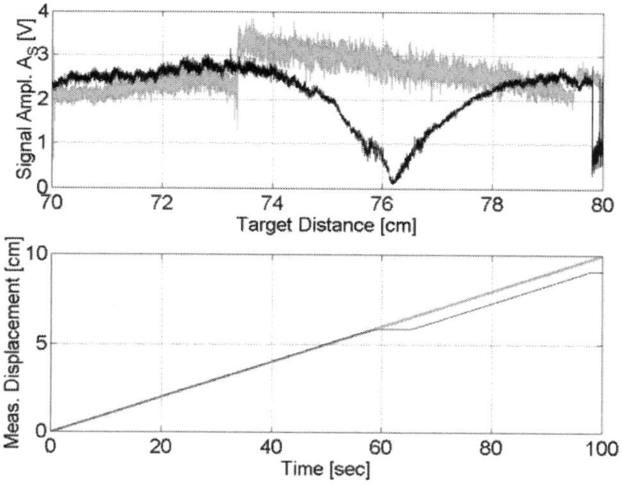

Fig.6-17 Top: comparison between the signal amplitude with (gray trace) and without (black) bright-speckle tracking. Bottom: the measured displacement measured by the interferometer displays an error near z=76 cm because of the speckle fading, whereas the error is removed with the bright-speckle tracking system. The target was moved from 70 to 80 cm at a 1-cm/sec speed (from Ref.[11], ©IEEE, reprinted by permission).

In the measurements, care was taken to avoid transversal movement of the target while actuating it by the z-axis motor driver. The transversal movement changes the speckle sample introducing extra fluctuation that increases the chance of fading in normal (non-tracked) operation, whereas the effect is readily tolerated when the bright-speckle tracker is working, as shown in Fig.6-18, left.

Another experimental result of interest, reported in Fig.6-18 (right) shows the general increase of signal field amplitude that we can obtain by speckle tracking with respect to a normal interferometer. Experimental results in Fig.6-18 are a sample, representative of the average behavior of the many statistical outcomes actually observed.

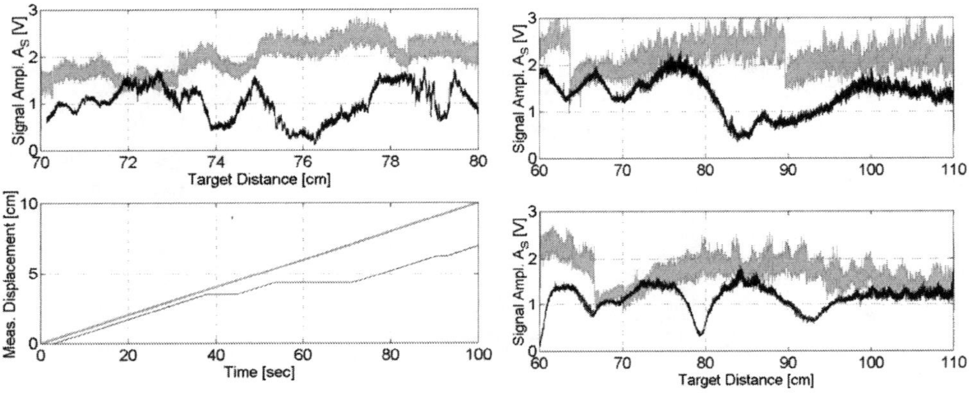

Fig.6-18 Left: transversal movement of the target worsens the fading problem in a normal interferometer, but can be tolerated when the bright-speckle tracker is added (top): loss of counts and error without tracker versus no error with the bright-speckle tracker. Right: typical experimental samples of signal amplitude with and without the bright-speckle tracker (from Ref.[11], ©IEEE, reprinted by permission).

Still another experimental result worth reporting is about the improvement we can obtain in interferometric measurements by a superdiffuser varnish or tape. In this case, the field amplitude level increases by the superdiffuser gain, typically G=20 to 40. Although this is good for small signals, when we may encounter a comparatively large signal, the amplitude may become so high to drive into saturation a preamplifier or another circuit. To avoid this, we may add an automatic gain control (AGC), conveniently implemented by a Liquid Crystal (LC) cell inserted in the outgoing laser-beam path as indicated in Fig.6-13. The LC is actuated by a control circuit that looks at the signal amplitude and keeps it constant.

In Fig.6-19, we report the probability density function of the field amplitude |E| for a target made of a Scotchlite™tape. Comparing it with the probability density function of a normal diffuser of Fig.6-15, we can appreciate the remarkable decrease of small-value amplitudes offered by the superdiffuser. Indeed, the probability of having a weak speckle with C<1 is in the range 10% for a normal diffuser without speckle tracking and 0.5% for a normal diffuser with speckle tracking. The same probability can go down to 2.2% and 0.15% for a superdiffuser without and with speckle tracking, respectively [11].

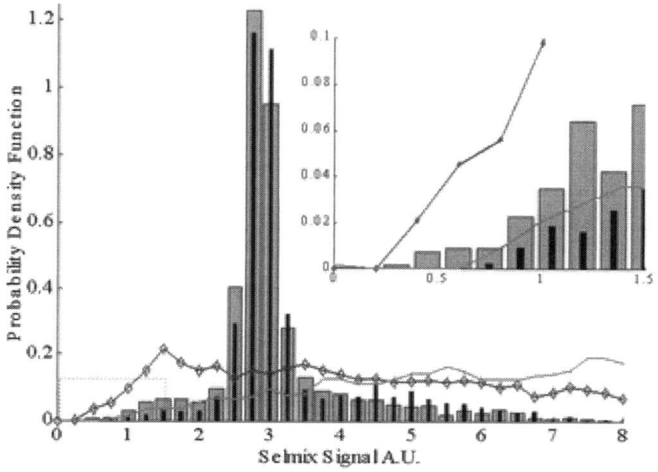

Fig.6-19 Probability density function of the field amplitude from a superdiffusing target (Scotchlite™ tape) with a superdiffuser gain G=20. Amplitude on the abscissa is normalized to the C=1 condition as in Fig.6-15. Line with aces is for the normal speckle statistics, and dotted line is for speckle with tracking on. Gray bars are for AGC on and speckle tracking off. Dark bars are for both AGC and speckle tracking on (from Ref.[11], ©IEEE, reprinted by permission).

6.2.3 Correction of the Speckle Phase Errors

After taking advantage of the methods discussed in Sect.6.2.2 to reduce the speckle errors, we may wonder if the residual statistical error may be corrected by any other method. We start from the availability of the amplitude signal $|E|$, and look for a relation connecting $|E|$ and φ, so that we may use the information in $|E|$ to correct the error in φ.

In general, the field given by Eq.6.4 can be written as $E_{st} = |E| \exp i(kz+\varphi_{sp})$, where kz is the propagation phase on distance z, and φ_{sp} is the speckle error of remaining terms. Usually, by beating with the local oscillator of the reference beam, we detect the real E_R and imaginary component E_I of the field, $E= E_R+iE_I$, and compute amplitude and phase as:

$$|E| = \sqrt{(E_R^2+E_I^2)} \qquad \varphi = \text{atan}\,(E_I/E_R) = kz+\varphi_{sp} \tag{6.32}$$

To be able to correct the phase error, we shall find a connection between φ_{sp} and $|E|$. In optics, such a connection is known as the Kramer-Kronig relation [12]. We use it when we calculate the phase shift $\phi(\nu)$ associated with the absorption/gain line $g(\nu)$ of a medium (see App. A.1.1), or when we connect real and imaginary parts of the index of refraction $n(\lambda)= n_R(\lambda)+i\, n_I(\lambda)$ (this effect is connected to the line-width enhancement factor α_{en} in Eq.5.12b).

In electronics, the Hilbert transform [13] relates the real and imaginary parts of a transfer function $F(\omega)$, ratio of output to input signals of a linear network in the frequency domain. It also relates the real and imaginary parts of a driving-point impedance, ratio of voltage to current. Specifically, if the network is physically realizable (that is, obeys the causality of effects) and has zero excess delay (with respect to the physical minimum), the real and the imaginary parts of $F(\omega)$ are Hilbert transforms of each other.

Mathematically, the condition for real and imaginary parts of a function $F(z)$ of a complex variable z to be Hilbert transforms is that the function $F(z)$ is analytic [13] or, written $z=u+iv$ and $F(z)=U+jV$, that the Cauchy-Riemann equations hold $\partial U/\partial u=\partial V/\partial v$, $\partial U/\partial v=-\partial V/\partial u$.

The Hilbert relation can be extended to amplitude and phase of the frequency response. For a zero-excess-delay network or system, the log of amplitude (or attenuation) and the phase are a Hilbert-transform pair. This statement easily follows from the previous, because by writing the attenuation as $\ln E = \ln [|E| \exp i\varphi]= \ln |E| + i\varphi$, we see that the real and imaginary parts of ln E are just the log-amplitude and the phase.

Moreover, for a system with excess phase $\exp i\psi$ with respect to the zero-excess-delay response, the Hilbert transform of the log-amplitude is given by the minimum-phase term plus the excess phase ψ.

The Hilbert transform $F(z)$ of a function $f(z)$ of the complex variable z is defined [12,13] as:

$$F(z) = H[f(z)] = \pi^{-1} \int_C f(\zeta)/(z-\zeta)\, d\zeta \qquad (6.33)$$

In this equation, the variables z and ζ are homogenous. The integral is a line integral around the origin of the complex plane, or, it is extended on $\zeta=-\infty, +\infty$ as a Cauchy principal value. The inverse Hilbert transformation is:

$$f(z) = H^{-1}[F(z)] = -\pi^{-1} \int_C F(\zeta)/(z-\zeta)\, d\zeta \qquad (6.34)$$

Interpreting Eqs.6.33 and 6.34 as convolution integrals, we can also write $F(z)=(1/\pi z)*f(z)$ and $f(z) = -(1/\pi z)*F(z)$.

Noting that the Fourier transform of $1/\pi t$ is i(sign ω), we can use Fourier numerical routines to compute the Hilbert transform. We first compute the Fourier transform of $f(z)$, $\Phi(\omega)$, and then change the sign of $\Phi(\omega)$ for $\omega<0$ and compute the inverse Fourier transform of the result, obtaining the Hilbert transform $F(z)$.

Examples of transforms. As an illustration, we list here a few Hilbert-transform (HT in the following) pairs. Sinusoidal functions cos z has the HT of sin z, and sin z has the HT of cos z. Thus, the complex exponential exp iz has real and imaginary parts that are HTs of each other. The impulse function (or Dirac delta) $\delta(t)$ has the HT $(\pi t)^{-1}$. A rectangular pulse centered at t=0 and of width 2T, that is, rect(t,T)=1 for $-T/2<t<+T/2$, =0 for $|t|>T/2$, has the HT: $\pi^{-1}\ln |(t-T)/(t+T)|$. A Gaussian pulse exp-

$t^2/2\sigma^2$ has the HT of Ei(t,T), where Ei is a special function (exponential-integral) [13]. The Lorentzian $1/[1+\omega^2T^2]^{1/2}$ has the HT: $\omega T/[1+\omega^2T^2]^{1/2}$, and these are the real and imaginary parts of a parallel-RC impedance.

Now we can go back to the field E(z) given by Eq.6.4. Under the integration sign, the diffuser function $\varphi(\eta,\xi)$ is real and well-behaved, and all other operations are linear and analytic. The terms out of the integral are analytic, too, so we may conclude that the field propagated at point P(z) is analytic with respect to the variable z. The same statement holds if we drop out the pure delay term exp ikz in Eq.6.4 and consider the remaining field E°(z), which is given by E(z)= exp ikz E°(z) and represents the reduced field carrying the speckle-induced phase error. It is easy see that E°(z) is analytic, too, with respect to the variable z. At this point the reader may wonder how z, the distance, is also a complex variable. It can be both at the same time: real when it's distance and complex when it's argument of the complex function, or is the *analytic continuation* of distance z.

In conclusion, the real and the imaginary parts of the total and reduced fields, E and E°, are Hilbert-transform pairs.

To test this statement, we have carried out numerical simulations of the reduced field E°(z). In the simulation [14], a square target, 100-μm per side, diffuses outward the illumination received by a coherent source (at λ=1 μm). The phase on each elemental area, a square of 1 μm per side, has been randomized on $-\pi$, $+\pi$. The real and imaginary parts of the field are calculated at several distances z according to Eq.6.4. An example of the results, representative of the actual statistics we have built up on a much larger sample size, is reported in Fig.6-20.

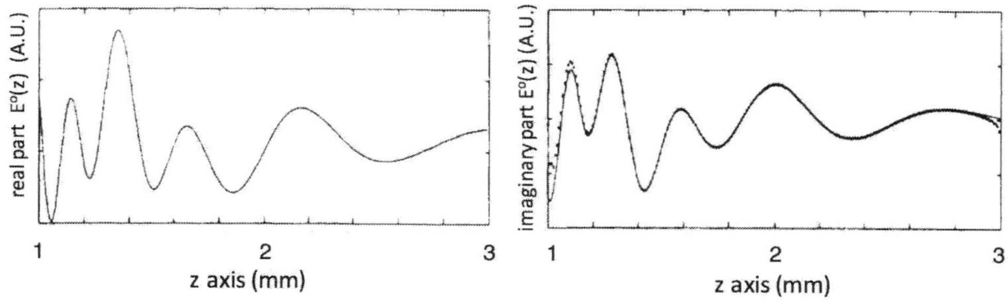

Fig.6-20 Real (left) and imaginary (right) part of the field E°(z) at distance z, for a speckle field generated by a square diffuser, 100 μm per side. Lines are results of simulation. Points in the right diagram are the results of a Hilbert transform of the real part.

As we can see from this figure, the agreement between the HT transform of the real part and the true imaginary part is very good.

The connection of real and imaginary parts does not help to correct the phase error, however. In fact, using the true E_R and either HT(E_R) or the true E_I to calculate the phase φ=atan (E_I /E_R)=kz+φ_{sp}, we end up with the same result, affected by the speckle error φ_{sp}.

On the other side, if we measure the intensity $I = E_R^2 + E_I^2$ and calculate the HT of ln I, we have φ=HT(ln I). If this expression is applied to the total field, we have $\varphi = kz + \varphi_{sp}$, and the measurement is affected by the speckle-error φ_{sp}. If we apply the expression to the reduced field, we get φ=HT(ln I) $=\varphi_{sp}$, and we get the error alone. The strategy for speckle error correction then follows: we shall subtract HT(ln I) from the true phase measured by the interferometer. With the same values given above, numerical simulations of amplitude and phase of the field from an ideal diffuser as a function of distance give the result shown in Fig.6-21.

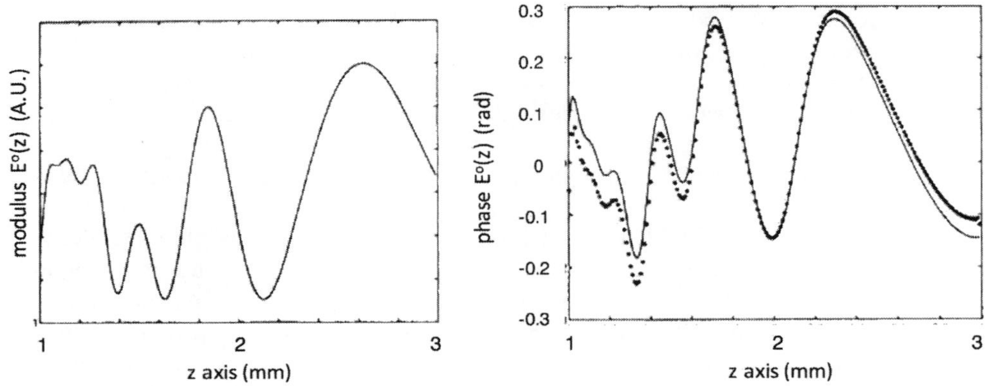

Fig.6-21 Intensity (left) and phase (right) of the speckle-pattern field from a diffuser as in Fig.6-20. Lines are the results of a numerical simulation, and dots in the right diagram are the result of the Hilbert transform of log-intensity data of the left diagram.

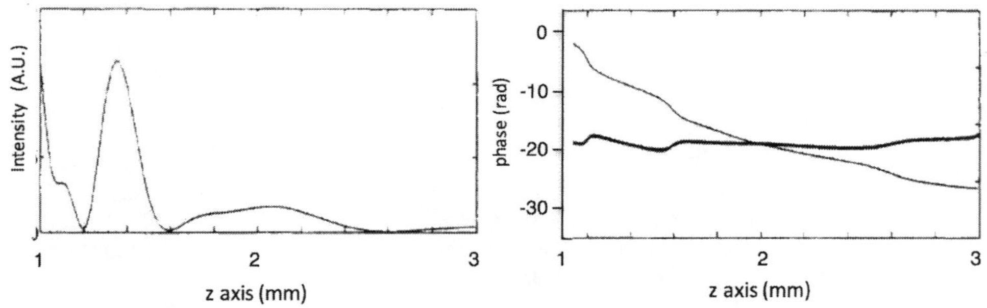

Fig.6-22 When intensity vs z falls to zero (left), the log has a singularity and the Hilbert transform of log intensity does not give the correct phase anymore, as shown at left by the thick line HT(ln I) compared to the thin line, the correct phase.

Fig.6-21 is a sample of good statistics, as we may expect when the intensity doesn't go to a zero on the distance considered. In Fig.6-21, the phase error φ_{sp} swings from –0.2 to +0.4 radians as distance is increased from 1 to 3 mm, and the HT(ln I) closely tracks φ_{sp}. Thus, if we subtract HT(ln I) from the measured phase, we cancel out a large fraction of the error, and a residue of $\approx \pm 0.01$ rad is left.
A serious limit of the HT method is, however, that on a dark speckle (with ≈0 intensity), ln I becomes infinity and the singularity of the complex function E°(z) makes HT(ln I) largely deviate from φ_{sp}. This is shown in Fig.6-22, where the case of two deep zeros in $E^2(z)$ is reported for illustration.
The HT correction has also a few other questionable features that we cannot consider here, yet the principle is intriguing and worth being considered.

6.3 Electronic Speckle Pattern Interferometry

In the previous section, speckle was evaluated as a source of error in interferometric measurements. Yet, we can also regard each speckle as an individual interferometric channel. Indeed, as coherence is maintained inside a speckle, each speckle is a spatial region within which a phase measurement can be carried out or, it is an individual pixel of an interferometric image. The only point to remember is that coherence is lost from speckle to speckle, so the measurement range cannot go outside the speckle. Also, as the phases in each speckle or pixel are uncorrelated, the interferometric image will be 'speckled' with a point like pattern as that of a normal diffuser shown in Fig.6-1.

Therefore, we can actually build an instrument measuring micrometer-displacements or vibrations pixel by pixel on an image, all simultaneously, and this instrument is called ESPI, the acronym for *Electronic Speckle Pattern Interferometry*.

The development of the ESPI concept dates back to the 1970s, when it was recognized as the electronic version of holography [15-18].
As such, ESPI is capable of supplying the same information without the burden of an anti-vibration table and of operation in darkness as required by the exposure of photographic plate, typical of holography, although with limited spatial resolution.

The basic setup of ESPI is shown in Fig.6-23. It comprises a laser emitting a power up to 20...50 mW in a single spatial mode. Traditionally, HeNe and Ar lasers have been used, with the HeNe being the best for low power, and later, two alternatives have become practicable: (i) quaternary (GaAlAsP) lasers emitting in the red (λ=620-670 nm) up to 100 mW and (ii), diode-pumped, frequency-doubled Nd-YAG lasers emitting in the green (λ=530 nm) up to several Watt.

Both sources require a beam quality control, e.g., through a pinhole spatial filter, to yield an illuminating field close to the single spatial mode.

The beam is projected on the object through a telescopic arrangement that eventually incorporates a spatial filter (Fig.6-23).

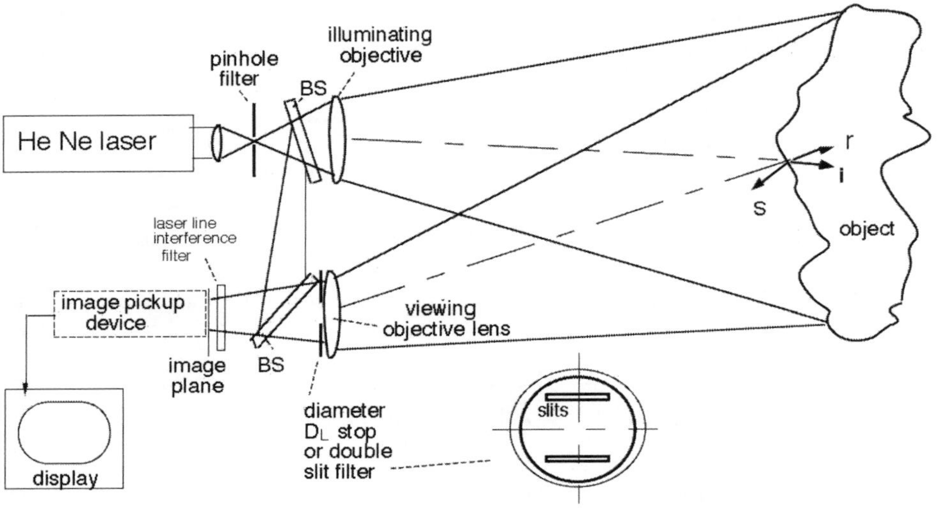

Fig.6-23 Basic setup of ESPI for pickup of displacement and vibration. Light from a He-Ne is shed on the object. By means of two beam splitters, part of the beam is superposed to the image formed by the viewing objective lens. Spatial filtering of the speckle is provided at the viewing objective lens by a circular stop or a double-slit stop.

A fraction of the beam is picked out by a beamsplitter and is diverted to the viewing objective lens. Here, the beam is superposed by a second beamsplitter to the image of the object. Thus, like in an interferometer, we recognize a measurement and a reference path. The measurement path is that going to a pixel on the object and back to the viewing lens, and the reference path is that of light coming from the beam splitters to the viewing lens.

Of course, a condition shall be met: that coherence length of the source shall be larger than the imbalance of reference and measurement paths. Eventually, we will adjust the balance with a dummy extra path inserted in the reference.

Power requirement may range from a few milliwatt for small-area objects (up to perhaps 10×10 cm^2) to hundreds of milliwatt and more for large-area testing. The required power depends, of course, also on the sensitivity of the detector (a CCD camera) and on the integration time allowed by the application.

In front of the photodetector, in Fig.6-23, we find an interference filter to get rid of ambient light. In this way, the ESPI instrument can operate without darkening, in the normal laboratory environment, differently from a holographic system.

Let us now consider the image formed on the display. Individual pixels that can be distinguished in the image are the transversal subjective speckles (Sect.6.1). The size of them, on the image, as given by (Eq.6.2), is $s_{t(trg)}=\lambda z/D_L$. By adjusting the observing lens diameter D_L, we can bring the speckle size to an optimal trade-off between fine (good resolution) and well visible (better contrast).

We may also recall the results of the speckle analysis (Sect. 6.1.6), by which the on-axis movement of the target drags along the speckle field. Then, the dragged speckle will interfere correctly (without phase error) with the reference field up to a pixel displacement given by the longitudinal size of the subjective speckle, given by $s_{l\,(trg)}=\lambda(z/D_L)^2$.

Thus, the longitudinal size $s_{l\,(trg)}$ is the allowed dynamic range in which we may perform the interferometer measurement of displacement, and the transversal size $s_{t\,(trg)}$ is the size of our pixel-level interferometer channel. The speckle image is then equivalent to a multiplicity of interferometers that work in parallel (but with uncorrelated initial phase) on the displacement s(x,y) of the target, from pixel to pixel in the x,y plane.

In each pixel of the image, the photocurrent generated by the detector is the usual interferometric signal (see for example Eq.4.3) made up of a constant term multiplying the phase dependent term $1+\cos\varphi$. For the in-line geometry considered so far, it was $\varphi=2ks$, but now we may consider a generalization. Looking at the geometry of illumination $\underline{i}$ and viewing $\underline{r}$ unit vectors (Fig.6-21), it is easy to write the phase generated for a displacement $\underline{s}$ as:

$$\varphi = k\,(\underline{i} + \underline{r})\cdot\underline{s} \tag{6.35}$$

At all pixel locations, the phase difference φ_M-φ_R of measurement and reference channels follows the speckle statistics and is randomly distributed on $-\pi..+\pi$ (Sect.6.1). Accordingly, the image of a still object viewed by the system of Fig.6-23 has the usual speckle appearance that carries no information.

In a particular pixel, we may initially find either a bright, dark, or gray speckle. Now, let the object vibrate, at a speckle position, with an amplitude Δs such that φ in Eq.6.33 is larger than 2π.

Then, we can see the pixel undergo a full cycle of brightness variation, e.g., from dark to bright and back, in analogy to a normal fringe of an interferometer subjected to a full 2π-phase cycle.

By an averaging of the detected signal followed by a high-pass filtering, we can cancel out the speckle constant field and obtain a map of vibrating modes.

Indeed, if the object under test is subjected to a vibration excitation, anti-nodal points have a large displacement and, in their pixels, the speckles appear smeared out by averaging. In nodal points, the vibration is small, and its speckle is nearly unchanged. We then obtain [17] the images displayed in Fig.6-24. This mode of operation of the ESPI instrument is called *time-averaging mode*.

With the setup shown in Fig.6-23, the time-averaging mode has an interferometric sensitivity to target movement, but the dynamic range is limited to $s_{l\,(trg)}$.

If we want to increase the dynamic range at the expense of sensitivity, we may simply remove the reference beam in the setup of Fig.6-23.

By doing so, we again obtain a signal averaging at an anti-nodal pixel location, but now smearing is due to the loss of correlation, which we incur when the displacement is larger than $s_{l\,(trg)}$. The longitudinal size $s_{l\,(trg)}$ can be made in the range of, say, 50 to 200 μm by an appropriate choice of lens diameter D_L and observation distance z. This mode is called time-averaging *without reference*, and using it we can observe large-amplitude vibration patterns even in direct sight by eye.

Returning to Fig.6-23, let us comment on the effect of the aperture stop of the objective lens. The stop affects the speckle size through the diameter D_L because it is a low-pass filter in the frequency (or Fourier) domain of the object. We can use other filtering functions to improve the speckle-image quality.

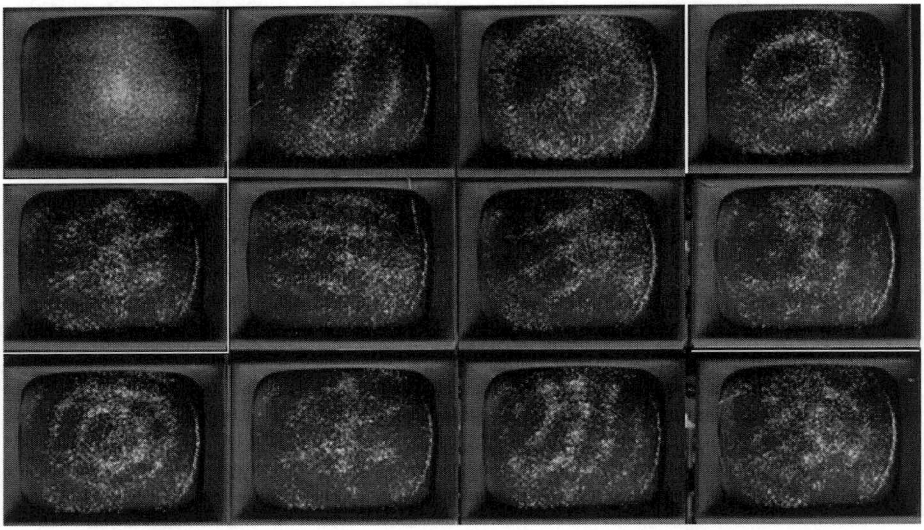

Fig.6-24 Typical ESPI images taken in the time-averaging mode. The target is a loudspeaker covered with a plain white sheet and is driven into vibration at increasing frequency (audio range). At nodes, the speckle is unchanged whereas at anti-nodes it is washed out. A few patterns of the many coming out from 200 to 2 kHz are shown. Upper left picture is the speckle image without excitation (from Ref.[17], ©AEIT, reprinted by permission).

For example, a double-slit stop [15,18] (Fig.6-23) allows increasing the signal amplitude while enhancing the high spatial frequency content of the speckle image picked up by an image photodetector scanned in a raster fashion.

When we combine a structured stop to filter the optical image with an electronic processing of the photodetected signal, the quality of the image is dramatically improved as shown in Fig.6-25. Here, the video output from the photodetector is high-pass filtered and peak-to-peak rectified. This processing yields a more detailed fringe pattern than with the simple high pass and averaging of Fig.6-23.

Now let us consider the pickup of interference images associated with a displacement distribution s(x,y). This is achieved by the *frame-subtraction mode* of operation of ESPI.
As previously noted, the image of a still object is a speckle image that carries no information. Each speckle is an individual interferometer channel, but all channels are uncorrelated because the initial phase φ(x,y) is randomly distributed from pixel to pixel.

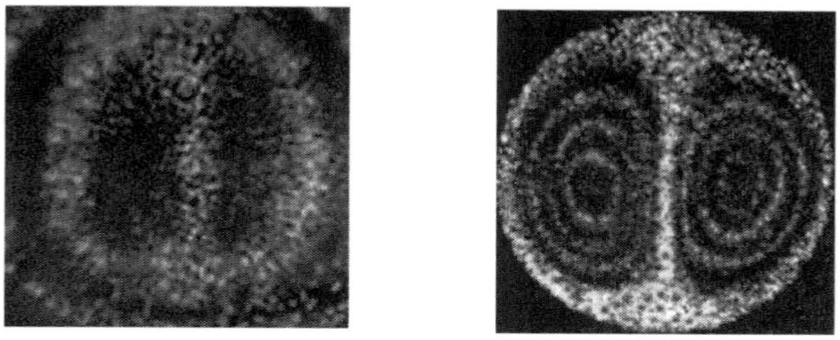

Fig.6-25 ESPI image of a loudspeaker taken with a circular stop and a video high-pass filter (left) and the same with the double-slit stop, high pass, and peak-to-peak rectifier (right) (from Ref.[17], ©AEIT, reprinted by permission).

Now, if the target is subjected to a displacement s(x,y), we get another speckle image, again with an intensity distribution uncorrelated from pixel to pixel. However, the pixel undergoes a brightness *variation* according to the interferometer phase $\varphi=k\ (\underline{i}+\underline{r})\cdot\underline{s}$. The variation is singled out by subtracting the initial image from the final image, pixel by pixel.
The result is exemplified in Fig.6-26 for an aluminum plate, observed at rest and after a deformation push at the center. Both initial and final images show no detail, but their difference unveils displacement fringes, each marking a $\varphi(x,y)=2\pi$ increment.

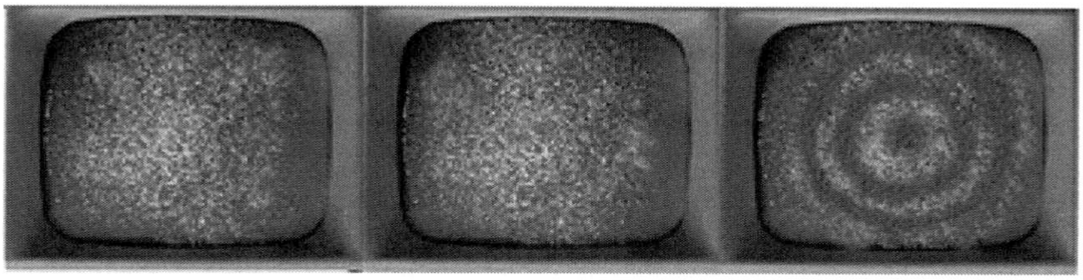

Fig.6-26 ESPI image of an aluminum plate before (left) and after (center) the application of a deformation in the center of it. The image at right is the difference of the two and unveils the strain-generated fringes (from Ref.[17], ©AEIT, reprinted by permission).

The apparatus for the frame-subtraction mode of operation of ESPI is shown in Fig.6-27 [17]. Besides the already described spatial filter in the objective lens stop and the electrical signal processing (high pass and peak-to-peak rectifier), we find a video image recorder.

In the recorder, we store the initial image of the object, corresponding to the quiescent state. The recorded image is read continuously, in synchronism with the photodetector that picks up the image of the object. Then, we may subtract pixel by pixel the initial intensity to the live image intensity. This generates an image containing the fringes associated with the displacement $\underline{s}$ as, $\varphi = k\,(\underline{i}+\underline{r})\cdot\underline{s}$. The optical correlation block indicated in Fig.6-27 can take several forms. In general, these forms are intended to develop fringes more complicated than the simple displacement measurement [17-23].

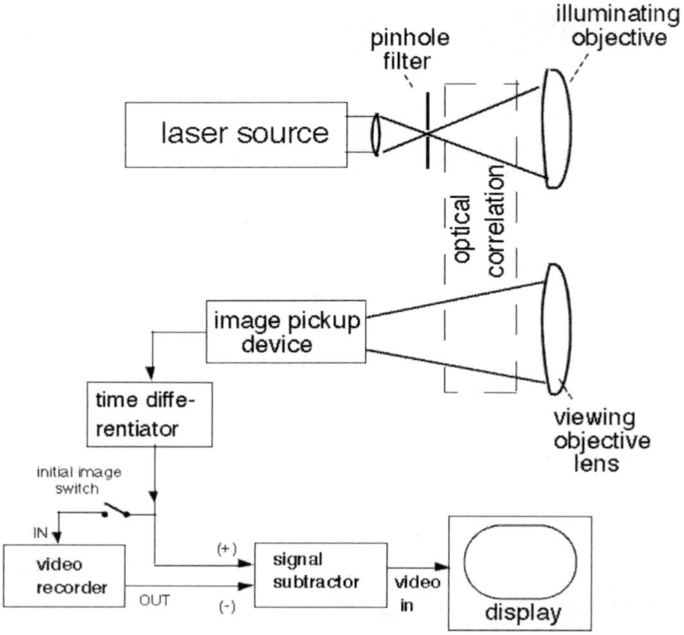

Fig.6-27 Operation of ESPI in frame subtraction mode. The high-frequency spatial content is enhanced with the appropriate lens stop and by time differentiation of the video signal. The initial speckle image is stored in the recorder so that we can subtract it from the live image presented on the display.

For example, instead of measuring the total displacement $Z=Z(x,y)$ of a pixel in the image, we may want to get the relative displacement $\Delta Z=Z(x,y) - \langle Z(x,y)\rangle$, where $\langle Z(x,y)\rangle$ is the average displacement of the target. This mode of operation removes the fringes due to rigid-body translation, a contribution that carries little information usually. We can obtain a speckle map of ΔZ by using, as a reference for superposition to the target image, the scattered light from a ground glass placed in front of the objective lens (Fig.6-28).

Second, in front of the objective lens we may place a beamsplitter with a slight tilt so that two images of the target are formed, laterally shifted by Δx (Fig.6-28). Assuming Δx is less than the transversal speckle size, we are correlating the fields received by a pair of pixels distant of Δx. The result is a measurement of the $\Delta Z/\Delta x$ strain component, with fringes indicating increment $\approx\lambda/2$ of the out-of-plane displacement ΔZ relative to points distant of Δx.

Another mode of operation of the ESPI is that of *shearing* measurement. In this interferometer, we want to measure the in-plane strain component $S_x=\Delta X/\Delta x$ following the application of a stress distribution. Here, the displacement X is that of a material point belonging to the target, and x is the laboratory coordinate in which the displacement is viewed. The ratio of their differentials is known as one component of the strain.

One of the several correlation methods developing shear fringes [17,18] is the double mirror [12] shown in Fig.6-28. The target is illuminated from a fixed direction $\underline{i}$ and viewed from two directions $\underline{r}_1$ and $\underline{r}_2$. In this way, the phase-sensitive term governing the speckle is $\varphi=(\underline{r}_1+\underline{i})\cdot\underline{s} -(\underline{r}_2+\underline{i})\cdot\underline{s} = (\underline{r}_1-\underline{r}_2)\cdot\underline{s}$, and it only depends on the difference of observation vectors $\underline{r}_1-\underline{r}_2$. As shown in Fig.6-28, the difference vector $\underline{r}_1-\underline{r}_2$ is perpendicular to the lens optical axis or, it is in the plane of the target surface.

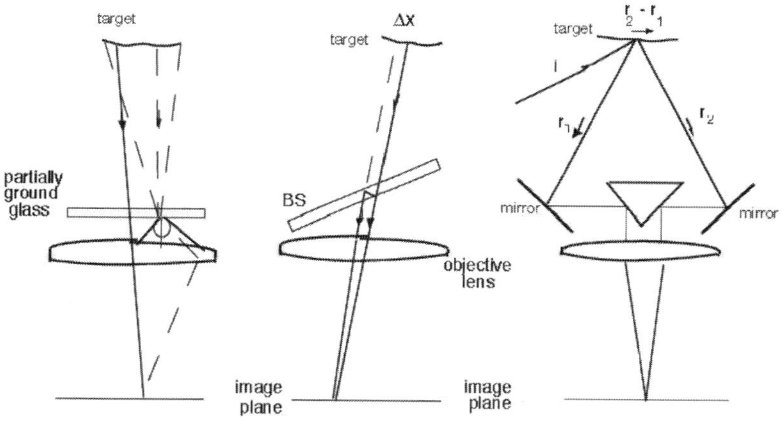

Fig.6-28 Optical correlation may take several forms. Left: a partially ground glass provides the reference for the average position of the target. Center: a tilted beamsplitter in front of the objective lens superposes contributions of image pixels shifted by Δx for strain measurement. Right: a mirror pair allows us to superpose images with different fields of view to get a shearing interferometer.

From the setup of Fig.6-28, one component of the in-plane strain, let us say $S_x=\Delta X/\Delta x$, is obtained. Of course, by rotating the setup by 90°, we can also get the other in-plane component of the strain, $S_y=\Delta Y/\Delta y$. We do not report here examples of measurement of strain components, which are rather obvious; the interested reader may consult Refs. [15-22].

Still another mode of operation of ESPI instruments is the *pulsed mode*. In it, we use a pulsed-laser source with adequate coherence to take snapshots of the target. Speckle fringes are formed by *frame subtraction* of the initial image off the live one. At each instant, we take a snapshot, and a speckle image is acquired. Thus, we can examine objects subjected to periodic stresses, like rotating machinery or turbine blades, and get information on the dynamical deformation undergone along a full cycle of the applied stress. This is valuable information, because it may reveal a pattern of strain evolution forewarning failure.

In processing the ESPI images, we face the problem of phase ambiguity like in any interferometer. Indeed, speckle fringes developed after a displacement s(x,y) obey the equation $\cos \varphi(x,y) = \cos k[(\underline{i}+\underline{r})\cdot\underline{s}(x,y)]$. As we know, a single trigonometric function of the type cosine leaves an ambiguity in the reconstruction of displacement from the phase $\varphi(x,y)$.

From the practical point of view, the ambiguity can be removed if we assume that the displacement distribution $\underline{s}(x,y)$ shall have the least peak-to-peak swing. Then, we may associate a large-width fringe with a local stationary point of $\underline{s}(x,y)$, that is, a minimum or a maximum and start from there to make the *phase unwrapping*. Also, if we know which point or points in the image are still, we can start from there to draw the paths of steepest variation of $\underline{s}(x,y)$.

Another clean approach to remove ambiguity is to make available a pair of orthogonal signals of the phase shift of the type sine and cosine. This can be easily obtained by applying a *phase dither* φ_{dith} to the reference channel [19,21].

To implement the dither, we add a PZT ceramic phase modulator in the reference beam path of Fig.6-23, so that the video signal at the photodetector output becomes $\cos[\varphi(x,y)+\varphi_{dith}]$ (Ref.24). If the ceramic is driven by a square wave voltage swinging between 0 and $V_{\pi/2}$, then the phase shift φ_{dith} swings between 0 and $\pi/2$.

By synchronizing the square wave to half the frame frequency of the image pickup, we obtain in one frame signal $\cos\ k(\underline{i}+\underline{r})\cdot\underline{s}(x,y)$ and, in the next frame, signal $\cos[k(\underline{i}+\underline{r})\cdot\underline{s}(x,y)+\pi/2] = -\sin k(\underline{i}+\underline{r})\cdot\underline{s}(x,y)$. Even and odd frames are stored separately and, after averaging, the reconstruction algorithm can be applied to obtain s(x,y).

REFERENCES

[1] J.W. Goodman: "*Statistical Properties of Laser Speckle Patterns*", in: Laser Speckle and Related Phenomena, ed. by J.C. Dainty, 2nd ed., Springer Verlag, Berlin, 1981, pp.9-74.

[2] A. Papoulis: "*Probability, Random Variables and Stochastic Processes*", 3rd edition, McGraw Hill: New York, 1991; see also: M.H. DeGroot, M.J. Schervish: "*Probability and Statistics*", 4th ed., Pearson Intl., London, 2011.

[3] D. Middleton: "*Introduction to Statistical Communication Theory*", McGraw Hill, New York, 1960; reprinted by IEEE Press: New York, 1996.

[4] J.D. Gaskill: "*Linear Systems, Fourier Transforms and Optics*", J. Wiley, New York, 1978.

[5] S. Donati: "*Photodetectors*", 2nd ed., Wiley IEEE Press, Hoboken 2021.

[6] S. Donati, G. Martini: "*Speckle-Pattern Intensity and Phase Second-Order Conditional Statistics*", J. Opt. Soc. of Am., vol. 69 (1979), pp.1690-1694.

[7] S. Donati, G. Martini, T. Tambosso: "*Speckle Pattern Errors in Self-Mixing Interferometry*", IEEE J. Select. Topics Quant. El. vol.49 (2013), pp.798-806.
[8] S. Donati, G. Martini: "*3D Profilometry with a Self-Mixing Interferometer: Analysis of the Speckle Error*", IEEE Phot Techn. Lett., vol.31 (2019), pp. 545-548.
[9] S. Donati, G. Martini: "*Systematic and random errors in Self-Mixing measurements: effect of the developing speckle statistics*", Appl. Opt. vol.53, (2014), pp.4873-4880.
[10] S. Donati, M. Norgia: "*Overview of self-mixing Interferometer Applications to Mechanical Engineering*", Opt. Engin., vol.57 (2018), doi:10.1117/1.OE.57.5.051506.
[11] M. Norgia, S. Donati, and D. D'Alessandro: "*Interferometric Measurements of Displacement on a Diffusing Target by a Speckle Tracking Technique*", IEEE J. Quant. Electr., vol. QE-37 (2001), pp.800-806; see also the companion paper: M. Norgia, S. Donati: "*A Displacement-Measuring Instrument Utilizing Self-Mixing Interferometry*", IEEE Trans. Meas. Instrum., vol. IM-52 (2003), pp.1765-1770.
[12] A. Yariv: "*Quantum Electronics*", 3rd ed., J. Wiley and Sons, New York, 2013, Appendix 4.
[13] S.L. Hahn: "*Hilbert Transforms in Signal Processing*", Artech House, Norwood, MA, 1996.
[14] M. Sorel, G. Martini, and S. Donati: "*Correlation between Intensity and Phase in Speckle Pattern Interferometry*", in Proc. ODIMAP II, 2nd Workshop on Distance Measurement and Applications, ed. by S. Donati, Pavia (Italy), 20-23 May 1999, pp.132-137.
[15] J.N. Butters, R. James, and C. Wykes: "*Electronic Speckle Pattern Interferometry*", in Speckle Metrology, ed. by R.K. Erf, Academic Press, New York, 1978, pp.111-158.
[16] R. Jones and C. Wykes: "*Holographic and Speckle Interferometry*", 2nd ed., Cambridge University Press, Oxford, 1989.
[17] S. Donati: "*A Speckle Pattern Instrument for Real Time Visualization of Vibration and Displacements*", Alta Freq., vol.44 (1975), pp.384-386.
[18] O.J. Lokberg and G.A. Slettemoen: "*Basic Electronic Speckle Pattern Interferometry*", in: Applied Optics and Optical Engineering, vol. X, ed. by R.R. Shannon and J.C. Wyant, Academic Press, New York, 1987, pp.455-503.
[19] K.J. Gasvik: "*Optical Metrology*", 3rd ed., J. Wiley, Chichester, 2002.
[20] R.S. Sirohi: (editor), "*Speckle Metrology*", Marcel Dekker, New York, 1993.
[21] P.K. Rastogi: "*Digital Speckle Pattern Interferometry and Related Techniques*", J. Wiley: New York, 2000.
[22] J. Burke, T. Bothe, H. Helmers, C. Kunze, R.S. Sirohi, and V. Wilkens: "*Spatial Phase Shifting in ESPI: Influence of Second-Order Speckle Statistics on Fringe Quality*", in Fringe '97, Automatic Processing of Fringe Patterns, ed. by W.J. Jüptner and W. Osten, Wiley Europe, London, 1997, pp. 111-119.
[23] D. Paoletti and G. Schirripa Spagnolo: "*Interferometric Methods for Artwork Diagnostics*", Progress in Optics, vol.35 (1996), pp.197-255.
[24] G. Martini, M. Facchini, and D. Parisi: "*Automatic Phase-Stepping in Fiber-Optic ESPI by Closed-Loop Gain Switching*", IEEE Trans. Meas. Instr., vol. IM-49 (2000), pp.823-828.

Problems and Questions

P6-1 *Calculate the transversal and longitudinal sizes of the speckle field generated at a distance of 50 cm, 2 m and 10 m by a white diffuser with spot size D=5 mm, using a laser wavelength λ=633 nm. How does the longitudinal size change if, at 2 m distance, we move transversally of 2 meters?*

P6-2 *Calculate the distance Z of the 'last speckle' with the data of Prob.P6-1. Comment on the relationship of Z to the Fresnel distance.*

P6-3 *Calculate the subjective transversal size of the speckle seen by the eye (D_{eye}=4 mm) when looking at a laser spot of diameter D= of 2 mm projected on a white screen at a distance z=1 m (assume λ=633 nm).*

P6-4 *How many speckles are contained in the spot imaged back on the photodetector of a triangulation telemeter? What about their effect on the measurement error?*

P6-5 *How frequent are the bright speckles in the speckle field? Calculate the probability of finding the intensity 2.5 times the average. Are dark speckles less frequent that bright ones?*

QP6-6 *Doesn't the bright speckle violate the principle of invariance of the brilliance in those points where a bright speckle is found?*

P6-7 *Consider a point, in a bright speckle, having 2.5 times the average intensity. What is the average intensity of a nearby point, correlated to the starting one with m=0.8? What is the phase variance?*

P6-8 *Consider a point, in a dark speckle, having 0.2 times the average intensity. What is the average intensity of a nearby point, correlated to the starting one with m=0.8? How about the phase variance?*

P6-9 *Starting with the Fresnel approximation to the field propagated to distance z, calculate rigorously the speckle transversal s_t and longitudinal s_l sizes, writing the correlation at an increment Δ of the coordinates ($x+\Delta$ for s_t and $z+\Delta$ for s_l).*

P6-10 *An interferometer works on a diffuse target to make a displacement measurement. Let the spot size be, like in Prob.P6-1 D=5 mm, and assume a distance of 50 cm, and a wavelength λ=633 nm. Calculate the amount of: (i) longitudinal displacement, (ii) transversal displacement; (iii) spot shift on the target; (iv) detector aperture that are allowed for a 1-λ (or 2π) phase error each.*

P6-11 *In an interferometer, amplitude fading due to speckle statistics is mitigated by operation in space diversity. We use two photodetectors, separated transversally of a speckle size, so as to have uncorrelated amplitudes. What is the probability of fading (and incorrect operation) in this case?*

P6-12 *We use the extension of Fig.4-16 for the operation of the interferometer on a diffuser. The focusing lens receives a collimated beam of waist w_0 and has a focal F and diameter D. What is the transversal dimension of the speckle exiting from the lens?*

P6-13 *I don't like the approximation to 1 of the multiplying factor in Prob.P6-12. How can you reconcile the results outlined in Prob.P6-12 without the need to introduce this approximation?*

P6-14 *How can we ensure that, on the distance swing Δ_{max} to be covered by an interferometer operating on a diffuse-surface target, the speckle phase error is a minimum? Specialize to the case Δ_{max}=1 m and calculate the attenuation of the returning signal.*

P6-15 *Try to mitigate the attenuation effects found in Prob.P6-14 while keeping the error at no more than a few λ's on the entire measurement range Δ_{max}=1m.*

P6-16 *In the bright-speckle tracking techniques, we not only oppose amplitude fading, but also have better phase statistics. Is there a simple physical reason that a bright speckle should have a smaller phase error whereas for a dark one the error is larger?*

CHAPTER 7

Velocimeters

Velocimeters were introduced in 1964 by Yeh and Cummins [1] and soon became another bright chapter of photonic instrumentation. The first commercial velocimeter, based on a crossing-beam interferometer and called the LDV (Laser Doppler Velocimeter) appeared in the market soon after the laser interferometer, that is, about the year 1970. Thereafter, several thousand LDV units were sold per year [2] and the LDV was recognized as an unparalleled tool for noncontact measurements of flow velocity in fluids [3]. The fluid can be either a liquid or a gas, and the velocity is desired for test, diagnostics and design applications to hydraulics, fluidics and wind tunnel. Because of the application, LDVs are also called Laser Doppler Anemometers (LDA). More recently, another approach has emerged, PIV (Particle Image Velocimetry), which takes advantage of the powerful computing resources available in consumer PCs to be able follow a sizeable number of particles in their individual movement along a fan-shaped illuminating beam, so as to build up the velocity field in a 2D or image format.

In this chapter, we start treating the well-established LDV, which is nicely representative of methods and technologies of photonics, then describe PIV and conclude with the SMI-based velocimetry and flow measuring instruments.

7.1 PRINCIPLE OF OPERATION OF THE LDV

Conceptually, a LDV is just an interferometer but, instead of looking at the usual out-of-plane component of displacement, given by the normal signal $2\underline{k} \cdot \underline{s}$, the LDV is designed to sense the *in-plane* component, perpendicular to the line of sight or $\underline{k}$. To obtain the in-plane component, two beams are used to illuminate the fluid, which contains small particles (μm-size) left back as residual particulate or intentionally dispersed in the medium.
Though there is no fundamental reason to identify the LDV as a Doppler-based instrument rather than a two-beam interferometer, the terminology has gained acceptance internationally, and we will therefore use it in the following.

The basic schematic of a velocimeter is shown in Fig.7-1. Historically, the preferred sources have been the He-Ne laser or the Ar-ion laser, for their good spatial quality.

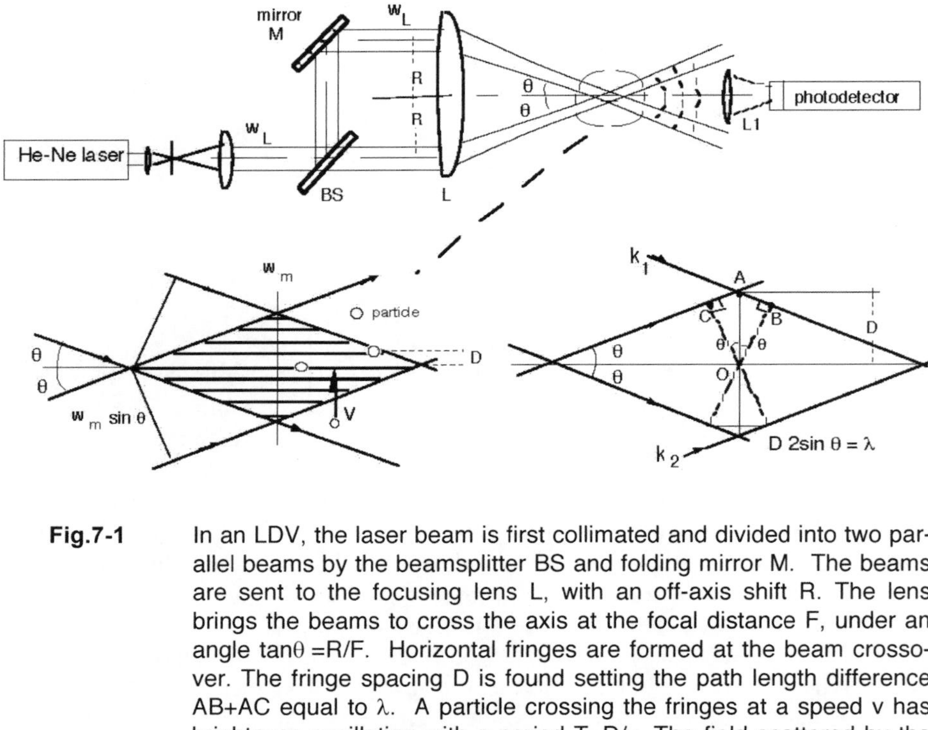

Fig.7-1 In an LDV, the laser beam is first collimated and divided into two parallel beams by the beamsplitter BS and folding mirror M. The beams are sent to the focusing lens L, with an off-axis shift R. The lens brings the beams to cross the axis at the focal distance F, under an angle tanθ =R/F. Horizontal fringes are formed at the beam crossover. The fringe spacing D is found setting the path length difference AB+AC equal to λ. A particle crossing the fringes at a speed v has brightness oscillating with a period T=D/v. The field scattered by the particle is converted by the photodetector in an electrical signal at a frequency v/D.

More recently, diode lasers with pinhole spatial filtering are used. The beam leaving the laser is collimated by a telescope to the desired size, and then it comes to a beamsplitter where it is divided in two parts. With a folding mirror, the two parts of the beam are brought parallel for off-axis incidence on a focusing lens. In this way, two beams are brought to cross each other on the axis of the lens at the focal distance F from it.

In the region of beam superposition (Fig.7-1, bottom left), stationary fringes are formed. In fact, the loci of equal phase difference between the two wave fronts are horizontal planes. To derive the fringe spacing D, let's move from point O on a fringe to point A in the next fringe (Fig.7-1, bottom right), and consider first the beam k_1. Along the wave front OB there is no phase shift, while along BA (perpendicular to the wave front) the path length delay is D sinθ.

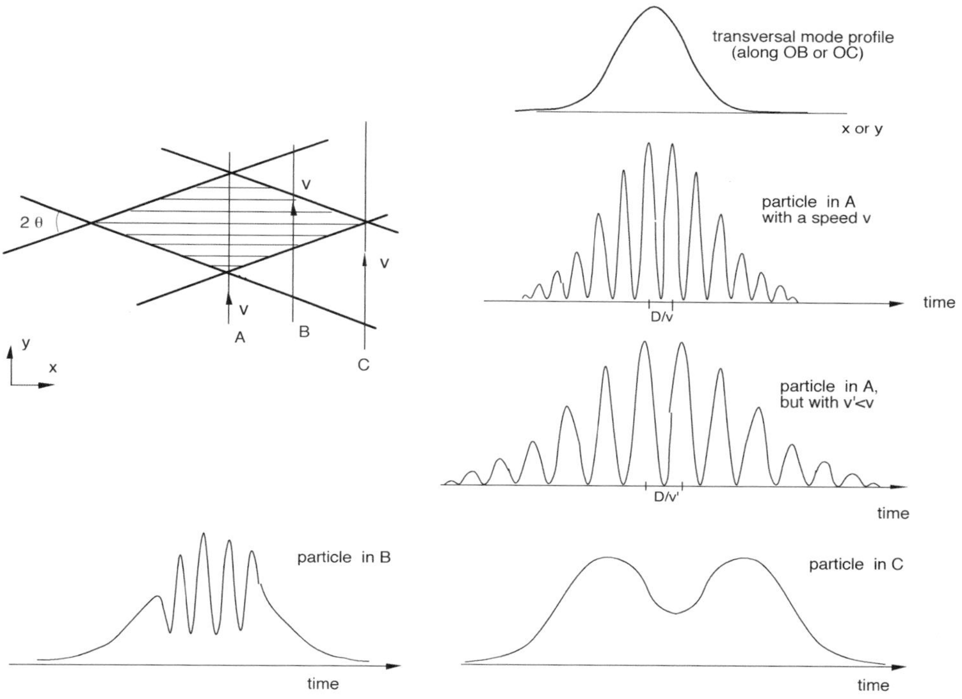

Fig.7-2 Particles crossing the fringes in the superposition region (top left) with a velocity v (the y-component) develop an oscillating signal at frequency f_D=v/ D, where D is the fringe spacing. The envelope of the oscillation replicates the Gauss distribution of the beam. Particle A, crossing the superposition region at the midpoint, develops the largest number of periods, and is the ideal response. Particle B gives fewer periods and its signal has less contrast. Particle C, outside the superposition region, only samples the beam distributions and its waveform carries no information.

For beam k_2, we get no phase shift along OC, and a path length delay -D sinθ along CA. The total path length difference of interfering beams is then 2 D sinθ. By equating this quantity to λ for adjacent fringes, we get the fringe spacing as:

$$D = \lambda/(2\sin\theta) \qquad (7.1)$$

A particle moving along the y-axis crosses the fringes and is illuminated by alternate bright and dark fringes (Fig.7-2), with a period T=D/v, where v is the velocity component along the y-axis. The particle scatters radiation outward (see Appendix A3), and scattered power is collected by a lens and photodetector combination (Fig.7-1), supplying an electrical signal for subsequent processing.

The electrical signal contains information on velocity in the form of frequency f_D of the oscillations developed by fringe crossing in the superposition region. The frequency is given by:

$$f_D = v/D = 2\sin\theta\, v/\lambda \qquad (7.2)$$

The envelope of the signal is a Gaussian distribution, a replica of the beam spatial profile of interfering beams (Fig.7-2). Not all the particle positions are equally good for developing the Doppler signal, however. Particle A at the middle of the superposition region will cross the largest number of fringes and develop a clean oscillation with a Gaussian envelope. Particle B is halfway between the middle and the border of the superposition region and crosses half as much fringes, and in addition passes through a portion of each beam alone.
Thus, particle B develops fewer cycles, has a long Gaussian tail, and its oscillation has non-unitary modulation depth because of beam intensity unbalance. Particle C is just outside the border of beam superposition and develops no fringes. Thus, particle C delivers a sample of the beam profile intensity instead of the desired Doppler signal. In the processing of signals, particle A has the most desired waveform, B is marginally useful, and C has to be discarded.

From Eq.7.2 we can see that, in the laser Doppler velocimeter, the fluid velocity measurement is brought to a frequency measurement on the electrical signal, with a scale factor given by $R = f_D/v = 2\sin\theta/\lambda$ [Hz/(m/s)].
This circumstance is very favorable, because the frequency measurement is the best we can perform in electronics. It may cover 9 to 12 decades, from millihertz or microhertz to gigahertz and above. In addition, a frequency meter can be easily traced to standards of frequency, and the result is that our measurement has also high *accuracy*, not only precision, with typical value of $\Delta f_D/f_D \approx 10^{-6}$ and better.

Returning to the frequency measurement, let us consider the typical values of the scale factor R and of the velocity range we can attain. Using λ=0.633 μm and θ= 0.1...0.5 rad (or 5°...30°), we get R = 0.3...1.6 MHz/(m/s).

Because the angle dependence is not that strong, let us take R = 1 MHz/(m/s) as a reference value. This figure means that we can go from very small velocity, ≈1 μm/s, in the low range of frequency (Hz's) to very high velocity, up to ≈100 m/s, in the high range of frequency (≈100 MHz). The wide span of frequency becomes translated in the wide *dynamic range* of velocity we can measure with the laser Doppler velocimeter.

7.1.1 The LDV as an Interferometer

We can now show that the following descriptions of the velocimeter signal are equivalent: (i) the fringe crossing; (ii) the Doppler effect; and (iii) the interferometric phase shift.

About fringe crossing, the result of the analysis is Eq.7.2 we have already considered.

About the Doppler effect, it is well known that the frequency deviation Δf_D experienced when the source moves with respect to the observer at a speed v is $\Delta f_D/f_D = v/c$, or $\Delta f_D = v/\lambda$. Around the direction of illumination (Fig.7-1), the particle is seen to approach by one beam ($\underline{k}_1$), and move away by the other beam ($\underline{k}_2$). Then, considering that the obliquity factor is $\sin\theta$, we have $\Delta f_D = v \sin\theta /\lambda$ and $\Delta f_D = -v \sin\theta /\lambda$ for the two beams. In conclusion, the total frequency difference is $2v \sin\theta /\lambda$, the result of Eq.7.2.

Third, we can use the interferometric phase shift to derive the equivalence to Eq.7.2. When a field E_0 impinges on the particle from direction $\underline{k}_1$ and we look at it from direction $\underline{k}_0$, the field is given by:

$$E_0 \exp i\,(\underline{k}_1 - \underline{k}_0) \cdot \underline{s}$$

This expression is recognized as the generalization of the usual term $E_0 \exp 2ks$, to the case of different directions of illumination and observation, where $\underline{s}$ is the displacement vector, and the dot " · " means scalar product. We have two fields of illumination in the LDV velocimeter, $\underline{k}_1$ and $\underline{k}_2$. Therefore, the total field observed from direction $\underline{k}_0$ is:

$$E = E_0 \exp i(\underline{k}_1-\underline{k}_0)\cdot\underline{s} + E_0 \exp i(\underline{k}_2-\underline{k}_0)\cdot\underline{s}$$

The signal found at the detector is proportional to the square modulus of the field, $I \propto |E|^2$. Developing the modulus, we easily get:

$$I \propto E_0^2\, 2 + E_0^2\, 2 \cos\,[(\underline{k}_1-\underline{k}_0)\cdot\underline{s} - (\underline{k}_2-\underline{k}_0)\cdot\underline{s}] = 2E_0^2\,[1+ \cos\,(\underline{k}_1-\underline{k}_2)\cdot\underline{s}]$$

Except for a constant multiplying term, the signal depends on the scalar product of the displacement $\underline{s}$ and the vector difference $\underline{k}_1-\underline{k}_2$ of the illuminating beam. Note that the result is independent from the observation vector $\underline{k}_0$.

With the aid of Fig.7-3, it is easy to see that the difference $\underline{k}_1-\underline{k}_2$ is parallel to the y-axis (the vertical axis in Fig.7-2), and its modulus is given by (see Fig.7-3):

$$|\underline{k}_1-\underline{k}_2| = 2k \sin\theta$$

For a particle making an in-plane displacement $\underline{s}$ parallel to the y-axis, the phase of the cosine function is then $\phi = (\underline{k}_1-\underline{k}_2)\cdot\underline{s} = 2ks \sin\theta$. By time differentiating the phase shift ϕ, we obtain angular frequency (or pulsation) $\omega=2\pi f$. Using the expression of ϕ, we can obtain frequency as:

$$f = (2\pi)^{-1}(d\phi/dt) = (2\pi)^{-1}2k(ds/dt)\sin\theta = 2 \sin\theta\, v/\lambda$$

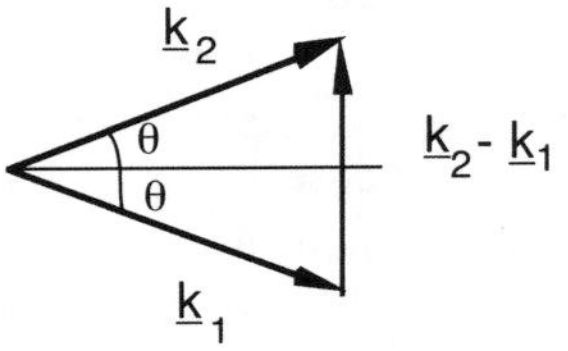

Fig.7-3 The vector difference of the illuminating wave vectors $\underline{k}_1$ and $\underline{k}_2$ is the argument of the phase shift $\phi = (\underline{k}_1 - \underline{k}_2)\cdot\underline{s}$ observed for a particle under a displacement $\underline{s}$.

Using the expression of ϕ, we can obtain frequency as:

$$f = (2\pi)^{-1}(d\phi/dt) = (2\pi)^{-1}2k(ds/dt)\sin\theta = 2\sin\theta\, v/\lambda$$

The result coincides with Eq.7.1 once more, and this demonstrates the equivalence of the Doppler effect and interferometric phase shift descriptions.

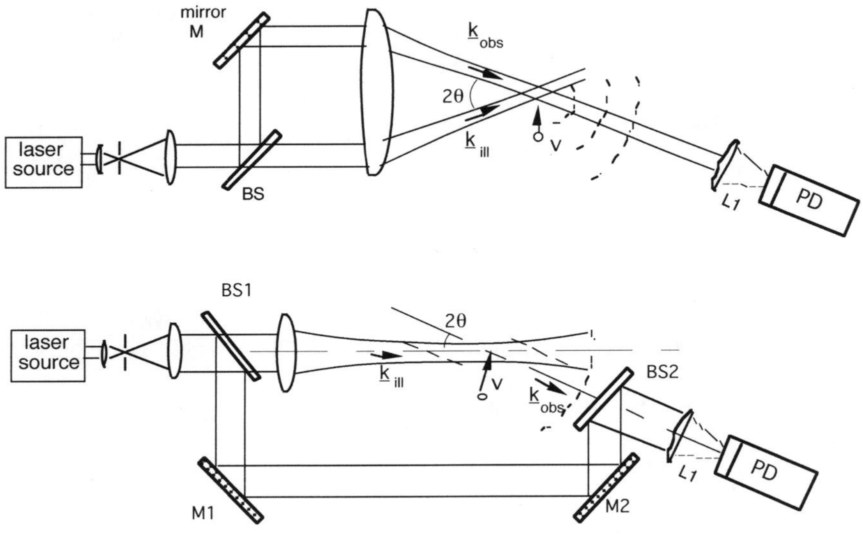

Fig.7-4 (Top): an LDV configuration named 'referenced' because the photodetector is positioned to collect one of the illuminating beams and thus acts as the local oscillator for the homodyne detection of the fringe signal. (Bottom): another referenced configuration uses a beam deviated from the laser to the detector. Though fringes are not formed, the vector difference $\underline{k}_{ill}$-$\underline{k}_{obs}$ determines the axis of sensitivity to the speed component.

Two other configurations related to interferometric arguments are reported in Fig.7-4. The first one (top in Fig.7-4) is called *referenced detection* and is obtained by moving the detector so that it intercepts the portion of the upper beam passing beyond the fringes. This beam acts as the local oscillator beam of a coherent homodyne detection [4], with the result that sensitivity is much increased compared to the placement of Fig.7-1, that operates in the direct detection regime.

The configuration is sensitive to the optical phase shift $(\underline{k}_{ill}-\underline{k}_{obs})\cdot\underline{s}$, that is, to the velocity component along the vertical axis in the drawing of Fig.7-4.

This circumstance is no surprise because it is the superposition region of the two beams that generates the usual horizontal fringes, like in the setup of Fig.7-1. The only basic difference with Fig.7-1 is the presence of the reference beam.

The second configuration in Fig.7-4 (bottom) is truly different from the previous ones, conceptually. The reference beam is again present, but illumination is provided by only one beam, and no fringes are formed. This configuration, one proposed and experimented in the early times of LDV [3], is itself sensitive to the optical phase shift $(\underline{k}_{ill}-\underline{k}_{obs})\cdot\underline{s}$.

A final comment is about the difference between velocimetry and vibrometry (Sect.4.5). As already pointed out, both are actually interferometers, and it makes no fundamental difference if we look at the frequency signal rather than the phase signal. The true difference is the direction along which the component of $\underline{s}$ or $\underline{v}$ is measured: parallel or perpendicular to the line of sight aimed at the measurement volume. Another difference is that the LDV looks at a velocity of a fluid moving in front of it, whereas the vibrometer looks at small periodic displacements.

7.2 Performance of the LDV

We now consider the parameters of operation of the LDV and the layout options affecting performance, such as accuracy of the velocity measurement, sampling volume, and effect of alignment errors.

7.2.1 Scale Factor Relative Error

The scale factor of the LDV measurement is the frequency-to-velocity ratio, or R = f/v= $2\sin\theta/\lambda$. The relative error of R, or $\Delta R/R$, is the sum of the relative errors $\Delta\theta/\theta$ and $\Delta\lambda/\lambda$ (absolute values). Usually, the wavelength error is not a problem because λ is known to a very good accuracy, at least $\Delta\lambda/\lambda\approx10^{-4}$ or better (see also App.A1).

The angle of beam crossing can be expressed as $\theta = \text{atan}\, R_{os}/F$, where R_{os} is the axial offset of the beams and F is the focal length of the objective lens in Fig.7-1.

Focal length F is usually known with a typical 10^{-4} accuracy, whereas R depends on the position of the beamsplitter and folder mirror. Quantity R_{os} can be measured to 10^{-4} accuracy as well, but the long-term stability of mechanical mounts may limit the repeatability to probably a few 10^{-3}. In conclusion, from the measured frequency, we may expect to go back to velocity $v=f_D/R$ with an accuracy of at least $\Delta R/R \approx 10^{-3}$.

7.2.2 Precision of the Doppler Frequency

Several sources limit the precision of the measured Doppler frequency f_D. Let us now consider the ultimate precision, the one set by the information content in the waveform being measured. Several types of waveforms are generated by particles crossing the fringe region in Fig.7-2, as already pointed out. The cleanest waveform is that generated by particle A at the fringe middle, and this waveform is likely the best for information content. Thus, we now take particle A as a reference for our considerations.

In crossing the beams, particle A is illuminated by a Gaussian distribution with spot size w_m (Fig.7-1). Therefore, the waveform s(t) of the signal scattered out to the detector is a sinusoidal function multiplied by a Gaussian envelope, or:

$$s(t) = \sin 2\pi f_D t \ \exp -t^2/2\sigma_t^2,$$

where $\sigma_t = w_m/v$ is the standard deviation of the Gaussian (or $1/e^2$ half-width), and v is the speed of the particle.

By going to the Fourier transform of s(t) [5], we find that the frequency spectrum of the Doppler signal is again a Gaussian distribution, centered at the Doppler frequency f_D, or:

$$s(f) = \exp -[(f-f_D)^2(2\pi \sigma_t)^2/2]$$

In this expression, the standard deviation σ_f of frequency f_D is $\sigma_f = 1/(2\pi \sigma_t)$. Substituting $\sigma_t = w_m/v$ in this expression gives $\sigma_f = [2\pi w_m/(f_D D)]^{-1}$, and by rearranging we get:

$$\sigma_f / f_D = 1/(2\pi N_f) \tag{7.3}$$

Here, we let $N_f = w_m/D$ for the number of fringes contained in w_m.

In conclusion, the relative accuracy σ_f/f_D of the Doppler measurement is proportional to the inverse of the number N_f of fringes crossed by the particle.

Of course, because it is $v=f/R$, the same result holds also for the relative precision of the speed measurement, $\sigma_v/v = 1/(2\pi N_f)$.

In point A, the height w_m of the fringe region (Fig.7-1) and the number N_f of fringes are maxima, and therefore the relative error σ_f/f_D is a minimum. This confirms the initial assumption that point A is the best for precision.

Now, let us consider those particles crossing the fringes off the optimal point A. Even if we are able to get rid of waveform artifacts (Fig.7-2), the number of fringes N_f' progressively decreases from the maximum N_f to zero at the crossing border (point C in Fig.7-2).

By repeating the above calculation, we find that the relative precision is still given by Eq.7.3, but with the actual number N_f' of fringes (or periods of the Doppler waveform) in place of N_f. Thus, as the particle crossing gets farther away from the middle of the superposition region, accuracy gets worse and worse.

Then, it will be appropriate to make a selection of 'good' particles and discard 'bad' ones. In the next section this selection is performed by means of an *a-posteriori* electronic validation, which is equivalent to the limit of the lateral offset of particles allowed in the measurement. If the selection restricts the particles to those crossing at least N/2 fringes, for example, the relative accuracy is intermediate between $(2\pi\, N_f)^{-1}$ and $(\pi\, N_f)^{-1}$.

The maximum number of fringes in the superposition region (particle A in Fig.7-2) is given by the ratio of radial offset R_{os} to the spot size w_L of the beams to be recombined:

$$N_f = (2/\pi)\, R_{os}/w_L \tag{7.4}$$

This result is derived in the next section in the case of a focusing lens, but has a general validity.

Last, let us consider what happens if the measurement is repeated on N_M particles. If we take the average value of N_M individual measurements, the relative precision of the measurement improves as $1/\sqrt{N_M}$.

Combining the $1/\sqrt{N_M}$ factor and the $\sigma_f/f_D \propto 1/N_f$ dependence, it is easy to see that the laser Doppler velocimeter can attain very low values of the relative error (in velocity), for example 10^{-3} or less. The caution is here that Eq.7.3 is an ultimate limit for precision that is approached only after properly curing other systematic errors, for example, the unequal spacing of fringes (see Sect.7.2.4).

7.2.3 Size of the Sensing Region

With the aid of Fig.7-5, we can calculate the size of the sensing region determined by the beam superposition where fringes are formed. This is the measurement region, also called the *sampling region* or volume of the LDV. A feature specific to the LDV is the small size (millimeter or less) of the sampling region. Together with the noncontact operation, this is a definite advantage of the LDV compared with other instruments.

Let w_L be the beam size produced by the beam expander and reaching the entrance of the focusing lens. In the focal plane, the beam is focused to a spot size w_f (Fig.7-5). We can find w_f by applying the invariance of the acceptance $\mathcal{a}$=aω [4], which is the product of the area and of the solid angle through which the bundle of rays is passing. For a single-mode spatial distribution, acceptance is equal to λ^2 [4]. Because $a = \pi w_f^2$ and $\omega = \pi(w_L/F)^2$, we may apply the previous statement and write for the spot size:

$$a\omega = \pi w_f^2\, \pi(w_L/F)^2 = \lambda^2$$

Solving this expression for w_f, we obtain:

$$w_f = \lambda F/\pi w_L \tag{7.5}$$

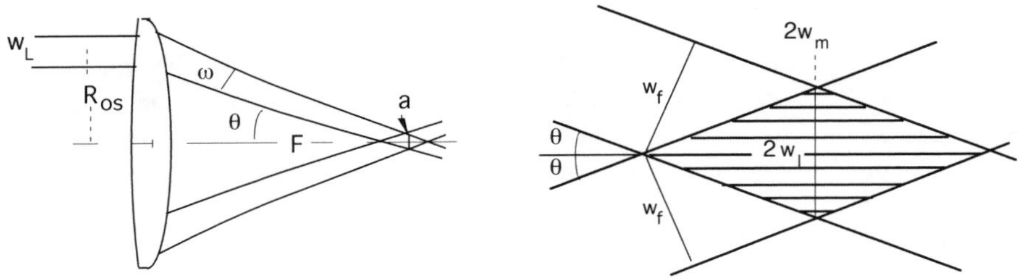

Fig.7-5 The sensing region is defined by the longitudinal and transversal sizes, w_m, and w_l. These quantities depend on beam size w_L and offset radius R_{os}.

Now, we can take into account the obliquity factor (Fig.7-5, right) connecting the spot size to the transversal size w_m (a factor $1/\cos\theta$) and to the longitudinal size w_l (a factor $1/\sin\theta$). Doing so, we can write the following results:

$$w_m = (\pi \cos\theta)^{-1} \lambda F/w_L, \quad \text{and} \quad w_l = (\pi \sin\theta)^{-1} \lambda F/w_L. \qquad (7.6)$$

In these expressions, the angle θ is determined by the offset radius R_{os} (Fig.7-5, left), and is given by $R_{os}/F= \tan\theta$. Recalling the fringe spacing $D=\lambda/2\sin\theta$ from Eq.7.1, we can work out the number of fringes as:

$$N_F = w_m/D = (\pi \cos\theta)^{-1} \lambda F/w_L [\lambda/2\sin\theta]^{-1} = (2/\pi) \tan\theta\, F/w_L = (2/\pi) R_{os}/w_L$$

The last term in the expression coincides with the result of Eq.7.4.

Example. To illustrate the typical values that can be found in a LDV, we may consider the case R_{os} = 50 mm and F = 200 mm, calling for a lens with a diameter $D_{lens}=2R_{os}=100$ mm (and with an F-number [4] F/ =0.5). The angle of superposition is found from $R_{os}/F = \tan\theta = 0.25$, as $\theta= 14°$.
With a laser spot size $w_L= 1$ mm on the focusing lens, we obtain a number of fringes $N_F = (2/\pi)50=31$. The accuracy of the single velocity measurement is therefore $\sigma_v/v = \sigma_f/f_D = (2\pi N_f)^{-1} = 0.5\%$.
At the He-Ne wavelength $\lambda=0.633$ μm, the fringe spacing is evaluated as D= (0.633 μm/2) sin 14° = 1.30 μm, and the scale factor is $R= 2\sin\theta/\lambda= 0.766$ Hz/(μm/s) or, if preferred, 0.766 kHz/(mm/s) or also MHz/(m/s). The focused spot size is w_f = (0.633μm) 200/π·1= 40 μm. The sizes of the sensing region are: $w_m = w_f/0.97= 41$ μm, and $w_l = w_f/0.24 = 166$ μm.
These values provide a good intrinsic accuracy and a small sampling volume. The sampling volume is not so far away from the focusing lens, however. If we need to sense the fluid farther away, we may use a larger focal length, for example F=500 mm, and the angle of superposition is $\tan\theta= 0.1$, or $\theta= 5.7°$. With the same laser beam size w_L, the number of fringes is unchanged, as well as the accuracy σ_v/v. Fringe spacing and scale factor are increased to D=3.16 μm, and $R= 0.316$ Hz/(μm/s). The focused spot size changes to w_f = (0.633 μm) 500/π·1= 100 μm, and the sizes of the sensing region become $w_m = w_f/0.995= 101$ μm, and $w_l = w_f/0.1= 1000$ μm. Should this large value of the longitudinal size w_l be inconvenient, we can always trim it by the a-posteriori validation of fringe (see Sect.7.3.1).

7.2.4 Alignment and Positioning Errors

Errors in alignment of the beams sent to the focusing objective have the consequence of distorting the fringes in the superposition region. As illustrated in Fig.7-6, an angular error of the parallelism of the beams results in an unequal spacing of fringes, whereas an unbalance of the lateral offset radius R tilts the fringes with respect to the optical axis.

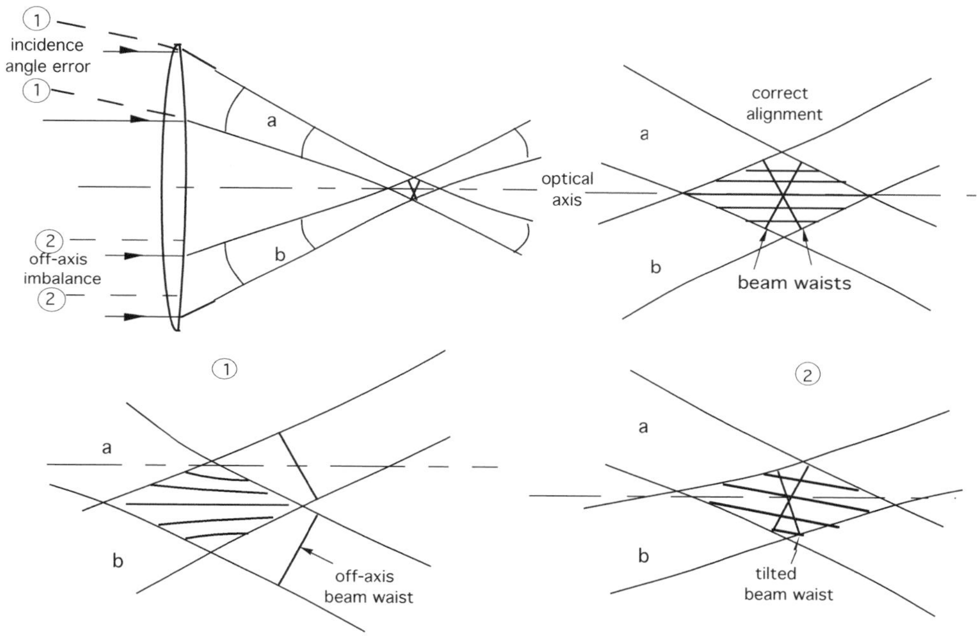

Fig.7-6 At the focusing objective, errors of the incidence angle or of the lateral offset of the two beams result in fringes with unequal spacing (bottom left) or with a tilt with respect to the optical axis (bottom right).

Of course, there are other possible types of alignment or positioning error, resulting in a specific distortion of the fringe pattern, which reduces the precision of the velocity measurement. To minimize these effects, we should employ an optical configuration that is easy to trim and mechanically stable in time. From this point of view, the configuration of Fig.7-1 is not the best because mirrors are affected by an angular error of their position, and each of them can go out of alignment.

A preferred configuration is shown in Fig.7-7 (left). Here, a cube beam splitter BS divides the incoming beam, and folding prisms are used in place of mirrors to derive beams b1 and b2. A prism, used as a mirror by the internal reflection at the 45° surface, is much sturdier and mechanically stable in time than a mirror on a flat. In addition, using the other two surfaces as a reference for the mechanical mount, the parallelism of P1, P2, and P3 is much easier to achieve. If desired, we could improve the configuration further with the use of Dove prisms as shown in Fig.7-7 (right).

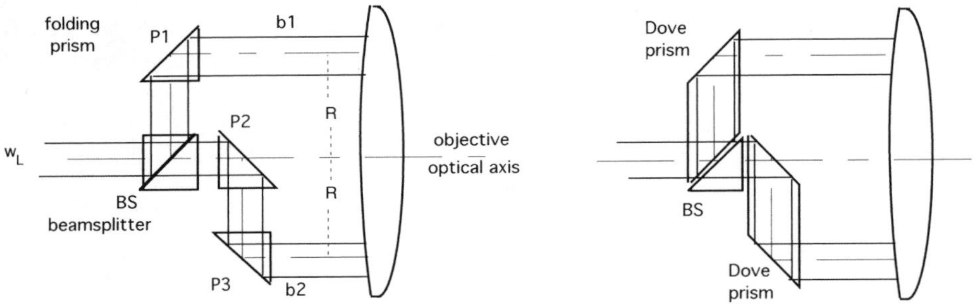

Fig.7-7 Using a beamsplitter prism BS and the folding prisms P1-3 as mirrors (left), the splitting configuration is more stable and easily trimmed as compared with that of Fig.7-1. Even better, we can employ Dove prisms to minimize parts count and improve stability further (right).

A last interesting feature of the prism configuration is optical the path-length balance. Looking at Fig.7-1, we can see that path lengths to the superposition region are different, because the lower beam reaches the lens straight, whereas the upper beam has the extra vertical path to cover. The difference is not so large, being of the order of 2R (typ. ≈100 mm). However, if the laser source is multimode in frequency, that is, emits several longitudinal modes, we may face a reduction of the coherence factor. In this case, the fringe visibility is not a maximum and we get a sort of pedestal in the Doppler signal, as illustrated in Fig.7-8.

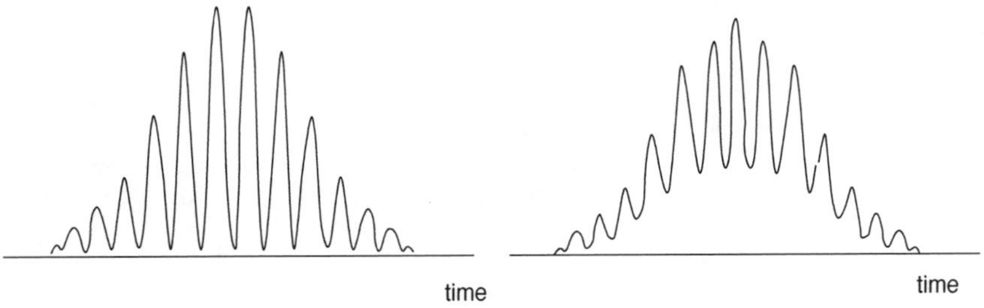

Fig.7-8 The Doppler waveform has 100% modulation (or V=1 fringe visibility) if the two superposed beams have a path-length difference much less than the coherence length of the source (left). If coherence is not unity, or beams have unequal intensity, waveforms exhibit a pedestal (right).

Using the prism configuration to generate the beam superposition, we enter parallel to the lens optical axis so that path lengths are balanced (Fig.7-7).

In addition to the configurations of Figs.7-1 and 7-7, several other variants have been proposed [3] and used in LVDs.

7.2.5 Placement of the Photodetector

The photodetector collects the signal scattered out by the particle crossing the fringes and converts it to an electrical signal. Because scattered light radiates in any direction, the photodetector could be, in principle, positioned everywhere around the beam-crossing region. But we want to collect a large signal and the fluid flows in front of the objective lens, and there are in practice just two options for the photodetector placement. As shown in Fig.7-9, we may place it beyond the fluid to look at the forward-scattered radiation or behind the objective lens to look at the backward-scattered radiation. In both cases, we will use an objective lens L' to improve light collection and to define a certain solid angle of collection Ω.

In the two cases, at equal distance from the scattering volume and equal aperture of the objective lens, the signal collected is different because it is proportional to the value of the scattering function $f(\theta)$ (App.A3.1) in the forward ($\theta \approx 0$) and in the backward ($\theta \approx \pi$) direction, respectively. The scattering function changes considerably when we go from the Rayleigh to the Mie regime. In the small-particle ($r << \lambda$) Rayleigh regime, the scattering function is nearly isotropic, and then the forward-scattering and backward-scattering photodetectors are equivalent. However, we do not prefer working with very small particles, because then the amplitude of the scattered signal, or the scattering cross-section Q_{ext} (App.A3.1), is very small. Indeed, Q_{ext} is proportional to $(r/\lambda)^4$ in the Rayleigh regime. If the particle radius is comparable to wavelength or larger, we are in the Mie regime, in which the cross-section factor is comparatively large ($Q_{ext} \approx 2$) and the scattered signal is much increased. In the Mie regime, however, the scattering function $f(\theta)$ is strongly peaked forward.

The ratio $p(\pi)/p(0)$, proportional to the relative photodetected signals in the backward and forward directions, is in general strongly dependent on the particle characteristics, but usually lies in the range 10^{-3} to 10^{-2}.

Clearly, in this case, we will prefer collecting the signal in the forward direction, if the placement of the detector beyond the fluid is accessible. If it is not, we shall work in the backward direction and will use a detector of increased sensitivity to compensate for the reduced signal amplitude.

In practice, both photodiodes (pin and avalanche) and photomultipliers are used according to the required sensitivity [4].

The amplitude of the signal supplied by the photodetector can be calculated as follows. Let P_L be the laser power, split in the two beams and recombined in the scattering volume. A particle crossing the scattering volume is illuminated with a power density P_L/w_l^2 and scatters out a total power $P_{ts} = (P_L/w_l^2)\, Q_{ext} \pi r^2$ (App.A3.1).

Denoting with $f(\theta_{meas})$ the scattering function evaluated at the measurement angle θ_{meas} ($=0$ or π), the signal collected in the solid angle Ω is written as $4\pi P_{ts}\, f(\theta_{meas})\, \Omega$. Last, recalling that the photogenerated current is σ times the detected power, σ being the spectral sensitivity [4], we get:

$$I_{ph} = \sigma\, (P_L/w_l^2)\, Q_{ext} \pi r^2\, 4\pi\, f(\theta_{meas})\, \Omega \tag{7.7}$$

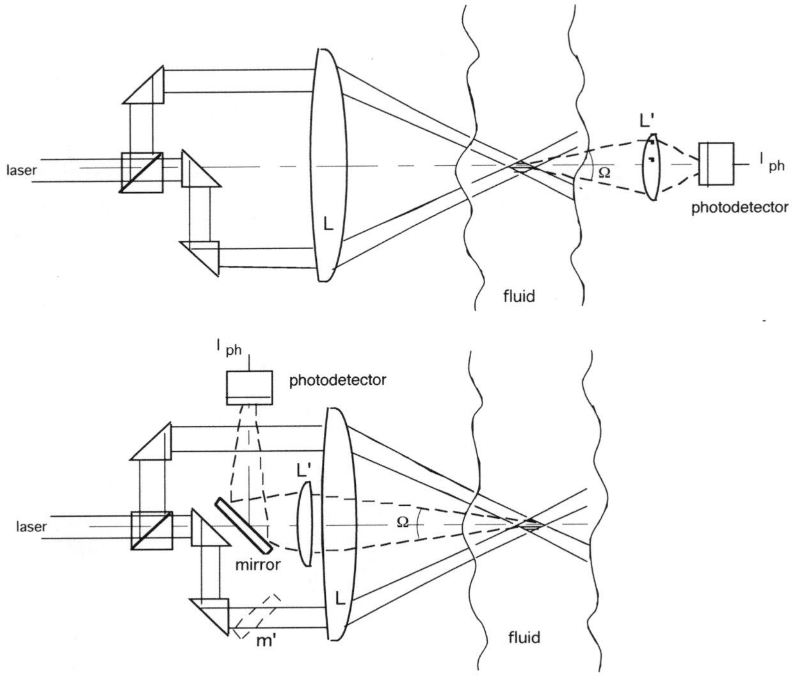

Fig.7-9 Light scattered by the particle in the measurement region can be collected by a photodetector looking at the forward scattering (top) or at the backward scattering (bottom).

In most schemes considered so far, we have assumed using direct detection to avoid unessential complication. Coherent detection can be employed as well, at the expense of some extra complexity in the optical setup, and it generally improves the sensitivity of photodiodes without internal gain. An example of this arrangement was provided in Fig.7-4.

In addition, most direct detection schemes of LDV can be easily converted to coherent ones. For example, with reference to Fig.7-9, we can convert to coherent detection by these three steps: (i) removing lens L' so that the returning beam comes out collimated from L, (ii) adding beamsplitter m' so that part of the reference beam is superposed on the photodetector, and (iii) changing the folding mirror in a beamsplitter.

7.2.6 Direction Discrimination

As in any interferometer, in a single-channel LDV we cannot distinguish the direction of the velocity, or tell v from -v. Actually, in a few cases the direction discrimination is absolutely necessary, for example, turbulence study.

In these cases, we may use the configuration of Fig.7-10 where a Bragg cell is inserted on one of the two recombining beams. The Bragg cell is a frequency shifter based on the

acousto-optical effect. If the frequency of the input beam is f, from the output of the Bragg cell we get a frequency $f+\Delta f$. The typical shift in the range $\Delta f = 20\text{-}100$ MHz.
Let us now consider the superposition of the two beams (Fig.7-10). We may repeat the reasoning of Sect.7.1.1, modified to take account of the different frequencies of the two beams.

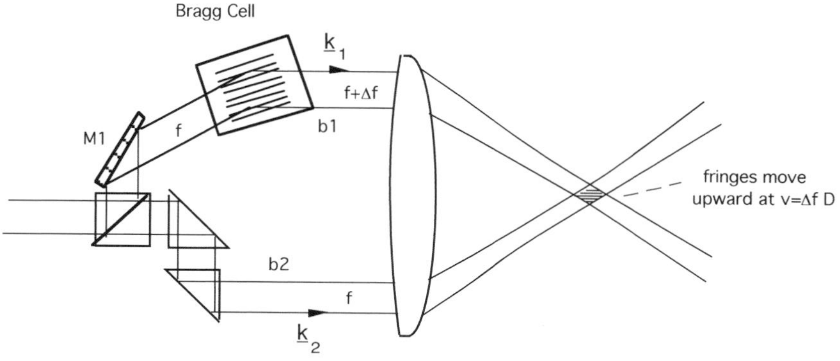

Fig.7-10 By adding a frequency shifter on the path of one of the beams of the velocimeter, we generate a moving fringe pattern in the measurement region and are able to discriminate the direction of velocity v.

The total field is given by:

$$E = E_0 \exp i\, [2\pi f t+(\underline{k}_1-\underline{k}_{obs})\cdot\underline{s}]+ E_0 \exp i\, [2\pi(f+\Delta f)t+(\underline{k}_2-\underline{k}_{obs})\cdot\underline{s}]$$

The signal at the detector, $I \propto |E|^2$ is given by:

$$I \propto E_0^{\,2}\, 2 + E_0^{\,2}\, 2 \cos\, [2\pi\Delta f t + (\underline{k}_1-\underline{k}_0)\cdot\underline{s} - (\underline{k}_2-\underline{k}_0)\cdot\underline{s}] = 2E_0^{\,2}\, [1+ \cos 2\pi\Delta f t+(\underline{k}_1-\underline{k}_2)\cdot\underline{s}]$$

By time differentiation of the phase in the cosine term, we obtain frequency as:

$$f = (2\pi)^{-1}(d\phi/dt) = (2\pi)^{-1}(d/dt)[2\pi\Delta f t +(\underline{k}_1-\underline{k}_2)\cdot\underline{s}] = \Delta f +2 \sin\theta\, v/\lambda \qquad (7.8)$$

The result tells us that the velocity-dependent frequency, $2\sin\theta\, v/\lambda$, is superposed to the frequency shift Δf. Thus, as long as $2\sin\theta\, v/\lambda<\Delta f$, we can determine the sign of v.
We can interpret the frequency bias introduced by the Bragg cell as an upward movement of the fringe pattern. Indeed, the shift Δf corresponds to a fringe velocity $v_{fr}= \lambda\Delta f/2 \sin\theta$.
The scheme with the Bragg shifter for direction discrimination has become increasingly popular in recent years, and most commercial LDV instruments incorporate it.

7.2.7 Particle Seeding

About particles generating the scattered signal, we can either have them already present in the fluid as unintentional contaminant or can seed them purposely in the fluid under measurement. The seeding operation is preferable if practicable, because we can choose the diameter best suited for generating a clean and strong signal.

Usually, seed particles are Latex (styrene butadiene) spheres of calibrated diameter, typically a few microns. These spheres are available from polymer companies in a range of sizes, from submicron to tens of micrometers in diameter.

The diameter is chosen large enough to give a large scatter signal in the Mie regime, but small enough compared to the fringe period to avoid averaging of the illumination spatial modulation (or fringe spacing).

The concentration of the spheres is readily calculated by requiring that no more than one particle at a time is present in the scattering volume, for the reasons seen in the next section. In view of the random spatial distribution the particles, we may take 0.1 as the design value of the average number. Then, we may write the concentration (in number per unit volume) as $C_N = 0.1/V_{meas} = 0.1 \tan\theta / w_l^3$. The corresponding relative volume concentration (or particle dilution) is $C_V = 0.4\, r^3 \tan\theta / w_l^3$. Typical design values of C_V turn out to be $\approx 10^{-5}$ to 10^{-4}, that is, very small.

These small values explain why we can easily find the fluid disseminated by a lot of scattering particles, already enough to perform the LDV measurement. When concentration is much larger than the desired one particle per scattering volume, we have two possibilities. One is to clean the fluid by filtering, the other is increasing the scattering volume and using signal processing techniques tolerant to multi-particle superposition.

7.3 Processing of the LDV Doppler Signal

There are two options for the electronic processing of the Doppler signal obtained from the photodetector: *time-domain* and *frequency-domain* processing.

We prefer time domain when there is, on the average, much less than one particle at the time in the measurement volume, as in the case represented in Fig.7-11, top. Then, the output of the detector is a sequence of Gaussian-envelope oscillations, well distinct and separated in time. Each oscillation indeed carries information.

As we have found in Sect.7.2.2, a waveform with N_f periods allows us to measure frequency or velocity with a relative error $\Delta f_D/f_D = \Delta v/v = 1/2\pi N_f$. There are also 'poor' waveforms with low N_f due to particles crossing near the edge of the fringe region, but these can be easily discarded by electronic processing in the time domain.

When the particle concentration increases, first we have the superposition of waveforms (Fig.7-11, middle), yet the Gaussian envelopes can be recognized. Then, we could still use time-domain processing, and wait for those waveforms that occasionally come isolated.

At a further increase of particle concentration (Fig.7-11, bottom), we cannot resolve individual waveforms anymore and must revert to a frequency-domain method. With this method, we will look at the average frequency contained in the random-like superposition of Gaussian-envelope oscillations.

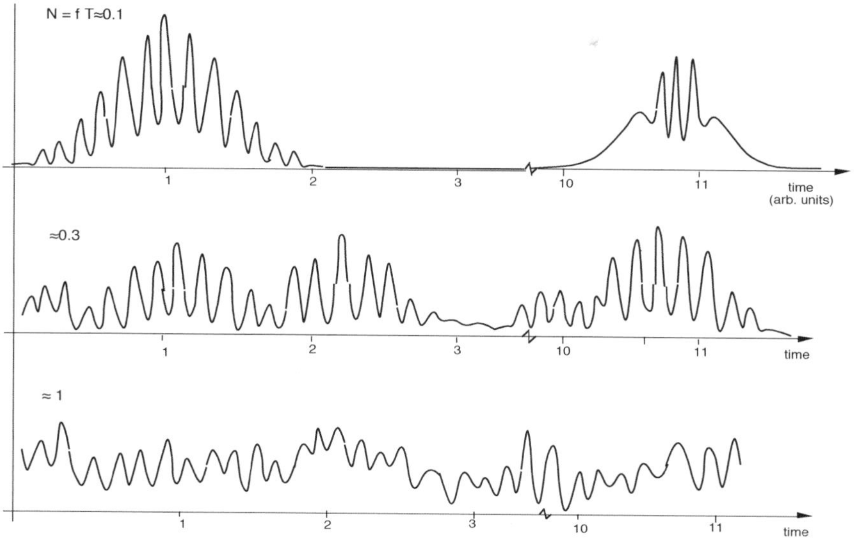

Fig.7-11 According to the average number N=fT of particles in the measurement volume, the waveform is best processed in the time domain (top) or in the frequency domain (bottom)

7.3.1 Time-Domain Processing of LDV Signal

We assume that waveforms are distinct in time and do not overlap, so the periods contained in the waveform are well correlated to the frequency f_D=v/D. Basically, the approach used to perform the velocity measurement follows the one outlined in Fig.7-12.

First, we perform a low-noise preamplification of the photodetector output signal and make available a clean Doppler waveform S(t). Then we time-differentiate S(t) so that eventual pedestal or slow drifts are canceled out. In the waveform S'(t) (Fig.7-13), the periods of oscillation are centered around the zero level and last $1/f_D$. Next, we pass the signal through an amplitude discriminator with a threshold S_0 close to zero and obtain a square-wave signal S_{SW} that contains a number of periods $N_f \approx w_L/D$ equal to the number of crossed fringes.

Frequency f_D is determined by two counters, enabled by the presence of the Doppler signal S(t). To form the enable command, we make the envelope of S(t), obtaining S_{env} (Fig.7-13) and pass it through a discriminator with threshold S_0 (Fig.7-12). The result is S_{enabl}, a signal lasting a time T_C, which enables the gates to open the counters.

During time T_C, a main counter counts the number of periods N_f, and, simultaneously, an auxiliary counter counts the N_C pulses of a clock running at a suitable frequency f_C (Fig.7-12). The counts of the signal are $N_f = f_D T_C$ and those of the clock are $N_C = f_C T_C$. By taking the ratio of the counts, we get a quantity $V = N_f / N_C = f_D / f_C$ proportional to $f_D = v/D$.

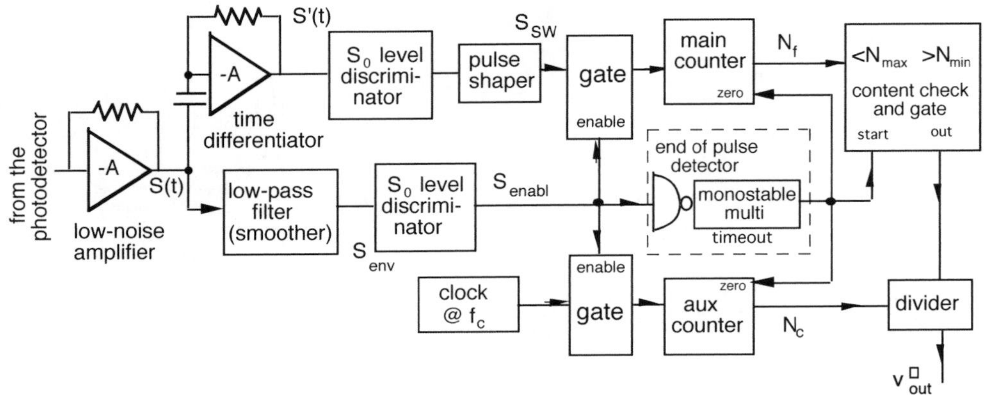

Fig. 7-12 In the time-domain processing, the Doppler signal is first time-differentiated and then squared by a discriminator with a threshold near zero. Another discriminator works on the signal envelope to determine the presence of the signal and to enable the counters. One counter is for the Doppler signal periods, and the other is for the clock. At the signal end, an inspection circuit compares the counter content to the reference minimum. If content is high enough, the result is validated and passed on to calculate velocity, whereas if it is low, it is discarded.

We can also adjust the clock frequency so that the ratio $v/f_C D$ is a decimal in the velocity v. At the end of the signal, after S_{enabl} has returned to zero, an end-of-pulse sequencer gates the division operation (Fig.7-12) and resets the counters for the next Doppler signal S(t). When a 'poor' Doppler signal is received, as exemplified by the second waveform in Fig.7-13, the number of counts N_f is low, and the content-check circuit of Fig.7-12 rejects the result at the end of the computing period. This operation is carried out by digital comparators and registers of minimum/maximum counts (Fig.7-12). If $N_f > N_{min}$ and $N_f < N_{max}$, the waveform is validated, and the content is passed on to the subsequent processing.

The minimum threshold N_{min} is usually set at half the maximum number of fringes N_f, as a good compromise between accuracy of the single measurement and waste of useful data. With the threshold placed at $N_f/2$, all particles crossing the fringe region farther than half the longitudinal width w_l (Fig.7-5) are discarded. Therefore, the actual measurement volume is determined by a longitudinal size $w_l/2$. About the upper threshold N_{max}, this is set to a little bit more than N_f (for example $1.2N_f$) to avoid taking two successive particles for one.

After the validation, the data processing may proceed in two ways. If we deal with a single point measurement, for best accuracy, we will integrate the results of N successive measurements and exploit the $1/\sqrt{N}$ dependence. If we are measuring a velocity profile, we will use one or a few measurements and then move to the next point of the scan, either along a line or in a raster pattern. A line scan is readily accomplished by translation of the focusing lens, which moves the measurement volume axially. A raster scan or a more complicated pattern will require a suitable moving mirror arrangement.

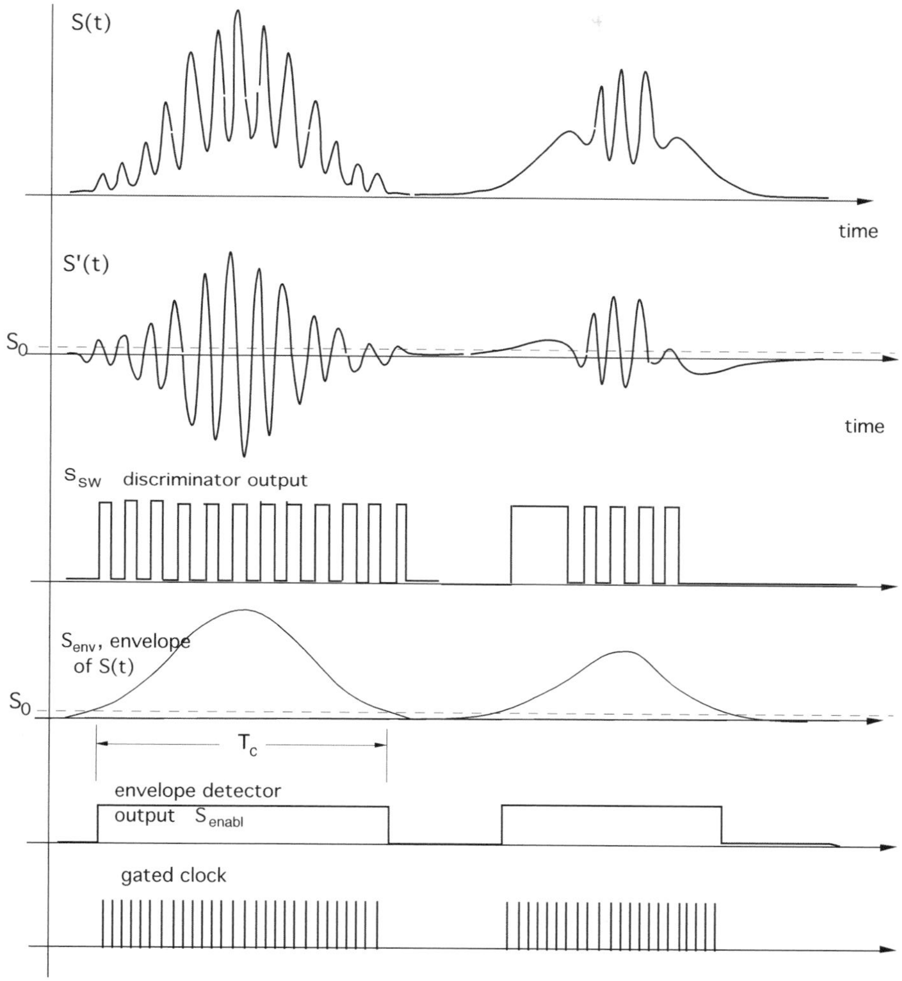

Fig.7-13 Waveforms in the time-domain processing. The first diagram is the Doppler signal S(t) at the photodetector output. Second: computing the time derivative S'(t), we are able to cancel out any pedestal and drift and get the Doppler frequency in the form of zero-crossings. Third: after discrimination with a threshold S_0, we get N_f pulses in the time duration T_C. Signal end is obtained by an envelope detector, and yields a gate square wave lasting a time T_C. When signal is over, N_f is validated by an inspection circuit to stay within the expected limits of good signals. Clock pulses (N_C) are counted during T_C. By computing the ratio N_f / N_C, velocity is obtained.

7.3.2 Frequency-Domain Processing

When particles crowd up the measuring volume as shown in Fig.7-11 bottom, time-domain processing can no longer be used. The signal oscillations carrying the Doppler frequency are damped out in amplitude and have a frequency jitter. In this case, given an average number N>>1 of particles in the measuring volume, the signal exhibits amplitude and phase fluctuations resembling those of the speckle pattern.

Indeed, the total signal is the summation of the contributions of N particles, and each particle yields a waveform of the type shown in Fig.7-11, top. Because the time of occurrence of each particle is randomly distributed, we have a summation of N uncorrelated contributions. The result is a field that replicates, for high N, the speckle pattern statistics, with zero mean value of the phase and quadrature components. Consequently, the intensity distribution is a negative exponential with a decay constant equal to the mean value.

It is then clear that, should we attempt a time-domain processing, the number of counts N_f per period T_c will be affected by a large error. More appropriately, we shall look directly at the frequency content of the signal S(t), as obtained at the output of the preamplifier stage following the photodetector.

There are two approaches for frequency-domain processing: *autocorrelation*, and *phase-locked loop*.

In the first approach, we compute the autocorrelation function $c(\tau) = \int_{0\text{-}\infty} S(t)S(t+\tau)\,dt$. The calculation can be performed by converting the signal S(t) to digital, with the aid of an Analog to Digital Converter (ADC), and then acquiring the data and computing $c(\tau)$ numerically in a computer unit. The ADC does not need a very high resolution (usually, 8 bits will be adequate) because the round-off error is usually small with respect to that of waveform noise. Rather, the ADC and the acquisition interface need to be fast if we are to cover the high-velocity range allowed by the LDV instrument. For example, if we get 100 kHz of conversion and acquisition rate, we are bound to a maximum velocity $v \approx 1\ \mu m \cdot 100$ kHz = 0.1 m/s.

From the autocorrelation function $c(\tau)$, we may already obtain the velocity by looking at the first zero of $c(\tau)$. If $c(\tau)=0$ for $\tau=\tau_z$, then the main frequency contained in S(t) is $f=1/\tau_z$. Alternatively, we may compute the Fourier transform of $c(\tau)$, for example by means of a FFT algorithm that will be easily available if we are using a personal computer for acquiring S(t) and computing $c(\tau)$. From the Wiener-Khinchin theorem [5,6], the Fourier transform of $c(\tau)$ is the power spectrum of the signal S(t). Thus, the power spectrum is a well-defined line at the Doppler frequency f_D if the fluid has a uniform velocity or is a distribution p(f) of spectral content if the fluid has a distribution of velocity.

When velocity is very high and the signal S(t) has a frequency content too high for the ADC and interface, we can resort to the PLL (phase-locked loop) technique. As shown in Fig.7-14, the signal S(t) at the preamplifier output is compared to the oscillation of a Voltage Controlled Oscillator (VCO). The phase error is used to correct the frequency of the VCO until it is dynamically locked to the frequency of the Doppler oscillation contained in S(t). The gain and filter sections following the phase discriminator act as the feedback loop of the frequency-tracking operation. With them, we can filter signal S(t), which may vary in frequency, yet keep a narrow band around the average frequency, which results in a very effective cleaning-up of the signal.

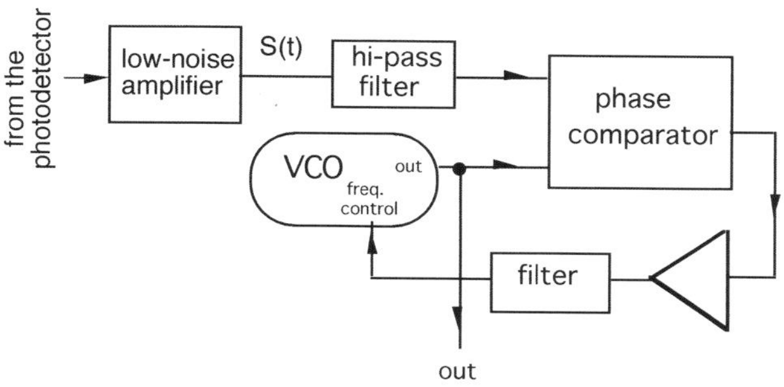

Fig.7-14 The Doppler signal of an LDV can be filtered to extract f_D by a simple PPL (phase locked loop) arrangement.

7.4 Particle Image Velocimetry

With the advent of powerful and cheap PCs, it has become technically feasible to pinpoint and individually follow a myriad of particles while they are flowing, so as to build up an image of the velocity distribution in the fluid.

This technique, called PIV (Particle Image Velocimetry) [9, 10] is quite different from LDV, which is more accurate and has a high dynamic range but measures a single point at a time, not an image.

In PIV, the particles are selected by a fan-shaped illuminating beam shed into the measuring volume (Fig.7-15), and the laser beam delivers two pulses separated by a short time Δt. A CCD acquires two frames of illumination.

Comparing the two frames, we can identify the displacement Δs undergone by each individual particle between the two successive pulses.

From the components Δx and Δy of displacement Δs, the velocity components $\Delta x/\Delta t$ and $\Delta y/\Delta t$ are calculated for each particle in the field of illumination.

In the basic scheme of the PIV (Fig. 7-15), the laser source is either a frequency-doubled Q-switched Nd crystal laser or a high-power quaternary diode laser emitting in the visible and pulsed through the bias current.

The time separation of the two pulses depends on the velocity of the fluid and on the selected size of the individual cell (Fig. 7-15) in which the total image is subdivided, and range typically between 10 and 500 ms.

The projecting optics is a collimating telescope that has a cylindrical objective lens, so that the output beam has a narrow divergence in one coordinate, and a large one in the other, or it's a fan-shaped beam.

To sort the particles from pulse to pulse, the entire image frame is subdivided into MxN individual cells, out of the available M_0xN_0 pixels of the CCD.
The cell size is chosen large enough to have the particle internal to it in both pulses, but small enough to have a large number MxN of cells, or a good spatial resolution of the velocity field measurement.

A frame grabber captures the image at time t and t+Δt, and then the images of individual cells at the two times are compared. For those cells that have a particle inside at both times, the distance Δs (and its components Δx and Δy) are computed.

This operation corresponds to calculating the cross-correlation of images $I_{t1}(x,y)$ and $I_{t2}(x,y)$, which is given by:

$$g(\Delta x,\Delta y) = \int\int_{x,y} I_{t1}(x,y)\, I_{t2}(x+\Delta x,y+\Delta y)\, dx\, dy \qquad (7.9)$$

For cells that have zero correlation, the data is discarded.
For fast enough processing, the frame sequence becomes a real-time movie of the kinematic field of velocity, a powerful tool in the study of hydrodynamics, and an example of it is offered by Ref. [11].

An extension of the PIV basic setup is 3D imaging, obtained by doubling the CCD imager into two units that look at the measurement volume from different directions, so as to develop a parallax, like in triangulation (see Sect.3.1), and consequently also allow the measurement of the Δz coordinate.

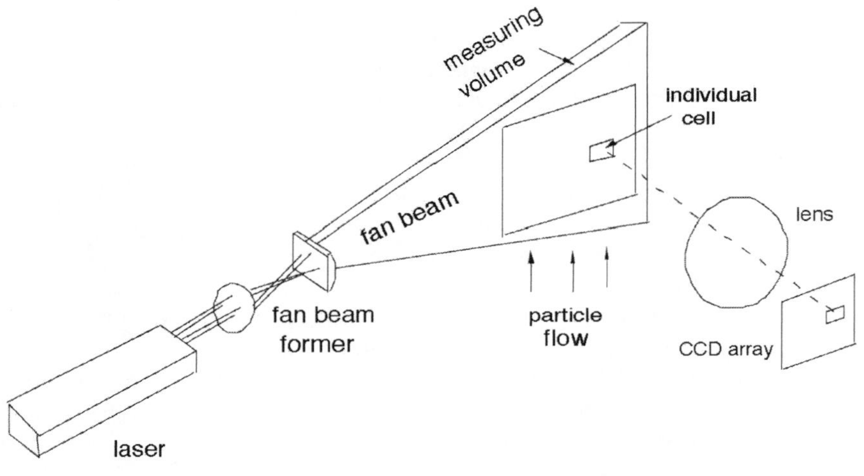

Fig.7-15 Schematic of the PIV (Particle Image Velocimetry). A fan-shaped illuminating beam shines on particles and a CCD detects their positions in the sampling volume. On two successive illuminating pulses, position changes of Δx and Δy, and velocity can be computed as Δx/Δt and Δy/Δt.

7.5 SMI VELOCIMETERS AND FLOWMETERS

The general expression for optical phase shift read by an interferometer is $\Phi=2\,\underline{k}\cdot\underline{s}$, as already discussed in Sect.7.1.1. To develop a phase signal for a *velocimeter*, we need that $\underline{s}$ makes an angle other than $\pi/2$ with respect to the wavevector $\underline{k}$.
Letting θ for such an angle (see Fig.7-16), we have $\underline{k}\cdot\underline{s}$ = ks cos θ. The frequency f_D generated by a target movement with velocity v= ds/dt is therefore given by:

$$f_D = (1/2\pi)\, d\Phi/dt = (1/2\pi)\, k \cos\theta\, ds/dt = (v \cos\theta)/\lambda \qquad (7.10)$$

and this is the well-known Doppler frequency. A lot of experiments have been reported on the measurement of f_D [13-16]. One application in the area of consumer products is the odometer already mentioned in Sect.5.7.1, and many more were performed with a rotating disk. As in this case the target is a diffuser, we observe an amplitude modulation of the SMI signal $I_0 \cos 2\pi f_D t$ due to the speckle pattern statistics. Indeed, the signal returning to the laser and photodiode combination is a speckle field that is dragged by the target movement (Sect.6.1.6), that is, translated transversally with respect to the laser receiving aperture (or laser spot).

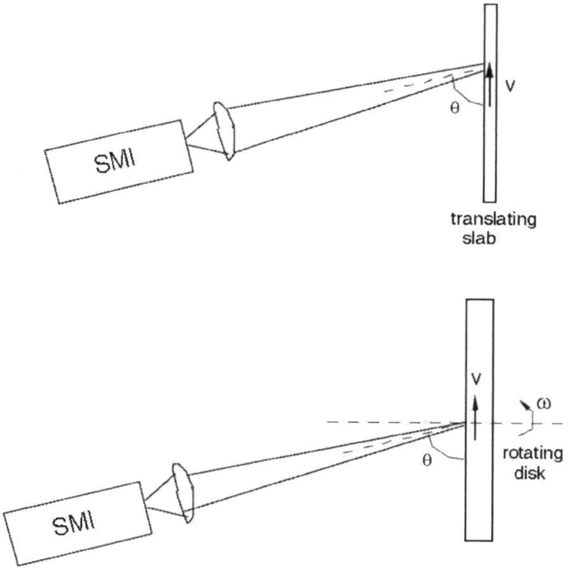

Fig.7-16 To measure the velocity of a translating or rotating body, we have to tilt the beam wavevector so as to develop a non-zero phaseshift Φ= 2 k s cos θ.

Thus, when the target movement is larger than the speckle transversal size s_t, we receive a new sample of the statistics, which is to say, the amplitude I_0 of the Doppler signal will change because it is uncorrelated to the previous one and exhibits a ripple. The result is an amplitude modulation with a period T= s_t/v or a frequency of the amplitude speckle-generated ripple, given by:

$$f_{sp} = v/s_t = v\, D/\lambda z \qquad (7.11)$$

where D is the beam diameter on the target. If the beam propagates at the diffraction limit, then $D=2\sqrt{(\lambda z/\pi)}$ and Eq.7.11 gives $f_{sp} = 2v/\sqrt{(\pi\, \lambda z)} \approx v/z_{Fr}$, where z_{Fr} is the Fresnel distance (see also Eq.2.4).

Note that frequency f_{sp} is independent from θ, and therefore a non-Doppler velocity signal, produced by the speckle dragging, is available also at normal incidence ($\theta=\pi/2$ in Fig,7-16) although of much less amplitude than the Doppler frequency given by Eq.7-10, but large enough to measure, as reported in Ref.[17].

Additional to the amplitude ripple, the speckle also brings about a phase error, as detailed in Sect.6.1.5. Both sources of error are minimized with a suitable filtering of the SMI waveform, as analyzed in Ref. [16].

About *flowmeters*, the schematic of Fig.7-16 (top) applies, with the only difference that now the target is a tube carrying the fluid to be measured. Zakian et al. [12] pioneered the SMI measurement of flow velocity, reporting the measurement of frequency spectra, ranging from 1 to 10 kHz in response to a laminar flow from 10 to 100 ml/h.
To have a minimum part-count SMI flowmeter, we can even remove the optical elements focusing the laser beam into the capillary carrying the liquid (Fig.7-17). In this case, the entrance beam has a distribution of incidence angles θ, and the SMI signal is the integral over θ of the Doppler frequency contributions.

Fig.7-17 A compact SMI flowmeter has the laser diode mounted directly on the wall of a capillary tube carrying the blood flow (from Ref. [18], ©IEEE, reproduced by permission).

Nevertheless, the average frequency is still proportional to velocity, and so a simple analysis of the SMI signal yields the desired flow velocity [18], as shown in Fig.7-18.

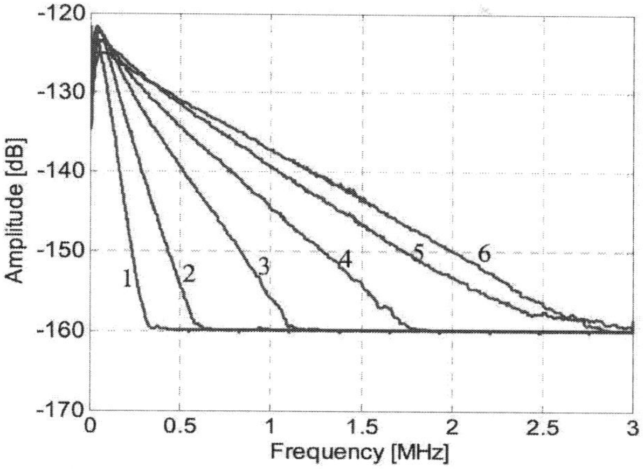

Fig.7-18 The measured frequency spectrum supplied by the flowmeter of Fig.7-17, for a blood flow of 1 to 6 l/min (from Ref. [18], ©IEEE, reproduced by permission).

REFERENCES

[1] Y. Yeh and H.Z. Cummins: "*Localized Fluid Flow Measurements with a He-Ne Laser*", Appl. Phys. Lett. vol.4 (1964), pp.176-178.

[2] Optical Industry Development Association (OIDA): "*OIDA Market Updates: 2020 Consolidated Report*", Optica Industry Report, vol.10 (2021), see https://opg.optica.org/abstract.cfm?URI=OIDA-2021-10.

[3] L.E. Drain: "*The Laser Doppler Technique*", J. Wiley and Sons, Chichester 1980.

[4] S. Donati: "*Photodetectors*", 2nd ed., Wiley IEEE Press, Hoboken 2021.

[5] J.D. Gaskill: "*Linear Systems, Fourier Transforms and Optics*", J. Wiley and Sons, New York 1978, Chapter 7.

[6] M.J. Buckingham: "*Noise in Electronic Devices and Systems*", Ellis Horwood, Chichester 1983.

[7] M.J. Rudd: "*A Laser Doppler Velocimeter Employing the Laser as a Mixer-Oscillator*", J. Physics E, vol.1 (1968), pp.723-726.

[8] P.J. de Groot, G. Gallatin, and S.H. Macomber: "*Ranging and Velocimetry Signal in a Backscattering Modulated Laser Diode*", Appl. Opt., vol.27 (1988), pp.4475-4480.

[9] R.J. Adrian: "*Particle-Imaging Techniques for Experimental Fluid Mechanics*", Ann. Rev. Fluid Mech., vol.23 (1991), pp.261-304.

[10] Y. Watanabe, Y. Hideshima, T. Shigematsu, K. Takehara: "*Application of three-dimensional hybrid stereoscopic particle image velocimetry to breaking waves*", Meas. Sci. Techn., vol. 17 (2006), pp.1456-1469.
[11] see a "*von Karman vortex*" sensed by PIV at https://www.dantecdynamics.com/solutions-applications/solutions/fluid-mechanics/particle-image-velocimetry-piv/.
[12] C. Zakian, M. Dickinson, T. King: "*Particle Sizing and Flow Measurement using Self-Mixing Interferometry with a Laser Diode*", J. Optics A, vol.7 (2005), pp. S445-S452.
[13] G. Plantier, N. Sarvagent, A. Sourice, T. Bosch: "*Real-Time Parametric Estimation of Velocity using Optical Mixing interferometry*", IEEE Trans. Instr. Meas., vol.50 (2001), pp. 919-926.
[14] X. Raoul, T. Bosch, G. Plantier, N. Servagent: "*A Double-Laser Diode Onboard Sensor for Velocity Measurement*", IEEE Trans Instr. Meas., vol.53 (2004), pp.95-100.
[15] J. Albert, M.C. Soriano, K. Panjotoff, J. Danckaert, P. Porta, D.P. Curtin, J.G. MacInerney: "*Laser Doppler Velocimetry with Polarization Bistable VCSELs*", IEEE J. Sel. Top. Quant. Electr., vol.10 (2004), pp.1006-1012.
[16] L. Scalise, G. Giuliani, G. Plantier, T. Bosch: "*Self-Mixing Laser Diode Velocimetry: Application to Vibration and Velocity*", IEEE Trans Instr. Meas., vol.53 (2004), pp.223-232.
[17] T. Shibata, S. Shinohara, H. Hikeda, H. Yoshida, T. Sawaki, M. Sumi: "*Laser Speckle Velocimeter using Self-Mixing Laser Diode*", IEEE Trans Instr. Meas., vol.45 (1996), pp.499-503.
[18] M. Norgia, A. Pesatori, S. Donati: "*Compact Laser Diode Instrument for Flow Measurement*", IEEE Trans. Instr. Meas., vol.64 (2016) DOI: 10.1109/TIM.2016.2526759.

Problems and Questions

P7-1 *To design the electronic part of a LDV, we start assuming a superposition angle θ ranging from 5 to 30 degrees, a wavelength λ ranging from 0.35 to 0.70 μm, and want to evaluate a speed measurement covering the range 1 μm/s to 200 m/s. Which frequency range has to be ensured?*
P7-2 *Consider an LDV that shall probe a fluid at a 200-mm depth, with a θ= 14° superposition angle. What is the size of the sampling region? What is the number of fringes and the precision?*
P7-3 *Repeat the calculation of Prob.P7-2 using a larger distance to the sampling spot, say 500 mm.*
P7-4 *With the LDV data of Prob.P7-2, calculate the amplitude of the signal supplied by the photodetector. Assume the detector is forward-looking at the sampling region, seeded by 0.25-μm particles, and the solid angle of collection by the detector objective is Ω=0.1 sr.*
P7-5 *Our laboratory has a 10,000 class-level of air cleanness. Calculate how many particles are found in the sampling volume of the LDV of Prob.P7-2.*
P7-6 *Changing the parameters of the LDV in Prob.P7-5 appropriately, estimate which class of cleanness can be measured by the Doppler technique.*
Q7-7 *In a fiberoptics version of the LDV, isn't the fiber detrimental to the beam wavefront quality needed in the superposition region?*

CHAPTER 8

Gyroscopes

The gyroscope is an inertial sensor that measures the angular rotation of a mobile vehicle. The rotation is measured with respect to the inertial reference frame, not the laboratory frame, and this gives the gyroscope a special place among sensors. Since the early times of aids to navigation, gyroscopes have been widely used in a number of applications, like inertial navigation, dead reckoning, attitude heading, horizon sensor, etc., all requiring the inertial-sensing property. Before the advent of the laser, mechanical gyroscopes (MG) were employed with success. Based on the measurement of the Coriolis force acting on the axis of a fast-rotating mass, after decades of improvements the MG worked nicely, and it was the only choice for inertial sensing up to about the 1970s. It is now abandoned because of drawbacks like: (i) the long (≈10 s) switch-on time, (ii) the dependence of performance and lifetime on fine rotational balancing of the mass, and (iii) the bulky and power-consuming structure.

With the advent of the laser, electro-optical gyroscopes based on the Sagnac effect were soon identified as a key application of lasers to avionics, soon superseding the MGs. In experiments dating back to 1962, the HeNe laser in the *internal configuration* of interferometry was used to read the very small Sagnac phase. Thereafter, much work was devoted to circumventing locking, and the final result was an inertial-grade (0.001 to 0.01 deg/h) sensor called the RLG (Ring Laser Gyroscope), a big success of photonics employed since the years

1980s in INUs (Inertial Navigation Unit) of virtually any new airplane. After the RLG, attempts followed in the years 1990s to make a more lightweight and modular version of the sensor, the FOG (Fiber Optic Gyroscope), based on a semiconductor source and the *external configuration* of readout, covering the sub-inertial (0.01 deg/h) performance. Recently, thermal perturbations were identified as the cause of residual drift and noise, and curing them, the FOG has reached the inertial grade performance.

For industrial and automotive applications, compact versions of the FOG were also developed, yet soon surpassed by the MEMS (Micro Electro Mechanical) gyroscope, a refurbishment of the old MG with the modern micro-fabrication technologies on a Si wafer. Like its predecessors, the MEM-gyro has also progressed fast in performances and, though unlikely to reach the inertial class, they have found extensive use in air bag control and robotics. Recently, the quest is not over: new approaches are sought (e.g., piezo, molecular beam, etc.) and R&D efforts (integrated technology, fibers and materials) continue to improve performance of existing devices.

8.1 Overview

When the first He-Ne laser was demonstrated in 1961, the idea of reading the Sagnac phase shift in a closed-cavity path was soon recognized as a promising approach for a new type of gyroscope, the *laser or electro-optical gyroscope*. Already in 1962 Macek and Davis [1] were able to demonstrate the detection of the earth rotation by the Sagnac phase shift induced in a square cavity, 1-m per side, made up of He-Ne laser tubes (Fig.1-7). They used what we today call an internal-readout configuration of interferometry (Sect.4.6).

From this encouraging result, a rush started to develop a compact, well-engineered electro-optical gyroscope with nearly quantum-limited sensitivity for use in avionics. Contrary to expectation, however, the road to such a device was paved with obstacles, not only technological but also conceptual. Indeed, the first experiments unveiled the problem of locking of counter propagating modes and the consequent washout of signal in the region most important to application, i.e., the low-speed rotations. After fifteen years of efforts by the international scientific community, evidenced in thousands of years of work and hundreds of papers published in scientific journals, the locking problem was finally circumvented [2]. Other technological improvements were also crucial in achieving the low-scatter mirrors and low thermal-expansion cavities required for a good sensitivity performance.

These progresses marked the coming of age of the modern *Ring-Laser-Gyroscope* (RLG), a sensor that since 1975 to 1980 has entered the mass-production stage and has been incorporated in the Inertial Navigation Unit (INU) of any newly fabricated aircraft (Fig.1-8).

The RLG then rapidly became an undisputed success of electro-optical instrumentation. This is true in terms of scientific sophistication incorporated in a product and also in terms of market – a rich niche totaling almost US$1 billion in sales per year since its inception and continuing to grow (at 3% per year) in the present years [3].

In a fitting comparison [2], if the MG is the like a mechanical clock, the RLG is the like an electronic clock, with the advantages and drawbacks that each of them may have.

While the RLG was maturing to a well-understood, producible and high-performance sensor as we know it today, another approach to the electro-optical gyroscope came out, the FOG (Fiber Optic Gyroscope). Proposed in 1976 by Vali and Shorthill [4], the FOG took advantage of the single-mode fiber, which just became available at that time, to develop a Sagnac fiber-interferometer based on the external-readout configuration as opposed to the internal of the RLG (Sect.4.6).

Initially, the hint was to exploit a long-fiber propagation path to substantially increase the Sagnac signal and hence the sensitivity. As another point of interest, the FOG presented a modular structure promising ease of fabrication, scalability, and low cost, as compared to the rather rigid structure RLG.

Like the RLG, the early FOG also started with performance far from those of interest to applications. And again, a decade of research efforts has been necessary to solve basic as well as technological problems [5-7] before the FOG at last has become an established device, though hardly approaching the ultimate sensitivity limit. After the reciprocity concept was clarified by Ulrich and Johnson [5], many small effects disturbing the balance of propagation path lengths in the fiber loop were identified and cured, yet without attaining the RLG performance.

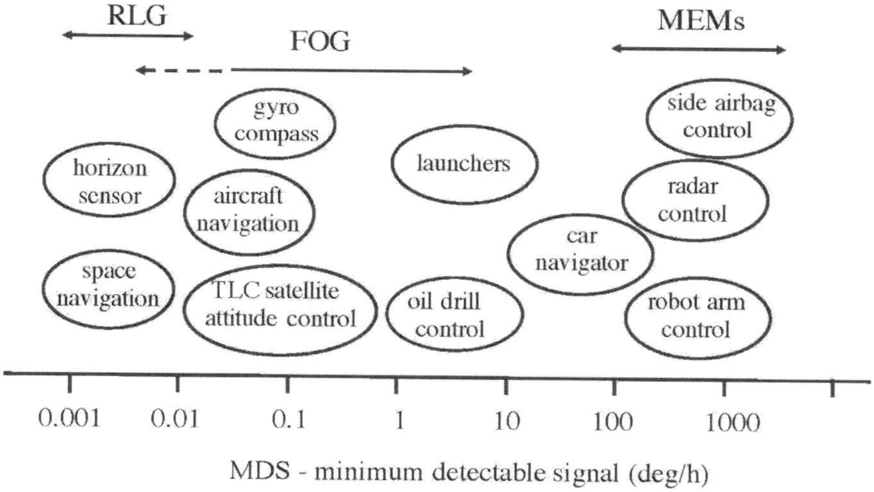

Fig.8-1 Typical applications of gyroscopes and minimum detectable signal of rotation speed (degree per hour) for the RLG (ring laser gyro), the FOG (fiber optic gyro) and the MEMS (micro-electro-mechanical-system).

Meanwhile, components were developed expressly for use in the FOG: high birefringence fibers, and super-luminescent LED (or SLED) [6,7]. Finally, it was understood that the excess of noise and zero baseline drift was due to thermal transient in the fiber coil [8], since then called the *Shupe effect*. Curing the Shupe effect by a specially wound coil has brought the FOG to the inertial performance.

Many parameters should be taken into account to assess a gyroscope. However, as a first-hand insight we may consider just the sensitivity, or minimum detectable signal of rotation (MDS), as in the diagram of Fig.8-1. Here, we can see that the RLG still marginally outperforms the FOG. Yet they are complementary devices, because the former attains better accuracy while the latter is lightweight, has a larger projected lifetime, and has better environmental characteristics (Table 8-1). The MEMS gyroscope has much less sensitivity, yet enough in many applications, like the automotive, where it is massively deployed for the airbag control.

A survey of gyroscope families is presented in Fig.8-2. We will discuss the details of the families in the next sections.

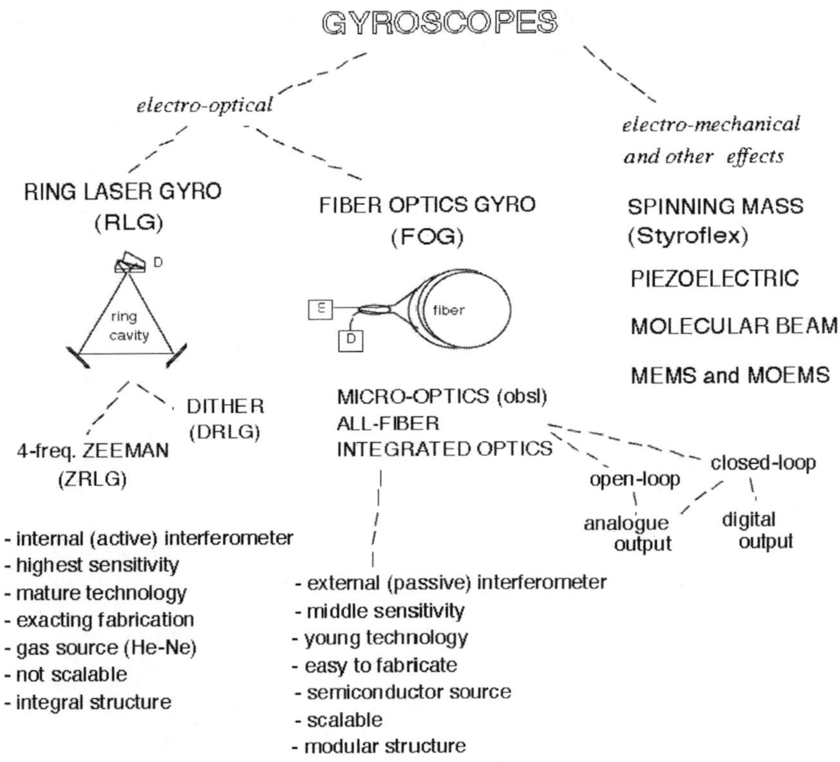

Fig.8-2 Classification of the gyroscope families and highlights of their main technical features. The inertial-grade gyros are the RLG and the FOG, while the spinning mass MG is not used in new designs.

While the performance of the standard gyroscopes has steadily improved through the years, thanks to the steady improvements of fabrication technologies, new ideas are proposed in the scientific literature and have undergone evaluation aiming at new devices with inertial grade performance. About the FOG, an effort has been devoted to shorten the fiber and to increase responsivity, and the result was the Resonator Fiber Optic Gyro (R-FOG), a device promising to surpass the traditional Sagnac-ring FOG. In the area of active RLG, attempts have dealt with the miniaturization of the laser cavity, using a semiconductor (typically GaAlAs) active waveguide, but no conclusive result has been yet obtained for these devices.

New solutions have been studied for application to the less-demanding robotics and automotive areas. These are aimed at a reduction of weight and size, even at a reduced sensitivity but at very low cost, for a number of applications (Fig.8-1). Thus, an attempt to miniaturize the FOG has been the so-called minimum-configuration or 3×3-coupler FOG. Another approach recently pursued has been the Integrated Optic Gyroscope (IOG), that aims to integrate on a suitable substrate, either active like a semiconductor chip or passive like Silica on Silicon (SOS), the functions of either the internal interferometer (RLG) or the external interferometer (FOG). The research is still underway but IOGs appear promising in the field of robotics and automotive sensors.

The MEMS gyro is more mature, and can be considered a refurbishment of the old mechanical gyro (MG), but fabricated in the modern technology of silicon micromachining, one that allows squeezing the device size down to 0.1 mm (but perhaps 10 mm when packaged) and cost to around US$1, but with sensitivity (or MDS) much less than the inertial grade (Fig.8-1), yet adequate for the automotive and some control applications.

A comparison of the basic performances of the three most important gyroscope classes is presented in Table 8-1.

Table 8-1 Comparison of RLG, FOG and MEMS gyroscopes

	RLG gyro	FOG gyro	MEMS gyro
Sensitivity, or MDS (deg/h)	0.001– 0.01	0.01–0.1	100–1000
Dynamic range (deg/s)	100–1000	100-1000	100–1000
Scale factor accuracy	10^{-5}	10^{-3}	10^{-3}
Typical size × height (cm)	15 (dia) × 6	8 (dia) × 4	0.1–1
Useful lifetime	typ. of He-Ne laser	typ. of laser diode	typ. of IC
Immunity to EM disturbance	high	very high	high
Immunity to microphonics	high	middle	middle
Structure	monolithic	modular	monolithic
Scalability	poor	good	poor
Manufacturing req'ment	very tight	middle	high
Clean room fabrication	yes	no	yes
Fab plant cost	very high	middle	very high

8.2 The Sagnac Effect

All electro-optical gyroscopes are based on the Sagnac phase shift: despite the different interferometer configuration used and the different signal obtained, the sensing mechanism is always the same.

The Sagnac effect is the nonreciprocal effect that takes place when light propagates on a curved path, as in Fig.8-3, and the plane π is subjected to a rotation with angular velocity Ω. Then, a small variation ϕ_S of the optical phase shift is observed, with respect to the normal path length measured at rest. The phase shift ϕ_S is proportional to Ω, sign included. Thus, the phase shift is nonreciprocal, and propagation in the clockwise and counterclockwise directions results in a variation $+\phi_S$ and $-\phi_S$, respectively.

We may evaluate the Sagnac phase shift ϕ_S by reference to Fig.8-3. Classically, let us assume that radiation undergoes the Doppler effect and write $\Delta\lambda/\lambda = v/c$. Changing to the wave number k and adding the vector dependence, we have:

$$\Delta \underline{k}/k = \underline{v}/c \tag{8.1}$$

where $\underline{v} = \underline{\Omega} \times \underline{r}$ is the tangential velocity. The corresponding optical phase shift for the elemental path $d\underline{s}$ is:

$$d\varphi = \Delta \underline{k} \cdot d\underline{s} \tag{8.2}$$

Inserting Eq.8.1 in Eq.8.2 yields:

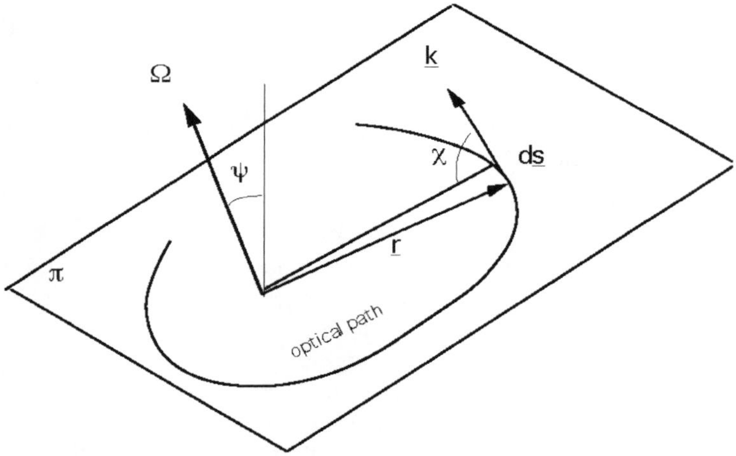

Fig.8-3 An optical path schematizing the propagation in a Sagnac-based gyroscope. The path (and $\underline{k}$) is contained in the plane π, and the rotation rate Ω is inclined of ψ with respect to the perpendicular to π.

$$d\varphi = k/c\ \underline{v} \cdot d\underline{s} = k/c\ (\underline{\Omega} \times \underline{r}) \cdot\ d\underline{s}$$
$$= \Omega\ (k/c)\ r \sin\chi\ \cos\psi\ ds \qquad (8.3)$$

The last expression follows by developing the vector and scalar products and taking into account angles ψ and χ (Fig.8-3). From Eq.8.3, we see that the Sagnac phase shift is sensitive to the rotation component perpendicular to the plane of propagation, or $\Omega_\psi = \Omega \cos\psi$. Keeping this in mind, we omit in the following the factor cos ψ to simplify notation.

Let us now consider the factor $r \sin\chi\ ds$ in Eq.8.3. From Fig.8-3, we can see that this factor is twice the area of the elemental triangle of base ds and height r. By integrating on a closed loop, we get $\int r \sin\chi\ ds = 2A$, where A is the area enclosed by the optical path. Therefore, we obtain for the Sagnac phase shift:

$$\phi_S = \Omega\ (k/c)\ 2A \qquad (8.4)$$

Noting that the total phase shift Φ_S between the two counter-propagating waves is $2\phi_S$, and substituting $k=2\pi/\lambda$, we finally get the total phase signal as:

$$\Phi_S = 8\pi\ A\Omega\ /\lambda c \qquad (8.5)$$

In this expression, A is the physical area of the RLG gyro, and the linked area in the FOG, also written as *N*A to put in evidence the number of turns *N* of the sensing coil of area A.

8.2.1 The Sagnac Effect and Relativity

Before proceeding, it is appropriate to comment about the method used to derive Eq.8.5. From the derivation, it would be reasonable to assume that wavelength λ and speed of light c in Eq.8.5 are those for the propagation medium. Thus, if n is the index of refraction of the medium, we should use values n times smaller than λ and c in vacuum, and have $\Phi_S = 8\pi n^2 A\Omega\ /\lambda c$. Actually, Eq.8.5 holds with λ and c for the vacuum at all times.

We may then wonder what is incorrect in the derivation based on the Doppler effect. The point is that a rotating system is an accelerated reference, for which a rigorous treatment requires the theory of general relativity. Thus, the simple classical vision based on the Doppler effect is not applicable, strictly speaking. It is just an approximation, not so bad after all, if it predicts correctly the $8\pi\ A\Omega\ /\lambda c$ dependence, although with λ and c to be interpreted.

The treatment of the Sagnac effect with general relativity is rather complex, and we refer the interested reader to specific works [9,10].

However, there exists a simple derivation, based on a conceptual experiment devised by Schultz DuBois [11], which is rigorous from the standpoint of relativity.

The derivation of Schultz DuBois starts considering an ideal, lossless toroidal cavity (Fig.8-4) with totally reflecting walls and vacuum inside. The cavity schematizes the sensor, whereas an observer integral with the cavity represents the detector. Now, suppose that light is somehow inserted in the cavity so that counter-propagating waves form a standing wave pattern with a spatial period $\lambda/2$ as in any real cavity.

At rest, the standing wave pattern is still, and the observer sees no movement of it. If the ring rotates at an angular Ω, the standing wave pattern is again still, because the waves do not exchange energy with the cavity. As the observer rotates integral with the cavity, he sees the standing wave pattern moving. At pattern speed ΩR, the observer sees $N= \Omega R/(\lambda/2)$ periods per second passing under him. He ascribes the movement to a difference of frequencies arisen in the two counter-propagating waves. The frequency difference, just the one observed in an RLG, is:

$$\Delta f_S = 2\,\Omega R /\lambda \tag{8.6}$$

To show the equivalence of this result and Eq.8.5, let us recall that the periodicity of the longitudinal modes (App.A1) of a ring cavity is $\Delta f = c/p = c/2\pi R$, with p being the perimeter. If the two counter-propagating modes are phase shifted by Φ_S, then the corresponding frequency difference Δf_S is related to Φ_S by the proportionality:

$$\Delta f_S /\Delta f = \Phi_S / 2\pi \tag{8.7}$$

Inserting $\Delta f=2\pi R/c$ in Eq.8.7 and using Eq.8.6, we obtain:

$$\Phi_S = 2\pi\ \Delta f_S / \Delta f = 4\pi\ (\Omega R /\lambda)\ (2\pi R/c) = 8\pi\ \ \Omega A /\lambda c \tag{8.8}$$

where we have used the expression $A= \pi R^2$.

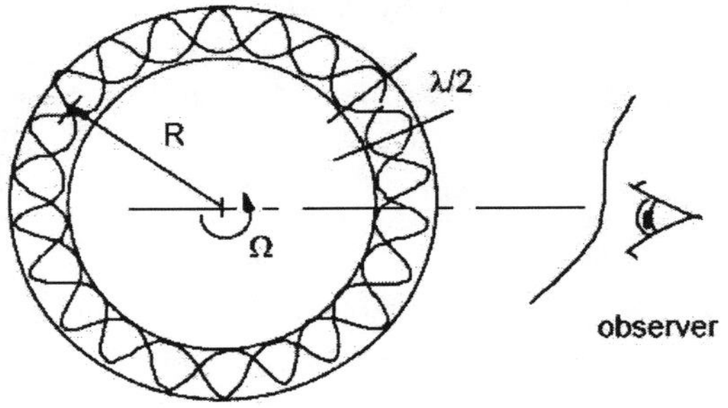

Fig.8-4 In the rigorous derivation of the Sagnac effect, we consider the standing wave formed in a toroidal cavity. An observer rotating at speed Ω sees the pattern running under him at a speed ΩR, and thus, a number of periods $\Omega R/(\lambda/2)$ per second, or, the Sagnac frequency $2\ \Omega R /\lambda$.

In conclusion, Eqs.8.5 and 8.7 are equivalent. We may regard the gyro as a device supplying a phase shift $\Phi_S=R_\Phi\,\Omega$ or a frequency difference $\Delta f_S =R_F\,\Omega$, where R_Φ and R_F are the frequency and phase responsivity of the conversion from rotation to optical quantity.

In closing, let us consider that the cavity of the Schultz DuBois ideal ring is filled with a material of index of refraction n.

Now, we have no reason to refuse changing λ to λ/n and c to c/n in Eq.8.5. This increases the responsivity to $n^2 R_\Phi$ (or $n^2 R_F$). This looks correct, if the standing wave pattern is still, because the wavelength in the material becomes shorter and we see more periods passing across.

Actually, relativity warns us that, in this case, light is partially dragged along by the material. This is the Fizeau effect [9-12], another effect known since the early times of relativity. The Fizeau effect develops a phase shift similar to the Sagnac and has a responsivity equal to $(n^2-1)R_\Phi$ [12].
Because the drag subtracts fringe crossings, we shall subtract the Fizeau phase shift $(n^2-1)R_\Phi$ from the Sagnac phase shift $n^2 R_\Phi$ to get the total phase shift resulting in the case of propagation through a material. Nicely, the result is $n^2 R_\Phi - (n^2-1) R_\Phi = R_\Phi$, or, it is independent from the index of refraction of the material, as of course it is well verified experimentally.

Other conceptual cases of deviation of the responsivity from the value given by Eqs.8.5 and 8.7 are those of (i) medium moving and detector still, and (ii) medium still and detector moving. From the previous arguments, the responsivity is found $(n^2-1) R_\Phi$ in case (i) and $n^2 R_\Phi$ in case (ii) [12].

In conclusion, we need relativity and not just the classical Doppler effect to treat the gyroscope correctly. When the cavity is filled with a medium, the phase shift Φ_S is still given by Eq.8.5, with λ and c in the vacuum. This is a consequence of the combination of Sagnac and Fizeau effects. The phase shift Φ_S also changes if the cavity or the observer (the photodetector) is still. This case is of little practical interest, however, because the gyroscope will invariably be used as a strap-down sensor, with all parts integral with the vehicle whose inertial rotation is being measured.

8.2.2 Sagnac Phase Signal and Phase Noise

Let us now discuss the typical values of the Sagnac signal we are going to measure with the gyroscope. The signal is a phase shift in the FOG and a frequency difference in the RLG. Because phase Φ_S and frequency Δf_S are related through Eq.8.7, we will consider only the phase shift in the following.

Plotting Φ_S against Ω for several values of the total area AN, we obtain the diagram of Fig.8-5. Dotted lines in the diagram are representative of typical RLG and FOG.
As we can see in the diagram, values of the phase shift are very small indeed. A big rotation speed, say Ω=1 rad/s, already gives a small Φ_S, $\approx 10^{-4}$ rad for the RLG and ≈ 0.1 rad for the FOG. If we go to the earth rotation rate, Ω=15 deg/h, the corresponding phase shift becomes $\approx 10^{-8}$ rad for the RLG and $\approx 10^{-5}$ rad for the FOG.

Important to note, the measurement of these small phase shifts must include the dc term, because the measurand Ω may well have a very low frequency content that prevents filtering. Thus, the measurement is much more difficult than that of a very small ac phase. We shall devise very robust techniques of phase measurement, and be extremely careful in avoiding even very small bias, either optical or electronic, that can enter in the setup and corrupt the minute phase shift Φ_S.

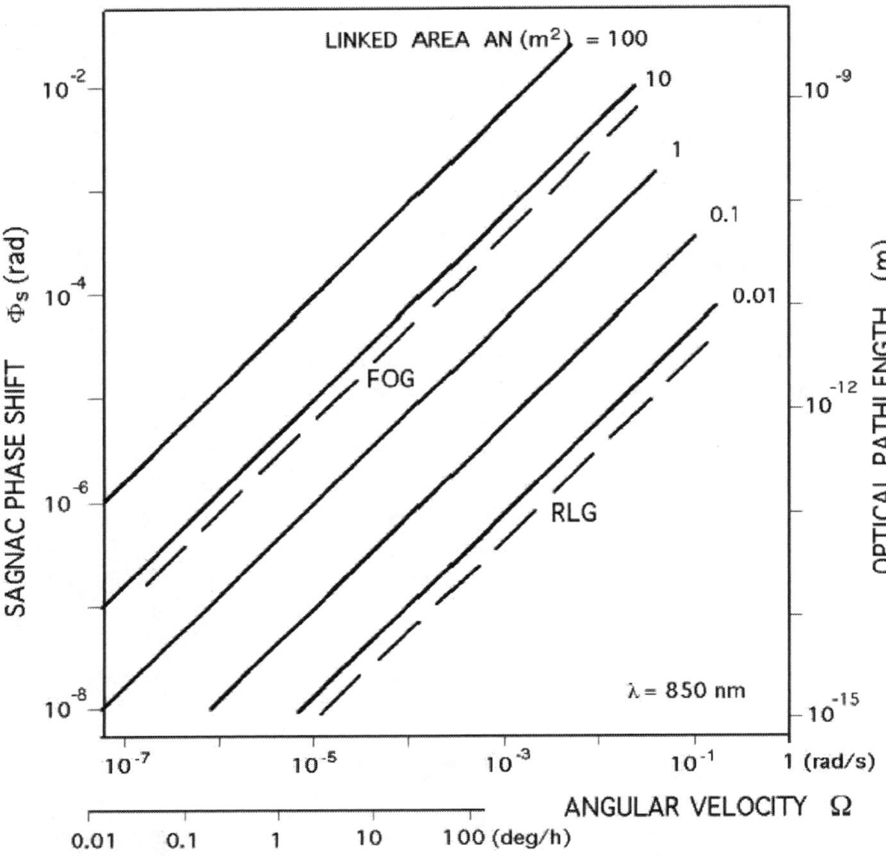

Fig.8-5 The Sagnac phase shift Φ_S developed in a gyroscope with linked area AN, as a function of angular velocity Ω. Dashed lines are for typical devices, a FOG using a 200-m length fiber coil, and RLG using a cavity with 10-cm per side. On the right-hand scale, the optical path change corresponding to Φ_S is indicated.

As a comment, measuring the previously quoted phase shift corresponds to detecting optical path length changes of the order of pico- and femto-meter (or, 10^{-12} m and 10^{-15} m), as we can see in Fig.8-5. This is a remarkable record, not only for interferometric measurements, but also in the area of instrumentation science.

Indeed, to obtain a resolution of $\approx 10^{-15}$m in a RLG with a 0.3-m perimeter requires that we are able to resolve 1 part out of $3 \cdot 10^{15}$. This performance is not so far from the sensitivity of the LIGO gravitational antenna (Fig.1-13) and other spatial gyro missions [14].

Returning to the Sagnac signal, we shall check that the phase noise permits the measurement we want to perform. The phase noise in interferometric measurement has been already treated in Sect.4.4.1 (see Eqs.4.22 and Fig.4-21). Recalling the result of Eqs.4.18 and 4.20, written in terms of the noise equivalent phase ϕ_n, and with R=V=1, we have for the phase noise:

$$\phi_n = k \cdot NED = (S/N)^{-1} = (2\ \kappa\ h\nu\ B/\eta\ P_0)^{1/2} \tag{8.9}$$

Here, B is the measurement bandwidth, η is the quantum efficiency of the photodetector, and P_0 is the power received by the photodetector.

The factor κ has been added to be able to write the same expression for the FOG and the RLG. It is κ=1 for the external configuration of the FOG (Eq.4.20) while for the internal configuration of the RLG is found [13] as:

$$\kappa = (1\text{-}R)^2/(1+\alpha_{en}^{\ 2}) \tag{8.10}$$

where R is the (power) reflectivity of the output mirror, the other two mirrors being assumed ideally reflecting, and α_{en} is the linewidth enhancement factor (see Sect.5.2.2).

The phase equivalent noise ϕ_n given by Eq.8.9 is plotted in Fig.8-6 versus bandwidth B and with the equivalent power $P=\eta P_0/\kappa$ as a parameter.

Typical equivalent powers of FOG and RLG are indicated by dotted lines in Fig.8-6. In the FOG, the equivalent power is appreciably less than the optical power available from the source (≈1 mW) because of the attenuation of the Sagnac interferometer and of the components used in it. In the RLG, the small $(1\text{-}r)^2$ factor brings up the equivalent power of 3-4 decades, with respect to the ≈1 mW power level that the source would be able to emit.

Thus, though the RLG starts from a smaller responsivity due to the smaller linked area with respect to the FOG, it readily recovers the disadvantage through the much larger equivalent power and ultimately surpasses the FOG in sensitivity by 2-3 decades (Fig.8-6). In addition, the RLG approaches the quantum limit more easily than the FOG because the mirror cavity has fewer residual non-idealities than a long fiber made of solid glass.

Comparing Figs.8-5 and 8-6, we can see that there is room to achieve the desired Minimum Detectable Signal (MDS) performance indicated in Sect.8.1 if we are not too far from the quantum limit.

The MDS is so called because it is made up of two terms: the phase noise ϕ_n and the zero-bias error ϕ_{ZB}. The zero-bias error is defined as the phase equivalent of the gyro output when Ω=0. Thus, we have:

$$MDS = \phi_n + \phi_{ZB} \tag{8.11}$$

The reason for considering ϕ_{ZB} is because the gyroscope shall work down to the zero-frequency, or dc term, included. Then, the zero-bias error ϕ_{ZB} is an integral part of the minimum detectable signal.

Usually, we will trim the zero-bias error to null with a calibration operation before using the gyroscope.

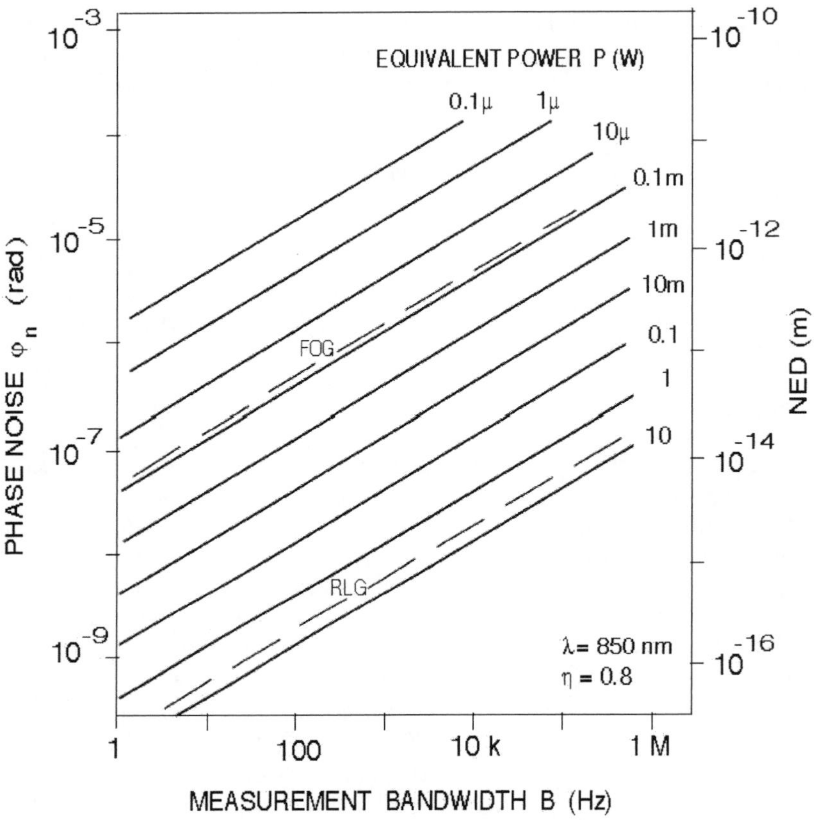

Fig.8-6 Phase noise ϕ_n at the quantum limit of detection, plotted as a function of the measurement bandwidth B and for several values of the equivalent power $\eta P_0/\kappa$. Dotted lines indicate the typical values for the RLG and the FOG.

After calibration, ϕ_{ZB} drifts away from null because of several residual fluctuations that cannot be controlled, like: dependence on temperature and other parameters, aging effects, switch-on/off, 1/f-noise components, etc.

In practice, while the phase-noise contribution scales with the bandwidth as $\sqrt{B}$ (see Eq.8.9), or with the integration time as $1/\sqrt{T}$ (because $B \approx 1/T$), the zero-bias error is about independent from time or slowly increases with T.

Thus, on long integration times, the ultimate performance of the gyroscope is likely be limited by ϕ_{ZB} rather than ϕ_n.

8.3 Basic Configurations of Gyroscopes

The basic schemes of RLG and FOG gyroscopes are shown in Fig.8-7. In both schemes, we find a propagation path, a triangle cavity in the RLG and a coiled fiber in the FOG. In the path, two counter-propagating waves CW (clockwise) and CCW (counter-clockwise) cumulate a phase difference Φ_S because of the Sagnac effect. At the output, with the prism and mirror combination in the RLG and the beamsplitter BS in the FOG, the CW and CCW waves are superposed onto a photodetector. The signal output from the photodetector carries the Sagnac phase Φ_S.

There are significant differences however, in the way the two configurations treat the phase signal, as already discussed in Section 4.6. The RLG is an active, internal configuration interferometer providing a frequency signal in response to a path length (or phase Φ_S) difference, whereas the FOG is an external configuration interferometer providing a phase modulated, baseband signal in response to Φ_S.

In the RLG, the readout of the cavity length is performed through the frequency difference of two counter-propagating modes of the triangular mirror cavity.
When the gyroscope is at rest (Ω=0), because of the resonant condition met by the laser oscillation, we find an exact number of wavelengths in the perimeter p of the cavity, and the CW and CCW frequencies are equal. When the gyroscope rotates, the Sagnac effect changes the perimeter length seen in the CW and CCW directions, by $\Delta p=+\phi_S/k$ and $\Delta p=-\phi_S/k$, where $k=2\pi/\lambda$ is the wave number. Accordingly, one frequency (f_{CCW}) increases, and the other (f_{CW}) decreases.

By differentiation of the resonance condition $kp=(2\pi f/c)p$=const., we get $f\Delta p+ p\Delta f=0$ for the frequency difference $\Delta f_S = |f_{CCW}-f_{CW}|$. By inserting $\Delta p=\phi_S/k$ in $\Delta p/p=\Delta f_S/f$, we obtain $\Delta f_S = (c/p)\phi_s/2\pi$, which is coincident with Eq.8.7.

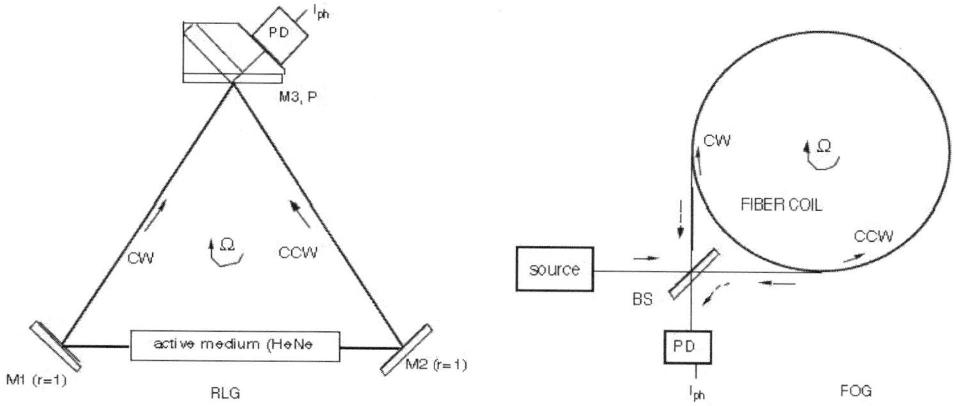

Fig.8-7 Basic configurations of the RLG (left) and of the FOG (right).

Using $\phi_S=4\pi A\Omega/\lambda c$, this expression can also be written $\Delta f_S = 2(A/p)\Omega/\lambda = 2R\Omega/\lambda$, where $R=A/p$ is the apothem of the triangle.

If we now recombine the CW and CCW waves at one of the cavity mirrors (M3 in Fig.8-7) by the prism arrangement, we get a photodetected signal given by:

$$I = I_0 [1+ \cos 2\pi\Delta f_S t] =$$

$$= I_0 [1+ \cos 2\pi\phi_S(c/p)t] = I_0 [1+ \cos 2\pi(2R\Omega/\lambda)t] \quad (8.12)$$

Thus, the RLG supplies the Sagnac signal in the form of a sinusoid whose frequency is proportional to the angular velocity of rotation Ω. This is an advantageous feature because frequency is measured easily and accurately by standard electronic circuits on a very wide dynamic range. In addition, the frequency signal $\Delta f_S=R_F\,\Omega$ has a scale factor $R_F=2R/\lambda$ that only depends on size R and wavelength λ and can be known with a very good precision and without the need of a calibration.

The scale factor is large, too. For example, a gyroscope with R=50 mm λ=1 µm, has $R_F=10^5$ Hz/(rad/s), and a rotation of 1 deg/h (or 4.7 µrad/s) is translated into a 0.47-Hz signal. By considering a measurement time interval of, say, 1000 s, the minimum observable frequency is 10^{-3} Hz, which corresponds to a minimum measurable rotation speed Ω_{min}=0.002 deg/h. On the opposite end, the maximum measurable frequency is that corresponding to the frequency spacing Δf_{Smax}= c/p of the longitudinal modes in the laser cavity. Using p= 30 cm, we get Δf_{Smax}= 1 GHz or also $\Omega_{max}=10^4$ rad/s, a very large theoretical limit.

A final advantage is that frequency is integrated without any low-frequency cutoff simply by counting the periods contained in the photodetected signal. This allows us to obtain numerically the quantity $\int_{0\text{-}t} \Omega\, dt = \Psi$, that is, the angle of (inertial) rotation Ψ of the gyroscope. The minimum angle of rotation that can be resolved with an RLG operating in the digital regime of signal processing is the angle Ψ_{min} corresponding to 1 count, or $\int_{0\text{-}t} \Delta f_S\, dt=1$. Using the expression $R_F\,\Omega$ for Δf_S, we get for the minimum resolved angle:

$$\Psi_{min} = 1/R_F = \lambda/2R \quad (8.13)$$

The value of Ψ_{min} is the same as the diffraction-angle from the gyro aperture 2R, with a typical value of 10^{-5}= 2 arc-sec with the numbers of the example. This value is of interest for attitude heading and gyrocompass applications in avionics [15].

Now, let's discuss the pitfalls of the basic scheme of the RLG. First, the signal supplied by the photodetector (Eq.8.12) is a sine or cosine function of a frequency proportional to Ω. If Ω changes to $-\Omega$, the Sagnac phase ϕ_S of each CW and CCW wave changes sign, and frequency Δf_S does too. But, as we cannot distinguish between positive and negative frequency, the sign of Ω is lost.

We may think of solving the sign problem by adding somehow a frequency-difference bias Δf_{bias} in the mode oscillations, to have $\Delta f_{bias}\pm\Delta f_S$ for $\pm\Omega$. This approach does not work, however, because the stability required to Δf_{bias} is far beyond practical feasibility.

Another serious problem is that of signal washout. The signal Δf_S is observed to disappear at small angular rotations Ω, and the reason for that is the frequency locking of the counter-propagating modes. The locking phenomenon is inherent to the internal configuration of interferometry, as already pointed out in Sect.4.6.1 and Fig.4-31, and is due to the even very minute coupling of the two oscillating modes. Any phenomenon producing an exchange of energy between the modes contributes to coupling. The most important sources of coupling are the scattering at the mirror surfaces and in the medium filling the cavity and the cross-saturation of gain provided by the active medium.

Let us briefly describe the locking regimes by considering two modes oscillating with an unperturbed frequency separation Δf_0. When coupling is negligible, we observe an unperturbed sinusoidal waveform of beating at frequency Δf_0. Then, at increasing coupling, we first observe frequency attraction of the two modes, whose separation becomes $\Delta f<\Delta f_0$ and the waveform of beating is a distorted sinusoid. This is the injection modulation region of Fig.4-31. By increasing the coupling further, attraction and distortion effects become very strong, and we finally observe the two modes collapse in one and the beating waveform disappear. This is the locking range or dead band in Fig.4-31.

An analysis of the locking regime can be carried out with the Lamb (or the Lang and Kobayashi) equations, written for the two modes and modified to include the coupling term (see Ref.[16], Sect.11.2). Solving these equations for the phase φ_1 and φ_2 of the modes, and with the notation used in Sect.5.2.2, the result is the following, known as Adler's locking equation [17]:

$$(d/dt)\,(\varphi_1-\varphi_2) = \Delta f_0\,[1+ K \sin(\varphi_1-\varphi_2+\zeta)] \qquad (8.14)$$

Here, K is the coupling factor. For $K<<1$, the oscillation frequency is Δf_0 and the beating waveform is sinusoidal. As K approaches unity, the beating waveform is more and more distorted [18]. At $K\geq 1$, we get locking because the solution of Eq.8.14 is $(d/dt)\,(\varphi_1-\varphi_2)=0$, or the phases of the two modes are locked at $(\varphi_1-\varphi_2)=\text{const}$. For modes of equal intensity, the expression of the coupling factor is [16,18]:

$$K = a_c\,(c/p)\,/\Delta f_0 \qquad (8.15)$$

where a_c is the (field) coupling coefficient between the modes. Eq.8.15 reveals why the locking problem is so hard. No matter how small a_c is, if the modes are closer and closer in frequency (that is, Δf_0 is small), we will ultimately get a large K.

Of course, to cure the locking problem, we will start trying to decrease coupling as much as possible. The most serious source of coupling is scattering at the mirrors, caused by minute residual irregularity of the surface. Scattering is described by the mirror diffusion coefficient δ, the fraction of power diffused according to Lambert cosine law [16] instead of being reflected. In a normal multilayer interference mirror, we may have $\delta\approx 0.3$-1%, but, with a special manufacture for the gyro application, we may achieve $\delta\approx 10^{-5}$. The coupling coefficient is easily related to δ using the acceptance invariance [16]. This yields $a_c^2= \delta\omega/\pi$, where ω is the solid angle associated with the oscillating mode, $\omega= \pi(\lambda/\pi w_0)^2$. Collecting the terms, we have:

$$a_c = (\lambda/\pi w_0)\sqrt{\delta} \qquad (8.16)$$

Using the condition K<1 in Eq.8.15, we get the minimum frequency separation that prevents locking as $\Delta f_0 < a_c(c/p) = (c/p)(\lambda/\pi w_0)\sqrt{\delta}$.

We can appreciate by a numerical example the technological effort in reducing δ, one of the key issues leading to the success of the RLG. Taking λ=1 μm and w_0= 1 mm and a normal mirror with δ=0.01, we have the typical value $a_c = 3\cdot 10^{-5}$. The corresponding de-locking frequency is $\Delta f_0 < 3\cdot 10^{-5}\cdot 1$ GHz = 3 kHz, which corresponds to a dead-band rotation speed of $\Omega_{db} = \Delta f_0/R_F$ =3 kHz/10^5 = 0.03 rad/s (or, ≈1.8 deg/s, a very large value indeed).
When a good low-scatter dielectric mirror is used, for example with $\delta=10^{-5}$, we improve the previous figures by a factor $\sqrt{(0.01/10^{-5})}$= 30, going down to a more reasonable value of dead-band rotation, Ω_{db}= 1.8/30 deg/s ≈ 200 deg/h, yet still far away from the inertial performance.
As an illustration, Fig.8-8 shows the typical response curve of the basic scheme of the RLG, that is, one without solution for locking and sign detection.
Last, we can circumvent locking by separating somehow the modes in frequency by more than Δf_0. The practical solutions following this hint, which also add the sign of Ω, will be discussed in Sect.8.4.

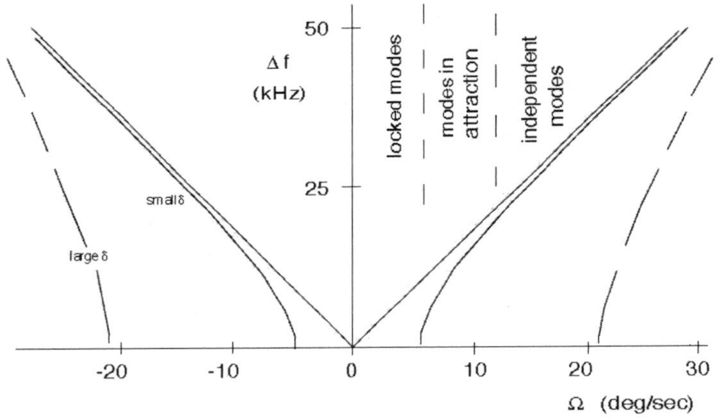

Fig.8-8 The response of the basic scheme of RLG. There is locking dead-band around Ω≈0, and no sign for Ω is supplied.

Now, let us turn to consider the basic scheme of the FOG (Fig.8-7). The FOG is developed around an external configuration called a Sagnac interferometer (see App.A2), using a relatively long single-mode fiber coil as the propagation medium.

Light from a semiconductor laser source is launched in the fiber pigtails of the coil with the aid of a beamsplitter (BS in Fig.8-9). Two modes, CW and CCW, travel down the fiber in opposite directions (full arrows) and, after completing the coil path length, they re-emerge at the beamsplitter (dashed arrows). After recombination of the modes on the photodetector PD, an output signal I_{ph} is generated.

The total phase of each two modes is the sum of the Sagnac phases, $+\phi_S$ and $-\phi_S$, and of the phase shift introduced by the beamsplitter ϕ_{BS}. Indeed, a beamsplitter delays the transmitted beam by $\phi_{BS}=\pi/2$ with respect to the reflected beam.

Considering the routes traveled by the CW and CCW modes (Fig.8-9), we can see that the CW is reflected twice by the beamsplitter and suffers no extra delay, while the CCW crosses the beamsplitter twice and suffers an extra delay π.

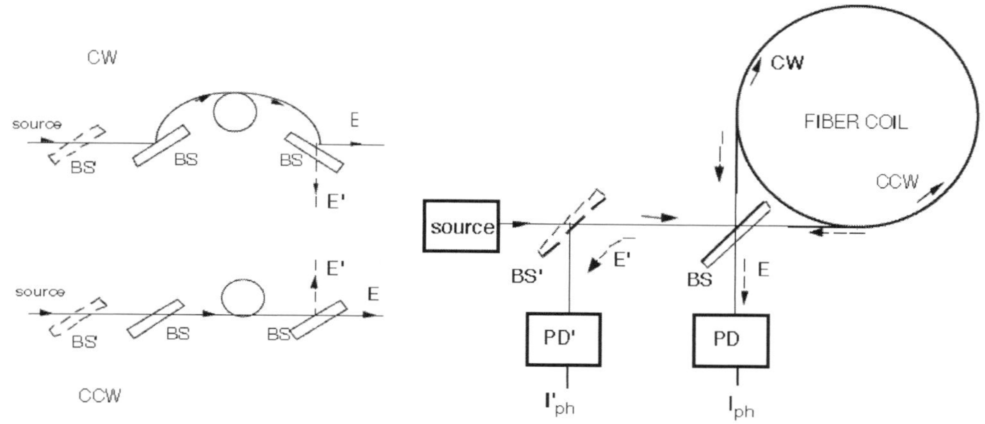

Fig.8-9 In the FOG, using the output E for the recombination of the counter-propagated beams, the BS is passed twice in reflection by the CW beam and twice in transmission by the CCW. To ensure reciprocity, it is better to use output E' detected by PD', in which both the CW and CCW beams make one reflection and one transmission at BS.

Thus, we may write the modes recombined on the photodetector as:

$$E_{CW} = E_0 \exp i(+\phi_S), \quad E_{CCW}=E_0 \exp i(-\phi_S+2\phi_{BS}) \tag{8.17}$$

Accordingly, the detected current is given by:

$$I_{ph} = | E_{CW}+E_{CCW} |^2 = E_0^2[1+\cos(2\phi_S+\pi)] = E_0^2 [1- \cos 2\phi_S] \tag{8.18}$$

As we can see from the last term in Eq.8.18, the FOG basic scheme supplies a phase signal $2\phi_S=R_\phi\,\Omega$ proportional to the rotation speed Ω, with a cosine dependence (Fig.8-10). The scale factor is given by Eq.8.8 as $R_\phi = 8\pi A/\lambda c$. With the typical values of $\lambda=1$ μm, and a 300-m long fiber wound on an 8-cm diameter coil, we have $R_\phi \approx 0.5$ rad/(rad/s). This value means that, to read a ≈1 deg/h rotation speed, we get a phase variation $2\phi_S \approx 1$ μrad and a very small signal, $\Delta I_{ph}/I_{ph} \approx 10^{-6}$, in the point of maximum sensitivity of the cosine.

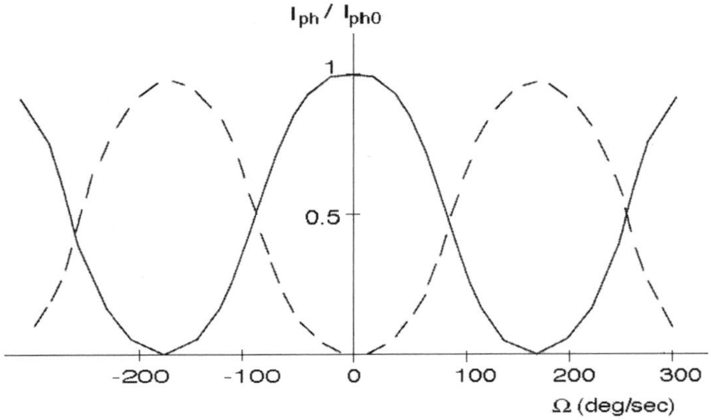

Fig.8-10 The response of the basic FOG scheme is a cosine function of the Sagnac phase, a functional dependence that is nonlinear for small Ω and cannot detect the sign of Ω. Moving from output E to E' in Fig.8-9 only changes the response in its complement to one (dotted line).

The point of maximum sensitivity cannot be reached so easily. In the cosine function, it is at a phase $=\pm\pi/2$, not 0, and we should be able to add a phase bias Φ_{bias} exactly equal to $\pi/2$ to reach the condition. However, the requirement of a bias $\Phi_{bias}=\pi/2$ with an accuracy of the order of microradian is far beyond the practical feasibility. In other words, with the cosine-type signal, we face again the problem of a vanishing sensitivity for small $2\phi_S$ and are unable to detect the sign of $2\phi_S$, as seen in Fig.8-10.

Another drawback of the FOG basic scheme (Fig.8-7) is the very noisy signal found when the output is obtained from the same beamsplitter used to launch in the coil. The excess noise is due to fluctuations of the phase shift ϕ_{BS} introduced by the beamsplitter.
Indeed, a real beamsplitter gives a phase shift $\phi_{BS}=\pi/2+\varepsilon$, where $\varepsilon=1-(t+r)$ is the loss, given by the complement to one of transmission plus reflection (see Ref.[16], p.294). As ϕ_{BS} directly impacts the signal (see Eq.8.17), even very minute fluctuations of ε (down to a few part per million) are of importance.

To avoid spoiling the ultimate quantum noise performance of the sensor, we need to use a reciprocal configuration for the propagation of CW and CCW waves, as first pointed out by Ulrich and Johnson [5]. To do so, we use the recombined wave E' going toward the source (Fig.8-9), deviated by BS' onto photodetector PD'. Now both CW and CCW waves pass the beamsplitter BS once in transmission and once in reflection, that is, the entire path is seen symmetrically by the counter-propagating waves. The output I'_{ph} is given by:

$$I'_{ph} = E_0^2\,[1+\cos 2\phi_S] \qquad (8.18')$$

The waveform of I'_{ph} is the complement to one of I_{ph} given by Eq.8.18, as shown in Fig.8-10. Unfortunately, the new signal does not help to circumvent the vanishing sensitivity and sign problems, yet it exhibits much improved noise performance because of reciprocity.
In the basic scheme of the FOG, there are several other issues to be improved before a practical configuration is developed. These issues will be discussed in Sect.8.5.

8.4 Development of the RLG

In the previous section, we have discussed the working principle of external and internal interferometer configuration for the readout of the Sagnac effect, and the problems inherent in both configurations. In this section, we present the solutions devised to develop the RLG concept in a commercial product and outline some technological hints that have been recognized as fundamental to achieve inertial-grade performance.

All practical configurations of the RLG employ a He-Ne ring laser, built around either a triangular or square cavity. The basic elements are shown in Fig.8-11. Despite the He-Ne may be regarded as an old technology, actually it's a perfect fit to the RLG and it's no surprise that the market of this gyroscope still keeps the US$ 1M per year revenue [3].

The cavity is machined from a single block of material, like Cervit or Zerodur, which are vitro-ceramic composites with very low thermal expansion (≈ ppm/°C typically).

The block is drilled to obtain the capillary bores for the He Ne mixture, and the edge surfaces are precisely lapped flat to accommodate the three (or four) mirrors of the cavity. The angular placement of the edge surfaces is made to coincide within arc-second to the nominal value required by the cavity. In this way, by leaning the mirrors on the block edges they are already aligned and no further adjustment is necessary.

Mirrors are not required to be cemented to the edge surface. By pressing them against the block, air is squeezed out from the interface (a joint known as anaerobic), and the mirror is kept firmly in place. Just a little stripe of vacuum-grade glue (like Tor-Seal) is adequate to protect the rim of the mirror and block interface.

The gas is the standard 10:1 He:Ne isotopic mixture using ^{20}Ne as the active atom. Natural Ne is not adequate because it is composed of two isotopes ^{20}Ne and ^{22}Ne with atomic lines slightly offset in frequency. Thus, natural Ne has an atomic line with a secondary peak disturbing frequency stabilization. Using the isotope, a smooth atomic line is obtained, allowing a clean frequency stabilization and multi-oscillator operation.

The pressure of the gas mixture is the usual 1-mbar value of linear He-He lasers. The discharge characteristic and the supply circuits are those described in Appendix A1.1.2. Typically, about 20 kV will be needed to switch the discharge on, and $V_{ak} \approx$ 1500 V will be the working voltage for an optimal discharge current $I_a \approx$ 5-7 mA. The cathode and the anode of the discharge tube are made by the usual Al-cylinder (a cold cathode for low thermal disturbance) and a W-wire, common to the standard He-Ne laser technology. Two anodes and two cathodes are provided in the block (Fig.8-11), arranged for opposite current flow to cancel the so-called Langmuir flow. The Langmuir flow is due to the ions flowing from anode to cathode and dragging light because of the Fizeau effect. The Langmuir flow contributes with a $\Omega_{BE}\approx 10$ deg/h zero-bias error per leg of discharge. Using two opposite discharges, the error is canceled out, provided the currents in the two legs are balanced within a $\approx\mu A$ difference (typical). Mirrors of the RLG are multilayer interference mirrors manufactured for very low scattering.

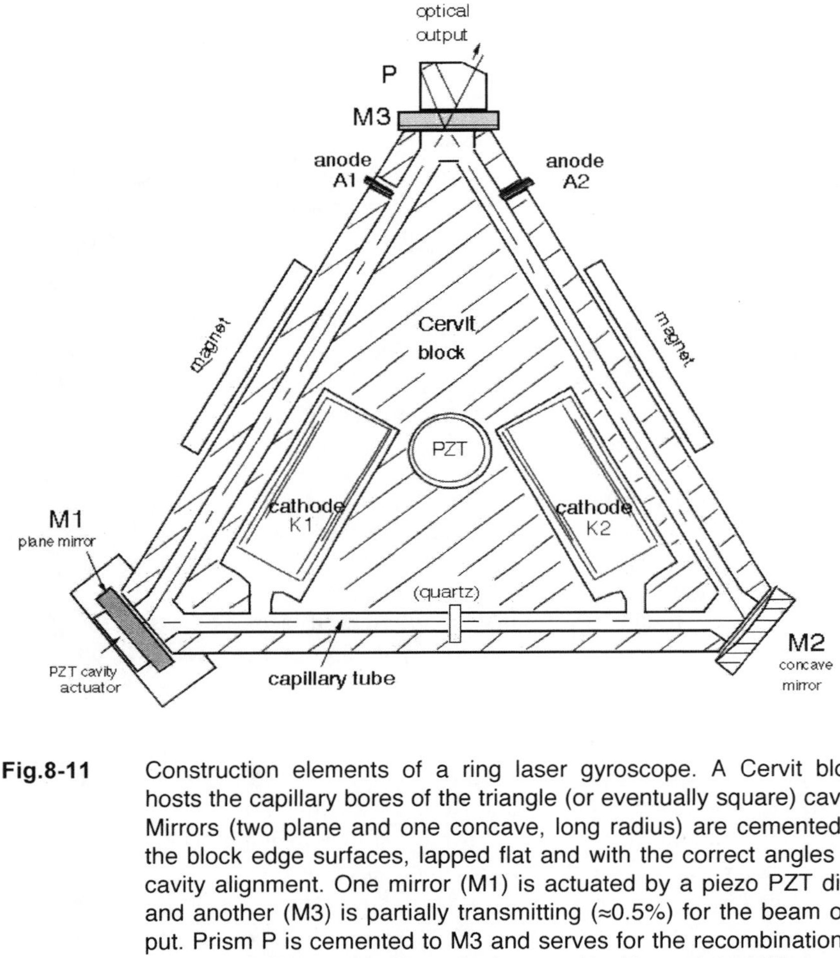

Fig.8-11 Construction elements of a ring laser gyroscope. A Cervit block hosts the capillary bores of the triangle (or eventually square) cavity. Mirrors (two plane and one concave, long radius) are cemented to the block edge surfaces, lapped flat and with the correct angles for cavity alignment. One mirror (M1) is actuated by a piezo PZT disk, and another (M3) is partially transmitting (≈0.5%) for the beam output. Prism P is cemented to M3 and serves for the recombination of CW and CCW mode. Flow discharges (A1-K1 and A2-K2) are arranged opposite to cancel out the Langmuir-flow error. Magnets are used in the ZRLG and a vibrating PZT in the DRLG.

Their curvature radii are chosen to ensure that the cavity stability condition is met and that the mirrors are easy to fabricate. Like with linear He-Ne units, the preferred choice is one long-radius concave mirror and plane mirrors for the remaining elements. The radius of the concave mirror is chosen a few times larger than the perimeter of the cavity, usually 1-3 m.

Of the two plane mirrors, one is maximum reflectivity (in practice 99.999%) and will be the mirror for cavity length control (M1 in Fig8-11), while the other will be the output mirror (M3). Transmission of M3 is chosen in the range t=1-r=0.3-1% to get an output power of ≈0.2-1 mW out of the cavity. It is not advisable to increase the output power beyond the minimum required for detection, because output power represents a loss for the cavity and affects ϕ_n (see Eqs.8.9 and 8.10). The concave mirror (M2) will have a maximum reflectivity too.

The frequency stabilization scheme may be one of those discussed in Appendix A1.2. In the Zeeman RLG, it is obvious to choose Zeeman splitting as the mechanism of marking the atomic line, whereas, in the dithered RLG, it is better to use Lamb's dip or the polarization-mode arrangement.

A practical RLG also contains additional items like (i) permanent magnets applying a longitudinal magnetic field of 300 to 500 Gauss, to split the atomic line in two; (ii) a quartz flat, to add a circular birefringence in the cavity; (iii) a PZT (lead-tin-zirconate) ceramic disk, to exert a rotational vibration to the device; (iv) magnetic shields (not shown in Fig.8-11); and (v) one or more temperature sensors (thermistors) to sense the cavity temperature and correct the zero bias drift. Items (i) and (ii) are specific of the ZRLG, whereas item (iii) is specific of the DRLG.

A final comment on the cavity shape is worth reporting. Of the two major companies that are active worldwide in the field of RLG for avionics and share a very wealthy market, one uses a square and the other a triangle cavity (Fig.8-12).
Both shapes have advantages and disadvantages. The triangle cavity saves one mirror with respect to the square, and works with a 30° incidence angle instead of 45°, so scatter is less and the lock-in range is smaller. The square cavity has a better surface-to-perimeter ratio A/p, and thus a larger responsivity (Eq.8.13) at equal size. Also, a square cavity is aligned more easily than a triangle, but it uses four mirrors instead of three.

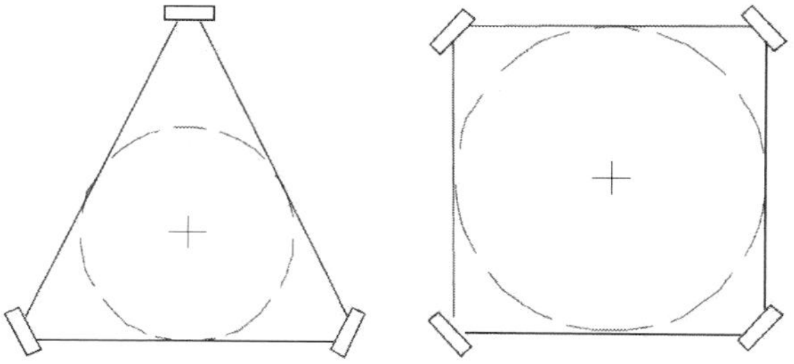

Fig.8-12 Both triangle and square cavities are used in the RLG by different manufacturers. There are minor differences in the two. The triangle saves a mirror and has less scattering, but the square has better responsivity at equal size and alignment is easier.

In good substance, all these differences are minor, but they are good enough to claim the superiority of one choice over the other. So, patents on both cavity shapes have been filed and accepted, and were apparently expected to waive one company from the other's charge of infringement. Actually, a lawsuit filed by one big manufacturer against the other big manufacturer, known in the field as the '*gyro dispute*' has lasted ten years in the United States, and was recently concluded with a compensation of about US$500 million [19].

8.4.1 The Dithered Laser Gyro

In the Dithered RLG (DRLG), we circumvent the problem of the lock-in dead band by applying an external small angular rotation to the device. The rotation is supplied by a piezo ceramic actuator mounted in the middle of the Cervit block (Fig.8-11). The piezo may be a single piece of ceramic, vibrating in the shear tangential mode (Fig.8-13 top) or a bimorph composite with the same function. The vibration consists of an angular movement of the top surface, integral with the gyro, with respect to the bottom surface, and integral with the reference body of the vehicle to which the gyro is strapped down.

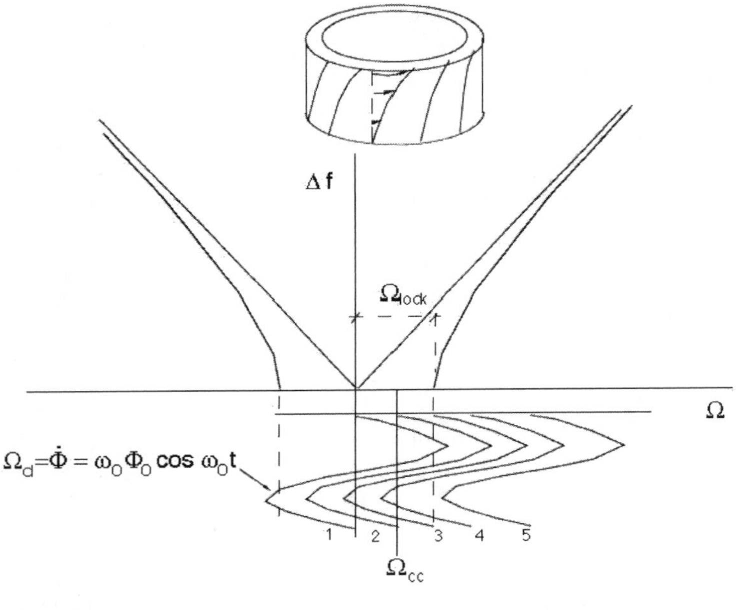

Fig.8-13 A ceramic tube vibrating in the tangential shear mode (top) is used to dither the RLG out of the locking dead-band (bottom). Even small vibration amplitude Φ_0 (≈ mrad) is enough to unlock the gyro at audio frequency ω_0.

The applied angular velocity Ω_d is such that the device exits from the dead band at each period of vibration (Fig.8-13 bottom), whence the name of *dither* given to the technique.

The required angular amplitude Φ_0 of vibration supplied by the piezo ceramic need not be so large to unlock the gyro. Indeed, if we use a reasonable frequency of excitation ω_0 (say $\approx$10 Hz), and thus have the instantaneous rotation $\Phi = \Phi_0 \sin \omega_0 t$, the corresponding angular velocity is $\Omega_d = d\Phi/dt = \omega_0 \Phi_0 \cos\omega_0 t$. Even with a small Φ_0=1 mrad, and ω_0 =60 rad/s, we get an amplitude of rotation speed $\Omega_{d0} = \omega_0 \Phi_0 = 0.06$ rad/s, which is largely in excess of the normal locking range (Sect.8.3).

The reason for a high frequency of dither is that ω_0 should be larger than the highest frequency contained in the signal. Otherwise, beating may occur, and the dithered signal may disappear. However, as the power required to actuate the gyro is $(1/2)\, I_{RLG}\, \omega_0^2$, where I_{RLG} is the momentum of inertia of the gyro, we shall limit ω_0 to the smallest value necessary.

Processing of the output signal of the DRLG follows the principle illustrated in Fig.8-14. We take the measurement of mode beating Δf for each semi-period of the dither waveform $\Omega_d = \Omega_{d0} \cos\omega_0 t$, let us call these signals C_1 (for $\Omega_d > 0$) and C_2 (for $\Omega_d < 0$).
First, we determine the sign of Ω by looking at which semi-period has the largest signal. When $C_1 > C_2$, Ω has the same direction of the dither rotation Ω_d (let's say $\Omega > 0$), whereas for $C_1 < C_2$, Ω has the opposite direction of Ω_d (and $\Omega < 0$).

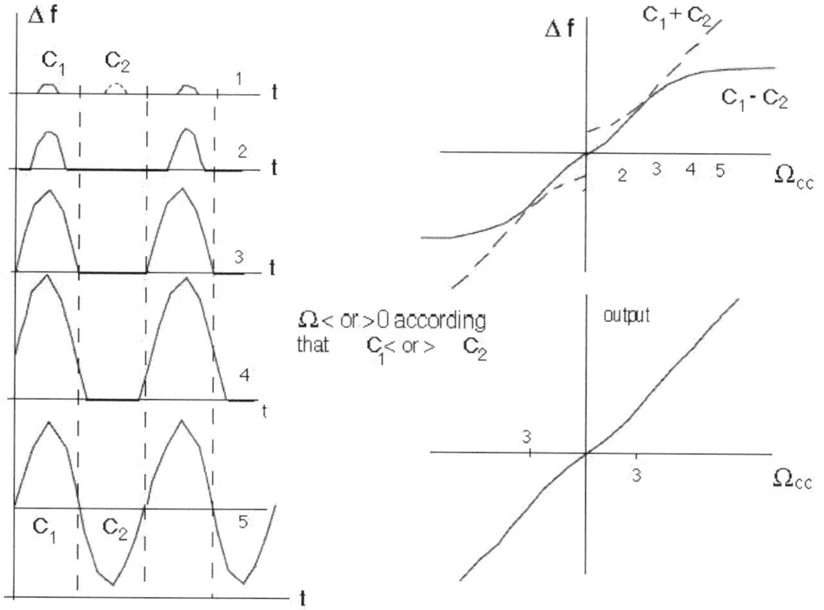

Fig.8-14 In the semiperiods of the dither drive cos ω_0t, the outputs C_1 and C_2 vary with the dc baseline Ω_{cc} as shown at left. By computing sum and difference of these signals, we get a good approximation for Ω_{cc} (right).

Second, let us consider the dependence of C_1 and C_2 from the baseline angular velocity Ω_{cc}, the signal now representing the relatively slow rotation velocity to be measured.
We can see that, as Ω_{cc} increases, C_1 increases and C_2 decreases. For a small Ω_{cc} internal to the lock-in range (cases 1 and 2 in Fig.8-14), a good approximation to Ω_{cc} is the signal difference C_1-C_2. When Ω_{cc} increases further and is external to the lock-in range (cases 3 and 4 in Fig.8-14), it finally converges to the average of C_1 and C_2, and we take their sum C_1+C_2 as the approximation of Ω_{cc}.

The break point between difference and sum is best placed at the amplitude marked 3 in Fig. 8-14, that is, where about half the dither peak comes out of the locking region. To correct the residual nonlinearity error, we may store the dependence of Ω_{cc} from the sum and difference signals in a look-up table and use the measured values of C_1-C_2 and C_1+C_2 to compute Δf and then Ω_{cc}.

The dither approach may appear disagreeable to the researcher because we start with an inherently static structure (the electro-optical gyroscope) and end up forced to apply a mechanical movement to have it work. The objection is founded, we shall admit. However, in engineering, we should never care about the beauty of our solutions. The best solution is definitely the one that works better, and this should be the concept of beauty for the engineer. When fabricating something, we should not prefer a conceptually appealing approach to one working better, because this could simply result in a loss for our company.

In actual products, manufacturers use the dither concept to realize the RLG with preference with respect to the more appealing ZRLG, because the DRLG is cheaper, uses fewer components, and is easier to fabricate.

8.4.2 The Ring Zeeman Laser Gyro

The Zeeman RLG (ZRLG) circumvents the lock-in problem in a very elegant way. It makes use of the combined effect of a double splitting of the oscillating modes to produce in the cavity four oscillations well separated in frequency [20,21]. The four oscillations are circularly polarized with a suitable order and are obtained by introducing appropriate circular birefringence [22] in the cavity.

Recalling circular birefringence. Circular birefringence is also called rotatory power and is described by rotation $\Psi=\rho L$ of the polarization plane of a linearly polarized wave propagated through a length L, where ρ is the rotatory power of the material. A material with $\rho>0$ rotates the polarization plane of a positive angle, or counter-clockwise, and is called *levogyrate*, whereas for $\rho<0$ the material is *dextrogyrate* and rotates a linear polarization in the clockwise direction.

We can attribute a rotating sense to the material because the rotatory power is independent from the sense along which we look through the material. Thus, if we take a flat of quartz as in Fig.8-11, we get the same rotation $\Psi=\rho L$ for both the left-bound and the right-bound wave passing through it.
This circumstance is also described by saying that the rotatory power of crystals is *reciprocal*. Usually, effects are reciprocal and, for example, no effect violating reciprocity is known in transmission optics or for linear birefringence. The only example of *nonreciprocal* birefringence in optics is the rotatory power associated with the Faraday or Zeeman effect.

The Faraday circular birefringence is again described by angle $\Psi=\rho L$, but now we have $\rho=V\underline{H}\cdot\xi$, where ξ is the unity vector associated with the propagation path through the material, and $\underline{H}$ is the vector of magnetic field intensity. Now, the material is levogyrate or dextrogyrate according to the sense we cross it (the sign of $\underline{H}\cdot\xi$,), or we deal with a *nonreciprocal* circular birefringence.
Another equivalent description of rotatory power is when considering circular polarization states.
Along a path with rotatory power, the LC and RC (left- and right-circular) states become phase shifted by $+\Psi/2$ and $-\Psi/2$, respectively. Thus, we have a circular birefringence with the two LC and RC states.
Now, this is the crucial point for the ZRLG working principle. When we have nonreciprocal circular birefringence, the path length is different for the two modes (in the same state) propagating in the $+\xi$ and $-\xi$ directions. Thus, we have the chance of distinguishing between the CW and CCW waves and will use the nonreciprocal birefringence effect to separate the modes in frequency, finally defeating the mode locking.

As a remark, the Faraday effect is strictly about the circular birefringence case, whereas the Zeeman effect is the splitting of the atomic line in two lines, one for the LC and one for the RC. It can be shown [23] that the tails of the lines split by the Zeeman effect are responsible, off resonance, for just the extra path length described by the Faraday birefringence.

Now, going to the schematic of Fig.8-11, the non-reciprocal circular birefringence (or Zeeman effect) is provided by the means of the permanent magnets. These are designed to apply a modest field (≈200 to 500 Gauss) to a portion (≈5 to 10 cm) of the capillary tube.
The reciprocal circular birefringence is provided by the means of the quartz flat. Normally, a ≈1-mm thick flat is used, for a zero-order delay of about ≈45°, the exact value being not so critical.

Because of the Zeeman effect, the atomic line is split in two lines (Fig. 8-15b) separated by ≈200 MHz, each sustaining a determined polarization state. As discussed in Appendix A1-2, when the magnetic field is applied longitudinally, the lines sustain circular polarization states, for example the LC of the lower-frequency line, and the RC of the upper-frequency. Thus, two modes with LC and RC polarization should be able to oscillate.

If we assume that the cavity line is somehow kept centered at the middle of the unperturbed atomic line, we should expect that oscillation is located right there in frequency, at ν_0 for both modes.
Actually, because of the pulling effect (App.A1.1), each polarization mode is slightly offset toward the center of the respective atomic line (Fig.8-15b). The frequency pulling $\Delta\nu_a$ depends on the cavity line width and on the detuning from the atomic line center [24]. With no detuning and with a typical Zeeman splitting of ≈200 MHz, we may have frequency pulling in the range $\Delta\nu_a \approx$20-50 MHz.

In these conditions, the ring cavity sustains two oscillations in opposite propagation directions, and each oscillation is degenerate in polarization, or it is double.
Indeed, in the CCW direction (Fig.8-15b), the oscillation with the LC polarization is at $\nu<\nu_0$ and that with the RC polarization is at $\nu>\nu_0$. The oscillation propagating in the CW direction has the LC mode at $\nu>\nu_0$ and the RC mode at $\nu<\nu_0$, contrary to the CCW.
This is no surprise, however, because a RC polarization propagating in the CCW direction is the same as a LC polarization propagating in the opposite CW direction, whatever is the convention used to define the handiness of a circular polarization. Because of this, the CCW-RC and CW-LC share the same population of atoms split by the Zeeman effect and oscillate at

the same frequency (Fig.8-15b).

Now, consider the effect of adding the quartz plate in the cavity, and with it a circular birefringence Δl, of a fraction of wavelength. Because of the birefringence, the ring cavity perimeter is seen to increase for one mode (say, the LC) and to decrease for the other mode (the RC). The quartz birefringence is reciprocal, so both the CW and CCW modes undergo the same Δl.

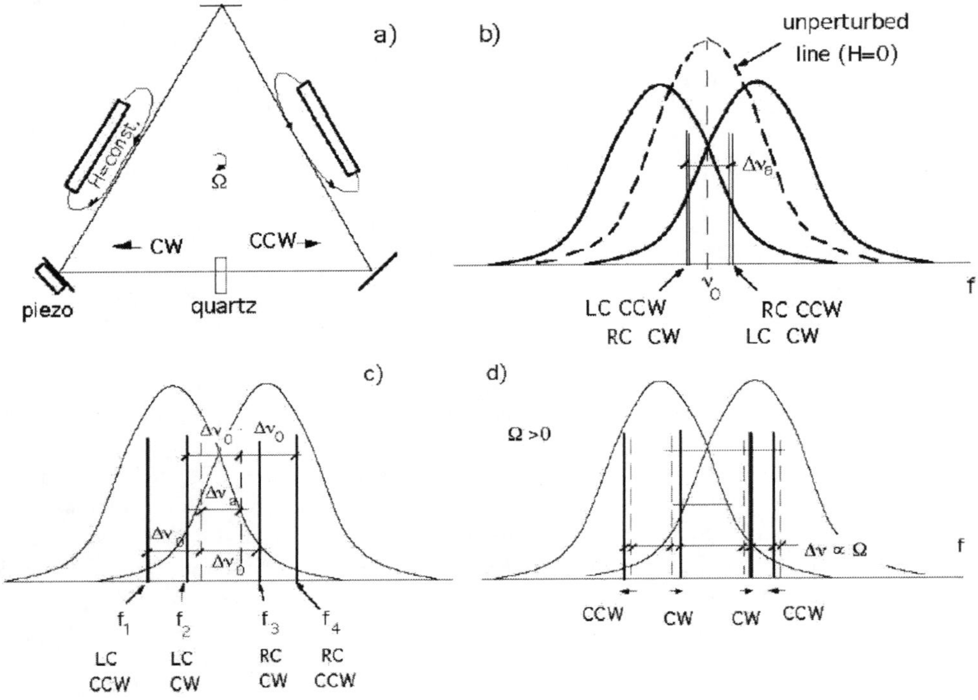

Fig.8-15 In the ZRLG, we get four oscillating modes by introducing (a) a reciprocal circular birefringence by the quartz flat and a nonreciprocal circular birefringence by the magnetic field. (b) The atomic line (dotted line) is split in two by the Zeeman effect produced by the axial magnetic field, and there is one line for LC (left circular) and one for the RC (right circular) polarization (dotted lines). Because of the frequency-pulling effect, the oscillations are moved apart by a frequency $\Delta\nu_a$, toward their atomic line center. (c) Introducing the quartz flat, the LC and RC waves are split further by $\Delta\nu_0$. (d) When the gyro rotates, frequency of the CW waves increases, and the frequency of the CCW decreases. Thus, f_1-f_2 and f_3-f_4 carry a frequency difference $\Delta\nu$ proportional to Ω and allow recovering its sign.

As a consequence of the added Δl, to maintain the zero phase-shift of the round-trip (as per the Barkhausen condition, App.A1.1.1), the oscillation frequency shall now change a little, by $-\Delta\nu_0$ for the LC mode and $+\Delta\nu_0$ for the RC mode (Fig.8-15c). The quantity $\Delta\nu_0$ is easily evaluated by recalling that $\Delta\nu=c/p$ is the frequency variation corresponding to $\Delta l=\lambda/2$, whence in the proportion $\Delta\nu_0=(2\Delta l/\lambda)(c/p)$.
We choose the quartz birefringence Δl so that $\Delta\nu_0\approx 50$-100 MHz, a value less than, but comparable to, the He-Ne atomic line width $\Delta\nu_{at}\approx 1500$ MHz.

Now, we have four oscillating modes in the cavity, as shown in Fig.8-15c. If splitting is symmetrical, the oscillation frequencies are so arranged that the differences of LC and RC modes are the same, or $|f_1-f_2| = |f_3-f_4|$ at rest ($\Omega=0$).

In addition, the frequency separation of the four modes is now large enough to prevent frequency attraction or locking between them. This assumes of course, that we have accurately cured all coupling effects, going down to a residual locking range of about 1 MHz or less.

Worth noting, the scheme lends itself to a method for frequency stabilization. In fact, when the mode pattern is symmetrical around the original atomic line ν_0 center, the amplitudes of oscillation E_1, E_2 at frequency f_1, f_2 are equal to amplitudes E_3, E_4 at frequency f_3, f_4. If a cavity drift causes a decrease in frequency of the mode pattern, then E_1, E_2 become larger than E_3, E_4. Thus, the difference $(E_1+E_2)-(E_3+E_4)$ is a good error signal to feed the piezo actuator of the cavity length control (mirror M1 in Fig.8-11).

We may now consider the dynamical mode oscillation when the gyroscope is put into rotation. Let $\Omega>0$ in the reference parallel to magnetic field $\underline{H}$. Then, the CCW waves decrease their frequency by an amount $\Delta\nu=R_F\Omega/2$, while the CW waves increase theirs by $\Delta\nu=R_F\Omega/2$, as indicated in Fig.8-15d.

The strategy to compute Ω now comes out easy. We will make the beating of the two CS and CD modes on two photodetectors PD1 and PD2, to obtain the two signals:

$$I_{PD1} = I_{01}\,(1+\cos 2\pi F_1 t), \qquad I_{PD2} = I_{02}\,(1+\cos 2\pi F_2 t)$$

$$\text{where} \qquad F_1 = |f_1-f_2|+\Delta\nu, \qquad F_2 = |f_3-f_4|-\Delta\nu \tag{8.19}$$

By counting the periods of signal I_{PD1} and I_{PD2}, we can process the signal in digital format. First, we compute the sign of Ω as >0 if $F_1>F_2$, and <0 if $F_1<F_2$. Then, the difference $F_1-F_2 = |f_1-f_2|-|f_3-f_4|+2\Delta\nu$ is the desired Sagnac inertial signal, and we get twice the responsivity of the normal ring (Sect.8.1), or $R_F=4R/\lambda$.

We may easily integrate the angular velocity signal and obtain the angle of rotation Ψ (see also Eq.8.13) by counting the zero crossings of I_{PD1} and I_{PD2} so that we integrate the frequency difference $F_1-F_2=2\Delta\nu$.

From the previous considerations, it is now appreciated why we wanted to match the mode frequency difference at rest.

Indeed, if it is $|f_1-f_2|=|f_3-f_4|$, we get $F_1-F_2= 2\Delta\nu$ exactly, whereas if we incur a residual small error $|f_1-f_2|-|f_3-f_4|= \Delta f_{offset}$ in the oscillation frequency pattern, the signal $F_1-F_2= 2\Delta f_S+\Delta f_{offset}$ is affected by an offset Δf_{offset}.

Last, let us describe the beam recombination arrangement (Fig.8-16). The CW and CCW waves, each carrying a LC and RC polarization mode, are brought together at the output mirror like in the basic scheme. A double-reflection 90° corner is used to fold one beam (the CCW in Fig.8-16) and to superpose it to the other beam (the CW).

In leaving the prism, the four-mode beam crosses a quarter-wave plate. The plate is oriented (its fast and slow axes) at 45° with respect to the incidence plane of the Glan cube that follows. The quarter-wave plate is a birefringence element that introduces a λ/4 (or 90°) phase shift in one linear polarization with respect to the other. The result is that circular polarization states are transformed in linear polarization states. For example, the RC becomes a HL (horizontal linear) and the LC becomes a VL (vertical linear), see Fig.8-16.

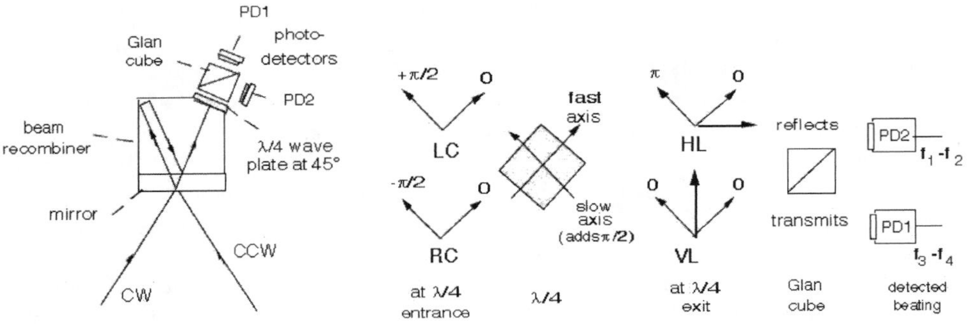

Fig.8-16 At the output mirror, total reflection at the prism corner is used to recombine the CW and CCW beams, each of which contains two LC and two RC polarizations. A quarter-wave plate oriented at 45° converts the polarization states to linear (HL and VL), and a Glan-cube polarizing beamsplitter transmits one linear component to PD1 and reflects the other to PD2. At photodetectors, optical signals at different frequency produce beating, and electrical signals at the frequency difference f_2-f_1 and f_4-f_3 are generated.

At the output of the quarter-wave plate, the Glan-cube beamsplitter cube divides the HL and VL components. These components are made available at the transmitted and reflected cube outputs; two photodetectors convert the HL and VL optical fields into electrical signals. Note that HL and VL components are correctly made up of the RC (CW plus CCW) and of the LC (CW plus CCW) modes, that is, those that we need to supply the F_1 and F_2 signals.

8.4.3 Performances of RLGs

Both the DRLG and the ZRLG are truly inertial-grade sensors. Let us briefly discuss performances and the way they are achieved, in this section.

Minimum Detectable Signal (MDS) In Sect.8.2.1, we have considered it in terms of the phase shift to be measured. More frequently, we may be interested in the minimum detectable angular velocity or MDΩ (deg/h). Using Eq.8.11, and dividing the phase MDS by the responsivity $R_\Phi = \Omega/\Phi$, we can write:

$$MD\Omega = \sigma_{\Omega n} + \Omega_{ZB} \tag{8.20}$$

The first term $\sigma_{\Omega n}$ is the random contribution coming from noise. At the quantum limit, using Eq.8.9, we can write $\sigma_{\Omega n}$ in the form:

$$\sigma_{\Omega n} = \phi_n/R_\Phi = (2\kappa\, h\nu/\eta P_0)^{1/2}/R_\Phi \sqrt{B} = S_\Omega \sqrt{B} \tag{8.20a}$$

where we have introduced quantity S_Ω with the meaning of spectral density of angular velocity fluctuation. S_Ω is measured in (deg/h)/$\sqrt{Hz}$ or also, more frequently, in (deg/$\sqrt{h}$). Here, let us note that 1 (deg/h)/$\sqrt{Hz}$ = 1 deg/($\sqrt{h}$ 60$\sqrt{s}$ $\sqrt{Hz}$)= (1/60) deg/$\sqrt{h}$, whence the spectral density S_Ω is a number 60 times smaller in the deg/$\sqrt{h}$ unit. So, for a gyro with $\sigma_{\Omega n}$ =1 deg/h/$\sqrt{Hz}$ we have S_Ω = 0.016 deg/$\sqrt{h}$, and thus the S_Ω figure looks more appealing than σ_Ω when specifying the MDΩ performance.

When the random term prevails in Eq.8.20, the minimum detectable Ω increases with bandwidth as indicated by Eq.8.20a. This holds at the quantum limit, strictly, but we may assume the dependence is $S_\Omega \sqrt{B}$ also for other source of excess noises. If we average or integrate Ω on a period T of time, then the measurement bandwidth is $B \propto 1/T$, and accordingly we get:

$$\sigma_{\Omega n} = S_\Omega/\sqrt{T} \tag{8.20b}$$

As Eq.8.20b tells us, the minimum detectable Ω decreases at increasing averaging time, and of course this statement holds until the second term Ω_{ZB} in Eq.8.20 is finally reached, or some nonwhite noise components of $\sigma_{\Omega n}$ come into play.

In Fig.8-17, we plot data representative of the MDΩ performance achieved in gyroscopes. As we can see there, the two RLG versions are not so far away from the quantum limit, with the ZRLG performing a little bit better than the DRLG, whereas the FOG has improved a lot in last decade and is only a factor ≈3 off. The maximum useful integration time differs as well. It is up to several hundred s in RLGs and just ≈10...50 s in the FOG.

The spectral density of angular velocity fluctuations can be read from the diagram of Fig.8-17 using the Nyquist value B=1/2T for the bandwidth. At T=0.5-s, we get the typical values $S_\Omega \approx 0.03$ deg/h$\sqrt{Hz}$= 0.0005 deg/$\sqrt{h}$ for the RLGs, and $S_\Omega \approx 0.06$ deg/h$\sqrt{Hz}$ ≈0.001 deg/$\sqrt{h}$ for the FOG. By increasing the integration time up to 100 s, we can attain a few 10^{-3} deg/h in the RLG, and perhaps 0.01-deg/h in the FOG.

Zero bias and null provision. The above figures of MDΩ are achieved if we are able to keep the second term in Eq.8.20, which is the zero-bias error Ω_{ZB}, consistently smaller than the random noise term.

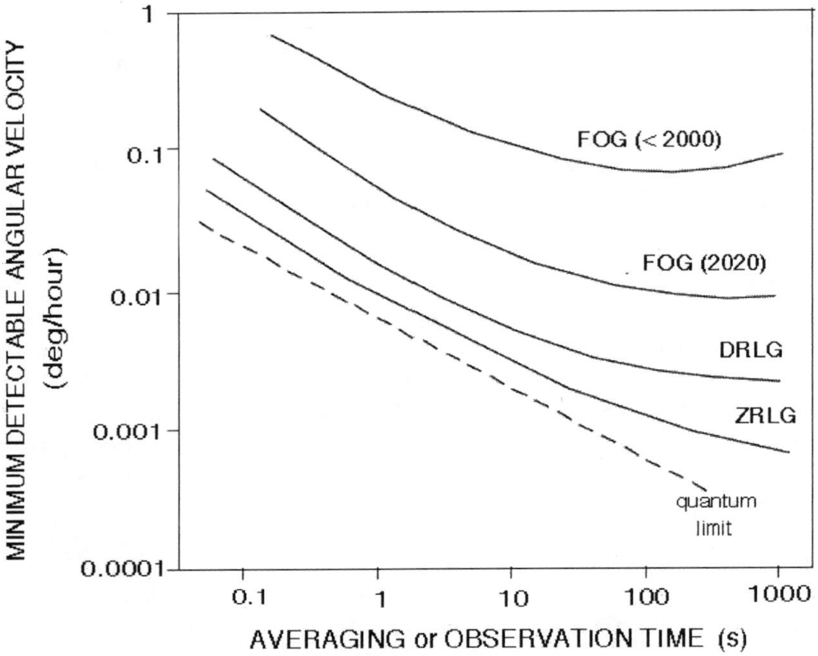

Fig.8-17 Typical minimum detectable angular velocity MDΩ of ZRLG, DRLG, and FOG. The FOF+G has progressed by about one decade in the last twenty years. Dotted line is the quantum limit (about the same for all).

A simple null operation of the Δf signal by a dc correction does not suffice, usually. We need a more sophisticated algorithm to be able reduce the initial zero bias error Ω_{ZB}, which usually may be in the range 1-10 deg/h, to the minute $\approx 10^{-3}$deg/h level.
The calibration operation is carried out in a test chamber well shielded from external, ambient-induced, mechanical disturbances. The gyro is oriented perpendicular to the earth rotation axis (Ω=0) and subjected to a well-defined set of physical quantities, or disturbances, M_k (k=1...n). For each M_k, we determine the deviation of the zero bias Ω_{ZB} from the initial value Ω_{ZB0}. In this way, we get the coefficients C_{ki} of the Ω_{ZB} expansion at the first-order:

$$\Omega_{ZB} = \Omega_{ZB0} + \Sigma_{k=1..n} C_{ki} (M_k - M_{k0}) \tag{8.21}$$

Here, Ω_{ZB0} is the zero offset in the standard conditions (M_k=M_{k0}). For each disturbance M_k, the measurement is repeated on a set of values (i=1...m) to determine if the linear regression (Eq.8.21) is adequate. Eventually, we need a higher-order expansion in M_k's, that easily follows as an extension of Eq.8.21.

In the device, the physical quantities in Eq.8.21 are measured by separate sensors (see also Sect.8.4). Typically, they are: (i) temperature of the cavity and temperature of the magnets (ZRLG only); (ii) currents in the discharge legs; (iii) magnetic field to the bore (ZRLG) or dither amplitude (DRLG); (iv) frequency detuning from the atomic line center. The measurement of coefficients C_{ki} is time-consuming, because each element requires a time of the order of the maximum useful integration time (Fig.8-17).

Following the calibration, the correction of the zero-bias error Ω_{ZB} through Eq.8.21 is performed by means of the on-board computer, in which the matrix C_{ki} is written. The correction is effective, and it typically brings down the initial value of Ω_{ZB} by a factor of $\approx 10^3$ or better. With aging however, the matrix C_{ki} changes, and the correction improvement may worsen to $\approx 10^2$ in a period of typically three years. Then, the sensor shall be recalibrated, and this operation requires disassembling the sensor and delivering it back for the factory.

Dynamic range and linearity. The dynamic range we need for inertial applications is 6 to 8 decades, spanning from the $\approx$0.001 deg/h of the best resolution to the maximum $\approx$100 deg/s required to cover fast jerks that the vehicle may experience. The scale factor should be calibrated to better than 10^{-4} as a specification for achieving a $\approx$0.1-km position accuracy after a $\approx 10^3$-km trip. For the same reason, the linearity error should be less than $\approx$10 ppm on the entire dynamic range (1 ppm = part-per-million, or 10^{-6}).

Both DRLG and ZRLG have a wide range of frequencies available for the signal Δf swing, about 100 MHz under the He-Ne atomic line of 1500 MHz width. However, a small linearity error arises, because of the frequency pulling effect. In addition, a large Δf swing may disturb the frequency stabilization, and hence result in a frequency error. Thus, in the end we get a $\approx 10^{-3}$ linearity error as the starting performance. This figure is improved and brought to the 10 ppm level again by calibration. Pendulum-based servos are employed to calibrate the measurement. Like for the zero correction, the linearity correction is also carried out by providing a matrix of coefficients to the on-board computer.

Conclusions. With all the provisions discussed above, the RLG becomes the heart of INU (inertial navigation units) and AHRS (attitude heading reference systems) extensively used in avionics [15]. Typical performance of an INU is $\approx$0.5-nm (nautical mile) error in dead reckoning after 1 hour of navigation, during which an aircraft flies approximately 500 nm. The INU is typically a box of 15” by 8” by 8” and includes three gyroscopes and the computer to handle all the signals, inertial and housekeeping, necessary to compute the actual position of the vehicle on the earth. Cost of an INU is typically in the US$2.5 to 5 million range.

8.5 Development of the Fiber Optic Gyro

The first problem encountered when developing the FOG from its basic configuration (Fig.8-7) has been that of the cosine-type dependence of the signal (Eq.8.18), which is unsuitable for measuring small Ω [4,7].

Several approaches to circumvent the cosine problem have been proposed. Despite the variety of solutions, all are based on moving the signal Φ_S from the baseband (f=0) to a carrier frequency f_m. In this way, looking at resulting signal $\cos(2\pi f_m t+\Phi_S)$ and comparing it to a reference $\cos 2\pi f_m t$, the Sagnac phase can be recovered.

Among the proposed solutions, we find those based on using two-frequency modes [7], a frequency shift, or a frequency modulation [25]. Over the years, however, the solution finally recognized as the most effective and easily interfaced to the basic scheme has been the one based on the *phase modulation* [7,26].

8.5.1 The Open-Loop Fiber Optic Gyro

The argument leading to phase modulation is understood looking at the cosine signal plotted in Fig.8-18. Usually, we are near $\Phi_S\approx 0$ and the photodetected current I_{ph} is about zero. If we are able to superpose a phase modulation to the Sagnac-induced phase Φ_S, let us say $\phi_m=\phi_{m0}\cos\omega_m t$, the photodetected current I_{ph} becomes modulated at frequency $f_m=\omega_m/2\pi$. As we can see from Fig.8-18, for $\Phi_S>0$, the waveform of I_{ph} is in phase with the modulation ϕ_m, whereas for $\Phi_S<0$, it is in antiphase. Thus, the sign of Φ_S is recovered.

Moreover, the amplitude of the modulated current increases at increasing Φ_S, removing the zero sensitivity near $\Phi_S\approx 0$. In the examples in Fig.8-18, note how the ac component at the modulation frequency f_m increases from *b* to *c*, reverses its phase in *d* and goes to zero in *a*, where a $2f_m$ component is left. Actually, as it will be shown later, the component of I_{ph} at the modulation frequency f_m is proportional to $\sin\Phi_S$ ($\approx\Phi_S$ for small Φ_S) and therefore to the measurand Ω.

A feature of the phase modulation measurement scheme is that we end up with the desired $\sin\Phi_S$ dependence starting from ta $\cos\Phi_S$ dependence. This tells us that the phase modulation readout yields the derivative of the initial response dependence.

We may also wonder if the working point of the interferometer, $1-\cos\Phi_S\approx 0$ for small Φ_S, is good from the point of view of noise. Indeed, the dc component is zero (or small at first order in Φ_S), but the signal is small at the second order, as $1-\cos\Phi_S \approx \Phi_S^2/2$. From an analysis of the S/N ratio [27], one can find that indeed the 'dark fringe' point of the interferometer is the best, and we work in the phase-noise limits of Sect.8.2.1.

To implement the phase modulation, we shall insert a suitable element in the FOG propagation path. We need a nonreciprocal element because the Sagnac signal we are dealing with is a nonreciprocal phase. To this end, we put the phase modulator in an asymmetrical position, as shown in Fig.8-18.

The phase modulator itself provides a reciprocal phase shift, that we may write as:

$$\Phi_m(t) = \Phi_{m0} \cos \omega_m t \tag{8.22}$$

yet, as it is placed at one of the fiber entrances (Fig.8-18), the CW and CCW beams cross it at different times, collecting different phase shifts. This is equivalent to nonreciprocity. Denoting the fiber coil length by L, the propagation phase shift is nkL, where $k=2\pi/\lambda$ and n is the effective index of refraction of the fiber. The propagation time through the coil is $T=nL/c$,

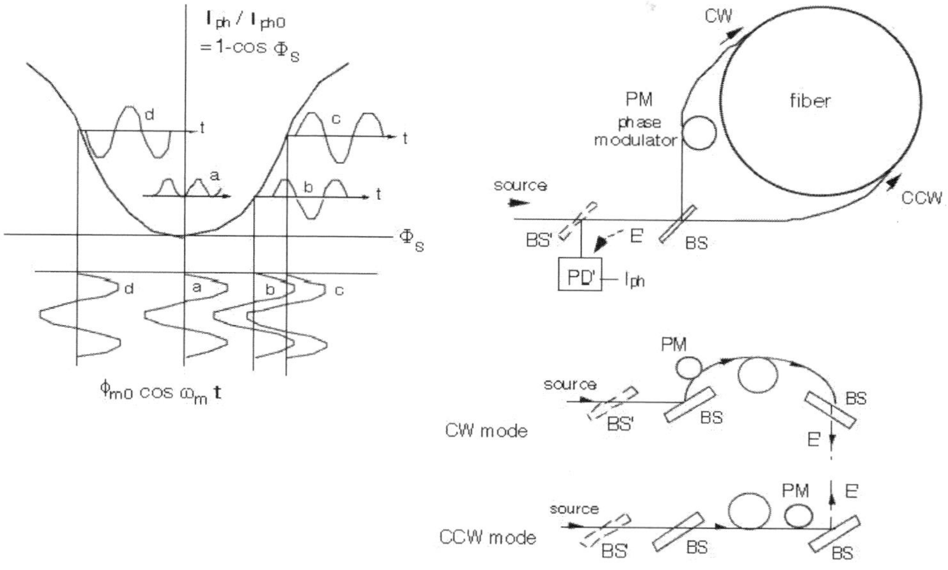

Fig.8-18 The cosine-type signal carrying the Sagnac phase Φ_S can be brought to linear with the aid of a phase modulator, adding a small modulated phase $\phi_m \cos \omega_m t$ to Φ_S (left). Then, the waveform of the photodetected current I_{ph} becomes modulated and carries the sign and the amplitude of Φ_S, as can be seen from the examples (a,b,c,d at left). The phase modulator shall be placed asymmetrical with respect to the coil (right, top) so that the modes pass through it with a delay and are able to collect a phase difference (right, bottom).

and this quantity is the delay that both the CW and CCW beams experience in crossing the phase modulator. After a complete propagation through the coil, the CW and CCW phases are written as:

$$\Phi_{CW} = \Phi_S/2 + \Phi_m(t\text{-}T/2) + n_{CW}kL + R^{BS}(t\text{-}T/2) + T^{BS}(t+T/2)$$

$$\Phi_{CCW} = -\Phi_S/2 + \Phi_m(t+T/2) + n_{CCW}kL + R^{BS}(t+T/2) + T^{BS}(t\text{-}T/2) \qquad (8.23)$$

Here, $\Phi_S/2$ and $-\Phi_S/2$ are the Sagnac phase shifts, $-T/2$ and $+T/2$ are the time delays in crossing the modulator, n_{CW} and n_{CCW} are the effective index of the modes, and R^{BS}, T^{BS} are (field) reflection and transmission of the launch beamsplitter BS.

At the photodetector, the beating of the recombined modes $E_{CCW}+E_{CW}$ yields a signal:

$$I_{ph} = |E_{CW}+E_{CCW}|^2 = E_0^2[1+\cos(\Phi_{CW}-\Phi_{CCW})] \qquad (8.24)$$

Let us now consider the phase difference Φ_{CW}-Φ_{CCW}. If the modes have the same effective index and the beamsplitter BS does not change appreciably with time, then we may drop the last three terms in the expressions of Φ_{CW} and Φ_{CCW} and get:

$$\Phi_{CW} - \Phi_{CCW} = \Phi_S + \Phi_{m0} [\cos \omega_m(t-T/2) - \cos \omega_m(t+T/2)]$$

$$= \Phi_S + \Phi_{m0}\, 2 \sin \omega_m t \sin \omega_m T/2 \qquad (8.25)$$

Now we let $\Phi_0 = 2\Phi_{m0} \sin \omega_m T/2$ for the effective phase deviation impressed by the modulator. Note that the maximum of phase deviation, $\Phi_{0max}=2\Phi_{m0}$ is achieved for $\omega_m T/2=\pi/2$ or also $2f_m T=1$. The most efficient frequency of modulation is then $f_m=1/2T$, the inverse of twice the time of propagation through the coil [for example, with L=200 m it is $T=200/3 \cdot 10^8=$ 0.66 μs, f_m= 0.75 MHz].

With this position, the photodetected signal follows from Eq.8.24 as:

$$I_{ph} / I_{ph0} = 1 + \cos (\Phi_S + \Phi_0 \sin \omega_m t) \qquad (8.26)$$

To see the frequency components at ω_m and multiples that are contained in the signal of Eq.8.26, we can expand the second-hand term in series of ω_m using Bessel functions [26, 28]. In this way, we get:

$$I_{ph} / I_{ph0} = 1 + [\, J_0(\Phi_0) + 2 \textstyle\sum_{k=1,\infty} J_{2k}(\Phi_0) \cos 2k\omega_m t] \cos \Phi_S +$$

$$+ [2 \textstyle\sum_{k=1,\infty} J_{2k-1}(\Phi_0) \cos(2k-1)\omega_m t] \sin \Phi_S \qquad (8.27)$$

Thus, the photodetected signal contains all the harmonic components of the modulation frequency $\omega_m/2\pi=f_m$. The amplitude of the odd-harmonic components is proportional to sin Φ_S, the desired dependence linear near $\Phi_S\approx 0$, whereas the even-harmonic components are proportional to the cosine term.

By analyzing the photodetected signal with the aid of a lock-in amplifier, driven by the phase modulator waveform, we can filter out the cos $\omega_m t$ component. The amplitude of this component yields the sin Φ_S signal.

We may wish to maximize the scale factor $2J_1(\Phi_0)$ of the sin Φ_S signal. The maximum is obtained at Φ_0=1.8 and the value is $2J_1$=1.16 [28].

As we have let $\Phi_0=2\Phi_{m0} \sin \omega_m T/2$, we need a phase amplitude Φ_{m0}=0.9-rad from the phase modulator, once the sine factor has been maximized (sin $\omega_m T/2=1$).

In Fig.8-19 we report the schematic of the most commonly used FOG configuration, called the *all-fiber FOG*. This schematic is the realization of the conceptual scheme of Fig. 8-18, and has been the first demonstrated as a practical device capable of close-to-inertial performance [6,26]. The name all-fiber is because all the components in the device are made of pieces of fiber. In particular, we use a coil of fiber wound on the piezo tube to realize the modulator [29], and fused-fiber couplers to realize the beamsplitters. Also shown in Fig.8-19 is a fiber-based device for polarization control and filtering that is intended for a fine trimming to improve the zero bias and noise.

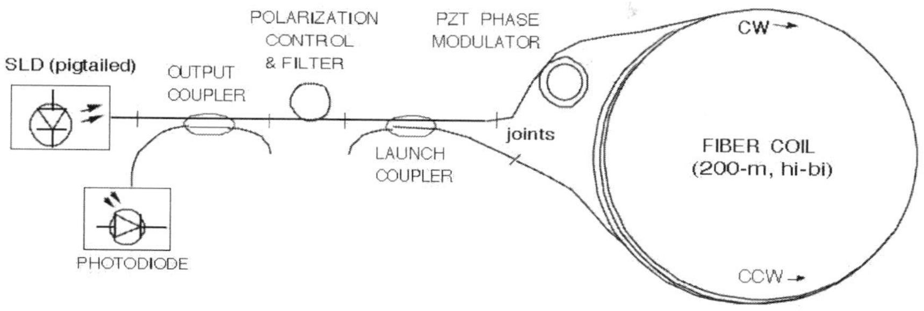

Fig.8-19 Basic scheme of the open-loop configuration of the FOG, implemented in all-fiber technology. Beam splitters are realized by fiber fused couplers, and the phase modulator by a coil wound on a PZT piezo ceramic tube. The fiber has high linear birefringence to minimize mode coupling. The source is a super-luminescent light emitting diode (SLED) with high radiance but low coherence length.

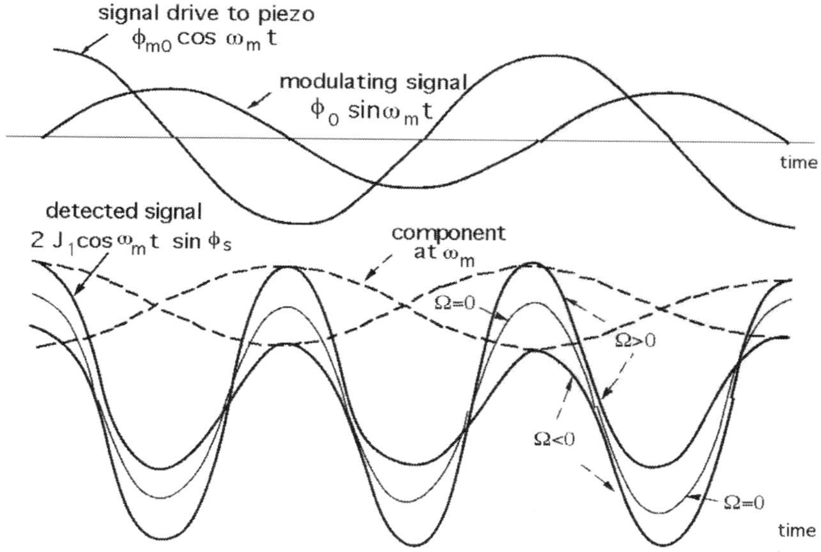

Fig.8-20 Waveforms of piezo drive and phase modulation (top), and of the detected signal I_{ph}/I_{ph0} (bottom). At rest (Ω=0), the signal is made up mostly of the second harmonic component, and its peaks have the same height. When the gyro rotates, for Ω>0, odd peaks are higher than the even, and vice versa for Ω<0. The signal at the fundamental frequency (dotted line) is in phase with the drive to the piezo.

The waveforms of the drive to the piezo, of the resulting phase modulation, and of the photodetected signal, are shown in Fig.8-20. At rest, only the second harmonic of the modulation is contained in the photodetected signal but, when $\Omega \neq 0$, the fundamental component shows up, in phase or antiphase with respect to the drive according that $\Omega<0$ or $\Omega>0$.

Fig.8-21 illustrates how the electrical sections of the open-loop FOG are interconnected. An oscillator provides the drive signal $V_{dr}=V_0 \cos\omega_m t$ to the piezo modulator, and it also feeds the reference-input of a lock-in amplifier. The measurement input is the photodetected signal, properly amplified by a transimpedance amplifier. The lock-in amplifier sorts out the components in-phase with respect to the reference. We retain that at the fundamental frequency, proportional to sin Φ_S. The second harmonic component is proportional to cos Φ_S and may also be used to introduce a linearity correction to signal sin Φ_S when Φ_S is not very small.

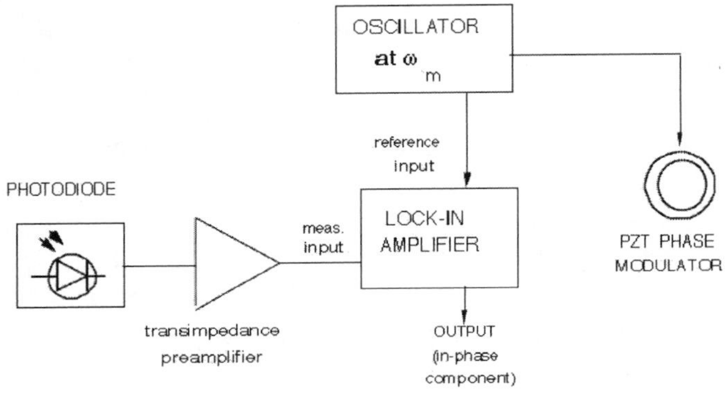

Fig.8-21 A lock-in amplifier is used to extract the output signal sin ϕ_S, proportional to the in-phase component at the modulator frequency ω_m.

8.5.2 Requirements on FOG Components

Let us now discuss the elements incorporated in the all-fiber FOG configuration of Fig.8-*P*19, and consider the requirements we need to achieve good performance.

Phase modulator. This device may be realized by winding a number (N=5-30 typ.) of fiber turns on a piezoceramic tube. The usual material is PZT (lead zirconate titanate), and typical size is a tube ≈30-mm diameter, ≈2-mm thick and ≈20-mm long [29]. When voltage is applied to the electrodes on tube walls, the tube strains by the piezoelectric effect, and the fiber is stretched to an increased length, say $L+\Delta L$. In addition, the optical path length changes because of the elastooptical effect, which changes the index of refraction by Δn. The two effects combine to give a total phase shift written as $\Phi_{m0}=k\Delta(nL)=k(L\Delta n+n\,\Delta L)$. Another expression is the form $\Phi_{m0}=\kappa NV$, showing that the phase modulation is proportional to the number of turns N and to the drive voltage V, besides a coefficient κ specific of the material. With the above size and N=10 turns, the phase shift is $\approx 2\pi$ (or $\approx\lambda$ in path length) for an applied voltage of a few volts [29].

The frequency response of the phase modulator is constant up to the first resonant frequency f_R of the mechanical structure. Typically, it is f_R=100-200 kHz (for the above size), but one can reach several tens of megahertz with linear (or slab) piezo elements [29], however at the expense of a reduced efficiency κ. To impart a non-reciprocal phase shift to the CW and CCW modes, the piezo phase modulator is placed asymmetrically with respect to the coil, as shown in Fig.8-19. Another possibility is to use an electro-optical material (typically lithium niobate, $LiNbO_3$) to build a chip that integrates a double modulator, as shown in Fig.8-19 (bottom right). The two sections of the modulator are connected to the two ends of the fiber coil, and are driven in phase opposition so as to double the resulting phase modulation. The chip can also integrate the input coupler, as shown in Fig.8-19 (bottom).

The fiber. This is the most critical component in a FOG. As already pointed out, we have the exacting requirement that the two counter-propagating waves shall see just the same optical path length or phase delay in opposite directions. Normally, in optical communications, nobody would doubt about the equality of path lengths in opposite directions, even through very long fiber trunks.

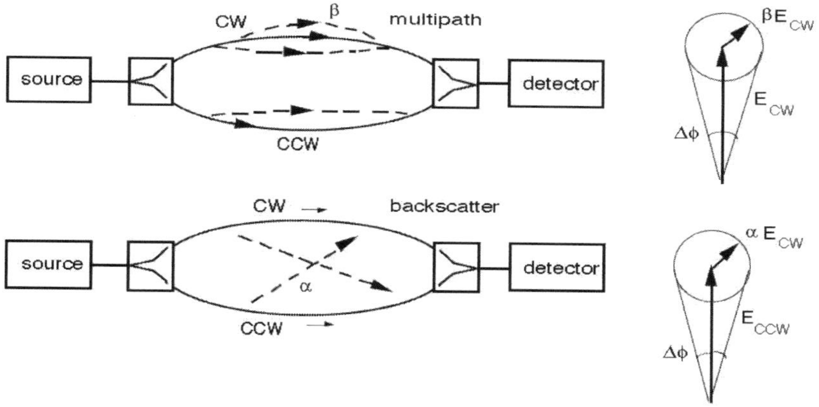

Fig.8-22 Phase errors are generated by multipath effects due to propagation with contributions from multimode or polarization diversity regimes (top). These effects are eliminated by working in single-mode, single polarization. Another phase error comes from very small back-reflections coupling the modes, as due to the Rayleigh scattering of the fiber (bottom).

However, when very minute differences become crucial, say the picometer or the microradiant as needed in a FOG, the matter changes. In other words, we need ensuring reciprocity of propagation among the two counter-propagating waves, as first pointed out by Ulrich [5].

In the fiber coil, reciprocity means that we shall avoid all sources of diversity in propagation, even minute. The first choice is a single mode fiber, to avoid any difference of mode mix in the two waves. In fact, different modes experience a different phase delay (Fig.8-22, top).

We can appreciate the consequence of this difference by considering a beam, mainly single mode, but with an added small fraction β of a higher-order mode. In this case, at the exit from the coil, the optical field would be the vector sum of a normally delayed contribution, E_{CW}, plus a minute fraction βE_{CW} with a different delay due to the higher-order mode.

On the fiber length of a FOG, this optical path length difference is very large (≈1000λ) and the delay spans the whole range 0-2π. We can represent the vector sum as in Fig.8-22, by a circle of uncertainty around the average position of the expected optical field E_{CW}.

The phase error is readily seen to be $\Delta\phi = \beta E_{CW}/E_{CW} = \beta$.

As an example of how severe the effect is, let us assume that our beam propagating down the coil has a spurious modal content of a bare 10^{-6} in power. Then, the field ratio is $\beta=\sqrt{10^{-6}}$ and the phase error is $\Delta\phi$=1 mrad, a very large value indeed, in our Sagnac-scale of phase shifts. Fortunately, not all the $\Delta\phi$ error contributes to noise. Should the modes have a very stable propagation constant, $\Delta\phi$ would also be a constant, and all would end with a zero-bias error, at most. Instead, because of fluctuations, $\Delta\phi$ has frequency components, and a fraction of it contributes to noise, actually. As a default estimate, about 10^{-2} of the total error may end up in noise, and this means that the phase noise is $\varphi_n=10^{-2}$ 1 mrad =10 μrad. Then, from Fig.8-5 we can read NEΩ≈5 deg/h in correspondence to the FOG dotted line.

The state of polarization also strongly affects the propagation constant, and we need a polarization holding fiber in the coil to get rid of phase fluctuations. This type of fiber, also called *polarization maintaining,* or hi-bi (high birefringence) fiber [31], is made by a structure similar to a normal fiber, but in the clad, stress elements are added. These elements run parallel to the core and their section is double circular (Panda fiber) or double wedge (bow-tie fiber) in shape. Because of the stress imparted to the core, a birefringence is induced by elas-to-optic effect. Mode propagation has then two different constants along the core slow and fast axis.

In consequence of this difference, phase matching between the two polarized modes is destroyed. This is also called mode decoupling, and it means that the effect of any distributed source of energy exchange is washed out.

Then, the factor β considered previously becomes much less than in an ordinary fiber (in practice, of a factor $\approx 10^{-3}...10^{-5}$).

From the discussion above, it is clear that, to avoid incurring a severe noise and zero bias spurious contribution, we shall carefully filter the beam launched into the propagation coil.

To this end, we use a filter (Fig.8-19) acting on the spatial distribution as well as on the state of polarization to approach the ideal condition of propagating the cleanest and conversion-resistant mode.

Actually, the fiber itself is a mode filter, when we use it at a normalized frequency V lower than the second-mode cutoff (V_{co}=2.405). Especially for low V (for example 1.3-1.7), we get a good rejection of higher modes eventually generated by geometrical and winding imperfections while keeping low the extra losses due to coil curvature.

The polarizer. Using the fiber itself as the spatial filter allows us to place a *polarizer* outside the coil for the polarization state filtering function. The requirements on spurious components not suppressed by the polarizer follow the same arguments as in the previous section.

A practical result is shown in Fig.8-23, and it clearly indicates that the best extinction ratio achieved in practice (≈50dB) is necessary to avoid degradation of the FOG performance.

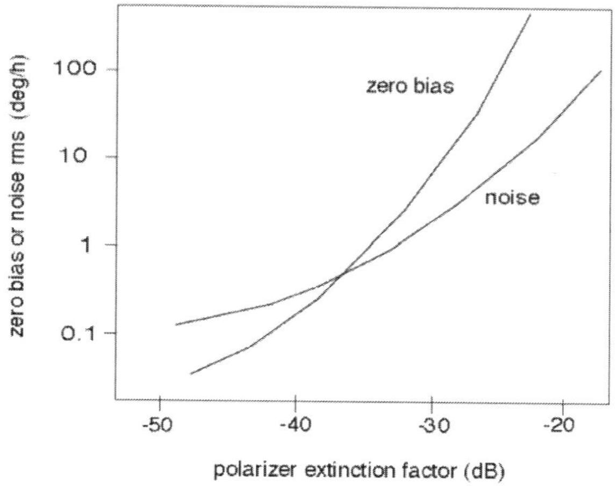

Fig.8-23 Typical experimental dependence, in an all-fiber FOG, of zero bias and noise from the extinction ratio of the polarizer as in Fig.8-19.

A fiber polarizer can be implemented [32] by winding a coil of a few turns of hi-bi fiber used near cutoff. Because the two polarizations have a different V parameter, the differential attenuation due to curvature can be exploited. On several turns of fiber, we can get a negligible loss (0.2 dB) for one polarization and a large loss (≈50 dB) for the other [32].

Backscattering effect. Another source of error associated with the fiber is cross-coupling between the counter-propagating waves (Fig.8-22 bottom). In the fiber coil, this is caused mainly by *Rayleigh scattering*. As in any material, also in the glass of the fiber, we find minute density fluctuations that act as a scattering center. At wavelengths outside absorption bands, scattering is the main source of loss in the fiber.

The Rayleigh loss is about 3 dB/km at λ=850 nm, the wavelength preferred for the all-fiber FOG operation. This figure amounts to a backward coupling α, between the CW and CCW modes, of about –55 dB/m. To bear the full extent of Rayleigh scattering on a 200-m (or +23 dB re 1-m) fiber, we should limit to (-55+23) =-32-dB or 1/1600 the power transferred from one mode to the other. Again, as calculated above for the multimode case, we would have a phase error of about $\Delta\phi=\alpha E_{CW}/E_{CCW}=1/\sqrt{1600}=25$ mrad, having taken $E_{CW}\approx E_{CCW}$. Even considering that the effective noise is only a fraction of the $\Delta\phi$ error, the value is so huge that it would make the FOG definitely useless.

The remedy is in avoiding that backscattered contributions can add up to the field in transit. The assumption by which they add is that superposition is coherent. As the backscattered contribution has run a path different from the path of the field on which it superposes, to have addition, the coherence length of the source shall be larger than this path difference. Now, let us decrease the coherence length to a small value, say of the order of l_c<<1 mm, instead of the large value 100-1000 m of frequency stabilized lasers.

Then, only the backscatter arising from locations in the fiber coil that are distant for less than l_c will contribute. The point where the path-length difference is zero is the midpoint of the fiber coil. With a coherence length l_c, only a length l_c of fiber contributes to backscattering. Thus, if we have l_c= 50 μm (=1/20.000-m or –43 dB re 1-m), we get as the Rayleigh scattering a value (-55-43) = -98 dB, now a safe value. Of course, the requirement of slashing the Rayleigh scattering is now transferred from the fiber to the source, which will be a low-coherence one.

Back-reflected contributions are also found in components external to the fiber. All joints (Fig.8-19) between the individual components are potential sources of back-reflection. The return loss (RL) is the power ratio of back-reflected power to input (forward going) power. Typical RL figures of good joints made by fusion splice are –50 dB for normal joint, -70 dB for angled joints. Considering that the field coupling factor is half the power factor, the previous figures would yield $\Delta\phi$=3 and 0.3 mrad, respectively. These are again unacceptable values but, fortunately, they are cured the same way the fiber backscatter is, i.e., by a low coherence source.

The source. Specially developed for the FOG is the *super-luminescent* LED or *SLED*. This device has a structure identical to that of a normal Fabry-Perot diode laser (App.A1.3) of comparable power level, but without the output mirror. In fact, after cleaving the chip out the wafer, the normal operation that creates the flat mirrors of the cavity like in a normal diode laser, for the SLED we coat the output facet with a single or a multiple anti-reflection layer. Thus, the reflectance of the facet may go down to a residual 10^{-4} from the initial value $(n-1)/(n+1) \approx 0.3$ of the Fresnel interface.

When the diode is driven by a bias current, the laser threshold is never reached. The emitted radiation is nevertheless different from a normal LED. Because the guiding structure provides confinement and the material is pumped to population inversion, the spontaneous emitted radiation is optically amplified. Thus, the radiance (power per unit surface and solid angle) of the emitted light is much larger than that of an LED and is comparable to that of a laser. This feature is very important in the FOG, to be able to launch power in a single mode fiber with a good efficiency. On the other hand, radiation is amplified spontaneous emission, not a coherent oscillation. The emission line width is about the same as the LED, that is $\Delta\lambda$=25–50 nm (Fig.8-24), and the corresponding coherence length $l_c = \lambda/\Delta\lambda$ is ≈ 20–40 nm, a nicely small value as we wish for the FOG.

A problem with the SLED is back-reflections fed back into the source. Back-reflections originate at components and down the coiled fiber. Those at components are minimized specifying an appropriately small return loss. Back-reflection from the coiled fiber is unavoidable, however, and not so small (≈ -32 dB for a 200-m fiber, as in the previous example).

As we can see in Fig.8-24, even small reflections distort the shape of the emission line, eventually resulting in a SLED breaking into oscillation. As the line shape is progressively distorted, coherence length increases back and an excess amplitude noise is found in the emitted power. Both effects contribute to increased noise and zero bias error of the FOG.

To prevent back-reflected light from reaching the SLED chip, the ideal remedy is to insert an optical isolator at the output, but optical isolators are not easily available at the wavelength ($\lambda \approx 800$ nm) of the GaAlAs source. A way out may be moving to another more favorable wavelength, 1500-nm for example, where optical isolators based on YIG crystal [33] are readily available. In this case, we may use as a source the output from an Er-doped EDFA.

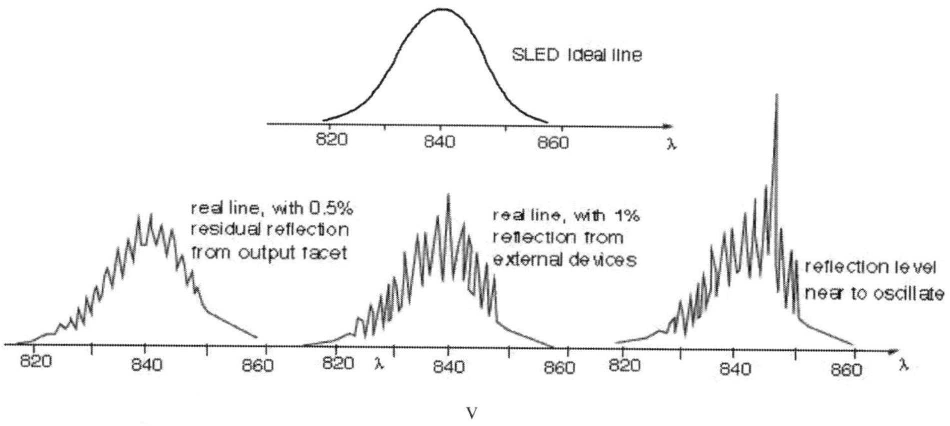

Fig.8-24 The source used in the FOG is the super-luminescent LED (or SLED), because it has a broad spectrum and thus a short coherent length, together with a small emitting area for high efficiency of coupling to fiber. If the output mirror has no residual reflectance R, the SLED has the ideal line (top), but at increasing R the spectral distribution becomes distorted (bottom). Distortion causes increased noise and zero bias error of the FOG.

At 800-nm, another way to reduce back reflection [34] is using a circular polarizer at the SLED output. The circular polarizer is in practice realized by cascading a linear polarizer and a $\lambda/4$ wave plate, both made of fiber pieces, in the all-fiber FOG. As circular polarization changes handiness when it returns upon reflection or from the Sagnac loop output, all contributions are blocked out when they return at the polarizer and $\lambda/4$-plate combination. Only scattering contributions undergoing a polarization change are unaffected, and this is a minor fraction (estimated in 10^{-3}) of the total back reflection.

Fiber couplers. For the all-fiber FOG, couplers can be fabricated with the fusion-splice or the lapping technology [35,36]. We may want to fabricate them out of the same hi-bi fiber in the coil, and then lapping technology is better suited because it disturbs the fiber structure less, whereas using a normal fiber, the fusion-splice technology is superior.

The fiber couplers of the FOG, and especially to the launch coupler, need a very stable splitting ratio, as close as possible to 50%, to get an interferometer signal with a nearly 100% visibility factor V (Eq.4.2). In addition, we need an exceptionally low excess loss ($\approx$ 0.01 dB) and the highest return loss (better than $\approx$60 dB) because these values impact the noise and zero bias error as discussed in the previous pages.

All these figures are rather demanding in the common practice of the field. However, they can be obtained with careful control of the fabrication process [36].

Photodiode receiver. This part presents no criticality. The standard amplifier based on the transimpedance circuit [30] is adequate as a front-end of the photodiode and provides quantum-limited performance already at low detected currents (say $\approx$ 0.01 to 1 μA).

Other sources of error. Initially, when noise and zero bias of an all-fiber FOG were measured and compared to the theoretical quantum limit values (Fig.8-6), a disappointing discrepancy of as much as 1.5 to 2 decades was found. Research has been actively carried out [7, 38] to identify the source of discrepancy, in the hope that removing or greatly reducing extraneous disturbances, an inertial grade performance could be finally achieved.

Three main effects about the fiber that have been studied are: (i) polarization mode dispersion (PMD); (ii) cross-phase modulation (XPM), and (iii) thermodynamic phase noise (TPN). All these contributions affect the term n_{CW}-n_{CCW} in Eq.8.23.

The *PMD* deals with the propagation time dispersion δ_P, a quantity that translates directly to phase delay dispersion. The PMD in a normal fiber is typically $\delta_P \approx 0.05$ ps/$\sqrt{}$km, a figure associated to a birefringence beat-length of $\approx$100 m. The corresponding phase dispersion is $\Delta\phi_P = (2\pi f)\, \delta_P \sqrt{L} = 2\pi\ 400\ 10^{12}\ 5\ 10^{-14} \sqrt{0.2} = 55$ rad for a L=200 m fiber length at λ=800 nm (at which f$\approx$400 THz). The main part of $\Delta\phi_P$ is deterministic, and we can estimate that only a fraction, $\approx 10^{-3}$ of it, or 0.05 rad, may contribute to the random fluctuation of phase. In addition, because $\Delta\phi_P$ is read out by two counter-propagating beams, a further reduction (estimated in $\approx 10^{-4}$) is obtained. The final phase noise that can be ascribed to PMD dispersion is $\approx$0.005 mrad, equivalent to a $\approx$3 deg/h noise in a typical FOG.

The *XPM* is due to the optical Kerr effect, by which the index of refraction weakly depends on the light intensity of the beam in transit through the medium. In a Sagnac interferometer, the $(n_{CW}-n_{CCW})L$ path length difference is proportional to the nonlinear $\chi^{(3)}$ coefficient of silica, and to the difference P_{CW}-P_{CCW} of counter propagating beams powers. For a L=200-m coil, the XPM contributes to an error of $\chi^{(3)}L(P_{CW}-P_{CCW}) \approx$ 10 (deg/h)/mW. This requires carefully balancing the power of beams launched in the opposite directions of the fiber coil. If we are able to keep the unbalance down to $P_{CW}-P_{CCW}<0.1$ μW, we get a negligible noise, down to $\approx 10^{-3}$(deg/h).

Thermodynamic phase noise (*TPN*) has already been discussed in Sect.4.4.5. The TPN contribution can introduce a large noise, up to $\approx$10 deg/h [36] in general. However, in the Sagnac interferometer, there is a strong cancellation effect due to the counter-propagation readout, so the effective error or noise can go down to $\approx 10^{-2}$ deg/h.

In conclusion, in the FOG there are several effects, each of which may well justify the excess noise found in experiments. However, after all the above causes were under control, with their effects seized down to the $\approx 10^{-3}$ deg/h level, a large unexplained disturbance of the zero bias and an excess noise remained, up to about the year 2000 (Fig.8-17), when the attention of researchers was attracted by an unassuming, 1-page Letter to the Editor of *Applied Optics* [7] by D.M. Shupe, who proposed the thermal transient in the fiber as the missing effect spoiling performance.

8.5.3 The Shupe Effect

We have seen that a reciprocal phase modulator, placed asymmetrically with respect to the coil midpoint, introduces a non-reciprocal phase shift because it is crossed by the CW and CCW modes in different times. The same happens for a thermal disturbance, and this is the Shupe effect [7]. Indeed, a temperature fluctuation ΔT affecting an edge of the fiber coil (see Fg.8-25) gives rise to a phase shift variation $\Delta\Phi = (2\pi/\lambda)L(\alpha + dn_c/dT)\,\Delta T$, where L is the fiber length, α the expansion coefficient of the fiber, and n_c the core index of refraction. If temperature changes in time as $\chi = \Delta T/\Delta t$, the apparently non-reciprocal phase shift induced by the thermal transient is:

$$\Phi_{CW} - \Phi_{CCW} = (\Delta\Phi/\Delta t)\,\chi\,\tau = (2\pi/\lambda)L(\alpha + dn_c/dT)\,\chi 2\,L/c \qquad (8.28)$$

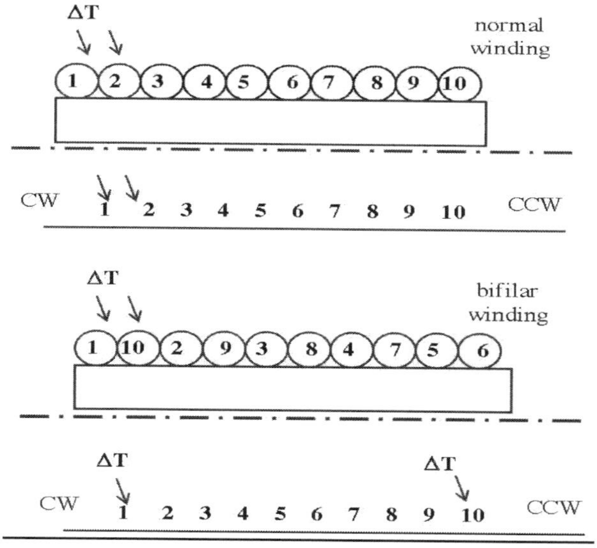

Fig.8-25 With a normal winding of a FOG (top), a thermal transient ΔT affects adjacent turns of the coil, in particular, the first ones seen by the CW and CCW modes at different times. By a bifilar winding (bottom), the transient ΔT equally affects the initial and final turns and is canceled out.

By putting numbers in Eq. 8.28, Shupe [7] evaluated that the maximum permissible ΔT has to be 0.006 °C to stay in the inertial performance grade, a requirement difficult to satisfy. Rather, we can wind the fiber coil in a new way, such that the extreme (first and last) fiber turns lay close to each other, and therefore minimize time τ in Eq.8.2.8. This is exemplified by the *bifilar winding* shown in Fig.8-25. By winding the spool with the fiber folded back on half of its total length, that is on a bifilar arrangement, we get the situation of Fig.8-24 (bottom), where the beginning and the end of the coil receive the same perturbation ΔT and cancel it out [38]. Even better, if we wind the fiber in a four-row (or quadripolar) fiber arrangement (not shown in Fig.8-24), it finally reaches the inertial grade performance (Fig.8-17) [39]. A special winding is also useful to remove another source of disturbance, acting similarly to the thermal disturbance, i.e., the microphonics (called Shupe mechanical disturbance) collected from the ambient and from transient mechanical stress imparted to the fiber coil through its own package. Both shearing and radial components can be counter-balanced by a suitable winding of the fiber coil, making the FOG nearly immune to mechanical stress.

8.5.4 Technology to Implement the FOG

The most popular technology for the FOG, the *all-fiber*, is used to designate a FOG in which all elements are made of pieces of fiber, properly worked or exposed to an external perturbation giving the desired functionality.

There are two more technologies to implement the FOG: the micro-optics and the integrated optics. The three technologies are also shared by all optical fiber sensors (Ch.9).

The *micro-optics* technology consists of using conventional optical components connected by free space propagation. This technology offers the widest choice of components to assemble in the device, but the free propagation in between components is the bottleneck. We need objective lenses to launch into the fiber, or relay lenses between components. As we attempt to develop a design, it soon becomes clear that launch and mechanical mountings are critical, bulky, and extremely sensitive to external mechanical disturbance.

This is a critical issue, because mechanical stability should be guaranteed over the full useful life of the sensor. Therefore, though used in the first developed prototypes of FOG, micro-optics is generally not a preferred approach for a device deployed in the field. It is just helpful to carry out experiments in the laboratory.

The *Integrated Optics* (IO) technology aims to put all required functions together, that is, to integrate the individual devices in a single chip. On the substrate of the chip, we may perform functions like (i) passive functions (SoS or silica on silicon), (ii) passive and modulation functions (lithium niobate or $LiNbO_3$), and (iii) active functions (compound semiconductors like GaAs and InP). Generally, the cost of facilities for an IO technology is high, so we may use it only for substantial volume of production to share the fixed cost on a large number of pieces. Unfortunately, applications with large volumes are scarce, both in the sensor and in the communication segments.

Another point of concern with IO is coupling of the chip to the fiber pigtails we shall make available to the user for accessing the device. This operation, referred to as *pigtailing*, is a critical manufacturing step, because we shall align the fiber core and the guide to within ≈0.1 μm and shall ensure that alignment remains good throughout the useful life of the device.

In addition, the fiber joint usually introduces a substantial pigtail loss. This is because of the differences in size, shape, and index of cores even when the residual alignment error is small. With a normal single-mode fiber, the loss may be as low as 1-3 dB per joint when the waveguide is not so different from the fiber (as with SoS and $LiNbO_3$).

On the other hand, the loss may go up to 6 to 10 dB per joint with compound semiconductors, which have a shape (a rectangle with high aspect ratio) and a high index of refraction much different from the fiber.

Examples of IO gyros have been reported in the literature [41]. The gyro closest to a fully integrated FOG is a 3×3 version (see Sect.8.7) entirely realized on an SoS (silica on silicon) substrate, a rectangle of about 30×50 mm^2 (Fig.8-26). The chip includes the coupler and a spiral-shaped propagation guide, totaling a few meters of optical path.

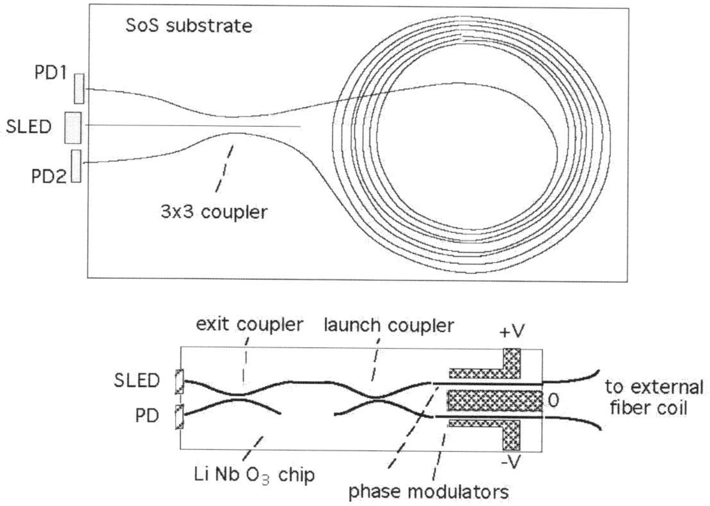

Fig.8-26 Examples of IO technology for FOG: (top) a fully integrated, minimum-configuration FOG with 3×3 coupler in SoS; (bottom) a lithium niobate chip implementing the splitting and phase modulation functions of a normal open-loop or closed-loop FOG.

The chip is butt-coupled to a SLED and two photodiodes. The device was intended for the low-performance end (Fig.8-1) of the gyro range.

Not a fully IO device, but one incorporating all the components external to the fiber, an electro-optical chip in lithium niobate has been used by several manufacturers of the FOG. The chip includes a coupler for source and detector, and two symmetric phase modulators

(Fig.8-26, bottom). Waveguides are obtained by photolithographic patterning of the chip, followed by a Ti in-diffusion to increase the index of refraction.

The phase modulator is simply realized by a waveguide with a pair of electrode stripes running parallel to it. Applying a voltage to the electrodes, the index of refraction of the waveguide is changed because of the electro-optical effect, and a phase $\Phi=\Delta nL$ is generated. The phase modulators outputs are pigtailed with the usual coil of ≈200-m long hi-bi fiber. To get the nonreciprocal phase modulation due to time-asymmetry, the phase modulators are driven with opposite-polarity waveforms.

Called the FOG with integrated chip or *iFOG*, the device has performances equaling those of the best all-fiber FOG. Performances are obtained with high (≈95%) yield on a production line, not just in the laboratory or after individual trimming. Noteworthy, the iFOG starts from a lower level of received power, typically -10...-15 dB less than in an all-fiber FOG because of the fiber-to-chip joints, but recovers the performance of zero bias and noise in virtue of the cleaner and stable setup.

Another approach based on IO technology aims to implement the RLG (ring-laser-gyro) concept on a semiconductor material like GaAlAs [42]. On a semiconductor chip, the enclosed area A shall be small (≈0.1 cm^2) if we aim to get an inexpensive sensor. Therefore, the potential applications of the *IO-RLG* are limited to the low-end sensitivity, like those required in the field of automotive and robotics (see Fig.8-1).

Work is in progress to demonstrate the feasibility of the concept, despite a variety of phenomena observed in semiconductor ring-lasers [43], like locking, self-pulsation and alternate oscillation, that have precluded so far to exploit the Sagnac effect.

8.5.5 The Closed-Loop FOG

The open-loop configuration of Fig.8-19 has a fair performance of linearity and dynamic range. Linearity is good at small ϕ_S, but at increasing ϕ_S is limited by the sin ϕ_S dependence of the signal (Eq.8.27).

Additionally, the maximum range of the Sagnac phase is $\phi_S<\pi/2$ if we are to avoid the sine-function ambiguity. By applying a post-distortion of the type y = arcsin x to the signal x= sin ϕ_S we can get a sizeable improvement in linearity, but the dynamic range is unchanged. To improve both linearity and dynamic range, we shall process the pair of signals sin ϕ_S and cos ϕ_S. The two signals are available in the open-loop gyro of Fig.8-19. As we have seen from Eq.8.27, sin ϕ_S and cos ϕ_S are contained in the amplitudes of the odd and even harmonics, respectively.

Thus, we can measure them with the lock-in amplifier as the amplitudes at frequency ω_m (proportional to sin ϕ_S) and at frequency $2\omega_m$ (proportional to cos ϕ_S), and process signals sin ϕ_S and cos ϕ_S in a number of ways, as seen in Chapter 4. Thus, we are able to compute ϕ_S also when it exceeds the $\pi/2$ ambiguity angle, and to reduce the linearity error. Typically, we may reach the 10^{-3} (or 1000 ppm) level of residual error on a range of several 2π.

Yet, this level may not be sufficient for the inertial applications, which call for a 10- to 100- ppm linearity error. In this case, we employ the *closed-loop* configuration.

The basic idea of the closed-loop configuration is that of keeping dynamically zero the total phase shift read by the gyro. We null the gyro by adding a phase shift Φ, equal and opposite of Sagnac phase ϕ_S (Fig.8-27). In this way, the lock-in works around zero at all times, and the linearity error of the sin ϕ_S dependence is eliminated. The element providing the phase shift Φ shall be a nonreciprocal phase modulator and is discussed below. The condition ϕ_S-Φ≈0 is ensured by the feedback loop, which is designed to have a large negative gain.

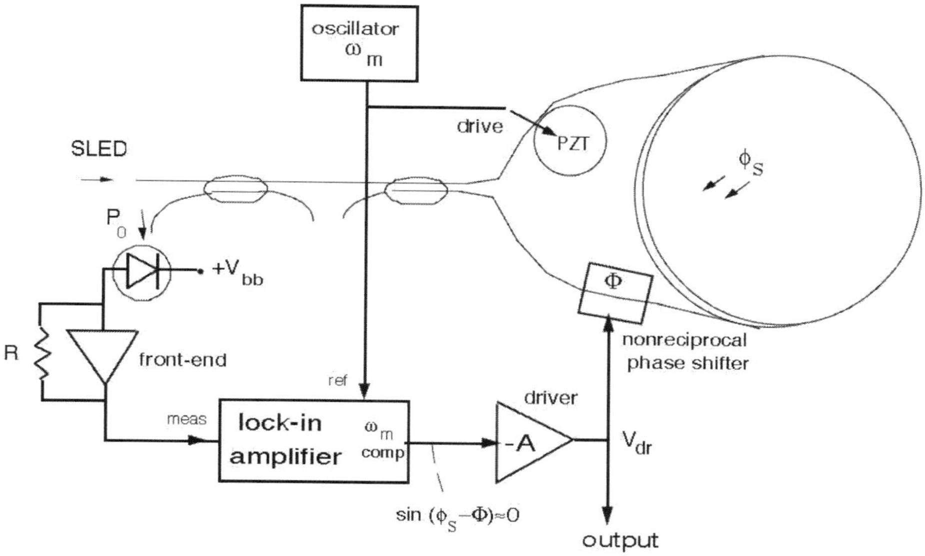

Fig.8-27 In the closed-loop FOG, the output of the lock-in amplifier is fed back to the optical path after amplification and conversion to a nonreciprocal phase shift Φ. The phase shift Φ is obtained by driving a suitable modulator with the amplified voltage V_{dr} of the lock-in output. Because the loop gain is large and negative, the phase difference is kept dynamically zero by the feedback loop, or $\phi_S-\Phi=0$. The output is then read from the drive voltage V_{dr} as $\phi_S=\Phi=f(V_{dr})$, where f is the transfer characteristic of the phase shifter.

We can check the negative feedback action by going along the loop in Fig.8-27. Let Φ suffer a negative variation, say, $-\varepsilon$. Then ϕ_S-Φ becomes positive ($=+\varepsilon$), the signal out of the front-end increases, and the lock-in output increases too, and finally we return to the starting point with a phase Φ increased by $+G\varepsilon$. As the counterpart of the electronic follower circuit, we get a feedback loop suppressing any deviations from the nominal ϕ_S-Φ=0 condition, by a factor equal to the loop gain G. By inspection of Fig.8-27, the loop gain is evaluated as:

$$G = 1.16\,\sigma\, P_0\, R\, A_{lock\text{-}in}\, A\, \kappa \qquad (8.28)$$

The symbols used in Eq.8.28 are as follows: σ=I/P = spectral sensitivity of the photodiode, P_0 =received power, R = transimpedance of the front-end, $A_{lock\text{-}in}$ = lock-in gain at the first harmonic, $\kappa=\Phi/V_{dr}$ is the modulator efficiency, and 1.16 = $2J_1$ is the modulation factor in Eq.8.28. In the closed-loop FOG, the output is voltage V_{dr}, and we go back to the Sagnac phase using the modulator transfer characteristic, $\phi_S = -\kappa V_{dr}$.

In doing so, we have moved from the sin ϕ_S dependence to the new dependence $-\kappa V_{dr}$. Now, we need a good linear transfer-characteristic of voltage to phase, or a constant κ on the whole range of Sagnac phase ϕ_S. Early attempts [25] employed a double-pass Bragg cell as the nonreciprocal modulator, but the setup was involved, and the linearity not satisfactory. The Faraday effect was not much better. Both approaches also suffered from criticality, in avoiding spurious nonreciprocal phase shifts that are inadvertently introduced.

Finally, it was just the asymmetrical phase modulator to be recognized as the best linear element capable of supplying the desired nonreciprocal phase shift. Because a PZT phase modulator is already there for the readout of the Sagnac phase, we may think of using it also for the phase null operation, adding the two signals sent to the PTZ drive.

Alternatively, we may introduce a second PZT located symmetrically with respect to the fiber coil, as in Fig.8-27. Having the two signals on separate PZTs, each with its own requirement of linearity to satisfy, leads to better linearity over a wider range of applied signals.

The phase-null PZT shall be driven by a periodic sawtooth signal. This waveform is our best effort to approximate a phase ψ=H t steadily increasing with time. As we cannot increase the phase of a modulator indefinitely, we use the sawtooth signal. During the linear ramp period, the signal has the form s(t) = H t (for t=0..τ), and then has a fast flyback to zero, in a time interval $\tau_{fb} << \tau$.

As the initial phase of the sawtooth is insignificant to the measurement, we get the equivalent of an indefinitely increasing ramp, provided we gate out the measurement during the flyback period. When we add the sawtooth phase, the nonreciprocal phase shift $\Phi=\Phi_{CW}-\Phi_{CCW}$ suffered by the two counter-propagating waves can be readily found by repeating the reasoning leading to Eq.8.27. If T=nL/c is the fiber coil transit time, the result is found as:

$$\Phi = H(t+T/2) - H(t-T/2) = H\,T \tag{8.29}$$

Eq.8.29 tells us that the quantity affecting the generated phase is the sawtooth slope. Therefore, we will use the voltage V_{dr} to drive a voltage-controlled ramp generator (not shown in Fig.8-27), whose output is the actual signal sent to the phase modulator. Thanks to the feedback loop, it is now the slope to be clamped at the correct value to null $\phi_S-\Phi$.

The above configuration is an *analogue* closed-loop FOG because of the analog processing of signals. The sawtooth arrangement is also referred to as a *serrodyne modulation* [45-47] and it was the first demonstration of the FOG potentiality in approaching inertial grade performance. The analogue processing is objectionable because, if we finally convert the FOG signal to digital for navigation calculations, we shall pass through the non-idealities of an A/D (analog to digital) converter.

Then, we look for an intrinsically digital readout of the Sagnac phase, while retaining the benefits of the closed-loop approach. The solution is shown in Fig.8-28 and is called the *digi-*

tal closed-loop FOG [7,47]. It employs a multistep sawtooth modulation, also called digital ramp because of its function.

Each step increases the phase by an increment $\Delta\phi$ equal to the least-significant-bit (LSB) we want in the FOG readout. Simply stated, if the phase modulation changes by M steps or $\Phi = M\,\Delta\phi$ in the transit time through the fiber coil, the Sagnac phase is read out as $\phi_S = M\Delta\phi$.

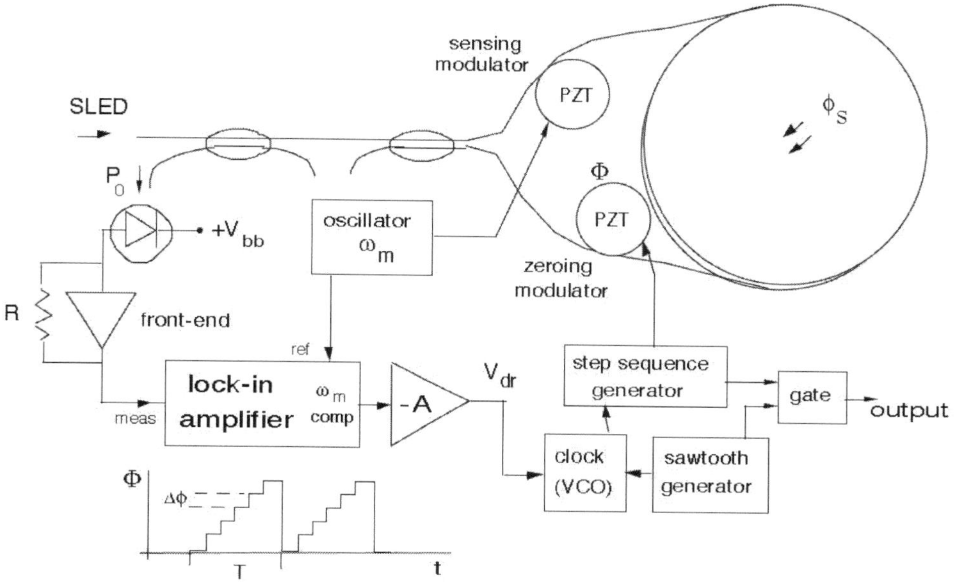

Fig.8-28 In the digital-readout closed-loop FOG, a digital (or multistep) ramp is used to drive the nonreciprocal Sagnac-zeroing modulator. Each phase step corresponds to one LSB of the readout.

To drive the PZT modulator, we shall convert the lock-in output voltage V_{dr} to a digital ramp. In the digital ramp, all the steps have the same amplitude, and their frequency (or inverse of the time spacing) shall increase as the lock-in output V_{dr} increases. These functions are readily implemented by the electronic circuits shown in Fig.8-28. A VCO clock fed by V_{dr} provides the pulse frequency, whereas the saw-tooth generator gives the underlying ramp waveform and provides the synchronism to validate the output through a gate.

The digital readout has the additional advantage of allowing an easy increase of resolution (App.A0) by averaging, as we now will illustrate. As a start, the LSB phase increment is chosen as a bare $\Delta\phi$=0.1 mrad, apparently a small value. Yet, the angular velocity that corresponds to ϕ_S=0.1 mrad is not at all small, typically about Ω=100 deg/h (see Fig.8-5).

Now, suppose that each measurement takes, say 0.1 ms. We can acquire 10^4 of them in a 1-s measurement time. Summing them all up, the result has a truncation error $\sqrt{10^4} = 10^2$

times smaller than the original truncation error of the individual measurement. For this statement to be true, we only need that the random noise superposed to the measurement is larger than 1 LSB. Thus, by averaging, we are able resolve Ω=1 deg/h.

As a concluding remark about the digital closed-loop configuration, the frequency pass band required for the phase modulator is not simply ≈1/T of the open-loop readout, but probably 20–50/T because of the ramp and multistep features that we need to preserve. This leads to preferring the integrated-optics approach based on the lithium niobate modulator (Fig.8-26) for implementing the digital closed-loop [7,47].

In conclusion, after many years of maturation, the FOG has evolved to probably its best level of technical appeal as a sensor with the *digital closed-loop* configuration discussed in the previous pages. The internal configuration counterpart of the FOG, that is, the RLG (Fig.1-8), has won an indisputable position in the field of application of INUs and AHRSs in avionics. On the other hand, the *digital closed-loop* FOG has reached good acceptance in some space applications (AHRS of low-orbit satellites, and gyro-compass) and other niches (oil drilling, rocket spin stabilization).

8.6 The Resonant FOG and Other Approaches

The research on the FOG is unsettled and has triggered a number of international R&D laboratories to pursue several alternative approaches, aimed to take to the stage of inertial-grade gyroscopes. A configuration that appears promising is the R-FOG, or *Resonant Fiber-Optic-Gyroscope* [48].

The basic idea is to use the fiber coil not just for propagation, but as an element to build a resonator. The resonator is the ring we obtain when input and output of the launch coupler of the basic FOG are interchanged.

The result is a long path-length resonator (Fig.8-29), which potentially can reach a high value of finesse F (or cavity Q-factor). As pointed out in Appendix A2.1, the ring resonator has a response curve much sharper than the normal Sagnac interferometer.

If we look at Fig.A2-2, we see that the slope of the relative power versus 2ks=ϕ curve is steeper than that of the normal sin 2ks response. We have an improvement of slope or responsivity by a factor F, the finesse of the resonator (Table A2-1). We can then achieve from the R-FOG the same responsivity of a normal FOG just using a fiber F times shorter than that of a FOG. As it is (Eq. A2.4) $F=2\sqrt{R}/(1-R)$, where R is the (power) through transmission of the coupler, with the reasonable choice R=0.99, we can easily get F=200. Thus, instead of L=200 m, we can work with a bare L=1 m in a R-FOG with the same signal.

The saving of fiber length and associated cost is not that important, however. Much more important is the strong decrease of non-idealities we get with the short fiber.
Paradoxically, we can get better performance by a shorter fiber in the ring than from the more responsive full-length fiber. In a certain sense, shortening the fiber moves the FOG near to the RLG and its performance closer to the quantum limit.

The readout of the R-FOG is performed by one or more photodetectors sensing the outputs tapped by the launch coupler or the result of their recombination.

The PZT modulator applies a sine phase shift like in the normal FOG. In addition, it also serves to lock the working point at the peak of the ring resonance, by applying a dc bias that dynamically tracks the resonance.

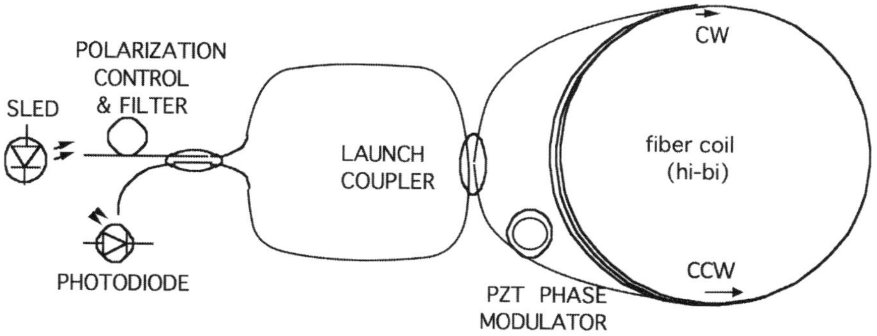

Fig.8-29 The R-FOG uses the fiber coil as a ring resonator. The PZT phase modulator keeps the total path length locked to the resonance peak, and senses the Sagnac phase shift by an added ac modulation of small amplitude.

Performance of the R-FOG started from a relatively unsatisfactory level, but rapidly caught and marginally surpassed the best figures provided by the conventional FOG.
Another interesting approach, following the advent of the EDFA (erbium doped fiber amplifier) was using this fiber as the source of the FOG. A coil of doped fiber, typically ≈10...30 m long, is optically pumped, using a ≈ 100-mW laser-diode, to population inversion.
The pump laser emits at an absorption band of the active Er ion, and the two most used wavelengths are 980 nm and 1480 nm.

The active transition of the Er ion sustains amplification, and eventually oscillation, in the well-known 3rd window of fibers, in the range λ=1520...1580 nm.
Thus, we may try to fabricate an EDFA-based version of the internal-configuration Sagnac-interferometer, that is, the fiber counterpart of the RLG. Once more, the experiment reveals that the fiber is a medium much dirtier than a He-Ne gas mixture, optically speaking.

Thus, instead of a clean narrow-line oscillation, the EDFA-based RLG has a broad line, jittering in wavelength because of longitudinal mode competition and of backscattering down the fiber. In addition, the counter-propagating modes are strongly locked by the much larger backscattering in the fiber. In view of these problems, the EDFA-based fiber laser is not likely to become the solid-state replacement of the He-Ne laser for an RLG configuration.

Yet the EDFA can be useful for application in the external-configuration FOG.

The pumped fiber is a source of *super-luminescent* radiation. This radiation is just the ASE (amplified spontaneous emission) supplied by the fiber [49], and we can use it as the source of the FOG in place of the normal semiconductor SLED.

The advantage of such *S-EDFA* is that the fiber is antireflection coated to a much better extent than a semiconductor facet, whence a smoother shape of the line. Second, we can protect our S-EDFA source from back reflections originated by the returned signal simply with the use of an optical isolator, readily available at $\lambda \approx 1550$ nm.
Actually, moving from 850 to 1550 nm about halves the response of the FOG, but the better control of the source more than compensates for the loss.

An improvement of about an order of magnitude in the NEΩ (down to a few 0.001 deg/h) has been reported for the S-EDFA i-FOG.

8.7 THE 3X3 FOG FOR THE AUTOMOTIVE

Another appealing market for the gyroscope is the automotive segment [50, 51]. Some requirements of the automotive are also common to the robotic market (Fig.8-1), and this circumstance adds value to a low-performance application with a potentially large volume.

A feature of the automotive application is that volumes are huge, but price shall be very low, even at a sacrifice of performance. Price comes first in the automotive segment, and only matching the price can you go ahead and talk of performance. Instead of the US\$ ≈5,000 to 10,000 of the best FOG, the automotive FOG calls for a price of just ≈ US\$ 10 to 20.

To try matching this severe price specification, we shall think of a very simplified configuration, with the minimum part-count, yet capable of reaching a respectable performance, say ≈ 0.1 deg/s or 360 deg/h.

Although we should abandon the reciprocal configuration because it requires too many components, we shall solve the cos ϕ_S ambiguity of the normal FOG, ensuring that the signal supplied by the sensor is of the form sin ϕ_S.

An alternative to the use of the phase modulator is that of employing a *minimum FOG* configuration based on a *3×3 coupler*.

This 3×3 configuration has been proposed [52] since the early times of the FOG as a solution to get the sin ϕ_S dependence, but it was soon abandoned because it is not inherently reciprocal. In the minimum configuration FOG, we clear the reciprocity requirement and take advantage of the drastic component-count reduction, so the 3×3 FOG becomes very appealing. About the 3×3 coupler, we rely on the phase shifts introduced between the input and output ports to obtain the sine dependence. To understand how, let us assume that the coupler is made by three fibers with the core sections symmetrically placed (Fig.8-30).

Then, coupling from one fiber to each of the other two fibers shall be identical, for symmetry reasons. We shall now apply the conservation of electric field vector as well as of the power (square of field amplitude), generalizing the well-known 2×2 coupler results (see for example Ref. [16], p.294).

Entering the coupler with E_0 at port 0 (Fig.8-30), the output vectors E_{1C} and E_{2C} at outputs 1 and 2 shall be parallel because of symmetry, and both shall be orthogonal to the output vector E_D of output 0.

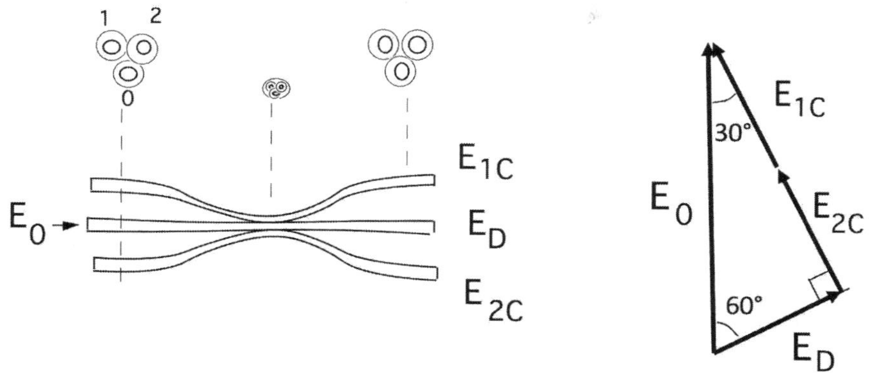

Fig.8-30 The 3×3 coupler splits the input field E_0 in three output ports. The vector at the direct output E_D is perpendicular to the crossed output vectors E_{1C} and E_{2C}, which are parallel to each other. If the splitting ratio is equal for all outputs (=1/√3), then angles at the hypotenuse are 30° and 60°.

The three vectors then form a right-angle triangle. As we can check, the vector sum gives E_0 as a result, and also the intensity E_0^2 is the sum of the squares E_D^2 and $(E_{1C}+E_{2C})^2$.

We may now assume that the splitting ratio is equal for all ports, that is 1/3 in power and 1/√3 in field amplitude. Then, the triangle of vector sum has one side that is double of the other, and the angles are 30° and 60°. We can write the above results as:

$$T_D = E_D / E_0 = (1\sqrt{3}) \exp -i\, 60°$$

$$T_{1C} = E_{1C}/E_0 = T_{2C} = E_{2C}/E_0 = (1\sqrt{3})\, E_0 \exp +i\, 30° \qquad (8.30)$$

Let us now turn to consider the 3×3 FOG configuration of Fig.8-31. As we can see, power from a SLED is launched from one fiber of the coupler, and the propagation coil is connected to the other two fibers of the coupler. When we come back after propagation, the two fields mix again at the coupler, and we collect them on the photodiodes PD1 and PD2.

Worth noting, the configuration uses the minimum possible number of components. The fiber coil is made of normal, telecommunication-grade, single mode fiber.

This fiber may cost ≈ 0.1 \$/m and we use a relatively short length, L=100 m, wound on a small coil (≈20...30-mm diameter), because increasing L does not improve the NEΩ in a non-reciprocal configuration. An SLED or even a high-radiance LED can be used, emitting at a wavelength anywhere in the broad range λ=1200 to 1600 nm of low fiber attenuation αL at the assumed length L.

The coupler is fabricated with the fused-coupler technology [36] out of the same fiber pigtails of the coil and of the SLED.

In this way, no joint is necessary, and we save a substantial cost of assembly. Photodiodes are normal, small area ternary (InGaAs) pn or pin PDs, with no special speed requirement. The only expensive component is the pigtailed SLED.

Let us now analyze the fields collected at the photodiodes PD1 and PD2, that is, E_{PD1} and E_{PD2}. They are made up of two contributions, coming from the same or the opposite-side fiber. As we can see from Figs.8-30 through 8-32, we have:

$$E_{PD1} = E_{1CC} + E_{2CD}, \quad \text{and} \quad E_{PD2} = E_{1CD} + E_{2CC} \tag{8.31}$$

Recalling Eq.8.30, we can write these terms as:

$$E_{1CC} = E_{1C}\exp{-i\phi_S}\, T_{1C},\; E_{2CC} = E_{2C}\exp{+i\phi_S}\, T_{2C},$$

$$E_{2CD} = E_{2C}\exp{+i\phi_S}\, T_D \text{ and } E_{1CD} = E_{1C}\exp{-i\phi_S}\, T_D \tag{8.32}$$

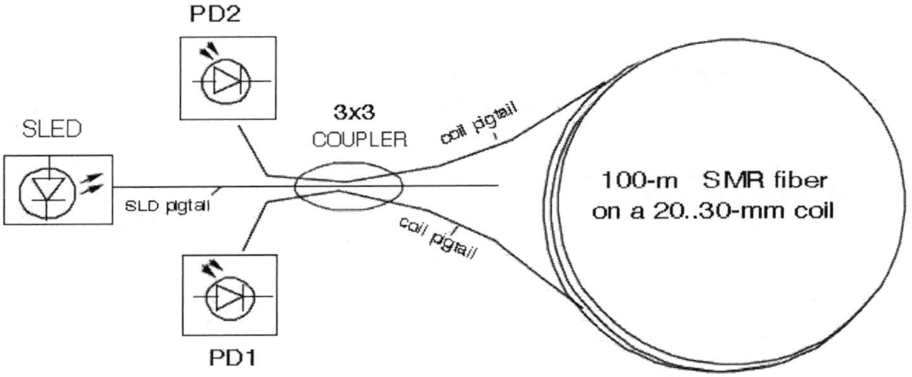

Fig.8-31 The 3×3 FOG is the gyroscope with the minimum part count. It requires just the fiber coil (a normal fiber, shorter than in a normal FOG), a SLED source, two photodiodes, and the 3×3 coupler. To dispense with fiber splices, the coupler is fabricated directly on the pigtail of the SLED and coil.

By inserting Eqs.8.32 in Eqs.8.31, we get:

$$E_{PD1} = E_{1C}\exp{-i\phi_S}\, T_{1C} + E_{2C}\exp{+i\phi_S}\, T_D = (E_0/3)\,[\exp i(+60° - \phi_S) + \exp i(-30° + \phi_S)]$$

$$E_{PD2} = E_{2C}\exp{+i\phi_S}\, T_{2C} + E_{1C}\exp{-i\phi_S}\, T_D = (E_0/3)\,[\exp i(+60° + \phi_S) + \exp i(-30° - \phi_S)] \tag{8.33}$$

The photodetected current follows [16] as the square mean value of the electric field, $I = \langle E^2 \rangle$. Because there are two terms in the fields of Eq.8.33, we shall develop the square. Doing so, we get: $I = \langle E^2 \rangle = \langle E_0^2(\exp i\xi + \exp i\zeta) \rangle = E_0^2[1+1+2\mathrm{Re}\,\{\exp i(\xi-\zeta)\}] = 2E_0^2[1+\cos(\xi-\zeta)]$. Applying this expression to the Eqs.8.33 gives:

$$I_{PD1}=2\,(E_0/3)^2\,[1+\cos(90°-2\phi_S)] = 2\,(E_0/3)^2\,(1+\sin 2\phi_S)$$

$$I_{PD2}=2\,(E_0/3)^2\,[1+\cos(90°+2\phi_S)] = 2\,(E_0/3)^2\,(1-\sin 2\phi_S) \qquad (8.34)$$

Now, computing the difference $I_\Delta=I_{PD1}-I_{PD2}$ of detected currents, we are able to suppress the dc term and obtain a signal proportional to the sine of the Sagnac phase ϕ_S for small ϕ_S:

$$I_\Delta = 4\,(E_0/3)^2 \sin 2\phi_S \qquad (8.34')$$

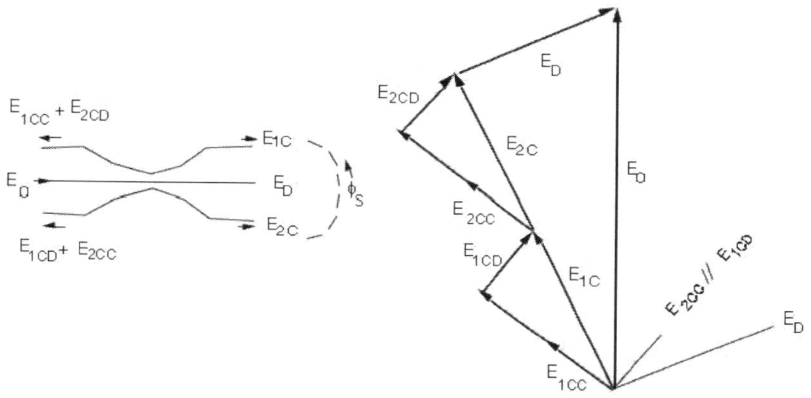

Fig.8-32 Propagation of signals through the 3×3 coupler of a minimum-configuration FOG. Each output port receives two contributions, propagated in opposite directions. Because of the phase shift suffered in the 3×3 coupler, the resulting beating are of the desired form ± sin $2\phi_S$.

A typical 3×3 FOG is built using a 30-mm diameter fiber coil with 100 m of single mode fiber [53], and performance of such a device is: baseline drift ≈0.05 deg/s, NEΩ≈0.01 deg/s (at B=10 Hz), and linearity error <0.5% up to a Ω≈ 200 deg/s dynamic range.

8.8 The MEMS Gyro and Other Approaches

The electro-optical gyroscope is the undisputed choice for inertial-grade applications, but in the automotive and robotic segments, other technical approaches have been developed, because they are competitive in terms of low cost at reduced performance (Fig.8-1).

In this section, we briefly describe these approaches to get an insight on the challenge coming from competing technologies.

All the approaches developed so far, based on an effect different from the Sagnac effect, can be traced to the Coriolis effect as the fundamental mechanism for sensing the inertial rotation speed Ω.

As we recall from physics, the Coriolis effect is about the apparent force F_C developed on a mass m moving with a velocity v, when the laboratory is subjected to an angular velocity Ω:

$$\underline{F_C} = 2m\, \underline{\Omega} \times \underline{v} \tag{8.35}$$

Here, $\underline{\Omega}$ and $\underline{v}$ are the vectors of angular velocity and spatial velocity, and $\times$ indicates the vector product.

In a schematization of the MG, the mass m is suspended with springs to the walls of the sensor box, as illustrated in Fig.8-33.

The constant of the x-axis spring is k_x, so that the displacement x produced in the spring by force F_x is $x= k_x F_x$. Also schematized in Fig.8-33, friction r_x is related to the force needed to keep a speed $x'=dx/dt$, as $F_x= r_x\, x'$. Last, we know that a force $F_x= m\, x''$ is required to accelerate the mass m at an acceleration $x''=d^2x/dt^2$.

By summing up the three terms, we can write the following balance equation for the external force F_X applied to the mechanical system:

$$m\, x'' + r_x\, x' + x/ k_x = F_x \tag{8.36}$$

For the two other axes, y and z, two more equations similar to Eq.8-36 should be written. Thus, our mechanical system is described by second-order responses for each axis.

To sense the Coriolis force F_C, the mass is put into vibration by means of a suitable actuator. Let the actuator move the mass along the x-axis, as shown in Fig.8-33 (left), so that $\underline{v}= v\, \underline{x}$, where $\underline{x}$ is the unit vector of the x-axis.

Then, a rotation $\underline{\Omega}$ parallel to the y-axis develops a force directed along the z-axis, because Eq.8.35 gives in this case $\underline{F_C}/2m= \underline{\Omega} \times \underline{v}= (\Omega \cdot v)\, \underline{y} \times \underline{x}= -\,(\Omega \cdot v)\, \underline{z}$. We will measure the displacement z produced by this force through a suitable sensor (Fig.8-33, left) to obtain the angular velocity component Ω along the y-axis.

Alternatively, if $\underline{\Omega}$ were parallel to the z-axis, we would get $\underline{\Omega} \times \underline{v}= (\Omega \cdot v)\, \underline{z} \times \underline{x}=-(\Omega \cdot v)\, \underline{y}$, and the signal is developed along the y-axis for the z-component of Ω.

If we are able to sense the vibrations transferred to both the z and y axes, we could in principle make a 2-axis measurement of the angular velocity vector $\underline{\Omega}$ with a single sensor. This was actually the case of the old mechanical gyros, for example, the Styroflex gyro family [15].

On the contrary, in all other gyros and in MEMS, the mechanical axes are not so well decoupled. Equations like $x=\eta_{xy}y$, $x=\eta_{xz}z$ and $z=\eta_{zy}y$ are to be added to Eq.8.35 to describe coupling. The result is a cross-sensitivity at small η's and more complicate chaotic effects at higher η's. To avoid all these errors, we shall make a single-axis measurement and strive to minimize the cross-sensitivity to other axes. Cross-coupling η coefficients can be minimized by design, and by appropriately increasing the stiffness k of the axes other than the measurement and excitation axes, so that the residual F/k is a small displacement.

The actuator indicated in Fig.8-33 provides a periodic excitation of the test mass M at a frequency ω_{exc}. This generates, in response to the measurand Ω, a periodic signal of the type $(\Omega v) \cos \omega_{exc} t$ at a carrier frequency ω_{exc}.

Of course, an ac signal is better than a dc one because the measurement is moved away from the low frequency range, affected by 1/f and slow drift components.

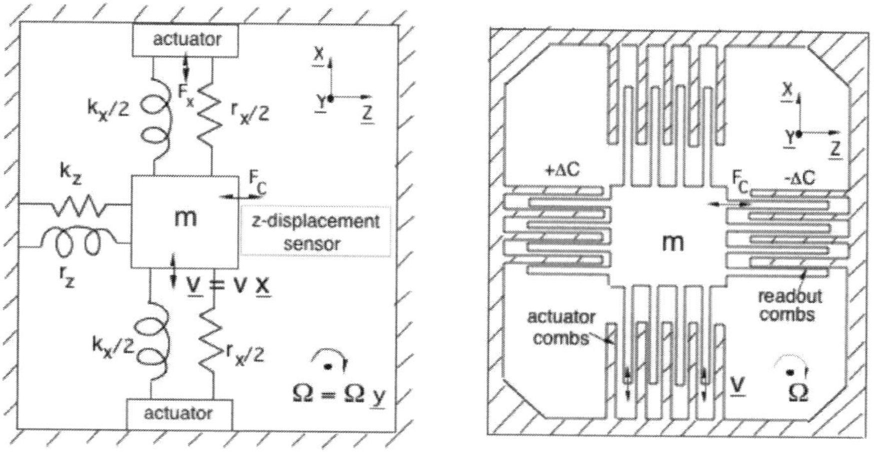

Fig.8-33 Left: a mechanical gyro is based on the Coriolis force acting on a sensing mass m. The mass is suspended by springs and is put into vibration along the x-axis by an actuator. By applying an angular velocity $\underline{\Omega}$ along the y-axis, a Coriolis force F_C is developed along the z-axis. A sensor measures the z-displacement, proportional to Ω. Right: in the MEMS gyro, the mass m is suspended above the substrate by a spring hinge (not shown) and has comb expansions along the x- and z-axis. The x-axis combs serve to actuate the mass by electrostatic force. The capacitance of z-axis combs varies with opposite sign as the mass moves along z, and this is the electrical readout of the z-displacement.

The frequency response of the sensor to the measurand Ω is a fraction 1/Q of ω_{exc}, the carrier frequency for the signal. The maximum frequency of excitation is determined by the mechanical resonance of the structure, expressed by Eq.8.35. The resonance is given by $\omega_{res}=\sqrt{(k/m)}$, m being the oscillating mass and k the stiffness (or spring constant) of the elastic structure bearing the mass.

Again from Eq.8.35, we find the fractional pass-band of the resonance as $\Delta\omega_{res}/\omega_{res}=1/Q$, where Q is the merit factor given by $Q=\omega_{exc}m/r$, and r is the friction coefficient (r_x in Fig.8-33). Along the excitation (x-axis) and the readout (z-axis) in Fig.8-33, the spring coefficients are k_x and k_z. For the best performance, we will require $k_x = k_z$ so that $\omega_{res(x)} = \omega_{res(z)}$.
Working with the excitation at resonance $\omega_{res(x)}$ gives a large velocity signal with a small drive power, reduced by a quantity equal to the Q_z-factor with respect to the power required off-resonance. Working with the readout at resonance $\omega_{res(x)}$ ensures an increase of response amplitude by a factor Q_x.

In general, we may assume that a displacement x is induced by the Coriolis force F_C, as a response to measurand Ω. The displacement is $x= k_z F_C$ in the example considered in Fig.8-33. Now, because we want a large response $1/k_z$ and a high frequency cutoff $\sqrt{(k_z/m)}$, it is clear that we need a mass m as small as possible. This is what we get in MEMS structures.

8.8.1 MEMS

A MEMS (Micro Electro Mechanical System) is just a micro-system fabricated in a silicon crystal with the technology of semiconductor devices [54-57]. As it is well known, silicon is the most used electronic material, and we know that devices and integrated circuits are fabricated by a well-established process, involving steps like epitaxy, lithography, etching, oxidation, dielectric and polysilicon deposition, metallization, etc.

When we look to properties other than electrical, silicon is recognized as a very good mechanical material too, with strength comparable to steel. Moreover, silicon has good thermal conductivity.

We can fabricate a miniaturized mechanical structure on a Si chip by combining the usual processing steps of selective growth and etching to one more specific step, the removal of sacrificial layers. These sacrificial layers can be etched from underneath to leave suspended structures in silicon [54]. As resolution is set by photolithography, the MEMS finest detail may be as small as ≈1 μm, and already complex structures may be fabricated with a typical size of only 50 to 100 μm.

Thus, the degree of miniaturization offered by MEMS is very good, and the integration may go up to a complexity only limited by the yield to be attained in the fabrication process.

The resulting technology is called *micromachining* and is currently pursued to develop new components, starting from machines like motors and actuators and including a wide range of devices for communication, displays, bioelectronics, and sensors.

The structure of MEMS gyros can take several different forms. We can use either a linear structure with mass suspended by springs (as in Fig.8-33) [55] or a rotor structure free of oscillating about a central hinge (Fig.8-34) [56]. Common to all structures is the excitation by means of electrostatic actuation and the readout of the displacement by means of capacitance variation.

Electrostatic actuation is very effective because, on the small scale of MEMS size, even a modest voltage V produces an appreciable force F=eV/d across the small gap d, more than adequate to move the small mass m.

The preferred actuation is a sinusoidal waveform at the resonant frequency of the structure, so that the oscillation amplitude of velocity v is high even with modest drive power.

Capacitance between the fixed part and the mobile part subjected to the Coriolis displacement is increased up to the desired level (≈pF) by appropriately increasing the surface of the comb structure (Fig.8-34). In this way, a large ΔC/C signal is developed following a transversal displacement Δx. If the gyro is put into vibration parallel to the x-axis (in the reference of Fig.8-33), and $\underline{\Omega}$ is directed along the y-axis, the displacement is along the z-axis and is in ac like the drive along the x-axis.

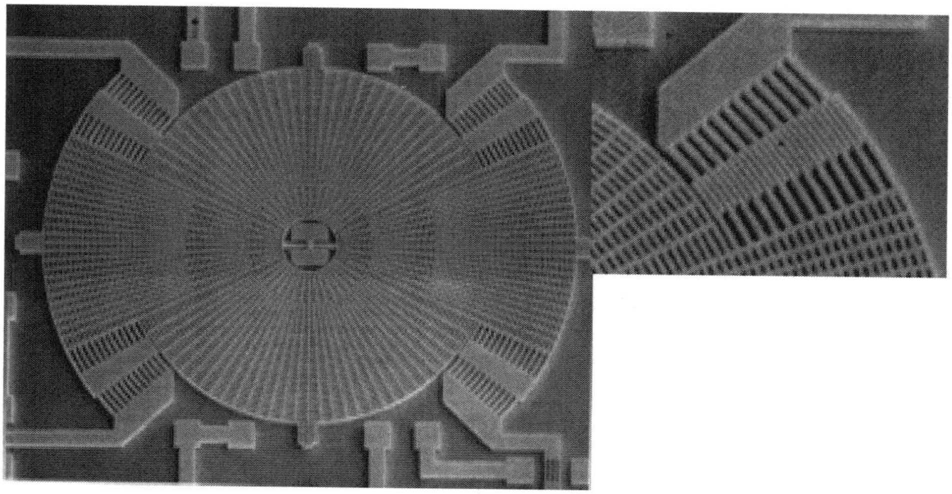

Fig.8-34 A typical MEMS gyro is made by a Si-rotor suspended above the substrate and free to oscillate around the central hinge. Comb fingers (magnified at right) located at the four corners allow electrical excitation of the oscillatory motion. An angular velocity Ω in the plane of the rotor displaces the rotor vertically (out of plane). The finger capacitance is measured to sense this displacement (from Ref. [56], ©IEEE, reproduced by permission).

We then need two small stiffnesses k_x and k_z, to let the structure move easily along the x- and z-axes, whereas the stiffness along the y-axis will be much larger than k_x and k_z.

The capacitance signal $\Delta C/C$ is read by arranging the readout comb capacitors in a bridge circuit. If the MEMS is designed so that two combs in it collect opposite signals ($+\Delta C$ and $-\Delta C$ as in Fig.8-33), we will place these elements at opposite diagonals in the bridge. The bridge is completed by two more dummy elements, similar in shape to the active combs.

By applying dc voltages to the dummy elements, the bridge can be balanced at rest, or the zero-bias is trimmed to zero.

The limit performance of the MEMS gyro is approached when we are able to cure a number of disturbances, like cross-axis sensitivity due to mechanical and/or electrical coupling, noise of the readout circuits, etc.

Then, the ultimate limit of NEΩ is set by mechanical-thermal noise because each degree of freedom of the mechanical structure (either rotational of translational) acquires by Boltzmann's theorem an energy (1/2)kT because of thermal agitation. By analyzing the effect of mechanical-thermal noise on a vibrating mass [57], it can be found that the corresponding NEΩ is given by:

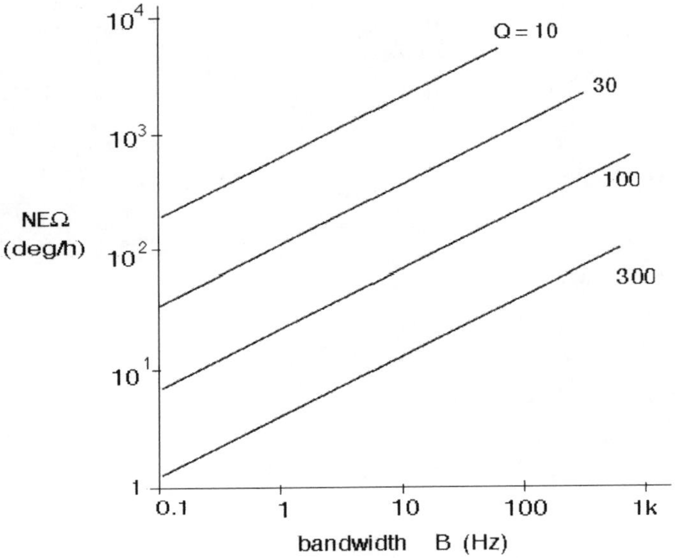

Fig.8-35 Noise equivalent angular velocity NEΩ versus measurement bandwidth B for a MEMS gyro as limited by the mechanical-thermal noise. Data is from Eq.8.36 with m=2 μg and ω=10 kHz.

$$NE\Omega = \omega^{3/2}(4kmTB)^{1/2}/2Q^{3/2}F_0 \tag{8.37}$$

Here, F_0 and ω are the peak force and frequency of excitation, $F= F_0 \cos \omega t$, Q is the quality factor along the sense axis, and m is the vibrating mass.

As we can see from Fig.8-35 where the NEΩ is plotted versus bandwidth [57], for Q=100…300 and a driving frequency $\omega/2\pi$= 5...20 kHz, appropriate to obtain a measurement bandwidth B≈100 Hz, we get a mechanical-thermal noise limit in the range NEΩ= 40-200 deg/h.

This is a satisfactory target for applications to automotive and robotics, but is obviously far away from the inertial-grade specifications, as all other non-optical approaches are.
In practical MEMS fabricated so far, performance differs from the mechanical-thermal noise limit by a factor ≈5...10.

Perhaps a better result can be obtained with a hybrid MEMS, combining the mechanical structure of a MEM gyro to the optical readout of Coriolis-induced vibration. The advantage of this approach is the separation of the mechanical structure from the readout, thus a release of design constraints.

As indicated in Fig.8-36, we can read the displacement along the z-axis, following an excitation along the x-axis and a rotation Ω parallel to the y-axis, by means of a self-mixing interferometer [58, 59] placed in front of the mass. A feasibility evaluation of the concept has indicated a potential performance of ≈0.01 deg/√h in terms of noise spectral density [58].

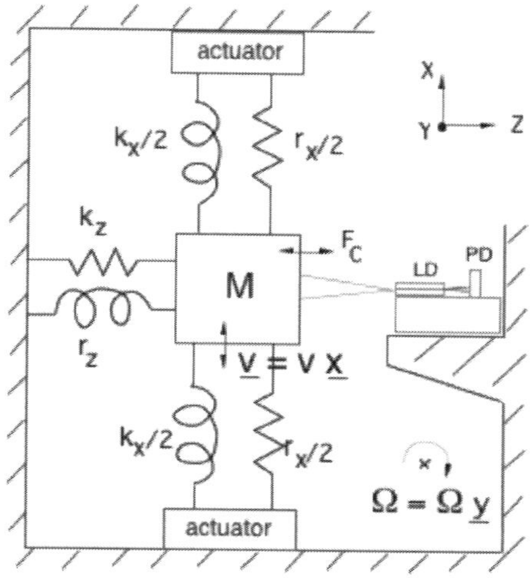

Fig.8-36 A hybrid MEMS with optical readout of the displacement along the z-axis, or MOEMS [58] can alleviate the design requirements and help improve NEΩ.

8.8.2 Piezoelectric Gyro

The piezoelectric gyro (PG) is another example of mechanical gyro (MG). In the PG, both excitation and readout of the mechanical oscillation are obtained through the piezoelectric effect. The piezoelectric effect is the transfer of energy from mechanical to electrical form and vice versa, taking place in piezoelectric materials like quartz, poled ceramics (like PZT), etc. The distinct advantage of PG, compared to other MG approaches, is that we have an easy access to the mechanical quantities of the sensor through electrical counterparts, simply by depositing a metal electrode at the chosen location.

The disadvantage is that bulk effects are used in devices reported so far. In consequence, the overall size of the PG is in the range of centimeters rather than millimeters as for Si-MEMS. Several different shapes have been proposed for the PG, all referable to the basic MG configuration of Fig.8-33.

Perhaps the first has been the tuning fork PG [60], shown in Fig.8-37. This configuration uses an inexpensive quartz crystal, similar to those of electronic watches. A feasibility evaluation of the concept has indicated a potential performance of ≈0.01 deg/√h in terms of noise spectral density [58].

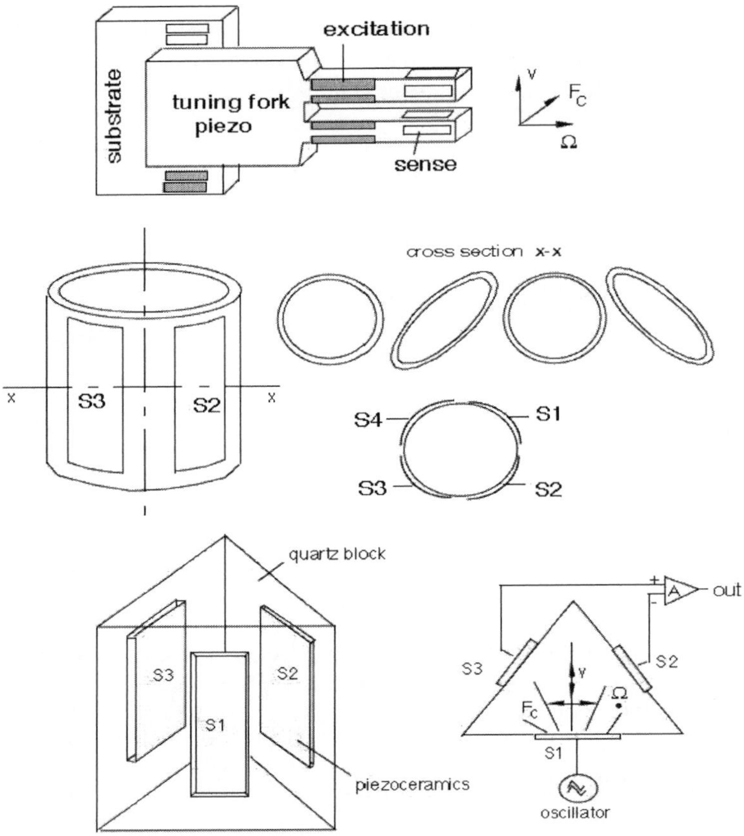

Fig.8-37 Examples of piezoelectric gyros configurations: tuning fork (top), wine-glass resonator (center), and triangle resonator (bottom).

Oscillation of the fork is along the axis labeled v, and sensing is along the axis labeled F_c in Fig.8-37. Then, the device is sensitive to angular velocity Ω in the plane of the fork.

Another popular configuration is that of the wine glass resonator (Fig.8-37, middle) [54]. In this resonator, a tube vibrates with an elliptical-shaped flexure. If the tube is well balanced mechanically, the major axes of the ellipse (oriented at 45 deg and -45 deg in Fig. 8-37) may take any inclination. Oscillation is excited by applying ac voltages at the resonant frequency of the tube to the electrodes. When an angular velocity ±Ω is applied along the tube axis, the inclination of the ellipse elongation moves clockwise or counterclockwise (according to the sign of Ω). Then, we get the measurement of Ω by comparing the difference of signals S1-S3 and S2-S4 collected at the electrodes.

Still another configuration widely employed is the resonating quartz triangle (Fig.8-37, bottom). In this device, we excite an acoustical wave by means of the transducer S1 (made of PZT) driven at the resonant frequency of the mechanical structure. Two more PZT transducers sense the amplitude of the incoming acoustical wave. When Ω is applied parallel to the prism axis, one transducer receives more power than the other due to the Coriolis force (Fig.8-36, bottom right). By computing the difference of outputs S2 and S3, a signal proportional to Ω is obtained.

About performances of the PGs, data are, of course, compliant with the thermal noise limitation of Fig.8-35. Typical sensitivity of commercial products is 0.2...1 deg/s and the dynamic range is 50 to 200 deg/s.

REFERENCES

[1] W.M. Macek and D.I.M. Davis: "*Rotation Rate Sensing with a Travelling Wave Laser*", Appl. Phys. Lett., vol.2 (1963), pp.67-69.

[2] V.E. Sanders and R.M. Kiehn: "*Dual Polarized Ring-Laser*", IEEE J. Quant. Electr., vol. QE-13 (1977), pp.739-745, see also: A. Mattews: "*The Laser Ring Gyro*", in: Proc. Rotation Sensing Conf., Paris 1979, AGARD vol.220, pp.1-21.

[3] Research and Markets: "*Ring Laser Gyroscopes - Global Market Trajectory 2022*", see:www.researchandmarkets.com/reports/5030627/ring-laser-gyroscope-global-market-trajectory.

[4] V. Vali and W. Shorthill: "*Fiber Ring Interferometer*", Appl. Opt., vol.15 (1976), pp.1099-1100.

[5] R. Ulrich and M. Johnson: "*Fiber Ring Polarization Analysis*", Opt. Lett., vol.4 (1979), pp.152-154.

[6] R.A. Bergh, H.C. Lefevre, and H.J. Shaw: "*All-single-mode Fiber Gyroscope*", Opt. Lett., vol.6 (1981), pp.198-200; see also: Opt. Lett., vol.6 (1981), pp. 502-504.

[7] H.C. Lefevre: "*The Fiber Optic Gyroscope*", Artech House: Boston, 1993, see Ch.11.4.

[8] D.M. Shupe: "*Thermally Induced Non-Reciprocity in the Fiber-Optic Interferometer*", Appl. Opt. vol.19 (1980), pp.654-6455; see also: R. Yang, S. Ge, B. Li, C. Guo: "*Analysis of Fiber-Optic Coil SHUPE Effect for Different Coil Winding Structure and Winding Process Optimization*", Proc. 2020 IEEE Int. Instr. Meas. Conf. (I2MTC), Dubrovnik, May 25-28, DoI 10.1109/I2MTC43012.2020.9128371.

[9] E.J. Post: "*The Sagnac Effect*", Rev. Mod. Phys., vol.39 (1967), pp.475-495.

[10] J. van Bladen: "*Relativity in Engineering*", Springer Verlag, Berlin 1984.

[11] E.O. Schultz DuBois: "*Alternative Interpretation of Fresnel-Fizeau Effect in a Rotating Optical Ring Interferometer*", IEEE J. Quant. Electr., vol. QE-2 (1966), pp.299-306.

[12] V. Vali and W. Shorthill: "*Fresnel-Fizeau Effect in a Rotating Optical Fiber Ring Interferometer*", Appl. Opt., vol.16, (1977), pp. 2605-2607.

[13] S. Donati and V. Annovazzi Lodi: "*Fiber Gyroscope with Dual Frequency Laser*", in: Fiber-Optic Rotation Sensors, ed. by S. Ezekiel and H.J. Arditty, Springer Verlag, Berlin 1982, pp.292-297.

[14] G.E. Stedman: "*Ring-Laser Tests of Fundamental Physics and Geophysics*", Progr. in Phys. vol.60 (1997), pp.615–688. See also: "*Ring Laser Gyroscope Receives its Centerpiece*", Europhot. 2001, p.15.

[15] R.P.G. Collinson: "*Introduction to Avionics*", 3rd ed., Springer, Berlin 2011.
[16] S. Donati: "*Photodetectors*", 2nd ed., Wiley IEEE Press, Hoboken 2021.
[17] S. Donati, S.-K. Hwang: "*Chaos and High-Level Dynamics in Coupled Lasers and their Applications*", Progr. Quant. Electr., vol.36 (2012), pp.293–341.
[18] see Ref. [16], p.312
[19] "*Gyroscope Battle is Settled*", Opto&Laser Europe, 92, (2002), p.15.
[20] V.E. Sanders and R.M. Kiehn: "*Dual Polarization Ring Laser*", IEEE J. Quant. Electr., vol. QE-13 (1977), pp.739-746, and: IEEE J. Quant. Electr., vol. QE-16 (1980), pp.918-935.
[21] W.W. Chow et al., "*Multioscillator Ring Laser Gyro*", Journal of Quantum Electronics, vol. QE-16 (1980) pp.918-935.
[22] M. Born and E. Wolfe: "*Principles of Optics*", 6th ed., Pergamon Press, Oxford 1983.
[23] B. Rossi: "*Optics*", Addison Wesley, Reading 1957.
[24] S. Donati: "*Laser Interferometry by Induced Modulation of the Cavity Field*", J. Appl. Phys., vol.49-2 (1978), pp.495-497, see also: O. Svelto: "*Principles of Lasers*", 5th ed., Springer, Berlin 2010.
[25] S. Ezekiel and H.J. Arditty: "*Fiber-Optic Rotation Sensors*", Springer Verlag: Berlin 1982, pp.292-297.
[26] R.A. Bergh, H.C. Lefevre, and H. J. Shaw: "*An Overview of Fiber Optics Gyroscopes*", IEEE J. Lightw. Techn., vol.LT-2 (1991), pp.91-107.
[27] D. D'Alessandro and S. Donati: "*Optimum Bias for Interferometers in the Quantum Noise Regime*", Alta Freq. vol.12 (2000), pp.72-75.
[28] A. Angot: "*Complements de Matematiques*", 6th ed., Masson et Cie, Paris 1997.
[29] G. Martini: "*Analysis of a Single Mode Fiber Piezoceramic Phase Modulator*", J. Opt. Quant. Electr., vol.19 (1987), p.179-190.
[30] see Ref. [16], Sect.5.3.
[31] T. Osaka, K. Okamoto, et al.: "*Low-loss Single-Polarization Fibres with Asymmetrical Strain Birefringence*", Electr. Lett., vol.17 (1981), pp.530-531.
[32] S. Donati, L. Faustini, and G. Martini: "*High-Extinction Coiled-Fiber Polarizers by Careful Control of Interface Reinjection*", IEEE Ph. Techn. Lett.vol.7, (1995), pp.1174-1177.
[33] G.P. Bendelli and S. Donati: "*Optical Isolators for Communications: Review and Current Trends*", Eur. Trans. on Telecomm., vol. 3 (1992), pp.63-69.
[34] V. Annovazzi Lodi, S. Donati, S. Merlo, L. Zucchelli, and F. Martinez: "*Protecting a Power-Laser Diode from Retroreflections by Means of a Fiber λ/4 Retarder*", IEEE Phot. Techn. Letters, vol.8, (1996), pp.485-488.
[35] V. Annovazzi Lodi and S. Donati: "*Technology of Lapped Optical-Fiber Couplers*", J. Opt. Comm., vol.11, (1990), pp.107-121.
[36] S. Donati, V. Annovazzi Lodi and F. Picchi, "*Ultra-low Insertion Loss Fused-Couplers Fabricated by a Long Furnace*", in Proc. LEOS Workshop on Passive Fiber Optics Components, ed. by S. Donati, Pavia, Sept.18-19, 1998, pp.161-164.
[37] E. Udd (editor): "*Fiber Optic Gyros 10th Anniversary Conference*", Proc. SPIE vol.719, (1988).
[38] W. Ling, X. Li, H. Yang, P. Liu, Z. Xu, Y. Wei: "*Reduction of the Shupe effect in interferometric fiber optic gyroscopes: the double cylinder-wound coil*", Opt. Comm., vol. 370 (2016), pp.62-67.
[39] See for example: https://www.gemrad.com/guidance-navigation-positioning/
[40] J. M. Lopez-Higuera (ed.): "*Optical Fibre Sensing Technology*", J. Wiley & Sons, Chichester 2002.

[41] O. Graydon: "*Integrated Gyro is set to Reduce the Cost of Navigation*", Opto-Laser Europe, Dec.1997, pp.23-25, see also: C. Wulf-Mathies, "*Integrated Optics for Fiberoptics Sensor*", Laser und Optoelektronik vol.21 (1989), pp.53-63.
[42] S. Donati, G. Giuliani, and M. Sorel: "*Proposal of a new Approach to the Electro-Optical Gyroscope: the GaAlAs Integrated Ring Laser*", Alta Freq. Riv. Elettron. vol.9 (1997), pp.61-63; see also: Alta Freq. Riv. Elettron. vol.10 (1998), pp.45-48.
[43] M. Sorel, P. Laybourn, G. Giuliani, and S. Donati: "*Unidirectional Bistability in Semiconductor Waveguide Ring Lasers*", Appl. Phys. Lett., vol.80 (2002), pp.3051-3053.
[44] B.Y. Kim and H.J. Shaw: "*Phase–Reading an All-Fiber-Optic-Gyroscope*", Opt. Lett., vol.9 (1984), pp.378-380.
[45] A. Edberg and G. Schiffner: "*Closed Loop Fiber-Optic-Gyroscope with a Sawtooth Phase Modulated Feedback*", Opt. Lett., vol.10 (1985), pp.300-302.
[46] C.J. Kay: "*Serrodyne Modulator in a Fiber-Optic-Gyroscope*", Proc. IEE part J Optoelectr., vol.132 (1985), pp.259-264.
[47] H.C. Lefevre, P. Graindorge, H.J. Arditty, et al.: "*Double Closed-Loop Hybrid Fiber Gyroscope with a Digital Ramp*", Proc. OFS'84, Munchen 1984, paper PSD-7.
[48] R. Carroll and J.E. Potter: "*Backscatter and Resonant Fiber-Optic-Gyro Scale-Factor*", IEEE Trans. Lightw. Techn., vol.7 (1989), pp.1895-1900.
[49] J.M. Lopez-Higuera: "*Superfluorescent Fiber Optic Sources*", Ch.10, Handbook of Optical Fibre Sens. Techn., J. Wiley & Sons: Chichester 2002.
[50] S. Oho, H. Kajioka, and T. Sasayama: "*Optical Fiber Gyroscope for Automotive Navigation*", IEEE Trans. Vehic. Techn., vol.44 (1995), pp.698-704.
[51] N. Barbour and G. Schmidt: "*Inertial Sensor Technology Trend*", IEEE Sens. J., vol.1 (2001), pp.332-339.
[52] S.K. Shyeem: "*Fiber Optic Gyroscope with a 3x3 Directional Coupler*", Appl. Phys. Lett., vol.37 (1980), pp.869-870.
[53] V. Annovazzi Lodi, S. Donati, and M. Musio: "*A Fiber Optic Gyroscope for Automotive Navigation*", Proc. Intl Conf. Adv. in Microsyst. for Automot. Applic., Berlin, 1996, pp.123-128.
[54] S.D. Senturia: "*Microsystem Design*", Kluwer, Boston 2000.
[55] K. Tanaka et al.: "*A Micromachined Vibrating Gyroscope*", Sens. Actuat., vol. A50 (1995), pp.111-115.
[56] D. Teegarden, G. Lorenz, and R. Neul: "*How to Model and Simulate Microgyroscope Systems*", IEEE Spectrum, July 1998, pp.66-75.
[57] V. Annovazzi Lodi and S. Merlo: "*Mechanical-Thermal Noise in Micromachined Gyro*", Microel. J., vol.30 (1999), pp.1227-1230, see also: V. Annovazzi Lodi, S. Donati, and S. Merlo: "*Thermodynamic Phase Noise in Fiber Interferometers*", J. Opt. Quant. Electr., vol.28 (1996), pp.43-49.
[58] S. Donati and M. Norgia: "*Hybrid Opto-Mechanical Gyroscope with Injection-Interferometer Readout*", Electr. Lett. vol.37 (2001), pp.756-757.
[59] O. Kilic, H. Ra, O. Can Akkaya, M.J.F. Digonnet, O. Solgaard: "*Haltere-Like Optoelectromechanical Gyroscope*", IEEE Sensor J., vol.16 (2016), pp. 4274-4280, see also: V. Apostolyuk: "*Theory and Design of Micromechanical Vibratory Gyroscopes*", in MEMS/NEMS Handbook, ed. by C.T. Cornelius, pp.173-195, Springer Berlin 2006.
[60] J. Soderkvist: "*Piezoelectric Beams and Vibrating Angular Rate Sensor*", IEEE Trans. Ultras., Ferroel. and Freq. Control, vol.38 (1991), pp.271-280.

Problems and Questions

P8-1 *When read as a phase difference, the signal out from an electro-optical gyro is proportional to the area or R^2 (Eq.8.5), whereas when read as a frequency difference it is proportional to the radius R (Eq.8.6). Isn't it a contradiction?*
Q8-2 *Why are no basic limitations other than the quantum noise considered to influence the gyro performance, like we have done for interferometers in Sect.4.4?*
P8-3 *Calculate the max/min angular velocity that can be measured at the limit of discretization with a He-Ne RLG gyroscope having an apothem R=6.33 cm. Find the associated dynamic range.*
P8-4 *Repeat the calculation of the minimum detectable angular velocity of Prob.P8-3 considering now the noise limits. Assume an output power P=1-mW from the laser, and a mirror reflectivity r_1=0.995.*
P8-5 *The He-Ne RLG gyroscope considered above is rigidly mounted (or, strapped-down) on a mobile vehicle. We get n=5 counts (or periods) of the difference frequency, in a certain unspecified time interval. What can be said about the rotation undergone by the vehicle?*
P8-6 *In a He-Ne RLG there is a residual diffusion $\delta=10^{-4}$ at one mirror. Does it produce locking?*
P8-7 *Calculate the responsivity of a FOG gyroscope made by L=300-m fiber wound on a 2R=8-cm diameter coil, when we read the coil with a source at λ=850 nm.*
P8-8 *Calculate the amplitude of the signal expected from a FOG, with the data as in Prob.P8-7, when the rotation is 1 deg/hour.*
P8-9 *Take into account a fluctuation of 0.1% of the loss of the launch coupler, and calculate the error generated in the output signal.*
Q8-10 *Since the FOG works on the Sagnac configuration, a balanced interferometer, the phase shift ϕ_{err} supposed in Prob.P8-9 should actually be cancelled out when the beating of the two counter-propagating beams is considered. Provide your comments on this issue.*
Q8-11 *Can any other waveform different from the sinusoidal be employed to drive the dither excitation (Fig.8-13 and 14) of the piezo actuator to unlock the gyro?*
P8-12 *Would you prefer a gyroscope tagged with a noise of 0.01 deg/h or one with 0.01 deg/√h?*
P8-13 *In the right-hand side of Fig.8-17, the levelling down to nearly constant of the diagrams is due to the zero-bias error coming into play or to other effects?*
P8-14 *We read the FOG gyro of Prob.P8-7 with the open-loop configuration based on a PZT phase modulator. What is the optimum value of the frequency modulation? What is the typical drive voltage required by the PZT? What is the signal amplitude obtained at the output?*
P8-15 *Should we wish to get the cosΦ in addition to the sinΦ signal in the FOG external configuration, how large shall we expect is it?*
P8-16 *How can we derive the CW/CCW mode coupling due to fiber backscattering (Sect.8.5.2)?*
P8-17 *Estimate the contribution, to the zero-bias error, of the Rayleigh backscattering from the fiber in a FOG gyro, when using a laser source with linewidth $\Delta\nu$=10 MHz and when using a SLED with spectral width $\Delta\lambda$=50 nm.*
P8-18 *An Integrated Optics FOG is made with N=10 turns of 4-cm=R radius waveguide on a SOS substrate. What is the responsivity of the FOG? What is the expected phase and NEΩ performance?*
Q8-19 *List the advantages of the closed-loop, digital readout FOG compared to the open-loop configuration. Consider: NEDΩ, linearity, dynamic range, signal swing, and bandwidth.*
P8-20 *A MEMS gyroscope has a mass m=2 μg and the resonant frequency is f_0=10 kHz. Find the spring constant and the amplitude of the comb displacement undergone by application of a constant force F=1 μN. Assuming then a readout comb made of 60 fingers, 2x300 μm in area and 2 μm in gap, calculate the capacitance variation obtained. Last, calculate the Coriolis force induced by a Ω=1 rad/s angular velocity.*

CHAPTER 9

Optical Fiber Sensors

The most important application of optical fibers is undoubtedly in optical communications. Yet, optical fibers can be used as sensors as well, if we look at how certain parameters of propagation are affected by external perturbations. This is an interesting niche from both the scientific and technical points of view.

In communications, we strive to get rid of sensitivity to external perturbations and make the fiber as immune as possible. In fiber sensors, we need to make the sensitivity reproducible, and eventually try to increase it. Because we are interested in its measurement, now the physical perturbation becomes the *measurand*.

In Optical Fiber Sensors (OFSs), the effect of a measurand may eventually be very minute, yet we are able to devise readout methods capable of responding with very high sensitivity. A very high sensitivity is actually a special feature of some configurations of OFSs, unequalled by any other approach. This is one of the reasons why OFSs have been actively studied and developed internationally by the scientific community as soon as high-quality fibers became available.

The most successful example of an OFS is the FOG, which was treated in Ch.8. Many other examples of viable OFSs have been reported in the literature, and prototypes of OFSs for measuring quantities such as temperature, pressure, strain, and pollutants have been proposed and/or developed into commercial products.

The revenues from OFSs other than the FOG have been rather disappointing, however. The sensor market has fierce competition among many different technologies. The market expectation is not for high-performance sensors only, but more commonly for cheap sensors and for sensors with a good price-to-performance ratio.

The new OFS technology can offer high performance, but is generally more expensive than conventional sensors. Thus, optical fiber sensors have not been accepted for general-purpose applications, the domain of conventional electronic sensors. Only very special circumstances (like for example chemically aggressive environment, strong electromagnetic interference, etc.) may render the OFS acceptable, but this is just a small percentage of the rich and appealing sensor market.

In this chapter, we will present the fundamentals of OFSs and discuss a few selected examples of application.

More exhaustive references may be found in the copious literature on the subject [1-5].

9.1 INTRODUCTION

The paradigm of an OFS is shown in Fig.9-1. The *measurand*, through the mechanism of interaction, affects an *optical parameter* of the fiber. Specifically, the optical parameter may be the intensity, or the state of polarization (SoP), or the phase of the field propagated through the fiber. Then, according to the parameter affected by the measurand, we arrange a *readout configuration* that transforms the variation of the parameter into an output electrical signal for the user.

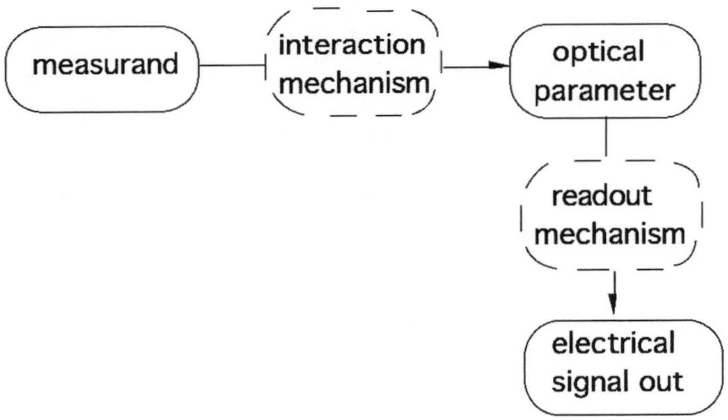

Fig.9-1 The paradigm of an OFS: the measurand, or quantity to be measured, affects an optical parameter, which is translated by a readout configuration into an electrical output.

9.1.1 OFS Classification

Sometimes it is suitable to convert the measurand into another, intermediate measurand, more conveniently measured by the fiber. According to this difference, OFSs are classified as *intrinsic* or *extrinsic*.

Another important classification is about the optical parameter involved in the interaction with the measurand. The most commonly used parameters are (i) intensity (or guided power), (ii) state of polarization (SoP), and (iii) optical phase of the propagated mode.
The corresponding readout (or OFS) is called an *intensity*, *polarimetric* or *interferometric* sensor (Fig.9-2).

This classification implies a type of fiber and source necessary to implement the readout. Intensity readout OFS can be implemented with multimode fibers, whereas polarimetric and interferometric OFSs require a single mode fiber to preserve the interaction without the errors originated by the mix up of the multimode regime.

In intensity-based OFSs, we do not look at the SoP nor need coherence. Therefore, we may use a cheap incoherent, unpolarized source like an LED.
In polarimetric and interferometric OFSs, on the other hand, we need to start with a clean input field, with well-defined SoP or coherence length. We shall therefore use a laser or a SLED (super luminescent LED).

READOUT (optical measurand)	*fiber type*	*source type*
INTENSITY (power)	multimode (occasionally monomode)	LED
POLARIMETRIC (state of polarization)	monomode	laser
INTERFEROMETRIC (optical phase)	monomode	laser, SLED

Fig.9-2 Classification of the three main readout principles and types of OFSs: intensity, polarimetric, and interferometric. Also indicated are the fibers and sources they use.

9.1.2 Outline of OFS

An outline of the most important OFSs, demonstrated in the laboratory or developed up to the level of units deployed in the field, is presented in Fig.9-3. As we can see, the range of applications is wide, spanning from mechanical to electrical to chemical quantities, and covering virtually all types of measurands.

Among the mechanical quantities, we find force, pressure, stress, strain, acoustical fields, and the OFSs developed for these quantities are mostly of the intrinsic type.
Some of them are known as the optical strain gage, the fiberoptic vibrometer, and the optical hydrophone. Transducer effects used include the elasto-optical effect, evanescent attenuation, and attenuation by proximity.

Optical Fiber Sensors

meas-urand	OFS type	*effect used, Intrinsic vs Extrinsic*
mechanical	STRAIN GAGE VIBROMETER HYDROPHONE	*elasto-optical I* *evanescent coupling I* *coupling attenuation E*
thermal	THERMOMETER	*birifringence temp coeff E* *index refraction temp coeff I*
inertial	GYROSCOPE ACCELEROMETER	*Sagnac effect I* *mass-loaded app force E*
electrical	ELECTRIC FIELD MAGNETIC FIELD	*Faraday and Pockels effects I* *electro-, magneto-striction E*
chemical	OPTRODES (pH, pCO, pollutants, etc.)	*fluorescence E* *chromatic reaction E*

Fig.9-3 Outline of the most developed OFSs, with the main effect used and type of interaction.

Temperature OFSs may be intrinsic as well as extrinsic. The temperature coefficient of the fiber is used for interferometric OFSs, whereas extrinsic OFSs have been developed using temperature-dependent birefringence in crystals and blackbody emission.

Electrical OFSs have been developed using the Faraday and Pockels effects, both intrinsic to the fiber, or occasionally the magnetostriction and electrostriction effects, both extrinsic to the fiber.

Inertial sensors include the FOG gyroscope (see Ch.8) which is an intrinsic OFS based on the Sagnac effect, and accelerometers, which are usually realized as extrinsic, mass-loaded strain gages.

Finally, OFSs have been demonstrated for the detection and measurement of chemical species (pollutants, hydrocarbons, etc.) and for chemical quantities like pH, pO_2, etc. Most chemical OFSs are extrinsic, and they are based on a color-developing reagent conveniently placed on the fiber tip, hence their name of *optrode* (optical electrode).

The advantages and disadvantages of OFSs with respect to other conventional sensors derive mainly from the fiber structure. Because of the electrically passive structure, their immunity to electromagnetic interference and to chemical species is good. They can withstand a high dose of ionizing radiation, in the range of Megarad compared to the 50-krad dose typical of electronic components. Their immunity to mechanical disturbance is also good because of the rugged structure of the fiber.

The overall size and weight of OFSs compare favorably to other sensors and flexibility of design in shape and size is good. Frequently, OFSs are able to operate without physically contacting the measurand, so they are *non-invasive*.

Finally, OFSs can attain sensitivity values unequalled by other types of sensors, especially when using the most sophisticated readout, i.e., the interferometric one (like in the FOG).

The list of disadvantages regarding OFSs is not short, unfortunately. To attain a high-sensitivity performance, OFSs are rather complex and expensive, whereas the simple intensity-based OFSs are cheap but not so appealing in performance.
Flexibility of reconfiguration is also modest. Ideally, in the user's wish list, an OFS should be composed of a readout electronic box and a probe. The electronic box should be capable of performing all three readouts (intensity, polarimetric, interferometric), with the appropriate fiber probe. In the user's view, there may be several probes available to perform all the measurements indicated in Fig.9-3 with a single instrument.

Unfortunately, no OFS is able to satisfy the wish list. More likely, after an OFS is finally completed for a specified measurement objective, if specifications are changed just a little (in sensitivity, dynamic range, or sensor placement), we need to redesign it in a completely different solution. Thus, while the performance capability offered by OFS technology is quite powerful and should always be considered a potential solution, the range of applications outside the specification is generally quite modest and allows little or no reconfiguration.

9.2 The Optical Strain Gage: A Case Study

The field of OFSs is characterized as being open to a variety of disparate approaches. Different from other segments of engineering, there are many possible solutions to a particular problem, that look at least technically plausible when first scrutinized.

To illustrate this point, we report in Fig.9-4 an example of OFS development, that is, the optical strain gage. This sensor has a well-known electronic counterpart, which is the resistive strain gage made of a thin film of conductive material deposited on a substrate, usually a polymer material. When the substrate is subjected to a stress, the film resistance changes. Arranging this resistance in a Wheatstone bridge, we can easily measure the stress or quantities connected to it (like strain, pressure, force, etc.).

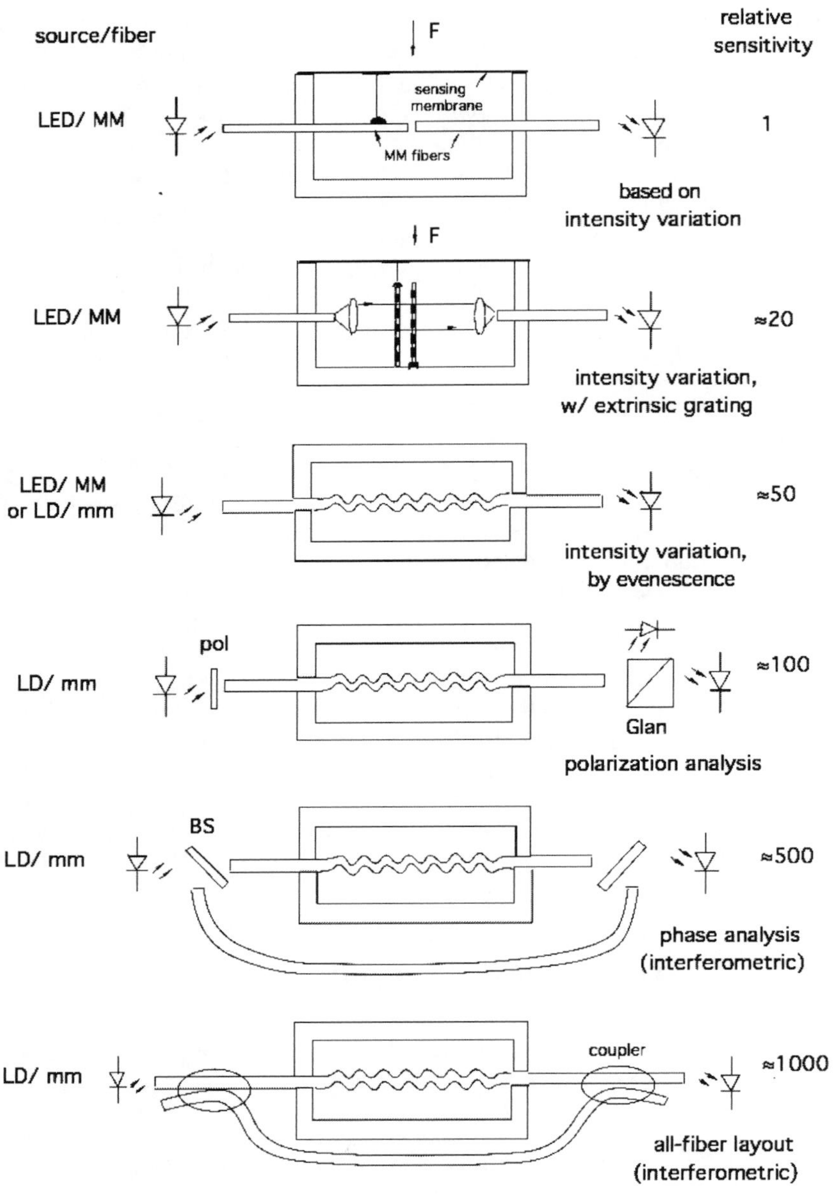

Fig.9-4 Evolution of different types of the optical strain gage, through intensity, polarimetric and interferometric readouts.

We can start by considering an intensity-based OFS (Fig.9-4, top), in which two fiber pigtails are aligned in the package and one of them is actuated by a thin membrane to which external force is applied. We can refer to this device as the benchmark of sensitivity (=1).

A second OFS easily follows as an extrinsic sensor when we put a grating inside the package and read the relative grating position through the fibers with collimating lenses. If the grating has black and white lines with thickness ≈ 2 μm, compared to the 50 μm of the fiber diameter, we get an improvement in sensitivity by a factor ≈ 25.

Further, we may think of sandwiching the fiber between two rippled jigs. If the jig is actuated in compression by the force to be measured, a curvature loss is induced in the fiber proportional to the force. This readout of loss is much more sensitive, perhaps by a factor ≈ 50 with respect to the reference scheme. Next, we may consider reading the birefringence induced by the jig. Doing so, we get a polarimetric readout of the effect. We can use a Glan cube to separate polarization, detect the two components, and compute the ratio of the two fields to get a quantity proportional to the applied force.

As a last step, we can measure the optical path length variation produced in the fiber by the applied force, i.e. convert the strain gage into an interferometric OFS. We need, as shown in Fig.9-4, two beamsplitters to build an interferometer around the cell. The sensitivity of the interferometer readout is increased, say about ≈ 500 times that of the reference configuration. The sensor works better by removing ambient-related vibration errors. Using two fiber couplers to build the interferometer (Fig.9-4), we are able to improve the stability of the setup and accordingly the working sensitivity.

9.3 Readout Configurations

We now consider sensor readout configurations in more detail, to assess the basic performance of each configuration and its limits, as well as showing examples of sensors developed from the concepts.

9.3.1 Intensity Readout

In intensity readout, the physical measurand affects power propagated down the fiber in a variety of mechanisms. The three cases used in fiber sensors are illustrated in Fig.9-5. The measurand M may act on the position of a fiber pigtail and thus affect the alignment loss of a fiber joint (Fig.9-5a). A variant is obtained when the measurand actuates a light stop inserted between two fiber pigtails (Fig.9-5b). In another mechanism, the measurand is related to the index of refraction n=n(M) of the medium surrounding the joint of two fiber pigtails (Fig.9-5c). A change of the index of refraction causes the variation of the joint loss and we can read the measurand from it.

Finally, we can take advantage of curvature-induced loss to exploit intensity readout. This can be done by mounting the fiber in a fixture with a serpentine profile of corrugation, as already considered in the fiber strain-gage of Fig.9-4.

The quantity of response obtained by these transducer mechanisms can be evaluated from the alignment losses plotted in Fig.9-6 (loss is defined as the attenuation A in dB, or a=20 $\log_{10}$A, where A is the ratio of output to input power).

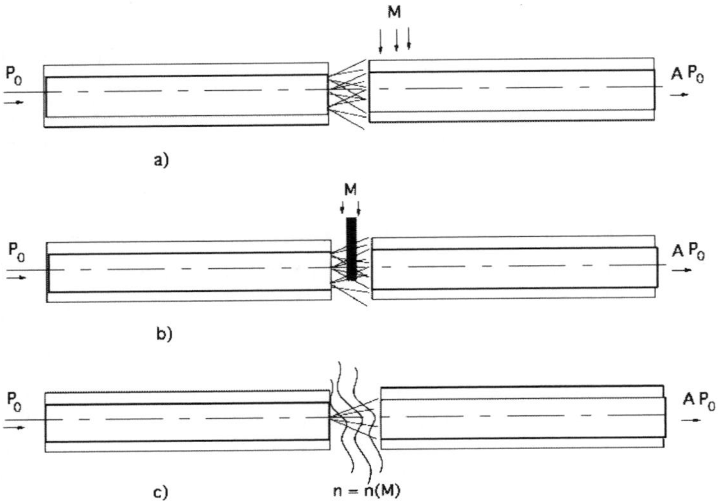

Fig.9-5 Intensity-based sensors at a fiber joint may use the measurand as: (a) the actuator of fiber pigtail position; (b), the attenuation by insertion of a light stop; (c) the variation of the index of refraction in the joint region.

The three basic contributions are considered in Fig.9-6: the longitudinal displacement z, the transversal displacement r, and the tilt angle α of the fiber axes. The fiber is assumed to be multimode with a core diameter 2a, and data are reported with the numerical aperture $NA=[\sqrt{(n_1^2-n_2^2)}]/n_0$ ranging from 0.1 to 0.3. In the loss data, the Fresnel reflection at input/output surfaces is not included. The data of Fig.9-6 also hold true for single mode fibers at first approximation. The case of n=n(M) can be treated, using the data in Fig.9-6, by considering that the external index of refraction n can affect the numerical aperture NA as n_0=n in this case.

The actuation by a stop can also be treated with the displacement-loss diagrams. Using a blade-like stop entering in the fiber gap, the total loss is the sum of a transversal displacement w/2a and a longitudinal displacement g/2a loss.

Fig.9-6 is the starting point for the design of a sensor, as we can determine the displacement leading to a full signal swing as well as the linearity of response.

However, as we start developing a practical sensor design, we soon realize that the conceptual configurations of Fig.9-5 put forth several undesirable features.

To illustrate this through an example, let us consider a pressure sensor developed from the blade-like stop concept (Fig.9-7). This sensor uses two pigtails of multimode fiber (with core diameter $2a = 50$ or 62.5 μm), aligned by sleeves to face each other precisely, typically with a gap $g = 20$ μm. A thin diaphragm membrane is deformed by the pressure under measurement. Integral with the diaphragm, a thin blade is slipped into the gap between the fibers.

From the data of Fig.9-6, we can calculate the attenuation $A=P_{out}/P_{in}$ of optical power passed though the sensor. This quantity is proportional to $I_{ph}=\sigma P_{out}$, the current supplied by the photodiode placed at the exit fiber, σ being the spectral sensitivity of the photodiode [6].

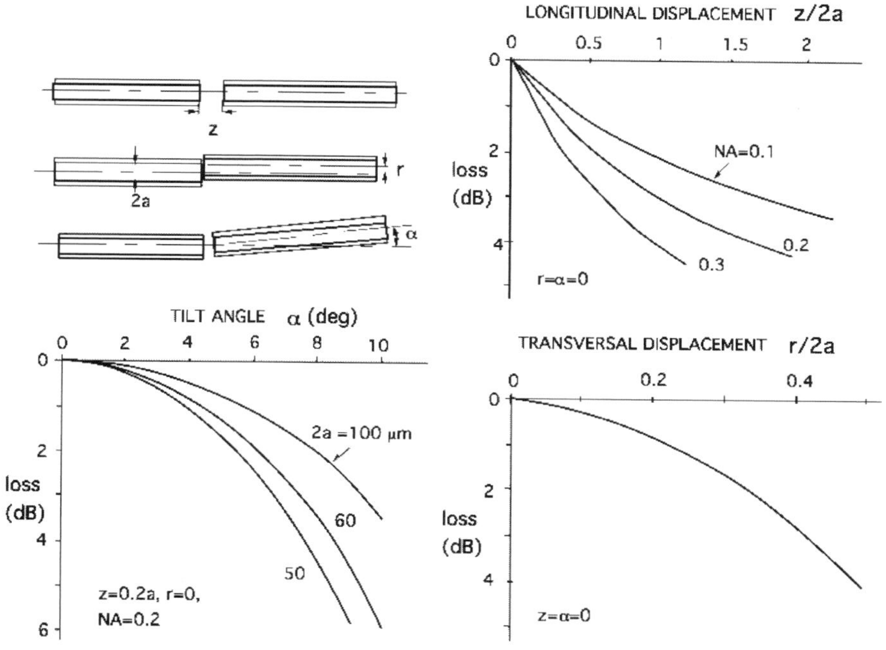

Fig.9-6 Losses due to misalignments in a joint between two fiber pigtails: longitudinal and transversal displacements and axis tilt. When more than one error is present, the total loss is the sum of individual contributions (for small losses).

Writing $P_{out} = AP_{in}$ and $I_{ph}=\sigma AP_{in}$, we can see that the photocurrent I_{ph} is a replica of attenuation A, provided the input power and spectral sensitivity are constant. Actually, both P_{in} and σ are likely to vary with temperature and aging. In particular, the input power will be supplied by an LED, and we should check temperature coefficient and long-term drift of input power by specific testing. We may as well correct the measurement from drifts in P_{in} and σ, adding a circuit to compute the ratio of I_{ph} and P_{in}/σ, at the expense of spoiling the inherent simplicity and low component count of the intensity-readout approach.

An additional error affecting the scale factor is the transversal misalignment of the two fiber pigtails. If this error drifts with time, it will seriously impair the sensor response calibration further.

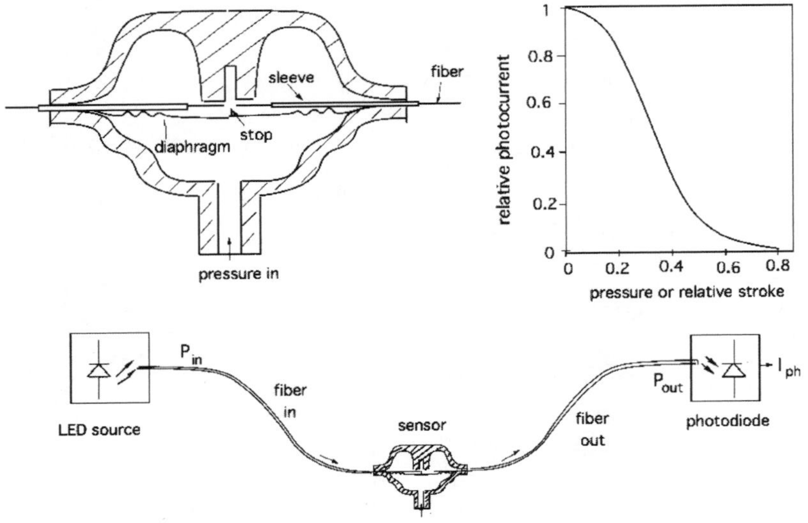

Fig.9-7 A pressure sensor (top left) uses a fiber joint with a stop actuated by a thin diaphragm. The response curve (top right) of detected current vs. stop displacement replicates the ratio of optical power P_{out}/P_{in}. The sensor is read by the LED/photodiode combination (bottom).

Another concern is the poor linearity of the pressure-to-current curve displayed in Fig.9-7 (right). We may improve linearity by limiting the range of the input pressure, but then we should require a calibration to set the working point at the middle of the dynamic range. In practice, the linearity error is not easily corrected unless we are ready to add substantial sophistication to the sensor. Thus, we would probably end up with accepting poor linearity for the least demanding application, i.e., an on/off alarm or the like.

In the design of Fig.9-7, we are using two fiber pigtails for sensor access and connection to the source (an LED) and the detector (a photodiode). Two fibers increase the cost of installation and are not really necessary. It would be better to access the sensor from just one fiber, used downstream to send optical power to the sensing region and upstream to collect the returning signal.

We may thus wish to modify the initial configuration of the intensity sensor to: (i) read in reflection; (ii) improve the linearity of response; (iii) have a self-aligned placement of the fiber. Two examples of the several possible solutions are shown in Fig.9-8. In the example, the end-face of the multimode fiber is placed at a suitable distance z_0 in front of the membrane subjected to external pressure. The membrane is a thin metal foil with enough reflectivity to act as a mirror for the light exiting from the fiber. In another implementation, the membrane is the cap of a bellow coaxial to the fiber (Fig.9-8 right) and overall size is shrunk to a minimum. The downstream and upstream optical signals can be separated with the aid of a 50% fiber coupler (Fig.9-9). The coupler is the all-fiber counterpart of the bulk-optics beamsplitter, and it transmits a fraction T and reflects a fraction R=1-T of incoming power. To minimize the total loss T(1-T) experienced by the go-and-return path to the sensor, we must take T=0.5 or, have a 50% coupler.

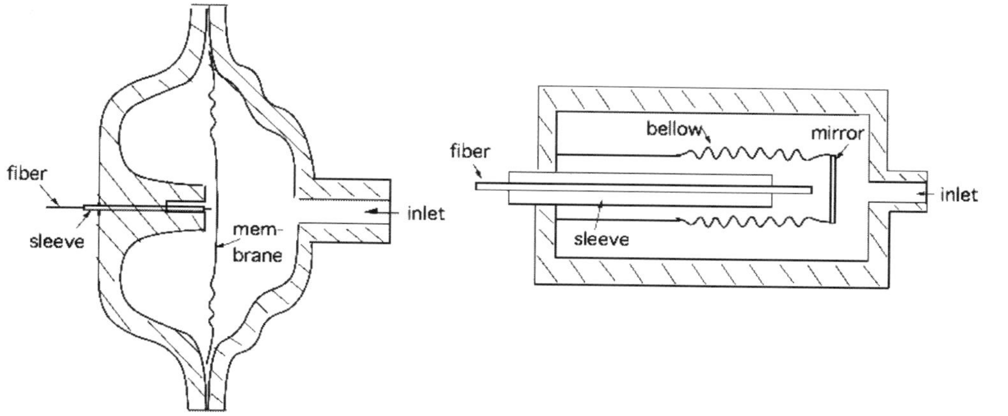

Fig.9-8 By using the membrane diaphragm as a reflector, the pressure sensor can be read in reflection by a single fiber (left). In place of the membrane, a bellow coaxial to the fiber can be used to reduce overall size (right).

The fiber and membrane combination is equivalent to a two-fiber joint with a gap twice the distance z_0, and the diagram of attenuation due to longitudinal displacement (Fig.9-6) is still applicable using $z= 2z_0$.

At the output of the measurement photodiode, the response of current I_{phm} is found close to a negative exponential (Fig.9-9) or, it is not at all linear. However, we can convert this response to an almost linear one by computing the logarithm of I_{phm}. In this way, we get the log-attenuation A (dB), a quantity which is about linear with longitudinal displacement or input pressure, as indicated by Figs.9-6 and 9-9.

We can make the conversion at little extra cost by using the photodiode itself as a logarithm converter (Fig.9-9). As it is well known [7], to perform this function we shall use an op-amp follower reading the photodiode in the open-circuit mode, so that the output is proportional to the logarithm of $I_{ph}=\sigma P =\sigma\ AP$, and hence to the logarithm of attenuation A. Working with a single measurement channel, however, the dependence of the log-signal from A would be readily masked by drifts in the power P emitted by the source as well as in the spectral sensitivity σ of the photodiode.

The logarithm circuit is important because it allows us to introduce a reference channel for the correction of drifts at a minor increase of circuit complexity and extra cost.

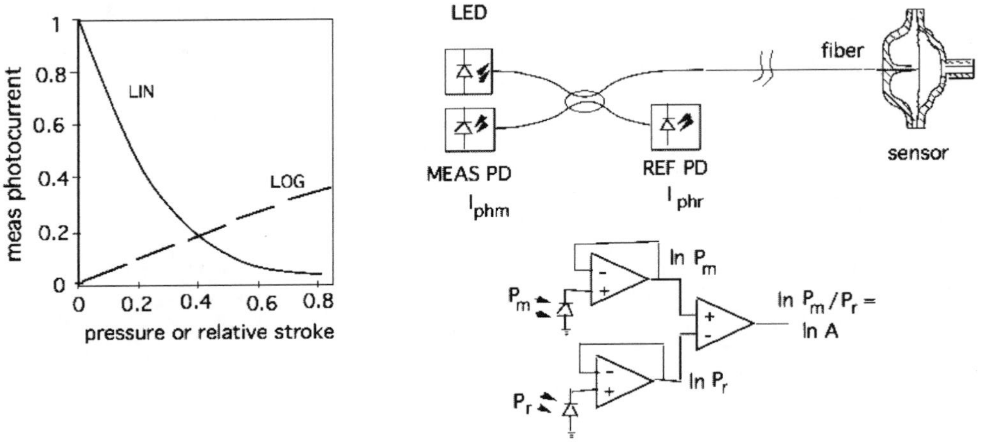

Fig.9-9 (Left) with linear processing, the response of the reflective read-out pressure sensor is a negative exponential, but we transform it into a linear one by a logarithm conversion performed by the photodiode itself. As the circuit performs the ratio of powers P_m and P_r (right), it also corrects the measurement from drifts in LED power and photodiode sensitivity.

As shown in Fig.9-9, we use a second photodiode at the unused port of the fiber coupler. This photodiode acts as a reference, and it yields a photocurrent $I_{phr}= \sigma_r P_r =(1-T)\sigma_r P_{LED}$ proportional to the power emitted by the LED and to the spectral sensitivity σ_r.

The voltages v_m and v_r at the outputs of the measurement and reference branches can be written as:

$$v_m = V_0 \log_{10} \sigma_m P_m/I_{m0} = V_0 \log_{10} \sigma_m \, A \, T \, P_{LED}/I_{m0}$$

$$v_r = V_0 \log_{10} \sigma_r P_r/I_{r0} = V_0 \log_{10} (1-T)\sigma_r P_{LED}/I_{r0} \qquad (9.1)$$

where V_0 is a scale factor (=59.6 mV, see [7]), $I_{m0,r0}$ are the dark currents, $\sigma_{m,r}$ are the sensitivity of the photodiodes, and $P_{m,r}$ are the received powers. By subtracting the outputs of the measurement and reference channels, we get a signal v_{out} given by:

$$v_{out} = v_m - v_r = V_0 \log_{10} \sigma_m P_m / \sigma_r P_r = V_0 \log_{10} A + V_0 \log_{10} T\sigma_m I_{m0}/(1-T)\sigma_r I_{r0}$$

$$= V_0 \log_{10} A + \text{const.} \qquad (9.2)$$

The last term of the first line has been assumed to be a constant because we use matched photodiodes in the measurement and reference channels, and coupler transmission T is reasonably stable with aging. In conclusion, with the log-conversion circuit we can obtain a response with good linearity and are able to introduce a reference channel that cancels the drifts of LED and photodiode [7]. With minor modifications to the basic scheme of Fig.9-9, the concept of double-channel readout with measurement and reference for drift cancellation is utilized in a number of intensity-based sensors.

A temperature OFS based on the fluorescence of a crystal excited by an LED source is illustrated in Fig.9-10. The crystal is a thin plate of GaAs cemented onto the fiber pigtail end-face, and we illuminate it with an LED emitting at a wavelength λ_{LED} below the photoelectric threshold λ_t of the GaAs material, so that the LED photons are absorbed by the crystal, generating electron-hole pairs. After a certain characteristic time, recombination of a pair takes place, with the emission of a fluorescence photon at wavelength λ_t.

Photons are radiated out of the crystal and a substantial fraction of them is collected by the fiber and brought back to the detector where we can measure the fluorescent power.

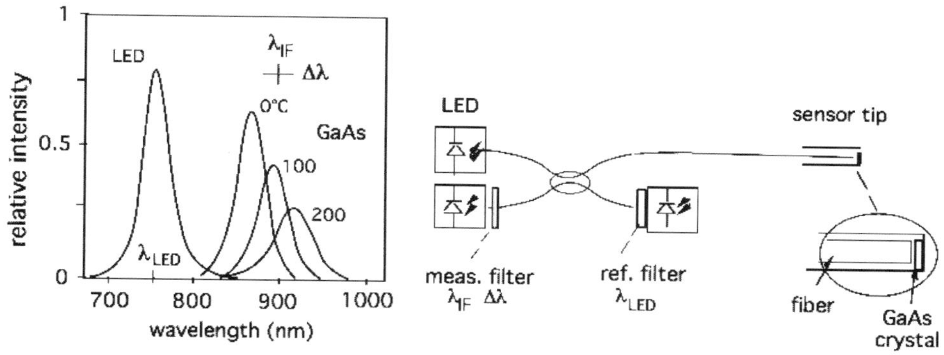

Fig.9-10 A temperature OFS uses the fluorescence of a GaAs crystal mounted on the tip of an optical multimode fiber. The LED illuminates the thin GaAs crystal at a wavelength (λ_{LED} ≈750 nm) of strong absorption. The crystal emits fluorescent light at a wavelength longer than λ_{LED}, with a precise temperature dependence of the peak wavelength (≈0.26 nm/°C). The fiber guides the light emitted by the crystal back to the photodiode.

Now, both the wavelength of emission λ_t and line width of fluorescence change with temperature, as illustrated in Fig.9-10. By inserting an interference filter in front of the photodiode, with a suitable central wavelength λ_{IF} and bandwidth $\Delta\lambda$, we can maximize the response as a function of temperature and render it approximately linear. Typical performance obtained by this approach is a ≈0.2°C accuracy on a measurement range extended from –50 to +200°C. An example of commercial temperature FOS is provided by LSENSU- Rugged Monitoring [8].

In another design aimed to cover a higher temperature range (0-800°C), we measure the temperature dependence of absorption $\alpha(\lambda)$ in a piece of special fiber (Fig.9-11). The special fiber is a modified version of the normal silica fiber, prepared to withstand high temperature and to exhibit a temperature-dependent $\alpha(\lambda)$.

In an ordinary fiber, as we increase temperature, we find several failure mechanisms to fix. The first is burnout of the secondary coating, a plastic material that can't exceed 150 to 180°C. We simply don't use coating in the high-temperature fiber, and protect the brittle fiber tip with an appropriate end-cap. Second, at about 400 to 500°C, the silica fiber itself starts softening. We can improve mechanical resistance of the silica fiber up to ≈800°C by using quartz cladding in place of the ordinary silica cladding.

Finally, we get the desired temperature dependence of $\alpha(\lambda)$ by suitably doping the silica core. There is a variety of choices available, but a good one is Nd, the same rare-earth metal used to fabricate an optical-amplifier fiber for the 1060-nm wavelength. At a ≈0.1% concentration in the core, the Nd-doped fiber has the absorption bands shown in Fig.9-11[3].
Of the two absorption bands, the one at λ_1 has attenuation strongly dependent on temperature and will be the measurement channel, whereas at λ_2 attenuation is nearly independent from temperature and will be used as the reference channel.

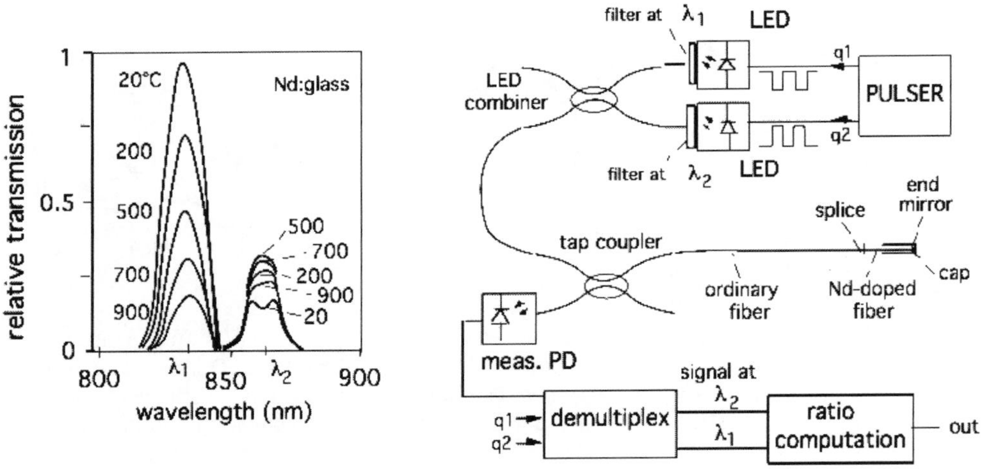

Fig.9-11 A high-temperature OFS uses a Nd-doped fiber that has two absorption bands at λ_1 and λ_2 (left): one with absorption $\alpha(\lambda)$ strongly dependent on T, and the other nearly independent from T. Outputs from LEDs are filtered at λ_1 and λ_2, combined in the 50/50 coupler and sent to the doped fiber (right). The fiber end-face is metal-coated and reflects the light signal back to the photodiode. LEDs are pulsed in alternate phases so that both λ_1 and λ_2 channels are read by a single PD. From the PD output, a demultiplexer returns the $\alpha(\lambda_1)$ and $\alpha(\lambda_2)$ signals, from which temperature is computed.

To perform the attenuation readout, we use two LEDs emitting at λ_1 and λ_2. The sensing fiber can be used in transmission or in reflection. In transmission, we need to fold back the fiber, and this operation is objectionable because the fiber without secondary coating is brittle. We may read the fiber in transmission if we metal-coat the fiber end so that it works as a mirror (as shown in Fig.9-11). In the receiver section, we can use two photodiodes with filters at λ_1 and λ_2, or a single photodiode with a multiplex of LED sources as shown in Fig.9-11. To perform multiplexing, we drive the LEDs in alternate phases by signals q_1 and q_2. Thus, only one LED is switched on at a time to feed the sensing fiber. When the returned signal is detected by the photodiode PD, we simply demultiplex it to the λ_1 and λ_2 channels through the phases q_1 and q_2. Last, we can compute the ratio of the signals, and obtain a signal related to temperature. Typical performance obtained by the doped fiber OFS is ≈1°C accuracy in a measurement range up to 800°C.

Still higher temperatures can be measured using fibers. We can look at the blackbody emission at two wavelengths, λ_1 and λ_2, and calculate temperature. We use a special fiber tip, made of a multimode sapphire core and coated with alumina black acting as the blackbody surface (Fig.9-12). The sapphire fiber is ≈10 cm long to withstand temperature, and is spliced to a normal silica fiber for the down-lead connection to the measuring instrument.

Principle of operation. The radiance per unit wavelength emitted by the fiber tip is given by the blackbody expression $r(\lambda,T)= (hc^2/\lambda^5)/(\exp hc/\lambda kT -1)$. Multiplying this quantity by the fiber acceptance $A\Omega$, A being the area and Ω the solid angle, we get the power per unit wavelength guided by the fiber down to the detection section. We use two photodetectors, looking at two bands of the emission at wavelengths λ_1 and λ_2, and compute the ratio of their currents as: $I_1/I_2 = (\lambda_2/\lambda_1)^5 \exp[hc/kT(1/\lambda_1-1/\lambda_2)]\ (\Delta\lambda_1/\Delta\lambda_2)$. Taking the inverse logarithm of this ratio, we get a quantity $[\ln I_1/I_2]^{-1}=C_a+C_bT$, where C_a and C_bT are instrumental constants. This quantity is linearly related to absolute temperature T and doesn't require any calibration, at least in principle.

The operating range of the radiant-emission OFS is 500 to 2000°C, the typical resolution is ≈0.05°C, and the accuracy is about 0.5°C [4].

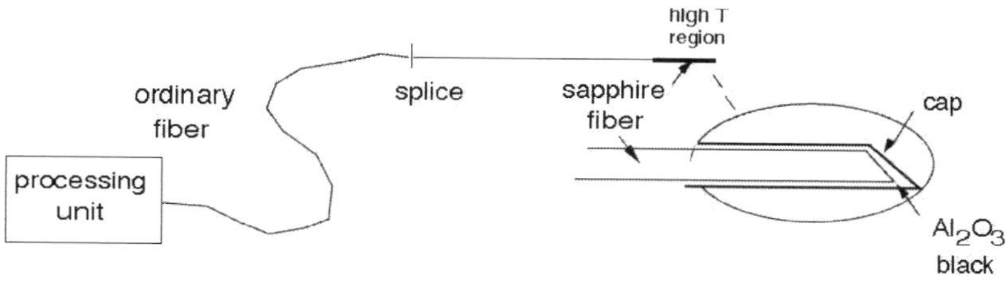

Fig.9-12 This OFS for temperature measurements in the range 500 to 2000 °C employs a temperature-resistant sapphire fiber coated with alumina black. The fiber tip emits blackbody radiation, and power collected by the fiber is sent to the two-color processing unit.

In the field of chemical OFSs, a simple sorter of chemical species is readily implemented by a multimode fiber exposed to the surrounding environment [9,11]. To accomplish this, we remove the clad from a piece of fiber of suitable length (usually ≈10 cm is enough). The external medium determines the local numerical aperture $NA=\sqrt{(n_1^2-n_{ext}^2)}/n_0$, n_{ext} being the index of refraction, and hence the extra attenuation experienced by the fiber tip.

The attenuation depends on the angle θ_l under which the source launches light in the fiber, the maximum value being that of fiber acceptance $\sin\theta_l = \sqrt{(n_1^2-n_2^2)}/n_0$. Light launched at angles θ between θ_l and $\theta_{ext} = \sin^{-1}NA$ is no longer guided and the corresponding attenuation is a function of n_{ext}. The typical attenuation read by a multimode OFS is shown in Fig.9-13 for several chemicals of interest in pollution monitoring. Attenuation can be read in reflection or transmission with the help of one of the several readout schemes discussed so far for intensity sensors.

Regarding this application, we actually get a large measurable signal from most species, as shown in Fig.9-13, but we have no means to discriminate one substance from another, or, the measurement has no selectivity to species.

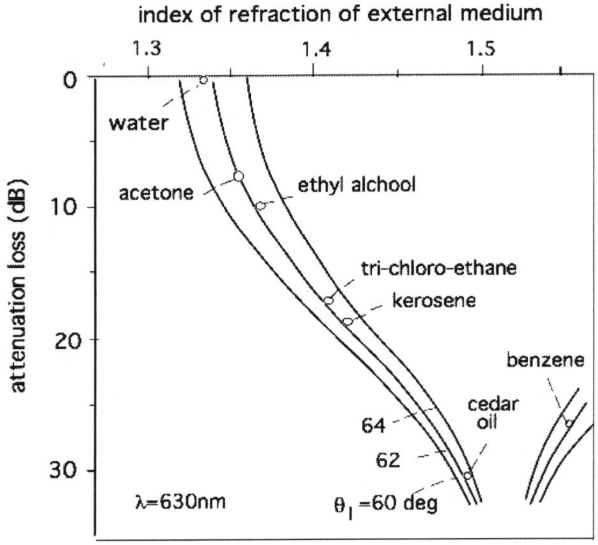

Fig.9-13 Transmission loss produced by an external liquid surrounding an uncoated silica fiber, as a function of the index of refraction and with the angle of launch θ_l as a parameter.

Another OFS used for measuring of chemical quantities is based on the optical change undergone by a special fiber tip, called an *optrode* (a contraction of *optical electrode*). As exemplified in Fig.9-14, the optrode is made of a thin layer of a polymer incorporating a reactant substance. The reactant may change either its spectral attenuation or the fluorescence response when exposed to the chemical species under measurement. For example, to measure the pH or acidity of a solution, the polymer layer is made of XAD-2 microspheres, treated for adsorption of phenol red.

Phenol red has a peak of absorption at λ=560 nm, which is strongly dependent on the pH, whereas at λ=475 nm we find a nearly constant attenuation (Fig.9-14).

Thus, we can measure the phenol-red optrode with the aid of a scheme similar to that of Fig.9-11, and obtain a quantity related to the relative attenuation. After a calibration, the pH is read with an accuracy of ≈ 0.02 in the range 6.8 to 7.9, and reproducibility is good. The measurement time is relatively long, about 60 s, because the external medium shall diffuse through the boundary membrane before coming in contact with the reactant.

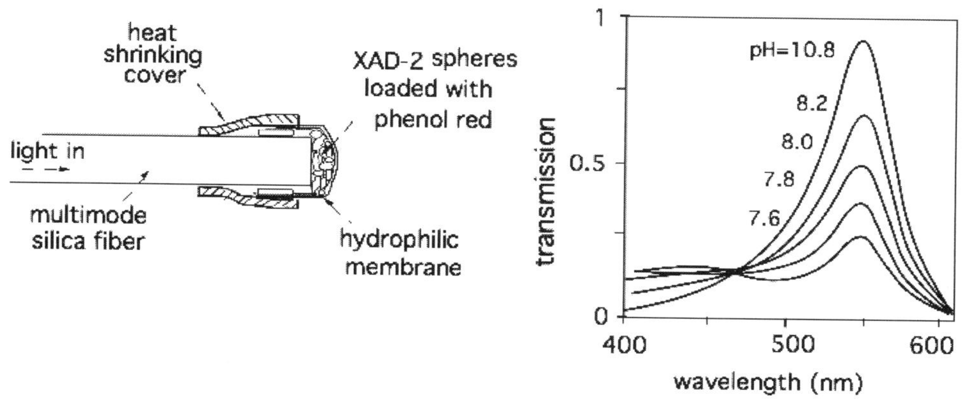

Fig.9-14 A pH optrode has its fiber tip covered with polymer microspheres (XAD-2) bearing the reactant phenol red (left). The absorption spectrum of phenol red shows peak sensitivity to pH at λ≈550 nm, and a reference at λ≈475 nm (nearly independent from pH).

Another reactant used for a somewhat wider range of measurement is bromothymol blue, a dye that has a maximum variation of absorption at λ=620 nm and a minimum at λ=500 nm. With it, a pH optrode has been developed for the measurement range 6.4 to 7.8.

A number of other chemical quantities can be measured with optrodes [11,12]. They are pO_2, pCO_2, and a number of water pollutants like pesticides, hydrocarbons, heavy metals, etc. For biomedical use, additional applications of OFS measurement include: glucose, metabolites, hemoglobin, breathing and blood gases, enzymes and co-enzymes, inhibitors, lipids, drugs, immunoproteins, etc. [12-14].

9.3.2 Polarimetric Readout

In a polarimetric OFS, the transducer mechanism affects the state of polarization and we perform a readout of the birefringence induced by the measurand.

Birefringence effects are classified as *linear* or *circular*.

We get linear birefringence when the measurand introduces an optical retardance Φ along a (slow) axis of the fiber with respect to the other, orthogonal (fast) axis. For example, in strain and force sensors, the axis parallel to the force is the slow axis, and retardance is proportional to the force.

Linear birefringence is described as a difference in the propagation constants β_s and β_f along the slow and fast axes.

Thus, down a propagation length L, we get a retardance $\Phi = (\beta_s - \beta_f)L$.

On the other hand, we get circular birefringence (also referred to as rotatory power) when the measurand affects the direction of a linear polarization state, making it rotate while it propagates through a piece L of fiber. The angle of rotation is then $\Psi = \Delta\beta\, L$.

Equivalently, circular birefringence can be described as retardance between right- and left-handed circular polarization states. We may then write $\Delta\beta = \beta_R - \beta_L$, where β_R and β_L are the propagation constants for right- and left-handed polarizations, respectively.

Different from linear birefringence, which can only be reciprocal, circular birefringence can also be non-reciprocal. Reciprocity has to do with the change of birefringence when the direction of propagation is reversed or, propagation vector $\underline{k}$ is changed to $-\underline{k}$.

Only two non-reciprocal effects of circular birefringence are known: the Faraday effect, which is due to a longitudinal magnetic field, and the Sagnac effect (see Sect.7.2), which is due to inertial rotation. All other effects (sugar rotatory power, crystal twist, etc.) are reciprocal, that is, rotation Ψ is unchanged when reversing the direction of observation (or the cell containing the material). Instead, reversing the direction changes rotation Ψ to $-\Psi$ in a non-reciprocal material or medium.

9.3.2.1 Circular Birefringence Readout

Among intrinsic polarimetric sensors, two examples have been developed to take advantage of circular birefringence effects. The first OFS is the torsion-bar balance, based on the reciprocal rotation induced by a twist of the fiber. The twist is applied by the small weight to be measured. Because we are able to resolve an angle of polarization rotation as small as $\approx 10^{-4}$ rad, the resulting OFS is a very sensitive balance, capable of appreciating a very small torque applied to the fiber (down to $\approx\mu g\cdot cm$).

The second OFS is a magnetic field or electrical current sensor, and is based on non-reciprocal rotation. In the following, we refer to this sensor to explain the readout of circular birefringence. However, the concept also holds for the readout of reciprocal circular birefringence as well as for extrinsic sensors.

Non-reciprocal rotation is also called the Faraday effect, and it is induced by the longitudinal component H_l of the magnetic field applied to the fiber. The angle of rotation Ψ is proportional to the line integral of the field H_l along the propagation path. The elemental contribution is $H_l\, dl = \underline{H}\cdot\underline{dl}$, where the dot $\cdot$ indicates the scalar product of vectors $\underline{H}$ and $\underline{dl}$. The constant of proportionality V is called the *Verdet constant*.

Thus, we may write the rotation angle as [15, 16]:

$$\Psi = V \int_L \underline{H} \cdot \underline{dl} \qquad (9.3)$$

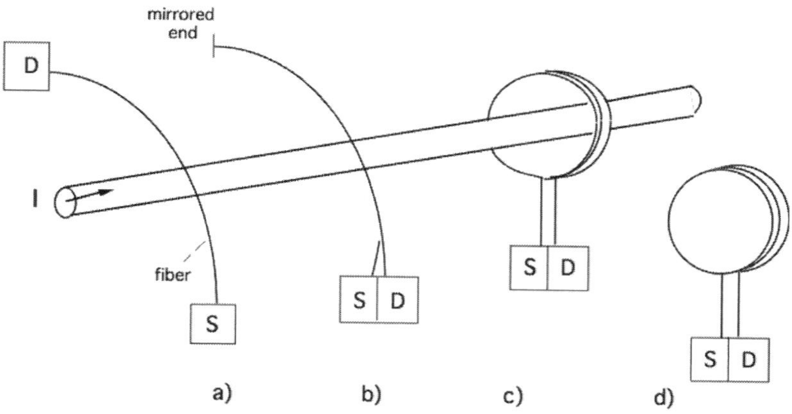

Fig.9-15 A piece of fiber or a coil can be used in connection with a polarized source S and a detector analyzer D to sense the Faraday rotation Ψ induced by a magnetic field. In a), with Ψ we read the field $\int \underline{H} \cdot \underline{dl}$, averaged along the fiber length. If the fiber has a mirrored end, the rotation doubles because it is non-reciprocal. With a fiber coil, rotation is proportional to the concatenated current I. Therefore, in c), Ψ is proportional to the current I in the concatenated wire, whereas in d) Ψ is exactly zero.

Using a piece of fiber to sense the Faraday rotation, we get a magnetic field OFS (Fig.9-15a). Indeed, in this case, rotation Ψ is proportional to the magnetic field $\int_L \underline{H} \cdot \underline{dl}$ averaged along the fiber length L.

With a mirrored end (Fig.9-15b) on the fiber, we fold back the propagation and get twice the rotation because the Faraday circular birefringence is non-reciprocal. By contrast, applying the fiber a twist-induced, reciprocal circular birefringence, we get rotations with opposite signs along a go-and-return path, and the total effect is zero.

With a fiber coil wound around the conductor (Fig.9-15c), the OFS is a true current sensor because the rotation becomes the circulation of magnetic field intensity.
This quantity is known from elementary magnetism to be equal to the magneto-motive force, that is, $\int_L \underline{H} \cdot \underline{dl} = N_f \, I$. Thus, if we arrange the sensor in the form of a winding of N_f turns of fiber wound around an electrical conductor carrying a current I, Eq.9.3 becomes:

$$\Psi = V \int \underline{H} \cdot \underline{dl} = V \, N_f \, I \qquad (9.3')$$

It is interesting to note that the result holds independently from coil size or shape, and only requires that the coil is linked to the conductor carrying the current I.

Also, if the coil is not linked to the conductor (Fig.9-15d), the rotation Ψ is exactly zero. This feature is very attractive for the application of the current OFS to three-phase power lines [16]. Here, the three wires are close and the sensor of each phase must be sensitive only to the current carried by its conductor and not to others.

For the readout of either reciprocal or non-reciprocal rotation, we can use the general scheme illustrated in Fig.9-16.

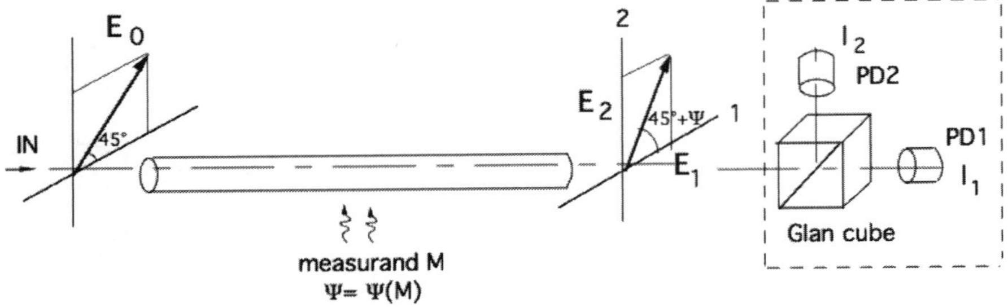

Fig.9-16 Schematic for the readout of circular birefringence. The input state of polarization E_0 is rotated by a small angle $\Psi=\Psi(M)$ by the measurand M. The Glan cube analyzer is oriented at 45° with respect to E_0, and splits the output field into components E_1 and E_2. The photodiodes provide output currents proportional to E_1^2 and E_2^2.

In this scheme, the fiber is read with a linearly polarized state, which can be shown [17] to be the optimal choice. As a reference system, let us take the axes (1 and 2 in Fig.9-16) of the output analyzer (the Glan cube), and assume that the input field E_0 is oriented at 45°. Along the fiber, the measurand M produces a rotation of the state of polarization, and the output field emerges from the fiber rotated by an extra angle $\Psi=\Psi(M)$ added to the initial 45°, ending up oriented at 45°+Ψ.

The Glan cube divides the output field in the components E_1 and E_2, and directs them on two separate photodiodes, PD1 and PD2. We may then write the fields E_1 and E_2 and the associated currents I_1 and I_2 as:

$$E_1 = E_0 \cos(45°+\Psi),\quad E_2 = E_0 \cos(45° -\Psi) = E_0 \sin(45° +\Psi)$$

$$I_1 = E_1^2 = E_0^2 \cos^2(45° +\Psi),\quad I_2 = E_0^2 \sin^2(45° +\Psi) \tag{9.4}$$

As we can see from Eq.9.4, neither the fields nor the currents are linearly related to the measurand. To obtain a better response, we compute the ratio S of difference I_1-I_2 to sum I_1+I_2 of the detected currents. Doing so, we obtain the following with easy algebra:

$$S = (I_1-I_2)/(I_1+I_2) = E_0^2[\cos^2(45°+\Psi)-\sin^2(45°+\Psi)]/E_0^2[\cos^2(45° +\Psi)+\sin^2(45°+\Psi)]$$

$$= \cos 2(45°+\Psi) = -\sin 2\Psi \tag{9.5}$$

Eq.9.5 is a good result because, with the difference-to-sum ratio, we have obtained a signal linearly related to the measurand Ψ [$\sin 2\Psi \approx 2\Psi$ for small Ψ]. In addition, the measurement is no longer dependent on E_0^2, the power of the input beam, or, common-mode fluctuations are eliminated thanks to the ratio-like structure of the signal S.

The range of linearity may also be improved further by computing the rotation angle as $2\Psi = \arcsin S \approx S + S^2/6$.

Next, we encounter a measurement ambiguity in the sine function when the rotation angle exceeds $\pi/2$. Similar to the treatment of interferometer signals, we need to supplement the measurement channel of Fig.9-16 with a second measurement channel to remove the ambiguity. This way, we make two orthogonal signals of the type $\sin 2\Psi$ and $\cos 2\Psi$ available. We can obtain the second channel by deriving a fraction of the output and analyzing it through a Glan cube oriented at 0° and 45° with respect to the direction of the input state of polarization, as illustrated in Fig.9-17.

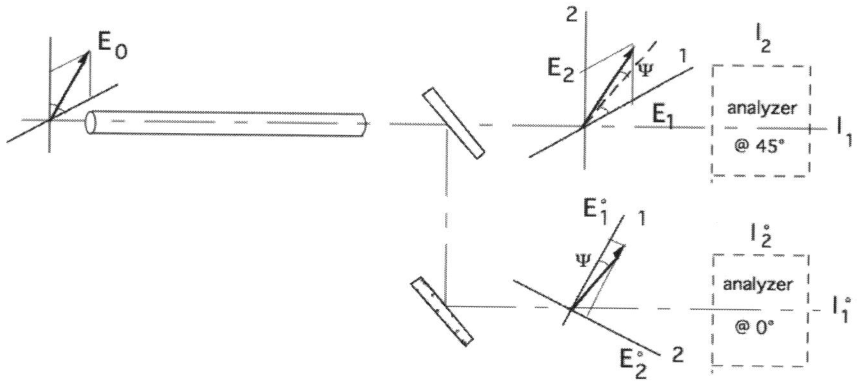

Fig.9-17 By using two Glan-cube analyzers of the output state of polarization, we can generate two signals, sin 2Ψ and cos 2Ψ, which allow us to compute the angle Ψ without ambiguity, also for $\Psi > \pi/2$.

Repeating the arguments leading to Eq.9.4 for signals $E_1°$ and $E_2°$ of the second channel, we can write:

$$E_1° = E_0 \cos \Psi, \quad E_2° = E_0 \cos(90° - \Psi) = E_0 \sin \Psi$$

$$I_1° = E_0^2 \cos^2 \Psi, \quad I_2° = E_0^2 \sin^2 \Psi$$

$$S° = (I_1° - I_2°)/(I_1° + I_2°) = \cos 2\Psi \tag{9.6}$$

The functions indicated in Fig.9-17 can be implemented with the all-fiber technology, as shown in the schematic of Fig.9-18. In it, we use a normal coupler C (polarization-insensitive) as the main divider of the output signal, and polarization-splitter couplers PS as the all-fiber versions of the Glan cube.

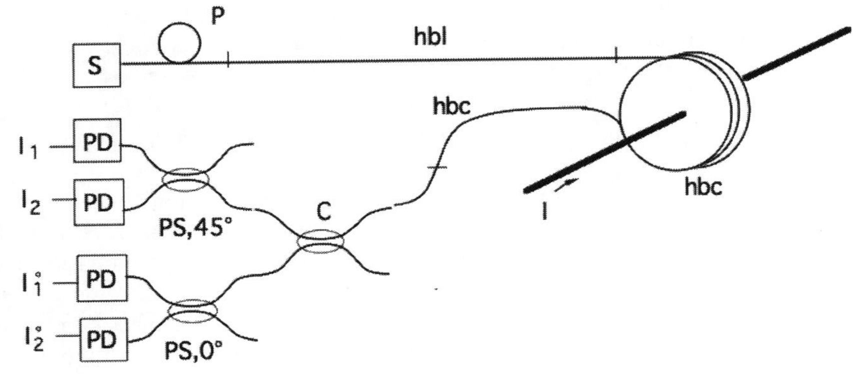

Fig.9-18 An all-fiber circuit for the measurement of circular birefringence. Light from source S is polarized by fiber polarizer P and sent up to the sensing coil through the high linear-birefringence (hbl) fiber to maintain polarization. The down-lead fiber is made of high circular-birefringence (hbc) fiber. Coupler C splits the output at 50% for analyzers PS. Analyzers are polarization-splitting couplers with slow axes oriented at 0° and 45°.

The fiber used in the coil and in the down-lead trunks is critical and requires brief discussion. In the coil, the bend of the fiber introduces linear birefringence that interferes with the small Faraday rotation. To quench the linear birefringence, a larger circular birefringence is added to the fiber of the sensing coil. This is done by twisting the fiber [16] with an appropriate period of twist (typically ≈1-3 cm). The obtained fiber becomes a high circular birefringence (hbc). It adds a reciprocal rotation Φ_{rec} to the useful Faraday rotation Ψ. This is not a problem with ac currents, however, because Φ_{rec} is a constant term and can be filtered out in the detected signal.

The same hbc fiber can also be used in the down-lead trunks connecting the coil to the instrument. Finally, in the upgoing lead, we can use a high linear birefringence fiber (hbl) in place of the hbc, as the hbl fiber is more insensitive to bends. [We can't use the hbl fiber in the downgoing lead, because the state of polarization out of light from the coil is unknown]. A final improvement to the all-fiber scheme of Fig.9-18 is to take advantage of the difference in reciprocity of the terms Φ_{rec} and Ψ. If we are able to cancel out Φ_{rec}, operation of the current sensor is extended down to the dc component. The reciprocal rotation Φ_{rec} of hbc fiber pieces is actually subtracted if we trace back propagation through the fiber, whereas the non-reciprocal rotation Ψ is doubled.

As shown in Fig.9-19, we obtain this condition by using a mirrored fiber coil end, reflecting back the radiation at the coil exit [16].

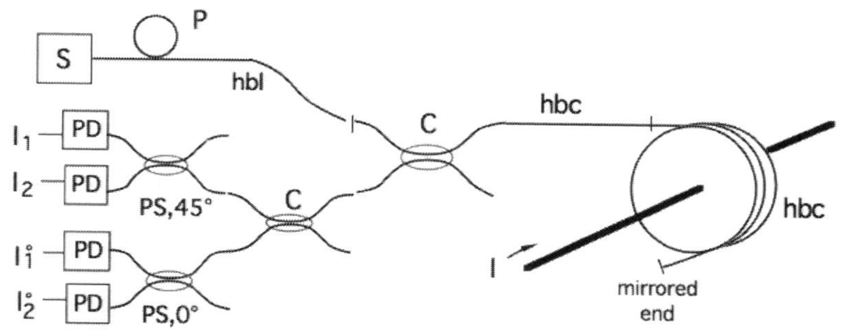

Fig.9-19 With a fiber coil ending on a mirrored end, we are able to cancel out the static rotation due to circular birefringence of the hbc fiber and extend the current measurement down to the dc component.

Developing the concept further, we might think of canceling the residual linear birefringence of the coil and access lead. This can indeed be done using a Faraday mirror at the exit of the coil, in place of the normal plane mirror obtained by metal coating the fiber end face (Fig.9-19). The Faraday mirror is a device combining a 45° Faraday rotator and a mirror. The rotator is simply made of a YIG crystal exposed to a magnetic field [16,18] that provides a non-reciprocal 45° rotation. Because of the total 90° rotation on reflection at the Faraday mirror, any input state of polarization is turned into its orthogonal on the Poincaré sphere [17]. Then, the go-and-return net change of state of polarization is zero, provided the effects generating the polarization changes are reciprocal. Therefore, all the linear and circular (reciprocal) birefringences are canceled out.

Unfortunately, cancellation does not affect the quenching effect discussed previously, and thus an eventual Faraday mirror can't free us from using the hbc fiber in the sensing coil to contrast bending birefringence [16].

Going back to consider the signals S and S°, we can reconstruct argument Ψ and current I also for large values of rotation (>2π). The processing of signals may be either analog or digital, as outlined in Fig.9-20.

In the analogue approach, we use the same method discussed in Sect.5.1.2 to process the sinΨ and cosΨ signals of an interferometer. We compute the derivatives of both signals, cross-multiply them with the other signal, and subtract the result. In this way (Eq.5.5), we obtain dΨ/dt and, after a final integration, the desired Ψ proportional to current I.

While relatively cheap to implement, the analog approach has some limitations at very small signals as the errors due to the analog multiplier become important. In addition, at low frequency (or in dc), the low-pass function of the derivative is not exactly canceled out by the final integration and a low frequency cutoff remains.

We may then prefer the digital processing (Fig.9-20), whose only limitations are associated with the offset and truncation errors introduced by the analog-to-digital converters (ADC). A reduced instruction set microcomputer (RISC), or even a fully programmable gate array (FPGA), may be adequate to carry out the modest computational load that is required.

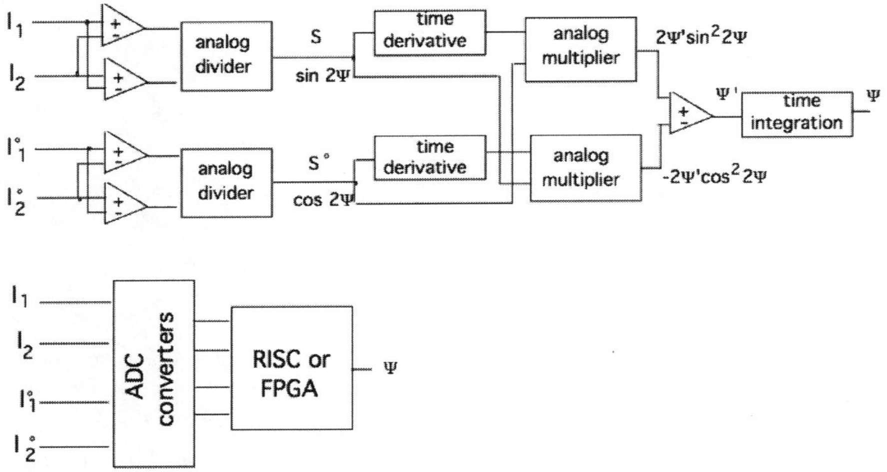

Fig.9-20 Analog (top) and digital (bottom) processing of photodetected currents to obtain rotation angle Ψ.

9.3.2.2 Performance of the Current OFS

After the measurement of currents I_1 and I_2 and the calculation of the angle 2Ψ (Eq.9.5), we shall now consider the minimum detectable signal NEI= σ_I (Noise-Equivalent Current) of the sensor. At the quantum noise limit, this quantity is given by the rms phase noise $(2eB/I_0)^{1/2}$ divided by the responsivity 2VN, or:

$$NEI = (1/2VN)\,(2eB/I_0)^{1/2} \tag{9.7}$$

Here, V= Ψ/NI is the Verdet constant of the material [15], expressing the specific Faraday rotation Ψ per unit magneto-motive force (NI), and I_0 is the photodetected current in each photodiode. The 3-dB bandwidth B of the sensor is primarily limited by the transit time τ through the coil and is given by:

$$B = 0.44/\tau = 0.44c/2\pi rN \tag{9.8}$$

where r is the coil radius and N the number of turns.

In Fig.9-21, we plot the NEI of the current OFS versus the bandwidth of measurement B, with the photodetected current I_0 as a parameter [16], as given by Eq.9.7. In the diagram, we also report the results obtained by a few research groups [19-23]. These are the dotted lines connecting the experimental results to the expected theoretical values.

As can be seen, the theoretical NEI limit due to the quantum noise associated with the readout beam is approached by a factor of 4...10, in practice.

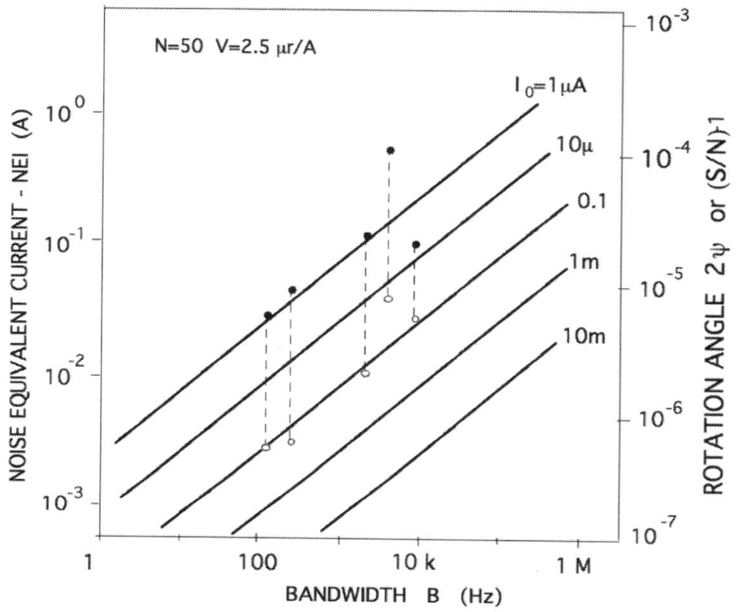

Fig.9-21 Minimum current (or NEI) that can be detected in a Faraday rotation OFS, as a function of bandwidth B and with photodetected current I_0 as a parameter. Experimental points are shown (full circle) along with the corresponding theoretical values (open circles).

These results tell us that there is space for further improvement, yet they are already satisfactory for applications to power lines, because currents of, let's say, 1 A can be detected with an adequate signal-to-noise ratio (10 or better), up to bandwidth of ≈10 kHz.
In Fig.9-22, we report the waveforms obtained by an experimental Faraday current OFS built at University of Pavia, along with the corresponding electrical waveforms, that are closely matched by the OFS optical transformer. In particular, for a sinusoidal waveform, the electrical one is delayed by 90° with respect to the optical, as expected because the secondary voltage $V_2=j\omega MI_1$ lags the primary current I_1 of 90°. Also shown in Fig.9-22 is the ability of the current OFS to reproduce fast transients, down to pulse widths in the range of 0.1 milliseconds or less.

Another approach to the current FOS has been using a closed-loop serrodyne interferometric technology, exactly like the one shown in Fig.8-28, to read the Faraday rotation. Experimental results reported this approach are in line with (but not outperforming) those of Fig.9-21, and the reason is that the minimum resolved rotation angle is again a few 10^{-6}.

Several optical current sensors have been announced by vendors, primarily ABB (Sweden) [25] but also Hitachi (J), NxtPhase Vancouver (Can), and SDO Arteche, Sidney (Aus) have units available.

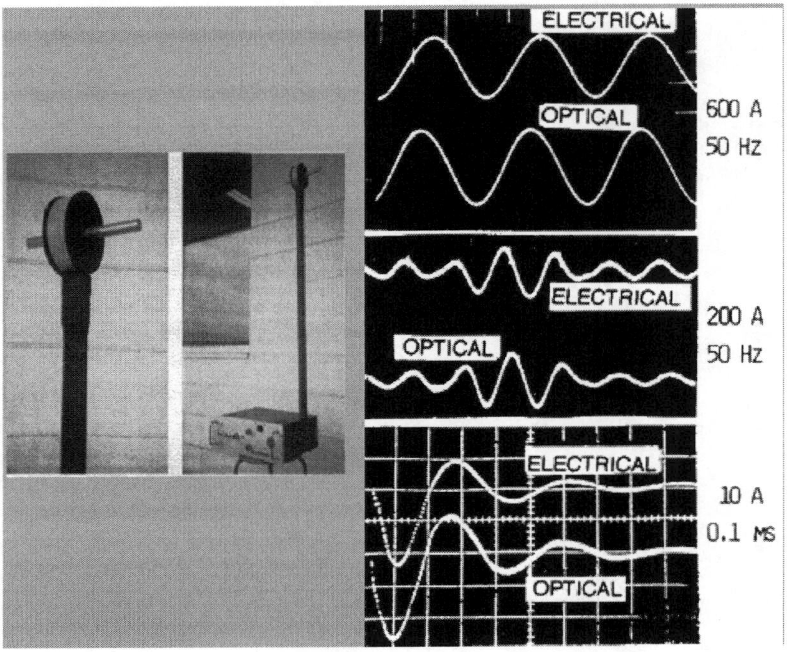

Fig.9-22 A laboratory prototype of the current sensor (left) and examples of detected waveforms (right). Top traces: currents measured by a normal electrical transformer, bottom traces: currents obtained with the OFS. Top: electrical mains current 600 A, 50 Hz; middle: a transient with 200 A peak current at 50 Hz; bottom: a fast disturbance with 10-A peak on a 0.1-ms time scale.

Another application of Faraday rotation is *magnetic field sensing*. In principle, we can use a single piece of fiber as in Fig.9-15a, to get a net non-zero rotation $\Psi = V \int_L H \cdot dl$, but the Verdet constant is small in ordinary silica fibers ($V \approx 2.5$ μr/A at 800 nm) and the rotation is 10^{-6} rad or less. To use a coil and cumulate the Faraday rotation on several turns, we shall circumvent the zero line-integral, $\int_{line} H \cdot dl = 0$. Taking advantage of bending birefringence, we may obtain a linear birefringence of 2π per turn and have a non-zero integral for our coil. To do so, we need a curvature radius R for the winding of about $R \approx 1$-cm [25], so that the state of polarization changes its orientation by π each half-turn. The Faraday elemental rotation, $+d\psi$ and $-d\psi$ on opposite half-turns, becomes summed with the same sign and we have a total rotation $\Psi = VH\pi r$, which is only half that corresponding to the total fiber length [25].

9.3.2.3 Linear Birefringence Readout

Several intrinsic and extrinsic OFSs are based on linear birefringence effects. Indeed, silica and other optical materials of fibers are sensitive to external mechanical perturbations, such as compression and bending. This circumstance leads to the development of OFSs for mechanical measurands (Fig.9-3). In addition, birefringent crystals like lithium niobate ($LiNbO_3$) exhibit a significant and reproducible temperature coefficient, and based on that, a temperature OFS can be developed from a linear birefringence readout.

Thanks to the elasto-optical effect, mechanical perturbations are translated into linear birefringence. As it is well known, linear birefringence consists of an optical delay Φ between the linear state of polarization propagating along fast and slow axes.

The optical delay or phase shift Φ is proportional to the measurand modulus, at least at small levels of perturbation, and the axes of birefringence are normally parallel and perpendicular to the axis along which the measurand is applied (Fig.9-23).

The basic scheme for the readout of linear birefringence is shown in Fig.9-23. Light from the source enters the fiber with a linear state of polarization, oriented at 45° with respect to the birefringence axes set by the measurand, that is, the x and y axes.

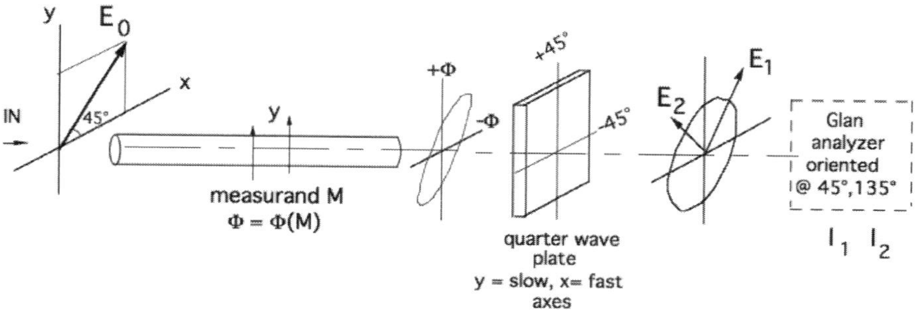

Fig.9-23 Schematic for the readout of linear birefringence. We enter the fiber with a linear polarization E_0 oriented at 45° so as to have two equal components on the x and y axes. At the fiber exit, we place a quarter wave plate for an additional +45° and –45° delay. We then analyze the field components E_1 and E_2 by means of a Glan cube, with the axes oriented at 45° and 135°.

At the fiber output, the two polarization components are phase shifted by +Φ and -Φ. Thus, light emerges from the fiber with an elliptical state of polarization. Before analyzing the state of polarization, we insert a quarter-wave (λ/4, yielding a 45° birefringence) plate with the principal axes oriented along x and y. This way, the total phase shift between the two components becomes +45°+Φ and -45°-Φ.

Next, we analyze the output field with a Glan cube beam splitter, with the axes oriented at 45° and 135° in the x-y reference, that is, parallel and perpendicular to input field E_0. To write the two field components E_1 and E_2 supplied to the Glan cube, we note that the input field feeds with two equal amplitudes $E_0/\sqrt{2}$ the x and y components. These components are phase-shifted by 45°+Φ and -45°-Φ as indicated in Fig.9-23, and then projected on the 45° and 135° axes. Adding the two contributions from the x and y axis, the fields of E_1 and E_2 are written as follows:

$$E_1 = (1/2)\, E_0 \exp i(-45°-\Phi) + (1/2)\, E_0 \exp i(45° + \Phi) = E_0 \cos(45° + \Phi),$$
$$E_2 = (1/2)\, E_0 \exp i(-45°-\Phi) - (1/2)\, E_0 \exp i(45° + \Phi) = -i\, E_0 \sin(45° + \Phi) \qquad (9.9)$$

In the above equations, the factor of 1/2 in the amplitude comes from the double projection on x and y, and on 45° and 135° of the initial component E_0. In addition, to simplify the algebra, we have written delays in a symmetrical format, using -45° and +45° in place of 0° and 90°, and -Φ and +Φ in place of 0 and 2Φ. Thus, the total phase introduced by the measurand is assumed to be 2Φ.

Associated with the fields, we get two signals, I_1 and I_2, from the photodetectors, given by the square of the field modulus or:

$$I_1 = |E_1|^2 = E_0^2 \cos^2(45° + \Phi),$$
$$I_2 = |E_2|^2 = E_0^2 \sin^2(45° + \Phi) \qquad (9.10)$$

It is now easy to proceed like in Sect.9.3.2.1 and derive a quantity linearly related to the measurand phase Φ. We compute the ratio S of difference I_1-I_2 to sum I_1+I_2 of the detected currents and obtain:

$$S = (I_1-I_2)/(I_1+I_2) = \cos 2\,(45°+\Phi) = -\sin 2\Phi \qquad (9.11)$$

As in the previous section, we may want to supplement the measurement by a second orthogonal signal, cos2Φ, to remove the ambiguity of the sine function. To accomplish this function, we take off the quarter-wave plate and analyze the output with the Glan cube, again oriented at 45° and 135° with respect to the x-y reference. We then obtain two field components, $E_1°$ and $E_2°$, given by:

$$E_1° = (1/2)\, E_0 \exp -i\Phi + (1/2)\, E_0 \exp i\Phi = E_0 \cos\Phi,$$
$$E_2° = (1/2)\, E_0 \exp -i\Phi - (1/2)\, E_0 \exp i\Phi = -i\, E_0 \sin\Phi \qquad (9.12)$$

Computing the signal $S° = (I_1°-I_2°)/(I_1°+I_2°)$ gives as a result:

$$S° = \cos 2\Phi \qquad (9.13)$$

Also for the linear birefringence OFS we can extend the range of linearity beyond 2π. The functions indicated in Fig.9-23 and the extensions required for S and S° can be implemented in all-fiber technology, as shown in the schematic of Fig.9-24.

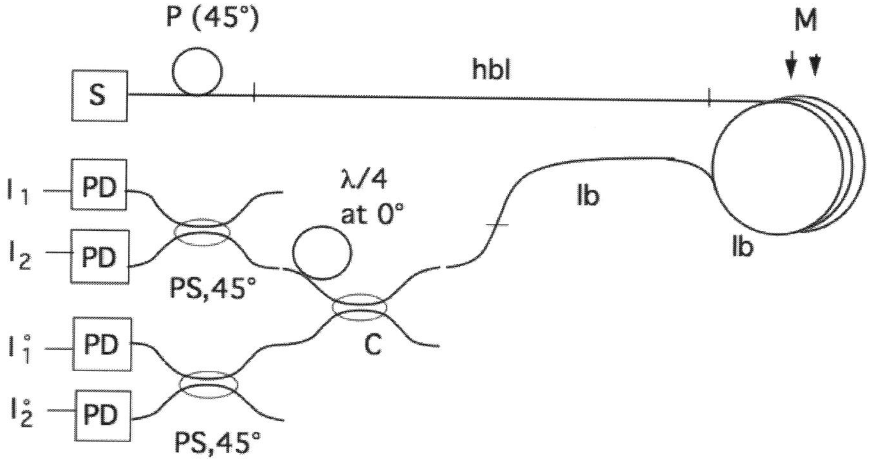

Fig.9-24 An all-fiber circuit for measuring linear birefringence. Light from source S is polarized by the fiber coil P at 45° with respect to the slow/fast axes of birefringence of the measurand M. Coupler C splits the output from the sensing region for the double-section analyzer. One section with the λ/4 plate and a Glan cube oriented at 45° gives the I_1 and I_2 signals for the sin2Φ term. The other section, without the λ/4 plate, gives the $I_1°$ and $I_2°$ signals for the cos2Φ term.

The schematic operates much like that for the readout of circular birefringence (Fig.9-18). We need two sections for deriving the I_1, I_2 pair of signals associated with S, and the $I_1°$, $I_2°$ pair associated with S°. In one section, the analysis is made after a quarter-wave phase shift, whereas in the other section, the quarter-wave is missing.

In principle, the adduction fiber should have high linear birefringence (with axes oriented at 45°and 135°) to preserve polarization, whereas the down-lead fiber is a low-birefringence (or spun) fiber and shall be shielded by external mechanical disturbances.
From the photodetected outputs, the processing to compute S and S° follows as outlined in Fig.9-20, and the same circuits apply.

It is interesting to note that the double-section analyzer can be simplified to a single-section analyzer if we replace the λ/4 plate with a controlled phase delay device. This can be realized by a PZT (lead zirconate titanate ceramic) element with a short piece of fiber cemented on it.

When the piezo is fed by the appropriate voltage, it squeezes the fiber and introduces a linear birefringence via the elastooptical effect. With ≈1 cm of fiber and a 1-mm thick PZT, we may need a voltage V_S≈10 V to impress the desired λ/4 birefringence.

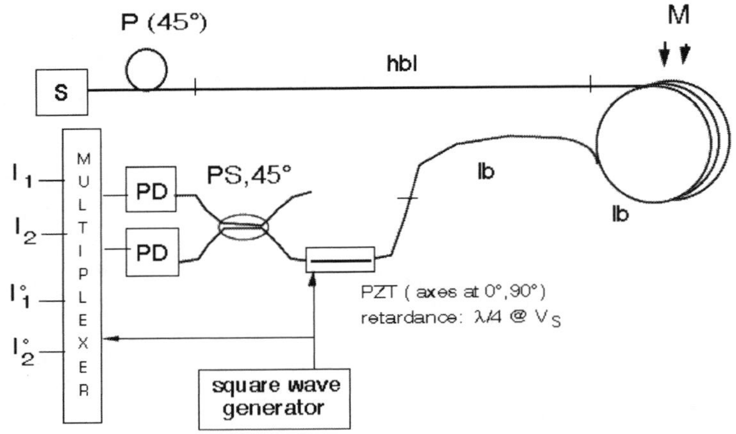

Fig.9-25 The linear birefringence can be read out with a single analyzer by using a controlled birefringence element PZT in place of the λ/4 in Fig.9-24. The PZT is driven by a square wave generator (amplitude V_S, frequency $f_S \approx$ kHz) and operates in combination with a multiplexer to recover the I_1, I_2 and $I_1°$, $I_2°$ signals.

At V_S=0, the birefringence is missing and we can therefore use just a single analyzer leg, as indicated in Fig.9-25.

The minimum amplitude of the measurand we can resolve with the linear birefringence readout depends on the responsivity $\Phi = \Phi(M)$ connecting M to the phase Φ. About the phase, the minimum detectable phase or NEΦ is ultimately determined by the shot noise associated with the detected currents I_1 and I_2. Thus, we have the same NEΦ as considered in the circular birefringence readout analyzed in Sect.9.3.2.2 and plotted in the diagram of Fig.9-21 (right-hand scale).

As we can see from the diagram, limiting resolutions or NEΦ as small as 10^{-5} rad are obtained experimentally and are actually representative of the state-of-the-art design. Values down to 10^{-7} to 10^{-6} rad are in the reach of the technique, and can be attained at the sacrifice of bandwidth.

9.3.2.4 Combined Birefringence Readout

In practice, it is uncommon to need a double measurement on two measurands, one inducing circular birefringence and the other inducing linear birefringence in the same fiber. Yet, it's interesting to consider this case, as it allows us to evaluate the effect of disturbance coming from the other type of birefringence, in the sensor fiber as well as in the down-lead piece of fiber accessing the sensor. So, let us assume a fiber with two birefringences acting on the fiber (Fig.9-26), a circular one with rotation Ψ and a linear one with retardance Φ. Then, let us employ the readouts illustrated in previous sections: one, the best suited for circular birefringence (Figs.9-16 and 9-17) and another, the best suited for linear birefringence (Figs.9-23 and 9-24), each fed by the outputs of a 50/50 beamsplitter.

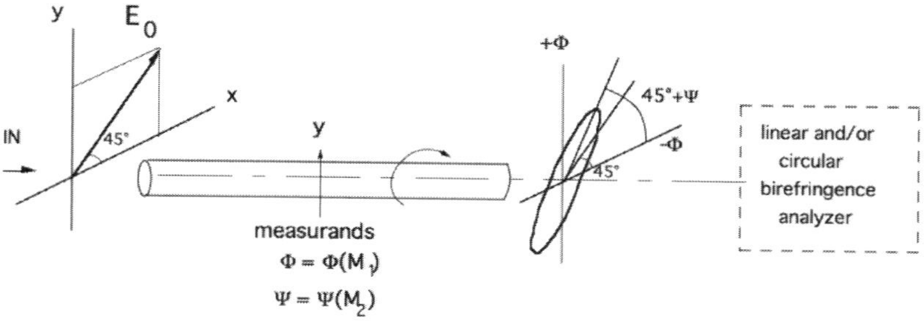

Fig.9-26 The case of simultaneous circular and linear birefringence in the fiber, read by both the linear and circular analyzers of Fig. 9-16 and 9-23.

To sort out the signals, let us call I_{1C} and I_{2C} the outputs of the circular birefringence scheme and I_{1L} and I_{2L} those of the linear birefringence scheme. Now, when Ψ and Φ are simultaneously present, the following relations hold in all cases:

$$I_{1C}+I_{2C} = I_{1L}+ I_{2L} = I_{1C}{}^{o}+I_{2C}{}^{o} = I_{1L}{}^{o}+I_{2L}{}^{o} = E_0^2 \qquad (9.14)$$

These equations follow from the conservation of energy in the readout schemes. In addition, with easy calculations similar to those leading to Eqs.9.6 and 9.9, and omitted here for saving space, we get:

$$I_{1C} - I_{2C} = - E_0^2 \sin 2\Psi \qquad (9.15a)$$

$$I_{1L} - I_{2L} = E_0^2 \sin 2\Phi \ \cos 2\Psi \qquad (9.15b)$$

$$I_{1C}{}^{o}-I_{2C}{}^{o}= I_{1L}{}^{o}-I_{2L}{}^{o}= E_0^2 \cos 2\Phi \ \cos 2\Psi \qquad (9.15c)$$

These expressions tell us that we can perform a simultaneous measurement of small-amplitude Ψ and Φ, putting together the readout schemes of both. In fact, for small values of the measurands, $\Psi, \Phi<<1$, we get $\cos 2\Psi\approx 1$ in Eq.9.15b and the difference signals are:

$$I_{1C}-I_{2C}\approx -2\Psi, \quad \text{and} \quad I_{1L}-I_{2L}\approx 2\Phi.$$

An example of application of the concept is provided by Ref. [25], in which two components, H_x and H_y of the magnetic field acting on a fiber coil, are simultaneously measured thanks to the interplay of Faraday (circular) and bending (linear) birefringence. To combine the schematic of linear and circular birefringence readout, we can take advantage of the PZT controlled-phase elements discussed above, and multiplex a single channel of polarization analysis for the entire setup, and a variety of useful configurations can be developed from this concept.

9.3.2.5 An Extrinsic Polarimetric Temperature OFS

Another example of an extrinsic polarimetric OFS is provided by the measurement of temperature in the range 0...500°C, as required in the application of geothermal mining [26].

Ordinary fibers do not work well at these high temperatures, and thus we may think of an extrinsic OFS. A suitable crystal is lithium niobate ($LiNbO_3$), chosen to withstand the required temperature and provide a linear birefringence with a sizeable and very reproducible temperature coefficient.

By measuring the phase shift $\Phi = \Delta\beta\, L$, where $\Delta\beta = \beta_x - \beta_y$ is the difference of propagation constants along the principal axes x and y, and L the crystal length, we obtain a $\Phi = \Phi(T)$ dependence for the measurand. Indeed, if $\alpha_{\Delta\beta} = d\Delta\beta/dT$ is the temperature coefficient of the birefringence, then we get $\Phi = \Delta\beta\, L = \alpha_{\Delta\beta}\, \Delta T\, L$.

As illustrated in Fig.9-27, the crystal is placed in a probe at the end of a long tube running parallel to the drill in search of the geothermal poll. The tube carries the adduction fiber and may be several km long. Thus, a requirement of the system is that the fiber shall be a rugged, multimode fiber with a large (50...100 μm) core diameter, so that mechanical tolerances of the mounting are minimized. We would also like to use a single fiber for both feeding the crystal with polarized light and bringing back the useful signal to be measured.

Because the multimode fiber does not maintain polarization, we must insert a polarizer (a calcite prism) in the gauge, in front of the sensing crystal. After the optical signal coming out of the $LiNbO_3$ crystal is properly polarized, we only need that the adduction fiber has little attenuation and a negligible polarization-dependent loss. This constraint can be amply satisfied with ordinary multimode fibers.

We now need to modify the double channel strategy illustrated in the last section and look for the $\sin\Phi$ and $\cos\Phi$ signals from a single amplitude measurement, performed external to the probe.

This can be done by taking advantage of the wavelength dependence of birefringence, $\Delta\beta = \Delta\beta(\lambda)$. Using a semiconductor laser diode, a wavelength sweep is easily achieved by applying a sweep to the drive current. Thus, in addition to the signal Φ, we have a time-dependent phase $\phi(t) = \Delta\beta(\lambda)\, L$. We manage $\phi(t)$ swings on a full 2π cycle.

The arrangement of Fig.9-27 has the crystal and the polarizer passed twice in the go-and-return path. With the arguments leading to Eq.9.9, it is easy to find the signal E_{1out} leaving the probe, and the result is:

$$E_{1out} = (1/2)\, E_0 \exp i[-\phi(t) - \Phi] + (1/2)\, E_0 \exp i[\phi(t) + \Phi] =$$

$$= E_0 \cos[\phi(t) + \Phi], \qquad (9.16)$$

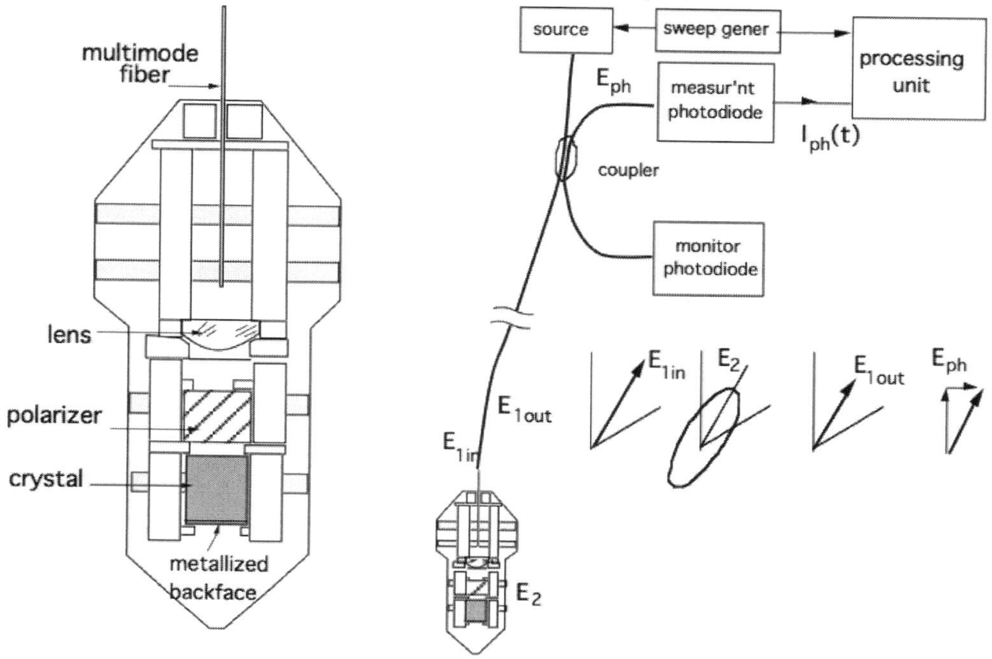

Fig.9-27 Left: layout of a temperature probe based on extrinsic linear birefringence. Light from the source is guided by the multimode fiber to the probe and passes through a polarizer oriented at 45° with respect to the $LiNbO_3$ principal axes. Light mirrored back from the crystal is analyzed by the polarizer itself and brought back to the detector location. Right: to recover the phase Φ of linear birefringence, the laser source is swept in frequency so that the optical path length $\Delta\beta L=\phi(t)$ undergoes a full 2π cycle. By calibrating the crystal response, we can trace back the phase $\Phi=\Phi(T)$ and hence the measurand T (from [26] ©Optica Publ., reproduced by permission).

After propagation through the multimode fiber, the state of polarization of E_{1out} will suffer a major change, yet the total power contained will be the same $E_{1out}{}^2$ value leaving the probe.

Then, after detection, we have a current $I_{1ph} = E_0{}^2 \cos^2[\phi(t)+\Phi] = E_0{}^2\{1+\cos 2[\phi(t)+\Phi]\}/2$. We may drop the constant term and consider the time-dependent signal $I_{1ph(td)} = \cos 2[\phi(t)+\Phi]$.

After calibrating the system (crystal, laser diode, and sweep amplitude), we know the times at which it is $\phi(t)=0$ and $\phi(t)=\pi/4$.

By sampling the output signal $I_{1ph(td)}$ at these times, we obtain the desired signals $\cos 2\Phi$ and $\cos 2[\pi/4+\Phi] = -\sin\Phi$, and we are brought back to the usual signal processing.

The measurement is incremental and we need to start with the FOS at a reference temperature (usually ambient temperature) and count the periods or fractions of them to get the current temperature. The incremental constraint is not a problem in geothermal gauging, because the drill starts from ground.

In the practical instrument [26], a resolution of 0.02°C has been achieved in the range of 0 to 500°C, while the accuracy has been estimated to be better than 0.1°C.

Modifications of the concept, with the use of a polarization-maintaining fiber in the adduction trunk, have been also reported [27].

9.3.3 Interferometric Readout

In an interferometric OFS, the measurand affects the optical path length of a piece of fiber exposed to it, and we obtain the path length or phase readout by means of a suitable configuration of interferometer.

In Chapter 7, we treated in detail the best-developed interferometric OFS, the FOG. Other interferometric OFSs share with the FOG excellent sensitivity, unparalleled by other electronic sensors. A feature common to all interferometric OFSs is the remarkable sophistication of the device structure.

Because of these features, interferometric OFSs have a very appealing potentiality of performance, but their rather high cost has been a serious obstacle to market penetration.

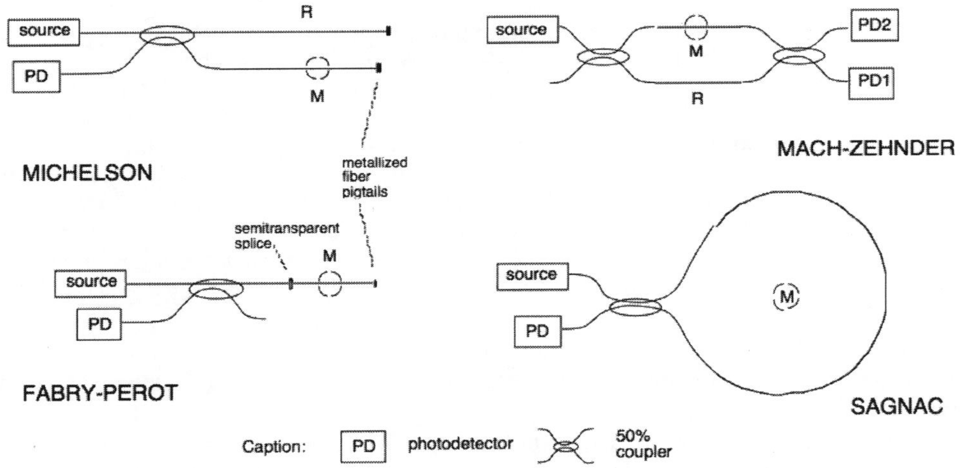

Fig.9-28 Commonly employed configurations of OFS interferometers are the all-fiber versions of conventional bulk-element interferometers. M and R indicate the measurement and reference arms.

In general, any type of optical interferometer can be virtually employed to provide a measurement and a reference arm, thus we are left to make the fiber counterpart out of the conventional bulk-optics interferometers described in Appendix A2 (see Fig.A2-1).

We obtain the interferometric OFS illustrated in Fig.9-28 by replacing free propagation with fiber-guided propagation, beam splitters with fiber-couplers, and mirrors with metal-coated pigtails.

Of course, if we need a well-reproducible phase of the optical field after propagation in the fiber, this shall be single mode. Then we may add some special feature, for example polarization control (the hi-bi fiber of the FOG), dispersion control, nonlinear effects, active-ion doping, and so on, yet the basic requirement is the single mode of the fiber. As a consequence, all the components in an interferometric OFS shall be single mode.

All the different configurations basically supply a signal of the type cos Φ. Interferometer performance differs from one configuration to another, however. As discussed in Appendix A2.1, the responsivity of the interferometer may vary from 1 (Mach-Zehnder and Sagnac) to 2 (Michelson) and to F (the finesse, >>1) in the Fabry-Perot interferometer (see Table A2-1).

In addition, a reference arm is available for the cancellation of extraneous disturbances in Michelson and Mach-Zehnder interferometers, whereas it is missing in the Fabry-Perot, and is not separately accessible in the Sagnac interferometer.

Other important features of the OFS are the balance or unbalance of the arms, the co-location of source and detectors on the same side of the fiber, and last but not least, the retro-reflection into the source.

In view of these features, it turns out that the Mach-Zehnder configuration is best for a general-purpose OFS, when we want an easy-to-implement sensor with a readily available reference arm and no disturbance effects from retro-reflections. Additionally, the differential output (PD1 and PD2 in Fig.9-28) is useful for common mode cancellation. The price for this configuration is the increase of component count (two couplers) and the placement of the laser and detector on opposite sides of the fiber.

If we don't want to fold the fiber to get a probe-like sensor with input and output on the same side, we may resort to the Michelson configuration. This uses a single coupler and has twice the responsivity of the Mach-Zehnder, but the dual detector can't be fitted in, and retro-reflection may disturb the source. If the source is a laser with a coherence length larger than the go-and-return optical path length, then we need to protect the laser with an optical isolator placed at the laser output before entering the OFS fiber.

Because the isolator is a rather expensive component, we generally try to avoid it and limit the coherence length to the value necessary for the measurement, that is, the path length difference between the arms.

The Fabry-Perot is an attractive configuration when we want an OFS with the highest responsivity. In fact, a finesse up to F=50-100 can be attained in metal-coated fibers, and the improvement in response may be quite significant. Disadvantages of the configuration are (i) the criticality of metal coating required to build the cavity, and (ii) the sensitivity of the access lead to external disturbances.

The Sagnac configuration is definitely ideal when sensing a non-reciprocal effect, as already discussed in connection with the inertial rotation in the FOG and the Faraday effect in current and magnetic field sensing. For reciprocal effects, on the other hand, one cannot separately access the reference and measurement paths for disturbance cancellation.

The ultimate sensitivity of interferometric OFS can be expressed in terms of a NEM (Noise Equivalent Measurand).

If $\Phi=\Phi(M)$ is the relationship connecting the measurand and the optical phase impressed to the fiber, we can write the NEM as NEM= ϕ_n/R_ϕ. Here, ϕ_n is the phase noise of the interferometric measurement (given by Fig.4-18), and $R_\phi=\Phi'(M)$ is the phase responsivity given by the derivative of the phase with respect to the measurand, calculated at the working point.

In practical OFS, we can attain phase noise ϕ_n in the range of 10^{-7} to 10^{-6} rad. Yet, these values require a careful design of both the optical layout and electronic processing. In addition, we should take into account the thermodynamic phase fluctuations of the fiber (Sect.4.4.5), as this limit is in the reach of OFSs.

9.3.3.1 Phase Responsivity to Measurands

The phase responsivity R_ϕ of a fiber is important to evaluate the signal that can be generated in an intrinsic OFS with interferometric readout.

Table 9-1 collects typical values that are illustrative of the responsivity of three classes of fibers: (i) standard single mode fiber with standard polymer secondary coating; (ii) fiber with special coating, for enhanced sensitivity to measurand; (iii) fiber with special coating, to desensitize it to the measurand. The reason for the three classes is that ordinary fibers are, surprisingly, appreciably sensitive to a lot of unintended measurands. We certainly will make our best effort to compensate for undesired measurands using a reference arm, which greatly helps to desensitize it. However, the ideal sensor, sensitive only to a specific measurand and immune to any other, is surely best approached if we use special fibers.

TABLE 9-1 Typical Phase Responsivity of Fibers

Measurand	Ordinary fiber	Coated for max response	Desensitized fiber	Unit
Temperature	300	5000	0.5	rad/°C·m
Pressure (isotropic)	36	600	0.1	μrad/Pa·m
Force (compression)	0.55		0.01	rad/N
Bending on radius R	0.6		0.01	rad·cm
Current	3	220		μrad/A

In particular, a special fiber designed for sensitivity to the measurand is useful in the OFS sensing section (the measurement arm), whereas another special fiber desensitized to the measurand is best suited for the reference section (or arm).

Finally, the piece of fiber adducting the sensing region should be desensitized to all extraneous measurands. As we can see from the typical data reported in Table 9-1, the desensitized fiber is about 2...3 decades less sensitive than the most responsive fiber. This is a good result, but perhaps not enough to seriously challenge conventional electronic sensors, which may put forward a typical cross-immunity of 10^5 to 10^7 to undesired measurands.

9.3.3.2 Examples of Interferometric OFS

Several interferometric OFSs have been proposed in the literature, and the most exhaustive reference on this topic is perhaps the recently prepared Collection of Optical Fiber Sensors Proceedings [28].

If we exclude the FOG, which is by far the unrivaled interferometric OFS, the most interesting OFS is the *hydrophone*, from the application point of view. The hydrophone is a device intended for hydrostatic pressure sensing in underwater environments.

Using a properly jacketed fiber sensitive to isotropic pressure, and laying a section of it coiled so as to cumulate the measurand on an appropriate length, we get the schematic of Fig.9-29. Here, the pressure-sensitized fiber is coiled at the end of the measurement arm of the Michelson configuration.

The reference arm brings a phase modulator (with a PZT) to adjust the quiescent working point in quadrature, so that the photodetected signal is of the form $I_{ph} = I_0 [1+\cos\{\Phi(M)-\phi_r\}] = I_0 [1+\sin\Phi(M)]$. Here $\Phi(M)$, is the phase induced by hydrostatic pressure, and ϕ_r is the reference arm phase shift whose low-frequency (or average) component shall be dynamically locked to $\pi/2$ (compare to Fig.4-23 and Sect.4.5.1).

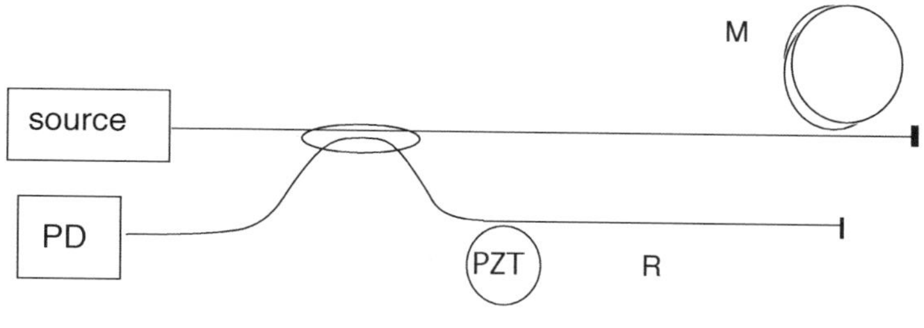

Fig.9-29 Schematic of an interferometric hydrophone using the Michelson readout configuration. The sensing fiber is a coil of pressure-sensitized fiber, whereas the adduction and reference arm fibers are desensitized.

With several centimeters of coiled fiber, the sensitivity of the OFS hydrophone can reach 0 dB_a corresponding to an acoustic pressure $p_a = 2 \cdot 10^{-5}$ Pa [29]. This is the reference level of acoustic pressure and corresponds to the threshold of audibility.

Another schematic similar to that of Fig.9-29 applies to the measurement of temperature. In this case, we take advantage of the PZT modulator to perform two functions: (i) removing the ambiguity of the cosine function, and (ii) extending the resolution to a small fraction of wavelength (or 2π angle). To do so, we sample the phase of the measurand $\phi=\Phi(M)-\phi_r$ by three or four values of the reference ϕ_r. Three values are those already illustrated in Fig.4-11; some researchers prefer using four values.

For example, we may switch ϕ_r on a sequence of three values: 0, ..$2\pi/3$...$-2\pi/3$. By sampling the $1+\cos[\Phi(M)-\phi_r]$ signal in correspondence, we triple the interferometric signal, as in the general scheme described in Sect.4.2.2.1. From the three signals, we can compute $\Phi(M)$ (see Eq.4.11) with $\approx$1 nm resolution (equivalent to $\approx\pi/200$ rad) and get the appropriate up/down signals required for a range exceeding 2π.

Similar to an interferometer for displacement measurement, when the phase induced by the measurand exceeds 2π, we count up/down transitions in a counter, and the measurement is an incremental one. To work correctly, the sensor shall be reset at an initial temperature T_0 and allowed to count the periods induced by temperature between T_0 and the current temperature T. Temperature resolutions down to 10^{-4} °C have been reported based on this configuration of interferometric OFS.

9.3.3.3 White-Light Interferometric OFS

The white-light interferometer described in Sect.4.7 is readily adapted to an OFS, as illustrated by the schematic in Fig.9-30.

In the Michelson version of the interferometer, the arm with in-fiber propagation is used to collect the measurand phase perturbation. In the other arm, we exit from the fiber with a collimating lens and project the beam onto the moveable mirror. When the mirror position matches the condition $s_m=s_r \pm l_c$, fringes appear and we can measure the deviation from the condition $s_m=s_r$ looking at the fringe envelope. The resolution of this measurement is of order of the coherence length l_c (Fig.9-30).

Using a SLED source, we may typically have $\Delta\lambda\approx 25$ nm at a central wavelength λ=800 nm, and the coherence length is $l_c = \lambda^2/\Delta\lambda = 0.8^2/0.025 = 26$ μm, a good starting value as resolution of a strain-measuring OFS.

If we scan the response curve a few times and average the results, we may go down to resolve typically $\approx 0.2\, l_c = 5$ μm.

This value of resolution is obtained with a very simple optical setup and needs only very limited electronic processing. Because it is cheap, the OFS lends itself to interesting application in the field of construction, called either *non-destructive testing* (NDT) or sometimes smart sensors. For example, we may start with a 1-m piece of fiber, with one end-face metal-coated and the other end-face ending on a connector.

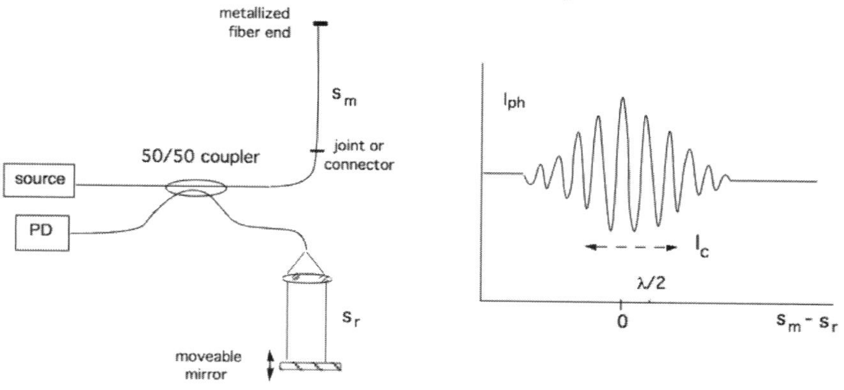

Fig.9-30 Schematic of a white-light interferometer OFS. Mirror M is mounted on a motorized stage, and when the arms are balanced ($s_m = s_r \pm l_c$), fringes appear.

The fiber is then threaded inside a steel tube and glued to it, and this constitutes an arm of the OFS. The arm duplicates the strain $\Delta l/l$ applied by the structure external to the tube.

Now we can cast the fiber inside a concrete pillar to monitor its state of strain during the useful life of the pillar [30]. The fiber tube is put in place while the concrete is poured to fabricate the pillar, leaving out just the connector for access [31].

After a period of time (e.g., months to years), we can go back to examine the sensor and check the strain applied to it and to the concrete structure. From this we get non-destructive diagnostics of the construction work. As a 10-μm resolution on a 1-m fiber length is a strain $\sigma = \Delta l/l = 10$ μm/1 m= 10 μstrain, we obtain a sensor amply capable of diagnosing incipient overload of the structure (usually found at 100...1000 μstrain).

About disturbances, temperature variations may actually affect the measurement of strain. Indeed, a typical sensitivity (Table 8-1) of 300 rad/°C·m amounts to about $300/2\pi \approx 50$ wavelength/°C. At $\lambda = 0.8$ μm, this quantity corresponds to a strain equivalent of $50 \cdot 0.8 =$ 40 μm/°C·1 m = 40 μstrain/°C. However, we can measure the temperature of the pillar and correct against thermal drift and associated errors if we characterize the response prior to installing the fiber.

The white-light approach is interesting and has led to successful products [31].

9.3.3.4 Coherence-Assisted Readout

A more elegant configuration of a white-light interferometer can be devised starting from the idea that, casting in the experiment a full interferometer instead of just the measurement arm, we make available a reference to subtract the undesired disturbances [32,33].

To do so, we shall read the path length difference s_{mS}-s_{rS}, between the measurement s_{mS} and reference s_{rS} paths contained in the sensing (S) section of the item under test.

For the subtraction, we secure the measurement fiber to the enclosure tube and leave the reference fiber loose. Thus, the difference s_{mS}-s_{rS} is sensitive to the measurand as before, whereas temperature and other common mode disturbances are canceled out. Now, the problem is how to read the s_m-s_r difference having already used both arms. With a white-light interferometer, we have one more degree of freedom available, which is coherence. Thus, we can develop a coherence-assisted readout as shown in Fig.9-31.

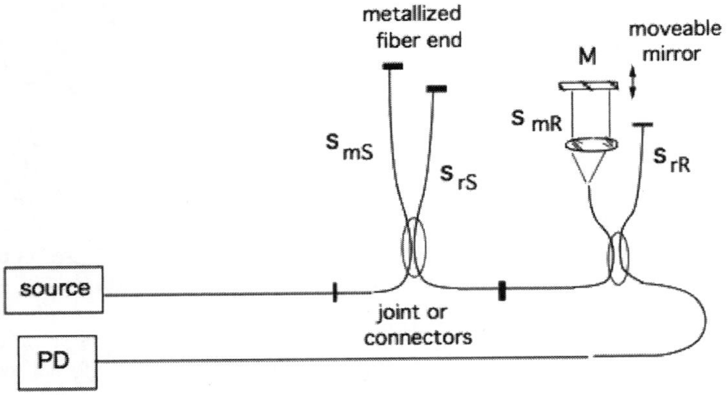

Fig.9-31 An optical fiber sensor using coherence-assisted readout employs two interferometer sections for sensing (S) and readout (R). The arm lengths of the interferometers differ by a quantity Δ much larger than the coherence length l_C. The moveable mirror is scanned, and when position $s_{mR} = s_{rR} + (s_{mS} - s_{rS})$ is reached, white-light fringes appear as in Fig.9-30.

In this scheme, we use two interferometers in series, in the sensing (S) and readout (R) section. We make the path lengths s_{mS} and s_{rS} differ by several coherence lengths l_c of the source so that the output fields from these arms will not beat at the photodetector. For example, we typically let s_{mS}-s_{rS} $=\Delta\approx 20\cdot l_c \approx 0.5$ mm, for arms s_{mS}, $s_{rS}\approx$1-m long.

In the readout section, we work with the same nominal lengths of the reference and measurement paths. For simplicity, let us take identical reference lengths $s_{rR}=s_{rS}$. Then, scanning the measurement path s_{mR} with the moveable mirror, we will find the readout interferometer in the same imbalance of the sensing interferometer when we reach $s_{mR}=s_{mS}$.

This condition corresponds to the balance of two special paths in the optical setup. One passes through the measurement arm of the sensing section and the reference arm of the readout section with a length $s_{mS}+s_{rR}$. The other passes through the reference arm of the sensing section and the measurement arm of the readout section with a length $s_{rS}+s_{mR}$. Along these two paths, the electric fields collect a phase delay $2k(s_{mS}+s_{rR})$ and $2k(s_{rS}+s_{mR})$, respectively. As the two fields are collected at the receiver, they generate the following photodetected signal:

$$(E_0^2/4)\{1+(1/2)\cos 2k[(s_{mS}+s_{rR})-(s_{rS}+s_{mR})]\} \quad (9.17)$$

With a coherence length l_c, interferometric fringes similar to those of Fig.9-30 are generated in the interval $s_{mR}= s_{mS}+(s_{rR}-s_{rS}) \pm l_c$ of moveable mirror position. When the reference paths are identical ($s_{rR}=s_{rS}$), this gives $s_{mR}= s_{mS}\pm l_c$.

As there are four possible paths in the double interferometer schematic of Fig.9-31, we shall also examine the additional terms beating at the photodetector. These terms are easily found by writing the amplitude and phase of the field propagating down the fiber circuit in Fig.9-31. Developing the calculation, we find the additional terms and write the total current at the photodetector as:

$$I_{ph} = (E_0^2/4)\{1+\cos 2k(s_{mS}-s_{rS}) + \cos 2k(s_{mR}-s_{rR}) +(1/2)\cos 2k[(s_{mS}+s_{rR})-(s_{rS}+s_{mR})]$$
$$+ (1/2)\cos 2k[(s_{mS}+s_{mR})-(s_{rS}+s_{rR})]\} \quad (9.18)$$

The first term in curl brackets has a zero mean-value because the argument, $s_{mS}-s_{rS}\approx\Delta$, is much larger than coherence length. The second term can give a fringe pattern, but only when $s_{mR}-s_{rR}\approx\pm l_c$. We never get it, because the difference $s_{mR}-s_{rR}$ is about equal to $s_{mS}-s_{rS}\approx\Delta$. Thus, the second term has a zero mean-value, too. The third term is the desired one, beating for $s_{mR}-s_{mS}\approx\pm l_c$. The last term has a beat only for $s_{mS}-s_{mR}\approx -2\Delta$, or it has zero mean value.
In conclusion, the extra terms may contribute with their beatings as well, but far away from the main one if we keep $\Delta >> l_c$, and we can easily reject them in the course of the scan.

Once we have been able to compensate for the interferometer from thermal and other undesired drifts, we can push the resolution further. Indeed, when the white-light fringes are still under the envelope waveform, we can look for the one with the largest amplitude.
This is the central fringe, defined by position to the usual $\lambda/2$-displacement per period. Therefore, if we look at the central fringe we get a resolution of about 0.5 μm, or 20 times better than the envelope (or coherence-based) position-sorting.

Important to note is that, near the balance condition, the white-light interferometer behaves exactly like a conventional interferometer, and its ultimate limits of performance are those already discussed in Sect.4.4. Thus, nanometer resolution is also achievable.

9.4 Multiplexed and Distributed OFSs

Despite the generally excellent technical performances of OFSs, their relatively high cost has so far prevented their widespread use in applications, especially in comparison with conventional electronic sensors. So, looking for ways to get around this problem, a possibility is to offer a number of measurement points instead of a single point. We can multiply the number of measurement points substantially, either by allowing the measurement to be distributed in space or by multiplexing a number of individual sensors on the same line served by a single readout unit.

Several approaches have been studied through the years to provide a viable approach to the multiplexed or distributed architecture, and copious literature has been published on the subject, see for example Refs. [34, 35].

9.4.1 Multiplexing

Multiplexing may be implemented by several methods and the most used are (i) space-division; (ii) time-division; (iii) wavelength-division; and (iv) coherence-division.

Space-division uses a switch connecting each individual sensor to the readout unit. Each sensor is read in sequence, one right after the other. Because of cost requirements, the switch is usually a mechanical rotary device, and the operation time (switching from one sensor to the other) is about a few milliseconds. For faster response, we may use an integrated approach: for example, we may start with a Mach-Zehnder waveguide interferometer realized in $LiNbO_3$ or in silica-on-silicon (SoS). These are 2-way switches, connecting the input guide to one of the two output guides, and are actuated by an electrical signal applied to the control electrode. By arranging N stages in cascade, we realize a 1×2^N switch array. Cost becomes prohibitively large at increasing N, however, and this solution is limited in practice to, say, $N\leq 4$.

Time-division uses short pulses to interrogate sensors located down a fiber at different lengths from the source. Fig.9-32 shows an example called a ladder configuration, a multiplexing scheme employing a pulsed light source and a series of sensors connected with couplers to common rails of fiber for feeding the sensors and for collecting the returned signals.

The time duration of the pulse τ is made shorter than the extra delay 2L/c experienced by a sensor with respect to the preceding one. Under this condition, each sensor responds with a pulse falling in a well-defined time slot (Fig.9-32) and thus we can demultiplex the responses at the receiver. Each coupler takes a fraction of the incoming power to its sensor. To equalize responses, the coupling factor at the n-th sensor should be chosen as $k_n=1/(N-n+1)$, where N is the total number of multiplexed sensors. By doing so, the first sensor has $k_1=1/N$ and leaves a through power of 1-1/N. The second sensor has $k_2=1/(N-1)$, takes a fraction $[1-1/N]/(N-1)=1/N$ of incoming power, and leaves a through power of [1-2/N]. The third sensor has $k_3=1/(N-2)$ and takes a fraction $[1-2/N]/(N-2)=1/N$. This continues up to the N-th sensor that has $k_N=1$ and takes a fraction 1/N of input power.

In the return path, the fraction of power deviated onto the common rail is the same value, 1/N of the power deviated from the feed rail. Thus, with respect to addressing individual sensors, the operation of multiplexing introduces an extra attenuation $1/N^2$.

With an appropriate design of source power and receiver sensitivity, attenuation limits the maximum number of sensors that can be multiplexed to $N\approx 15...20$. The time-division scheme has been employed for polarimetric as well as interferometric sensing [36] of temperature, strain, and acoustical emission, especially in connection with marine applications.

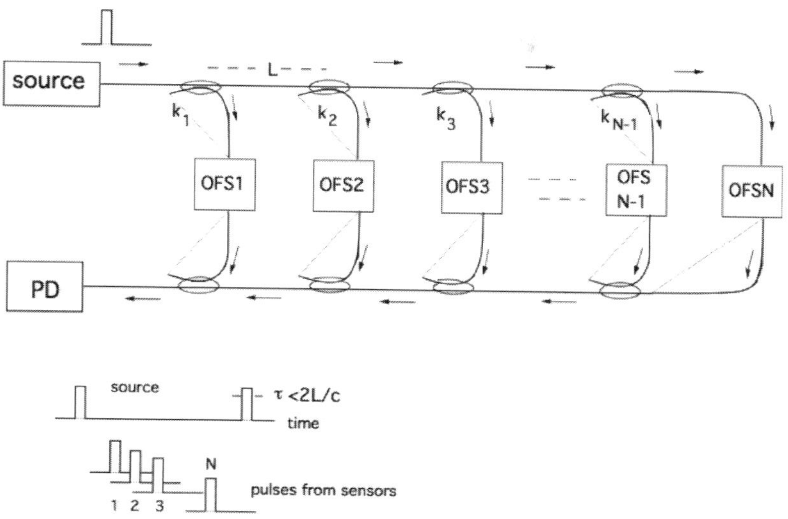

Fig.9-32 Time-division multiplexing of an OFS, using a pulsed source and individual sensors arranged in a ladder array. To distinguish the pulses returning from individual sensors, the pulse width τ should be shorter than the delay 2L/c added at each additional stage. Extra fiber length can be added between sensors to match the condition.

Wavelength-division multiplexing consists of assigning a wavelength slot to each individual sensor and reading the composite output with a narrow-band source tunable in frequency. To be suitable for cascade connection, the sensor should be frequency-selective, respond in transmission or reflection, and let the out-of-band components pass unaltered in reflection or transmission.

These requirements are satisfied by a Fiber Bragg-Gratings (FBGs), a passive component well known in optical fiber communications. An FBG is a longitudinal grating, written along the axis of the fiber as a periodic variation Δn of the refraction index, produced by exposure to UV radiation. If Λ is the spatial period of the grating, the Bragg condition $2n\,\Lambda=\lambda_B$ determines the wavelength of resonance λ_B. For wavelengths close to λ_B, the device reflects back the incoming radiation, whereas radiation off-resonance ($\lambda\neq\lambda_B$) passes unaltered through the FBG. We summarize this behavior by saying that the FBG is a band-reject filter in transmission or a band-pass filter in reflection.

The FBG can be used as a sensor because the resonant wavelength λ_B depends, in a reproducible way, on temperature and strain, the two quantities most commonly measured with FBGs. The relative change of resonance wavelength is written as:

$$\Delta\lambda_B/\lambda_B = \Delta n/n + \Delta\Lambda/\Lambda = (1-p_e)\,\varepsilon + (\alpha+\psi)\Delta T \qquad (9.19)$$

Here, p_e is the elastooptic coefficient of the fiber (typically $p_e \approx 0.22$ for silica), $\kappa=\alpha+\psi$ is the total thermooptic coefficient (typ. $\kappa \approx 5 \cdot 10^{-6}$ in silica), and $\varepsilon=\Delta l/l$ is the strain applied to the FBG.

Connecting a number of FBGs in series, as indicated in Fig.9-33, we get a wavelength-multiplexed sensor.

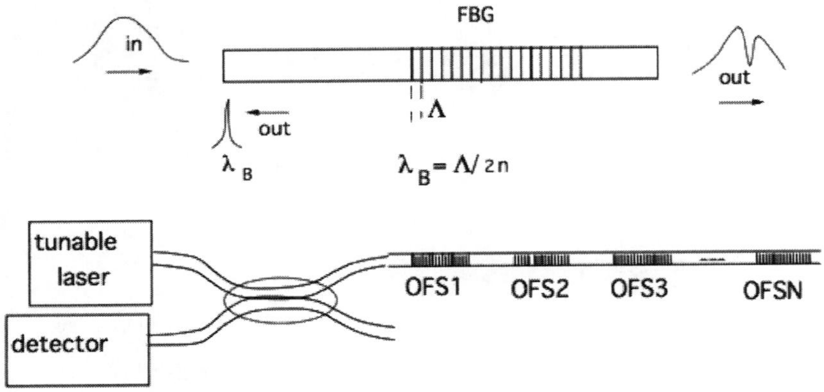

Fig.9-33 An FBG reflects the wavelengths close to the resonant wavelength λ_B and transmits the rest of the spectrum (top). A wavelength-multiplexed OFS is arranged by cascading a number of FBGs at different wavelengths λ_B and using a tunable laser source for reading the individual sensors.

The array can be read by a wavelength-tunable laser source because each FBG other than the selected one is transparent, whereas the sensor interrogated by the laser source will respond with a peak reflection in correspondence to its resonant frequency.

The shift in wavelength resonance $\Delta\lambda_B$ is dependent on temperature T and strain ε applied to the FBG. If we are to avoid cross-sensitivity to T when measuring ε, or vice versa, we should use a pair of FBGs, arranged in the experiment to have different sensitivity to T and ε. By combining the results of the measurements in a set of two equations, we can solve for both T and ε measurands.

A feature that is unsatisfactory in the λ-multiplexed sensor of Fig.9-33 is the use of a wavelength-tunable laser source. This is an expensive item and, if it is the only possibility, it is likely that an OFS incorporating it will hardly leave the laboratory.

Thus, we may want to modify the concept of λ-multiplexed readout. One of the several solutions reported in the literature is shown in Fig.9-34. Here, we use a broadband source to illuminate the sensors and duplicate each FBG in the sensing chain with another FBG in the measurement chain, with nominally the same resonant wavelength λ_B.
Upon application of the measurands, the k-th sensor changes its resonance wavelength by, say, a deviation $\Delta\lambda_{Bk}$. To measure the set of all $\Delta\lambda_{Bk}$ simultaneously, we apply to each k-th

FBG of the reference chain a stress appropriate to produce the same $\Delta\lambda_{Bk}$ shift of the corresponding FBG in the measurement chain. This is done by the piezo actuators, the PZT slabs stretching the fiber of each FBG (Fig.9-34).

To operate all sensors independently, each k-th piezo is actuated with a different electrical modulation frequency f_k. The photodetector output is filtered in k bands, centered at the f_k resonance, and in each band, we look for the maximum response of the k-th FBG sensor.

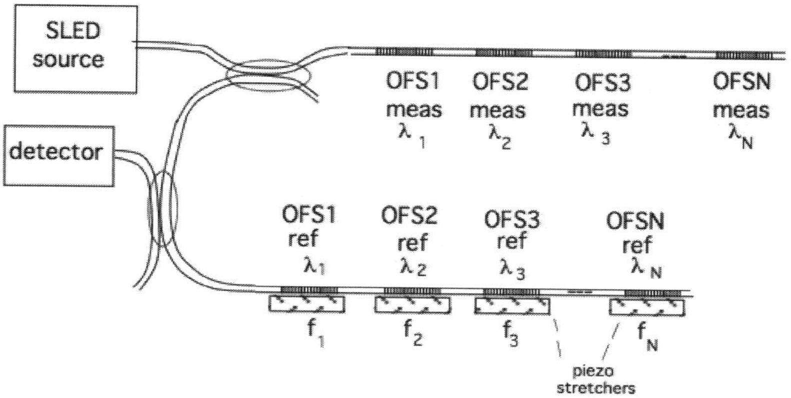

Fig.9-34 Schematic of a λ-multiplexed sensor that uses two sets of FBGs in the measurement and reference chain. Corresponding OFSs have nominally identical resonant wavelengths λ_B. The signal returning from the measurement OFS is brought to the reference OFS for response comparison. Modulating the reference λ_B with the piezo actuator allows us to demultiply the responses of individual sensors.

Finally, *coherence-multiplexing* is an interesting technique that has been recently proposed. We may think of it as a generalization of the concepts discussed in Sect.9.3.3.4.
As illustrated in Fig.9-35, we can multiplex several interferometers (for example, Mach-Zehnder interferometers) by using a different arm imbalance in each OFS, and by arranging all of them in a double chain, one for sensing and the other for readout.

In the interferometers, we let the imbalance be much larger than the coherence length l_c, that is, L_k-l_k >>l_c. In addition, we let the imbalance of consecutive sensors be larger than the coherence length L_k -L_{k-1}>>l_c.

At each detector, say the k-th of the sensing section, we find the superposition of many field contributions propagated through different paths across the interferometers. All of them are unbalanced by more than the coherence length l_c, except the two that go through the corresponding k, OFSk and Rk, in a crossed fashion, that is, $l_{k(OFS)}+L_{k(R)}$ and $L_{k(OFS)}+l_{k(R)}$, and share the same path in the other interferometers.

Thus, we can simultaneously read all the multiplexed sensors because the beating is returned only at the appropriate interferometer.
Coherence is an elegant solution for multiplexing, but the drawback is a large attenuation.

First, we shall equalize the channel amplitudes by choosing k_n=1/(N-n+1) as in the time-division scheme (Fig.9-32), and this causes an attenuation of 1/N.

In addition, at each interferometer of the sensing section, we waste a factor of 2 because of the unused ports, and this is an extra 2^N attenuation. In practice, we can hardly go beyond N≈4 unless we introduce a costly optical amplifier.

In conclusion, when designing a new FOS, we shall use the concept of multiplexing whenever possible to reduce the cost per point.

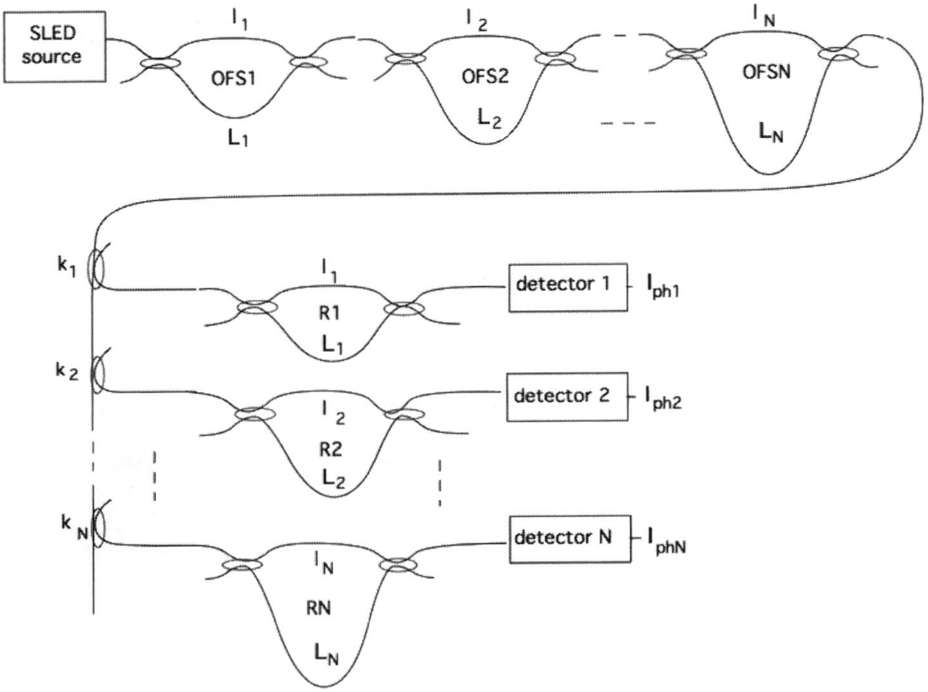

Fig.9-35 Schematic of a coherence-multiplexed OFS. Light from a low-coherence source is passed through a cascade of interferometers with different arm-length imbalance. In the readout section, light is split by means of couplers to a corresponding number of interferometers, duplicating the arm imbalance of the sensing section. The cross-path through the sensing and readout interferometers returns the useful signal.

But, the total cost of the array remains high anyway. Because complexity increases with N, a practical limit of maximum number of points comes out. Thus, we are not really gaining a new segment of the sensor market by multiplexing.

9.4.2 Distributed Sensors

When the measurement of an optical parameter is resolved in distance, each interval we resolve defines an individual sensor and a measurement point of a *distributed sensor* [37-40]. Most of the optical parameters we use for sensing are position-dependent, but in a single-point FOS we only make an averaged or cumulative reading of the optical parameter affected by the measurand. Distributed sensors offer a decisive advantage with respect to single-point sensors, because with a single readout box we can measure a large number N of individual sensing points. The information supplied by the distributed sensor is then increased by N, while the unit cost per point decreases by N. Even more important, we can develop a sensor with a very large number of points, many more than is reasonable with multiplexed sensors. This feature is unequalled by conventional technology and is a great advantage in applications.

In a distributed sensor, we should be able to remotely interrogate the fiber and have a parameter telling us the distance of the fiber piece being measured. Then we should read the measurand through an optical parameter and a readout scheme (Sect.9.3).

Common measurands are strain and temperature, and common optical parameters are intensity and SOP of light scattered at the individual sensing point.

Several approaches have been reported in the literature [1-4] for the interrogation of distributed sensors, namely time-domain, frequency-domain, and coherence-domain.

The time-domain approach is readily adapted to sensing because it takes advantage of instrumentation and measurement techniques [32] developed in optical fiber communication, for the remote testing of losses along fiber lines. The instrument for such a distributed measurement of attenuation is known as Optical Time-Domain Reflectometer (OTDR).

The OTDR has been used to sense stress and temperature in a variety of fibers since the early times of OFSs [37]. The interaction mechanisms first tested were attenuation by microbends to measure local stress, and temperature coefficient of the Rayleigh scattering to measure temperature. Both effects are small in ordinary silica fibers, however, and therefore special fibers (with a liquid core or doped with rare-earth elements) were tested in an attempt to improve sensitivity and resolution [33,34].

Later, variants of the basic OTDR scheme rather than fiber were able to provide the improvement. One is the Polarization OTDR (POTDR), by which we get a distance-resolved measurement of the SoP and hence of the birefringence of the fiber [36]. As birefringence is sensibly related to the stress imparted to the fiber, the POTDR is a powerful tool to remotely test and localize mechanical stresses in the fiber or in a pipeline, for example, to which the fiber is cemented.

The basic scheme of the OTDR readout for the distributed sensing of attenuation is shown in Fig.9-36. Similar in concept to the LIDAR discussed in Ch.3.5, the OTDR uses a short pulse of power at a suitable wavelength to interrogate the fiber. Wavelength is chosen in the windows of optical fibers (850, 1300, or 1500 nm), so that attenuation is at a minimum and the range covered is at a maximum.

The source is a pulsed laser, usually a semiconductor laser driven by a fast current-pulse and emitting peak powers of a few tens of milliwatts. The source is eventually followed by a booster optical amplifier and protected by reflections from the fiber with an optical isolator.

The time duration τ_p of the pulse determines, as discussed in Sect.3.5, the spatial resolution of the OTDR measurement. Assuming that the response time T of the detector is faster than the pulse duration $T \ll \tau_p$, the resolution is given by $\Delta z = c\tau_p/2n$, where $n \approx 1.5$ is the effective index of refraction of the fiber.

Resolutions of $\Delta z \approx 1$ m are readily achieved with pulses of ≈10-ns duration. The quantity Δz also represents the length on which the returning signal is averaged, and thus the equivalent length of the elemental sensor provided by the distributed OFS.

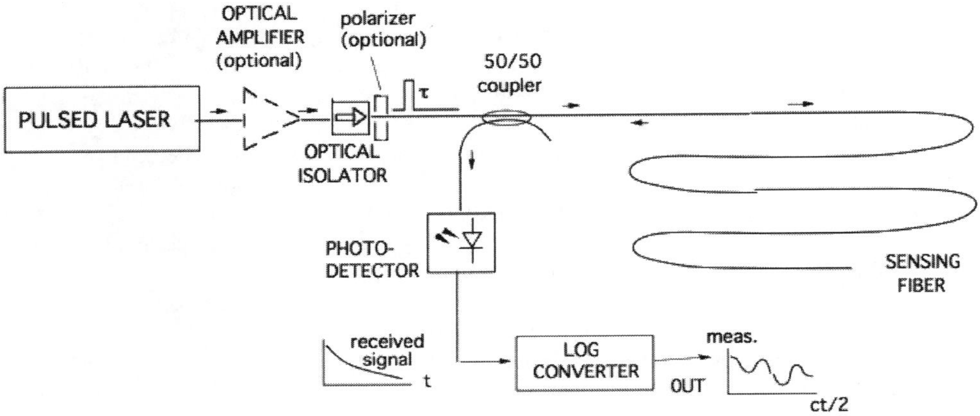

Fig.9-36 In a distributed fiber sensor, the local back-scattered power is read by an OTDR. The duration τ_p of the optical pulse used for interrogation determines the spatial resolution $c\tau_p/2n$ of the measurement, that is, the size of the virtual sensing element. The back-scattered signal looks the same as it does in Fig.3-31. After detection, the logarithm of the signal is computed, yielding the dependence from distributed attenuation and back-scattering.

The dependence from fiber attenuation α_f and back-scattering σ_{bs} are described by the same equations as for a LIDAR (Eqs.3.37 through 3.40), if we change the acceptance term $(\pi D^2/4z^2)$ into the solid angle of the fiber πNA^2, and the Neper-base attenuation in the base-10 attenuation, $\alpha = 2.3\ \alpha_f$.

The power in transit down the fiber is attenuated exponentially with distance z (as $10^{-\alpha z}$, where α is the attenuation in dB/km), and a fraction $\sigma_{bs}\,dz$ of it is back-scattered by the elemental length dz of fiber at z. The back-scatter coefficient $\sigma_{bs}(z)$ is dependent on the fiber material (and fiber imperfections) and on the interaction of the measurand with the fiber.

The back-scattered power returning to the source reaches the fiber coupler, and half of it is deviated to the photodiode (Fig.9-36).

The detected output signal is log-converted and the result L(t) supplies the time dependence of attenuation α and the back-scattering σ_{bs} (Eq.3.40).

With typical powers and measurement time, an OTDR may resolve attenuation of about 0.1 dB, but we need a special fiber if such a resolution is to correspond to a meaningful temperature and strain range.

Alternatively, we need a different optical parameter such as polarization. The setup of Fig.9-36 becomes a POTDR by adding a polarizer in the launch section and a Stokes parameter analyzer in the detector section [37]. Using the POTDR with a normal single mode fiber at λ=1500 nm, a typical sensitivity to $\approx$100-μstrain with 100-m resolution can be obtained.

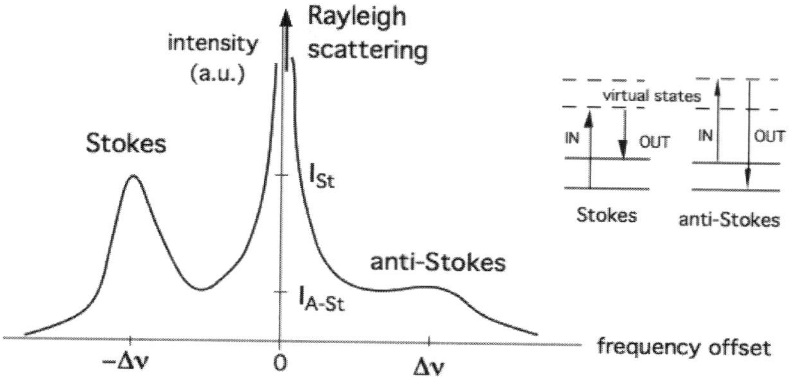

Fig.9-37 Spectrum of anelastic scattering with Stokes and anti-Stokes peaks located at the sides of the much larger Rayleigh peak of elastic scattering. The frequency shift $\Delta\nu$ depends on the type of scattering.

Another approach to improve performance is using a scattering different from the normal Rayleigh-Glan elastic scattering related to attenuation. Raman and Brillouin anelastic scatterings provide the desired mechanism.

These have to deal with phonon-assisted scattering, which introduces a virtual energy level in the interaction (Fig.9-37). Because of the new level, scattered radiation gains or loses a small amount of energy, $E_{sc}=h\Delta\nu$, and thus the line corresponding to scattered energy splits into lines, separated by E_{sc} from the central Rayleigh-Gans (or elastic scattering) line. The upper and lower frequency lines are called Stokes and anti-Stokes lines, respectively.

Raman and Brillouin scattering differ by the energy of the phonon they interact with. The Raman effect deals with the optical phonon, relatively energetic and such that, in silica, $\Delta\nu_R\approx$13.2 THz (or $\Delta\lambda_R\approx$100 nm).

This large frequency separation makes it relatively easy to separate the weak Raman lines from the much larger Rayleigh scatter line, a clear advantage in developing the readout instrument.

The Brillouin effect deals with the acoustic phonon and has a much smaller frequency separation ($\Delta\nu_B\approx$11 GHz, or $\Delta\lambda_R\approx$0.08 nm). Thus, narrow line lasers and very selective filters are required to resolve the lines.

An interesting feature of anelastic scattering is the clean and repeatable dependence from temperature of the relative amplitudes of Stokes and anti-Stokes lines. This is given by:

$$I_{AS} / I_{St} = (\lambda_{St}/\lambda_{AS})^4 \exp -hc/\lambda kT \qquad (9.19)$$

A Raman OTDR can be developed starting from the wavelength-resolved measurement of back-scattered power done at the photodetector (Fig.9-36) and by inserting suitable optical filters to sort out the Stokes lines [39]. By measuring the amplitudes of the Raman lines, temperature T can then be computed. A typical distributed sensor based on the measurement of Raman lines ratio can resolve 1°C on individual 1-m elements out of a 10-km total length of a multimode fiber, using ≈500 mW of peak power and τ_p=10 ns.

In applications, the Raman-based approach is the undisputed choice when we need distributed temperature only, because the signal is stronger and more easily sorted out.

With Brillouin lines, however, in addition to the temperature dependence given by Eq.9.19, we can also measure the frequency dependence of the separation $\Delta\nu_B$ from strain $\Delta l/l$, a small but measurable quantity because the unperturbed $\Delta\nu_B$ is small, too [39].

The dependence is written as $\Delta\nu_B \approx \Delta\nu_{B0} + \zeta\Delta l/l$. In silica fibers, the strain coefficient is about $\zeta \approx$ 50 (kHz/με), and $\Delta l/l$ is in microstrain (με) units. An experimental setup based on the Brillouin scattering has demonstrated [40] the simultaneous distributed measurement of strain and temperature, resolving about 4°C and $\Delta l/l \approx$300 με on a 15-km long fiber.

REFERENCES

[1] J.M. Lopez-Higuera (editor): "*Handbook of Fibre Sensing Technology*", J. Wiley & Sons, Chichester 2002.
[2] B. Culshaw and J. Dakin: "*Optical Fiber Sensors: Principles and Components*" vol.1, Artech House, Norwood 1988.
[3] B. Culshaw and J. Dakin: "*Optical Fiber Sensors: Systems and Applications*", vol.2, Artech House, Norwood 1989.
[4] B. Culshaw and J. Dakin: "*Optical Fiber Sensors*", vol.3, Artech House: Norwood 1996.
[5] B. Culshaw and J. Dakin: "*Optical Fiber Sensors: Applications, Analysis and Future Trends*", vol.4, Artech House: Norwood 1997.
[6] S. Donati: "*Photodetectors*", 2nd ed., Wiley IEEE Press, Hoboken 2021, Sect.5.2.
[7] see Ref. [6], Sect.5.3.1.3.
[8] LSENSU *Temperature Sensor*, -200 +250 °C, 2020, see https://www.ruggedmonitoring.com/product-details/lsensu/5cf240d4fec70f0001d1a59e
[9] see Ref. [1], Sect 32-34.
[10] M. Nagai, M. Shimitzu, and N. Ohgi: "*Sensitive Liquid Sensor for Long Distance Leak Detection*", Proc. 4th Intl. Conf. on Fiber Sensors, OFS-4: Stuttgart, 1984, pp.207-210.
[11] A. Harmer and A.M. Scheggi: "*Chemical, Biochemical, and Medical Sensors*", in Ref. [3], pp.599-651.
[12] G. Boisde' and A. Harmer (ed.): "*Chemical and Biochemical Sensing with Optical Fibers and Waveguides*", Artech House: Norwood 1996.

[13] O.S. Wolfbeis: "*Fiber Optics Chemical Sensors and Biosensors*", Anal. Chem., vol.76 (2004), pp.3269-3284.
[14] A.G. Mignani and F. Baldini: "*In-vivo Medical Sensors*", in Ref. [5], pp.289-326.
[15] P.W. Milonni, J.H. Eberly: "*Laser Physics*", Wiley, Hoboken 2010.
[16] S. Donati, V. Annovazzi Lodi, and T. Tambosso: "*Magnetooptical Fibre Sensors for the Electrical Industry: Analysis of Performances*", IEE Proc. J, Optoel., vol.135 (1988), pp.372-382.
[17] B.U. Billings (ed.): "*Polarization*" SPIE Milestone Series, vol. MS-23 (2006), pp.388-390.
[18] K. Bohnert, P. Gabus, H. Brändle: "*Fiber-Optic Current and Voltage Sensors for High-Voltage Substations*" (Invited Paper), 16th Intl Conf. OFS, Oct.13-17, 2003, Nara Japan, pp.752-754.
[19] A. Papp and H. Harms: "*Magneto-Optical Current Transformers*", Appl. Opt., vol.19 (1980), pp.3729-3745.
[20] S.C. Rashleigh and R. Ulrich: "*Magneto-Optical Current Sensing with Birefringent Fiber*", Appl. Phys. Lett., vol.34 (1979), pp.768-770.
[21] V. Annovazzi Lodi, S. Donati: "*Fiber Current Sensors for HV Lines*", in Proc. SPIE Symp. Fiber Optics Sensors II, vol.798 (1987), pp.270-274; see also: Alta Freq., vol.53 (1984), pp.310-314.
[22] M. Berwick, J.D.C. Jones, and D.A. Jackson: "*Alternate Current Measurement and Noninvasive Data Ring Utilizing Faraday Effect in a Closed-Loop Magnetometer*", Opt. Lett., vol.12 (1987), pp.293-295.
[23] H. Ahlers and T. Bosselman: "*Complete Polarization Analysis of a Magnetooptic Current Transformer with a new Polarimeter*", Proc. OFS-9, (1993), pp.81-84.
[24] V. Annovazzi Lodi, S. Donati, and S. Merlo: "*Vectorial Magnetic-Field Fiberoptic Sensor based on Accurate Birefringence Control*", Proc. OFS-9 (1993), pp. 293-302, see also: "*Coiled-Fiber Sensor for Vectorial Measurement of Magnetic Field*", J. of Lightw. Techn., LT-10 (1992), pp.2006-2010.
[25] ABB announces (2014) the launch of its latest generation of *Fiber Optic Current Sensor*, see https://www.cablinginstall.com/testing/article/16477034/abb-says-fiberoptic-current-sensors-boost-smart-grids-iq
[26] L. Fiorina, S. Mezzetti, and P.L. Pizzolati: "*Thermometry in Geothermal Wells: an Optical Approach*", Appl. Opt., vol.24 (1985), pp.402-406.
[27] M. Corke, D. Kersey, K. Lin, and D.A. Jackson: "*Remote Temperature Sensing using a Polarization Preserving Fiber*", Electr. Lett., vol.57 (1984), pp.77-80.
[28] The Collection of Early "*OFS Conference Proceedings*" from OFS-1 to OFS-13, a CD published by SPIE, vol.CDP-01 SPIE: Bellingham, 1997; the Conf. Proc. of OFS-13 to OFS-27 (2022) are published by Optica (formerly OSA).
[29] J.H. Cole, J.A. Bucaro, C.K. Kirkendall, A. Dandridge: "*The Origin, History and Future of Fiber-Optic Interferometric Acoustic Sensors for US Navy Applications*", Proc. 21st Conf. on Fiber Optics Sensors, Ottawa 2011, paper 775303.
[30] C.K.Y. Leung, K.T. Wan, D. Inaudi, X. Bao, W. Habel, Z. Zhou, J. Ou: "*Optical Fiber Sensors for Civil Engineering Applications*", Mater. Struct., vol.48 (2015), pp. 871-904.
[31] "*Geotechnical and Structural Monitoring FOS*", Smartec, Manno (CH), see https://smartec.ch/en.
[32] R. Kist, "*Point Sensor Multiplexing Principles*", in Ref. [1], vol.2, pp.511-574.

[33] A. Dandridge and C. Kirkendall: "*Passive Fiber Optic Sensor Networks*", in [4], pp. 433-449.
[34] J.P. Dakin: "*Distributed Optical Fiber Sensor Systems*", in Ref. [1], vol.2, pp.575-598.
[35] A.J. Rogers: "*Distributed Optical Fiber Sensing*", in Ref. [4], pp.271-309.
[36] B.Y. Kim and S.S. Choi: "*Backscattering Measurement of Bending-Induced Birefringence in Single-Mode Fibers*", Opt. Lett., vol.17 (1981), pp.193-194, see also "*Single-End Polarization Mode Dispersion Measurement using Backreflected Spectra and a Linear Polarizer*", J. Light. Techn., vol.17 (1999), pp.1835-42.
[37] B. Hutter, B. Gisin, and N. Gisin: "*Distributed PMD Measurement with a P-OTDR in Optical Fibers*", IEEE J. Light. Techn., vol.17 (1999), pp.1843-48.
[38] G.P. Lees at al.: "*Advances in Optical Fiber Distributed Temperature Sensing using the Landau-Placzek Ratio*", IEEE Phot. Techn. Lett., vol.10 (1998), pp.126-128.
[39] H.H. Kee, G.P. Lees, T.P. Newson: "*Low-loss, Low-cost Brillouin-based System for Simultaneous Strain and Temperature Measurement*", Proc. CLEO 2000, paper CthI4, p.432.
[40] J. Hu, X. Zhang, Y. Yao, X. Zhao: "*A Brillouin OTD Analyzer with Break Interrogation Function over 72 km Sensing Length*", Opr. Expr. vol. 21 (2012), pp.145-153.

Problems and Questions

Q9-1 *How would you classify a hypothetical optical-fiber-sensor (OFS) for SARS detection, based on the color change induced in an organic tracer cemented to the end-face of the fiber?*
P9-2 *How can I design a scale for weighting small weights, say in the range 50-500 g, based on an intensity readout OFS?*
P9-3 *Can an additional measurand other than force be measured with the arrangement of fiber misalignment?*
P9-4 *Can the intensity readout OFS considered in Prob.P9-1 and P9-2 be implemented by a single-ended access, using just one fiber pigtail?*
Q9-5 *Two common problems of intensity-readout OFS are the lack of a reference and the poor linearity. How can we mitigate them with a minimum of added complexity?*
P9-6 *How large is the circular-birefringence rotation induced in a fiber subjected to a 1-Ampere-turn magnetic field intensity?*
P9-7 *How can we increase the small value of the rotation angle of Prob. P9-6?*
P9-8 *Can we sense a magnetic field produced by a normal magnet, say supplying B=1000 Gauss, with the following: (i) a short piece (10 cm) of fiber; (ii) a coil of N=100, 1-cm dia. fiber turns; (iii) a 1-cm long YIG-crystal?*
P9-9 *Calculate the minimum current we are able to measure (with a S/N=1) using a fiber coil linked to the current wire, with* N_f *=25 turns, using an optical power* P_0*=1 mW and a measurement bandwidth B=10 Hz.*
P9-10 *A 1-cm long fiber is loaded with a small weight F=0.01 N. Knowing that the photoelastic birefringence coefficient is q=0.3 rad/N, calculate the signal amplitude that is read with the basic scheme (Fig.9-23) using a 1-mW power.*
P9-11 *Can we develop an OFS hydrophone more sensitive than human hearing?*

CHAPTER 10

Quantum Sensors

Till now we have considered sensors read by a laser or by an LED or an incoherent source. LED and incoherent sources emit a mixture of modes and needn't be considered separately. The laser emission is a mode or a *coherent state*, that is, a field carrying a number of photons n with mean <n> and a Poisson distribution, whereas the electric field E has mean value <E> and a fluctuation ΔE obeying the Gauss distribution. This is the classical result that follows from the *first quantization* of the field, introduced by Planck's law [1] stating that the propagating field carries along quanta $h\nu$ of energy, the photons. About twenty-five years later [2], Dirac proposed the second quantization, introducing operators for both modulus $|E|$ and phase arg(E) of the optical field E that [3] became a theoretically standard treatment in quantum optics.

While the second quantization was asymptotically in full agreement with classical results, it also unveiled unexpected and somehow bizarre aspects of the quantum theory, especially non-locality, that were strongly criticized by Einstein [4] with his famous saying "God doesn't play dices" and generated the EPR (Einstein-Podolski-Rosen) paradox arguing the description of physical reality by quantum mechanics is incomplete, and hinting at the existence of hidden variables to justify non-locality.

Several years later Bell [5] proposed a theorem experimentally testable to verify (or falsify) the quantum theory, and thereafter Aspect et al. [6] were able to follow up with the conclusive experimental evidence that quantum theory is correct.

But, apart from these fundamental theoretical issues, two quantum tools, totally new and escaping a classical explanation, have been already developed and can be used in photonic measurement instrumentation: they are the *squeezed* states, and the *entangled* states we describe in next sections.

Squeezed states [7] offer an uncertainty in the measurement of amplitude or phase which is smaller by a factor F (the squeezing factor) than the classical limit set by the well-known quantum noise 2hνPB. This is a very welcome improvement (generally of $F^{1/2}$) of the minimum detectable measurand, and is directly obtained by inserting the new (squeezed) source in the experiment, although this operation is not at all a simple matter nor is the source easy to make, and last, F is not that small (see next section).

Entangled states offer the bizarre pairing of measurement outcomes of two (entangled) photons, even very distant from each other, and their main and already demonstrated field of application is quantum cryptography [8] followed by the nascent quantum computing technology. In measurement science, entangled states are a hint for remote measurement, also known as teleportation of results.

From the engineering point of view, however, no conclusive evidence of quantum-based instruments outperforming classical schemes has been demonstrated so far.
In another caveat, the new quantum states are obviously *not* a new measurement mechanism, but a new type of source that can be used as a reduced-fluctuation source, or eventually hint at new strategies of measurement.

Despite these limitations, the field of quantum sensing is undoubtedly interesting and may lead in the near future to breakthrough results.

10.1 SQUEEZED STATES SENSING

Before introducing squeezed states, which is a new kind of radiation, let us briefly recall the ultimate performances of sensors based on classical radiation, the ones seen so far using coherent states limited by (classical) quantum noise.

10.1.1 Classical Quantum Noise Performances

Let's briefly review the classical quantum limits found in the instruments developed so far.

In Sect. 3.2, we have seen that the quantum-noise limited accuracy of a time-of-flight telemeter is given by (Eq.3-16) $\sigma_t = \tau/\sqrt{N_r}$ for pulsed modulation and time domain processing, and by (Eq.3.29') $\sigma_t = (1/2\pi f_m)/\sqrt{N_r}$ for the sine wave modulation telemeter.

Multiplied by 2c, these quantities translate into an uncertainty of the distance L=2cT measurement as $\sigma_L = 2c\tau/\sqrt{N_r}$ and $\sigma_L = (c/\pi f_m)/\sqrt{N_r}$, respectively, for the two telemeters.

In interferometers, we found the quantum noise limit as NED = $(\lambda/2\pi)\,(2eB/I_0)^{1/2}/RV$ (Sect.4.4.1 and Eq .4.22), an expression that can be brought to the form NED $=\lambda_r/\sqrt{N_{ph}}$, in which $\lambda_r = \lambda/(2\pi RV)$ and the number of detected photons is $N_{ph} = (I_0/2B)\,h\nu/\eta$, η being the quantum efficiency.

Similarly, the precision of the absolute distance meter (Sect. 5.5) at the quantum limit is found as $\sigma_l = (L_{res}/2\pi)/\sqrt{N_{ph}}$ [9].

In the gyroscope, the minimum detectable angular velocity is $\Omega_{min} = \phi_n/R$, where R is the responsivity, and it is (Eq.8.9) $\phi_n = (2\kappa\, h\nu\, B/\eta\, P_0)^{1/2} = (\kappa/N_{ph})^{1/2}$.

Other precisions have been calculated in Ref. [9]. In the position sensing photodiode (and also in the quadrant photodiode), the rms error σ_m of the position coordinate has been found as $\sigma_m = (w\sqrt{\pi/2})/\sqrt{N_{ph}}$, where w is the spot size of the beam. In the same way, the precision of the optical rule (described in Sect.12.3.4 of Ref. [10]) has been analyzed and the result reads $\sigma_m = (1/2\pi p)/\sqrt{N_{ph}}$, p being the period of the rule grating. Still another result, the precision of the distance measurement (Sect.3.1) by means of triangulation, is found in [9] as $\sigma_L = (\pi/2)^{1/2}(wL^2/DF)/\sqrt{N_{ph}}$, where L=distance, D=parallax base, w= spot size at the detector, and F =focal length of the viewing objective.

In conclusion, all the measurements of length-related quantities share the same dependence on the number of detected photons when they reach the *quantum limit* of performance, and the rms square error σ_L is:

$$\sigma_L = m_f\, L_c/\sqrt{N_{ph}}, \tag{10.1}$$

where L_c is a characteristic length, and m_f is a multiplicative factor of the order of unity, specific of the particular measurement at hand.

We shall warn the reader that the result expressed by Eq.10.1, although a nice generalization, is an ideal target that can only be approached in practice. Rather than limited by the quantum regime, we will more frequently find our experiment or instrument limited by the *thermal regime* of detection, see Ref. [10], Sect. 3.2 and, to cover this case, the error σ_L of Eq.10.1 shall be multiplied by $(1+I_{eq}/I_{ph})^{1/2}$, where I_{eq} is the noise equivalent current of the receiver or detector. Usually, we have to struggle to reduce noise contribution I_{eq} below I_{ph} and approach the quantum limit by say a few dB, already a very good result from the engineering point of view.

The $N_{ph}^{-1/2}$ dependence comes from the statistics of radiation, in which photons are emitted completely *at random* as schematized in Fig.10-1, and act independently from each other to probe the experiment. Indeed, by repeating the measurement N_m times the denominator of Eq.10.1 would become $\sqrt{(N_m N_{ph})}$, just the same as a single measurement made with $N_m N_{ph}$ detected photons.

In the language of quantum mechanics, Eq.10.1 or more in general the $N_{ph}^{-1/2}$ dependence of a classical measurement is called the Standard Quantum Limit (SQL) or Cramer-Rao bound (CRB) [13]. Using quantum states, the SQL limit can be surpassed and the ultimate dependence becomes the Heisenberg limit (HL) or Quantum Cramer-Rao bound (QCRB) given by N_{ph}^{-1}. In both cases, for a measurement repeated N_m times, the uncertainty decreases as $N_m^{-1/2}$.

Returning to Fig.10-1, the randomness of photons time of occurrence is described by the Fano factor of the number n of photons collected in an arbitrary time interval T. In physics, the Fano factor FF is defined as the ratio of variance $\sigma^2{}_n = \langle[n-\langle n\rangle]^2\rangle$ to mean value:

$$FF = \langle[n-\langle n\rangle]^2\rangle / \langle n\rangle$$

Compete randomness of photon occurrence has a Poisson distribution, $p(n)=e^{-\langle n\rangle}\langle n\rangle^n/n!$ for which mean value $\langle n\rangle$ and variance coincide, so that FF=1. For a squeezed state, the Fano factor can be less that one and can reach the zero variance or FF=0 for a number squeezed state (Fig.10-1, bottom).

The Poisson distribution is well matched by natural sources (e.g., incandescent lamps, photoluminescence devices, LEDs) as well as by lasers above threshold or, most of the sources commonly found in applications. In the few other cases (e.g., laser near threshold, superluminescence, gas lamps), the statistical distribution is invariably broader than the Poissonian, and the variance is larger than the mean value, $\sigma^2{}_n > \langle n\rangle$ (or FF>1) and there is excess noise.

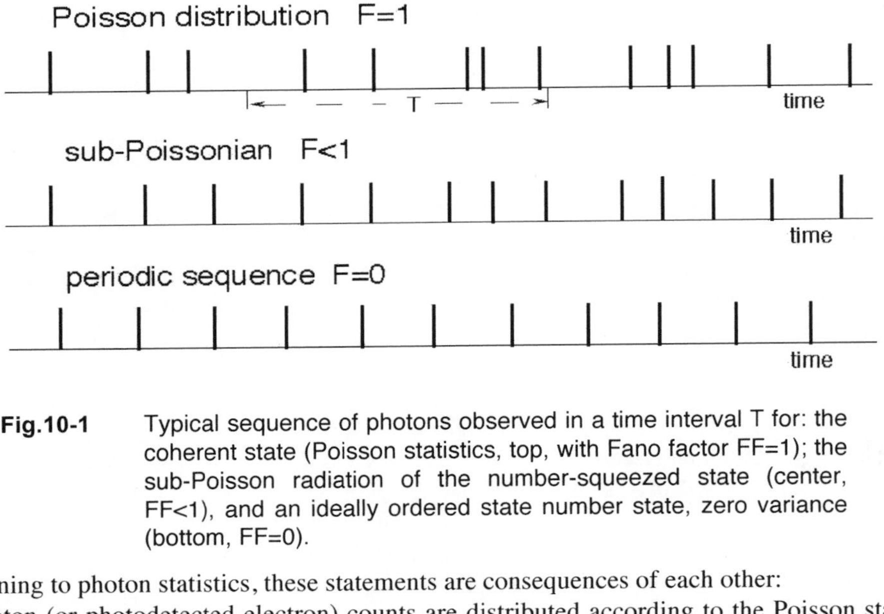

Fig.10-1 Typical sequence of photons observed in a time interval T for: the coherent state (Poisson statistics, top, with Fano factor FF=1); the sub-Poisson radiation of the number-squeezed state (center, FF<1), and an ideally ordered state number state, zero variance (bottom, FF=0).

Returning to photon statistics, these statements are consequences of each other:

- photon (or photodetected electron) counts are distributed according to the Poisson statistics;
- shot (or quantum) noise has a white spectral density $2h\nu P$ (in power) or $2eI$ (in photodetected current) and variance $2h\nu PB$ or $2eIB$, respectively,
- the SNR is given by $\sqrt{N_{ph}}$ for photons and $\sqrt{\eta N_{ph}}$ for detected electrons.

Now, to decrease the variance below the quantum limit, we shall build an unconventional radiation source with a sub-Poisson statistics of radiation and $\sigma^2_n<\langle n\rangle$ (or F<1), that is, a source with a regular time distribution of emitted photons, as illustrated in Fig.10-1 bottom.

10.1.2 Squeezed States

For a rigorous analysis of sub-Poissonian sources, we need to employ the *quantum mechanics* treatment of the radiation field [11], see also some introductory publications [12-13]. However, a quantum mechanics treatment is outside the scope of this book, and therefore we present in the following a shortcut, that is, a *semiclassical description* that retains the key results of quantum mechanics as starting postulates, and develops the correct results in a consistent classical framework.

To start, let us represent the electric field associated with radiation by a rotating vector at optical frequency, $\underline{E}$=E exp i(ωt+φ), and let the uncertainty in $\underline{E}$ be a random vector $\Delta\underline{E}$ added to $\underline{E}$, as shown in Fig.10-2 (top).

For a coherent state obeying the Poisson statistics, the result from quantum mechanics is that $\underline{E}+\Delta\underline{E}$ lies along a circle as indicated in Fig.10-2, and the components E_1 and E_2 of $\Delta\underline{E}$, projected on the Cartesian axes a_1 and a_2, have the same Gaussian distribution, with zero average, $\langle\Delta E_1\rangle = \langle\Delta E_2\rangle = 0$ and a quadratic mean value $\langle\Delta E^2\rangle$ given by:

$$\langle\Delta E^2\rangle = h\nu /2\varepsilon_0 V \qquad (10.2)$$

where V is the volume of observation and ε_0 the vacuum permittivity. Alternatively, the same result in terms of the power spectral density of field fluctuations reads:

$$d(\Delta E^2 A/2Z_0)/d\nu = {}^1/_2\, h\nu \qquad (10.3)$$

or, the spectral density of the field E fluctuations is *half photon per Hertz*.
Also, standardizing the electric field E to $h\nu/({}^1/_2\varepsilon_0 V)$, the resulting dimensionless field amplitude a has a fluctuation given by:

$$\langle\Delta a^2\rangle = {}^1/_2\varepsilon_0 V\langle\Delta E^2\rangle/h\nu = {}^1/_4 \qquad (10.4)$$

Semiclassical description. By representing the instantaneous electric field as the sum E+ΔE of an average E and a fluctuation ΔE, the associated power it carries is $P=(A/2Z_0)|E+\Delta E|^2$. By developing the square modulus, we get: $P=(A/2Z_0)[|E|^2+2\mathrm{Re}(E\Delta E^*)+|\Delta E^2|]$. The first term in square brackets is the average power $P=(A/2Z_0)|E|^2$, the last term is small and can be ignored, and the second has a zero mean-value and a variance:

$$\sigma^2_P = \langle\Delta P^2\rangle=(A/2Z_0)^2 4|E|^2\langle\Delta E^2\rangle= 4P(A/2Z_0)\langle\Delta E^2\rangle$$

If we require that this expression is coincident to that of the shot noise, $\sigma^2_P=2h\nu PB$, we get the condition $2(A/2Z_0)\langle\Delta E^2\rangle=h\nu B$, and dividing it by B for a white spectrum, we obtain exactly Eq.(10.3).

In addition, letting $Z_0=\sqrt{(\mu_0/\varepsilon_0)}$ and B=1/2T (from the Nyquist condition) where T=L/c is the time of observation and L=V/A in terms of volume V and cross-section A, we readily get Eq.(10.4). In Eq. (10.4) a^2 can be interpreted as the mean number of photons, and $\langle\Delta a^2\rangle$ its fluctuation, because $(1/2)\varepsilon_0 VE^2$ is the energy contained in the observation volume V, and dividing it by hν we get a photon number.

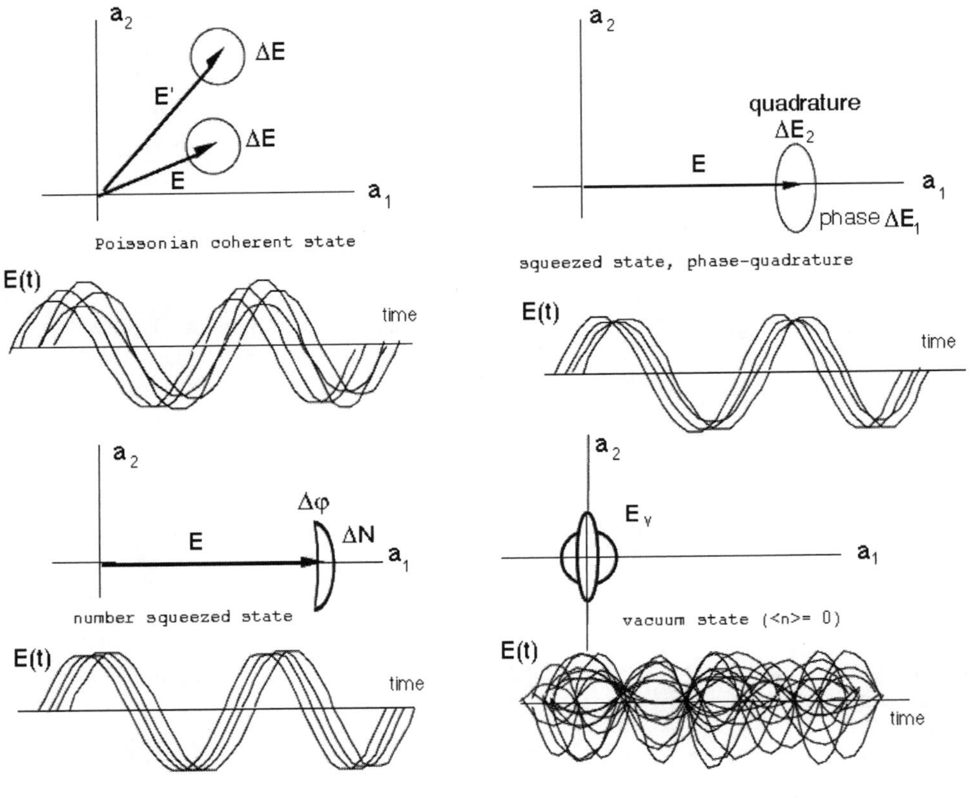

Fig.10-2 The electric field E of the optical field is represented as a rotating vector in the frequency domain, and with its time waveform, shown for several types of radiation. For a coherent state (Poisson distributed source), the uncertainty ΔE is represented by the circle (top left) and is independent from the amplitude of E (or square root of the mean photon number). In the *quadrature* squeezed state (top right) it is $\Delta E_1<\Delta E_{coh}<\Delta E_2$, and in *the number-phase* squeezed state (3rd line, left) it is $\Delta N<\Delta N_{coh}$ and $\Delta\phi >\Delta \phi_{coh}$. The vacuum state has the fluctuation ΔE too, and can be squeezed as well.

Thus, from a semiclassical standpoint, the shot noise is explained as the *coherent beating* of the average field E and its fluctuation ΔE, whose power spectral density is *half photon* (hν/2) *per Hertz* of bandwidth, Eq.(10.3).

Also, the fluctuation in the number of observed photons is 1/2 photon [1/4 as mean square, see Eq.(10.4)] and comes from the (hν/2) uncertainty of the electric energy $(1/2)\varepsilon_0 VE^2$ in the observation volume V.

As a corollary of the above, the field fluctuations ΔE shall have a white frequency spectrum, and be of course uncorrelated to the signal E, whence the correctness of the random vector addition representation of Fig.10-1.

It is important to note that the fluctuation ΔE always has the same amplitude, *independent of the field amplitude E* (and of the photon number N).

Surprisingly, this is true also for E=0 or n=0, that is, for the *zero-field* or *vacuum state*, and for this reason the fluctuation ΔE is frequently referred to as the *coherent-state* or the *vacuum-state* fluctuation.

The very small, yet finite energy $^1/_2$ hνB of the vacuum is another paradoxical, yet incontrovertible consequence of quantum mechanics.

The phase fluctuation associated with the field E can be expressed as the ratio of fluctuation ΔE to mean amplitude E, i.e., $\Delta\phi = \Delta E/E$. By squaring and averaging and using $\langle \Delta E^2\rangle = (1/2)h\nu B/(A/2Z_0) = (1/2)\, h\nu B|E|^2/P$, we get:

$$\langle \Delta\phi^2\rangle = (1/2)\, h\nu B\, /P = h\nu/4TP = 1/(4N) \qquad (10.5)$$

where $N = \langle n\rangle = PT/h\nu$ is the mean number of photons.

Finally, the Poisson fluctuation of the photon number $\langle \Delta n^2\rangle$ is calculated as $\langle \Delta P^2\rangle T^2/(h\nu)^2 = 4P(A/2Z_0)\langle \Delta E^2\rangle T^2/(h\nu)^2 = 4P(1/2)\, h\nu BT^2/(h\nu)^2 = PT/h\nu = N$, the mean value as expected from the Poisson distribution.

In conclusion, from both second quantization and its semiclassical description, the quadratic fluctuations of the photon number N and of the phase ϕ for a coherent state (or Poisson-statistics source) are:

$$\langle \Delta n^2\rangle = N$$
$$\langle \Delta\phi^2\rangle = 1/(4N) \qquad (10.6)$$

These expressions satisfy the Heisenberg *uncertainty principle*, which holds for canonically conjugated variables such as: (i) the number N and phase ϕ pair, and (ii) the in-phase a_1 and in-quadrature a_2 field components pair. The Heisenberg relations are written as:

$$\langle \Delta n^2\rangle\langle \Delta\phi^2\rangle \geq {}^1/_4$$
$$\langle \Delta a^2{}_1\rangle\langle \Delta a^2{}_2\rangle \geq {}^1/_{16} \qquad (10.7)$$

The Heisenberg principle does not forbid a measurement with an uncertainty smaller than the limit given by Eq.(10.6) for one of the two conjugated variables, but not for both at the same time because Eq.(10.7) must always be satisfied.

Thus, quantum mechanics theoretically allows one to *squeeze* the uncertainty of one phase (N or ϕ and a_1 or a_2) provided the other phase is left to fluctuate so that the product satisfies Eq.(10.7).

In a squeezed-state source, we will have therefore (Fig.10-1) one fluctuation (e.g., that of N or of a_1) decreased by a factor F compared to the quantum limit, at the expense of an increase by a factor 1/F of the fluctuation of the other phase (e.g., that of ϕ or of a^2).

The factor F is called the *squeezing factor*, and, apart from a subtle difference for very small $\langle n \rangle$, is coincident with the Fano factor. Thus, it is possible to improve the SNR ratio of a signal beyond the quantum limit, provided another conjugate signal is available to carry an equal amount of worsening. But then, the total information content per unit of energy is unchanged, because what is gained on one phase is lost on the other [Eq.10.7].

An interesting and surprising corollary of the quantum treatment is that the vacuum state or the E=0 field, with a mean value $\langle E_{vac} \rangle = 0$ and zero number of photons $\langle N_{vac} \rangle = 0$, also has the same fluctuation as a coherent state (or Poisson-statistics source) $\Delta E_{vac} = \Delta E$ and can be squeezed in its turn [14-16].

Squeezed-state generators are rather complex and critical to implement in practice. Two examples of them are reported in Fig.10-3.

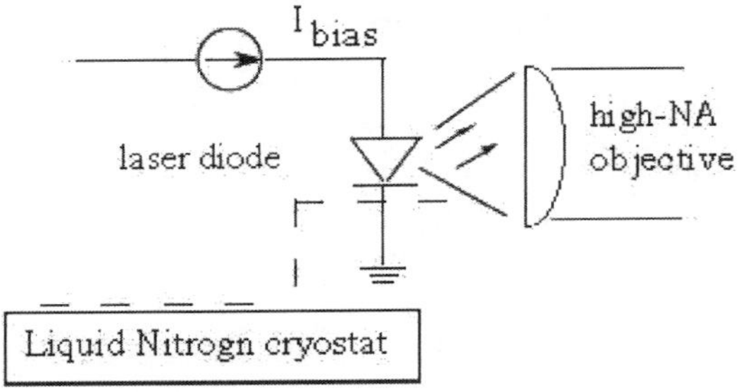

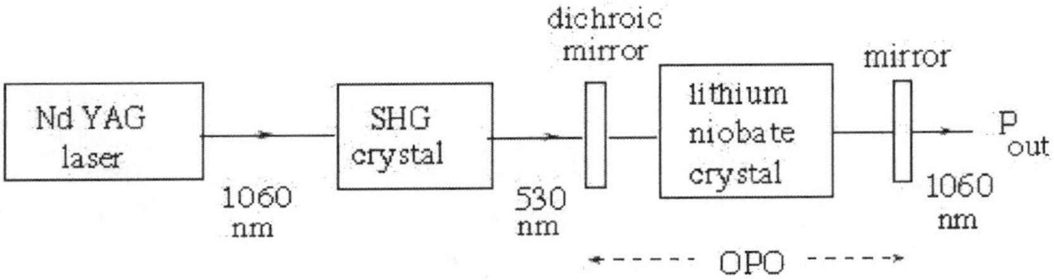

Figure 10-3 Squeezed-state sources: (top) number-phase, obtained by biasing through a constant current a high-efficiency laser (cooled at 77 K); (bottom) phase-quadrature, obtained by pumping an optical parametric oscillator OPO (a Lithium Niobate crystal, like in figure, or a ring in Si nitride as in [17]).

For the number-squeezed state, one can take advantage of the high-efficiency conversion (η>0.95) from electrons to photons found in a semiconductor laser working at low temperatures: if the laser is fed by a constant current, i.e., an ordered flux of electrons, the photon flux replicates the electron statistics and is sub-Poissonian (achieving squeezing factors down to F=0.2 or 7 dB).

For the phase-quadrature squeezing, a source that has been demonstrated since the early times is the parametric laser based on an OPO (*optical parametric oscillator*), see Fig.10-3, where the gain diversity for the two phases due to the pump is used to generate the squeezing.

More recently, the same function integrated in a silicon nitride microring chip pumped by a tunable laser followed by an optical amplifier (EDFA) has been reported [17], with a net squeezing of 1.7 dB (5 dB before losses).

Impact of squeezing on the SNR. For the direct detection of a squeezed-state signal with squeezing factor F, by means of an ideal detector with η=1, the signal-to-noise ratio is:

$$SNR^2 = \langle n\rangle^2/\sigma^2_n = N / F \tag{10.8}$$

The same result is found for the squeezed phase of a phase-quadrature. For the other phase, as a conjugate variable we have $SNR^2 = F N$.

In practice, the improvement of SNR offered by a squeezed-state source is difficult to obtain, because any attenuation suffered in the propagation from source to detector will spoil the squeezing factor.

Indeed, the attenuation ε encountered by a squeezed-state radiation with an initial factor F changes the squeezing factor to a new value F ', given by [15,16]:

$$F' = 1 - \varepsilon\,(1-F) \tag{10.9}$$

and the result, plotted in Fig.10-4, shows that F is degraded severely at the smaller values of ε [13]. Additionally, the finite photodetector quantum efficiency η has the same effect as attenuation, and can be modeled according to Eq.10.9 with ε changed into $\eta\varepsilon$, as shown in Fig.10-4.

Physically, the attenuation corresponds to a random absorption of photons in the sequence shown in Fig.10-1, with a survival probability ε for each photon; when ε<<1 from the ordered sequence we go back to the maximum disorder, which is the Poisson distribution.

This special sensitivity is called *lability* to attenuation of the squeezed-state radiation, and, of course, is a serious drawback in applications where the attenuation is intrinsic and cannot be eliminated, as in optical fiber communications. Nor does using coherent detection schemes give a substantial advantage in SNR for the squeezed-state radiation.

Indeed, by using homodyne detection with a balanced detector, the signal, noise, and SNR ratio of the photocurrent are found as [15]:

$$SNR^2_{hom,ss} = I^2/\langle\Delta I\rangle^2 = 4\mu^2 I\, I_0 / [2eB\,(IF_0+I_0F+I_b) + 4kTB/R] \tag{10.10}$$

where F and F_0 are the squeezing factors of signal I and local oscillator I_0.

Note that the squeezing factors multiply each other's signal, so the squeezing factor ($F_0<1$) of the local oscillator multiplies the (small) signal to be detected, while the dominating term I_0 (the local oscillator intensity) is multiplied by the squeezing factor of the signal (presumably a coherent state, so F=1). So, using a squeezed-state local oscillator is useless.

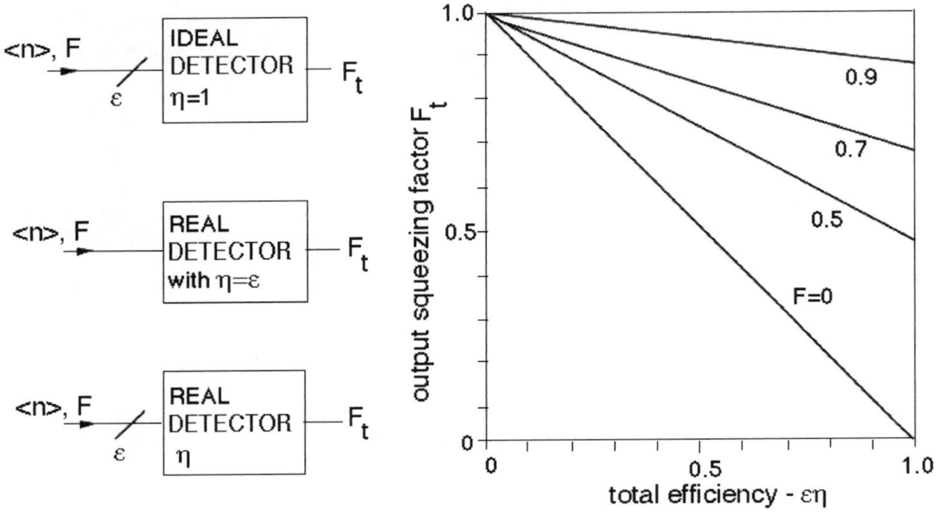

Figure 10-4 Left: detection of squeezed states: a non-ideal photodiode with η <1 is equivalent to an ideal one preceded by a beamsplitter with transmission $\varepsilon=\eta$; right: plot of the output squeezing-factor F_t versus total finite efficiency $\varepsilon\eta$, and with the input factor F as a parameter.

Instead, if we squeeze the signal, from Eq.10.10 we see that, for $I_0>>I+I_b+I_R$, the SNR ratio is that of the classical homodyne divided by the signal squeezing factor:

$$SNR^2_{hom,ss} = SNR^2_{hom,cl} / F \tag{10.11}$$

But, if the signal has undergone a substantial attenuation (e.g., 10 dB, see Fig.10-4), F will have returned close to 1 and no improvement will be obtained, nor will a squeezing of the local oscillator help.

Squeezing states in phase measurements. A quite different situation luckily occurs in interferometers, because propagation attenuation can be made negligible and we can fully exploit the improvement expected from the squeezing factor on the variance $\langle\Delta\Phi^2\rangle$ of the phase Φ under measurement.

For example, let us consider a Mach-Zehnder interferometer reading phase Φ as in Fig.10-5, by means of a coherent radiation source of power P entering at an input port of the coupler beamsplitter.
The output at the balanced detector is $V_u = R\sigma P \cos\Psi \approx RI\Phi$, where $\Psi = nk(l_1 - l_2) = \pi/2 + \Phi$ is the optical pathlength difference between the two arms of the interferometer, adjusted in quadrature so as to read Φ [15].

If we now add, at the normally unused input port of the interferometer, a source of squeezed radiation as shown in Fig.10-5, the phase variance can be found as [13,15]:

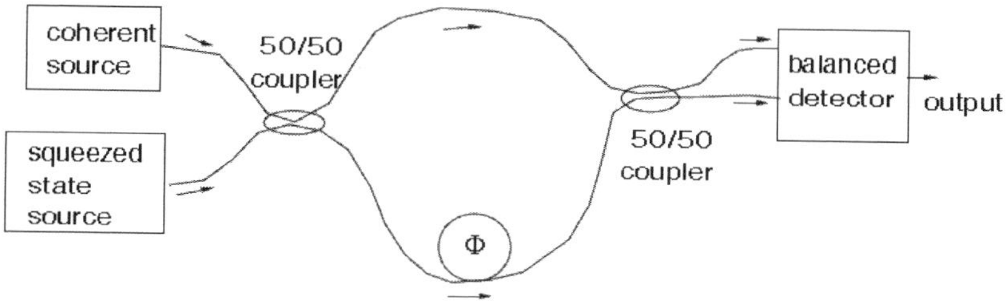

Figure 10-5 In a Mach-Zehnder interferometer, applying a squeezed-state radiation to the normally unused port improves by F_0 the sensitivity of the phase Φ measurement with respect to the classical quantum limit.

$$\langle\Delta\Phi^2\rangle_{ss} = F_0 \langle\Delta\Phi^2\rangle_{cl} = F_0\, 2h\nu B / P \qquad (10.12)$$

where P is the power of the coherent-state source and F_0 is the squeezing factor of the squeezed-state source; the power of the squeezed-state source is uninfluential, so we can use for it a squeezed vacuum as well.

In conclusion, adding a squeezed state (or vacuum) in a leg of the interferometer, the phase uncertainty $\sqrt{\langle\Delta\Phi^2\rangle_{ss}}$ is improved by a factor $\sqrt{F_0}$ with respect to the classical limit.

This method of increasing the sensitivity in a phase measurement has been proposed in the context of gravitational wave interferometers, and a factor $F_0 = 10$ has been predicted as feasible in near future experiments [18].

10.2 ENTANGLED STATES

Entanglement is another important peculiarity of quantum mechanics, and is the phenomenon by which two photons that are created simultaneously (for example, by an OPO halving the energy of an input photon like in Fig.10-3) are entangled in the same quantum state $|\psi\rangle$. In general, the quantum state is described by the equation:

$$|\psi> = \alpha\,|0> + \beta\,|1> \quad (10.13)$$

and represents the superposition of the two states |0> and |1> with probability amplitudes α and β (such that $\alpha^2+\beta^2=1$). With polarization states, for example, state |0> is for H, the horizontal polarization, and state |1> is for V, the vertical polarization.

For quantum mechanics, different from the classical view, the photon is not in a defined polarization state |0> or |1>, but is undecided on the two until it is interrogated with a measurement, that is, it is passed through a polarizer. The measurement makes it collapse to H or V (that is, to |0> or |1>) with a probability α and β.

Thereafter, the state of the photon becomes $|\psi> = |0>$ [or $|\psi> = |1>$] as the outcome of the polarizer. But then, the quantum mechanical properties inherent in Eq.10.13 are lost. This totally counter-intuitive statement of quantum mechanics, a blatant violation of locality, has been debated for a long time as already pointed out in the previous section.

As we make two photons entangled in the same state $|\psi>$ described by Eq. (10.13), any interrogation or measurement carried out on *one of them* (with e.g. the result |0>) will also make the *other one* collapse to the same state |0>, even if meanwhile the two photons have flown far apart from each other, so that no information can have been exchanged between them.

The crucial experiment was carried out by Aspect et al. [6] on entangled photons from a Na-vapour source. They demonstrated that Bell inequality is satisfied, and nonlocality is fully confirmed.

Since then, quantum theory and nonlocality have become the base of the nascent field of quantum-related technologies, in particular *quantum computing* and *quantum communication*. Both these fields are appealing and promising, yet outside the scope of this book.

The best developed to date is quantum communication and in particular *quantum key distribution*.

Entangled states are central to secure communication because, once information or a secure key is exchanged between the partners, it enables traditional digital cryptography to be applied with high security, just requiring the key is sufficiently long, for example 64 bits (corresponding to a $1{:}10^{19}$ probability of detection by chance).

The interested reader may find more details in Ref. [10, Sect.11.5.]. Quantum computing is also based on qubits, and an introduction to the state-of-the-art is provided by Ref. [19].

About *quantum sensing*, entangled states allow implementation of quantum teleportation, that is, to transmit the information necessary to prepare a remote state in the same quantum state as a source state, without apparent information exchange between the two.

The example reported above of a quantum state $|\psi>$ based in polarization is the simplest case of teleportation, and in general one may expect that the result of a measurement by one actor of a couple can be read at distance by the other actor if both are using an ensemble of entangled states, in much the way the secret key is transmitted in quantum cryptography.

In another interesting experiment, hinting at quantum-based Lidar, has been carried out with entangled single photons generated by spontaneous parametric down-conversion in a GaAs waveguide pumped by a 783-nm laser and supplying a continuous-wave flux of 1566-nm entangled photons. Photons are emitted in pairs, and in the pair, the two photons are temporally coincident [20]. One photon is detected as the reference, and the other is left to propagate to a distant target and back, and then detected in the measurement channel.

The detection of photons is performed by a super-conducting nanowire single-photon detector (SNSPD, cooled at liquid He temperature) [21] that supplies the necessary high detection efficiency (η>95%) and the fast response to single photons ($\sigma_t \approx$100 ps), that translates itself into a distance resolution of 1.5 cm.

With a 0.5-mW source power, the generated photon pair had a frequency of 4 MHz (corresponding to a pair-to-pump efficiency of 2.1 10^{-8}) and the detection rate of photons, returning from a target placed at 0.86 m, was 26 Hz.
Yet, thanks to the entanglement, the measurement could tolerate a stray photon flux of 46.1 kHz and had a range extended to about 1 m distance [20].

This example demonstrates that quantum sensing requires considerable hardware and has great potentialities, although presently not yet approaching the performance of instruments described in other chapters of this book and based on "classical" approaches.

REFERENCES

[1] M. Planck: *"On the Law of the Energy Distribution in the Normal Spectrum"*, Annalen der Phys., vol.4 (1901), pp. 553-563.
[2] P.A.M. Dirac: *"The Quantum Theory of Emission and Absorption of Radiation"*, Proc. Royal Soc. London, vol.114 (1927), pp.243-265.
[3] L. Mandel, E. Wolf: *"Optical Coherence and Quantum Optics"*, Cambridge University Press, Cambridge 1995.
[4] A. Einstein, B. Podolski, N. Rosen: *"Can Quantum Mechanical Description of Physical Reality be Considered Complete?"*, Phys. Rev. vol.47 (1935), pp.777-780.
[5] J.S. Bell: *"On the EPR Paradox"*, Physics, vol.1 (1964), pp.195-200.
[6] A. Aspect, P. Grangier, G. Roger: *"Experimental Test of Realistic Local Theories via Bell's Theorem"*, Phys. Rev., Lett. vol. 47 (1981), pp.460-463.
[7] D.F. Wall: *"Squeezed States of Light"*, Nature, vol.306 (1983), pp.5939-5947.
[8] H.M. Wiseman, S.J. Jones, A.C. Doherty: *"Steering, Entanglement, Nonlocality, and the EPR Paradox"*, Phys. Rev. vol.98 (2007), pp.1402-06.
[9] S. Donati, C.-Y. Chen, C-C. Yang: *"Uncertainty of Displacement Measurements in Quantum/Thermal Regimes"*, IEEE Trans Instr. & Meas., vol.IM-56 (2007), pp.1658-1665.
[10] S. Donati: *"Photodetectors"*, 2nd ed., Wiley IEEE Press, Hoboken 2021.
[11] R. Loudon: *"The Quantum Theory of Light"*, Oxford University Press, Oxford 2000.
[12] G.M. D'Ariano, G. Chiribella, P. Perinotti: *"Quantum Theory from First Principles: An Informational Approach"*, Cambridge University Press, Cambridge 2017.
[13] E. Polino, M. Valeri, N. Spagnolo, F. Sciarrino: *"Photonic Quantum Technology"*, AVS Quant. Sci., vol.2 (2020), doi 10.1116/5.0007577.
[14] V. Braginski, F. Khalili: *"Quantum Measurements"*, Cambridge University Press, Cambridge 1992.
[15] V. Annovazzi Lodi, S. Donati, S. Merlo: *"Squeezed States in Direct and Coherent Detection"*, J. Opt. Quant. Electr., vol.24 (1992), pp.285-301.
[16] R. Slusher, B. Yurke: *"Squeezed Light for Coherent Communications"*, IEEE J. Lightw. Techn. vol. LT-8, (1990), pp. 466-477.

[17] A. Dutt, K. Luke, S. Manipatruni, A.L. Gaeta, P. Nussenzveig, M. Lipson: "*On-Chip Optical Squeezing*", Phys. Rev. Appl., vol.3 (2015).
[18] R. Schnabel, M. Mavalvala, D.E.Mc Clelland, P.K. Lam: "*Quantum Metrology for Gravitational Wave Astronomy*", Nature Comm., vol.1 (2010), art. number 121.
[19] N.D. Mermin: "*Quantum Computer Science: History, Theories and Engineering, An Introduction*", Cambridge University Press, Cambridge 2007.
[20] H. Liu, D. Giovannini, H. He, D. England, B.J. Sussman, B. Balaji, A.S. Helmi: "*Enhancing LIDAR Performance Metrics using Continuous-Wave Photon Pair Sources*", Optica, vol.6 (2019), pp. 1359-1355.
[21] see Ref. [10] Sect.7.3.

APPENDIX A0

Nomenclature

In this book, we frequently talk about the ultimate limits of sensitivity of a specific instrument or measuring method. Terms like *sensitivity*, *quantity of response*, *ultimate limit of performance*, etc. are not devoid of ambiguity. Indeed, in electro-optical instrumentation, we use results coming from different disciplines, like optics, electronics, and measurement science, where terms are not always intended in the same way. For the sake of clarity, in this appendix we review the terms used in the text.

A0.1 Responsivity and Sensitivity

In a classical paper, Jones [1] considered the response of a generic sensor (actually, it was a photodetector, but the concepts had a more general applicability) and introduced the following quantities:

- The *responsivity* R, ratio of the output signal (let's say, a voltage signal V_u) to the input physical quantity M (or measurand) $R = V_u/M$. Clearly, this is not a sensitivity limit, because using an amplifier we can increase it at will. However, the responsivity is an interesting quantity because it supplies the scale factor of the conversion performed by the sensor.
- The *NEI*, or *noise-equivalent-input*, of the sensor. This is given by the output noise, usually taken as the rms value v_n of the fluctuation found at the sensor output, divided by the responsivity, or $NEI= v_n/R$.
- The *dynamic range* DR, which is defined as the ratio of the maximum signal V_{max} supplied at the output before the sensor saturates (or its response becomes appreciably distorted) to the noise seen at the output, that is, $DR= V_{max}/v_n$.
- The *detectivity* D*, figure of merit of the sensor's ability to detect small signals [2].

With these definitions, the term *sensitivity* is no longer necessary. We think it helps to alleviate ambiguity because some researchers understand it as the NEI, whereas especially in electronic measurements [3,4], *sensitivity* is used with the meaning of responsivity.

A0.2 Uncertainty and Resolution

The uncertainty of the result of a measurement generally consists of several contributions, which have been traditionally classified as *accidental* (or random) and *systematic*. Recently, however, the NIST and other International Committees have adopted [5] the more correct definition of uncertainty as *Type A* (those evaluated by statistical methods) and *Type B* (those evaluated by other methods). The correspondence between the old and new classifications is not one-to-one, as is explained by the example below.

In the old classification, uncertainty is systematic when the deviation from the true value is repeatedly the same in successive measurements. Incorrect calibration and bias or offset of measurement are systematic. Systematic uncertainty cannot be reduced by averaging. To remove it, we must introduce data correction or act on the measurement method.

Uncertainty is accidental when the deviation from the true value randomly varies from measurement to measurement. Because of the random nature, this uncertainty can be reduced by averaging the results of several successive measurements. Usually, measurements affected by accidental effects obey Gaussian statistics and their rms deviation decreases as $1/\sqrt{N}$ with the number of measurements N being averaged.

In the previous classification, *accuracy* was the qualitative term for systematic uncertainty of an instrument, whereas *precision* referred to the accidental uncertainty. Frequently, these terms are still used, rather loosely. Either of the two may prevail in a sensor.

One last quantity of interest is resolution. *Resolution* is defined as the minimum increment of response that can be perceived by a sensor. If the sensor has a digital readout, resolution is just one unit of the least significant digit (LSD) and represents the effect of the truncation (or round off error). Truncation is a Type B effect if the measurement randomness is much less than 1-LSD, whereas it is Type A if the randomness is much larger than 1-LSD (or, a type A effect is added). In this case, we can improve resolution by averaging.

Example. Let a 99.4-mm stick be measured with a sensor with 1-mm resolution. If the accidental uncertainty is e_a<0.1 mm, every measurement will be rounded to the same result, 99 mm. But, if e_a=1 mm, we will get different outcomes from the measurement, for example, 99, 100, 99, 99, 101, etc. By averaging, we will get a result approaching the true 99.4 at increasing N.

REFERENCES

[1] R.C. Jones: "*Performances of Detectors for Visible and Infrared Radiation*", in Advances in Electronics and Electron Physics, vol.4, pp.2-88, Academic Press: New York, 1952.

[2] S. Donati: "*Photodetectors*", 2nd ed., Wiley IEEE Press, Cambridge 2021. Ch.3.

[3] H. K.P. Neubert: "*Instrument Transducers*", 2nd ed., Oxford Univ. Press: London 1976.

[4] "*ISO Intl. Vocabulary of Basic and General Terms in Metrology*", ISO: Geneva 1993.

[5] B.N. Taylor and C.E. Kuyatt: "*Guidelines for Evaluating and Expressing the Uncertainty of NIST Measurement Results*", NIST Technical Note 1297, 1994.

APPENDIX A1

Lasers for Instrumentation

In instrumentation applications, we want a low-cost, compact laser source capable of supplying the necessary power at the wavelength of operation with the quality of emission required by the application.

The quality of emission may be substantiated by parameters like modal composition, spatial and temporal coherence, wavelength accuracy and repeatability, immunity of emission to external disturbances, etc.

In this Appendix, we will review the basic features of a few laser sources relevant to the instrumentation described in the text, that is, He-Ne lasers, semiconductor laser diodes, and diode-pumped, solid-state lasers. Here, we have no pretense of completeness, but will discuss parameters of the source affecting performance of the instrument, and special solutions specific to instrumentation applications.

Readers interested in a more thorough discussion of lasers can find it treated in a number of excellent textbooks see, for example, Refs. [1-3].

In photonic instrumentation, lasers are central to the development and design of new, clever measurement concepts. The special emission characteristics they offer are unparalleled by conventional light sources, and lead to instruments outperforming the previous electronics counterparts. To attempt classification, we may list, as in Table A1-1, the most important characteristics exploited in each particular instrument.

Instruments for alignment and sizing applications require moderate power preferably in the visible wavelength range. The beam is projected on the measuring target by a collimating telescope, so the beam should have a single-mode spatial distribution. Additionally, for use in alignment instruments, we require good pointing stability of the beam.

These specifications are satisfied by the red He-Ne (helium-neon) lasers, and partly by LDs (laser diodes). Units commercially available from several vendors readily provide several milliwatts of power in the red (λ=633 nm) with a single spatial mode (TEM_{00}) distribution.

Table A1-1 Classes of electro-optical instruments and the lasers they use

Application	Typical Laser	Main Characteristic
Alignment, Pointing and Tracking	He-Ne LD	collimation, beam quality
Diameter Sensors and Particle Sizing	He-Ne	collimation, beam quality
Laser Telemeters - geodimeters (d>1 km) - topography (d<1 km)	 solid-state (Nd, etc.) GaAlAs LD	 high peak-power (Q-switched) high-frequency modulation
Interferometers for dimensional metrology	He-Ne, DFB LD	temporal coherence, 6-digit wavelength accuracy
Doppler velocimeters	He-Ne, quat LD	spatial coherence, beam quality
ESPI vibration analyzers	He-Ne, quat LD	spatial coherence, beam quality
FOG gyroscopes and Fiber Optics Sensors	GaAlAs diode and SLED	high radiance, spatial coherence, controlled temporal coherence

Pointing stability is excellent in internal mirrors, side-arm He-Ne tubes (Sect. A1.1.2), and may reach an angular stability of ≈0.1 μrad, whereas in a LD (laser diode) it is poor (may range from ≈0.1 to 1 mrad).

In interferometers, coherence and wavelength accuracy are the most demanding performance. Though the general discussion is valid also for other applications, in the next sections we will describe frequency and wavelength stabilization.

Temporal coherence allows us to operate the laser interferometer on a substantial target distance, that is, with a large arm-length difference. Wavelength accuracy and stabilization imply that the measurement we are performing is inherently calibrated and that several digits of the measurement are indeed meaningful. In the early times of lasers, the most popular choice to satisfy these requisites at low cost has been the He-Ne laser at λ=633 nm.

When frequency-stabilized, it easily supplies coherence lengths well in excess of kilometers and a wavelength precision easily going to 6 or 7 decimal digits.

Despite being a gas laser, He-Ne has a very satisfactory useful lifetime (in excess of 10,000 hours) demonstrated by use, it is cheap and readily available in quantities by vendors all over the world. Decades of progress have rendered the He-Ne laser a well-reproducible device, with plain and controlled behavior, ideal for safe design of instruments with a long MTTF (mean time to failure).

Emission in the visible and moderate power (0.5 to 2mW), adequate in most applications, add the benefit of a generally eye-safe device (see Sect.A1.5) requiring no special effort to comply with laser safety standards.

Drawbacks of He-Ne lasers are:

- a relatively bulky size (typically the bore is 15 to 25 cm in length and 2 to 5 cm in diameter),
- a high-voltage supply (however, not dangerous from the shock hazard point of view, as the level 5 to 10 mA is rarely exceeded),
- a relatively low optical power ($\approx$ 1 mW) available with reasonable tube length (power is proportional to the length of the discharge).

On the other hand, semiconductor LDs are very compact and work with a low-voltage supply, but they start with a wavelength accuracy and stability that are a few orders of magnitude worse than those of He-Ne lasers.

In recent years, however, LDs are catching up in these performance areas, especially the MQW (multi-quantum-well) and the DFB, and the stabilized external grating may eventually also become competitive in interferometers.

LDs based on the ternary compound GaAlAs (gallium aluminum arsenide) have been used as an optical source since the early years of laser telemetry, both in sine-wave modulated and pulsed versions for medium-distance and middle-accuracy applications.

To push up the attainable range, power was increased using a stack of several diodes connected in series. The largest peak powers of solid-state lasers operating in the Q-switching regime have never been surpassed, however. Thus, very long-range telemeters have traditionally used solid-state lasers (glass and YAG-doped Nd, Er, Yb) optically pumped by flash lamps.

A1.1 Laser Basics

Let's start illustrating the basic laser properties with reference to the He-Ne laser, one of the first lasers developed, dating back to 1961. Its production volume is second only to semiconductor lasers, reaching about 300,000 units per year. The He-Ne laser can oscillate on several lines, the most commonly used being the red at λ=633 nm, followed by the green at λ=543 nm, and by a few infrared lines at λ=1.15 μm, 1.52 μm, and 3.39 μm.

The medium is a low-pressure (a few Torr) gas mixture of He and Ne, in $\approx$5:1 proportion. Neon atoms provide the active transitions through a number of energy levels, spaced in energy by $\Delta E = h\nu = hc/\lambda$, where λ is the wavelength of the line. The helium atoms are excited by collision with the electrons of the discharge, and transfer excitation to neon atoms by collision between ions. The capillary tube is carefully sealed to attain a loss of <0.01 torr/year, and the wall thickness is kept ≈5 mm to limit He leakage through the glass.

The He-Ne active medium is a low-gain one, with a typical gain of γ=0.5-1%/cm. Correspondingly, the optical gain per pass (that is, exp γL) is just ≈1.1-1.2 in a typical tube of 20-cm length. The end-surfaces of the capillary tube are worked flat and parallel. The surfaces accommodate the mirrors or the Brewster-angle windows, in laser units with internal or external mirrors, respectively.

Mirrors of the optical cavity are made by deposition of dielectric multilayers working on interference. This allows us to limit the optical loss to <0.1% or less, compared to several % of metal layers. To get the best cavity properties, usually the output mirror is made flat (radius of curvature $r_1=\infty$) and the other (rear mirror) is concave, with a radius of curvature larger than the tube length (typically r_2=1 m). Mirror reflectivity values in this example are R_1=0.95 (to optimize the output power) and R_2=1 nominally (and 0.998 in practice).

The mirror cavity is a frequency-selective element because it is a Fabry-Perot resonator (see also Appendix A2). Radiation in the cavity bounces back and forth between the mirrors. The resonance wavelengths λ are those at which the round-trip optical phase shift k2L is a multiple of 2π, so that the electrical field of the optical signal continues to add in phase in successive round-trips. Writing the condition k2L= N2π in terms of k=2π/λ, we get the condition that the cavity length is an integer multiple of half-wavelength:

$$N\,(\lambda/2) = L \tag{A1.1}$$

Here, the integer N is the order of the resonance, also indicated in the mode designation as TEM_{00N}. By writing the above condition for the orders N and N+1 and subtracting, we obtain the frequency *spacing* $\Delta\nu$ of the resonance as:

$$\Delta\nu = \nu_{N+1}-\nu_N = c[1/\lambda_{N+1} - 1/\lambda_N] = c[(N+1)/2L - N/2L] = c/2L \tag{A1.2}$$

The c/2L spacing of resonance is typical of the so-called *longitudinal modes*, or TEM_{00} modes, characterized by an electric field distribution given by:

$$E(r) = E_0 \exp -r^2/w_0^2 \tag{A1.3}$$

In the Gaussian field distribution given by Eq. A1.3, the parameter w_0 has the meaning of a characteristic radius, called the *spot size* of the laser beam. In particular, w_0 is the radius at which the field amplitude drops off to 1/e=0.37 and the power density (proportional to E^2) drops off to $1/e^2$=0.13 of the maximum value.

In addition, by integrating $E^2(r)$ on r, we find that the relative power contained within w_0 is $1-1/e^2$=0.86 of the total beam power.

Each longitudinal mode satisfying Eqs. A1.1-A1.3 is accompanied by a set of *transversal modes*. These modes are the distributions of the electric field that satisfy Eq. A1.1 by virtue of a non-vanishing transversal component of the wave vector.

The field distribution of the transversal mode TEM_{pqN} of order p, q can be written in polar coordinates (r,ϕ) as [3]:

$$E_{pq}(r, \phi) = E_0\, \Pi_p(r/w_0) \cos(q\, \phi) \exp -r^2/w_0^2 \qquad \text{(A1.3')}$$

Π_p being a polynomial of order p [1-3] and q in an integer. From Eq. A1.3', we can see that the main spatial dependence is the same Gaussian of the longitudinal mode (Eq. A3.1). However, because of the multiplication by a polynomial of order p, the resulting distribution is considerably broader than the Gaussian, and wavier because of the zeroes of Π_p.

Transversal modes have a spot size, again defined as the radius containing 86% of the beam power, larger than the fundamental longitudinal mode w_0 and increasing with mode order. In frequency, TEM_{pqN} transversal modes are close to the lowest of them TEM_{00N}, the separation being much less than c/2L. Losses experienced by the transversal mode are larger than the fundamental mode (see Fig. A1-1), and this circumstance is useful for preventing modes from oscillation, so that the beam quality of a purely Gaussian mode is preserved.

Propagation [1-3] of Gaussian modes is such that the mode distribution (Eq. A1.3) keeps itself unaltered in free propagation of the beam and in imaging of it by lenses and mirrors. Only the *spot size* w(z) changes, and its relation to the beam waist or spot-size w_0 originated by the laser at z=0 is:

$$w^2(z) = w_0^2 + (\lambda z / \pi w_0)^2 \qquad \text{(A1.4)}$$

In Eq. A1.4, when z=0 we get $w=w_0$, the beam waist of the laser. For z>0, w increases as the quadratic sum of two terms: the beam waist w_0 and a propagation term proportional to distance z. Writing this term as $\theta_0 z$, we can see that $\theta_0=\lambda/\pi w_0$, and this is the angular divergence of the beam, equal to the diffraction angle of the aperture w_0.

When z is large enough ($z>\pi w_0^2/\lambda$), the propagation term prevails and $w(z) = \theta_0 z$. Then, at the angle $\theta=r/z$ off the beam axis, the electric field given by Eq. A1.3 becomes $E(\theta) = E_0 \exp -\theta^2/\theta_0^2$. This tells us that θ_0 duplicates in angle the spot-size meaning w_0 has in the radius. In a real laser with spot size w_0, the 86%-power angle θ_{eff} is larger than $\theta_0=\lambda/\pi w_0$, the ideal single mode value.

Then we define an *M-squared factor* as:

$$M^2 = \theta_{eff}^2 / \theta_0^2 \qquad \text{(A1.4')}$$

The M^2 factor is close to unity as emission approaches a single mode, and is therefore used to describe the angular (or single-mode) quality of a generic laser beam.

By analyzing propagation in the mirror cavity [1-3], we calculate the *beam waist* dependence from the curvature radii of mirrors and distance L, as well as the position of the waist (or, z=0 origin) with respect to mirror position. The result reads:

$$w_0^2 = (\lambda/\pi)\, mL \qquad \text{(A1.5)}$$

In this expression, m is a numerical factor that depends on the ratio of the mirrors' curvature radii to distance L. For example, a plano-concave mirror cavity ($r_1=\infty$, $r_2=KL$) has $m=\sqrt{(K-1)}$, a confocal cavity ($r_1=r_2=KL$) has $m=\sqrt{(K/2-1)}$, etc. The beam waist is located on the plane mirror in a plano-concave cavity, whereas in the confocal cavity, it lays at the midpoint of the mirrors.

When $M^2>1$, considering the *acceptance* $A\Omega=\pi w_0^2\pi\theta_{eff}^2$ and using Eq.A1.4', we can also write $M^2 = w_{eff}^2 / w_0^2$ in terms of the effective spot size w_{eff}, to show that it is larger than the TEM_{00} mode w_0 of a factor M.

A1.1.1 Conditions of Oscillation

Illustration of the conditions leading to laser oscillation in a He-Ne medium is provided in Fig.A1-1. The drawing shows the atomic line providing optical gain, as well as the resonance of the mirror cavity. The width of the atomic line is $\Delta\nu_{at}$ =1.5 GHz in the He-Ne mixture, and the spacing of the longitudinal modes is $\Delta\nu$ =c/2L in frequency (or $\Delta\nu$ =500 MHz for L=30 cm).

As described above, besides the main peaks corresponding to the TEM_{00N} longitudinal modes, we find smaller resonance due to the higher-order spatial modes.

The medium provides a round-trip gain exp2γL, where γ (≈0.01 cm^{-1}) is the gain per unit length and L the length of the active medium.

As in any oscillator, we shall apply the Barkhausen's conditions to find which frequency can oscillate and how amplitude (or gain) will adjust in the permanent regime of oscillation.

The first Barkhausen's condition is about the onset of oscillation and states that the round-trip gain shall have a modulus larger than 1. Once oscillations set in, the second Barkhausen's condition states that the round-trip gain shall become exactly equal to 1 and the phase shift equal to 0. The necessary gain reduction is achieved by saturation of the medium gain.

With reference to Fig.A1-1, when the gain exp2γL is large enough to overcome the losses 1/p [mainly due to the mirrors' reflectivity, $p=r_1r_2$], the pattern of cavity modes determines the oscillating frequencies. Those modes falling within the exp2γL>1/p dotted line in Fig.A1-1 are allowed to oscillate, whereas the others are not. In the example, modes 2 and 3 will oscillate. With oscillation, the mode diminishes the gain (dotted line), preventing the oscillation of further modes (including the undesired higher-order modes).

Ideally, the oscillation frequency is centered around the cavity resonance peak. The medium itself introduces a phase shift, however. Then, the actual frequency moves a little under the cavity line toward the atomic line center. In this way, the atomic line phase shift is compensated for by an equal and opposite amount of phase shift due to the cavity line (Fig.A1-1). Because of this small shift, longitudinal modes are not spaced exactly by 500 MHz, but deviate a modest quantity from c/2L, equal to a fraction of the cavity line width, 50-500 kHz, typically. This effect is called *frequency pulling*.

In a short-length (L=20 to 30 cm) He-Ne laser, we usually find 2 or 3 oscillating modes, the exact number depending on the actual position of the cavity pattern with respect to the atomic line center. However, the cavity pattern is far from being still in a non-stabilized laser. Any perturbation (mechanical, thermal, etc.) will cause ample drifts.

This can be appreciated by considering that a $\lambda/2 = 0.3$ μm variation of the mirror distance (on a L=30 cm) will shift the mode pattern of one full c/2L period, and eventually produce a change in oscillating modes.

Also, when the main mode is right under the atomic line center ν_{lc}, the gain dip and its symmetrical dip become superposed, and there is a decrease in power with respect to frequencies slightly detuned off ν_{lc}. This effect is the so-called Lamb's dip, and is useful for frequency stabilization.

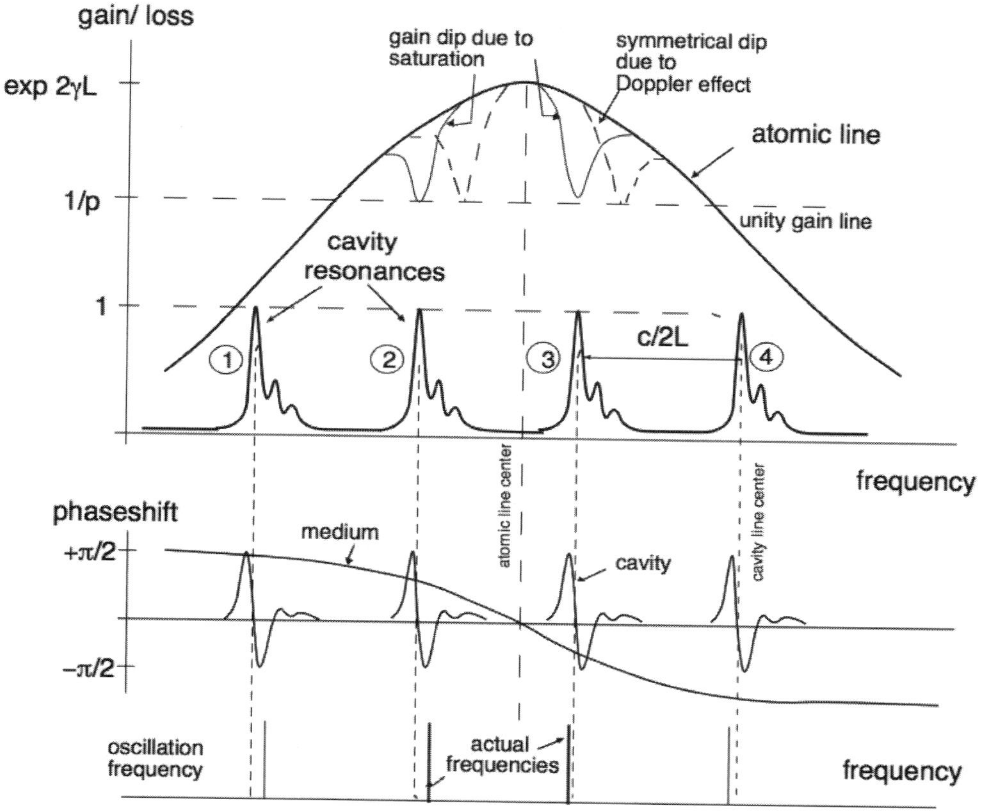

Fig.A1-1 Oscillations in a He-Ne laser. Under the atomic line, the modes labeled 2 and 3 can oscillate because their gain is larger than the loss. Because of oscillation, holes are burned in the atomic line by saturation and the Doppler effect. Mode-pulling makes the longitudinal mode spacing slightly different from c/2L.

A1.1.2 Coherence

Coherence is a very important feature of laser sources, widely used in instrumentation (Table A1-1). In the following, we will recall some basic quantities about it.
In general, the term *coherence* is used to deal with the property of correlation between the optical fields in two points $P_1(x_1,y_1,z_1)$ and $P_2(x_2,y_2,z_2)$.
Correlation is expressed by the coherence factor μ_c, already written in Eq.6.10. When μ_c is high (e.g., $\mu_c>0.5$), we say fields E_1 and E_2 are well correlated, or that there is coherence between them. When μ_c is small or negligible, we say fields E_1 and E_2 are not coherent.
Usually, we distinguish two aspects of coherence, spatial and temporal.

Spatial coherence is when points P_1 and P_2 are taken on the wavefront, or transversal to the propagation direction ($z_1=z_2$). The region inside which μ_c is high is defined as the *coherence area*, and its size $r=\sqrt{[(x_1-x_2)^2+(y_1-y_2)^2]}$ is called the *coherence radius*. Along a fundamental spatial mode of a laser, we have $\mu_c=1$, and coherence dimension is as large as the spot size w_0. Instead, if some power is shared by higher-order modes, we use the factor M^2 to describe the laser emission. Then, at increasing number of modes N, the coherence radius becomes smaller and smaller, and $r\approx w_0/\sqrt{N}$.

Temporal coherence is when points P_1 and P_2 are taken along the wavefront, say at a distance z. Then, the coherence factor is a function of z, $\mu_c=\mu_c(z)$, and the distance z at which μ_c is dropped to 0.5 is defined as the *coherence length* L_c. So, the coherence length is the length of the wave packet inside which the beating term $\langle E(P_1)E^*(P_2)\rangle$ is not less than half the maximum value, given by $[\langle |E(P_1)|^2\rangle\langle |E^*(P_2)|^2\rangle]^{1/2}$ (see Eq.6.10), or a good interference signal is developed.
Writing $L_c=cT_c$, where c is the speed of light and T_c is the time to cover L_c, unveils the *coherence time* associated with the source. The coherence time is interpreted as the time during which the phases of fields $E(P_1)$ and $E(P_2)$ maintain a definite difference or are coherent.

Time T_c is connected to the *linewidth* $\Delta\nu$ of the laser oscillator, and is conventionally assumed [3] $T_c\approx 1/2\pi\,\Delta\nu$.

If the laser emission is multimode (in frequency), the atomic line is likely be filled with oscillating lines (Fig.A1-1), and the line width is about the whole atomic line value, $\Delta\nu\approx\Delta\nu_a$. For a He-Ne, $\Delta\nu_a\approx 1$GHz and then T_c=0.16 ns and L_c=5 cm.

By contrast, if the laser oscillates on a single longitudinal mode, the line width of oscillation is not much different from the cavity resonance width (Fig.A1-1), with typical values of $\Delta\nu\approx$ 1...10 MHz in a He-Ne laser.

This corresponds to a coherence time $T_c\approx$160...16 ns and to a corresponding coherence length $L_c\approx$ 48...4.8 m.

Lastly, if the laser is frequency-stabilized (Sect.A1.2) line width may go down to ≈ 1-10 kHz, coherence time up to ≈ 1-5 ms, and corresponding coherence length may readily be in excess of ≈ 50 km.

A1.1.3 Types of He-Ne Lasers

A number of typical, commercial He-Ne tubes are reported in Fig.A1-2. Depending on the application, we may choose an easy-to-mount, compact internal-mirror tube, or we may prefer to have access to the cavity, like in a Brewster's window tube with external mirrors. The state of polarization of the emitted beam is different in the two cases. It is randomly polarized in internal mirror units (more precisely, it is a random mixture of linear states apparently behaving as unpolarized) and is linearly polarized in Brewster's window tubes, the polarization being parallel to the window's incidence plane.

Windows are tilted at Brewster's angle α_B= atan n_{glass} to have zero reflection loss for the parallel polarization (while perpendicular polarization has a substantial loss and can't oscillate).

All the tubes in Fig.A1-2 are 15 to 30 cm long and their longitudinal mode spacing is accordingly c/2L=1000 to 500 MHz, a value suitable for near-single mode operation in a medium with an atomic line width of $\Delta\nu$ =1.5 GHz.

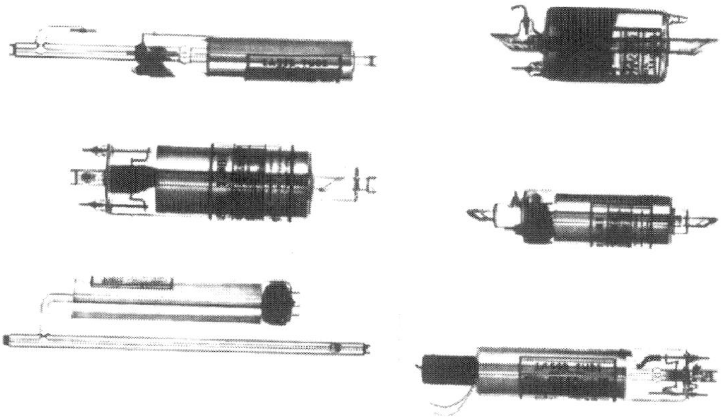

Fig.A1-2 Typical He-Ne lasers for instrumentation. Left, with internal mirrors cemented to the capillary bore (unpolarized output). Right, with Brewster's windows, those with external mirrors, that yield a linear polarization output.

Their typical gain per pass exp2γL is about 1.05 to 1.10, that is, barely in excess of unity but enough to tolerate a mirror loss of 0.95 to 0.98, respectively. These values call for good (multilayer interference) mirrors. Usually, one is chosen with maximum reflectivity (r_1=0.995, the rear mirror), the other with r_2=0.95 to 0.98 reflectivity (the output mirror) adjusted to maximize the output power.

The radius of curvature of the mirrors is as follows: one is flat (usually the output mirror), the other has R=2L to 5L. This choice matches the so-called stability condition of the Fabry-

Perot cavity formed by the two mirrors, by which no extra loss in excess of the mirrors' reflectivity is found because of the back-and-forth propagation of the beam.

The power obtained by the specimen in Fig.A1-2 is typically 0.5 to 2 mW, increasing with tube length. Additional details of tube construction are supplied in Fig.A1-3. In coaxial tubes, the transversal size is substantially reduced with respect to side-arm types, at the expense, however, of a questionable self-alignment and beam pointing stability.

Indeed, in a side-arm tube, mirrors are glued against the tube end-faces, which can be figured flat and parallel with very good accuracy (typically, a few arcsec) by normal optical tooling. As the capillary is kept rather thick (10 mm for a 1-mm bore) to ensure a low loss rate of He, a gas that can appreciably leak through glass in the long term, the structure is inherently stable and alignment-tight. This is a better solution as compared to the coaxial tube, which requires a two-section capillary with a gap in between, to allow the discharge going from cathode to anode.

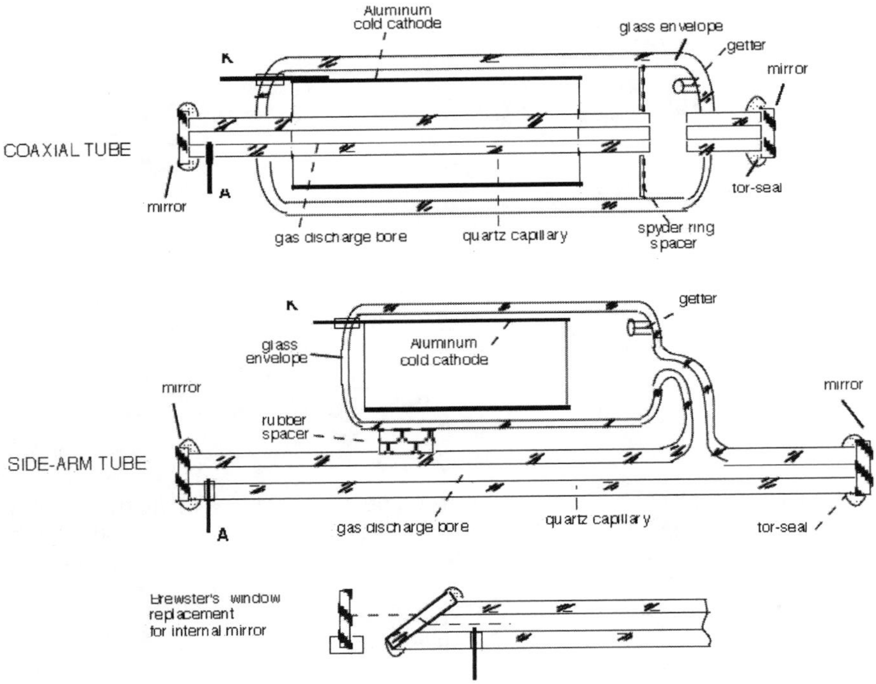

Fig.A1-3 Coaxial and side-arm construction of He-Ne laser tubes with internal-mirrors. Detail of a Brewster's window is shown at bottom.

In coaxial tubes, pointing stability is left to the strength of a relatively thinner outer envelope. To save filament-heating power, cold cathodes are used throughout He-Ne lasers, in the form of an Al tube. This means a ≈2 W savings with respect to a thermionic cathode.

Because the current density obtained by a cold cathode is rather low (<0.1 mA/cm^2), we need a relatively bulky Al cylinder to get the 5 to 15 mA cathode current normally required by the discharge. Regarding the tube supply, the I-V characteristic of the He-Ne tube is that typical of a low-pressure gas discharge (Fig.A1-4). When we start applying a voltage to the tube, initially current is very low. We need a V_T=+15 to 20 kV to trigger the switch-on of the tube and get I_a ≈5 to 10 mA to pass through it. Then, the voltage is lowered to the value (typically V_A=1500 V) required by the gas discharge. We can apply the correct sequence of supply voltage by means of the circuit shown in Fig.A1-4. Here, the rectifier DP is the main diode giving the V_A on-voltage. To also get the initial V_T trigger overdrive, other diodes are used in a diode Marx pump circuit, supplying high voltage when the current is low and being cleared by the current passing through them when the tube is on.

Further, to prevent rectifier ripple from reaching the discharge current, and affecting the emitted power with a mains ripple, a transistor is added in series to the tube so as to feed the discharge by a constant-current generator (equal to $I_a=V_Z/R_e$, see the schematic in Fig.A1-4).

The ballast resistance R_B put in series to the tube is because the tube differential resistance is negative (Fig.A1-4) at the quiescent point of operation (e.g., ≈ 5mA, 1600 V). To avoid spurious oscillations in the supply circuit, we allow for an R_B larger than the negative resistance of the discharge.

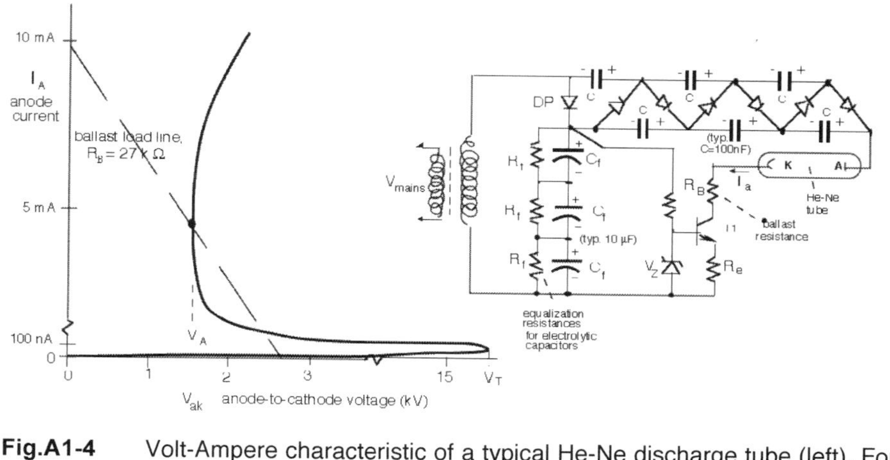

Fig.A1-4 Volt-Ampere characteristic of a typical He-Ne discharge tube (left). For discharge switch-on, we need a high-voltage V_T (typically +15 to 25 kV, depending on tube length L), but with a very small current (less than 1 μA). When the tube starts conducting, the anode voltage drops to about 1500 V. In the drive circuit (right), a diode-pump circuit supplies the trigger voltage peak, while transistor T1 serves to stabilize the tube current.

Values of R_B in the range 10 to 50 kΩ are adequate for the purpose, and we shall minimize the parasitic capacitance of R_B by proper wiring.

A1.2 FREQUENCY STABILIZATION OF THE HE-NE LASER

To stabilize the frequency of emission, three basic functions must be performed:

- a frequency reference (i.e., a frequency marking)
- a mechanism for generating a signal error
- an actuator to change the frequency (through the cavity length)

The *reference* determines how good the ultimate frequency stability of our laser will be. After choosing the reference, we get an *error signal* indicating how far from the reference the actual frequency is, best if it is proportional to it. This signal will be used in a *control loop* to actuate the cavity length, thus changing the frequency (for a $\lambda/2$ change in length, we get a c/2L shift in frequency).

Several methods are available for implementing the frequency reference, and the most performant and used are the following:

- Lamb's dip
- Two-mode, cross-polarized
- Zeeman splitting
- Iodine (external cell) signature

In the section to follow we will review the basic features and performance of each of them.

A1.2.1 Frequency Reference and Error Signal

The *Lamb's dip reference* consists of looking at the power amplitude P_m of the mode as its frequency f_m sweeps under the atomic line (Fig.A1-5). It works well with a single longitudinal mode regime of oscillation, as obtained with a not-too-long cavity length (say 15 to 20 cm), either with internal or external mirror units. When the detuning $\Delta\nu$ is large, the active medium supplies less gain and power is small, while near to the line's center we get maximum power. However, because of the self-saturation (Sect. A1.1.1), right at the line's center we find a small dip, both in gain and in emitted power. The waveform of P_m versus detuning $\Delta\nu$ is readily seen experimentally, during the laser warm-up following the tube switch-on (Fig.A1-5).

Now, if we control the cavity length by adjusting the power to be locked at the Lamb's dip minimum, frequency stabilization is achieved.

To generate the error signal, we cannot directly use the power, because near the peak $P_m(\Delta\nu)$ is about quadratic in $\Delta\nu$ and does not tell us the sign of detuning. But, recalling a well-known technique used in control systems engineering, we can add a small ac modulation, ΔL, to the cavity actuator, and look to the corresponding power modulation ΔP_m. Now, the ac signal ΔP_m is in-phase with $\Delta\nu$ (or with the ΔL drive signal) for $\Delta\nu<0$, is in-phase opposition for $\Delta\nu>0$, and for $\Delta\nu=0$ carries only a second harmonic component.

Thus, by phase detection of ΔP_m with respect to the ΔL drive signal used as a reference, we can obtain an error signal adequate for control.

Another popular technique, again taking advantage of the power dependence $P_m(\Delta\nu)$ on detuning, is that of the *two-mode, crossed-polarization regime* of oscillation.

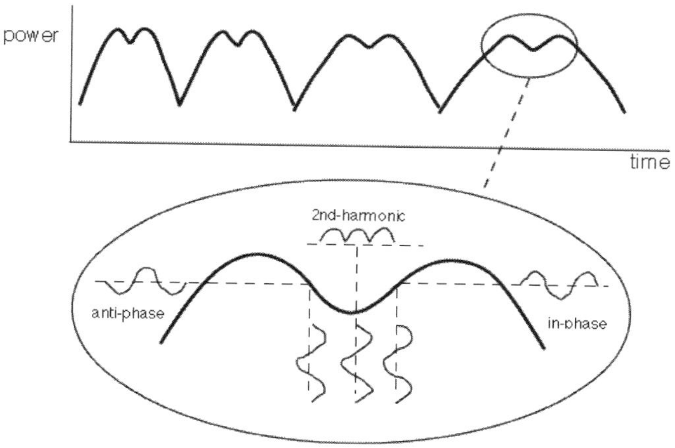

Fig.A1-5 At switch-on, the cavity warms up and power from the He-Ne laser undergoes cycles of variation, replicating the P_m vs. detuning waveform (top). Adding an ac modulation to the cavity length actuator allows us to get a signal proportional (with sign) to detuning from the Lamb's dip center.

The underlying principle is known as spatial hole-burning. When a mode breaks into oscillation, a standing wave pattern is established in the medium, subtracting energy from the active atoms aligned with its particular polarization. If another mode is about to break into oscillation, the gain for the state of polarization orthogonal to that already running is at its maximum, and is at its minimum for the same state of polarization.

The result is such that, when two adjacent longitudinal modes oscillate simultaneously, they put themselves spontaneously in orthogonal states of polarization. In internal-mirror lasers, there is no theoretically preferred polarization to start with, but even minute deviations from ideal symmetry in a practical structure will privilege a specific linear polarization.
Thus, in a two-mode laser tube ($L\approx 15$ to 30 cm for best operation), two adjacent longitudinal modes oscillate with linear orthogonal polarization states. If P_s and P_p are the powers they carry, we will find replicas of the normal $P_m(\Delta\nu)$ curves for each of them, and the diagram of powers as a function of detuning is that shown in Fig.A1-6.

Note that the power difference P_s-P_p is adequate as an error signal marking the symmetry in frequency of the two modes with respect to the atomic line center.

The signal P_s-P_p is easily obtained by placing two photodiodes at the output of a Glan cube polarizing beamsplitter receiving a fraction of the emitted power. A convenient location for the Glan cube is at the laser's rear mirror (the one normally unused), where a small (typ. a few microwatts) but sufficient power is available.

The advantage of the two-mode cross-polarized method is that two polarization modes are readily made available, separated by c/2L = 500 to 1000 MHz, typically.

The frequency stabilization performances obtained by the Lamb's dip and by the two-mode, cross-polarized methods are comparable.

The frequency rms deviation from the average, $\sigma_f(T)$, depends on the observation time T. Typical values relatively easy to obtain are in the range $\sigma_f \approx 1...5$ MHz for short times (T=1 ms), but the best results reported may be 10 to 50 times better.

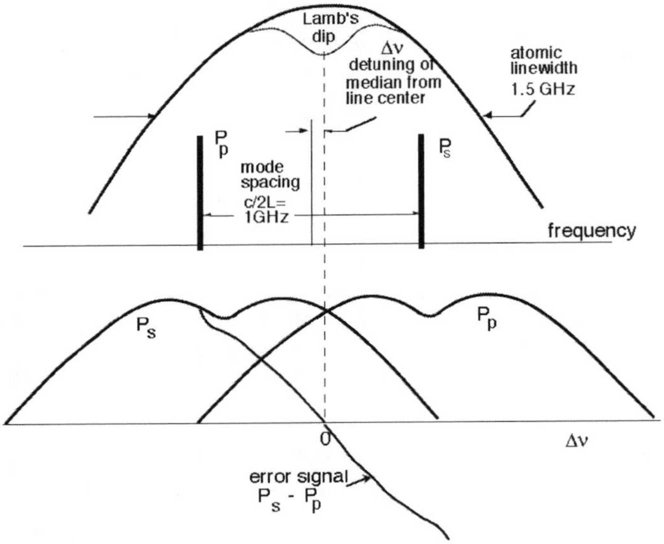

Fig.A1-6 In a two-mode laser, the difference in power amplitude is an easy marking of detuning with respect to the atomic line center. The two modes can be separated and detected because their polarizations are orthogonal.

At λ=600 nm, an rms stability $\sigma_f \approx 2$ MHz means a relative accuracy of frequency $\sigma_f/f = 2$ MHz/500 THz = $4 \cdot 10^{-9}$, and a coherence length $L_c = c/\sigma_f = 24$ m. These are very good figures indeed, largely adequate for industrial applications of interferometry. To get still better results, and also precision of the frequency reference, many efforts can be pursued. Perhaps the first point to care about is the composition of the active Ne gas.

Natural Ne is mixture of ^{20}Ne and ^{22}Ne isotopes, which have slightly different atomic lines giving rise to a waveform distortion (not indicated in Fig.A1-1 to A1-6) that disturbs accuracy and repeatability. The remedy is to use the pure isotope ^{20}Ne.

The *Zeeman-splitting* method of stabilization is based on the application of a magnetic field to the atomic medium. If the magnetic field is applied *parallel* to the tube axis (or along the propagation direction of the oscillating mode), the atomic line becomes split in two frequency shifted lines, one supporting the right-handed circular state of polarization, the other supporting the left-handed circular state of polarization (Fig.A1-7).

The amount of frequency splitting is proportional to magnetic flux density B, and is given by $\Delta\nu_Z = C_Z B$, where $C_Z = e/4\pi m = 1.4$ MHz/Gauss is the gyromagnetic ratio or Zeeman constant.

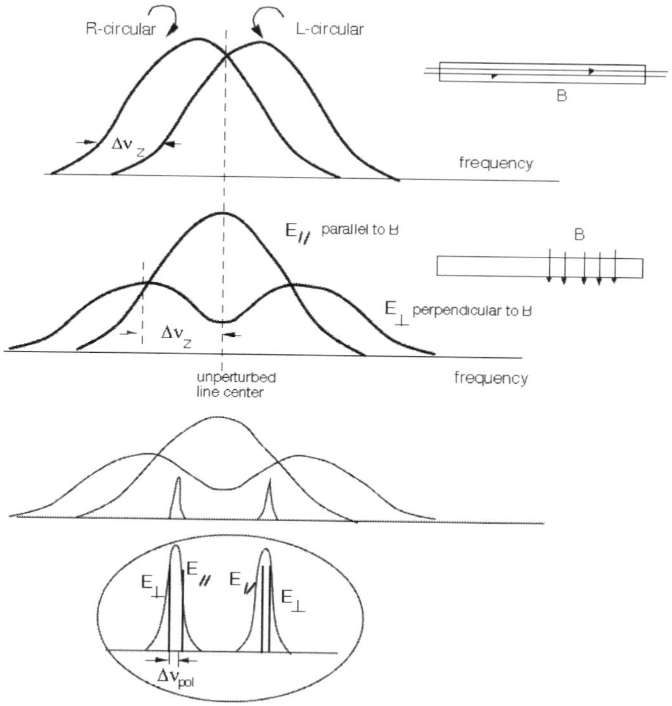

Fig.A1-7 Zeeman splitting of the atomic line, for a magnetic field applied parallel (top) and perpendicular (center) to the tube axis. Because of mode-pulling, the two polarization modes oscillate at slightly different frequencies (bottom), typically $\Delta\nu_{pol}$=50-200 kHz, dependent on the detuning of the cavity line from the original atomic line center. The difference $\Delta\nu_{pol}$ provides the error signal for cavity control, and can be recovered by a photodiode and polarizer combination detecting the mode beating at the rear mirror.

Applying a magnetic field *perpendicular* to the tube axis leaves the atomic line unperturbed for the linear polarization parallel to the magnetic field (say, the horizontal), while the other polarization (vertical) line becomes double-peaked, as indicated in Fig.A1-7.

The consequence of splitting is that the active medium now can support two oscillating modes with orthogonal polarizations, each sustained by the atomic population matched to that polarization.

With an internal mirror cavity adding no extra polarization selectivity, the cavity line is the same for the two polarization modes.

But, because of frequency pulling (Sect.A1.1), the modes oscillate at a slight deviation from the cavity line's center (Fig.A1-7) to compensate for the medium phase shift, each being attracted toward the respective atomic line center.

Thus, we have a frequency difference $\Delta\nu_{pol}$ between the two polarization modes, which changes as detuning $\Delta\nu$ changes.

This is a frequency marking of the atomic line and readily provides the error signal for cavity length control. To get the error signal, it is customary to use the rear mirror of the laser (opposite of the main output mirror), where a small but significant power is emitted. Here we place a polarizer and photodiode.

The polarizer is oriented at 45° with respect to linearly polarized modes so that they beat on the photodiode.

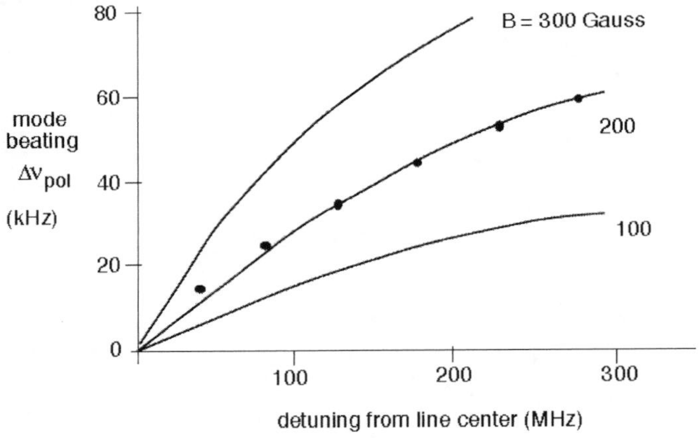

Fig.A1-8 The frequency difference between linearly polarized orthogonal modes in a He-Ne Zeeman laser with a transverse magnetic field, as a function of detuning and for several applied magnetic fields. Lines are the theoretical results, and points are the experimental values for B=200 Gauss.

The photodetected current is then a sinusoidal signal at frequency $\Delta\nu_{pol}$ and, after a frequency-to-voltage conversion, the error signal is ready for the actuator. Both longitudinal and transversal Zeeman effects are used in practice.

An example of the dependence of $\Delta\nu_{pol}$ from the magnetic flux density and detuning is reported in Fig.A1-8, which shows that moderate fields (hundreds of Gauss) are sufficient.

This magnetic field is easily provided, either by a coil wound on the capillary tube (B parallel) or by Al-Ni-Co permanent magnets (B perpendicular).

By actuation of the cavity length (see below), the typical frequency stability of the beating note obtained using a commercial laser He-Ne tube [4] is reported in Fig.A1-9. Translated to optical frequency, this amounts to a relative stability of $\sigma_f/f \approx 1 \cdot 10^{-11}$. However, this is not the actual precision of the wavelength because the beat note $\Delta\nu_{pol}$ depends on B and its eventual drifts.

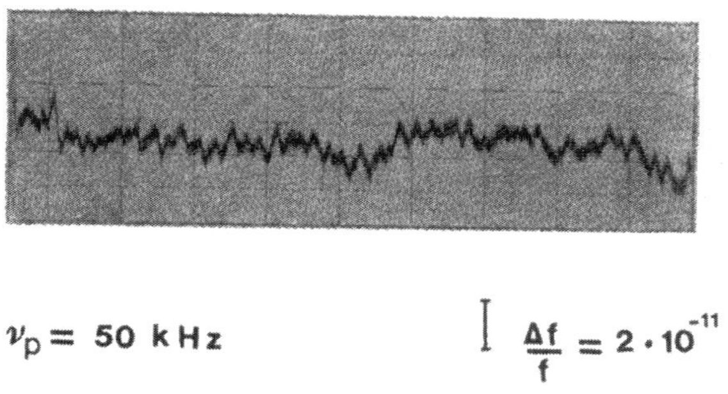

Fig.A1-9 Typical stability of frequency difference in a He-Ne Zeeman laser with transverse magnetic field, on a time period of 1 hour.

With the *iodine (external cell) stabilization method*, we take advantage of the several lines of absorption that the $^{127}I_2$ vapor has under the Ne atomic line [1,4]. These lines are due to the hyperfine structure of iodine and are very narrow dips, typically ≈100 kHz wide.

We can probe the $^{127}I_2$ cell with the laser beam, sweeping its frequency by means of the actuator. If we use the same modulation technique explained for the Lamb's dip method, we can lock the frequency at one of the fine dips. Frequency stability down to ≈100 Hz has been reported with an external cell, the best achievable and of relevance for metrology [5]. Despite that, the absorption cell method is seldom used for interferometers, because so many lines falling under the atomic line require a sophisticated control strategy to lock on a particular one, adding complexity to the final source layout.

A1.2.2 Actuation of the Cavity Length

The two methods most frequently used for actuation of the cavity length are (i) thermal expansion of the capillary tube, best suited for internal mirror lasers, and (ii) piezoelectric movement of the mirror, best suited for external mirrors lasers. Both can be used, of course, with any of the frequency reference schemes described previously.

Thermal expansion is accomplished by Joule dissipation in a resistive wire (e.g., Ni-Cr), directly wound on a substantial fraction of the capillary tube length so as to minimize the thermal power required to expand it and shorten response time. Usually, a few Watts of dissipated power is enough to ensure a $\approx$100-λ expansion from 0 to full control. The response time of thermal actuation is rather slow, typically τ=20-50 ms, yet fast enough to compensate for slow drifts and thermal transients at switch-on and from external temperature, but not enough to reduce vibration-induced fluctuations picked from the external environment (which may have components up to $\approx$100 Hz).

To optimize the response time, the amplifier feeding the resistive winding will include a PID (Proportional-Integral-Derivative) controller. The amount of each PID action will be carefully trimmed to obtain the maximum open loop gain and the fastest closed-loop response.

Piezo actuation is implemented by mounting one of the external mirrors on a PZT piezoceramic element, on its turn secured to the mounting basement carrying the other mirror and the tube. A few-millimiter-thick ceramic disk will usually provide a few micrometers of mirror movement when driven by a 500-1500 V supply (however, with small current demand), and will present capacitance load (typically 1000 pF) to the drive amplifier.

This amount of movement is enough for frequency stabilization when the basement has limited thermal expansion, for example, when we are using an Invar bearing structure or enclose the structure in a thermostat. If not, we have to wait for thermal equilibrium at switch-on and at large ambient temperature changes, before the frequency regulation stops skipping from one mode to the next (for typically 30 minutes or more in a normal He-Ne tube).

The advantage of the piezo compared to thermal actuation is its faster response time, down to 1–10 μs, one capable of canceling ambient-related vibration disturbances as well.

A1.2.3 Ultimate Frequency Stability Limits

We can analyze the performance of frequency-stabilized lasers with the aid of the block-scheme of Fig.A1-10, describing the functions implemented in the feedback control loop.
The frequency error signal is transferred into an electrical signal by a conversion factor G(ν), the power actuator amplifier has a voltage gain A, and the cavity control makes a ΔL variation for unit voltage with a factor K. Lastly, the ΔL variation turns into a $\Delta\nu$ change with a ratio of $\Delta\nu /\Delta L=(c/2L)/(\lambda/2)=c/\lambda L$.

The loop gain of the control circuit is accordingly $G_{loop}= G(\nu)AK(c/\lambda L)$. Thus, following control theory basic results, any disturbance σ_ν altering the frequency of oscillation is reduced by a factor $1+G_{loop}$. Usually, in the above-quoted examples, we have a typical $G_{loop}\approx 10^3\text{-}10^4$. The resulting closed-loop rms frequency fluctuation is 1-10 MHz, typically.

The ultimate limit to frequency stability, when $\sigma_\nu /(1+G_{loop})$ is made negligible, is set by the quantum limit [1] as $\sigma_{\nu(limit)} =[2h\nu B/P]^{1/2}/\tau_c$, where P is the power in the laser cavity and $\tau_c =(1-r_1r_2)c/2L$ is the decay time of radiation in the mirror cavity.
Values for $\sigma_{\nu(limit)}$ are down to a few Hertz for $P\approx 1$ mW, but are seldom approached in practice.

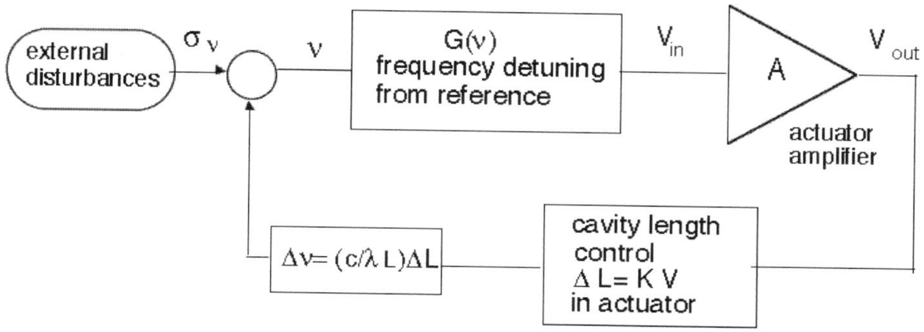

Fig.A1-10 Block scheme of frequency stabilization.

A final remark on He-Ne lasers. He-Ne lasers continue to be preferred as the safe engineering choice in instrumentation, and especially in interferometers. Stabilization not only gives a narrow-line, long coherence length source, but also provides inherent frequency calibration, and not explicitly noted above, amplitude stabilization.

Although laser diodes are catching up, yet He-Ne lasers maintain a significant role in precision applications like interferometers for mechanical engineering because of the favorable performance-to-price ratio.

A1.3 Narrow-Line and Frequency Stabilized LDs

Semiconductor LDs (laser diodes) are the best choice for compact size, low-drive voltage, and small power requirement. However, they have poor mode quality (or, a large M^2), and a less precise wavelength.

Yet, if a plain, cheap LD can be found with sufficient performance for an application, it is undoubtedly the ideal choice for the instrumentation application.

On the other hand, performances of LDs are not satisfactory for high-precision interferometry like calibration of a tool machine on meter-size displacements with 10^{-6} accuracy. However, the situation is slowly improving and LDs incorporating gratings like DFB (distributed feedback laser) or DBR (distributed Bragg reflectors) already offer narrow-line, long coherence length sources with reasonable wavelength accuracy to be considered in sub-parts per million precision interferometer applications.

In this section, we discuss the performances of LDs relevant to applications in electro-optical instruments. In view of applications to measurements, the focus is primarily on engineering issues rather than on device physics. A more detailed treatment can be found in a number of excellent textbooks, see Refs. [1-3,6].

A1.3.1 Types and Parameters of LDs

For ease of use and safety requirements (see Sect.A1.5), a laser with emission at a visible or near-infrared wavelength is desirable. For the range λ=620-800 nm, materials include the ternary semiconductors GaAlAs (the 780 nm being the best choice for high-yield CD lasers), InGaP (going down to 620 nm), and the quaternary InGaAsP.

For the junction, both planar and vertical structures are used. The planar structure incorporates a waveguide and thus the active region is long (typ. 100 to 300 μm) and high gain, whereas in the vertical structure, called VCSEL (vertical cavity surface emitting laser), the cavity is only a few micrometers long.

As the volume of the active region is small, the VCSEL requires much less bias current (typ. a 1–10mA) and the threshold is low, but emitted power is also much less than in a planar structure laser.

The pn-junction structure is usually a heterojunction for optical and electrical confinements and, in the active layer, quantum wells are preferred for band confinement. Band confinement greatly reduces the threshold current, increases the output power, and improves the single-mode operation by narrowing the line width and increasing the side mode suppression, all key points for applications in interferometers and electro-optical instruments.

The I/V characteristic of a planar LD is that of a normal diode (Fig.A1-11), and the P/I curve has a step increase in power in correspondence to the threshold current I_{th}; a VCSEL has the same trend but currents are scaled down by a factor ≈10–20.

The planar LD chip has a length L=100 to 300 μm, typically. The corresponding mode spacing (for λ=800 nm and L=150 μm) is c/2nL=300 GHz (or 0.64 nm in wavelength). The chip ends with cleaved facets, used as mirrors of a plain parallel cavity without additional treatment. The Fresnel reflection at the facet gives a reflectance ≈30% (being n ≈ 3.0...3.5 in most semiconductors).

A point of concern in the use of LDs for instrumentation is that emitted power (Fig.A1-11), as well as modal composition and wavelength (Fig.A1-12), depend markedly on temperature and bias current.

To keep power constant, we should apply a substantial change of bias current to the LD as temperature varies, see Fig.A1-11 (right).

For example, if bias is unregulated and set at I=45 mA to have 5 mW emitted power at 25°C, when temperature becomes 0°C, power will increase to 8 mW, and when it reaches 50°C power will decrease to 2 mW.

Frequently, the power swing is even larger than in this example and poses a risk of burning the device.

The remedy is using a photodiode to monitor the emitted power, and add a feedback loop to keep the power constant.

To this end, a monitor photodiode is usually mounted facing the back-mirror side of the LD chip (Fig.A1-14) to sense power.

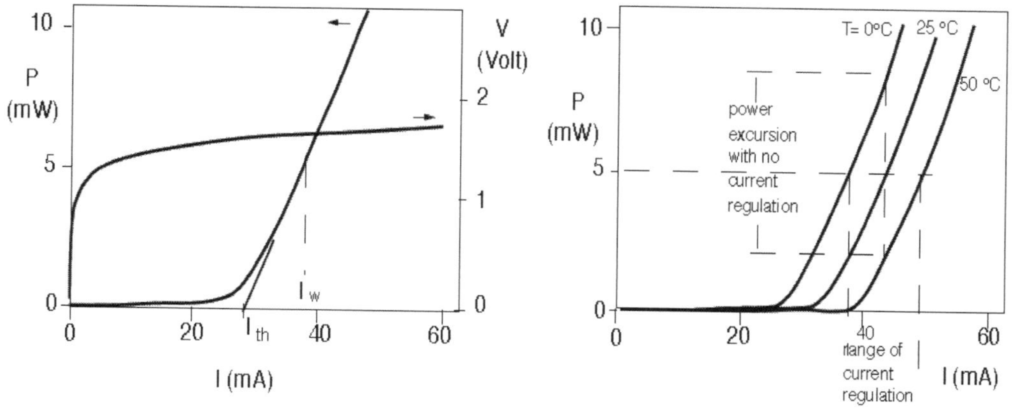

Fig.A1-11 Left: typical I/V and P/I characteristics of a small-power LD (laser diode) at 780 nm; right: a regulation of bias current is required to keep the emitted power at the desired value if temperature changes, e.g., from 0°C to 50°C. The monitor photodiode usually provided by the manufacturer in the same package of the LD (Fig.A1-14) serves this function.

As shown in Fig.A1-12, when current is gradually increased from threshold, first the laser emits a multimode spectrum, then the mode with the largest gain increases faster than the adjacent and finally dominates over the others' side modes.

This is because in LDs the ASE (Amplified Spontaneous Emission) is comparatively larger than in other lasers. Filtered by the cavity mode resonances, the ASE looks like a multimode spectrum. Only when most of the power available is sunk by the main mode, the spectrum looks single mode, hinting that the laser is better driven with the largest permissible current.

A very disturbing feature of LDs is the sensitivity of wavelength to temperature and bias current. In a single-mode Fabry-Perot structure with plane cavity mirrors, at λ=800 nm we may have

$$\alpha_\lambda = d\lambda/dT \approx 0.02 \text{ nm/°C (or, } \approx 10 \text{ GHz/°C)},$$

and

$$\alpha_I = d\lambda/dI \approx 0.004 \text{ nm/mA (or, } \approx 2 \text{ GHz/mA)}$$

In addition, we have *mode hopping* (Fig.A1-12 right) due to the drift of the longitudinal mode pattern under the gain line (Fig.A1-1) as temperature is changed.

Mode hopping comes from the concurrent drift of the atomic line and the cavity line pattern, which move at different rates. When the oscillating mode drifts out the edge of the gain curve (see the representation in Fig.A1-1) it will switch off, while another mode close to the

gain peak wavelength will switch on, with an abrupt change in the emission wavelength as shown in Fig.A1-12, right, typ. of ≈0.5 nm (equivalent to ≈250 GHz in frequency).
The same hopping trend is found versus the LD bias current I_{bias} (Fig.A1-12 bottom), because a current variation ΔI_{bias} changes the dissipated power of $P_{diss} = V_{ak}\, \Delta I_{bias}$ and hence the junction temperature of $\Delta T = K_{ja}\, P_{diss}$, where K_{ja} is the thermal resistance of the junction to ambient.
In a DFB of DBR planar structure, thanks to the selectivity of the grating mirrors, mode hopping is less severe (Fig.A1-12), yet still disturbing in applications.

Thus, it can be concluded that LD can't offer a wavelength precision better than, say $\Delta\lambda/\lambda \approx 10^{-3}$.

The active region has a cross-section $w_{y0} \times w_{x0} = 0.5 \times 1.5$ µm (typically), and the spot size of the mode emerging from the output facet has the same dimensions (Fig.A1-13).

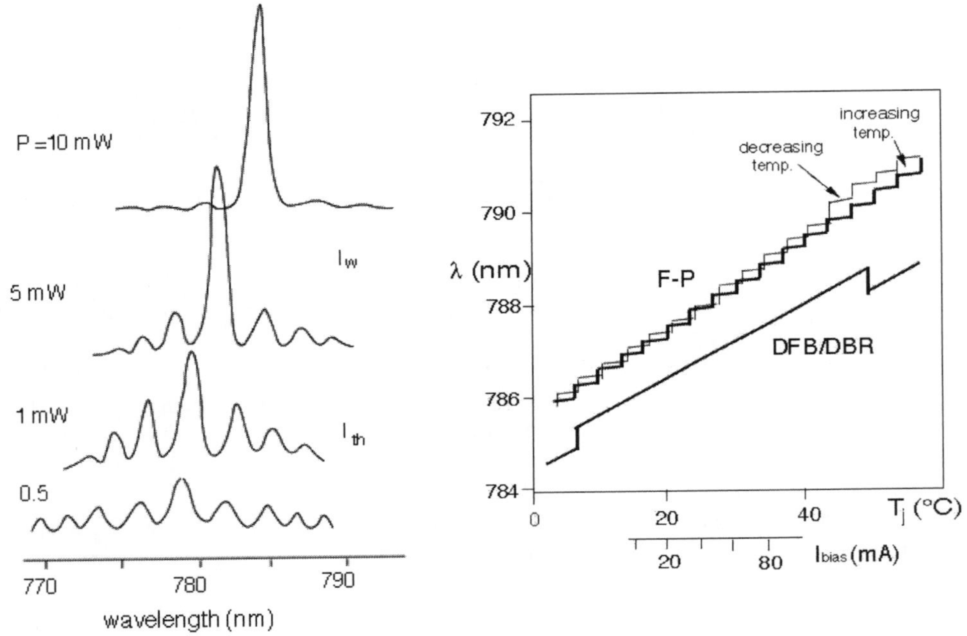

Fig.A1-12 Left: the mode spectrum as a function of drive current at several levels of emitted power, starting from under-threshold to a clean single-mode regime (at 10 mW); right: the λ–dependence on temperature of planar structures. The Fabry-Perot (FP) may exhibit as many as 10-15 mode hops in a temperature swing of 50°C on a total wavelength drift of 5-6 nm. Bias current induces mode hops as well, to a lesser extent. DBR and DFB structures have a comparable drift but fewer mode hops thanks to the filtering effect of the gratings.

Spot size is the parameter of the Gaussian approximation to mode field distribution, that is, $E(x,y) = E_0 \exp -(x^2/w_{x0}^2+y^2/w_{y0}^2)$. A fraction 0.86 of the total power is contained within an ellipse of semi-axes w_{x0} and w_{y0}.

In propagating out of the laser chip, the laser beam expands with divergence angles $\theta_x=\lambda/\pi w_{x0}$ and $\theta_y=\lambda/\pi w_{y0}$ (typically $\theta_x\approx10°$ and $\theta_y\approx30°$). At a distance z from the output facet, the spot-size has dimensions given by: $w_x^2=w_{x0}^2+(\theta_x z)^2$ and $w_y^2=w_{y0}^2+(\theta_y z)^2$.
The elliptical shape of the spot may require a correction, in several measurement applications.

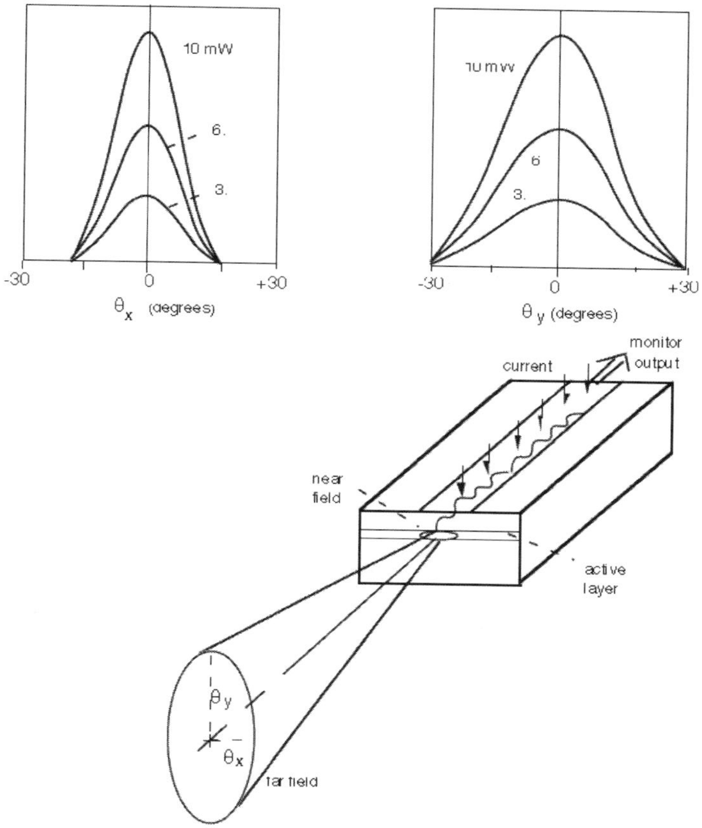

Fig.A1-13 Top: the far-field angular distributions θ_x and θ_y of the LD power (at 10, 6 and 3 mW) along the junction axes. Bottom: the major axis of the beam spot changes from horizontal to vertical in going from the near to the far field.

To transform the elliptical spot into a circular symmetry spot, anamorphic prisms or cylindrical (stigmatic) lenses are commonly used.

The laser chip is welded on a sub-mount (Fig.A1-14) of the package, which incorporates an access window or a collimating lens and, already mentioned, a monitor photodiode mounted on the rear to detect the emission from the back mirror.

To add flexibility in laser operation and reduce thermal requirements, a *Peltier thermo-electric cooler* (or TEC) may be used, sandwiched between sub-mount and case. A *thermistor* is also added in this case to monitor chip temperature.

With a typical 1-W power spent to control the Peltier, we can stabilize the chip temperature against a ±30°C swing from ambient temperature.

A last option, more frequently found in LDs for optical fiber communications than in units intended for instrumentation, is the optical isolator used to prevent back-reflections from returning in the active cavity.

Several packages for LDs are available from manufacturers. Of course, chip and sub-mount versions of laser diodes allow the OEM (original equipment manufacturer) to optimize the key aspects of source design, as well as reduce costs (up to 50% is due to packaging), at the expense of a further effort being required on the component.

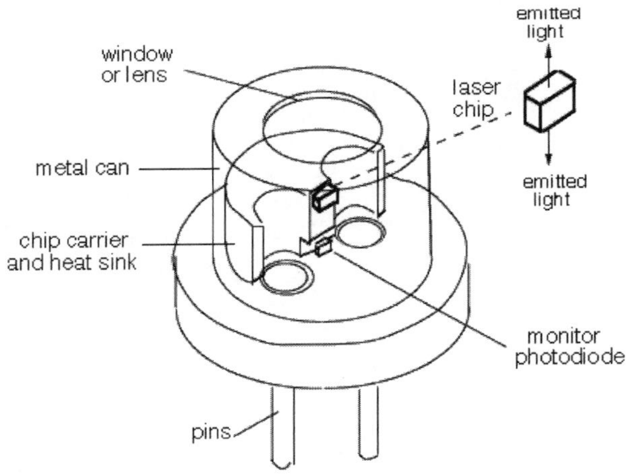

Fig.A1-14 In the typical package of a LD for instrumentation, the laser chip is mounted on a chip carrier that also accommodates the monitor photodiode at the bottom. A Peltier cell (not shown) is eventually mounted between the chip carrier and the package basement. The top contact to the chip is by thermo-compression. A window (or lens) seals the metal package.

About frequency response, LDs are intrinsically very fast and easily exceed the gigahertz cutoff frequency even in plain devices unintended for high frequency. This is a consequence of the heterojunction structure being very thin, and of the fast removal of carriers crowding the junction through stimulated emission.

Yet normally the frequency response is limited to megahertz because of the parasitics of the electrical access circuit.

If it were not for the mode-hopping and the excessive α_λ and α_I dependence, Fabry-Perot lasers would be good for interferometers as they have long coherence length, typically >10 m, when single-mode units are used well above threshold.

A better λ–behavior is obtained by other types of structures, such as:

- DBR (distributed Bragg reflector)
- external FBG (fiber Bragg grating)
- external bulk-optics grating

A1.3.2 Narrow-Line and Tunable LDs

Frequency-stabilized and narrow-line lasers are based on the use of a frequency-selective element possessing just one resonance corresponding to the active medium wavelength.

The *DBR laser* incorporates a Bragg-grating reflector as one of its mirrors (see Fig. A1-15). The grating is obtained by etching a periodic corrugation in the semiconductor at one mirror location.

Because of the corrugation, a small index of refraction step Δn is seen by the propagating mode at each grating period, giving a field contribution reflected back to the active region. When the number of periods N is of the order of $1/\Delta n$, a large reflectivity is found, provided the individual contributions come back and add all in phase.

This is the Bragg condition, written as $2n\Lambda=k\lambda$, where Λ is the spatial period and k is the order of the grating. For k=1, we need a grating period $\Lambda=\lambda/2n$, or Λ=120 nm for λ=800 nm, being n=3.3. Such a short period is a challenge to fabrication by photolithography, and it increases the cost of the device substantially, compared to a normal Fabry-Perot laser.

For visible wavelength DBR lasers, the required period is even smaller, and we shall resort to the second order k=2 to keep Λ a reasonable value.

With k=2, we get $\Lambda=\lambda/n$ and it is Λ=200 nm for λ=680 nm and n=3.4.

Because the DBR structure has a single resonance in spectral range of the active medium gain, mode hopping is suppressed and wavelength is no longer dependent on the current injected in the active region.

Yet, the temperature dependence remains as in the Fabry-Perot diode because in the grating, n is affected by temperature and Λ by thermal expansion. Using a temperature control, we can keep this drift as low as desired, at least in principle. As an example, using a Peltier cell, a $\Delta\lambda/\lambda$ stability of a few ppm has been reported over a 1-year period [7], at λ=1500 nm.

Another interesting feature of the DBR is that wavelength can be finely tuned by a current injected in the Bragg grating region.

The general drawback of DBR lasers is that the device is difficult to fabricate at wavelengths in the visible range.

The *external-FBG laser* is the affordable version of the DBR. It uses a normal chip like that of Fabry-Perot diodes, but one facet is now ARC (anti-reflection coated) and the other is treated for total reflection (TR). Residual reflection may be <0.01% on the ARC facet, while the TR facet may have a reflectivity R=99.9%. The FBG (fiber Bragg grating), that is, the Bragg grating written in the fiber core, of widespread use in fiber communications, is butt-coupled to the ARC facet.

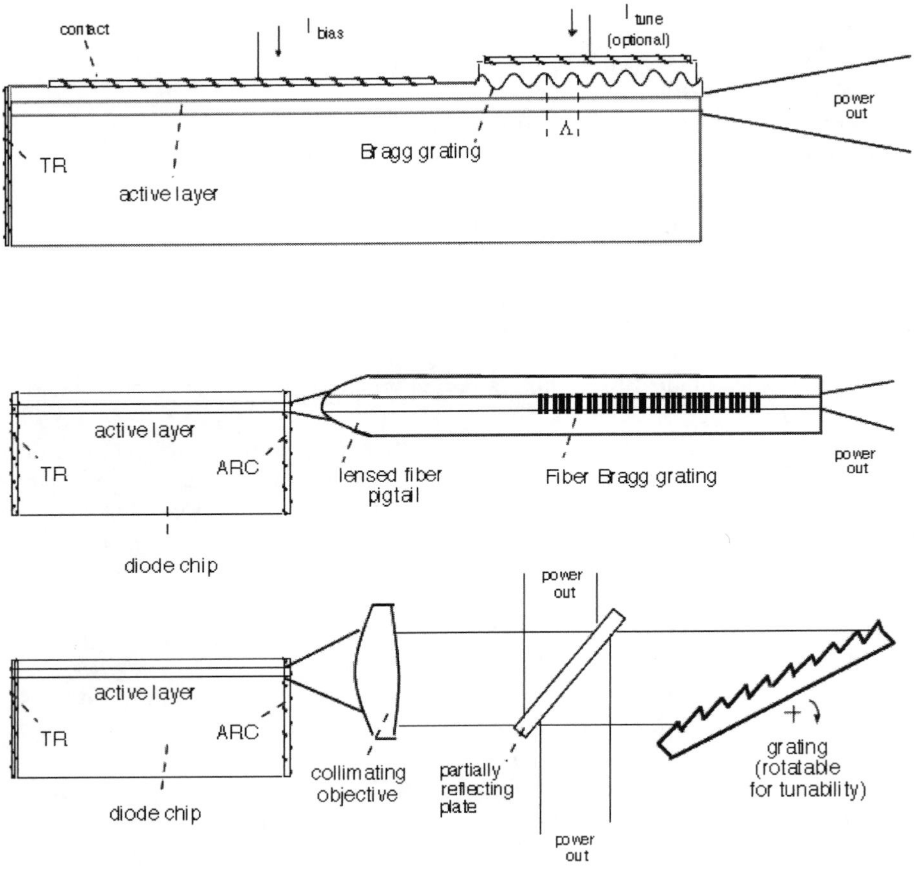

Fig.A1-15 DBR laser chip (top), external DBG laser (center), and external bulk optics (bottom). Drawings are not to same scale.

Thus, except for the longer cavity (requiring a few cm of fiber in the practical implementation), this source is identical to the DBR, sharing the same narrow line width (100 kHz typically) and the modest λ-temperature dependence (of the glass fiber).

Another variant uses bulk-optics, *external grating* as the end-mirror of a cavity. The cavity is made of a Fabry-Perot chip with ARC/TR treatment on the facets and a collimating objective to fill the grating's useful area.

Turning the grating, the wavelength changes according to the grating equation $2p \sin\alpha = n\lambda$, where α is the incidence angle and p the period of the grating.
The wavelength span is limited by the linewidth of the active medium, and is typically $\Delta\lambda$= 30...50 nm using compound semiconductors for the 600...1500 nm central wavelength range. For a wider range of λ-tuning, it's better to use a crystal laser, for example Ti:sapphire, that yields a $\Delta\lambda$ =180 nm wavelength span, from 700 to 880 nm [8].

In some applications, like SS-OCT (Sect.4.7.2) and SMI distance measurement (Sect. 5.5) we need a source with a wide $\Delta\lambda$ together with the possibility of *sweeping* the wavelength. Just rotating the grating, as in the scheme Fig.A1-15 (middle), isn't good, because a constant cavity length L_{cav} is affected by mode hopping. Indeed, as we sweep the wavelength, we should at the same time change the cavity length so that it accommodates the same number N of half wavelength, or $L_{cav}=N\lambda$, and the condition of resonance (Eq. A1.1) is satisfied. Instead, when the cavity length is constant, starting from a particular wavelength $\lambda_1=L_{cav}/N$, only when the wavelength changes to a new $\lambda_2=L_{cav}/(N+1)$ we will have a new oscillation, that is, we suffer a mode hop of $\lambda_2-\lambda_1=L_{cavv}/N^2=\lambda^2\ L_{cavv}$.

The way out to the mode hop is the *Littman-Metcalf* configuration [9] (Fig.A1-16), which consists of rotating the end mirror in such a way that both grating condition $2p \sin\alpha = n\lambda$ and cavity resonance $L_{cav}=N\lambda$ are dynamically satisfied.

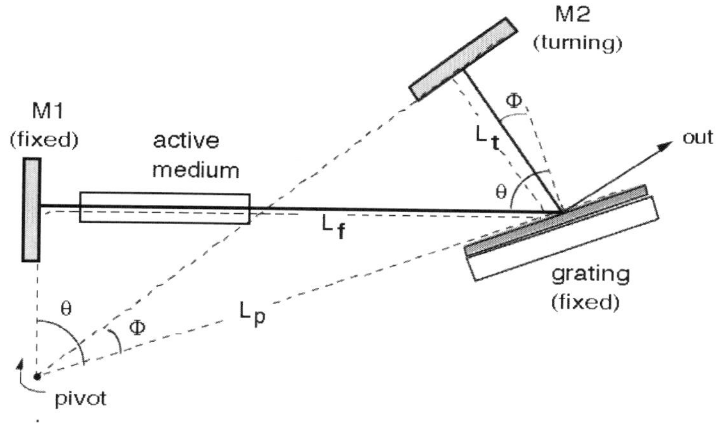

Fig.A1-16 The Littman-Metcalf configuration allows us to scan the wavelength without any mode hop, because the rotation around the pivot point changes the cavity length of a constant multiple N of λ.

As we can see in Fig.A1-16, we shall rotate the end-cavity mirror around a pivot point, at the intersection of front-mirror surface and grating surface, so that $L_p \sin\theta = L_f$.
By substituting in the resonance equation $L_f+L_t=N\lambda/2$, we get $(L_p \sin\theta + L_t)(2/N)= \lambda$, and comparing with the grating equation $p \sin\theta+p \sin\Phi =\lambda$, we have an identity for $L_p(2/N)=p$ and $L_t(2/N)= p \sin\Phi$, independent from (the rotating) θ.

Accordingly, the cavity length is $L_{cav}= L_f+L_t= L_p \sin\theta+ L_p \sin\Phi =(N/2)\lambda$, with N=const., or there is no mode hop because cavity length is dynamically made a multiple of λ.

The Littman-Metcalf configuration works fine also on large $\Delta\lambda$'s, but as the tuning is mechanical on bulky parts, its response time is rather slow (typ. 10...100 ms). In addition, a precise and backlash-free movement of the rotating parts, with micrometer-accuracy of positioning is mandatory, and this is, of course, a constraint in many applications.

Another recent approach to swept-λ sources is offered by the *MEMS-actuated VCSEL* [9]. As the VCSEL native cavity is short, we can use an external mirror actuated by a MEMS to change the cavity length appreciably. As the wavelength follows the cavity length dynamically, Eq. A1-1 is always satisfied and there is no mode hop. The speed of response of this solution, thanks to the very lightweight structure of the MEMS, goes up to 5...10 μs (or 100...200 kHz in frequency). With a quaternary material for the VCSEL, we can obtain $\Delta\lambda$ =10 nm at 1550 nm and 30...50 nm at 1060 nm (Fig.A1-17).

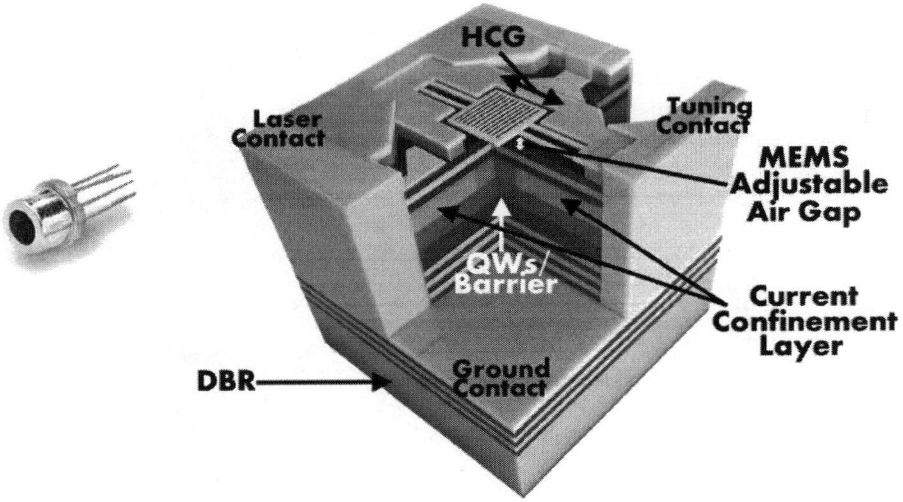

Fig.A1-17 The tunable laser diode BW-10-1550 (left) incorporates a quantum well VCSEL with a DBR bottom mirror and tuning over $\Delta\lambda$=10 nm is achieved (right) by moving a high-contrast grating (HCG) top mirror actuated by a MEMS (by courtesy of Bandwidth10, USA).

A1.4 DIODE-PUMPED, SOLID-STATE LASERS

In pulsed telemeters, we need laser sources capable of emitting short pulses with high peak power so that the timing (and distance) accuracy is good and the maximum distance of operation is large. The typical pulse duration may range from a few nanoseconds to perhaps 10 ns and the peak power we require may go from kilowatts to megawatts, while the pulse repetition frequency is generally low (a few Hertz to a kilohertz).

These figures are typical of Q-switched solid-state lasers. Solid-state lasers [2,3] employ a rare earth element such as Nd, Yb, Er, Cr as the active material. The active atom is embedded in an optically transparent host, a glass matrix, or a garnet-like YAG (yttrium aluminum garnet) with a 0.01-0.5% concentration in weight, typically. The wavelengths of active-level transition are located in the red and near-infrared, depending on the active atom. For operation through the atmosphere, the wavelength is chosen in regions of good transparency (see Fig.A3-3). Useful absorption bands with a fast decay to the active level, are located in the visible or near-infrared, blue-shifted with respect to emission wavelength, and have a spectral width which determines pump efficiency.

Pumping is performed optically, and traditionally the choice for pump sources was a high-pressure flash lamp, however, its bulkiness, limited efficiency, and poor reliability were because of the strong electrical stress on the lamp.

With the availability of low-cost, high-power, high-efficiency LD structures, the stack or array of semiconductor LD has become the ideal source for pumping solid-state lasers.
Stacks are made by piling up several bars, each filled with many individual diodes placed side by side (Fig.A1-18). One bar may contain typically n_d=500 diodes on a 1-cm wide chip. Each diode may emit p_1=10-50 mW, for a total power of 5-25 W.

Up to n_b=5-20 bars may be piled one above the other, with a limit only due to thermal dissipation. In the bar, the LD has a comparatively long (400-800 μm) cavity and broad emitting area (4×6 μm^2) to maximize the active volume [6]. It is designed for low threshold current and high saturation, for example using the QW-GRINSCH structure.

The QW (quantum well) ensures confinement in the band structure, whereas the graded-index, separate confinement hetero-structure (GRINSCH) optimizes optical confinement and mode size. The wavelength of emission is determined by the composition of the active QW.

Using GaAs substrates and elements like Al and In in the active well and surrounding buffer layer, the wavelength of emission can go from 700 to 1000 nm, with λ=808 nm as the best value for pumping Nd doped or co-doped crystals.

A diode stack may typically deliver from 1 up to 300W of optical power in the CW regime. The actual power is proportional, of course, to the number of bars n_b in the stack and to the number of diodes n_d per bar.

The limit to the total number of diodes $n_b\, n_d$ is set by thermal dissipation, whereas the limit to individual diode power is set by saturation and catastrophic optical damage (COD) of the output mirrors due to excessive power density. Neither limit is abrupt, but as we approach them, the useful life of the device is severely shortened.

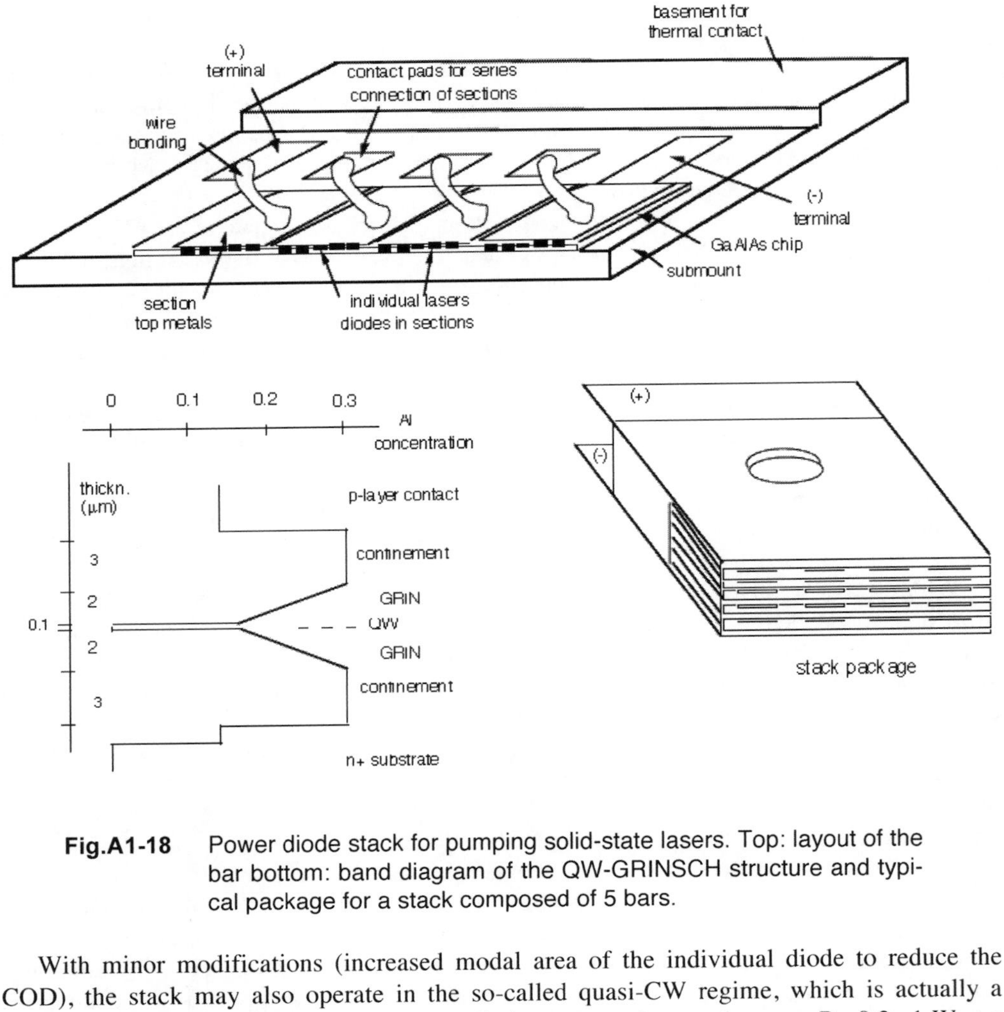

Fig.A1-18 Power diode stack for pumping solid-state lasers. Top: layout of the bar bottom: band diagram of the QW-GRINSCH structure and typical package for a stack composed of 5 bars.

With minor modifications (increased modal area of the individual diode to reduce the COD), the stack may also operate in the so-called quasi-CW regime, which is actually a pulsed regime. When pulsed, a single laser diode may supply a peak power P_p=0.2...1 W at a duty cycle of η=2%.

A typical specified pulse width is τ_p=100 μs and the corresponding pulse period is $T=\tau_p/\eta$= 0.1 m/0.02= 5 ms (or f=200 Hz). From these figures, the average power emitted by the diode is $P_p\eta$=(0.2..1)·0.02= 4..20 mW, nearly the same as the power from a CW laser diode, whereas the peak power is $1/\eta$=50 times as much, at least for pulse duration shorter than the thermal time-constant τ_{th} of the device (milliseconds).

Thus, a diode stack may typically deliver P_p=50 W to 15 kW of peak optical power, with 2% duty-cycle and pulse repetition frequency of a few hundred Hertz.

Because of power dissipation, we cannot use pulse duration greater than about $\tau_{th}\eta$ (at constant duty cycle).

At the opposite end, a pulse duration much shorter than ≈100 μs (or B≈3.5 MHz) is difficult to achieve because of the high current handling of the device.

A1.5 LASER SAFETY ISSUES

All instruments incorporating a laser source must comply with laser safety standards. The actual legal regulations that enforce laser safety issues may vary from country to country, according to the applicable laws.

Below we introduce a few concepts of laser safety, warning the reader that for the obvious legal implications of the matter, it's advisable to seek advice from a professional to make a product compliant and certified to the safety standards.

Safety rules always make reference to standards issued by international organizations like IEC (International Electrotechnical Commission), ISO (International Organization for Standardization), or other national or learned societies like ANSI (American National Standardization Institute) and to the MIL-STD (Military Standards).

Both manufacturers and users shall take the appropriate actions in order to ensure that a laser-based product (or instrument) is prevented from being harmful. In particular, manufacturers must classify their products according to classes expressing the hazard about the laser used in the product.

Users must use or install the product so that the physical regions where harm may occur are prevented from access.

Of course, the subject is much involved and requires a careful study of the applicable standards [10] for a correct answer, because of the many parameters to be evaluated. Here, we simply report a brief overview to elucidate the general issues that we shall consider when dealing with laser safety.

Laser sources and laser-based equipment are categorized in *classes*, from Class 1 to Class 4, according to their potential hazards, which include eye (sight) damage and skin (burning) damage, mainly. Worth noting, the original and worldwide applied IEC 825 standard has been corrected and reissued, after about twenty years of application, with the introduced in the new IEC 60825 (particularly, the new classes 1C, 1E, 2M and 3R).

However, as many laser instruments manufactured in the past are still on the market have been classified according to the IEC 825, we start considering the original version of the standard and will afterwards report the changes recently published.

One fundamental distinction for classification is between CW lasers and pulsed lasers. The first parameter to be taken into account is the optical accessible power (CW or peak).

In addition, other parameters that influence classification are important and include: (i) wavelength, beam area, and beam divergence for all lasers, and (ii) pulse duration, pulse optical energy, and duty cycle for pulsed lasers.

As an illustration, we report in Fig.A1-19 the power versus wavelength diagram defining the past IEC 825 safety classes for CW laser sources [10,11]. Though the last version of the IEC standard has modified the diagram, the guidelines are unchanged.

Class 1 was that of intrinsically safe sources. Safety is because the emitted power is below the threshold of harm, or because the laser is equipped with an automatic shutdown of emission, preventing the operator from being exposed to a dangerous level.

Of the newly introduced classes, Class 1M is for lasers that can be viewed by the naked eye for an indefinitely long time without exceeding the MPE (maximum permissible exposure), but become unsafe when viewed with a magnifying lens or a telescope, and Class 1C is for the lasers intended for treatment of skin, tissues and the like, for medical and/or cosmetic purposes, and they can also emit well in excess of the MPE provided the eye exposure is prevented by means of appropriate shielding of the beam, thus avoiding the eye's access to it.

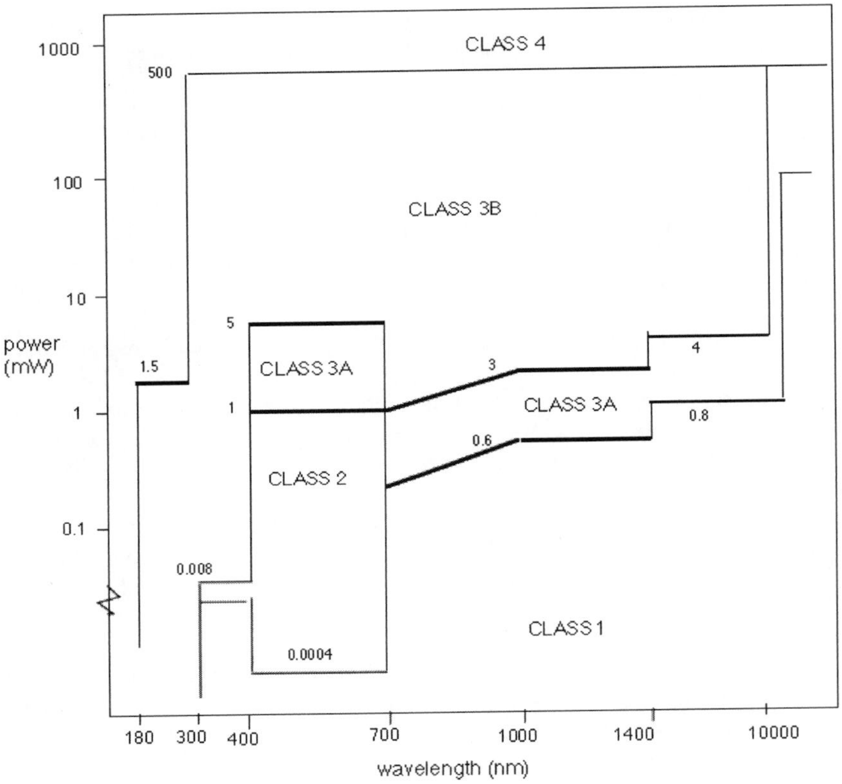

Fig.A1-19 The IEC laser safety classification for CW sources, as a function of wavelength, according to IEC 825 laser safety standard [10].

Class 2 was that of lasers emitting in the visible range (400 to 700 nm), where the blink reflex of the eye allows tolerating a larger CW power (typ. 1 mW).

No special safety measures were required for Class 1 and 2 laser equipment, other than just a label indicating the aperture from which radiation is emitted and a warning not to stare into the laser beam.

In view of this circumstance and from the point of view of engineering design, in all instruments where a power of $\leq$1 mW is technically sufficient, we shall absolutely prefer to work with such a low value to clear the safety problems.

Examples of Class 1 or 2 products are: bar-code laser scanners, alignment and pointing systems, laser interferometers, and sine-wave modulated telemeters.

Class 2M lasers are those safe for short-term exposure of the naked eye, but surpass the MPE limit when viewed by a magnifier of a telescope, therefore they require that instruction labels warn the user against continued or lens-aided viewing of the beam.

Class 3R lasers and equipment are those that have low risk of injury for short time exposure direct vision. In the visible, the CW power limit of Class 3R is twice the Class 2 limit, and five times the power limit of Class 1.

In Class 3B, now abolished with the new IEC 60825 but still found in laser products fabricated years ago, the CW power in the visible was from 5 mW to 500 mW, and this type of laser was always dangerous for direct vision, but still safe for view of the return from a diffuse reflection. Class 3B lasers required a turn-on key for operation by authorized personnel only and a warning sign in the area of operation.

Last, Class 4 lasers are the most dangerous. For a CW source, we get Class 4 when power is just $\geq$0.5 W. Even unintentional reflections (from rings, metallic objects, etc.) and diffuse reflections may scatter to the eyes a power well in excess of the Maximum Permissible Exposure (MPE).

Therefore, operation of a Class 4 laser requires a restricted access area, equipped with warning signs and with acoustic and red-light alarms indicating laser switch-on.

This situation is typical of a CO_2 power laser for welding and other mechanical works, but is surely a big hindrance also for a measurement instrument like a telemeter. Pulsed telemeters are easily Class 4 lasers, and their operation falls under specifications for open-air and propagation through the atmosphere of laser-based equipment.

In this case, the standards prescribe that the Ocular Risk Zone (ORZ) be shielded by appropriate barriers from the reach of the public, which shall have a safety distance of 3 to 6 m (IEC 825) from the reach of the beam.

Another quantity of importance is the DOR (Distance of Ocular Risk). This is defined as the distance at which the power or other related quantity falls below the level, and operation is safe.

In conclusion, it has to be stressed that laser safety shall be seriously considered in any product containing a laser or a LED of non-negligible optical power.

Therefore, the source shall be classified according to the applicable standards (for example, the presently in force IEC 60825) and measures shall be enforced to reduce the hazard risk, like labels and instructions, laser goggles, turnkeys, warning signals and barriers.
In exploiting these tasks, the assistance of a certified professional is advised, up to the settling of appropriate measures and the release of a certificate of compliance.

REFERENCES

[1] A.E. Siegman: "*Lasers*", University Science Books: Mill Valley 1990.
[2] O. Svelto: "*Principles of Lasers*", 4th ed., Springer, Berlin 2009.
[3] P.W. Milonni, J.H. Eberly: "*Laser Physics*", Wiley & Sons Inc., Hoboken 2010.
[4] S. Donati: "*Laser Interferometry by Induced Modulation of the Cavity Field*"; see also Opt Engineer., vol.57 (2018), doi:10.1117/1.OE.57.5.051506.
[5] F. Bertinetto et al.: "*International Comparison of He-Ne lasers stabilized with* $^{127}I_2$ *at 633 (July 1995 to Sept. 1995)*", Metrologia, vol.36 (1999), pp.199-208.
[6] S.-L. Chuang: "*Physics of Photonic Devices*", J. Wiley and Sons, Hoboken, 2009.
[7] R.S. Vodhanel et al.: "*Long-term* λ*-drift of 0.01nm/yr for 15 running DFB lasers*", Opt. Fiber Conf. Techn. Digest, San Jose, Feb. 20-25, 1994, pp.103-104.
[8] T.T. Yang, Y.-I. Yang, R. Soundararajan, P.S. Yeh, C.Y. Kuo, S.-L. Huang, S. Donati: "*Widely Tunable, 25-mW Power, Ti:sapphire Crystal-Fiber Laser*", IEEE Phot. Techn. Lett. vol.31 (2019), DOI 10.1109/LPT.2019.2950020.
[9] K. Li, C. Chase, P. Qiao, C.J. Chang-Hasnain: "*Widely Tunable 1060-nm VCSEL with High-Contrast Grating Mirror*", Opt. Expr., vol.25 (2017), pp.11844-11854.
[10] IEC "*Laser Safety*", Standards 60825-12, Geneva, 2022.
[11] R. Henderson, K. Schulmeister: "*Laser Safety*", CRC Press, Boca Raton 2003.

APPENDIX **A2**

Optical Interferometers

In this appendix, we review the properties of basic optical interferometers, namely Michelson, Fabry-Perot, Mach-Zehnder, and Sagnac (Fig.A2-1). All these configurations are basic and can be improved by the addition of extra components or by suitable variants, like exemplified by the Michelson interferometer that becomes the Twyman-Green interferometer when corner cubes are substituted with mirrors (Sect.4.2.1).

A2.1 Configurations and Performances

The Mach-Zehnder interferometer (Fig.A2-1) is used in fiber sensors, and compared to the Michelson, has two separate arms and makes two complementary outputs available. The Fabry-Perot interferometer has no reference path, and interference is between the back-and-forth contributions reflected by the two mirrors. The multipath propagation adds an increased sensitivity compared to the other configurations. Last, the Sagnac has two counter-propagating beams (clockwise and counterclockwise) in a single propagation path, and is sensitive to non-reciprocal phase shifts only, as required in the gyroscope and in magnetic field sensors.

Now, let's consider the signal obtained at the output of the interferometer. In general, the fields returned on the photodetector from the measurement and reference arms can be written in the form of rotating vectors, or $E_m = \mathrm{E_m} \exp i(\omega t+\phi_m)$ and $E_r = \mathrm{E_r} \exp i(\omega t+\phi_r)$.

The photodetected current is the average of the square modulus of the total field [1], or $\mathrm{I_{ph}} = \sigma \langle |(E_m + E_r)|^2 \rangle$, where σ is a conversion factor and the brackets $\langle .. \rangle$ denote averaging.

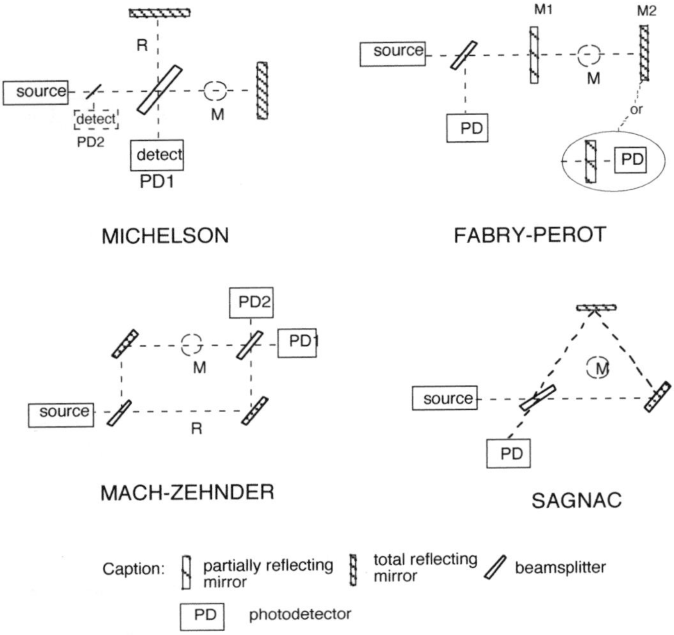

Fig.A2-1 Common optical interferometers and their bulk optics configuration. M and R are the measurand and reference arms.

Inserting the fields E_m and E_r in this expression and developing the modulus, we obtain $I_{ph}= \sigma \langle E_m{}^2+E_r{}^2+2E_mE_r\mathrm{Re}\{\exp i(\phi_m-\phi_r)\}\rangle = I_m+I_r+2(I_mI_r)^{1/2}\langle\cos(\phi_m-\phi_r)\rangle$, where I_m and I_r are the currents produced by the measurement and reference fields, respectively. For $I_m \neq I_r$, we can write the detected current as $I_{ph}=(I_m+I_r)\{1+[2C^{1/2}/(1+C)]\langle\cos(\phi_m-\phi_r)\rangle\}$, where $C= I_m/I_r$ is the signal contrast.

Further, by developing the phase term $(\phi_m-\phi_r)$ in an average part $\langle\phi_m-\phi_r\rangle=\phi_m-\phi_r$ and a random deviation $\Delta\phi$, as $\phi_m-\phi_r = \phi_m-\phi_r+\Delta\phi$, the averaged cosine becomes $\langle\cos(\phi_m-\phi_r)\rangle= \mu \cos(\phi_m-\phi_r)$ [1]. The term $\mu= \langle\cos \Delta\phi\rangle$ has the meaning of the coherence factor of the measurement and reference fields, whereas $\phi_m-\phi_r$ can be written as $R\, k(s_m-s_r)$ in terms of the measurement and reference arm lengths s_m and s_r. In summary, we may write I_{ph} in the form:

$$I_{ph} = I_0 \{1+V \cos[R\, k(s_m-s_r)] \} \qquad \text{(A2.1)}$$

where $V= 2\mu C^{1/2}/(1+C)$ is the fringe visibility, or contrast factor [3], and R is the responsivity (or response factor) of the interferometer. Ideally, we should have V=1, but if the coherence factor μ is not unity, polarization is not matched, the beam powers are unequal (C<1) or some stray light is collected, V<1. The factor V is also called the contrast factor because it is coincident with the difference divided by the sum of the maximum and minimum signal amplitudes:

$$V = (I_{phMAX} - I_{phMIN})/(I_{phMAX}+I_{phMIN})$$

The *responsivity* R of the interferometer is a measure of the ability to supply a relative variation of the photocurrent I_{ph} to a variation of the optical path length in the measurement arm. In terms of the photocurrent signal, the responsivity is the maximum value of the relative derivative of I_{ph} with respect to ks when V=1:

$$R = \max_s \{(I_{ph}k)^{-1}\, dI_{ph}/ds\}$$

The reason for this definition is that, given the transfer ratio Φ=dks=HdM for the measurand M, the relative signal variation obtained in the photocurrent is readily found as:

$$dI/I= VR\, H\, dM \qquad (A2.2)$$

In the Michelson interferometer, from Eq.4.3 of the text, the responsivity is seen to be R = 2, a value that is the simple consequence of the go-and-return path.

By inspection of the scheme in Fig.A2-1, it is easy to see that R=1 for a Mach-Zehnder interferometer. In the Sagnac, we find a non-reciprocal dependence on +HM and -HM for the two counter-propagating waves, and thus we take R=2 (see Table A2-1).

In the Mach-Zehnder we have two available outputs from PD1 and PD2 (Fig.A2-1), and they are complementary to unity. For V=1 they are written as:

$$I_{ph1} = I_{ph0}\,[1+ \cos 2k(s_m-s_r)], \qquad I_{ph2} = I_{ph0}\,[1- \cos 2k(s_m-s_r)] \qquad (A2.3)$$

The two outputs are present also in the Michelson and Sagnac interferometers. Signal I_{ph1} is the output from PD in the scheme of Fig.A2-1, while I_{ph2} (not shown in Fig.A2-1) could be recovered from the beamsplitter output directed toward the source.

Last, in the Fabry-Perot, the photocurrent detected at either the front or rear mirror has a dependence of the type:

$$I_{ph}/I_{ph0} \propto 1/(1+F^2 \sin^2 ks) \qquad (A2.4)$$

where $F=2R^{1/2}/(1-R)$ is the finesse of the Fabry-Perot resonator and R is the mirror (power) reflectivity [2]. Looking for the maximum of the relative slope $d(I_{ph}/I_{ph0})/dks$ in Eq.A2.4, one can find that the responsivity of the Fabry-Perot is just equal to F.

Derivation of Eq.(A2.4). Let R_1 and R_2 be the (power) reflectivity of M1 and M2, and A=exp-α2s the (power) attenuation suffered in the cavity of length 2s.

Just beyond the first mirror M1, the cavity field E_{cav} is the sum of two terms: one coming from the input field E_0 and transmitted with $\sqrt{(1-R_1)}$ through M1, and one coming after the round-trip of the cavity, with reflectance $(R_1R_2)^{1/2}$ attenuation A and delay 2ks. Thus, we can write $E_{cav} = E_0\sqrt{(1-R_1)} + E_{cav}\, A(R_1R_2)^{1/2} \exp i2ks$. Solving for E_{cav}, we get:

$$E_{cav} \approx E_0\sqrt{(1-R_1)}\, /\, [1- A(R_1R_2)^{1/2}\exp i2ks] \qquad (A2.5)$$

TABLE A2-1 Comparison of basic optical interferometers

	Michelson	Fabry-Perot	Mach-Zehnder	Sagnac
Responsivity R	2	F	1	2
Balanced/Unbalanced configuration	B or U	U	B or U	B
No. of channels of basic setup	1	1	2	1
Reference available	yes	no	yes	no
Minimum number of beamsplitters (or partially reflecting mirrors)	1	1	2	1
Source and detector from the same side	yes	yes/no	no	yes
Retro-reflection to the source	yes	yes	no	yes
Metallization required (for fiber versions)	2	2	0	0
Cascadeability	no	yes	yes	no

Notes: Michelson and Mach-Zehnder may be operated as either unbalanced or balanced (the latter being the preferred choice). The Michelson and the Fabry-Perot can have source and detector on the same side if the output is taken by a beamsplitter added on the input path (Fig.A2-1). Responsivity is R=2 for the Sagnac, assuming equal and opposite effects on the two counter-propagating waves. Retro-reflection is for the basic configuration with mirrors; using corner cubes, it is avoided in the Michelson.

Both the outputs from M1 and M2 are proportional to E_{cav} and we can write them as:

$$E_{M1} = E_{cav}\, A\, [\exp i2ks]\, \sqrt{R_2}\, \sqrt{(1-R_1)}, \text{ and } E_{M2} = E_{cav}\, \sqrt{A}\, [\exp iks]\, \sqrt{(1-R_2)}.$$

Therefore, the outputs have the same selectivity, versus ks or λ, as the cavity field E_{cav}, and we can limit ourselves to study just E_{cav}. By taking the square modulus of the field, we obtain the power. Letting $R^2=A^2R_1R_2$, the power in the cavity is:

$$P_{cav} = P_0\,(1-R_1)/\,|1-R\exp i2ks\,|^2 = P_0(1-R_1)/[1+R^2-2R\cos 2ks]$$

$$= P_0\,(1-R_1)/[(1-R)^2+4R\sin^2 ks] \qquad \text{(A2.6)}$$

where P_0 is the incident power, and the last passage is obtained by noting that $\cos 2ks = 1-2\sin^2 ks$. The output powers from the mirrors M1 and M2 are found as:

$$P_{M1}/P_0 = 4\,A^2\,R_1\,R_2\,\sin^2 ks\,/\,[(1-R)^2+4R\sin^2 ks], \quad \text{and}$$

$$P_{M2}/P_0 = (1-R_1)(1-R_2)A\,/\,[(1-R)^2+4R\sin^2 ks]$$

The diagram of P_{cav}/P_0 is shown in Fig.A2-2. It has a periodicity 2π in the argument 2ks, or it exhibits periodic resonance at $2(2\pi/\lambda)s=N2\pi$ or at $s=N\lambda/2$ (that is, at multiples of half-wave-length).
In frequency, the periodicity of resonance $\Delta\nu=\nu_{N+1}-\nu_N$ is obtained by letting $2(2\pi\nu/c)s=N2\pi$, and it reads: $\Delta\nu=c/2s$. For $2ks=N2\pi$, the resonance has a peak value $P_{cav}/P_0=(1-R_1)/(1-R)^2$, even larger than one because of the buildup of power in the resonator cavity, while off-resonance, the response has a minimum at $2ks=\pi$, where it is $P_{out}/P_0=(1-R_1)/(1+R)^2$.

The half-width of the resonance curve is calculated from the condition: $P_{cav}/P_0=1/2[P_{cav}/P_0]_{peak}$, which from Eq.A2.6, becomes $(1-R)^2=4R\sin^2 ks$ and is solved as $\sin ks \approx ks=(1-R)/2\sqrt{R}$.
In frequency, the half-height line width is: $\Delta\nu_{HW}=(c/2s)(1-R)/\pi\sqrt{R}$.

Last, an important parameter of the Fabry-Perot is the finesse [2], defined as the ratio of half-height line width to period, or $F=(2/\pi)\Delta\nu/\Delta\nu_{HW}$. From the above results, the finesse is given by $F=2R^{1/2}/(1-R)$. Thus, we may substitute F for the factor at the denominator of Eq.A2.6 to obtain the more elegant expression Eq.A2-4.

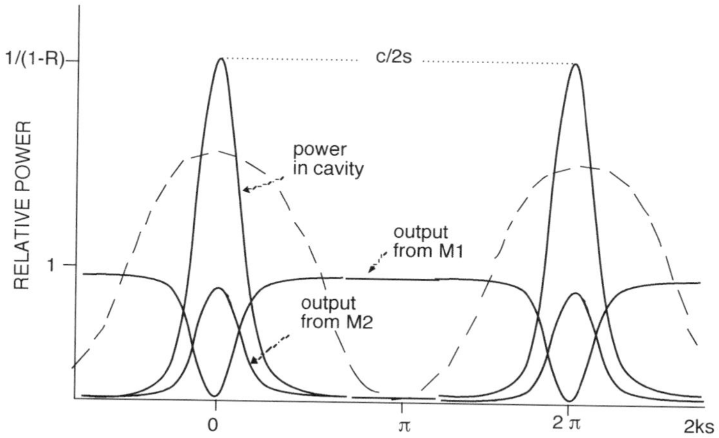

Fig.A2-2 The response of the Fabry-Perot is periodic by 2π in phase shift 2ks, or by c/2s in frequency. The selectivity is F (=finesse) times better than in other interferometers with cosine-type response (dotted line).

When using interferometers, further details relevant to performance are the following [2,3] (see also Table A2-1):

- the availability of a reference path: the reference is important as it allows compensation for the measurement against external disturbances other than the measurand (e.g., temperature, vibrations, EMI). The reference path is laid close to the measurement path but shielded from the measurand so as to collect the same disturbance in both arms;
- the number of channels available: basically, the Mach-Zehnder has two outputs readily available for detection, while the other interferometers have one. By an additional beamsplitter in the path going back to the source, a second channel can be made available;
- the position of the detector with respect to the source: it is easier to have both on the same side to accommodate them in a single box;

- the retro-reflection from the optical interferometer into the source: retro-reflections may severely impair the quality of the emitted field, adding amplitude fluctuations as well as widening the spectral line of the laser, especially in frequency stabilized sources;
- the balanced versus unbalanced paths: if reference and measurement paths are nearly the same, requirements on the coherence length of the source are mitigated;
- the number of beamsplitters and mirrors: this clearly affects the component count and cost of the final optical assembly;
- cascadeability: this feature has to deal with multiple-sensor operation (as in fiberoptic sensors) or when using the optical interferometer as a filter (frequency channel separation in fiberoptic communications).

A last point is alignment of the mirrors. All interferometers considered above are intended for operation with a laser source and basically work with a single-mode spatial distribution. Thus, their mirrors require an angle alignment accuracy within the diffraction limit $\theta_d=\lambda/\pi w_0$ associated with the Gaussian beam spot size w_0, otherwise incomplete superposition (Sect.4.5.3) and reduction of fringe visibility will occur. If we can use a corner cube in place of the mirror, like in the Michelson interferometer, the angular alignment problem is solved (provided that the corner cube dihedral error is less than θ_d).

Stray light suppression may also be of concern in the interferometer operation. When the laser line width is narrow, an interference filter in front of the detector will usually be effective to suppress light from the environment.

Alternatively, we may use a spatial filter (see Fig.A2-3), consisting of a lens-and-pinhole combination placed in front of the detector. If the pinhole has a radius r and is placed in the focal plane of the lens with focal length F, only the rays contained in an angle θ=r/F can pass through the filter. Because the minimum θ is the diffraction value, this filtering is very effective.

A2.2 Choice of Optical Components

When we move from conceptual schemes with ideal components as presented in the text, to practical implementation with real components, a number of non-idealities need to be taken into account, which if overlooked, may impair or severely degrade the performance we expect in principle.

To illustrate the problems that may be encountered, we will briefly discuss some issues concerning the choice of optical components with reference to a Michelson-based laser interferometer where the beamsplitter and the corner cube are the most critical components.

Beamsplitters. A plain glass-flat model of the popular BK7 glass can be used, provided that it has a reasonable flatness and few surface defects. But, unless we need just the R=4-8% reflection obtained at a low-incidence angle (<10°), we shall treat one surface by a single- or multiple-layer reflection coating to get the desired R. Also, as both front and rear surfaces contribute to reflection, we inadvertently have a sort of parasitic Fabry-Perot interferometer inserted in the propagation path. To avoid this effect, in a glass-flat beamsplitter we shall have one surface coated for the intended reflection, and the other be anti-reflection coated (ARC). Alternatively, we may consider using wedge-flat and pellicle beamsplitter versions. As for any optical surface crossed by the wavefront, the surface quality of the beamsplitter should be specified at least as $\lambda/4$ distortion and 60-40 scratch and dig, as a rule of thumb.

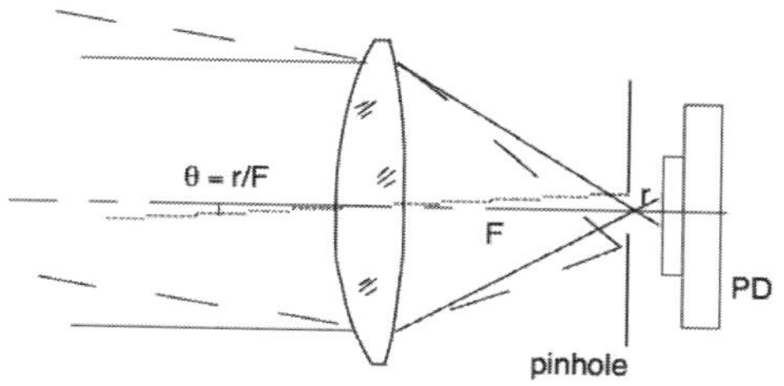

Fig.A2-3 A pinhole objective combination can be used for spatial filtering of the field reaching the photodetector. The acceptance angle is r/F.

Polarizing beamsplitters. We may either choose from multireflection or birefringent versions. A multireflection cube is based on the reflection property at Brewster's angle incidence, by which $R_{//}$=0 for the polarization parallel to the incidence plane, while $R_{\perp}\approx$15%(for n=1.5) for the polarization perpendicular to the incidence plane. Adding several layers of alternatively high and low n, we can obtain a reflection $R_{\perp}\approx$99%, while $R_{//}$ stays low (typ. $R_{//}$<1%). The layers are deposited on the diagonal surface of a square prism, and the input/output surfaces will be treated ARC.

A better extinction ratio $R_{\perp}/R_{//}$ (up to 10^5-10^7) is obtained by birefringent cube polarizing beamsplitters that use calcite as the material of the prisms. The two halves of the cube are cut along properly chosen directions so as to obtain an index of reflection difference at the diagonal surface boundary, adequate for total reflection of one polarization and transmission for the other. A pair of prisms is normally used to compensate for beam deviation, so that one beam is delivered parallel to the input and the other is about perpendicular. Several cube versions are available under the names of Glan, Glan-Thomson, Wollaston, Nicol, etc.

We can also use a polarizer beamsplitter as a polarizer, but if we only need to select a (linear) polarization state while blocking the other, we may prefer cheaper dichroic-sheet polarizers. Discovered by Kodak's researchers and widespread in applications, these sheets offer a reasonable extinction ratio (typically 10^2-10^3) in the visible and near-infrared, which is often adequate for use in front of the photodetector.

Corner cubes and retro-reflectors These components can be understood by starting from the in-plane version of the 90° prism (Fig.A2-4).

If the vertex angle is β, the angular deviation between incident and reflected rays is found to be 2β, or 180° for a 90° prism. Incidentally, the doubling is general rule, and thus the pentaprism, which has a vertex angle of 45°, deflects the beam of exactly 90° (Fig.A2-4). The angular deviation is irrespective of the incidence angle α, provided α is within an acceptance angle α_{max} determined by the total reflection angle internal to the cube. To extend the concept in three dimensions, we need two dihedrals, or the corner cube structure depicted in Fig.A2-4, bottom, in which three reflections are used to deflect the incident beam in two orthogonal planes.

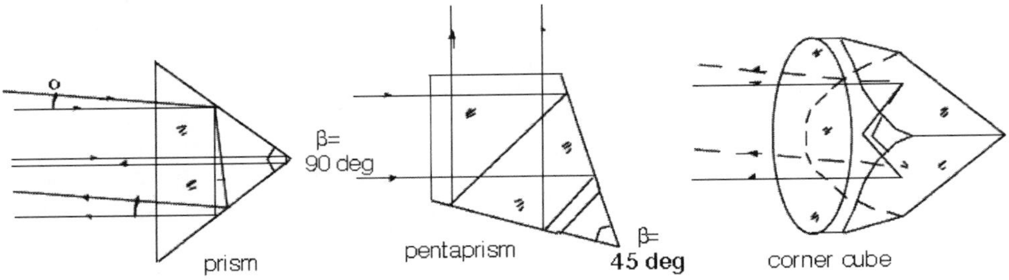

Fig.A2-4 Retro-reflection in a 90° prism and in a pentaprism with a 45° vertex angle (top), and in a corner cube retro-reflector (bottom).

Basically, we have solid-glass and hollow versions of corner cubes. A solid-glass corner cube (as shown in Fig.A2-4) has an ARC front surface and may reach a reflection efficiency of 95% or better. The points of concern are (i) the fragility of the sharp edges and vertex, easily damaged even during handling of the device; and (ii) the polarization state change that radiation undergoes because of the phase shift at total internal reflection. Because of this effect, for an input linear state of polarization, the output is a mixture of elliptical states, dependent on the cube sector. This is disturbing if we use the state of polarization for the beam handling in our instrument. Yet, in a Michelson laser interferometer with two retro-reflectors, if they have the same edge orientation, the beams superpose with the same state of polarization in different sectors, and fringe visibility is unity.

To make the corner cube less sensitive to polarization and protect the total reflection surfaces, these may be metallized (with a thin Au or Al film) and covered by a protection coating. Metallization slightly decreases reflection efficiency (to 90% typically), but increases the field-of-view or acceptance of the device.

Alternatively, we may use a hollow-type corner cube, made of three mirrors mounted in a trihedral arrangement at 90° angles. The hollow corner cube is cheaper and has the same performance as the metallized solid-glass cube, but the stability of dihedral angles (to be kept typ. below ≈ 1 arcsecond for the useful device's life) may be questionable because of the demanding mechanical stability.

REFERENCES

[1] S. Donati: "*Photodetectors*", 2nd ed., Wiley IEEE Press, Hoboiken 2021, Sect. 10.1.2.
[2] B.E.A. Saleh, M.C. Teich:"*Fundamentals of Photonics*", 2nd ed., Wiley 2007.
[3] W. Lauterborn, T. Kurz, and M. Wiesenfeld: "*Coherent Optics*", 2nd ed., Springer Verlag: Berlin, 2003.

APPENDIX A3

Propagation through the Atmosphere

When propagated through a substantial path length in the atmosphere, a number of phenomena affect the optical beam and distort the wavefront. Usually, atmospheric effects are classified in two categories: (i) those associated with the non-ideal transparency of the atmosphere, notably absorption and scattering, called *turbidity effects*, and (ii) those coming from non-homogeneity of the index of refraction of the atmosphere, caused by thermal exchange and air mass motion, called *turbulence effects*.

A vast amount of literature is available on the subject (see, for example, Refs. [1,2]), and here we will report a few results to supply basic information useful for evaluating performance of instruments treated in the text.

A3.1 TURBIDITY

A beam propagated through a medium on a path length L (Fig.A3-1) suffers a power loss, with respect to the initial value P_0, described by the Lambert-Beer law of attenuation:

$$P = P_0 \exp -(\alpha+s)L \tag{A3.1}$$

Quantities α and s are called the *absorption* and *scattering* coefficients of the medium, and their sum, $a=\alpha+s$, is the (total) *attenuation* coefficient. Units of a, α, and s are m^{-1} or cm^{-1}. Their inverse values represent the propagation length for the initial power to decrease to 1/e=37% and are called the absorption, scattering, and attenuation lengths, respectively.

The quantity $T= \exp -(\alpha+s)L$ is referred to as the *transmittance*.

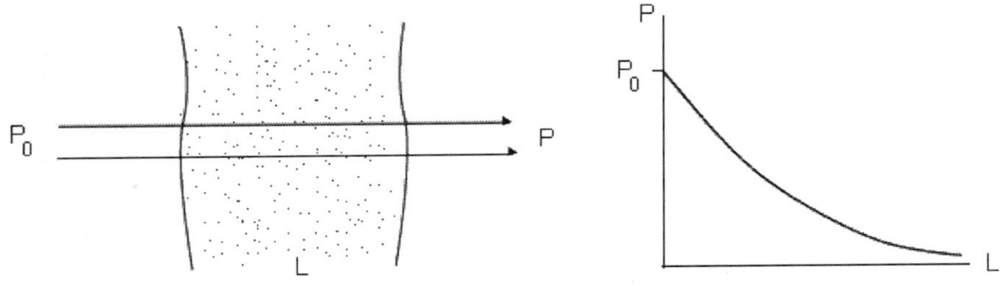

Figure A3-1 The attenuation of power propagated through a path length L follows the negative-exponential Lambert-Beer law.

In the atmosphere, absorption is due to constituent species (nitrogen, oxygen, carbon dioxide, etc.) and water vapor, while scattering is dominated by small-size (<1 μm) particulate and eventually water droplets in the case of haze or fog.

The absorption contribution is fairly constant, whereas the scattering contribution strongly depends on weather conditions. In Fig.A3-2 we report what is the most representative and famous diagram [3] of atmospheric transmission T versus wavelength, measured at ground level on a normal, clear day on a path length of L=1800 m.

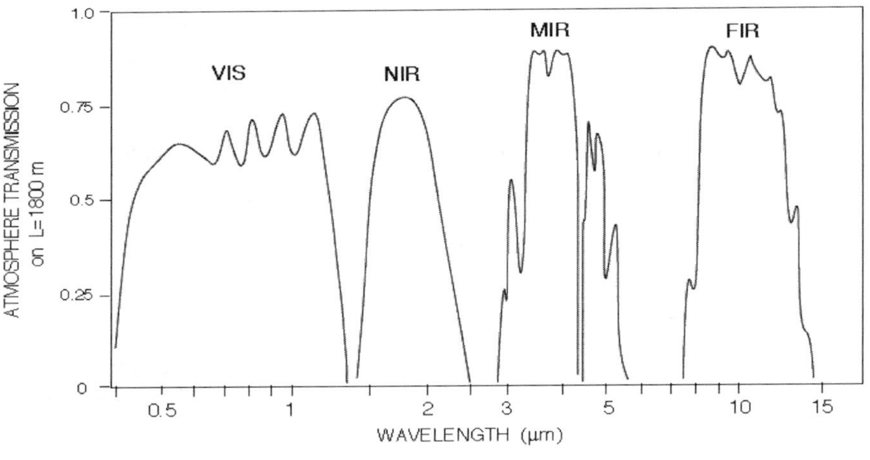

Figure A3-2 The standard atmospheric transmission (%) for a propagation on an L=1800 m length path at ground level.

As we can see from Fig.A3-2, the atmosphere has a few *transmission windows* of reasonable transparency, useful for operating laser instrumentation on medium distances (say, 100 to 1000 m).

Another diagram, useful for evaluating atmospheric transmission on large distances (tens of km) is shown in Fig.A3-3. The diagram is the plot of solar spectral irradiance received at ground level on a clear day at different elevation angles θ of the sun on the horizon. The quantity AM=$(\cos\theta)^{-1}$ is the air mass number, and represents the equivalent number of atmospheres crossed by the rays at elevation θ. The curve with AM0 in Fig.A3-3 is the irradiance outside the atmosphere. The ratios of the values read on the curves at different AM# and at AM0 are the transmissions of the atmosphere for different elevations or air masses.

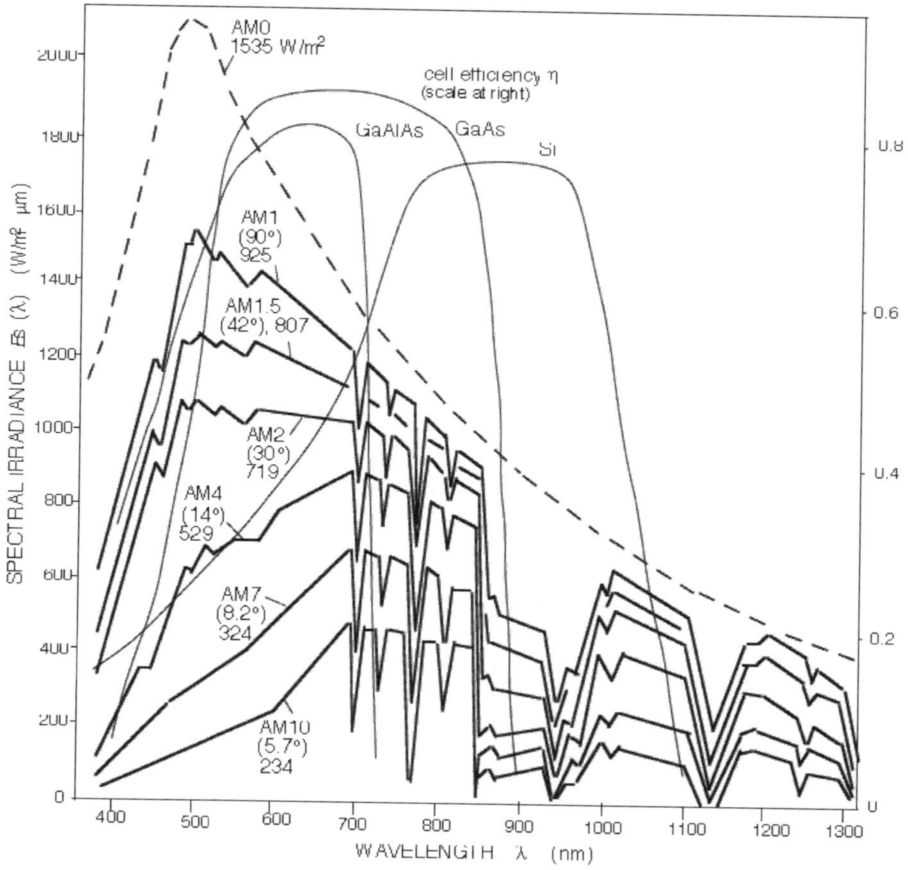

Figure A3-3 Spectral solar irradiance (W/m^{2}μm) received at ground level for several elevation angles or air masses AM.

In addition to the VIS (visible) window, we find the three well-known IR (infrared) windows of the:

- near-infrared (or NIR, λ=1.5-2.3 μm),
- middle-infrared (or MIR, λ=3-5 μm), and

- far-infrared (or FIR, λ=8-14 μm).

Beyond the FIR window, no other window is found up to millimeter waves. For a very dry atmosphere (RH<5%), there is actually a modest window at λ=330±10 μm with a peak transmission T=5% (L=300 m), but T rapidly decreases with increasing relative humidity RH, going down to T<0.1% at RH=70%.

In the UV, the atmospheric transmission gradually decreases at shorter wavelengths until we reach VUV (vacuum-ultra-violet), requiring vacuum for propagation because of an intolerable loss, even for a path that is only a few centimeters in length.

In the visible range, the attenuation coefficient of a standard atmosphere has been computed from constituents as the sum $a_{st} = \alpha_{oz}+s_R+s_a$ of the following: ozone contribution α_{oz} (important for λ<290 nm), Rayleigh scattering s_R (important for λ<400 nm), and aerosol scattering s_a, giving a value a=0.18 km^{-1} at λ=500 nm.

The term s_a is dominant for λ>400 nm and gives the trend $a/a_o=\sqrt{(\lambda/\lambda_o)}$ for total attenuation, a good approximation for λ= 0.5...5 μm, except for the narrow molecular absorption lines α_{mol} that must be added to the background value.

Data shown in Fig.A3-2 represent the transmission of the standard atmosphere and include the s_R and s_a terms. In addition to that, a much larger term, s_{par}, comes from scatters extraneous to the clear atmosphere, such as particulate matter and small droplets.

Scatters are small-size particles that can be schematized as spheres randomly distributed in space, with a certain radius r_p and an index of refraction n_p. The radius is found to range from < 0.1 μm to over 100 μm, and the index of refraction is always well in excess of n_{air}=1 (e.g., n_p =1.33 for water droplets), the index of the surrounding medium.

Small size and large index of refraction difference is the peculiarity of *turbidity*. Conversely, *turbulence* is the regime where scatters are large in size (from 1 mm to 1 m) and their index of refraction difference is very minute (10^{-6} to 10^{-3}) and may also fluctuate spatially and temporally.

Considering a small particle with a substantial Δn, the outgoing ray will largely deviate in angle on refraction inside and outside the particle (Fig.A3-4), whereas the deviation (or beam wandering) is very little if Δn is small. Also, as a large angular deviation brings the outgoing ray out of the beam, a scattered ray is equivalent to a lost ray, and thus scattering does contribute to attenuation. By contrast, the small deviation introduced by turbulence can be recovered by increasing the acceptance area of the detector and does not contribute to attenuation, at least in principle.

But, with turbulence, we will additionally have a random spatial mixing of contributions inside the beam, an effect called scintillation, which destroys spatial coherence.

The quantity used to describe the angular scattering process is the *scattering function*, $f(\theta)d\Omega=dP(\theta)/4\pi P$, as first introduced in a seminal book of van der Hulst [4]. It is defined as the relative power per unit solid angle Ω that we collect at the angle θ with respect to the θ=0 incidence (Fig.A3-4), 4π being a normalization factor.

Regarding the shape of the scattering function $f(\theta)$, the critical parameter is the ratio r/λ, and we can distinguish between two limit cases:

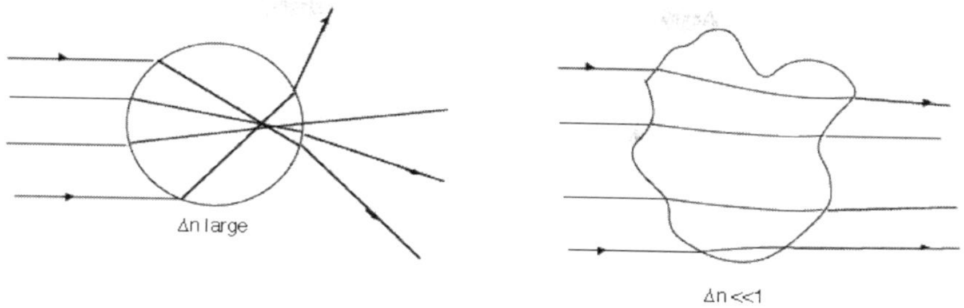

Figure A3-4 Ray tracing through scatterers with a large Δn (turbidity, left) and a small Δn (turbulence, right) reveals the difference in angle deviation.

(i) for small particles, i.e., r<<λ, we are in the so-called Rayleigh scattering regime and the function f(θ) is nearly isotropic. This means that we collect nearly as much power in the forward (θ≈0) as in the backward direction (θ≈π), the minimum being found at the right angle (θ≈π/2);

(ii) for very large particles, or r>>λ, we are in the Mie regime and the scattering function is strongly peaked in the forward direction, and an appreciable contribution in the backward direction is left (Fig.A3-5).

Another important parameter to describe the scattering process is the cross-section A_{sc} of the scatterer, intended as the effective area subtended to incoming rays.

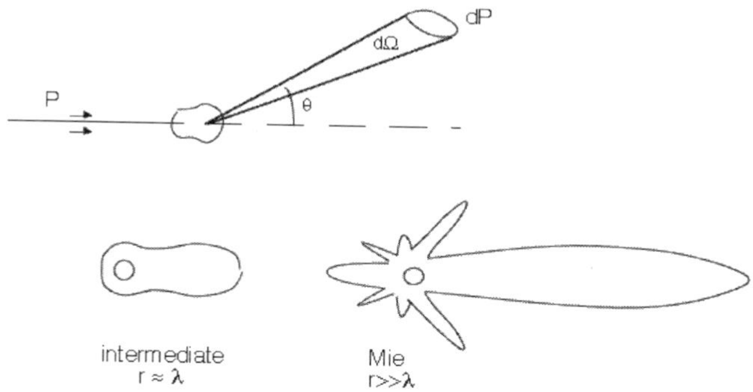

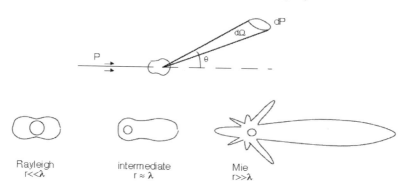

Figure A3-5 Top: definition of the scattering function f(θ). Bottom: the scattering function f(θ) in the Rayleigh and Mie regimes.

Like the scattering function f(θ), this quantity is calculated by an electromagnetic field analysis of the scattering process. From the analysis, it is found that:

$$A_{sc} = Q_{ext}\, \pi r^2 \tag{A3.2}$$

The factor Q_{ext} is called the extinction efficiency and is plotted in Fig.A3-6 versus the normalized radius r/λ. In the Rayleigh regime for small particles, the extinction efficiency Q_{ext} varies as $(r/\lambda)^{-4}$. The particles are seen from the electromagnetic radiation as if they were much smaller than their actual physical size. In the Mie regime for large particles, Q_{ext} =const ≈2 and particles are seen as twice as large as their actual radius. This apparently odd result is correct, because the diffraction contribution is found to be exactly equal to the physical obstruction produced by the particle.

Now we can sum up the effects of scattering. If we have a medium containing a concentration of c (cm^{-3}) particles per unit volume, each with an effective area A_{sc}, then the power remaining in the beam after propagation on a path length L will be:

$$P = P_0 \exp -c\, A_{sc}\, L \qquad \text{(A3.3)}$$

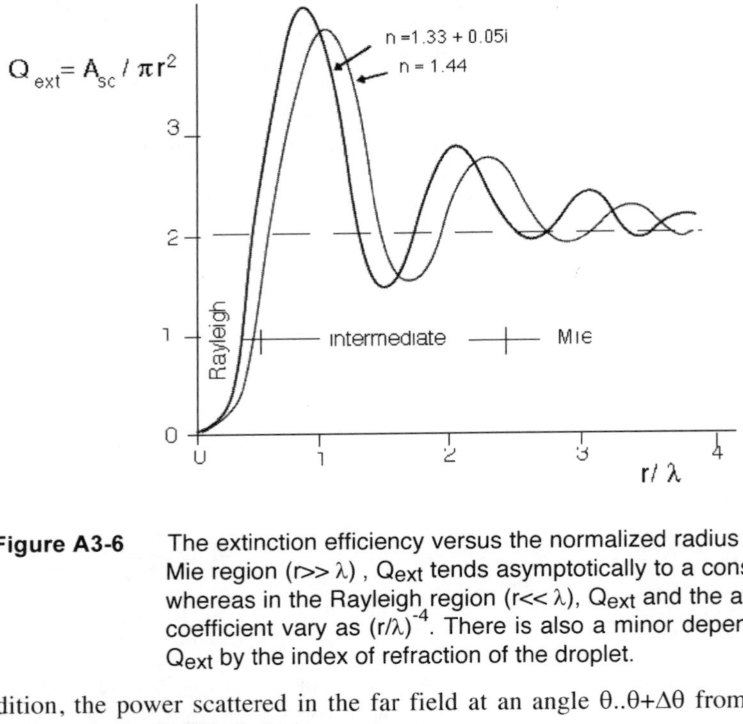

Figure A3-6 The extinction efficiency versus the normalized radius r/λ. In the Mie region ($r >> \lambda$), Q_{ext} tends asymptotically to a constant (≈2), whereas in the Rayleigh region ($r << \lambda$), Q_{ext} and the attenuation coefficient vary as $(r/\lambda)^{-4}$. There is also a minor dependence on Q_{ext} by the index of refraction of the droplet.

In addition, the power scattered in the far field at an angle $\theta..\theta+\Delta\theta$ from the incidence is $P_{sc} = P_0\,[1 - \exp -cA_{sc}L]\, f(\theta)\, 2\pi\theta\Delta\theta$.

A quantity of importance, related to the attenuation coefficient, is the *visibility range*. Conventionally, the visibility range for vision through the atmosphere and other scattering media is taken as the value L_{vis} at which scene contrast is reduced to 2% of the initial value.

This definition is related to the visual acuity of the eye, which can resolve a ≈2% contrast in good viewing condition [6]. As the contrast reduction is the same as the useful non-attenuated power, using Eq.A3.1 we find the visibility range as:

$$L_{vis} = (\ln P/P_0)/a = 3.92\ /a \tag{A3.4}$$

Typical values of attenuation a, in km^{-1} are the following:

0.18 for a clear atmosphere at sea level, (L_{vis} =22 km),
0.5-1 for a light haze (L_{vis} =7.8-3.9 km),
2-5 for thick haze (L_{vis} =1.9-0.78 km),
50-200 for a medium fog (L_{vis} =78-20 m),
300 for a thick fog (L_{vis} =13 m),
0.02-0.1 for an exceptionally clear atmosphere, with Rayleigh diffusion limit, seldom reached, of s_R=0.012 (L_{vis} =326 km).

Looking at the dependence of Q_{ext} from r/λ in Fig.A3-6, we can see that visibility in fog and haze can be improved by moving the wavelength of operation to the infrared.

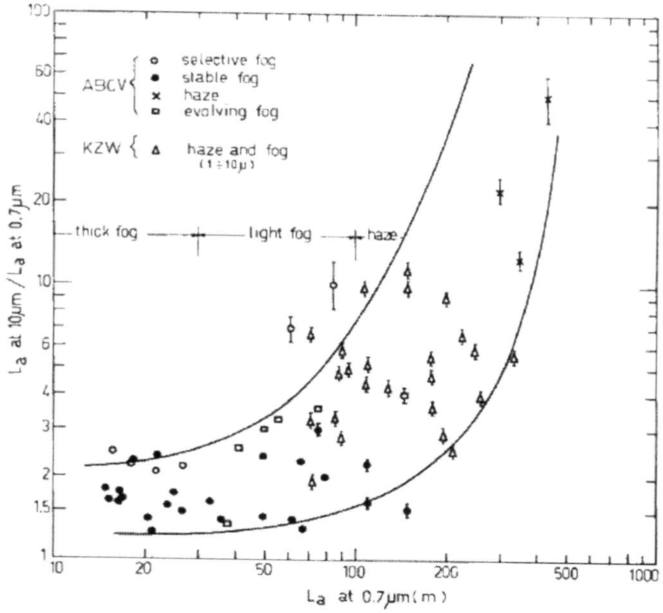

Figure A3-7 The ratio $L_{a10}/L_{a0.7}$ of attenuation lengths at 10 μm (FIR) and at 0.7 μm (visible) vs the attenuation length in the visible $L_{a0.7}$.

With a smaller r/λ, we may go from the Mie regime with Q_{ext}≈2 to the Rayleigh regime with Q_{ext} and attenuation proportional to $(r/\lambda)^{-4}$. In Fig.A3-7, we report experimental results on the FIR-to-visible ratio of attenuation length $L_{a10}/L_{a0.7}$, as measured by several researchers [7].

Other things being equal, this ratio is the expected visibility improvement. The ratio depends on $L_{a0.7}$, because the average radius of a particle changes with the type of turbidity, being small (r≈0.5 μm) in hazes and fairly large (r≈0.5-10 μm) in fogs.

In usual conditions, the ratio $L_{a10}/L_{a0.7}$ may range from ≈1.5 for $L_{a0.7}$=20 m (L_{vis} = 78m) to ≈5 for $L_{a0.7}$=150 m (L_{vis} = 600 m), up to perhaps ≈10 for $L_{a0.7}$=100 m (L_{vis} = 390 m).

These values are very attractive in several applications, including navigation aids in automotive, avionics, and military systems.

Unfortunately, the improvement is the smallest at the short visibility range, where it would be the most desirable. In addition, the actual visibility improvement is appreciably less than is expected from the $L_{a10}/L_{a0.7}$ ratio. The reason is that the scene appearance changes appreciably from the visible to the FIR regions. In the visible, the scene contrast is dominated by the surface diffusion coefficient, whereas in the FIR, it is determined by temperature differences [8]. Thus, an infrared scene has details correlated to the visible image, but with significant deviations introducing a task of identification to the observer, whereas detection and recognition [9] are easier.

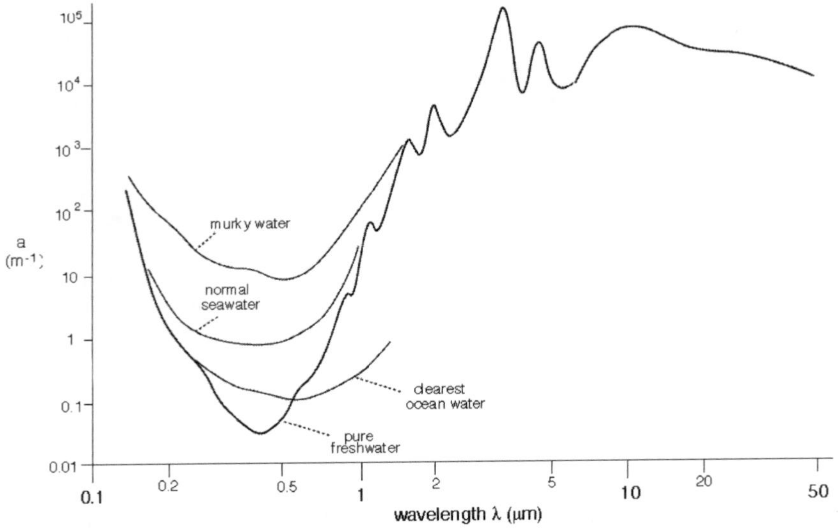

Figure A3-8 Typical dependence of attenuation vs. wavelength in water.

In detection, improvements of visibility up to a factor 50 have been actually observed [7] on 10-km ranges with FIR imaging. For identification tasks, however, we should take into account a difficulty factor [9]. This can be roughly estimated to decrease, by a factor of 2–3, the actual improvement with respect to the attenuation ratio $L_{a10}/L_{a0.7}$.

Another natural medium of importance for electro-optical instrumentation is water. As shown in Fig.A3-8, there is just one window in water, centered at 450 nm (blue-green), where the attenuation coefficient reaches the ultimate value of a=0.025 m^{-1} for distilled water (L_{vis} =156 m). In very clear deep ocean waters, attenuation may reach, at λ=440 to 510 nm, a=0.05-0.1 m^{-1} (L_{vis} = 78-39 m, near Madeira island) [10], but usually it is a≈1 m^{-1} and up to a≈10 m^{-1} depending on particulates. In the maximum transparency window, the relative weight of diffusion s and absorption in attenuation a=s+α are 0.6 and 0.4, respectively, whereas outside (Fig.A3-8), absorption is dominant (α>>s).

A3.2 TURBULENCE

Propagation over a substantial distance (say, >30 m) in the atmosphere is affected by turbulence. Thermal exchanges with the ground and the sun's heating are responsible for a daily temperature excursion producing convection and vortex in the air.

Because of the associated density fluctuations, the local index of refraction Δn has a small randomness superposed on the average value, and the propagated beam suffers a small angular deviation (Fig.A3-4).

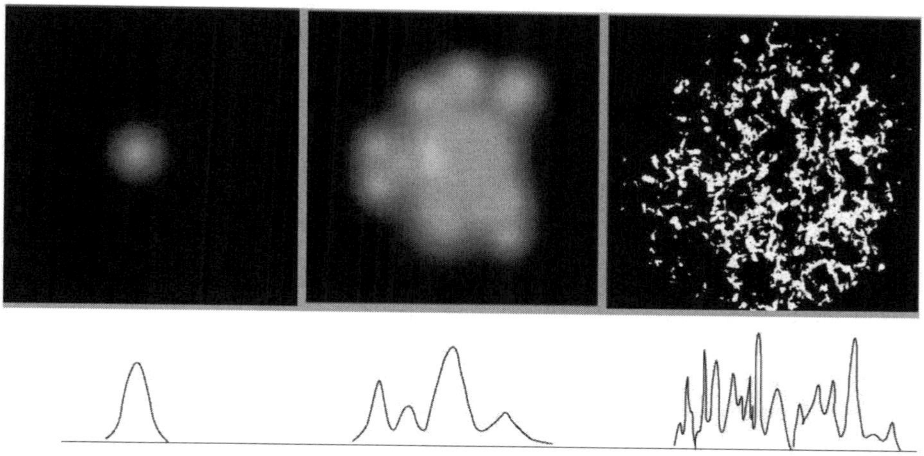

Fig.A3-9 Scintillation of a 20 mW, w_0=1 mm He-Ne beam propagated through air at H=5 m height from ground. Top: the beam spatial distributions (not to same scale) as they appear at L=0 (left), 40 m (center), and 200 m (right); exposure is 0.1-s. Bright spots change their position randomly in the beam cross sections with a typical 0.1 to 1-s time scale. Bottom: profiles of intensity along a diameter exhibit an increasing number of spikes with L.

If this randomness Δn were constant across the beam, there would be just a deflection, or a random *wandering* of the beam around the unperturbed direction. But, most frequently, the Δn scale of correlation is smaller than the size of the propagated beam after a substantial distance, and different portions of the beam are deflected differently. As the result, the wavefront is distorted and the spatial distribution deviates from the Gaussian emitted from the laser, with dark and bright spots known as *scintillation* (Fig.A3-9).

For such a distribution, the coherence factor μ_{sp} of superposition at the detector is decreased. Also, when the wandering deviation is about the same as the spot size (Fig.A3-10), in addition to intensity fading, we face a serious speckle phase error.

The study of atmospheric turbulence is segmented into three steps: (i) assessing the statistical properties of Δn fluctuations, (ii) translating them into the statistics of the propagated optical field, and (iii) describing the consequent effect on the detected signal.

The details of the theoretical development are very complex, and here we will report just a few basic results, derived under a number of assumptions.

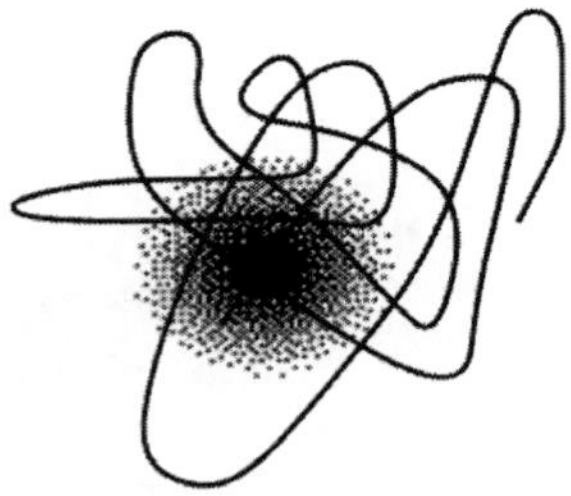

Fig.A3-10 Wandering of the propagated beam for a He-Ne laser (as in Fig.A3-8). The centroid of the beam cross section (dotted circle) moves randomly in time around the average line-of-sight position.

An excellent review of turbulence is provided in the seminal work of Hufnagel [11] and by other authors, see for example Ref. [12].

From the engineering standpoint, nearly all the quantities of interest are related to $C_n^2 \propto \langle \Delta n^2 \rangle$, called the *structure constant*, and associated with the correlation function defined as $D_n(\rho)=\langle [n_1(0)-n_2(\rho)]^2 \rangle$, where ρ is the distance between two points P_1 and P_2. Tatarski, in a famous paper [13], calculated that $D_n(\rho)= C_n^2(\rho)^{2/3}$, so the quantity C_n^2 is measured in $m^{-2/3}$.

The constant C_n^2 is connected to the temperature variance $C_T^2 \propto \langle \Delta T^2 \rangle$ by the error propagation, as $C_n^2= (dn/dT)^2 C_T^2$.

The temperature coefficient is $|dn/dT| \approx 1$ ppm at sea level [see Eq.4.33] and decreases exponentially with height as $|dn/dT| \approx \exp-(H_{(km)}/12.6)$ [11].

According to the model of Kolmogorov [13], the kinetic energy of a vortex is dissipated by friction on an inertial sub-range scale, spanning from a few millimeters to a few meters. In this range, temperature fluctuations have a correlation $D_T(\rho)= C_T^2(\rho)^{2/3}$, and Δn fluctuations have $D_n(\rho)= C_n^2(\rho)^{2/3}$, provided that $l_0<\rho<L_0$, where l_0 and L_0 are called the inner and outer scales of turbulence.

Experimentally, the constant C_n^2 can range from $\approx 10^{-15}$ to $\approx 10^{-13}$ $m^{-2/3}$ for weak to strong turbulence, respectively, at sea level.
The lowest values are found in the early morning and at dusk, in the absence of wind. With height, C_n^2 varies as $H^{-2/3}$ (low H or high wind) or $H^{-4/3}$ (high H or still air), with a break point at H≈1-10 m.

To assess the effect of C_n^2 on the propagating optical beam, following Rytov [14], we can represent the electric field $E= \langle E\rangle \exp[i(kr-\omega t)+\chi(r,t)+i\phi(r,t)]$ as composed of an average part $\langle E\rangle$ with the usual propagation term $\exp i(kr-\omega t)$, and of a random exponential attenuation $\chi(r,t)$ and phase disturbance $i\phi(r,t)$, both fluctuating and due to turbulence.
From the Rytov analysis, it is found that these random variables have zero mean values, $\langle\chi\rangle=0$, $\langle\phi\rangle=0$, and a normal distribution.

Thus, the distribution of the attenuation $\exp \chi$ is *log-normal*.

In the simple case of a horizontal path propagation ($C_n^2\approx$const) of a plane wave on a distance L, the variance σ_χ^2 of the log-normal attenuation χ has been calculated [11] as $\sigma_\chi^2=0.131\cdot C_n^2 k^{7/6} L^{11/6}$, with $k=2\pi/\lambda$ being the wave number. For a spherical wave, the numerical factor is $\sigma_\chi^2=0.546$. Experimentally, the predicted values nicely match the measurements for $\sigma_\chi^2<0.3$, while for larger values a saturation is found at about $\sigma_\chi^2\approx 0.5$, which is typical for L≈ 300 m or larger.

Correspondingly, the variance σ_E^2 of amplitude fluctuations is $\sigma_E^2=\langle E\rangle^2[(\exp 4\sigma_\chi^2)-1]$ and the variance of the intensity is $I=\langle E\rangle^2$ is $\sigma_I^2=4I^2(\exp\sigma_\chi^2-1)$.

About the phase variance σ_ϕ^2, it has been reported [14] that it is equal to σ_χ^2 or twice as much in the two cases $l_c<<\sqrt{\lambda L}$ and $l_c>>\sqrt{\lambda L}$, i.e., in the case of a turbulence scale of correlation l_c much smaller or much larger than the Fresnel-zone size $\sqrt{\lambda L}$, respectively.
Conservatively, we may take $\sigma_\phi^2\approx C_n^2 k^{7/6} L^{11/6}$ and find the maximum distance L at which the turbulence effect is negligible by equating σ_ϕ^2 to the value of accuracy required in our experiment.

For example, letting $\sigma_\phi^2=0.01(\mathrm{rad}^2)$, at $\lambda=1$ µm we get L=47 m for a weak turbulence ($C_n^2=10^{-15}$), but just L=3.8 m for a strong turbulence ($C_n^2=10^{-13}$).

Beam wandering can be characterized by a radius ρ_w, defined as the rms value of the beam centroid deviation off the expected position. An approximate expression for a collimated beam is: $\rho_w= w/(1 - 2.45\, C_n^2 k^{1/3} L^{11/6} w^{-5/3})$, where w is the unperturbed beam size at distance L, or explicitly $w^2=w_0^2+(\lambda/\pi w_0)^2L^2$ for a Gaussian beam.
At a relatively short distance, the beam shape is substantially unchanged and its centroid wanders randomly by a rms deviation ρ_w.

The wandering has a characteristic time τ_w (given by $\approx w/v_w$, v_w being the wind velocity), and σ_χ^2, σ_ϕ^2 are negligible. On a time-scale slower than wandering time, the apparent beam size is $w_{turb}^2=w^2+\rho_w^2$. At a large distance, where $\rho_w>>w$, the beam no longer wanders appreciably, appears widened to a size w_{turb} with scintillation effects now showing up, and σ_χ^2, σ_ϕ^2 are no longer small.

A quantity of interest for the operation of interferometers through a turbulent atmosphere is the *transverse coherence length* S_t of the propagated beam, which corresponds to the speckle transversal size discussed in Sect.6.1.1 and is roughly the size of the individual grains found in both the scintillation and beam wandering regimes.

For propagation on a length L, the transverse coherence length is given by an approximate relation [14] as: $S_t=(0.5\, k^2\, C_n^{\;2} L)^{-3/5}$.
For example, at $\lambda=1$ μm and L=200 m, we get $S_t=40$ cm for a weak turbulence ($C_n^{\;2}=10^{-15}$) and $S_t=2.6$ cm for a strong turbulence ($C_n^{\;2}=10^{-13}$).

All these statistical results describe the point-like detection of amplitude and phase of the propagated field.

Now, we can consider the effect of detector size. When we increase the detector diameter from $d_{det} \ll S_t$ up to the coherence size S_t, the average signal increases and the relative variance improves, such as $\sigma_I^{\;2}/I^2 = 4\,(\exp\sigma_\chi^{\;2}-1)$ for the intensity.

For $d_{det} > S_t$, we collect uncorrelated samples of the χ and ϕ random variables, and the relative variance of intensity weakly improves, only as the logarithm of the number of coherence areas, or $\ln(1+d_{det}/S_t)$, while the phase variance is substantially unchanged.

Finally, regarding the frequency distribution of fluctuations, the power spectrum S(f) is related to the variance σ^2 by the Wiener-Khintchin theorem as $\int_{0-\infty} S(f)df = \sigma^2$.
If the spectrum falls off at a corner frequency f_{tur}, we can assume $S(f) = \sigma^2/f_{tur}$ as a reasonable first-order approximation. The corner frequency f_{tur} for both amplitude and phase fluctuations depends on the wind transversal velocity v_t as $f_{tur}=l_c/v_t$, where l_c is the turbulence scale of correlation. Typical values of the frequency corner are in the range of $f_{tur}\approx 10$ to 200 Hz.

REFERENCES

[1] W.L Wolfe: "*The Infrared Handbook*", Environ. Res. Inst. Michigan, 1978, a volume of 1725 pages, https://play.google.com/books/reader?id=3J-xjNousrsC&hl=en&pg =GBS.PP1.

[2] R.D. Hudson: "*Infrared System Engineering*", paperback by J. Wiley, New York, 2006.

[3] H.A. Gebbie et al.: "*Atmospheric Transmission in the 1-14 micrometer Region*", Proc. Royal Soc., vol. A206 (1951), pp. 87-96.

[4] H.C. Van de Hulst: "*Light Scattering by Small Particles*", J. Wiley, New York, 1957.

[5] OSA Staff: "*Optics Handbook*", vol. 1, Opt. Soc. of Am., Washington 1992.

[6] S. Donati: "*Photodetectors*", 2nd ed., Wiley IEEE Press, Hoboken 2021, Sect.A2.1.

[7] S. Donati, "*Thermal Imaging Through Hazes and Fog: Experimental Results*", Alta Frequenza, vol.42 (1973), pp.101-105.

[8] See Ref. [6], Sect.8.4.

[9] See Ref. [6], Sect.A2.1.

[10] S. Duntley: "*Light in the Sea*", J. Opt. Soc. of Am., vol.53 (1963), pp. 214-221.

[11] R.E. Hafnagel: "*Propagation through Atmospheric Turbulence*", in Ref. [1].

[12] J.W. Strohbehn: "*Laser Beam Propagation in the Atmosphere*", Springer, Berlin 2014.

[13] V.I. Tatarski: "*Propagation of Waves in a Turbulent Medium*", McGraw Hill: New York 1961.

[14] V. Razier: "*Remote Sensing of Turbulence*", CRC Press, Boca Raton 2021.

APPENDIX A 4

Propagation and Diffraction

In this appendix, we derive some useful relationships used in the text about propagation and diffraction when we have dealt with size measurements (Ch.2) and speckle pattern distributions (Ch.6). The cases considered here are a useful reference, yet not at all exhaustive. A more complete treatment can be found in several well-known textbooks (see for example Ref. [1-3]).

A4.1 Propagation

Let's start with a fundamental equation relating the amplitudes of the optical field in the propagation from a source plane (ξ,η) to a receiving plane (x,y), as shown in Fig.A4-1. The source is divided into elemental areas $d\xi d\eta$, each with an amplitude $A_1(\xi,\eta)\, d\xi d\eta$. In view of the Huygens-Fresnel principle, each contribution is a source radiating as a spherical wave, and thus comes to the point (x,y) in the image plane with a phase delay kr_{12}, where $k=2\pi/\lambda$ is the wave number and r_{12} is the distance between point $P1(\xi,\eta)$ in the source plane and point $P2(x,y)$ in the receiving plane. The component perpendicular to the (x,y) plane is $A \cos\theta$, where $\cos\theta= z/r_{12}$ is called the obliquity factor. Summing all the elemental contributions, we have:

$$A_2(x,y) = (1/\lambda z) \int\int_{-\infty,+\infty} A_1(\xi,\eta)\, (z/r_{12})\, \exp(ikr_{12})\, d\xi d\eta \qquad (A4.1)$$

This is the Rayleigh-Sommerfeld diffraction formula, valid with little loss of generality in the case $z>>\lambda$ [1,2].

We can write the explicit expression for r_{12} as:

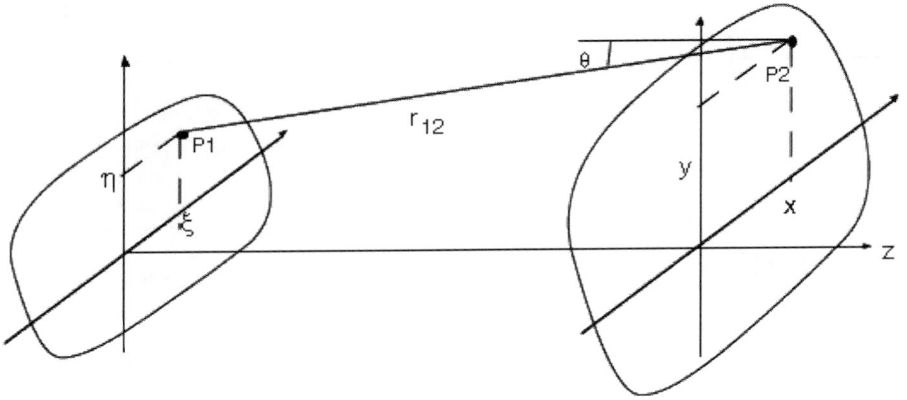

Fig.A4-1 Geometry for analyzing the propagation from a source plane (ξ,η) to a receiving plane (x,y).

$$r_{12} = [z^2+(x-\xi)^2+(y-\eta)^2]^{1/2} = z\{1+[(x-\xi)/z]^2+[(y-\eta)/z]^2\}^{1/2} \quad \text{(A4.2)}$$

By inserting in Eq. A4.1, we obtain:

$$A_2(x,y) = (1/\lambda z) \iint_{-\infty,+\infty} A_1(\xi,\eta) \frac{\exp ikz\{1+[(x-\xi)/z]^2+[(y-\eta)/z]^2\}^{1/2}}{\{1+[(x-\xi)/z]\}^2+[(y-\eta)/z]^2\}^{1/2}} d\xi d\eta \quad \text{(A4.3)}$$

Eq. A4.3 is a convolution integral of the form $A_2(x,y) = \iint A_1(\xi,\eta)H(x-\xi,y-\eta)d\xi d\eta$, and we can then write it in a compact form as:

$$A_2(x,y) = A_1(x,y) ** H(x,y) \quad \text{(A4.4)}$$

Thus, the function $H(x,y) = (1/\lambda z) \exp ikz\{1+(x/z)^2+(y/z)^2\}^{1/2}/\{1+(x/z)^2+(y/z)^2\}^{1/2}$ assumes the meaning of the impulse response of the propagation between input (source) and output (receiving) planes.

The Fourier transform $H_F(k_x,k_y)$ of the impulse response $H(x,y)$ in the spatial frequency k_x and k_y domain, conjugate to coordinates x and y is written as:

$$H_F(k_x,k_y) = \iint_{-\infty,+\infty} H(x,y) \exp i(xk_x+yk_y)\, dx\, dy \quad \text{(A4.5)}$$

The function $H_F(k_x,k_y)$ is called the transfer function between input and output planes, and it is found to be given by the following expression [1]:

$$H_F(k_x,k_y) = \exp ikz[1-\lambda^2(k_x{}^2+k_y{}^2)]^{1/2} \quad \text{(A4.6)}$$

Now, if we wish to compute the propagation problem in the frequency domain, in terms of the Fourier transforms $A_1(k_x,k_y)$ and $A_2(k_x,k_y)$ of the spatial domain distributions $A_1(x,y)$ and $A_2(x,y)$, we can take advantage of the convolution in Eq.A4.4 becoming the product of the transforms in the frequency domain, or:

$$A_2(k_x,k_y) = A_1(k_x,k_y)\, H_F(k_x,k_y) \qquad \text{(A4.7)}$$

A4.2 The Fresnel Approximation

Eqs.A4.1 to A4.3 are insightful from a physical point of view, but are difficult to treat in most cases. A very helpful approximation is provided by the well-known Fresnel condition $L_{\xi,\eta}$, $L_{x,y} \ll z$, in which the transverse dimensions $L_{\xi,\eta}$ and $L_{x,y}$ of the source and receiving apertures are small as compared to distance z.

With this condition, we can approximate the obliquity factor to unity, $\cos\theta \approx 1$, and write the distance as:

$$r_{12} = z\{1+[(x-\xi)/z]^2+[(y-\eta)/z]^2\}^{1/2} \approx z\{1+{}^1/_2[(x-\xi)/z]^2+{}^1/_2[(y-\eta)/z]^2\}$$
$$\approx z+ (x-\xi)^2/2z+(y-\eta)^2/2z \qquad \text{(A4.8)}$$

Further, we may develop the square factor and obtain:

$$r_{12} = z + (x^2+y^2)/2z + (\xi^2+\eta^2)/2z - (x\xi+y\eta)/z \qquad \text{(A4.9)}$$

Going back to the propagation equation (A4.3), we can see that, in the Fresnel approximation, the impulse response H becomes:

$$H(x,y) = (\lambda z)^{-1} \exp ikz \times$$
$$\times \exp i[k(x^2+y^2)/2z] \exp i[k(\xi^2+\eta^2)/2z] \exp -i[k(x\xi+y\eta)/z] \qquad \text{(A4.10)}$$

The first exponential term in this equation gives the propagation delay on the distance z, a constant term independent from x,y. The second is a field curvature term and comes out of the integral (A4.3). The third term is a field curvature of the source field, and can eventually be incorporated in the source distribution $A(\xi,\eta)$. The last term is the kernel of a Fourier double-transform integral that conjugates the spatial variables ξ and η to the transform-domain variables kx/z and ky/z.

By rewriting Eq.A4.3 with the previous terms, we have in the Fresnel approximation:

$$A_2(x,y) = (\lambda z)^{-1} \exp i[k(x^2+y^2)/2z]\; FT\; \{A_1(\xi,\eta)\; \exp i[k(\xi^2+\eta^2)/2z]\} \qquad \text{(A4.11)}$$

Note on terminology. In the Fourier transform (abbreviated above with *FT*), the variables conjugated to ξ and η are the spatial frequencies (Eqs.A4.5-A4.7) k_x=kx/z and k_y=ky/z, and so we should write the left-hand term in Eq.A4.11 as $A_2(k_x,k_y)$. However, as $x=zk_x/k$ and

$y=zk_y/k$ are connected to k_x and k_y, the notation $A_2(x,y)$ is also correct and can be used when we regard the propagation process as one conjugating the spatial domains (ξ,η) and (x,y). Additionally, we may interpret $x/z=\psi_x$ and $y/z=\psi_y$ as angular coordinates of diffraction, and write the left-hand term in Eq. A4.11 as $A_2(\psi_x, \psi_y)$ to indicate the Fourier conjugation from the spatial to the angular domain.

When distance is large, so that $z>>kL_{\xi,\eta}{}^2$, $kL_{x,y}{}^2$, the curvature terms become negligible and we are in the Fraunhofer region (or far field). Eq. A4.11 can be written as the Fourier transform integral relating the source and image fields:

$$A_2(x,y) = \int\int_{-\infty,+\infty} A_1(\xi,\eta)\exp -i[k(x\xi+y\eta)/z]\,d\xi\,d\eta$$
$$A_2(x,y) = FT\ [A_1(x,y)] \qquad \text{(A4.12)}$$

Last, it is useful to recall that the irradiance, or power per unit surface, $I(x,y)$, associated with the electrical field is in general given by $I(x,y)= E^2(x,y)/Z_0$, where Z_0 is the free-space impedance. In the following, we will consider the irradiance as proportional to the square of the electric field amplitude, or simply by $I \propto E^2$.

A4.3 EXAMPLES

Rectangular aperture of width D across x, and infinite along y. In this case, we have $A_1=A_0$ rect(-D/2,+D/2) and Eq.A4.12 gives:

$$A_2(x,y) = FT\,[A_1(x,y)] = A_0\int_{-D/2\,..\,+D/2} A_1 \exp -ik(x\xi)/z\,d\xi$$
$$= A_0(\sin kxD/2z)/(kxD/2z) = A_0 \operatorname{sinc} xD/\lambda z \qquad \text{(A4.13)}$$

In the last equation, we used $k=2\pi/\lambda$ and have introduced the sinc function:

$$\operatorname{sinc} x = (\sin \pi x) / \pi x \qquad \text{(A4.14)}$$

When defined with the normalization of Eq. A4.14, both the sinc function and its square, $\operatorname{sinc}^2 x$, have unity area [1] when integrated on x from $-\infty$ to $+\infty$.

Letting $x/r_{12} \approx x/z = \sin\theta$ for the direction of diffraction, we have for the diffraction distribution in the far field:

$$E(\theta) = \operatorname{sinc}(\sin\theta D/\lambda).$$

For small angle θ, it is $\sin\theta\approx\theta$, and then we have:

$$E(\theta) \approx \operatorname{sinc}(\theta D/\lambda) \qquad \text{(A4.15)}$$

This is the well-known formula for the diffraction from a slit.

In addition, if P_0 is the radiant power intercepted by the slit and rediffused, and in view of the normalization to 1 of the $\text{sinc}^2 x$ area, the density $I(\theta)$ of radiant power diffracted at the angle θ is written as:

$$I(\theta) = P_0 \, \text{sinc}^2 (\theta D/\lambda) \tag{A4.15'}$$

The functions sinc x and $\text{sinc}^2 x$ are plotted in Fig.A4-2. The HWHM (half-width at half-maximum) of the field (sinc x) is at x=0.6, whereas that of irradiance ($\text{sinc}^2 x$) is at x=0.44. As it is $x=\theta D/\lambda$, we can solve for the angle of diffraction at half-power as $\theta=0.44\ \lambda/D$.

Other features of the sinc^2 function are: a first zero located at x=±1 (whence $\theta_{zero}=\pm\lambda/D$) and secondary maxima located at x=±1.43, ±2.47 and ±3.47, with amplitudes of 4.7, 1.6 and 0.8% of the maximum, respectively.

Comment on notation. We follow the notation of Gaskill [1] for the functions sinc and somb (see below). Other authors prefer incorporating the factor π inside the argument.

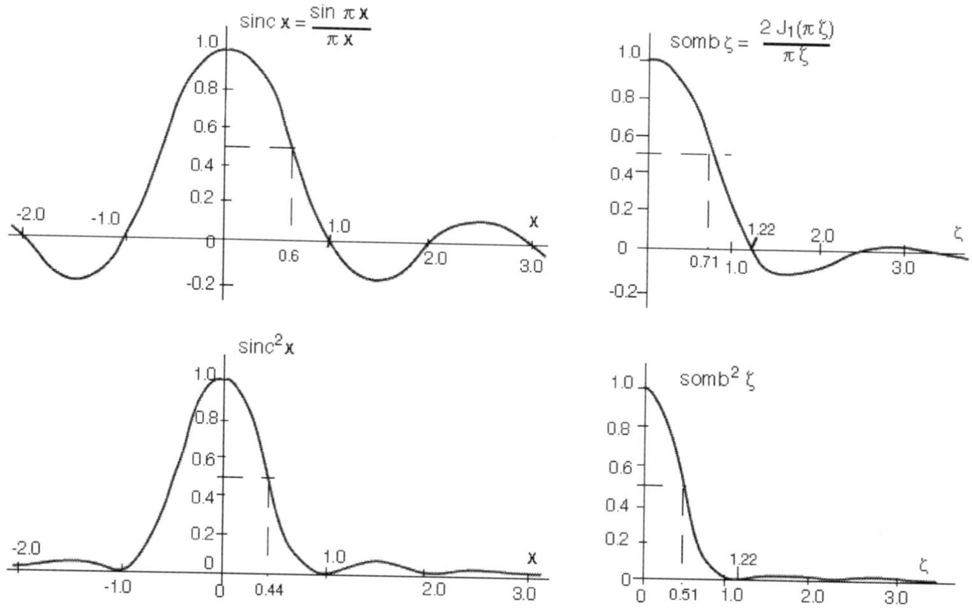

Fig.A4-2 Left: the sinc x function describes the relative amplitude of the field diffracted at $x=(D/\lambda)\sin\theta$ from a slit aperture of width D. Its square, sinc^2 x, gives the associated radiant power density. Right: the somb and somb^2 functions describe the relative amplitude of the field and the power density diffracted at $\zeta=(D/\lambda)\sin\theta$ from a sphere of diameter D.

Circular aperture of diameter D

In Eq. A4.12, Cartesian variables $d\xi d\eta$ are changed to the radial variables $\rho d\rho d\psi$, and the integral is evaluated on the limits ρ=0-D/2 and ψ=0-2π. Letting $r^2=x^2+y^2$, the result reads:

$$A_2(r) = A_0 \text{ somb } r\, D/\lambda z \qquad \text{(A4.16)}$$

In this expression, we have used the notation somb (from sombrero, the bidimensional generalization of the sinc function to radial coordinates) [1], defined as:

$$\text{somb}(\zeta) = 2J_1(\pi\zeta)/(\pi\zeta) \qquad \text{(A4.17)}$$

here, J_1 is the Bessel function of the first kind. Again, letting $r/r_{12} \approx r/z = \sin\theta$ for the direction of diffraction and squaring A_2, we obtain the well-known Airy diffraction formula of the radiant power density $I_2(\theta)$ diffracted at the angle θ:

$$I_2(\theta) = I_0 \text{ somb}^2 \sin\theta\, D/\lambda \qquad \text{(A4.16')}$$

The functions somb and somb^2 are plotted in Fig.A4-2. They are standardized so that the maximum value is unity and the volume over the x-y plane, from $-\infty$ to $+\infty$, is $4/\pi$ [1].

The half-width at half maximum (HWHM) of somb^2 is x=0.51, which corresponds to a diffraction angle θ =0.51 λ/D.

The first zero of the relative radiant power is at x =1.22, leading to the well-known diffraction formula θ_{zero}=1.22 λ/D.

Secondary maxima are found at x=±1.63, ±2.67, and ±3.69, with amplitudes of 1.7, 0.4 and 0.15% of the maximum, respectively.

Gaussian-distributed spot

This is the case of a real mode representing the field distribution E(r) of the laser beam along the radial coordinate $r=(x^2+y^2)^{1/2}$:

$$E(r) = E_0\, [(2\pi)^{1/2} w_0]^{-1} \exp -r^2/w_0^2 \qquad \text{(A4.18)}$$

The spot size parameter of the Gaussian beam is w_0, the rms deviation. This quantity is the radius at which the field has dropped at 1/e=0.37 of the maximum value, and the intensity $I \propto E^2$ at $1/e^2$=0.13 of the maximum.

The HWHM of the Gaussian is 1.18w_0 for the field and 0.84w_0 for the intensity. Additionally, w_0 is the radius carrying 86% of the total power.

Using Eq.A4-12, the Fourier transform E(r) is found as [1]:

$$E_2(r) = E_0 \exp -w_0^2 k^2 r^2/4z^2 \qquad \text{(A4.18')}$$

As a function of the angular coordinate of observation θ=r/z, this equation can be written as:

$$E_2(\theta) = E_0 \exp -\theta^2/\theta_{w0}^2 \qquad \text{(A4.18'')}$$

The parameter $\theta_{w0} = \lambda/\pi w_0$ introduced in this equation represents the diffraction angle associated with the spot size w_0.

As we can see from Eq. A4-18", the diffracted field in the Fraunhofer region has a Gaussian distribution with angular (spot) size θ_{w0}. This is the angular radius containing 86% of the total power.

For comparison to the Airy disk $\theta = 0.51\lambda/D$, we can use the spot diameter $D_{w0} = 2w_0$ in θ_{w0} and obtain $\theta_{w0} = 2\lambda/\pi D_{w0} = 2\lambda/e = 0.63\lambda/D_{w0}$.

Sinusoidal amplitude grating of spatial period Λ
In this case we get $A_1 = \sin 2\pi\xi/\Lambda$. By substitution in Eq. A4.12, we get:

$$A_2(x,y) = FT\,[A_1(x,y)] = (1/2i)\,FT\,[\exp i(2\pi\xi/\Lambda) - \exp -i(2\pi\xi/\Lambda)]$$
$$= [\delta(kx/z+2\pi/\Lambda) - \delta(kx/z-2\pi/\Lambda)]/2i \qquad \text{(A4.20)}$$

The result shows that we have two diffracted beams at $kx/z = \pm 2\pi/\Lambda$, or at angles $\sin\theta = x/z = \pm 2\pi/\Lambda k = \pm\lambda/\Lambda$, the first order of diffraction, a well-known result in optics.

Sinusoidal phase grating of spatial period Λ
It is $A_1 = \exp i[\phi_0 \cos 2\pi\xi/\Lambda]$, and by developing the exponential in series, we get the field in the output plane as:

$$A_2(x,y) = [1+J_0(\phi_0)]/2 + \Sigma_{n=1..+\infty} (i^n/2) J_n(\phi_0)[\delta(kx/z+n2\pi/\Lambda) - \delta(kx/z-n2\pi/\Lambda)] \qquad \text{(A4.21)}$$

From this expression, we can see that for $\phi_0 \ll 1$, there is only the first order of diffraction as in Eq.A4.20, whereas when ϕ_0 is not so small, all the orders of diffraction are generated.

Grating of period Λ and a finite width D
We have $A_{1f} = A_1(x,y)\ \mathrm{rect}(-D/2,D/2)$, where D represents the width of the reticle. Using Eq.A4.12 gives:

$$A_{2f}(x,y) = A_2(x,y) * FT\ \mathrm{rect}(-D/2,D/2) \qquad \text{(A4.22)}$$

If $A_2(x,y)$ is the output of an amplitude grating (Eq.A4.16), we get:

$$A_{2f}(x,y) = \mathrm{sinc}\,(kx/z+2\pi/\Lambda)D/2 - \mathrm{sinc}\,(kx/z-2\pi/\Lambda)D/2. \qquad \text{(A4.21')}$$

Arbitrary profile illuminated with a divergence Θ
This is the case of a target positioned not exactly in the beam waist of the illuminating laser. Let the illuminating beam arrive at the ξ,η plane with an angle Θ or, given a diameter D of the target, with a curvature of the wavefront $R = D/\Theta$.

At point ξ,η, the phase due to the spherical wavefront is:

$$\varphi = k(R^2+\xi^2+\eta^2)^{1/2} - kR = kR\,[1+(\xi^2+\eta^2)/2R^2] - kR \approx k(\xi^2+\eta^2)/2R$$

Thus, in addition to the function $A_1(\xi,\eta)$ describing the target, in Eq.A4.12 we have a multiplying term $A_1^{\#}=\exp ik(\xi^2+\eta^2)/2R$.

We know that the Fourier transform of a product is the convolution of the transforms. Thus, our result is $A_2(x,y)^{**}FT\,[\exp ik(\xi^2+\eta^2)/2R]$.

The transform of $A_1^{\#}$ is [1]:

$$FT\,[A_1^{\#}]= \exp\text{-}i[kR(x^2+y^2)/2z^2]=\exp\text{-}i[\theta^2/(2/kR)]$$

where $\theta= (x^2+y^2)^{1/2}/z$.

The argument of the transform is small enough to be safely neglected as long as the angular variance 2/kR is small with respect to the diffraction angle θ_2 associated with the distribution $A_2(x,y)$.

This condition is written as $2/kR<<\theta_2^2=(\lambda/D)^2$ in the case of a particle of diameter D. Solving for R, we get $R>>(2/k)(D/\lambda)^2=D^2/\pi\lambda$, the well-known Fresnel distance [1].

REFERENCES

[1] J.D. Gaskill: "*Linear Systems, Fourier Transforms and Optics*", J. Wiley and Sons: New York 1978, see Chapter 10.

[2] C. Scott: "*Introduction to Optics and Optical Imaging*", J. Wiley & Sons, Chichester 1998.

[3] J.W. Goodman: "*Introduction to Fourier Optics*", 3rd ed., Robert & Company Publ., New York 2005.

APPENDIX A 5

Source of Information on Photonic Instrumentation

Readers who want to keep abreast of the latest news and scientific and technical developments in the field of electro-optical instrumentation may consult the journals and magazines listed in the following.

Scientific Journals

IEEE Publications, 445 Hoes Lane, Piscataway, NJ (see web site at *ieee.org*):

IEEE Transaction on Instrumentation and Measurements (monthly)
IEEE Sensors Journal (bimonthly)
IEEE Journal of Quantum Electronics (monthly)
IEEE Photonics Technology Letters (monthly)
IEEE Selected Topics in Quantum Electronics (monthly)
IEEE Proceedings of the IEEE (monthly)
IEEE Spectrum (monthly)

The IEEE also provides access and download privileges to a large library ($\approx 7\ 10^6$ titles) of PDF documents, including journal papers, conference proceedings, and standards published by IEEE and IEE since 1980 (see *ieeexplore.com*).

Optica (formerly OSA) Publications (see web site at: *osa.org*):

Optics Letters (semi-monthly)
Journal of the Optical Society of America (monthly, issued in two parts)
Applied Optics (monthly, issued in three parts)

Optics and Photonics News (monthly)
Optics Express (monthly, web only)

IEE, the Institution of Electrical Engineers (London, UK) publications:
Electronics Letters (weekly)
IEE Proceeding part J: Optoelectronics (bimonthly)
IEE Proceeding part E: Instrumentation (bimonthly)

IOP, the Institute of Physics (Bristol, UK) publications (search engine *iop.org/EJ/welcome*):
Journal of Optics, part A and B (monthly)
Journal of Measurement Science (monthly)

SPIE (Bellingham, WA) publications (search engine *spie.org*):

Optical Engineering (monthly)

AIP, the American Institute of Physics publications (search engine *aip.org/japo*):

Journal of Applied Physics (monthly)
Review of Scientific Instruments (monthly)
Applied Physics Letters (semi-monthly)

Journal of Optical and Quantum Electronics (Kluwer, Dodrecht, Holland, bimonthly)
Sensors and Actuators (North Holland, Dodrecht, Holland, monthly)

Magazines

IEEE Photonics Society Newsletters (IEEE, bimonthly)
Optics &Photonics News (Optica, monthly)
IEEE Instrumentation and Measurement Magazine (IEEE, 9 issues a year)
Photonics Spectra (Laurin Publications, monthly) search engine *photonics.com*
Laser Focus World (Pennwell, Nashua, NH, monthly) search engine *lfw.pennNET.com*
Europhotonics (Laurin Publications, NL, monthly) search engine *photonicsonline.com*

Societies

Several societies deal with laser and photonics instrumentation as topics of interest:
IEEE - IM Society (Instrumentation and Measurement Society), see *im.ieee.org*
IEEE -PhoS Society (Photonics Society), see *photonicssociety.ieee.org*
Optica (Optical Society of America), see *optica.com*
SPIE (Society of Photonic Instrumentation Engineers), see *spie.org*

Meetings

Annual international meetings and conferences on electro-optical instrumentation are:
CLEO and CLEO Europe (Conference on Laser and Electro-Optics),
promoted by IEEE and OSA
IMTC (International Measurement Techniques Conference), promoted by IEEE
IEEE PhoS Annual Meeting, promoted by IEEE
IMEKO (International Measurement Conference), promoted by IMEKO

Index

M

N

O

P

Q

V

W

X

Y

Z

About the Author

Silvano DONATI is Emeritus Professor of Optoelectronics at University of Pavia (Italy), Faculty of Engineering, Department of Electronics.

Upon earning a PhD in Physics with honours from University of Milano, he joined CISE, a famous research institution now merged with the Electrical Generating Board of Italy, where he worked on photodetectors and optoelectronic instrumentation. At University of Pavia he has first a full-time lecturer, holding courses in Electronics Circuits Design, Electronic Materials and Technologies, and Electro-Optical Systems, then became a full professor in Optoelectronics introducing novel courses on Photodetectors, Optoelectronic Instrumentation, and Optical Communications. Under his guidance, 120 Masters students and 20 PhDs have graduated so far. At University of Pavia, he was proactive in starting the curriculum in Optoelectronics for electronic engineers in 1980, and there he has created and headed since then a research group in Optoelectronics, totalling 12 researchers.

Through the years, he has contributed to the fields of electronics (noise, CCD, coupled oscillators) and optoelectronic instrumentation (laser interferometers, fiber gyros, fiberoptic current sensors). More recently, his interests have been in all-fiber passive components for communications, noise in photodetection, self-mixing interferometry, optical chaos and cryptography.

He has helped in several R&D programmes with national Companies active in the areas of communications, instrumentation and avionics. He also made a successful spin-off on fiber-couplers with an Italian company and promoted another on self-mixing technology.

He has organised several national and international meetings and schools, as a member of the Steering and Programme Committees, and as the Chairman of 'Fotonica' (Roma, 1997) and 'Elettroottica' (Pavia, 1994 and Roma, 2004). In 2019, at the National Taiwan University (Taipei), he has served as the Program Chair of the 4th IEEE International Conference on Biophotonics.

From 1986 to 1992 he has been the Director of the Italian scientific review in electronics 'Alta Frequenza - Rivista di Elettronica' of the Italian Electronics Association (AEI).

He and his group have been awarded 7 prizes, one from Philips Morris and six from AEI, including the G. Marconi gold medal to the scientific career. In 2015, he was awarded the IEEE A. Kressel prize for his pioneering research on self-mixing interferometry.

He was the founder and first Chairman of the IEEE-LEOS Italian Chapter in 1997.

He started and chaired the WFOPC (Fiber Optics Passive Components) IEEE International Conference in 1998, and chaired ODIMAP II and III, the International Conference on Interferometry held in Pavia. In 2010, he started COMPENG, the IEEE Conference on Complexity in Engineering, now on its 5th edition.

Prof. Donati has authored or co-authored about 350 papers, and holds 12 patents.

He is the author of the book 'Photodetectors', published by Prentice Hall, 2000 and with a 2nd edition by Wiley IEEE Press, 2020.

He is a Life Fellow Member of IEEE, Life Fellow Member of Optica, and Meritorious Member of AEI. He is also a member of several other Societies, including SPIE, IMAPS, and IoP.